Quantum Electrodynamics

- annotated sources

Volume II

by

Trevor G. Underwood

Quantum physics is the study of matter and energy at the most fundamental level.

Quantum mechanics is the branch of physics that deals with the behavior of matter and light on a subatomic and atomic level.

Quantum electrodynamics deals with the electromagnetic field and its interaction with electrically charged particles.

Relativistic quantum mechanics, the synthesis of quantum mechanics with relativity, is quantum field theory.

Published by Trevor G. Underwood
18 SE 10th Ave.
Fort Lauderdale, FL 33301

ISBN: 979-8-218-18286-1 (hardcover)

Library of Congress Control Number: 2023905784

Printed and distributed by Lulu Press, Inc.
627 Davis Dr.
Ste. 300
Morrisville, NC 27560
http://www.lulu.com/shop

CONTENTS of Volume II. (1930-65)

Page no.

33 **PREFACE**

63 **Werner Heisenberg – 1932 Nobel Lecture, December 11, 1933.** ***The development of quantum mechanics.*** https://www.nobelprize.org/prizes/physics/1932/heisenberg/lecture/. Heisenberg received the 1932 Nobel Prize in Physics "for the creation of quantum mechanics, the application of which has, inter alia, led to the discovery of the allotropic forms of hydrogen". In 1925, Heisenberg formulated a type of quantum mechanics based on matrices. In 1927 he proposed the "uncertainty relation", setting limits for how precisely the position and velocity of a particle can be simultaneously determined. In his Nobel Prize lecture, he noted that the impossibility of harmonizing the Maxwellian theory with the pronouncedly visual concepts expressed in the hypothesis of light quanta subsequently compelled research workers to the conclusion that *radiation phenomena can only be understood by largely renouncing their immediate visualization.* Classical physics seemed the limiting case of visualization of a fundamentally unvisualizable microphysics, the more accurately realizable the more Planck's constant vanishes relative to the parameters of the system. After describing the development of the current theory of quantum mechanics, Heisenberg concluded that a visual description for the atomic events is possible only within certain limits of accuracy - but within these limits the laws of classical physics also still apply. Owing to these limits of accuracy as defined by the uncertainty relations, moreover, a visual picture of the atom free from ambiguity has not been determined. On the contrary the corpuscular and the wave concepts were equally serviceable as a basis for visual interpretation. The laws of quantum mechanics were basically statistical. He noted that the attention of the research workers was now primarily directed to *the problem of reconciling the claims of the special relativity theory with those of the quantum theory.* The attempts made hitherto to achieve a *relativistic* formulation of the quantum theory were all based on visual concepts so close to those of classical physics that it seemed impossible to determine the fine-structure constant within this system of concepts.

73 **Erwin Schrödinger – 1933 Nobel Lecture, December 12, 1933.** ***The fundamental idea of wave mechanics.*** https://www.nobelprize.org/prizes/physics/1933/schrödinger/lecture/.] Schrödinger shared the 1933 Nobel Prize in Physics with Paul Dirac "for the discovery of new productive forms of atomic theory". In Niels Bohr's theory of the atom, electrons absorb and emit radiation of fixed wavelengths when jumping between fixed orbits around a nucleus. The theory provided a good description of the spectrum created by the hydrogen atom, but needed to be developed to suit more complicated atoms and molecules. Assuming that matter (e.g., electrons) could be regarded as both particles and waves, in 1926 Schrödinger formulated a wave equation that accurately calculated the energy levels of electrons in atoms. In his Nobel Prize lecture, Schrödinger described a fascinating analogy between the path of a ray of light and the path of a mass point. In optics the old system of mechanics corresponds to operating with isolated

mutually independent light rays. The new wave mechanics corresponds to the wave theory of light. We are never in a position to say what really is or what really happens, but we can only say what will be observed in any concrete individual case. He concluded that *the ray or the particle path corresponded to a longitudinal relationship of the propagation process (i.e. in the direction of propagation), the wave surface on the other hand to a transversal relationship (i.e. normal to it). Both relationships were without doubt real;* one is proved by photographed particle paths, the other by interference experiments. *To combine both in a uniform system has proved impossible so far. Only in extreme cases does either the transversal, shell-shaped or the radial, longitudinal relationship predominate to such an extent that we think we can make do with the wave theory alone or with the particle theory alone.*

83 **Paul Dirac – 1933 Nobel Lecture, December 12, 1933. *Theory of electrons and positrons.*** https://www.nobelprize.org/prizes/physics/1933/dirac/lecture/. Dirac shared the 1933 Nobel Prize in Physics with Schrödinger "for the discovery of new productive forms of atomic theory". During the intense period of 1925-26 quantum theories were proposed that accurately described the energy levels of electrons in atoms. These equations needed to be adapted to Einstein's theory of relativity, however. In 1928 Dirac formulated a fully *relativistic* quantum theory. The equation gave solutions that he interpreted as being caused by a particle equivalent to the electron, but with a positive charge. This particle, the positron, was later confirmed through experiments. In his Nobel Lecture, Dirac described the current state of his theory of electrons and positrons. He focused on electrons and the positrons on the grounds that the theory has been developed further, and hardly anything could be inferred theoretically about the properties of the others. He noted that photons are so simple that they can easily be fitted into any theoretical scheme, and protons and neutrons seem to be too complicated and no reliable basis for a theory of them had yet been discovered. He noted that the question that must first be considered is *how theory can give any information at all about the properties of elementary particles.* There existed at the present time a general quantum mechanics which could be used to describe the motion of any kind of particle, no matter what its properties are. The general quantum mechanics, however, was valid only when the particles have small velocities and failed for velocities comparable with the velocity of light, when the effects of *relativity* come in. There existed no *relativistic* quantum mechanics for particles with large velocities which could be applied to particles with arbitrary properties. But, *subjecting quantum mechanics to relativistic requirements imposes restrictions on the properties of the particle, and in this way information about the particles can be deduced from purely theoretical considerations.* Dirac described how the spin properties of the electron could be deduced, and the existence of positrons with similar spin properties and with the possibility of being annihilated in collisions with electrons could be inferred. He started with the equation connecting the *kinetic energy* W and *momentum* p_r ($r = 1, 2, 3$) of a particle in *relativistic* classical mechanics. From this he obtained a *wave equation* of quantum mechanics, by letting it operate on the wave function ψ, where W and p_r were the operators $ih\partial/\partial t$ and $-ih\partial/\partial x_r$. In order to satisfy the general requirement of quantum mechanics that its *wave equations* be linear in the operator

W or $\partial/\partial t$, and the requirement of *relativistic invariance* that they must also be linear in the p's, he replaced this with a *wave equation* linear in kinetic energy W and momentum p_r. This involved four new variables α_r and α_0, which were operators that could operate on ψ. The new variables α, which he introduced to get a *relativistic* wave equation linear in W, gave rise to the spin of the electron. The variables α also gave rise to some unexpected phenomena concerning the motion of the electron. *It was found that an electron which appears to be moving slowly, must actually have a very high frequency oscillatory motion of small amplitude superposed on the regular motion, which resulted in the velocity of the electron at any time being equal to the velocity of light.* This prediction could not be directly verified by experiment, but other consequences of the theory which are inseparably bound up with this one, such as the law of scattering of light by an electron, were confirmed by experiment. A feature of these equations also led to the prediction of the positron. These quantum equations were such that they allowed as the possible results of a measurement of W either something greater than mc^2 or something less than $-mc^2$. An examination of the behavior of these negative-energy states in an electromagnetic field showed that they *corresponded to the motion of an electron with a positive charge instead of the usual negative one*, i.e. a positron. Dirac solved *the difficulty of negative energies* by identifying the positron with a "hole" using his "hole theory". On this view, the positron was just a mirror-image of the electron, having exactly the same mass and opposite charge.

89 **Dirac, P. A. M. (January, 1930). A theory of electrons and protons.** *Roy. Soc. Proc., A*, 126, 801, 360-65; https://doi.org/10.1098/rspa.1930.0013; in *relativistic* quantum theory in which the *electromagnetic field* is subjected to quantum laws the *wave equation* refers equally well to an electron with charge + e with *negative kinetic energy*, the difficulty is not restricted to the quantum theory of the electron but appears in all relativity theories, it arises because there is an ambiguity in the sign of W, or rather $W + eA_0$, in the *relativity* Hamiltonian equation of the classical theory, in the quantum theory transitions can take place in which the *energy* of the electron changes from a positive to a negative value, Dirac uses his "*hole theory*" to address the problem by assuming that in a vacuum all negative-energy electron eigenstates are occupied, if negative-energy *eigenstates* are incompletely filled each unoccupied *eigenstate* – called a "hole" – would behave like a positively charged particle, which Dirac initially thought might be a proton, for scattering of radiation by an electron the *exclusion principle* forbids the electron to jump into a state of *negative energy* so Dirac assumes a double transition process in which first one of the negative-energy electrons jumps up into the *final state* for the electron with *absorption* (or *emission*) of a photon, and then the original positive-energy electron drops into hole formed by first transition with *emission* (or *absorption*) of a photon.

95 **Heisenberg, W. (November, 1930). Die Selbstenergie des Elektrons. (The self-energy of the electron.)** *Zeit. Phys.*, 65, 1-2, 4-13; https://doi.org/10.1007/BF01397404; (translation in Miller, A. I. (1994). *Early quantum electrodynamics. A sourcebook.* Selected papers, I); https://www.cambridge.org/core/books/abs/early-quantum-electrodynamics/selfenergy-of-the-electron/D1B89DCC1B534F4

CA24A23EA2915DCA2); *point-like* electron results in infinite energy density, but *finite radius for electron is inconsistent with special relativity*, *chooses to represents electron by point charge*, conditions are investigated under which the self-energy of electron vanishes, shows that one-electron problem could be treated correctly without an infinite *self-energy* if there were solutions of vacuum electrodynamics without a *zero-point energy* but *such solutions do not exist*, difficulties of *field theory* do not come directly from infinite *self-energy* of electron, *a solution of the basic equations has therefore not been found for the time being*, it also is not probable that one will achieve a solution without substantial modification of the quantum theory of wave fields.

102 **Dirac, P. A. M. (May, 1932). Relativistic Quantum Mechanics.** *Roy. Soc. Proc., A*, 136, 829, 453-64; http://www.jstor.org/stable/95782; alternative to Heisenberg and Pauli's approach to *relativistic* quantum mechanics which regards the field itself as a dynamical system amenable to Hamiltonian treatment and its interaction with the particles as describable by an *interaction energy*, there are serious objections to these views, if we wish to make an observation on a system of interacting particles the only effective method of procedure is to subject them to a field of electromagnetic radiation and see how they react, the role of the *field* is to provide a means for making observations, the nature of an observation requires an interplay between the *field* and the *particles*, we cannot suppose the *field* to be a dynamical system on the same footing as the *particles* and thus something to be observed in the same way as the *particles*, the *field* should appear in the theory as something more elementary and fundamental, in this paper an *interaction representation* is proposed which gives interplay between particles and field, translates *equations of motion* of *relativistic* classical theory directly into equations expressible entirely in terms of *probability amplitudes* referring to one *ingoing field* and one *outgoing field*, assumes passage from the field of ingoing waves to the field of outgoing waves is a quantum jump performed by one field composed of waves passing undisturbed through the electron and satisfying Maxwell's equations, *relativistic* observable quantities always *transition probabilities*, *probability amplitudes* analogous to Heisenberg's matrix elements, quantization assumes *intensities* and *phases* are operators satisfying usual quantum conditions governing the *intensities* and *phases* of Fourier components of electromagnetic field in empty space, determines matrix elements associated with electron jumps, assumes interaction of each electron with the *field* can be described by an *interaction energy* equal to its charge multiplied by the *potential* at the point where it is situated, *wave equation* of interactions of two electrons due to motions of each being connected with same field
$\{ih\, \delta/\delta t + h^2/2m_1\, \delta^2/\delta x_1^2 + h^2/2m_2\, \delta^2/\delta x_2^2 - \varepsilon_1 V(x_1 t) - \varepsilon_2 V(x_2 t)\}\, \psi = 0$.

113 **Dirac, P. A. M., Fock, V. A., & Podolsky, B. (1932). On quantum electrodynamics.** *Phys. Zeit. Sowjetunion*, 2, 468; *relativistic* model in which a fixed number of electrons interact through a second-quantized electromagnetic field, applies Dirac's *interaction representation* formulation of quantum field theory to full electrodynamics, formulated with the help of a *multi-time* wave function $\psi(t_1,\mathbf{x}_1,\ldots,t_N,\mathbf{x}_N)\psi(t_1,x_1,\ldots,t_N,x_N)$ that generalizes the Schrödinger's

multiparticle wave function to allow for a manifestly *relativistic* formulation of wave mechanics, all non-trivial dynamics due to interaction between charged matter and electromagnetic field relegated to time evolution of state vector, second Maxwell equation - the Ampère-Maxwell law - satisfied only through the action of the field operators on the wave function, implies that electromagnetic field operators still obey the free field equations and consequently the covariant commutation relations of Jordan and Pauli, but method of quantizing by imposing covariant commutation relations still an isolated technique applied only to Maxwell field, shows relationship between *Maxwell's equations* for empty space, the *wave equation* of Heisenberg-Pauli's theory, and the *wave equation of* Dirac's *interaction representation.*

118 **Dirac, P. A. M. (1933). The Lagrangian in Quantum Mechanics.** *Phys. Zeit. Sowjetunion*, 3, 1, 64-72; alternative formulation of quantum mechanics in terms of Lagrangian in place of Hamiltonian, *coordinates* and *velocities* instead of *coordinates* and *momenta*, allows *equations of motion* to be expressed as stationary property of *action function* which is the time-integral of the Lagrangian and *relativistic invariant*, there is no corresponding *action principle* in terms of *coordinates and momenta* in Hamiltonian theory, Lagrangian closely connected with theory of *contact transformations*, transformation functions have classical analogues expressible in terms of the Lagrangian, function of coordinates at time t and time t + dt rather than of the coordinates and velocities, applies to *field dynamics* using suitable field quantities or potentials as coordinates, "*many-time*" theory, each coordinate a function of four *space-time* variables instead of one time variable in particle theory, *generalized transformation function.*

126 **Dirac, P. A. M. (March, 1934). Discussion of the infinite distribution of electrons in the theory of the positron.** *Proc. Camb. Phil. Soc.*, 30, 2, 150-63; doi:10.1017/S030500410001656X; attempt by Dirac to address problem with his *relativistic* 'hole' theory which implies an infinite number of negative-energy electrons (per unit volume) with energies extending continuously from $-mc^2$ to $-\infty$, when an electromagnetic field is present positive- and negative-energy states cannot be distinguished in *relativistically* invariant way, need to set up assumptions for production of *electromagnetic field* by the electron distribution such that any finite change in distribution produces a change in the field in agreement with Maxwell's equations and such that the infinite field which would be required by Maxwell's equations from an infinite density of electrons is in some way cut out, *assumes each electron has its own individual wave function in space-time* and each electron moves in an *electromagnetic field* which is the same for all electrons part coming from external causes and part from the electron distribution itself, introduces *relativistic density matrix* $\sum_{oc} \psi_{k'}(x' t') \psi^*_{k''}(x'' t'')$ referring to two points in space and two times, separates density distribution into two parts where one contains the singularities, and the other describes the *electric* and *current densities* physically present.

134 **Heisenberg, W. (March, 1934). Bemerkungen zur Diracschen Theorie des Positrons. (Remarks on the Dirac theory of positron.)** *Zeit. Phys.*, 90, 3-4, 209-31; https://doi.org/10.1007/BF01333516; (translated by D. H. Delphenich; https://neo-classical-physics.info/ electromagnetism. html); Heisenberg's reconstruction of Dirac's *theory of the positron* in the formalism of quantum electrodynamics, demands that symmetry between positive and negative charge should be expressed in the basic equations from the outset, in addition to the well-known difficulties with the divergences no new infinities should appear in the formalism, theory should provide an approximation for the treatment of problems that have been treated by quantum electrodynamics up to now, Dirac [(1934). Discussion of the infinite distribution of electrons in the theory of the positron.] showed that a quantum mechanical system of many electrons that fulfill the Pauli principle and move in a given force field without back-reaction can be characterized by a *density matrix* $(x', t', k' \mid R \mid x'', t'', k'') = \sum n\ \psi^{*}n\ (x', t', k')\ \psi_{n}\ (x'', t'', k'')$ when $\psi_{n}(x', t', k')$ means the normalized eigenfunctions of the states that possess one electron, and x′, t′, k′ (x″, t″, k″, resp.) are position, time, and spin variables, all physically-important properties of quantum-mechanical systems like *charge density*, *current density*, etc., can be read off from the *density matrix*, the temporal change in the density matrix is determined by the Dirac differential equation, Dirac made different choice of *density matrix* representing external field resulting in a different energy and impulse density, by restricting oneself to an *intuitive analogue theory of matter fields the negative energy levels in the Dirac theory could be avoided by replacing the homogeneous Dirac differential equation with an inhomogeneous equation* where the inhomogeneity is indicative of pair creation, for most practical applications e.g. pair creation, annihilation, Compton scattering, etc. *the theory described here does not yield anything new compared to the formulation of the Dirac theory*, in the Maxwell theory a continuous charge distribution also leads to a finite self-energy; it is the "quantization" that leads to the infinite self-energy, *if one represents the quantization of the electromagnetic field by point-like light quanta then the infinitude of the self-energy also emerges in the intuitive theory of matter waves.*

146 **Victor Frederick Weisskopf (September 19, 1908 – April 22, 2002).**

149 **Weisskopf, V. (January, 1934). Über die Selbstenergie des Elektrons. (The Self-energy of the Electron.)** *Zeit. Phys.*, 89, 27-39; https://doi.org/10.1007/BF01333228; (translation in A. I. Miller. (1994). *Early Quantum Electrodynamics. A Sourcebook*, Cambridge University Press, 157-68; https://doi.org/10.1017/CBO9780511608223.015; the *self-energy* of the electron is derived in close formal connection to classical radiation theory without direct application of quantum electrodynamics, the radiation field is calculated classically from *current and charge densities* of atom, divides electromagnetic field into rotation-free *electrostatic* part and divergence free *electrodynamic* part, *self-energy* derives from electrostatic part, self-energy of electron at occupied negative energy states diverges logarithmically as in Dirac's "hole theory" in contrast to the linear divergence of the classical theory and the quadratic divergence of the one particle Dirac theory.

154 **Pauli, W. & Weisskopf, V. (1934). Über die Quantisierung der skalaren relativistischen Wellengleichung. (The Quantization of the Scalar Relativistic Wave Equation.)** *Helv. Phys. Acta*, 7, 709-31; https://doi.org/10.1007/978-3-322-90270-2_36; (translation in A. I. Miller. (1994). *Early Quantum Electrodynamics. A Sourcebook*, Cambridge University Press, 188-205; https://doi.org/10.1017/CBO9780511608223.017; the *scalar relativistic wave equation* $E^2/c^2 - \sum_{k=1}^3 p_k^2 - m^2c^2 = 0$ has generally been relinquished in favor of *Dirac's four-component wave equation* $\{p_0 + \rho_1 (\boldsymbol{\sigma}, \mathbf{p}) + \rho_3 mc\} \psi = 0$ because the former does not yield the spin of the particles, shows that Dirac's *a priori* arguments based on limitation to a *single-body problem* and *particle density* being a meaningful observable need to be revised, shows that *charge density* is a meaningful observable, no reason for special form of *charge density* $\sum_r \psi^*_r \psi_r$, in *scalar relativistic wave equation theory* the energy of material particles is always positive after wave fields have been quantized, no new hypothesis such as the "hole theory" is required, applies Heisenberg-Pauli formalism of quantization of wave fields to *scalar relativistic wave equation* for matter fields for particles without spin and with Bose-Einstein statistics, shows that quantization of the *Klein-Gordon relativistic wave equation* for *scalar particles* gives rise to particles with opposite charge but with same rest mass which can be created or destroyed with absorption or emission of electromagnetic radiation, frequency of pair creation and annihilation processes of same order of magnitude as frequency for particles of same charge and mass which follows from Dirac "hole theory", *also leads to infinite self-energy not only of the particles but also to an infinite polarizability of the vacuum.*

164 **Ernst Carl Gerlach Stueckelberg (February 1, 1905 – September 4, 1984).**

169 **Stueckelberg, E. C. G. (September, 1934). Relativistisch invariante Störungstheorie des Diracschen Elektrons I. Teil: Streustrahlung und Bremsstrahlung. (Relativistically invariant perturbation theory of Dirac's electron Part I: scattered radiation and Bremsstrahlung.)** *Ann. Phys.*, 413, 4, 367-389; https://doi.org/10.1002/andp.19344130403; also at https://archive-ouverte.unige.ch/unige:161855; the main innovation of Stueckelberg's paper is the introduction of a new perturbative scheme yielding relativistic expressions for the *matrix elements* which are manifestly *gauge invariant,* this is achieved by performing a four-dimensional Fourier transformation of the *wave-function* eliminating space and time variables based on the interaction picture of Dirac, Fock and Podolsky, the starting point is the Dirac equation for the *spinor wave-function* $[1/i\ (\gamma, \partial/\partial x) + M + eV(x)]\ \Psi(x) = 0$, the contributions of positive and negative energies (corresponding to virtual electrons and positrons) are contained in a single propagation function which corresponds to what later was called the *Feynman propagator*, his method for calculating the cross-section starts with the definition of *on mass-shell wave-functions* and uses integration over the *complex energy-plane*, this paper introduces *Stueckelberg diagrams* later adopted by Feynman and subsequently renamed *Feynman diagrams*, all of Stueckelberg's expressions for matrix elements are identical to those obtained nowadays from Feynman diagrams.

186 **Einstein, A., Podolsky, B. & Rosen, N. (May, 1935). Can Quantum-Mechanical Description of Physical Reality Be Considered Complete?** *Phys. Rev.*, 47, 777-80; the description of reality as given by a wave function in quantum mechanics is not complete.

187 **Bloch, F. & Nordsieck, A. (July, 1937). Note on the Radiation Field of the Electron.** *Phys. Rev.*, 52, 54; previous methods of treating radiative corrections in non-stationary processes such as the scattering of an electron in an atomic field or the emission of a β-ray, by an expansion in powers of $e^2/\hbar c$, are defective in that they predict infinite low frequency corrections to the transition probabilities, difficulty can be avoided by a method developed here which is based on the alternative assumption that $e^2\omega/mc^3$, $\hbar\omega/mc^2$ and $\hbar\omega/c\Delta p$ (ω = angular frequency of radiation, Δp = change in momentum of electron) are small compared to unity, in contrast to the expansion in powers of $e^2/\hbar c$ this permits the transition to the classical limit $\hbar$=0, external perturbations on the electron are treated in the Born approximation, it is shown that for frequencies such that the above three parameters are negligible the quantum mechanical calculation yields just the directly reinterpreted results of the classical formulae, namely that the total probability of a given change in the motion of the electron is unaffected by the interaction with radiation, and that the mean number of emitted quanta is infinite in such a way that the mean radiated energy is equal to the energy radiated classically in the corresponding trajectory.

188 **John Archibald Wheeler (July 9, 1911 – April 13, 2008).**

191 **Wheeler, J. A. (December, 1937). On the Mathematical Description of Light Nuclei by the Method of Resonating Group Structure.** *Phys. Rev.*, 52, 11, 1107-22; https://doi.org/10.1103/physrev.52.1107; introduction of S-matrix.

192 **Pauli, W. & Fierz, M. (March, 1938). Zur Theorie der Emission langwelliger Lichtquanten. (On the theory of the emission of long-wave light quanta.)**; *Nuovo Cimento*, 15, 3, 167-88; https://doi.org/10.1007/BF02958939; the usual radiation theory provides an infinitely large value for the effective cross-section dq of a charged particle when passing through a force field when deflected to a given angular range ("*ultrared catastrophe*"). If one prescribes that the energy loss of the particle should lie between E and E + dE, then for small E according to this theory dq = const. dE/E, which diverges logarithmically at the point E = 0 when integrated. The present study investigates in more detail what quantum electrodynamics yields for this cross-section when a finite expansion is attributed to the charged body. It turns out that then infinity is always eliminated and that the distractions considered radiationless in the ordinary theory appear here as those with finite, albeit very small energy loss. On the other hand, according to the exact theory, the closer course of dq for very small energy losses E depends so much on the expansion of the charged body that a direct application of the result to real electrons is possible. *It must therefore be concluded that the problem in question is essentially linked to the fundamental difficulties of quantum electrodynamics which have not yet been resolved.*

193 **Dirac, P. A. M. (August, 1938). Classical Theory of Radiating Electrons.** *Roy. Soc. Proc., A*, 167, 929, 148-69; https://royalsocietypublishing.org/doi/pdf/10.1098/rspa.1938.0124; Dirac referenced Frenkel (1925) in describing *relativistic* form of classical theory of radiating electrons assuming electron to be a *point charge* with no volume, *an extended electron inconceivable in special theory of relativity due to intrinsic connection between space and time*, Lorentz *model of electron as possessing mass on account of the electromagnetic field around it fails* without further assumptions if the electromagnetic field varies too rapidly or if the acceleration of the electrons is too great, no natural way of introducing further assumptions has been discovered, discovery of neutron introduced a form of mass that is difficult to believe is of electromagnetic nature, theory of the positron in which positive and negative values of an electron play symmetrical roles cannot be fitted into idea of electromagnetic idea of mass which requires all mass to be positive, *departure from electromagnetic theory of mass removes main reason for believing in finite size of the electron*, results in difficulty that the *field* in the immediate neighborhood of the electron has an infinite mass, in quantum mechanics results in divergence in the solution of the equations that describe the interaction of an electron with an electromagnetic field and prevents its application to high-energy radiation processes, *Dirac chose to represent electron by point charge and to avoid the difficulties with the infinite energy by direct omission or subtraction of unwanted terms* as had been used in the *theory of the positron. new theory must be in agreement with special relativity* and the conservation of energy and momentum, first addressed problem of a *single electron moving in an electromagnetic field*, used equations for electromagnetic *potentials* in terms of the *charge-current density* vector, *assumed that the charge-current density vector vanishes everywhere except on the world-line of the electron where it is infinitely great*, derived *field* quantities from electromagnetic *potentials*, obtained solutions in terms of *retarded* and *advanced* potentials and potentials representing the incoming and outgoing radiation, calculated field of radiation produced by the electron, in agreement with classical theory but provided a value for the field of radiation throughout space-time, obtained exact *equations of motion* of the electron in a specified incident *field* from the laws of conservation of energy and momentum within limits of classical theory, imposed condition that solutions that occur in Nature when there is no incident field are those for which velocity in terms of proper time is constant, requires solutions of equations of motion *for which the final acceleration as well as the initial position and velocity are prescribed*, electron responds to pulse of electromagnetic radiation before it reaches the center of the electron, behaves as if it has a radius of order $1/a$ where $a = 3m/e^2$, implies that signal can be transmitted *faster than the speed of light* through the interior of an electron, *interior of electron is a region of failure of some of the elementary properties of space-time*, showed how theory can be extended to any number of electrons interacting with each other and with a field of radiation.

208 **Dancoff, S. M. (May, 1939). On Radiative Corrections for Electron Scattering.** *Phys. Rev.*, 55, 959; https://doi.org/10.1103/PhysRev.55.959; a *relativistic*

treatment of the radiative correction of order $e^2/\hbar c$ to the elastic scattering cross section leads to the following results: (a) for the scattering in an electrostatic field of a particle described by the Pauli-Weisskopf theory, the correction is finite;
(b) for the scattering of a Dirac electron in an electrostatic field, the correction diverges logarithmically and is positive; (c) the convergence or divergence of the correction depends critically on the type of scattering potential considered.

209 **Dirac, P. A. M. (July, 1939). A new notation for quantum mechanics.** *Proc. Camb. Phil. Soc.*, 35, 416-8; https://doi.org/10.1017/S0305004100021162; introduction of *bra ket* notation, quantum mechanics deals with *vectors in Hilbert space*, representing the *states* of a dynamical system, and with *linear operators*, representing dynamical *variables*, sometimes one makes calculations using the *vectors and linear operators* directly treating them as *abstract quantities*, at other times one works with *coordinates* (or *representatives*) of these *quantities*, the *bra ket* notation provides concise way of writing the *abstract quantities* themselves and their *coordinates* in a single scheme, leads to a unification of ideas.

212 **Weisskopf, V. F. (July, 1939). On the Self-energy and Electromagnetic Field of the Electron.** *Phys. R*ev., 56, 1, 72-85; https://doi.org/10.1103/PhysRev.56.72; *the main purpose of this paper is to show the physical significance of the logarithmic divergence of the self-energy of the electron and to demonstrate the reason for its occurrence*, the *self-energy of the electron is its total energy in free space when isolated from other particles or light quanta*, $W = T + (1/8\pi) \int (H^2 + E^2)\, dr$ where T is the *kinetic energy* of the electron and H and E are the *magnetic* and *electric field strengths* at point r, identifies three reasons why quantum theory of the electron results in infinite *self-energy* of the electron, claims that *quantum kinematics shows that the radius of the electron must be assumed to be zero*, resulting in infinite energy of the *electrostatic field*, the contributions of the electric and magnetic fields of the spin to the *self-energy* of the electron cancel one another, *quantum theory of the electromagnetic field postulates the existence of field strength fluctuations in empty space*, gives rise to an additional energy which diverges more strongly that the electrostatic self-energy, induces the electron to perform vibrations with energy that diverges quadratically *for an infinitely small radius*, Dirac's positron theory implies that the charge and magnetic dipole of the electron are extended over a finite region, explains why the *self-energy* is only *logarithmically infinite*, *divergences are consequence of assumption of point electron.*

216 **Hideki Yukawa (January 23, 1907 – September 8, 1981).** 1949 Nobel Prize in Physics.

218 **Yukawa, H. (1942).** *Kagaku* (Science), 7, 249; 8, 282; 9, 322 (in Japanese); unavailable online; attempt at extended space-time description of quantum field theory, non-separability of cause and effect, density matrix, field equation with additional conditions, Lagrangian and Hamiltonian, numerable elementary particles, point-like space-time.

219 **Dirac, P. A. M. (March, 1942.) Bakerian Lecture - The physical interpretation of quantum mechanics.** *Roy. Soc. Proc., A*, 180, 980, 1-40, https://doi.org/10.1098/rspa.1942.0023; describes how a satisfactory *non-relativistic* quantum

mechanics had been established, *Heisenberg method* focuses on quantities which enter into experimental results, connects together in one calculation probability coefficients from all initial states to all final states, *Schrodinger method* connects together in one calculation probability coefficients for transitions from one particular initial state to any final state, both methods rested on same mathematical formalism, generalization of Hamilton form of classical dynamics involving linear operators instead of ordinary algebraic variables, methods of physical interpretation differ, probably are still not finally settled, *the theory is not in agreement with the theory of special relativity* as is evident by the special role played by the time t, works very well in the *non-relativistic* region of low velocities where it appears to be in complete agreement with experiment, can be considered only as an approximation, setting up the mathematical formalism for a *relativistic* theory is fairly straightforward, the classical mechanics needs to be put into a *relativistic* Hamiltonian form taking into account that that the various particles comprising the dynamical system interact through the medium of the electromagnetic field, use Lorentz's *equations of motion* including the dampening terms which express the reaction of radiation, this Hamitonian formulation can then be made into a quantum theory following the procedure from *non-relativistic* quantum mechanics, appears satisfactory mathematically but meets serious difficulties in its physical interpretation, *it results in states of negative energy and negative probability*, it appears that whether one is dealing with particles of integral spin or of half-odd integral spin the mathematical methods at present in use in quantum mechanics *are capable of direct interpretation only in terms of a hypothetical world differing very markedly from the actual one*, all that *relativistic* theory does is provide a consistent means of calculating experimental results.

244 **Heisenberg, W. (July, 1943). Die beobachtbaren Größen in der Theorie der Elementarteilchen. I. (The "observable quantities" in the theory of elementary particles. I.)** *Zeit. Phys.*, 120, 7–10, 513-38; https://doi.org/10.1007/ bf01329800; describes S-matrix theory as a principle of particle interactions, the present work attempts to extract from the conceptual structure of the quantum theory of wave fields those terms that are unlikely to be affected by the future change and which will therefore also form a part of the future theory, it avoids the notion of space and time by replacing it with abstract mathematical properties of the S-matrix, the S-matrix relates the infinite past to the infinite future in one step, without being decomposable into intermediate steps corresponding to time-slices.

246 **Yoshio Nishina (December 6, 1890 – January 10, 1951).**

248 **Shinichiro Tomonaga (March 31, 1906 – July 8, 1979).** 1965 Nobel Prize in Physics.

253 **Tomonaga, S. (1943). On a Relativistically Invariant Formulation of the Quantum Theory of Wave Fields.** *Bull. I. P. C. R. Riken-iho*, 22, 545 (in Japanese); translation Tomonaga, S. (August, 1946). *Prog. Theor. Phys.* 1, 2, 27-42; *existing formalism of quantum field theory not perfectly relativistic*, commutation relations refer to points in space at different times, Schrodinger equation for the vector representing the state of the system is a function of time,

time variable plays a different role than space variables, *probability amplitude* not *relativistically invariant* in the space-time world, follows Dirac (1932) [Relativistic Quantum Mechanics] in *generalizing the notion of probability amplitude as far as is required by the theory of special relativity*, substitutes four-dimensional form of the commutation relations, *generalizes the Schrodinger equation* following the Dirac (1933) [The Lagrangian in Quantum Mechanics] *many-time* formalism, then introduces his *super many-time theory* $[\{H_{12}(P) + h/i\ \partial/\partial C_P\}\ \Psi[C] = 0$ at point P on surface C with infinitely many time variables representing the local time for each position in the space, results in *relativistic interaction representation*, three-dimensional manifold (space-like "surface") in four-dimensional space-time world, not necessary to also assume time-like surfaces for the variable surface as was required by Dirac, previous formalism built up in way too analogous to ordinary *non-relativistic* mechanics, theory was divided into one section giving the kinematical relations between various quantities at the same instant of time and another section determining the causal relations between quantities at different instants of time with the *commutation relations* belong to the first section and the *Schrodinger equation* to the second, this way of separating the theory into two sections very *unrelativistic*, "same instant of time" plays a distinct role, *new formalism consists of one section giving laws of behavior of the fields when they are left alone and the other giving the laws determining the deviation from this behavior due to interactions*, can be carried out *relativistically*, although theory in more satisfactory form no new contents added, *divergence difficulties inherited*, fundamental equations admit only catastrophic solutions *due to non-vanishing zero-point amplitudes of the fields which inheres in the operator $H_{12}(P)$, a more profound modification of the theory is required in order to remove this fundamental difficulty*.

269 **Dirac, P. A. M. (April, 1945). On the Analogy Between Classical and Quantum Mechanics.** *Rev. Mod. Phys*., 17, 195; https://doi.org/10.1103/RevModPhys.17.195; mathematical methods available for working with *non-commuting quantities* much weaker than those available for *commuting quantities* owing to the fact that the only functions of *non-commuting variables* that one has been able to define are those expressible algebraically, shows how this difficulty can be avoided in the case when the *non-commuting quantities* are *observables*, can set up a theory of functions of them (a generalization of the concept known as *well-ordered functions*) of almost the same degree of generality as the usual functions of *commuting variables*, can use this theory to make closer analogy between classical and quantum mechanics, enables discussion of trajectories for motion of a particle in quantum mechanics, method is given for defining *general functions of non-commuting observables*, method is developed to provide formal *probability* for *non-commuting observables* to have numerical values (in general a complex number), also enables the analogy between classical and quantum *contact transformations* to be set up on a more general basis.

276 **Richard Phillips Feynman (May 11, 1918 – February 15, 1988).** 1965 Nobel Prize in Physics.

287 **Wheeler, J. A. & Feynman R. P. (April, 1945). Interaction with the Absorber as the Mechanism of Radiation.** *Rev. Mod. Phys.*, 17, 157-81; https://doi.org/10.1103/RevModPhys.17.157; the motive of the analysis was to clear the present quantum theory of interacting particles of those of its difficulties which have a purely classical origin, method was to define as closely as one can within the bounds of classical theory the proper use of the *field* concept in the description of nature, this paper represents the third part of the survey, an analysis of the *mechanism of radiation* believed to complete the last tie between *action at a distance* and *field theory*, only section now finished, difficulties to obtain a satisfactory account of the *field* generated by an accelerated charge at a remote point and to understand the source of the force experienced by the charge itself as a result of its motion, takes up suggestion by Tetrode (1922) that *the act of radiation should have some connection with the presence of an absorber*, develops this idea into the thesis that the force of radiative reaction arises from the action on the source owing to the half-advanced fields of the particles of the absorber, absorber response as the mechanism of radiative reaction, expresses in terms of *action at a distance*, *Wheeler–Feynman absorber theory, assumes solutions of the electromagnetic field equations* must be invariant under time-reversal transformation as are the field equations themselves, considers an accelerated charge located in the absorbing system as the source of radiation, *elementary particles not self-interacting*, a complete correspondence is established between *action at a distance* and the usual formulation of *field theory* in the case of a completely absorbing system.

302 **Lamb, Jr., W. E. & Retherford, R. C. (August, 1947). Fine Structure of the Hydrogen Atom by a Microwave Method.** *Phys. Rev.*, 72, 3, 241-3; https://doi.org/10.1103/PhysRev.72.241; the spectrum of hydrogen has a fine structure of the energy levels which according to the *Dirac wave equation* for an electron moving in a Coulomb field is due to the combined effects of *relativistic* variation of mass with velocity and spin-orbit coupling, according to this theory the $2^2S_{1/2}$ state should exactly coincide in energy with the $2^2P_{1/2}$ state which is the lower of the two P states, previous attempts at measurement have alternated between finding confirmation and discrepancies of as much as eight percent, using microwave method depending on a novel property of the $2^2S_{1/2}$ level results indicate that contrary to the *Dirac wave equation* the $2^2S_{1/2}$ state is higher than the $2^2P_{1/2}$ by about 1000 Mc/sec.

306 **Hans Albrecht Bethe (July 2, 1906 – March 6, 2005).** 1967 Nobel Prize in Physics.

314 **Bethe, H. A. (August, 1947). The Electromagnetic Shift of Energy Levels.** *Phys. Rev.*, 72, 339-41; https://doi.org/10.1103/PhysRev.72.339; Lamb and Retherford results show that fine structure of second quantum state of hydrogen does not agree with *Dirac wave equation*, Schwinger, Weisskopf, and Oppenheimer suggest might be due to *shift of energy levels by interaction of electron with the radiation field*, this shift comes out infinite in all existing theories and has therefore always been ignored, possible to identify the most strongly (linearly) divergent term in the level shift with an *electromagnetic mass effect* which must exist for a bound as well as for a free electron, already included in the *observed mass* of the electron so should

be subtracted, assumes *relativistic cut-off* in quantum energies (frequencies) of included atomic states, then calculation of Lamb shift for hydrogen atom using *non-relativistic* ordinary radiation theory gives shift of the levels due to *radiation interaction* in close agreement with observed value, removes discrepancy with Dirac theory, did not carried out *relativistic* calculations.

318 **Tati, T. & Tomonaga, S. (December, 1948). A Self-Consistent Subtraction Method in the Quantum Field Theory, I.** *Prog. Theor. Phys.*, 3, 4, 391-406; https://doi.org/10.1143/ptp/3.4.391; notes procedure previously used by Bloch and Nordsieck and then by Pauli and Fierz in the treatment of problems with the self-field of an electron by first separating the radiation field into part bound to the electron and part of unbound photons by means of a canonical transformation, they then obtained a term in the Hamiltonian that could be interpreted as the *interaction energy* of electron with the radiation field bound to it, this term though infinite gave rise to a modification of mass of electron so that it could be amalgamated into the mass term in the Hamiltonian for the free electron, the term was dropped off to obtain *observed mass* when electron mass reinterpreted as already including it, same result as Bethe for level-shift of bound electron in *non-relativistic* treatment, Tati and Tomonaga considered it desirable to obtain *relativistic* generalization of this treatment using a canonical transformation, paper addresses *field reaction problem when no external field* to e^2 approximation, does not address infinity related to *vacuum polarization* effect that occurs in *relativistic* treatment, *omitted as cannot be amalgamated into equation for free radiation*, starts from *relativistic* formalism of quantum field theory proposed by Tomonaga [August, 1946. On a Relativistically Invariant Formulation of the Quantum Theory of Wave Fields] in which Schrodinger equation with Hamiltonian for interaction density between radiation and electron fields is $\{H(P) - i\ \partial/\partial C_P\}\ \Psi[C] = 0$ where H(P) is the *interaction energy density* between radiation and electron fields at the world point P, Ψ[C] is the generalized Schrodinger functional which is a functional of the space-like variable surface C in the four-dimensional world and $\partial/\partial C_P$ its partial functional differentiation at the point P, the point P being considered as lying on C, decomposes fields into parts oscillating with positive and negative frequencies corresponding to the electron and positron, transforms Schrodinger functional retaining terms up to the order e^2, expands integrand into Fourier integral and defines 4-vector to obtain Hamiltonian for transformed equation that enables use of first order calculation, allows processes through intermediate states such as emission and reabsorption of virtual particles (self-energy of an electron) to be treated as direct processes, calculates the commutator and rearranges factors in each term into the correct order, identifies (1) terms to represent *interactions between electrons* caused by exchange of virtual photons including ordinary M¢ller interaction between two electrons (2) terms connected with *self-energy* of an electron (3) terms responsible for the *scattering of a photon* by a free electron, terms for *creation of a pair* by two photons (4) variables describing the radiation field only representing modification of the radiation field in vacuo due to the *polarization of the vacuum* that cause an infinite energy level shift of the vacuum itself and an infinite *self-energy* of a photon, and (5) a *mass-modifying term* with a logarithmic diverging quantity representing the electromagnetic mass of an electron, assumes

mass-modifying term already included in free field equation so no level-shift caused by the interaction between the electron and radiation, subtraction method "*self-consistent*" in this sense, gives *relativistic* generalization of transformation that separates *radiation field* into field of "unbound" photons and field of photons bound to the electron, procedure similar to Hartree method of the *self-consistent* field in which *interaction* between electrons is considered as a perturbation but some part of its effect is already included.

331 **Koba, Z. & Tomonaga, S. (September, 1948). On Radiation Reactions in Collision Processes. I: Application of the "Self-Consistent" Subtraction Method to the Elastic Scattering of an Electron.** *Progr. Theor. Phys.*, 3, 3, 290-303; https://doi.org/10.1143/ptp/3.3.290; applies Tati and Tomonaga's "self-consistent" subtraction method to the elastic scattering of an electron by a fixed electrostatic potential, *formal infinity associated with mass-modifying term attributed to defect of current theory so empirical value substituted for theoretical value* (assuming it was already included in the free field equation), maintains total Hamilton describing the interaction of the electron and electromagnetic fields unaltered, requires interaction-energy part of the Hamiltonian to undergo corresponding change by inclusion of "counter-self-energy" term ("mass-type" correction to scattering cross-section), results in finite value for self-energy of electron in the e^2 approximation, when new formalism applied to elastic scattering of electron *effective cross-section* for scattering by a fixed potential in zeroth approximation has value in good agreement with experiment but as soon as reaction of electromagnetic field with electron is taken into account the correction becomes infinite, first infinite term in modified Hamiltonian eliminated using *subtraction hypothesis of positron theory*, second difficulty disappears by applying *"self-consistent" subtraction method* using modified Hamiltonian, first term can also be eliminated if one interprets the sum of the external potential and its infinite correction due the *vacuum polarization* effect as the physically observable potential and *substitute the empirical value for it* whilst interaction part of the Hamiltonian supplemented by an additional "counter-vacuum polarization" term, *method by no means give the real solution of the fundamental difficulty of the quantum electrodynamics* but reveals nature of various diverging terms and reduces them to two quantities - the *self-energy* and the *vacuum polarization*, in this way *it becomes possible in an unambiguous and consistent manner to treat the field reaction problem without touching the fundamental difficulty by employing the finite empirical values instead of the infinite "theoretical" values for these two quantities.*

340 **Koba, Z. & Takeda, G. (December, 1948). Radiation Reaction in Collision Process, II: Radiative Corrections for Compton Scattering.** *Prog. Theor. Phys.*, 3, 4, 407-21; https://doi.org/10.1143/ptp/3.4.407; previous paper applied Tati & Tomonaga's *"self-consistent" subtraction method* to *elastic scattering of an electron*, all divergences were eliminated by counter-terms in modified form of the Hamilton function, but this was a simple and particularly favorable case, need to examine whether method is still effective in fundamental processes between elementary particles, applies method to calculate e^2-corrections to *Klein-Nishina formula for Compton scattering* using *perturbation method,* two types of diverging

terms, one related to *polarization of vacuum* and other to *self-energy* of electron, sufficient to add two new terms to Hamiltonian function to cancel out these infinite terms, *self-energy* logarithmic divergence eliminated by introducing new term into interaction Hamiltonian derived in a plausible way as a "counter-term" which compensates change in free field Hamiltonian and conserves total Hamiltonian unaltered, one of *vacuum polarization* terms could be treated in similar way, while other term could not be foisted into the theory without radically changing the Maxwell equation for the free electromagnetic field, *far from final settlement of the fundamental problem in theory of elementary particles.*

348 **Koba, Z. & Takeda, G. (June, 1948). Radiative Corrections in e^2 for an Arbitrary Process Involving Electrons. Positrons, and Light Quanta.** *Prog. Theor. Phys.*, 3, 2, 203-5, letter to the editor [Referenced in Dyson (February, 1949). The Radiation Theories of Tomonaga, Schwinger, and Feynman, but full account in later issue of this journal not found online, nor were any references found]; https://doi.org/10.1143/ptp/3.2.203; investigated further the *radiation correction in collision processes between electrons, positrons and photons* in a general manner and confirmed modified Hamiltonian sufficient to eliminate divergence difficulty *as long as we are concerned with the first radiative correction, introduces "transition diagram method"* [similar to the Stueckelberg (1934) diagrams, later introduced by Feynman at Ponoco in spring 1948 and published in September 1949, and subsequently renamed *Feynman diagrams*] *to analyze complicated connection between initial and final states through a number of intermediate ones*, expresses *electrons* and *positrons* as *world lines* in *momentum space*, while *emission* and *absorption* of *photons* are "*leaps*" of these *world lines*, diagram acquires two additional *leaps* (the *emission* and *reabsorption* of a *virtual photon*) when *radiative corrections* are taken into account, two distinct ways of attaching them, existing *world lines* gain new *leaps* or additional closed *world line* is introduced and coupled with existing ones through *virtual photon*, first case implies "*mass-type*" or "*self-energy-type*" divergences when *emission* and *reabsorption* of the *virtual photon* take place in succession which can be eliminated by *counter-self-energy term*, second case represents *vacuum polarization* corrections which can be subdivided according to number of *leaps* in additional closed *world line*, which afford quadratic and logarithmic divergences that are cancelled by the "*counter-terms*" and found not to contribute to divergence.

352 **Julian Seymour Schwinger (February 12, 1918 – July 16, 1994).** 1965 Nobel Prize in Physics.

357 **Schwinger, J. (February, 1948). On quantum-electrodynamics and the magnetic moment of the electron.** *Phys. Rev.*, 73, 4, 416-7; https://doi.org/10.1103/PhysRev.73.416; *electrodynamics unquestionably requires revision at ultra-relativistic energies,* desirable to isolate those aspects of the current theory that essentially involve high energies and are subject to modification by a more satisfactory theory, this goal has been achieved by transforming the Hamiltonian of current *hole theory* electrodynamics to exhibit explicitly the logarithmically divergent self-energy of a free electron which arises from the virtual emission and absorption of light quanta, the electromagnetic *self-energy* of a free electron can be

ascribed to an *electromagnetic mass* which must be added to the mechanical mass of the electron, new Hamiltonian involves experimental *electron mass* rather than unobservable *mechanical mass*, electron now interacts with radiation field only in presence of external field such that only an accelerated electron can emit or absorb a light quantum, *interaction energy* of electron with external field subject to *finite* radiative correction, *polarization of the vacuum* still produces logarithmically divergent term proportional to *interaction energy* of electron in an external field, such term equivalent to altering value of *electron charge* by constant factor with only final value being identified with experimental charge, interaction between matter and radiation produces *renormalization* of electron charge and mass, all divergences contained in *renormalization* factors, radiative correction for energy of electron in external magnetic field corresponds to *additional magnetic moment associated with electron spin* of magnitude $\delta\mu/\mu = (½\ \pi)e^2/hc = 0.001162$, experimental measurements on hyperfine splitting of ground states of atomic hydrogen and deuterium larger than expected from directly measured nuclear moments, finds additional *electron spin magnetic moment* to account for measured hydrogen and deuterium *hyperfine structures* to be $\delta\mu/\mu = 0.00126$ and $\delta\mu/\mu = 0.00131$ respectively, these discrepancies accounted for by additional spin magnetic moment to the electron of $\delta\mu/\mu = 0.0018 \pm 0.00003$, *values yielded by relativistic calculation of Lamb shift differ only slightly from those conjectured by Bethe on basis of non-relativistic calculation and are in good accord with experiment.*

361 **Feynman, R. P. (April, 1948). Space-Time Approach to Non-Relativistic Quantum Mechanics.** *Rev. Mod. Phys.*, 20, 2, 367-87; https://doi.org/10.1103/RevModPhys.20.367; third formulation of *non-relativistic* quantum theory in addition to *differential equation of Schroedinger* and *matrix algebra of Heisenberg*, *path integral formulation* utilizing the *action* principle as suggested in Dirac (1933) [The Lagrangian in Quantum Mechanics] and Dirac (1945) [On the Analogy Between Classical and Quantum Mechanics], in quantum mechanics the probability of an event which can happen in several different ways is the absolute square of a sum of complex contributions one from each alternative way $\varphi_{ac} = \sum_b \varphi_{ab}\varphi_{bc}$ where φ_{ab}, φ_{bc}, φ_{ac} are complex numbers such that $P_{ab} = |\varphi_{ab}|^2$, $P_{bc} = |\varphi_{bc}|^2$, and $P^q{}_{ac} = |\varphi_{ac}|^2$, and P_{ab} as the probability that if measurement A gave the result *a*, then measurement B will give the result *b*, and $P^q{}_{ac}$ is the quantum mechanical probability that a measurement of C results in c when it follows a measurement of A giving a, the probability that a particle will be found to have a path lying somewhere within a region of space time is the absolute square of a sum of contributions - one from each path in the region, the contribution from a single path is postulated to be an exponential whose (imaginary) phase is the classical *action* for the path in question where *action* refers to time integral of Lagrangian along a path, restriction to finite time interval, the total contribution from all paths reaching x, t from the past is the *wave function* $\psi(x, t)$, shown to satisfy Schroedinger's equation, *probability amplitude for a space-time path associated with entire motion of particle as function of time rather than with position of particle at particular time*, establishes postulates that describe *non-relativistic quantum mechanics neglecting spin*, mathematically equivalent to Heisenberg and Schroedinger formulations, no fundamentally new results, *suffers serious*

drawbacks, requires unnatural and cumbersome division of the time interval, not formulated so that it is physically obvious that it is invariant under unitary transformations, improvements could be made through use of notation and concepts of mathematics of functionals.

391 **Dirac, P. A. M. (May, 1948). Quantum Theory of Localizable Dynamical Systems.** *Phys. Rev.*, 73, 9, 1092-103; https://doi.org/10.1103/PhysRev.73.1092; a dynamical system is called *localizable* if its *wave functions* can be expressed in terms of variables, each referring to physical conditions at only one point in *space-time*. These variables may be at points on any three-dimensional space-like surface in space-time. A general investigation is made of how the *wave function* varies when the surface is varied in any way. The variation of the *wave function* is given by equations of the Schrödinger type involving certain operators which play the role of Hamiltonians. The *commutation relations* for these operators are obtained. The theory works entirely with *relativistic* concepts and it provides the general pattern which any *relativistic* quantum theory must conform to, provided the dynamical system is *localizable*.

392 **Feynman, R. P. (October, 1948). A Relativistic Cut-Off for Classical Electrodynamics.** *Phys. Rev.*, 74, 939-946; https://doi.org/10.1103/PhysRev.74.939; *in this paper a consistent classical theory is described which the author believes can be quantized*, previous attempts to address *problem of infinite self-energy* that results from assuming *point electron* in *relativistic* theory met with considerable difficulties when attempt made to quantize them, the *potential* at a point in space at a given time depends on the charge at a distance r from the point at a time previous by t = r (taking the speed of light as unity), *relativistically* interaction occurs between events whose four-dimensional interval s defined by $s^2 = t^2 - r^2$ vanishes, this results in *infinite action of a point electron on itself, the present theory modifies this idea by assuming that substantial interaction exists as long as the interval s is time-like and less than some small length a of order of the electron radius*, reduces infinite *self-energy* to a finite value for accelerations which are not extreme, *action* of an electron on itself appears as *electromagnetic mass*, formulates in terms of *action at a distance*, satisfies Maxwell's equations, not the usual *retarded* solution for which there is no *self-force* but *half the retarded plus half the advanced solution*, effect of modification is to change slightly the field of one particle on another when they are very close and to add a *self-force*, *little reason to believe that the ideas used here to solve the divergences of classical electrodynamics will prove fruitful for quantum electrodynamics*.

399 **Feynman, R. P. (November, 1948). Relativistic Cut-Off for Quantum Electrodynamics.** *Phys. Rev.*, 74, 1430-8; https://doi.org/10.1103/PhysRev.74.1430; describes model based on quantization of a classical theory for which all quantities automatically come out finite described in his previous paper [Feynman (October, 1948). A Relativistic Cut-Off for Classical Electrodynamics], contains an arbitrary function on which numerical results depend, only term that depends significantly (logarithmically) on the cut-off frequency is the *self-energy* which can be used to *renormalize* the electron mass, remaining terms are nearly independent of the function, applies only to results for processes in which virtual quanta are

emitted and absorbed, *terms representing processes involving a pair production followed by annihilation of the same pair are infinite and not made convergent by the present scheme,* problems of *permanent emission* and the position of *positron theory* need to be addressed, *the present paper may be looked upon as presenting an arbitrary rule to cut off at high frequencies in a relativistically invariant manner the otherwise divergent integrals appearing in quantum field theories*, produces finite invariant *self-energy* for a free electron, but problem of *polarization of the vacuum* not solved, alternative cut-off procedure which eliminates high frequency intermediate states offers to solve *vacuum polarization* problems as well.

410 **Schwinger, J. (November, 1948). Quantum Electrodynamics. I. A Covariant Formulation.** *Phys. Rev.*, 74, 10, 1439-61; https://doi.org/10.1103/PhysRev.74.1439; *lack of convergence in current formulations of quantum electrodynamics indicates that revision of electrodynamic concepts at ultra-relativistic energies is necessary*, elementary phenomenon in which divergences occur as a result of virtual transitions involving particles with unlimited energy are *polarization of the vacuum* and *self-energy of the electron* which express *the interaction of the electromagnetic and matter fields with their own vacuum fluctuations*, this alters the constants characterizing the properties of the individual fields and their mutual coupling by infinite factors, the question is whether all divergencies can be isolated in such unobservable *renormalization* factors, *this paper is occupied with the formulation of a completely covariant electrodynamics*, manifest covariance with respect to Lorentz and gauge transformations essential in a divergent theory, customary *canonical commutation relations* fail to exhibit the desired covariance since they refer to field variables at equal times and different points of space, *can be put in covariant form by replacing the four-dimensional surface t = const. by a space-like surface*, offers the advantage over the Schrodinger representation in which all operators refer to the same time providing distinct separation between *kinematical* and *dynamical* aspects, formulation that retains evident covariance of the Heisenberg representation but offers something akin to Schrodinger representation can be based on distinction between the properties of *non-interacting fields*, and the effects of *coupling between fields*, constructs a *canonical transformation* that changes the *field equations* in the *Heisenberg representation* into those of *non-interacting fields*, *supplementary condition* restricting the admissible states of the system and the *commutation relations* must be added to the *equations of motion*, describes the coupling between fields in terms of a varying state vector, then simple matter to evaluate commutators of *field* quantities at arbitrary *space-time* points, one thus obtains an obviously covariant and practical form of quantum electrodynamics expressed in a mixed Heisenberg-Schrodinger representation called the *interaction representation*, discusses *covariant* elimination of longitudinal field in which customary distinction between longitudinal and transverse fields is replaced by a suitable *covariant* definition, describes collision processes in terms of an invariant *collision operator* which is the unitary operator that determines the over-all change in state of a system as the result of interaction, notes that a *second paper* treats the problems of electron and photon *self-energy* together with the *polarization of the vacuum* and a *third paper* is concerned with the determination of the *radiative corrections* to the properties of

an electron and the comparison with experiment [this was not addressed, it stated that "*radiative corrections to energy levels* will be treated in the next paper of the series" but *this did not appear, nor are there any references to it*].

437 **Freeman John Dyson (December 15, 1923 – February 28, 2020).**

441 **Dyson, F. J. (February, 1949). The Radiation Theories of Tomonaga, Schwinger, and Feynman.** *Phys. Rev.*, 75, 3, 486-502; https://doi.org/10.1103/PhysRev.75.486; the recent and independent formulations of quantum electrodynamics by Tomonaga, Schwinger, and Feynman have made two notable advances, the foundations and applications of the theory have been simplified by being presented in a *completely relativistic way* and the divergence difficulties *have been at least partially overcome*, the advantages of the Feynman formulation are simplicity and ease of application while those of Tomonaga-Schwinger are generality and theoretical completeness, this paper presents a unified development of quantum electrodynamics embodying the main features of Tomonaga-Schwinger and Feynman radiation theories, *emphasis on application of the theory*, aims to show how the Schwinger theory can be applied to specific problems in such a way as to incorporate the ideas of Feynman, the main results are general formulas from which radiative reactions on motions of electrons can be calculated, divides *energy-density* into two parts, *energy of interaction* of two fields with each other and energy produced by external forces, *interaction energy* alone is treated as a perturbation, important results of the paper are the *equation of motion* $ihc[\partial\Omega/\partial\sigma(x_0)] = \{S(\sigma)\}^{-1}H^e(x_0)S(\sigma)\Omega$ for the *state vector* $\Omega(\sigma)$ and the interpretation of the *state vector* Ω, simplifies Schwinger theory for using it for calculations, demonstrates equivalence of the theories within their common domain of applicability, in the *Schwinger theory* the aim is to calculate the matrix elements of the "*effective external potential energy*" between *states* specified by their *state vectors*, in the *Feynman theory* the basic principle is to "preserve symmetry between past and future" so the matrix elements of the operator are evaluated in a "*mixed representation*" in which the matrix elements are calculated between an *initial state* specified by its *state vector* and a *final state* specified by its *state vector,* a *graph* corresponding to a particular matrix element is used not merely as an aid to calculation but as a picture of the physical process which gives rise to that matrix element, derives fundamental formulas for the operator in the *equation of motion* for the *state vector* $\Omega(\sigma)$ which represents the interaction of a physical particle with an external field for both the Schwinger and the Feynman theories, derives set of rules by which matrix element of Feynman operator may be written down in form suitable for numerical evaluation, shows equivalence of the two theories, develops graphical representation of matrix elements, *the theory as a whole cannot be put into a finally satisfactory form so long as divergencies occur in it however skillfully these divergencies are circumvented*, present treatment should be regarded as justified by its success in applications rather than by its theoretical derivation, *paper suffers from a series of significant errors*.

470 **Fukuda, H., Miyamoto, Y. & Tomonaga, S. (March, 1949). A Self-Consistent Subtraction Method in the Quantum Field Theory. II-1.** *Prog. Theor. Phys. (Kyoto)*, 4, 1, 47-59; https://doi.org/10.1143/PTP.4.47; (June, 1949). *Idem.*, II-2, 24, 2, 121-9; https://doi.org/10.1143/ptp/4.2.121; Tati and Tomonaga (December 1948) proposed a *subtraction procedure* to be used in the treatment of quantum-theoretical problems involving infinite field reactions, this consists in generalizing the method of *canonical transformation*, and by means of this transformation the electron and the radiation fields are separated into two parts, one bound with each other and the other unbound and propagating freely in the absence of an external field, then the infinite energies related to the infinite *self-energy* of the free electron and to the *vacuum polarization* of the radiation field are separated as the interaction energies between bound parts of the fields, then *these infinite terms in the energy are dropped off considering that only the remaining finite terms have physical meaning*, when an *external field is present* the unbound fields no longer propagate freely and transitions take place, the electron wave is able to change its state of propagation not only elastically by the external field but also by emitting or absorbing unbound photons, thus *an interaction appears between electron and radiation* which were free from interaction with each other in the absence of the external field, this interaction causes a radiation reaction upon the electron so that the motion of the electron will be modified, *in this paper we give an example of how one can calculate this reaction and obtained a finite result as the consequence of our subtraction procedure*, we are able to obtain a *finite radiative level-shift* of a bound electron in the external field and a *finite e^2-correction* to the *scattering cross section*, estimated Lamb shift for hydrogen atom as 1076 Mcycles by adding additional terms to Bethe's value, we have thus successfully obtained finite answers for these field reaction problems and they are of the magnitude agreeing with experimental results, but *we must nevertheless confess that the calculation carried out in this paper is still unsatisfactory because we had to make a non-relativistic treatment in the evaluation of the effective energies,* our work is therefore only of a provisionary character, *still problematic whether this procedure corresponds to the correct prescription.*

483 **Kanesawa, S, & Koba, Z. (September, 1949). A Remark on Relativistically Invariant Formulation of the Quantum Field Theory.** *Prog. Theor. Phys.*, 4, 3, 297-311; https://doi.org/10.1143/ptp/4.3.297; new formalism of quantum field theory more satisfactory from stand-point of *relativistic invariance*, formulated in terms of *invariant space-time* concepts but employed *canonical formalism* in which time variable unnecessarily distinguished from space variables, *the purpose of the present paper is to show that the same formalism can be reached without referring to the canonical formulation*, in current theory the *interaction Hamilton density* coincides with *interaction Lagrange density* except the sign if surface-dependent part neglected, suggests to connect *generalized interaction Lagrange density* directly to generalized Schrodinger equation, guiding principle the integrability of the Tomonaga-Schwinger equation, shows how to apply new method to the general case when canonical description is impossible, *problematic aspects of theory such as universal length or ultraviolet divergencies were not addressed*, aim of paper

lies in demonstrating the possibility of including more general kinds of interactions in field theory than those which allow the canonical description.

489 **Kroll, N. M. & Lamb, Jr., W. E. (February, 1949). On the self-energy of a bound electron.** *Phys. Rev.*, 75, 3, 388-98; https://doi.org/10.1103/PhysRev.75.388; calculation of electromagnetic shift of energy levels of bound electron based on usual formulation of *relativistic* quantum electrodynamics and positron theory gave 1052 megacycles per second for $^2 2S_{1/2} - ^2 2P_{1/2}$ shift in hydrogen in close agreement with the *non-relativistic* calculation of 1040 megacycles per second by Bethe.

490 **French, J. B. & Weisskopf, V. F. (April, 1949). The Electromagnetic Shift of Energy Levels.** *Phys. Rev.*, 75, 8, 1240-8; https://doi.org/10.1103/PhysRev.75.1240; *relativistic* calculation of Lamb shift using conventional form of perturbation theory and removing infinite *self-energy* of the electron by subtracting "mass operator" from Hamiltonian gives 1051 megacycles per second for $2s_{1/2} - 2p_{1/2}$ separation in hydrogen, compared with *non-relativistic* calculation of 1040 megacycles per second by Bethe, and *relativistic* calculation of 1052 megacycles per second by Kroll & Lamb, results of the different calculations suggest that they are not dependent on whether they are *relativistic* or *non-relativistic*.

492 **Schwinger, J. (February, 1949). Quantum Electrodynamics. II. Vacuum Polarization and Self-Energy.** *Phys. Rev.*, 75, 4, 651-79; https://doi.org/10.1103/PhysRev.75.651; *interaction representation* is applied to *polarization of the vacuum* and the *self-energies of the electron and photon*, in *first section* the vacuum of the non-interacting *electromagnetic* and *matter* fields is *covariantly* defined as *state* for which the eigenvalue of an arbitrary time-like component of the *energy-momentum four-vector* is an absolute minimum, covariant decomposition of field operators into positive and negative frequency components introduced to characterize vacuum *state vector*, shows that *state vector for electromagnetic vacuum* annihilated by positive frequency part of transverse four-vector potential and *state vector for matter vacuum* annihilated by positive frequency part of Dirac spinor and its charge conjugate, these properties of vacuum *state vector* employed in the calculation of the *vacuum expectation* values of quadratic field quantities, specifically the *energy-momentum tensors* of the independent *electromagnetic* and *matter fields* and the *current four-vector*, infers that the *electromagnetic energy-momentum tensor* and *current vector* must vanish in the vacuum, while the *matter field energy-momentum tensor* vanishes in the vacuum only by the addition of a suitable multiple of the unit tensor, *second section treats the induction of a current in the vacuum by an external electromagnetic field*, supposes that *external electromagnetic field* does not produce actual electron-positron pairs, considers only phenomenon of *virtual pair creation*, restriction is introduced by requiring that establishment and subsequent removal of the external field produce no net change in state for the *matter field*, *demonstrates that the induced current at a given space-time point involves the external current in the vicinity of that point and not the electromagnetic potentials*, this *gauge invariant* result shows that a light wave propagating at remote distances from its source induces no current in the vacuum and is therefore undisturbed in its passage through space, indicates *absence of a*

light quantum self-energy effect, *current* induced at a point consists of two parts, a logarithmically divergent multiple of the *external current* at that point *which produces an unobservable renormalization of charge* and a more involved finite contribution which is the physically significant *induced current*, *third section considers the modification of the matter field properties arising from interaction with the vacuum fluctuations of the electromagnetic field*, analysis carried out with two alternative formulations, one employing the complete *electromagnetic potential* together with a *supplementary condition*, the other using the *transverse potential* with the variables of the *supplementary condition* eliminated, no real processes produced by first order coupling between the fields, alternative *equations of motion* for the *state vector* are constructed from which the first order interaction term has been eliminated and replaced by the *second order coupling* which it generates, this includes the *self-action* of individual particles and light quanta, the *interaction* of different particles, and a *coupling* between particles and light quanta which produces such effects as Compton scattering and two quantum pair annihilation, concludes from comparison of the alternative procedures that *for the treatment of virtual light quantum processes the separate consideration of longitudinal and transverse fields is an inadvisable complication*, light quantum *self-energy* term is shown to vanish, while that for a particle has the form for a change in *proper mass* but is logarithmically divergent in agreement with previous calculations, identification of *self-energy* effect with a change in *proper mass* is confirmed by removing this term from the state vector *equation of motion*, alters the *matter field equations of motion* in the expected manner, verifies that *the energy and momentum modifications produced by self-interaction effects are entirely accounted for by the addition of the electromagnetic proper mass to the mechanical proper mass—an unobservable mass renormalization*, an appendix is devoted to the construction of several invariant functions associated with the *electromagnetic* and *matter* fields.

495 **Dyson, F. J. (June, 1949). The S Matrix in Quantum Electrodynamics.** *Phys. Rev.*, 75, 1736-55; https://doi.org/10.1103/PhysRev.75.1736; the covariant *quantum electrodynamics* of Tomonaga, Schwinger and Feynman is used as basis for a general treatment of scattering problems involving electrons, positrons, and photons, addresses relation between Schwinger and Feynman theories when restriction to one-electron problems is removed, in these more general circumstances the two theories appear as complementary rather than identical, *the Feynman method is essentially a set of rules for the calculation of the elements of the Heisenberg S matrix* corresponding to any physical process and can be applied directly to all kinds of scattering problems, *the Schwinger method evaluates radiative corrections by exhibiting them as extra terms appearing in the Schrodinger equation*, *practical usefulness of S matrix as connecting link between the Feynman technique of calculation and Hamiltonian formulation of quantum electrodynamics*, the *Feynman radiation theory* provides a set of rules for the calculation of matrix elements between *states* composed of any number of ingoing and outgoing free particles, it is thus an S matrix theory, scattering processes including the creation and annihilation of particles are completely described by Heisenberg's S matrix, the elements of this matrix are calculated by a consistent

use of perturbation theory to any desired order, detailed rules given for carrying out such calculations, divergences arising from higher order radiative corrections removed from S matrix by consistent use of *mass and charge renormalization, the operators so calculated are divergence-free, the divergent parts at every stage of the calculation being explicitly dropped after being separated from the finite parts*, must be justified *a posteriori* by the fact that they ultimately lead to a clear separation of finite from infinite expressions, *such an a posteriori justification of dubious manipulations is an inevitable feature of any theory which aims to extract meaningful results from not completely consistent premises*, the perturbation theory of this paper is *applicable only to a restricted class of problems* when not only the radiation interaction but also the external potential is small enough to be treated as a perturbation, *does not give a satisfactory approximation either in problems involving bound states or in scattering problems at low energies,* in other situations the Schwinger theory will have to be used in its original form, *problems of extending treatment to include bound-state phenomena and of proving convergence of the theory as the order of perturbation itself tends to infinity not addressed*, analysis suggests that *the divergences of electrodynamics are directly attributable to the fact that the Hamiltonian formalism is based upon an idealized conception of measurability*, now no longer a compelling necessity for a future theory to abandon some essential features of the present electrodynamics, *the present electrodynamics is certainly incomplete, but is no longer certainly incorrect*.

517 **Richard John Eden (July 2, 1922 –September 25, 2021).**

518 **Eden, R. J. (September, 1949). Heisenberg's S Matrix for a System of Many Particles.** *Proc. Roy. Soc., Series A*, 198, 1055, 540-559; https://doi.org/10.1098/rspa.1949.0118; this vignette is included as Dr. Eden provided lectures in Nuclear Physics and Quantum Theory to the author in 1963 as part of his M.A. Cantab. degree. Eden's thesis advisors were Dirac and Heisenberg. The author attended a joint lecture on May 23, 1963, by Heisenberg and Dirac (both speaking whilst writing and erasing on dueling pairs of blackboards at the same time) at the old Cavendish Laboratory.

522 **Feynman, R. P. (September, 1949). The Theory of Positrons.** *Phys. Rev.*, 76, 749-59; https://doi.org/10.1103/PhysRev.76.749; this is the first of a set of papers dealing with the solution of problems in *quantum electrodynamics*, main principle is to deal directly with *solutions* of the Hamiltonian differential equations rather than with equations themselves, this paper *addresses motion of electrons and positrons in given external potentials*, second paper considers *interactions* (*quantum electrodynamics*), problem of charges in a fixed potential usually treated by method of second quantization of the electron field using the *theory of holes*, here we replace Dirac's "hole theory" by reinterpreting *solutions* of the Dirac equation, results simplified by *following the charge rather than the particles* because *the number of particles is not conserved* whereas *charge is conserved*, in the approximation of classical *relativistic* theory the creation of an electron pair (electron and positron) represented by the start of two world lines from the point of creation, *following the charge rather than the particles corresponds to considering continuous world line as a whole rather than breaking it up into its pieces*, over-all

space-time point of view, quantum mechanically the direction of the world lines is replaced by the direction of propagation of waves, quite different from Hamiltonian method which considers the future as developing continuously from the past, in a scattering problem this over-all view of the complete scattering process is similar to the *S-matrix view-point of Heisenberg*, temporal order of events during scattering which is analyzed in such detail by the Hamiltonian differential equation is irrelevant, development stemmed from idea that in *non-relativistic* quantum mechanics the *amplitude* for a given process can be considered as the sum of an *amplitude* for each *space-time* path available, in *relativistic* case the restriction that the paths must proceed always in one direction in time is removed, results more easily understood from more familiar physical viewpoint of *scattered waves* used in this paper, after the equations were worked out physically the proof of the equivalence to the *second quantization theory* was found, solution in terms of boundary conditions on *wave function* contains all possibilities of pair formation and annihilation together with the ordinary scattering processes, negative energy states appear in space-time as waves traveling away from the external potential backwards in time (as suggested in Stückelberg (December, 1942)), such a wave corresponds to a positron approaching the potential and annihilating the electron, a particle moving forward in time (electron) in a potential may be scattered forward in time (ordinary scattering) or backward (pair annihilation), when moving backward (positron) it may be scattered backward in time (positron scattering) or forward (pair production), *amplitude* for transition from an initial to a final state analyzed to any order in the *potential* by considering it to undergo a sequence of such scatterings, *amplitude* for a process involving many such particles is the product of the *transition amplitudes* for each particle, vacuum problems do not arise for charges which do not interact with one another, shows equivalent to the *theory of holes* in second quantization.

538 **Feynman, R. P. (September, 1949). Space-Time Approach to Quantum Electrodynamics.** *Phys. Rev.*, 76, 6, 769-89; https://doi.org/10.1103/PhysRev.76.769; in previous paper motion of electrons *neglecting interaction* was analyzed by dealing directly with the *solutions* of the Hamiltonian time differential equations rather than the equations, here the same technique is applied to include *interactions* to express in simple terms the solution of problems in *quantum electrodynamics* rather than the differential equations from which they come, (1) it is shown that *a considerable simplification can be attained by writing down matrix elements for complex processes in electrodynamics*, a physical point of view is available which permitted them to be written down directly. The simplification results from the fact that previous methods separated into individual terms processes that were closely related physically, (2) *electrodynamics is modified by altering the interaction of electrons at short distances*, all matrix elements now finite with the exception of those relating to *vacuum polarization*, latter evaluated in manner suggested by Pauli and Bethe to give finite results, phenomena directly observable insensitive to the details of the modification, electrodynamics viewed as direct *interaction* at a distance between *charges* rather than behavior of a *field* (Maxwell's equations), *field point of view*, which separates the production and absorption of light, most practical for problems involving *real quanta* while *interaction view* best for

discussion of *virtual quanta* when dealing with close collisions of particles or their actions on themselves, Hamiltonian method not well adapted to represent direct *action at a distance* between *charges* because action is delayed, forces use of field viewpoint rather than interaction viewpoint, for *collisions* much easier to treat the process as a whole, effects of longitudinal and transverse waves can be combined, begins with solution in *space* and *time* of Schrodinger equation for particles interacting instantaneously, solution is expressed in terms of *matrix elements, derived from Lagrangian form of quantum mechanics*, then modifies in accordance with the requirements of the Dirac equation and phenomenon of pair creation, made easier by reinterpreting the *theory of holes*, generalizes to *delayed interactions* of *relativistic* electrons, for practical calculations expressions developed in a power series of $e^2/\hbar c$, *derivation will appear in a separate paper*, by forsaking Hamiltonian method, the wedding of *relativity* and *quantum mechanics* accomplished most naturally, *relativistic* invariance becomes self-evident but the *matrix elements for complex processes* and the *self-energy diverges* then can see how the matrix elements can be written down directly, then introduces modification in interaction between charges at short distances, *assumes substantial interaction exists as long as the four-dimensional interval is time-like and less than some small length of order of the electron radius* [Feynman (October, 1948). A Relativistic Cut-Off for Classical Electrodynamics.], *convergence factors* then introduced such that the integrals with their convergence factors now converge, *self-energy* now convergent, corresponds to a correction to the electron mass, *all matrix elements now finite* with exception of those relating to *vacuum polarization*, *strict physical basis for rules of convergence not known*, after *mass and charge renormalization* results are equivalent to those of Schwinger in which the terms corresponding to corrections in mass and charge are identified and removed from the expressions for real processes, *although in the limit the two methods agree neither method appears to be completely satisfactory theoretically*, practical advantage of new method is that ambiguities can be more easily resolved by direct calculation of otherwise divergent integrals, complete method therefore available for calculations of all processes involving electrons and photons, shows how matrix element for any process can be written down directly in *Feynman diagrams*, paper includes first published "*Feynman diagram*", *attempt to find a consistent modification of quantum electrodynamics is incomplete, not at all clear that the convergence factors do not upset the physical consistency of the theory.*

569 **Schwinger, J. (September, 1949). Quantum Electrodynamics. III. The Electromagnetic Properties of the Electron—Radiative Corrections to Scattering.** *Phys. Rev.*, 76, 6, 790-817; https://doi.org/10.1103/PhysRev.76.790; a covariant form of *quantum electrodynamics* has been developed and applied in the previous articles of this series to two elementary phenomena that are produced by the *vacuum fluctuations of the electromagnetic field*, these applications were the *polarization of the vacuum* expressing the modifications in the properties of an electromagnetic field arising from its interaction with the *matter field* vacuum fluctuations, and the *electromagnetic mass of the electron* embodying the corrections to the mechanical properties of the *matter field* in its single particle aspect, in these problems *the divergences that mar the theory are found to be*

concealed in unobservable charge and mass renormalization factors, the previous paper was confined to consideration of *vacuum polarization* produced by the field of a prescribed *current* distribution, *we now consider how induction of current in a vacuum by an electron results in an alteration in its electromagnetic properties* revealed by scattering in Coulomb field and energy level displacements, *this paper is concerned with the computation of the second-order corrections to the current operator as modified by the coupling with the vacuum electromagnetic field and its application to electron scattering* by a Coulomb field, applies canonical transformation to *renormalize electron mass*, correction to *current operator* produced by coupling with electromagnetic field developed in power series, first- and second-order terms retained, results in second-order modifications in *current operator* of same general nature as the previously treated *vacuum polarization current* apart from contribution in form of *dipole current*, the latter implies fractional increase of $\alpha/2\pi$ in the *spin magnetic moment* of electron, the only flaw in second-order current correction is *logarithmic divergence attributable to infra-red catastrophe,* in the presence of an *external field* the first-order *current* correction will introduce a compensating divergence, thus the second-order corrections to particle electromagnetic properties cannot be completely stated without regard for the manner of exhibiting them by an *external field*, accordingly in the second section we consider the interaction of three systems - the *matter field*, the *electromagnetic field*, and a given *current distribution*, shows that this can be described in terms of an *external potential* coupled to the *current* operator as modified by the interaction with the *vacuum electromagnetic field*, applies to *scattering of an electron* by an *external field* which is regarded as a small perturbation, convenient to calculate the total rate at which collisions occur and then identify the *cross sections* for individual events, correction to the *cross section* for radiation-less scattering is determined by the second-order correction to the *current* operator, scattering that is accompanied by single quantum emission is a consequence of the first-order *current* correction, the final object of calculation is the *differential cross section* for scattering through a given angle with a prescribed maximum energy loss which is completely free of divergences, an *Appendix* is devoted to an alternative treatment of the *polarization of the vacuum* by an external field, *radiative corrections to energy levels will be treated in the next paper of the series* [*but this did not appear*].

586 **Feynman, R. P. (November, 1950). Mathematical Formulation of the Quantum Theory of Electromagnetic Interaction.** *Phys. Rev.*, 80, 3, 440; https://doi.org/10.1103/PhysRev.80.440; in two previous papers rules were given for calculation of matrix element for any process in electrodynamics, no complete proof of equivalence of these rules to conventional electrodynamics was given, nor was a closed expression given valid to all orders in $e^2/\hbar c$, this paper addresses these formal omissions, *gives the derivations of the formulas of Feynman (September 1949).* [*Space-Time Approach to Quantum Electrodynamics.*] *by means of the form of quantum mechanics given in Feynman (April 1948).* [*Space-Time Approach to Non-Relativistic Quantum Mechanics*], derivation of rules using Lagrangian form of quantum mechanics permits motion of any part to be solved first and the results to be used in solution of motion of other parts, *electromagnetic field* a simple system,

interaction of matter (electrons and positrons) and field analyzed by first solving for the behavior of the field in terms of the coordinates of the matter, integration over field oscillator coordinates eliminates field variables from *equations of motion* of electrons, then address behavior of electrons, all of the rules given in second paper derived, *restricted to cases in which the particle's motion is non-relativistic* but the transition of the final formulas to the relativistic case is direct and the proof could have been kept relativistic throughout, *generalized formulation unsatisfactory because for situations of importance it gives divergent results, problems of divergences are not discussed, present theory assumes that substantial interaction exists as long as the interval is time-like and less than some small length of order of the electron radius.*

600 **Feynman, R. P. (October, 1951). An Operator Calculus Having Applications in Quantum Electrodynamics.** *Phys. Rev.*, 84, 1, 108; https://doi.org/10.1103/PhysRev.84.108; suggests alteration in mathematical notation for handling operators, new notation permits considerable increase in ease of manipulation of complicated expressions involving operators, *no results which are new are obtained in this way, the mathematics is not completely satisfactory, no attempt has been made to maintain mathematical rigor*, it is believed that to put the present methods on a rigorous basis may be quite a difficult task, beyond the abilities of the author.

602 **Dirac, P. A. M. (October, 1962). The Conditions for a Quantum Field Theory to be Relativistic.** *Rev. Mod. Phys.*,34, 592; a quantum field theory in agreement with *special relativity* can be built up from the infinitesimal operators of translation and rotation. These operators are expressible in terms of a momentum density and an energy density. The momentum density is determined by the geometrical properties of the fields concerned. The energy density has to satisfy commutation relations for which certain conditions hold.

603 **Tomonaga, S. (May 6, 1966). Nobel Lecture. *Development of Quantum Electrodynamics - 1966: Personal recollections*.** https://www.nobelprize.org/prizes/physics/1965/tomonaga/lecture/; Tomonaga's Nobel Prize lecture describes the evolution of his work and that of others since 1932 when he started his research career up until 1948.

616 **Feynman, R. P. (December, 1965). Nobel Lecture. *The Development of the Space-Time View of Quantum Electrodynamics*.** https://www.nobelprize.org/prizes/physics/1965/feynman/lecture/; Feynman's Nobel Prize lecture describes the sequence of ideas which occurred and by which he finally came out the other end with an unsolved problem for which he ultimately received the Nobel prize, he represented conventional electrodynamics with *retarded interaction*, not his half-advanced and half-retarded theory, the *action* expression was not used, the idea that *charges* do not act on themselves was abandoned, the *path-integral formulation* of quantum mechanics was useful for guessing at final expressions and at formulating the general theory of electrodynamics in new ways though was not absolutely necessary, the rest of his work was simply to improve the techniques then available for calculations, *making diagrams to help analyze perturbation theory quicker*, he

concluded: "*I don't think we have a completely satisfactory relativistic quantum-mechanical model*, ... Therefore, I think that the *renormalization theory* is simply a way to sweep the difficulties of the divergences of electrodynamics under the rug".

637 **Schwinger, J. (December 11, 1965). Nobel Lecture. *Relativistic Quantum Field Theory.*** https://www.nobelprize.org/prizes/physics/1965/schwinger/lecture/; Schwinger does not mention any of his 1948 and 1949 papers in his Nobel Prize lecture for which he was awarded the Nobel Prize, instead he focuses on papers he wrote between 1951 and 1965, he starts by describing the logical foundations of *quantum field theory* or *relativistic quantum mechanics* which he defines *as the synthesis of quantum mechanics with relativity*, he notes improvements in the formal presentation of quantum mechanical principles by himself described in a series of six papers on the theory of quantized fields published in *Physical Review* between June 1951 and June 1954 and by Feynman described in his April 1948 Space-Time Approach to Non-Relativistic Quantum Mechanics utilizing the concept of *action* based on a study of Dirac concerning the correspondence between the quantum *transformation function* and the classical *action*, he identifies two distinct formulations of quantum mechanics - a *differential formulation* utilizing a *differential* version of the *action* principle and the *path integral formulation* of Feynman, he claims that his *differential* version transcends the *correspondence principle* and incorporates on the same footing the two different kinds of quantum dynamical variable that are demanded empirically by the two known varieties of particles obeying Bose-Einstein or Fermi-Dirac statistics, the *quantum action principle is a differential statement about time transformation functions*, *all quantum-dynamical aspects of the system are derived from a single dynamical principle*, he also claims that *quantum field theory* had failed no significant test nor could any decisive confrontation be anticipated in the near future contrary to Feynman's reservations and Schwinger's statements in the preface to his 1958 book, *classical mechanics is a determinate theory*, knowledge of the *state* at a given time permits precise prediction of the result of measuring any property of the system, *quantum mechanics is only statistically determinate*, it is the *probability* of attaining a particular result on measuring any property of the system not the outcome of an individual microscopic observation that is predictable from knowledge of the *state*, *relativistic* structure of the *action principle* is completed by demanding that it present the same form independently of the particular partitioning of *space-time* into space and time, facilitated by the appearance of the *action operator* - the time integral of the Lagrangian - as the *space-time* integral of the Lagrange function, he also discusses a further eleven papers that he had published in *Physical Review* between July 1962 and October 1965 which attempted to extend his theory to gravitational fields and develop a *field theory of matter*.

PREFACE to Volume II

> Wikipedia: *"In particle physics, quantum electrodynamics (QED) is the relativistic quantum field theory of electrodynamics. In essence, it describes how light and matter interact and is the first theory where full agreement between quantum mechanics and special relativity is achieved. QED mathematically describes all phenomena involving electrically charged particles interacting by means of exchange of photons and represents the quantum counterpart of classical electromagnetism giving a complete account of matter and light interaction"*. [???]

These two volumes were spun off from a larger project involving a review of the current state of our scientific understanding of electromagnetic radiation and gravity[1].

> [1] For the first spin-off, see Underwood, T. G. (2021). *Urbain Le Verrier on the Movement of Mercury - annotated translations*. Lulu Press, Morrisville, NC.

I decided that in my self-imposed lockdown to avoid COVID, I could usefully pick up where I had left off in 1965, after graduating in theoretical physics from Cambridge University. As the volume of papers relevant to understanding the current state of *quantum electrodynamics* was extremely large and some were not easily obtainable, and a few had not been translated into English, I decided to produce an annotated sourcebook from which it would be easier to understand the current state of our knowledge about the electromagnetic field and its interaction with electrically charged particles. This would not have been possible in 1965, before the advent of personal computers in 1971, the origin of internet in 1969, the increase in disc capacity, and the availability of much of this material on the internet in the last two decades.

This is not a sourcebook in the conventional sense. It is a working document which brings together annotated extracts from 107 primary sources, or translations of them, of the development of quantum electrodynamics so that it is easier for a researcher to deal with the large volume of relevant sources. Papers hidden behind pay walls for which no alternative internet source or abstract is available have been omitted, but fortunately these are of little consequence. Links to internet copies of the primary material or alternative sources are provided where available to enable these to be consulted. A summary is provided at the head of each paper and in the Contents. The references in each paper are expanded to include the title (and its translation where relevant), and a copy of the summary where this is available and helpful to avoid unnecessary cross referencing. Biographies of the main contributors are also provided, of which 18 received Nobel Prizes in Physics. Explanations of terminology (which is highlighted in italics) and biographical information are largely drawn from Expedia, unless otherwise referenced. The chronological development of the theories of quantum mechanics and quantum electrodynamics constituted an interesting interplay between theory and experiment.

Volume I, covering the period from 1896 to 1931 is primarily focused on the development of the largely successful *non-relativistic* theory of quantum mechanics and quantum electrodynamics. Volume II, covering the period from 1930 up until 1965, when

Tomonaga, Feynman, and Schwinger received their Nobel prizes, addresses the attempts to formulate a *relativistic* quantum electrodynamics or quantum field theory for the electron when the energy of the electron is *relativistic*, and in particular to address, through a process of *renormalization*, the still unresolved divergencies arising largely, if not entirely, from the assumption of a point electron.

Despite the claims to the contrary in modern textbooks, there have been no significant developments in the quantum electrodynamics or quantum field theory since 1965 to resolve the underlying occurrence of divergencies recognized by Dirac, Tomonaga, Schwinger and Feynman. Witness to this is the very comprehensive 2018 *An Introduction to Quantum Field Theory* by Michael Peskin and Daniel Schroeder, first published in 1995, which replaced the 1965 two-volume text by James Bjorken and Sidney Drell, *Relativistic Quantum Fields*, which focuses on the application of Feynman diagrams. The former claims that "Quantum Electrodynamics (QED) is perhaps the best fundamental physical theory we have" then devotes Part II (pages 265-470), nearly one third of the book, to *renormalization.*

Born, in his 1954 Nobel prize speech, noted that "Planck, himself, belonged to the sceptics until he died. Einstein, De Broglie, and Schrodinger have unceasingly stressed the unsatisfactory features of quantum mechanics and called for a return to the concepts of classical, Newtonian physics while proposing ways in which this could be done without contradicting experimental facts".

Schwinger, in the Preface of his 1958 book [*Selected Papers on Quantum electrodynamics.*], "questioned whether *renormalization* simply corrected a mathematical error that causes the divergencies, or whether *there is a serious flaw in the structure of field theory*". He concluded that "the observational basis of quantum electrodynamics is self-contradictory" and that "a convergent theory cannot be formulated consistently within the framework of present space-time concepts" … "It can never explain the observed value of the dimensionless coupling constant measuring the electron charge … a full understanding of the electron charge can exist only when the theory of elementary particles has come to a stage of perfection that is presently unimaginable".

Tomonaga, in his 1965 Nobel prize speech, note that "In order to overcome the difficulty of an infinitely large *electromagnetic mass, Lorentz considered the electron not to be point-like but to have a finite size. It is very difficult, however, to incorporate a finite sized electron into the framework of relativistic quantum theory*. Many people tried various means to overcome this problem of infinite quantities, but nobody succeeded".

Feynman, in his 1965 Nobel prize speech, described *renormalization* as "simply a way to sweep the difficulties of the divergences of electrodynamics under the rug".

Dirac's final judgment on *quantum field theory*, in his last paper published in 1987 [The inadequacies of quantum field theory.], was that "These rules of *renormalization* give surprisingly, excessively good agreement with experiments. Most physicists say that these working rules are, therefore, correct. I feel that is not an adequate reason. Just because the results happen to be in agreement with observation does not prove that one's theory is correct."

I was tempted to include my own initial analysis of the problem but decided against this so that others were free to draw their own conclusions and spot the error or errors, and, ideally, provide a solution free of the divergencies, for which they might be awarded a Nobel prize.

Volume II.

On December 11, 1933, Werner Heisenberg received the 1932 Nobel Prize in Physics "for the creation of quantum mechanics, the application of which has, inter alia, led to the discovery of the allotropic forms of hydrogen". In 1925, when he was 24 years old, Heisenberg formulated a type of quantum mechanics based on matrices [Heisenberg, W. (July, 1925). Über quantentheoretische Umdeutung kinematischer und mechanischer Beziehungen. (On the quantum-theoretical re-interpretation of kinematic and mechanical relations.)]. In 1927 he proposed the "uncertainty relation", setting limits for how precisely the position and velocity of a particle can be simultaneously determined. In his Nobel Prize lecture [Werner Heisenberg – 1932 Nobel Lecture, December 11, 1933. *The development of quantum mechanics*.], he noted that "the impossibility of harmonizing the Maxwellian theory with the pronouncedly visual concepts expressed in the hypothesis of light quanta subsequently compelled research workers to the conclusion that *radiation phenomena can only be understood by largely renouncing their immediate visualization. ...* Classical physics seemed the limiting case of visualization of a fundamentally unvisualizable microphysics, the more accurately realizable the more Planck's constant vanishes relative to the parameters of the system". After describing the development of the current theory of quantum mechanics, he concluded that "a visual description for the atomic events is possible only within certain limits of accuracy - but within these limits the laws of classical physics also still apply. Owing to these limits of accuracy as defined by the uncertainty relations, moreover, a visual picture of the atom free from ambiguity has not been determined. On the contrary the corpuscular and the wave concepts are equally serviceable as a basis for visual interpretation. The laws of quantum mechanics are basically statistical". He noted that "the attention of the research workers was now primarily directed to *the problem of reconciling the claims of the special relativity theory with those of the quantum theory. ...* The attempts made hitherto to achieve a *relativistic* formulation of the quantum theory are all based on visual concepts so close to those of classical physics that it seems impossible to determine the fine-structure constant within this system of concepts".

On December 12, 1933, Erwin Schrödinger received the 1933 Nobel Prize in Physics, together with Paul Dirac, "for the discovery of new productive forms of atomic theory". In Niels Bohr's theory of the atom, electrons absorb and emit radiation of fixed wavelengths when jumping between fixed orbits around a nucleus. The theory provided a good description of the spectrum created by the hydrogen atom, but needed to be developed to suit more complicated atoms and molecules. Assuming that matter (e.g., electrons) could be regarded as both particles and waves, in 1926, when he was 39 years old, Schrödinger formulated a wave equation that accurately calculated the energy levels of electrons in atoms [Schrödinger, E. (December, 1926). A Wave Theory of the Mechanics of Atoms and Molecules.]. In his Nobel Prize lecture, Schrödinger described a fascinating analogy between the path of a ray of light and the path of a mass point. In optics the old system of mechanics corresponded to operating with isolated mutually independent light rays. The new wave mechanics corresponds to the wave theory of light. We are never in a position to say what really is or what really happens, but we can only say what will be observed in any concrete individual case. He concluded that *the ray or the particle path corresponded to a longitudinal relationship of the propagation process (i.e. in the direction of propagation), the wave surface on the other hand to a transversal relationship (i.e. normal to it). Both relationships were without doubt real;* one is proved by photographed particle paths, the other by interference experiments. *To combine both in a uniform system had proven impossible so far. Only in extreme cases did either the transversal, shell-shaped or the radial longitudinal relationship predominate to such an extent that we think we can make do with the wave theory alone or with the particle theory alone.*

On December 12, 1933, Paul Dirac received the 1933 Nobel Prize in Physics, together with Erwin Schrödinger "for the discovery of new productive forms of atomic theory". During the intense period of 1925-26 quantum theories were proposed that accurately described the energy levels of electrons in atoms. These equations needed to be adapted to Einstein's theory of relativity, however. In 1928 Dirac, when he was 26 years old, formulated a fully *relativistic* quantum theory [Dirac, P. A. M. (February, 1928). The Quantum Theory of the Electron.]. The equation gave solutions that he interpreted as being caused by a particle equivalent to the electron, but with a positive charge. This particle, the positron, was later confirmed through experiments. In his Nobel Lecture, Dirac described the current state of his theory of electrons and positrons. He focused on electrons and the positrons on the grounds that the theory has been developed further, and hardly anything could be inferred theoretically about the properties of the others. He noted that the question that must first be considered is *how theory can give any information at all about the properties of elementary particles*. There existed at the present time a general quantum mechanics which could be used to describe the motion of any kind of particle, no matter what its properties were. The general quantum mechanics, however, was valid only when the particles had small velocities and failed for velocities comparable with the velocity of light, when the effects of *relativity* come in. There existed no *relativistic* quantum mechanics for particles with large velocities which could be applied to particles with arbitrary properties. But, *subjecting quantum mechanics to relativistic requirements imposed restrictions on the properties of the particle, and in this way information about the particles could be deduced from purely theoretical considerations*. Dirac described how the spin properties of the electron could be deduced, and the existence of positrons with similar spin properties and with the possibility of being annihilated in collisions with electrons could be inferred. The new variables which he introduced to get a *relativistic* wave equation linear in *kinetic energy* W gave rise to the spin of the electron. These variables also gave rise to some unexpected phenomena concerning the motion of the electron. *It was found that an electron which appears to be moving slowly, must actually have a very high frequency oscillatory motion of small amplitude superposed on the regular motion, which resulted in the velocity of the electron at any time being equal to the velocity of light.* The wave equation also allowed negative-energy states in an electromagnetic field which *corresponded to the motion of an electron with a positive charge instead of the usual negative one*, i.e. a positron, which Dirac identified with a "hole" in his "hole theory". On this view, the positron was just a mirror-image of the electron, having exactly the same mass and opposite charge.

In January 1930, Dirac published a paper [A theory of electrons and protons.], on which his 1933 Nobel Lecture was based, in which he formulated a fully *relativistic* quantum theory. He noted that in *relativistic* quantum theory, in which the *electromagnetic field* is subjected to quantum laws, the *wave equation* refers equally well to an electron with charge + e with *negative kinetic energy*, a difficulty not restricted to the quantum theory of the electron but one which appears in all relativity theories. It arises because there is an ambiguity in the sign of W, or rather $W + eA_0$, in the *relativity* Hamiltonian equation of the classical theory. In the quantum theory transitions can take place in which the *energy* of the electron changes from a positive to a negative value. Dirac used his "*hole theory*" to address the problem by assuming that in a vacuum all negative-energy electron eigenstates were occupied. He noted that if negative-energy *eigenstates* are incompletely filled each unoccupied *eigenstate* – called a "hole" – would behave like a positively charged particle, which he initially thought might be a proton. He noted that for the scattering of radiation by an electron the *exclusion principle* forbids the electron to jump into a state of *negative energy*, so he assumed a double transition process in which first one of the negative-energy electrons jumps up into the *final state* for the electron with the *absorption* (or *emission*) of a photon, and then the original positive-energy electron drops into hole formed by first transition with *emission* (or *absorption*) of a photon.

In November 1930, Heisenberg published a paper [Die Selbstenergie des Elektrons (The self-energy of the electron)] in which he noted that a *point-like* electron results in infinite energy density, but a *finite radius* for the electron is inconsistent with Special Relativity. *He chose to represent an electron by a point charge*, and investigated the conditions under which the self-energy of the electron vanished. He showed that the one-electron problem could be treated correctly without an infinite *self-energy* if there were solutions of vacuum electrodynamics without a *zero-point energy*, but *such solutions did not exist*. He showed that the difficulties of *field theory* did not come directly from the infinite *self-energy* of electron, and concluded that *a solution of the basic equations had therefore not been found for the time being,* and that it was also not probable that one would achieve a solution without substantial modification of the quantum theory of wave fields.

In May 1932, Dirac published a paper [Relativistic Quantum Mechanics] describing an alternative to Heisenberg and Pauli's approach to *relativistic* quantum mechanics which regarded the *field* itself as a dynamical system amenable to Hamiltonian treatment and its interaction with the particles as describable by an *interaction energy*. He noted that there were serious objections to these views. *If we wish to make an observation on a system of interacting particles the only effective method of procedure is to subject them to a field of electromagnetic radiation and see how they react.* The role of the *field* is to provide a means for making observations. The nature of an observation requires an interplay between the *field* and the *particles.* We cannot suppose the *field* to be a dynamical system on the same footing as the *particles* and thus something to be observed in the same way as the *particles.* The *field* should appear in the theory as something more elementary and fundamental. In this paper he proposed an *interaction representation* which gave interplay between particles and the field, and he translated the *equations of motion* of *relativistic* classical theory directly into equations expressible entirely in terms of *probability amplitudes* referring to one *ingoing field* and one *outgoing field.* He assumed that passage from the field of ingoing waves to the field of outgoing waves was a quantum jump performed by one field composed of waves passing undisturbed through the electron and satisfying Maxwell's equations. The *relativistic* observable quantities were *transition probabilities*, with *probability amplitudes* analogous to Heisenberg's matrix elements. For quantization he assumed that *intensities* and *phases* were operators satisfying the usual quantum conditions governing the *intensities* and *phases* of Fourier components of electromagnetic field in empty space, which determined the matrix elements associated with electron jumps. He assumed that the interaction of each electron with the *field* could be described by an *interaction energy* equal to its charge multiplied by the *potential* at the point where it was situated, and derived the *wave equation* for interactions of two electrons due to motions of each being connected with same field to be $\{ih\ \delta/\delta t + h^2/2m_1\ \delta^2/\ \delta x_1^2 + h^2/2m_2\ \delta^2/\ \delta x_2^2 - \varepsilon_1 V(x_1 t) - \varepsilon_2 V(x_2 t)\}\ \psi = 0$.

In late 1932, Dirac, Fock, and Podolsky published a paper [On quantum electrodynamics] in the *Phys. Zeit. Sowjetunion* which described a *relativistic* model in which a fixed number of electrons interacted through a second-quantized electromagnetic field. They applied Dirac's *interaction representation* formulation of quantum field theory to the full electrodynamics. This was formulated with the help of a *multi-time* wave function $\psi(t_1,\mathbf{x}_1,\ldots,t_N,\mathbf{x}_N)\psi(t_1,x_1,\ldots,t_N,x_N)$ that generalized the Schrödinger's multiparticle wave function to allow for a manifestly *relativistic* formulation of wave mechanics. All non-trivial dynamics due to interaction between charged matter and electromagnetic field were relegated to the time evolution of a state vector. The second Maxwell equation - the Ampère-Maxwell law – was satisfied only through the action of the field operators on the wave function. This implied that electromagnetic field operators still obeyed the free field equations and consequently the covariant commutation relations of Jordan and Pauli, but the method of quantizing by imposing covariant commutation relations was still an isolated technique applied only to Maxwell field. This showed relationship between *Maxwell's equations* for empty space, the *wave equation* of Heisenberg-Pauli's theory, and the *wave equation of* Dirac's *interaction representation.*

In 1933, Dirac published a paper [The Lagrangian in Quantum Mechanics], in the same journal, which provided an alternative formulation of quantum mechanics in terms of Lagrangian in place of Hamiltonian. This used *coordinates* and *velocities* instead of *coordinates* and *momenta*, and allowed the *equations of motion* to be expressed as stationary property of the *action function*, the time-integral of the Lagrangian, which was a *relativistic invariant*. There is no corresponding *action principle* in terms of *coordinates and momenta* in Hamiltonian theory. Lagrangian theory is closely connected with theory of *contact transformations*. Dirac derived *transformation functions* that had classical analogues expressible in terms of the Lagrangian. The classical Lagrangian is a function of coordinates at time t and time t + dt rather than of the coordinates and velocities. He introduced his "*many-time*" theory and applied it to *field dynamics* using suitable field quantities or potentials as coordinates, in which each coordinate was a function of four *space-time* variables instead of one time variable as in particle theory. This resulted in a quantum analogue of the classical *transformation function* between dynamical variables that Dirac described as a sort of *generalized transformation function*.

In March 1934, Dirac published a paper [Discussion of the infinite distribution of electrons in the theory of the positron] which attempted to address the problem with his *relativistic* 'hole' theory that implied an infinite number of negative-energy electrons (per unit volume) with energies extending continuously from $-mc^2$ to $-\infty$. When an electromagnetic field was present positive- and negative-energy states could not be distinguished in a *relativistically* invariant way. It is necessary to set up assumptions for production of *electromagnetic field* by the electron distribution such that any finite change in distribution produced a change in the field in agreement with Maxwell's equations and such that the infinite field which would be required by Maxwell's equations from an infinite density of electrons is in some way cut out. The exact treatment is very complicated; the present paper gave an approximate treatment. Dirac assumed that *each electron had its own individual wave function in space-time* and *each electron moved in an electromagnetic field which was the same for all electrons*, part coming from external causes and part from the electron distribution itself. He defined the distribution of electrons in terms of the *relativistic density matrix* $\sum_{oc} \psi_{k'}(x'\, t')\, \psi^*_{k''}(x''\, t'')$ referring to two points in space and two times, and separated the distribution into two parts, where one contained the singularities, and the other described the *electric* and *current densities* physically present.

In March 1934, Heisenberg published a paper [Bemerkungen zur Diracschen Theorie des Positrons (Remarks on the Dirac theory of positron)] with the purpose of reconstructing Dirac's *theory of the positron* in the formalism of quantum electrodynamics. It demanded that the symmetry in nature between positive and negative charge should be expressed in the basic equations from the outset. In addition to the well-known difficulties with the divergences, no new infinities should appear in the formalism, moreover the theory should provide an approximation for the treatment of the circle of problems that have been treated by quantum electrodynamics up to now. Dirac [(1934). Discussion of the infinite distribution of electrons in the theory of the positron.] showed that a quantum mechanical system of many electrons that fulfilled the Pauli principle and moved in a given force field without back-reaction could be characterized by a *density matrix*
$(x', t', k' \mid R \mid x'', t'', k'') = \sum n\, \psi^* n\, (x', t', k')\, \psi_n\, (x'', t'', k'')$ where $\psi_n(x', t', k')$ meant the normalized eigenfunctions of the states that possessed one electron, and x′, t′, k′ (x″, t″, k″, resp.) were the position, time, and spin variables. All physically-important properties of

quantum-mechanical systems like *charge density*, *current density*, etc., could be read off from the *density matrix*. The temporal change in the *density matrix* was determined by the Dirac differential equation. Dirac made different choice of *density matrix* representing the external field, resulting in a different energy and impulse density. By restricting oneself to an *intuitive analogue theory of matter fields, the negative energy levels in the Dirac theory could be avoided by replacing the homogeneous Dirac differential equation with an inhomogeneous equation*, where the inhomogeneity was indicative of pair creation. For the most practical applications e.g., pair creation, annihilation, Compton scattering, etc. *this theory described did not yield anything new compared to the formulation of the Dirac theory*. In the Maxwell theory, a continuous charge distribution also led to a finite self-energy; it is the "quantization" that leads to the infinite self-energy. *If one represents the quantization of the electromagnetic field by point-like light quanta then the infinitude of the self-energy also emerges in the intuitive theory of matter waves.*

In January 1934, Victor Weisskopf published a paper [Über die Selbstenergie des Elektrons. (The Self-energy of the Electron.)] in which the *self-energy* of the electron was derived in close formal connection to classical radiation theory without direct application of quantum electrodynamics. The radiation field was calculated classically from the *current and charge densities* of the atom. The electromagnetic field was divided into a rotation-free *electrostatic* part and a divergence free *electrodynamic* part. The *self-energy* derived from the electrostatic part. A correction to the paper showed that the *self-energy* of electron at occupied negative energy states diverged logarithmically as in Dirac's "hole theory", in contrast to the linear divergence of the classical theory and the quadratic divergence of the one particle Dirac theory.

The *scalar relativistic wave equation* $E^2/c^2 - \sum_{k=1}^{3} p_k^2 - m^2c^2 = 0$ had generally been relinquished in favor of *Dirac's four-component wave equation* $\{p_0 + \rho_1 (\boldsymbol{\sigma}, \mathbf{p}) + \rho_3 mc\} \psi = 0$ [Dirac, P. A. M. (February, 1928). The Quantum Theory of the Electron.] because the former did not yield the spin of the particles.

In July 1934, Pauli and Weisskopf sent a paper for publication [Über die Quantisierung der skalaren relativistischen Wellengleichung. (The Quantization of the Scalar Relativistic Wave Equation.)] which argued that Dirac's *a priori* arguments based on limitation to a *single-body problem* and *particle density* being a meaningful observable needed to be revised. It showed that *charge density* is a meaningful observable, that there was no reason for special form of *charge density* $\sum_r \psi^*_r \psi_r$, in the *scalar relativistic wave equation theory*, and that the energy of material particles is always positive after wave fields have been quantized. No new hypothesis such as the "hole theory" was required. It applied Heisenberg-Pauli formalism of quantization of wave fields to the *scalar relativistic wave equation* for matter fields for particles without spin and with Bose-Einstein statistics, and showed that quantization of the *Klein-Gordon relativistic wave equation* for *scalar particles* gave rise to particles with opposite charge but with same rest mass which can be created or destroyed with absorption or emission of electromagnetic radiation. The frequency of pair creation and annihilation processes was of same order of magnitude as the frequency for particles of same charge and mass which followed from Dirac's "hole theory". It *also led to infinite self-energy not only of the particles but also to an infinite polarizability of the vacuum.*

In September 1934, Ernst Stueckelberg, published a paper [Relativistisch invariante Störungstheorie des Diracschen Elektrons I. Teil: Streustrahlung und Bremsstrahlung. (Relativistically invariant perturbation theory of Dirac's electron Part I: scattered radiation and Bremsstrahlung.)] in which the main innovation was the introduction of a new perturbative scheme yielding relativistic expressions for the *matrix elements* which were manifestly *gauge invariant*. This was achieved by performing a four-dimensional Fourier transformation of the *wave-function* eliminating space and time variables based on the interaction picture of Dirac, Fock and Podolsky. The starting point was the Dirac equation for the *spinor wave-function* $[1/i\ (\gamma, \partial/\partial x) + M + eV(x)]\ \Psi(x) = 0$. The contributions of positive and negative energies (corresponding to virtual electrons and positrons) were contained in a single propagation function which corresponded to what later was called the *Feynman propagator*. His method for calculating the cross-section started with the definition of *on mass-shell wave-functions* and used integration over the *complex energy-plane*. This paper introduced *Stueckelberg diagrams* later adopted by Feynman and subsequently renamed *Feynman diagrams*. All of Stueckelberg's expressions for matrix elements were identical to those obtained nowadays from Feynman diagrams.

In August 1938, Dirac published a paper [Classical Theory of Radiating Electrons] describing a *relativistic* form of classical theory of radiating electrons assuming the electron to be a *point charge* with no volume. He noted that an extended electron was inconceivable in the *special theory of relativity* due to intrinsic connection between space and time, and that the Lorentz *model of the electron as possessing mass on account of the electromagnetic field around it failed* without further assumptions if the electromagnetic field varied too rapidly or if the acceleration of the electrons was too great. No natural way of introducing further assumptions had been discovered. The discovery of the neutron introduced a form of mass for which it was difficult to believe that it was of electromagnetic nature. Also, the theory of the positron in which positive and negative values of an electron play symmetrical roles could not be fitted into the electromagnetic idea of mass, which required all mass to be positive. The *departure from the electromagnetic theory of mass removed the main reason for believing in a finite size of the electron.* This resulted in the difficulty that the *field* in the immediate neighborhood of the electron had an infinite mass. In quantum mechanics, this resulted in a *divergence* in the solution of the equations that described the interaction of an electron with an electromagnetic field and prevented its application to high-energy radiation processes.

In this paper Dirac chose to represent electron by point charge and to avoid the difficulties with the infinite energy by direct omission or subtraction of unwanted terms, as had been used in the theory of the positron. *The new theory had to be in agreement with special relativity* and the conservation of energy and momentum. Dirac first addressed the problem of a *single electron moving in an electromagnetic field.* He used the equations for the electromagnetic *potentials* in terms of the *charge-current density* vector, and a*ssumed that the charge-current density vector vanished everywhere except on the world-line of the electron, where it was infinitely great*. He derived the *field* quantities from the electromagnetic *potentials*, and obtained solutions in terms of *retarded* and *advanced potentials* and *potentials* representing the incoming and outgoing radiation, from which he calculated the field of radiation produced by the electron. These were in agreement with classical theory but provided a value for the field of radiation throughout space-time. He obtained exact *equations of motion* of the electron in a specified incident *field* from the laws of conservation of energy and momentum within the limits of classical theory, and imposed the condition that solutions which occur in Nature when there is no incident field were those for which the velocity in terms of proper time is constant. This required solutions of the equations of motion *for which the final acceleration as well as the initial position and velocity were prescribed*. The electron responded to a pulse of electromagnetic radiation before it reaches the center of the electron; behaving as if it had a radius of order $1/a$ where $a = 3m/e^2$. This implied that a signal could be transmitted *faster than the speed of light* through the interior of an electron, and that the *interior of electron was a region of failure of some of the elementary properties of space-time*. He showed how the theory could be extended to any number of electrons interacting with each other and with a field of radiation.

In July 1939, Dirac published a paper [A new notation for quantum mechanics] introducing the *bra ket* notation. He noted that quantum mechanics dealt with *vectors in Hilbert space*, representing the *states* of a dynamical system, and with *linear operators*, representing dynamical *variables*, and that sometimes one made calculations using the *vectors and linear operators* directly, treating them as *abstract quantities*, at other times one worked with *coordinates* (or *representatives*) of these *quantities*. The *bra ket* notation provided a concise way of writing the *abstract quantities* themselves and their *coordinates* in a single scheme, which led to a unification of ideas.

In July 1939, Weisskopf published a paper [On the Self-energy and Electromagnetic Field of the Electron] of which t*he main purpose was to show the physical significance of the logarithmic divergence of the self-energy of the electron and to demonstrate the reason for its occurrence*. The self-energy of the electron is its total energy in free space when isolated from other particles or light quanta, $W = T + (1/8\pi) \int (H^2 + E^2)\, dr$ where T is the *kinetic energy* of the electron and H and E are the *magnetic* and *electric field strengths* at point r. He identified three reasons why the quantum theory of the electron resulted in the infinite *self-energy* of the electron. *Quantum kinematics required that the radius of the electron must be assumed to be zero*, resulting in infinite energy of the *electrostatic field*. The contributions of the electric and magnetic fields of the spin to the *self-energy* of the electron canceled one another. *Quantum theory of the electromagnetic field postulated the existence of field strength fluctuations in empty space*, which gave rise to an additional energy that diverged more strongly that the electrostatic self-energy It induced the electron to perform vibrations with energy that diverged quadratically *for an infinitely small radius*. Dirac's positron theory implied that the charge and magnetic dipole of the electron were extended over a finite region. Weisskopf explained why the *self-energy* was only *logarithmically infinite*. *Divergences were a consequence of the assumption of a point electron.*

On June 19, 1941, Dirac delivered the Bakerian Lecture [The physical interpretation of quantum mechanics, published March 1942] at Burlington House, whilst London was being bombed. He described how a satisfactory *non-relativistic* quantum mechanics had been established, in which the *Heisenberg method* focused on quantities which enter into experimental results, and connected together in one calculation probability coefficients from all initial states to all final states; and in which the *Schrodinger method* connected together in one calculation probability coefficients for transitions from one particular initial state to any final state. Both methods rested on same mathematical formalism, a generalization of the Hamilton form of classical dynamics, involving linear operators instead of ordinary algebraic variables. The methods of physical interpretation differed, and was probably still not finally settled. *The theory was not in agreement with the theory of special relativity*, as was evident by the special role played by the time t. While it worked very well in the *non-relativistic* region of low velocities, where it appeared to be in complete agreement with experiment, it could be considered only as an approximation. Setting up the mathematical formalism was fairly straightforward. Firstly, the classical mechanics needed to be put into *relativistic* Hamiltonian form, taking into account that that the various particles comprising the dynamical system interact through the medium of the electromagnetic field, using Lorentz's *equations of motion*, including the dampening terms which expressed the reaction of radiation. This Hamitonian formulation could then be made into a quantum theory following the procedure from *non-relativistic* quantum mechanics. Although this appeared satisfactory mathematically, it met serious difficulties in its physical interpretation. In particular, *it resulted in states of negative energy and negative probability*. It appeared that, whether one was dealing with particles of integral spin or of half-odd integral spin, the mathematical methods at present in use in quantum mechanics *were capable of direct interpretation only in terms of a hypothetical world differing very markedly from the actual one*. All that *relativistic* theory did was provide a consistent means of calculating experimental results.

S-matrix theory was proposed as a principle of particle interactions by Heisenberg in July 1943 [Heisenberg (July 1943). Die beobachtbaren Größen in der Theorie der Elementarteilchen. I. (The "observable quantities" in the theory of elementary particles. I.)], following Wheeler's 1937 introduction of the S-matrix. [Wheeler (December 1937). On the Mathematical Description of Light Nuclei by the Method of Resonating Group Structure.]

Heisenberg's paper attempted to extract from the conceptual structure of the quantum theory of wave fields those terms that were unlikely to be affected by the future change and which would therefore also form a part of the future theory. It avoided the notion of space and time by replacing it with abstract mathematical properties of the S-matrix. The S-matrix related the infinite past to the infinite future in one step, without being decomposable into intermediate steps corresponding to time-slices.

In 1943 Sin-Itiro Tomonaga published a paper in Japanese, of which a translation was published in August 1946, [On a Relativistically Invariant Formulation of the Quantum Theory of Wave Fields] that drew heavily on Dirac (1932) (Relativistic Quantum Mechanics), and Dirac (1933) (The Lagrangian in Quantum Mechanics). It noted that *existing quantum field theory was not relativistic*. Commutation relations referred to points in space at different times. The Schrodinger equation for the vector representing the state of the system was a function of time. Time variable played a different role than space variables. The *probability amplitude* was not *relativistically invariant* in the space-time world. Tomonaga followed Dirac (1932) in *generalizing the notion of probability amplitude as far as was required by the theory of special relativity*. He substituted a four-dimensional form of the commutation relations, then generalized the Schrodinger equation following the Dirac (1933) *many-time* formalism. Tomonaga then introduced his *super many-time theory* in which $[\{H_{12}(P) + h/i\ \partial/\partial C_P\}\ \Psi[C] = 0$ at point P on surface C with infinitely many time variables which represented the local time for each position in the space. This resulted in the *relativistic interaction representation*, in terms of a three-dimensional manifold (space-like "surface") in the four-dimensional space-time world. It was not necessary to also assume time-like surfaces for the variable surface as was required by Dirac. The previous formalism was built up in way too analogous to ordinary *non-relativistic* mechanics. The theory was divided into one section giving the kinematical relations between various quantities at the same instant of time and another section determining the causal relations between quantities at different instants of time; the *commutation relations* belonging to the first section and the *Schrodinger equation* to the second. This way of separating the theory into two sections was very *unrelativistic*, in which the "same instant of time" played a distinct role. The *new formalism consisted of one section giving the laws of behavior of the fields when they were left alone and the other giving the laws determining the deviation from this behavior due to interactions*. This could be carried out *relativistically*. Although the theory was brought into a more satisfactory form no new contents were added. The *divergence difficulties were inherited*, and the fundamental equations admitted only catastrophic solutions *due to non-vanishing zero-point amplitudes of the fields which inhered in the operator $H_{12}(P)$. A more profound modification of theory was required in order to remove this fundamental difficulty*.

In April 1945, Dirac published a paper [On the Analogy Between Classical and Quantum Mechanics] which noted that mathematical methods available for working with *non-commuting quantities* were much weaker than those available for *commuting quantities* owing to the fact that the only functions of *non-commuting variables* that one had been able to define are those expressible algebraically. He showed how this difficulty could be avoided in the case when the *non-commuting quantities* were *observables*, for which it was possible to set up a theory of functions of them (a generalization of the concept known as *well-ordered functions*) of almost the same degree of generality as the usual functions of *commuting variables*. He showed how this theory could be used to make a closer analogy between classical and quantum mechanics, enabling the discussion of trajectories for the motion of a particle in quantum mechanics. A method was given for defining *general functions of non-commuting observables* in quantum mechanics, and developed to provide the formal *probability* for *non-commuting observables* to have numerical values (in general a complex number). This method also enabled the analogy between classical and quantum *contact transformations* to be set up on a more general basis.

The spectrum of hydrogen has a fine structure of the energy levels which according to the *Dirac wave equation* for an electron moving in a Coulomb field was due to the combined effects of *relativistic* variation of mass with velocity and spin-orbit coupling. According to this theory the $2^2S_{1/2}$ state should exactly coincide in energy with the $2^2P_{1/2}$ state which is the lower of the two P states. Previous attempts at measurement had alternated between finding confirmation and discrepancies of as much as eight percent.

In August 1947, Willis Lamb and Robert Retherford published a paper [Fine Structure of the Hydrogen Atom by a Microwave Method] containing the results of their work at the Columbia Radiation Laboratory at Columbia University using a microwave method depending on a novel property of the $2^2S_{1/2}$ level. This showed that, contrary to the *Dirac wave equation*, the $2^2S_{1/2}$ state was higher than the $2^2P_{1/2}$ by about 1000 Mc/sec.

The Lamb and Retherford results showed that the fine structure of the second quantum state of hydrogen did not agree with the *Dirac wave equation.* Schwinger, Weisskopf, and Oppenheimer suggested that this might be due to a *shift of energy levels by interaction of the electron with the radiation field.* This shift came out as infinite in all existing theories and had therefore always been ignored.

Hans Bethe published a response 15 days later [(August 1947). The Electromagnetic Shift of Energy Levels] which noted that it was possible to identify that the most strongly (linearly) divergent term in the level shift was due to an *electromagnetic mass effect* which must exist for a bound as well as for a free electron, and was therefore already included in the *observed mass* of the electron so should be subtracted. He assumed a *relativistic cut-off* in quantum energies (frequencies) of included atomic states. Then calculation of Lamb shift for hydrogen atom using *non-relativistic* ordinary radiation theory gave a shift of the levels due to *radiation interaction* in close agreement with the observed value, and removed the discrepancy with the Dirac theory. He did not carry out *relativistic* calculations.

In December 1948, Takao Tati and Tomonaga published a paper [A Self-Consistent Subtraction Method in the Quantum Field Theory, I] which noted that the procedure previously used by Bloch and Nordsieck and then by Pauli and Fierz in the treatment of problems with the self-field of an electron first separated the radiation field into a part bound to the electron and a part of unbound photons by means of a canonical transformation, then obtained a term in the Hamiltonian that could be interpreted as the *interaction energy* of electron with the radiation field bound to it. This term though infinite gave rise to a modification of mass of electron so that it could be amalgamated into the mass term in the Hamiltonian for the free electron. They dropped off this term to obtain the *observed mass* by reinterpreting the electron mass as already including it, and obtained the same result as Bethe for level-shift of bound electron in his *non-relativistic* treatment.

Tati and Tomonaga considered it to be desirable to obtain a *relativistic* generalization of this treatment of the self-field of an electron using a canonical transformation. This paper addressed *the field reaction problem when there was no external field to the* e^2 *approximation.* It did not address the infinity related to *vacuum polarization* effect that occurs in the *relativistic* treatment. This was simply omitted *as it could not be amalgamated into the equation for free radiation.* They started from the *relativistic* formalism of quantum field theory proposed by Tomanaga [August 1946. On a Relativistically Invariant Formulation of the Quantum Theory of Wave Fields] in which the Schrodinger equation with the Hamiltonian for the interaction density between radiation and electron fields was $\{H(P) - i\ \partial/\partial C_P\}\ \Psi[C] = 0$, where H(P) was the *interaction energy density* between radiation and electron fields at the world point P, Ψ[C] was the generalized Schrodinger functional which was a functional of the space-like variable surface C in the four-dimensional world, and $\partial/\partial C_P$ was its partial functional differentiation at the point P, the point P being considered as lying on C. They decomposed the fields into parts oscillating with positive and negative frequencies corresponding to the electron and the positron, and transformed the Schrodinger functional, retaining terms up to the order e^2. They then expanded the integrand into a Fourier integral and defined a 4-vector to obtain the Hamiltonian for the transformed equation. This enabled use of a first order calculation, and allowed processes through intermediate states such as emission and reabsorption of virtual particles (self-energy of an electron) to be treated as direct processes. They then calculated the commutator and rearranged the factors in each term into the correct order. From this they identified (1) terms to represent *interactions between electrons* caused by exchange of virtual photons including ordinary M¢ller interaction between two electrons, (2) terms connected with the *self-energy* of an electron, (3) terms responsible for the *scattering of a photon* by a free electron, terms for *creation of a pair* by two photons, (4) variables describing the radiation field only representing modification of the radiation field in vacuo due to the *polarization of the vacuum* that cause an infinite energy level shift of the vacuum itself and an infinite *self-energy* of a photon, and (5) a *mass-modifying term* with a logarithmic diverging quantity representing the electromagnetic mass of an electron. Finally, they assumed that the *mass-modifying term* was already included in the free field equation so no level-shift was caused by the interaction between the electron and radiation. Their subtraction method was "*self-consistent*" in this sense. It gave a *relativistic* generalization of the transformation that separated the *radiation field* into a field of "unbound" photons and a field of photons bound to the electron, similar to the Hartree method of the *self-consistent* field in which *interaction* between electrons were considered as a perturbation but some part of its effect was already included.

In September, 1948, Ziro Koba and Tomonaga published a paper [On Radiation Reactions in Collision Processes. I: Application of the "Self-Consistent" Subtraction Method to the Elastic Scattering of an Electron] which applied Tati and Tomonaga's "self-consistent" subtraction method to the *elastic scattering of an electron by a fixed electrostatic potential.* The formal infinity associated with *mass-modifying term* was attributed to a defect of current theory so *the empirical value was substituted for the theoretical value* (assuming it was already included in the free field equation). This maintained the total Hamilton function describing the interaction of the electron and electromagnetic fields unaltered. It required the *interaction-energy* part of the Hamiltonian to undergo a corresponding change by inclusion of a "counter-self-energy" term ("mass-type" correction to the scattering cross-section), which resulted in a finite value for the *self-energy* of electron in the e^2 approximation. When they applied the new formalism to the *elastic scattering of electron*, the effective cross-section for scattering by a fixed potential in zeroth approximation had a value in good agreement with experiment, but as soon as the reaction of electromagnetic field with the electron was taken into account the correction became infinite. The first infinite term in the modified Hamiltonian was eliminated using the *subtraction hypothesis of positron theory*. The second difficulty disappeared by applying the *self-consistent" subtraction method* using a modified Hamiltonian. The first term can also be eliminated if one interprets the sum of the external potential and its infinite correction due the vacuum polarization effect as the physically observable potential and *substitutes the empirical value* for it, while the interaction part of the Hamiltonian is supplemented by an additional "counter-vacuum polarization" term. *This method by no means gave the real solution of the fundamental difficulty of quantum electrodynamics* but revealed the nature of various diverging terms and reduced them to two quantities - the *self-energy* and the *vacuum polarization*. In this way *it became possible in an unambiguous and consistent manner to treat the field reaction problem without touching the fundamental difficulty by employing the finite empirical values instead of the infinite "theoretical" values for these two quantities*.

All divergences were eliminated by counter-terms in a modified form of the Hamilton function, but this was a simple and particularly favorable case. It was necessary to examine whether this method was still effective in fundamental processes between elementary particles.

In December 1948, Koba and Gyo Takeda published a paper [Radiation Reaction in Collision Process, II: Radiative Corrections for Compton Scattering] which applied this method to calculate e^2-corrections to *Klein-Nishina formula for Compton scattering* using the *perturbation method*. It noted two types of diverging terms, one related to *polarization of vacuum* and other to *self-energy* of electron, and showed that it is sufficient to add two new terms to Hamiltonian function to cancel out these infinite terms. The *self-energy* logarithmic divergence was eliminated by introducing a new term into interaction Hamiltonian derived in a plausible way as a "counter-term" which compensates the change in the free field Hamiltonian and conserves the total Hamiltonian unaltered. One of the *vacuum polarization* terms could be treated in similar way, while other term could not be foisted into the theory without radically changing the Maxwell equation for the free electromagnetic field. This was *far from the final settlement of the fundamental problem in theory of elementary particles*.

In June 1948, Koba and Takeda published a letter [Radiative Corrections in e^2 for an Arbitrary Process Involving Electrons. Positrons, and Light Quanta] which investigated further the *radiation correction in collision processes between electrons, positrons and photons* in a general manner and confirmed that a modified Hamiltonian was sufficient to eliminate the divergence difficulty *in the case of the first radiative correction.* They *introduced the "transition diagram method"* [similar to the Stueckelberg (1934) diagrams, later introduced by Feynman at Ponoco in spring 1948 and published in September 1949, and subsequently renamed *Feynman diagrams*] *to analyze the complicated connection between initial and final states through a number of intermediate ones. Electrons* and *positrons* were expressed as *world lines* in *momentum space*, while *emission* and *absorption* of *photons* were described as "*leaps*" of these *world lines*. The diagram acquired two additional *leaps* (the *emission* and *reabsorption* of a *virtual photon*) when *radiative corrections* were taken into account. There were two distinct ways of attaching them, existing *world lines* gained new *leaps* or additional closed *world line* were introduced and coupled with existing ones through *virtual photon*. The first case implied "*mass-type*" or "*self-energy-type*" divergences when the *emission* and *reabsorption* of the *virtual photon* take place in succession, which can be eliminated by a *counter-self-energy term.* The second case represented *vacuum polarization* corrections, which can be subdivided according to the number of *leaps* in additional closed *world lines*, which afford quadratic and logarithmic divergences that are cancelled by the "*counter-terms*" and found not to contribute to the divergence.

In February 1948, Julian Schwinger published a paper [On quantum-electrodynamics and the magnetic moment of the electron] which noted that *electrodynamics unquestionably required revision at ultra-relativistic energies.* It was desirable to isolate those aspects of the current theory that essentially involve high energies and are subject to modification by a more satisfactory theory. He claimed that this goal had been achieved by transforming the Hamiltonian of current *hole theory* to exhibit explicitly the logarithmically divergent *self-energy* of a free electron which arises from the virtual emission and absorption of light quanta. The electromagnetic *self-energy* of a free electron could be ascribed to an *electromagnetic mass* which must be added to the mechanical mass of the electron. The new Hamiltonian involved the *experimental electron mass* rather than the *unobservable mechanical mass*; the electron then interacted with the radiation field only in the presence of an external field such that only an accelerated electron could emit or absorb a light quantum. The *interaction energy* of an electron with an external field was now subject to a *finite* radiative correction. *Polarization of the vacuum* still produced a logarithmically divergent term proportional to the *interaction energy* of an electron in an external field, which was equivalent to altering the value of the *electron charge* by a constant factor with only the final value being identified with the experimental charge. This resulted in the interaction between matter and radiation producing a *renormalization* of the electron charge and mass, with all divergences contained in the *renormalization* factors. The radiative correction for the energy of an electron in an external magnetic field corresponded to an *additional magnetic moment associated with electron spin* of magnitude
$\delta\mu/\mu = (\frac{1}{2}\pi)e^2/hc = 0.001162$. The experimental measurements on the hyperfine splitting of the ground states of atomic hydrogen and deuterium were larger than expected from directly measured nuclear moments. It was found that the additional *electron spin magnetic moment* accounted for measured hydrogen and deuterium *hyperfine structures* to be

$\delta\mu/\mu = 0.00126$ and $\delta\mu/\mu = 0.00131$ respectively, with these discrepancies accounted for by additional spin magnetic moment to the electron of $\delta\mu/\mu = 0.0018 \pm 0.00003$. *The values yielded by this relativistic calculation of the Lamb shift differed only slightly from those conjectured by Bethe on the basis of non-relativistic calculation and were in good accord with experiment.*

In April 1948 Richard Feynman published a paper [Space-Time Approach to Non-Relativistic Quantum Mechanics] which described a third formulation of *non-relativistic* quantum theory in addition to the *differential equation of Schroedinger* and the *matrix algebra of Heisenberg*. This was the *path integral formulation*, utilizing the *action* principle as suggested in Dirac (1933) [The Lagrangian in Quantum Mechanics] and Dirac (1945) [On the Analogy Between Classical and Quantum Mechanics]. In quantum mechanics the probability of an event which can happen in several different ways is the absolute square of a sum of the complex contributions, one from each alternative way, $\varphi_{ac} = \sum_b \varphi_{ab}\varphi_{bc}$ where φ_{ab}, φ_{bc}, φ_{ac} are complex numbers such that $P_{ab} = |\varphi_{ab}|^2$, $P_{bc} = |\varphi_{bc}|^2$, and $P^q_{ac} = |\varphi_{ac}|^2$, where P_{ab} is the probability that if measurement A gave the result *a*, then measurement B will give the result *b*, and P^q_{ac} is the quantum mechanical probability that a measurement of C results in c when it follows a measurement of A giving a. The probability that a particle will be found to have a path lying somewhere within a region of space time is the absolute square of a sum of contributions, one from each path in the region. The contribution from a single path was postulated to be an exponential whose (imaginary) phase is the classical *action* for the path in question where *action* refers to time integral of Lagrangian along a path. This was restricted to a finite time interval. *The probability amplitude for a space-time path was associated with the entire motion of the particle as a function of time rather than with the position of the particle at a particular time.* The total contribution from all paths reaching x, t from the past was the wave function $\psi(x, t)$. This was shown to satisfy Schroedinger's equation. Postulates were established that described *non-relativistic quantum mechanics neglecting spin*, which were mathematically equivalent to Heisenberg and Schroedinger formulations. There were no fundamentally new results, and this formulation *suffered serious drawbacks*. It required an unnatural and cumbersome division of the time interval, and was not formulated so that it was physically obvious that it was invariant under unitary transformations, though improvements could be made through the use of the notation and concepts of mathematics of functionals.

Quantum electrodynamics was built from a classical counterpart that already contained many difficulties which remained upon quantization. It had been hoped that if a classical electrodynamics could be devised which did not contain the difficulty of *infinite self-energy*, and this theory could be quantized, then the problem of a self-consistent quantum electrodynamics would be solved. Previous attempts to address the problem of *infinite self-energy* that resulted from assuming point electron in *relativistic* theory were met with considerable difficulties when attempts were made to quantize them.

In October 1948, Feynman published a paper [A Relativistic Cut-Off for Classical Electrodynamics] which described *a consistent classical theory which he believed could be quantized.* The *potential* at a point in space at a given time depended on the charge at a distance r from the point at a time previous by t = r (taking the speed of light as unity). *Relativistically*, interaction occurred between events whose four-dimensional interval, s, defined by $s^2 = t^2 - r^2$, vanished. It was formulated in terms of *action at a distance*. This resulted in an infinite *action* of a *point electron* on itself. This theory was essentially that of Friedrich Bopp (1942). It modified this idea by assuming that substantial interaction exists as long as the interval s was time-like and less than some small length, *a*, of order of the electron radius. This reduced the infinite *self-energy* to a finite value for accelerations which were not extreme, in which the *action* of an electron on itself appeared as *electromagnetic mass*. It satisfied Maxwell's equations; not the usual *retarded* solution for which there was no *self-force* but the half the *retarded* plus half the *advanced* solution in Wheeler and Feynman (April, 1945) [Interaction with the Absorber as the Mechanism of Radiation]. The effect of the modification was to change slightly the field of one particle on another when they are very close, and to add a *self-force*. Feynman concluded that there was *little reason to believe that the ideas used here to solve the divergences of classical electrodynamics would prove fruitful for quantum electrodynamics*.

In November 1948, Feynman published a paper [Relativistic Cut-Off for Quantum Electrodynamics] which described a model based on the quantization of the classical theory for which all quantities automatically come out finite described in his previous paper. This contained an arbitrary function on which numerical results depend. The only term that depended significantly (logarithmically) on the cut-off frequency was the *self-energy* which could be used to renormalize the electron mass. The remaining terms were nearly independent of the function. This model applied only to results for processes in which virtual quanta were emitted and absorbed. *Terms representing processes involving a pair production followed by annihilation of the same pair were infinite and not made convergent by this scheme*. Problems of *permanent emission* and the position of *positron theory* still need to be addressed. *This paper may be looked upon as presenting an arbitrary rule to cut off at high frequencies in a relativistically invariant manner the otherwise divergent integrals appearing in quantum field theories*. It produced finite invariant *self-energy* for a free electron, but the problem of *polarization of the vacuum* was not solved. An alternative cut-off procedure which eliminated high frequency intermediate states offered to solve the *vacuum polarization* problems as well.

Lack of convergence in current formulations of *quantum electrodynamics* indicated that a revision of electrodynamic concepts at *ultra-relativistic* energies was necessary. Elementary phenomenon in which divergences occurred as a result of virtual transitions involving particles with unlimited energy were *polarization of the vacuum* and *self-energy of the electron* which expressed *the interaction of the electromagnetic and matter fields with their own vacuum fluctuations*. This altered the constants characterizing the properties of the individual fields and their mutual coupling by infinite factors. The question was whether all divergencies could be isolated in such unobservable *renormalization* factors.

In November 1948, Schwinger published a paper [Quantum Electrodynamics. I. A Covariant Formulation], which was occupied with *the formulation of a completely covariant electrodynamics*. He asserted that manifest covariance with respect to Lorentz and gauge transformations was essential in a divergent theory. Customary *canonical commutation relations* failed to exhibit the desired covariance since they referred to field variables at equal times and different points of space. They could be put in a covariant form by replacing the four-dimensional surface t = const. by a space-like surface. This offered the advantage over the Schrodinger representation, in which all operators refer to the same time, by providing a distinct separation between *kinematical* and *dynamical* aspects. A formulation that retained the evident covariance of the Heisenberg representation but offered something akin to the Schrodinger representation could be based on the distinction between the properties of *non-interacting fields*, and the effects of *coupling between fields*. In the second section, he constructed a *canonical transformation* that changed the *field equations* in the *Heisenberg representation* into those of *non-interacting fields*. He added *supplementary condition* restricting the admissible states of the system and the *commutation relations* to the *equations of motion* to obtain a description of the coupling between fields in terms of a varying *state vector*. Then it was a simple matter to evaluate commutators of *field* quantities at arbitrary *space-time* points. He thus obtained an obviously covariant and practical form of quantum electrodynamics expressed in a mixed Heisenberg-Schrodinger representation, called the *interaction representation*. The third section discussed the *covariant* elimination of the longitudinal field in which the customary distinction between longitudinal and transverse fields was replaced by a suitable *covariant* definition. The fourth section described collision processes in terms of an invariant *collision operator*, which was the unitary operator that determined the over-all change in state of a system as the result of interaction. He noted that a *second paper* would treat the problems of the electron and photon *self-energy* together with the *polarization of the vacuum*, and a *third paper* was concerned with the determination of the *radiative corrections* to the properties of an electron and the comparison with experiment. This was not addressed in that paper; it stated that "*radiative corrections to energy levels* will be treated in the next paper of the series" but *this did not appear nor are there any references to it*.

The recent and independent formulations of quantum electrodynamics by Tomonaga, Schwinger, and Feynman have made two notable advances, the foundations and applications of the theory have been simplified by being presented in *a completely relativistic way* and the *divergence difficulties have been at least partially overcome*. The advantages of the Feynman formulation were simplicity and ease of application while those of Tomonaga-Schwinger were generality and theoretical completeness,

In February 1949, Freeman Dyson, at age 24, published a paper [The Radiation Theories of Tomonaga, Schwinger, and Feynman] which presented a unified development of quantum electrodynamics embodying the main features of the Tomonaga-Schwinger and the Feynman radiation theories. It aimed to show how the Schwinger theory could be applied to specific problems in such a way as to incorporate the ideas of Feynman. The emphasis was on the application of the theory. The main results were general formulas from which radiative reactions on the motions of electrons could be calculated. It divided *energy-density* into two parts, the *energy of interaction* of two fields with each other and the energy produced by external forces. *Interaction energy* alone was treated as a perturbation. Important results of the paper were the *equation of motion* $i\hbar c[\partial\Omega/\partial\sigma(x_0)] = \{S(\sigma)\}^{-1}H^e(x_0)S(\sigma)\Omega$ for the *state vector* $\Omega(\sigma)$, and the interpretation of the *state vector* Ω. It simplified the Schwinger theory for using it for calculations, and demonstrated the equivalence of the theories within their common domain of applicability. In the *Schwinger theory* the aim was to calculate the matrix elements of the "*effective external potential energy*" between *states* specified by their *state vectors*. In the *Feynman theory* the basic principle was to "preserve symmetry between past and future" so the matrix elements of the operator were evaluated in a "*mixed representation*" in which the matrix elements were calculated between an *initial state* specified by its *state vector* and a *final state* specified by its *state vector*. A graph corresponding to a particular matrix element was used not merely as an aid to calculation but as a picture of the physical process which gave rise to that matrix element. Dyson derived fundamental formulas for the operator in the *equation of motion* for the *state vector* $\Omega(\sigma)$ which represented the interaction of a physical particle with an external field for both the Schwinger and the Feynman theories, and a set of rules by which matrix element of Feynman operator might be written down in a form suitable for numerical evaluation. He showed the equivalence of two theories, and developed graphical representations of the matrix elements. He noted that *the theory as a whole could not be put into a finally satisfactory form so long as divergencies occurred in it however skillfully these divergencies were circumvented.* The present treatment should be regarded as justified by its success in applications rather than by its theoretical derivation. The *paper suffered from a series of significant errors*.

In March and June 1949, Hiroshi Fukuda, Yoneji Miyamoto, and Tomonaga published a sequel [A Self-Consistent Subtraction Method in the Quantum Field Theory] to Tati and Tomonaga (December 1948) which addressed the case when an *external field is present*. The unbound fields no longer propagate freely and transitions took place. The electron wave was able to change its state of propagation not only elastically by the external field but also by emitting or absorbing unbound photons. Thus, *an interaction appeared between an electron and radiation* which were free from interaction with each other in the absence of the external field. This interaction caused a radiation reaction upon the electron so that the motion of the electron would be modified. *In this paper they gave an example of how one could calculate this reaction and obtained a finite result as the consequence of the subtraction procedure*. They were able to obtain a *finite radiative level-shift* of a bound electron in the external field and a *finite e^2-correction* to the *scattering cross section*, and estimated the Lamb shift for hydrogen atom as 1076 Mcycles by adding additional terms to Bethe's value. They thus successfully obtained finite answers for these field reaction problems of the magnitude agreeing with experimental results. But they also noted "*we must nevertheless confess that the calculation carried out in this paper is still unsatisfactory because we had to make a non-relativistic treatment in the evaluation of the* *effective energies* ... Our work is therefore only of a provisionary character. ... It is *still problematic whether this procedure corresponds to the correct prescription*".

The new formalism of quantum field theory was more satisfactory from the stand-point of *relativistic invariance* as it was formulated in terms of *invariant space-time* concepts, but it employed *canonical formalism* in which the time variable was unnecessarily distinguished from the space variables.

In September 1949, Suteo Kanesawa and Koba published a paper [A Remark on Relativistically Invariant Formulation of the Quantum Field Theory] which showed that *the same formalism could be reached without referring to the canonical formulation.* In the current theory the *interaction Hamilton density* coincided with the *interaction Lagrange density*, except the sign, if the surface-dependent part was neglected. This suggested the idea of connecting the *generalized interaction Lagrange density* directly to generalized Schrodinger equation. The guiding principle was the integrability of the Tomonaga-Schwinger equation. This paper showed how to apply the new method to the general case when a canonical description was impossible. *Problematic aspects of theory such as universal length or ultraviolet divergencies were not addressed.* The aim of the paper lay in demonstrating the possibility of including more general kinds of interactions in field theory other than those which allowed a canonical description.

The initial calculation of the correct Lamb shift by Bethe (August, 1947) was *non-relativistic.* Various *relativistic* calculations followed the next year. A comedy of errors ensued: *both Feynman and Schwinger made an incorrect patch between hard and soft photon processes, and so obtained identical, but incorrect, predictions for the Lamb shift,* and the weight of their reputations delayed the publication of the correct, if pedestrian, calculations by Kroll & Lamb and French & Weisskopf until February 1949.

In February 1949, Norman Kroll and Willis Lamb published a paper [On the self-energy of a bound electron] which contained their calculation of the electromagnetic shift of energy levels of a bound electron based on the usual formulation of *relativistic* quantum electrodynamics and *positron theory.* This gave 1052 megacycles per second for the $^2 2S_{1/2} - {}^2 2P_{1/2}$ shift in hydrogen in close agreement with the *non-relativistic* calculation of 1040 megacycles per second by Bethe.

In April 1949, James Bruce French and Victor Weisskopf published a paper [The Electromagnetic Shift of Energy Levels] which contained their *relativistic* calculation of Lamb shift using the conventional form of perturbation theory and removing the infinite *self-energy* of the electron by subtracting a "mass operator" from the Hamiltonian. This gave 1051 megacycles per second for $2s_{1/2} - 2p_{1/2}$ separation in hydrogen, compared with the *non-relativistic* calculation of 1040 megacycles per second by Bethe, and the *relativistic* calculation of 1052 megacycles per second by Kroll & Lamb. The results of the different calculations suggested that they were not dependent on whether they are *relativistic* or *non-relativistic*.

In February 1949, Schwinger published a paper [Quantum Electrodynamics. II. Vacuum Polarization and Self-Energy] which applied the *interaction representation* to the *polarization of the vacuum* and the *self-energies of the electron and photon*. In the *first section* the vacuum of the non-interacting *electromagnetic* and *matter* fields was *covariantly* defined as the *state* for which the eigenvalue of an arbitrary time-like component of the *energy-momentum four-vector* was an absolute minimum. The covariant decomposition of field operators into positive and negative frequency components was introduced to characterize the *vacuum state vector*. He showed that the *state vector for the electromagnetic vacuum* was annihilated by the positive frequency part of transverse four-vector potential and *state vector for matter vacuum* was annihilated by positive frequency part of Dirac spinor and its charge conjugate. These properties of *vacuum state vector* were employed in the calculation of the *vacuum expectation* values of quadratic field quantities, specifically the *energy-momentum tensors* of the independent *electromagnetic* and *matter fields* and the *current four-vector*. It was inferred that the *electromagnetic energy-momentum tensor* and *current vector* must vanish in the vacuum, while the *matter field energy-momentum tensor* vanished in the vacuum only by the addition of a suitable multiple of the unit tensor. The *second section* treated the *induction of a current in the vacuum by an external electromagnetic field*. It was supposed that the *external electromagnetic field* did not produce actual electron-positron pairs, and considered only the phenomenon of *virtual pair creation*. This restriction was introduced by requiring that the establishment and subsequent removal of the external field produced no net change in state for the *matter field. He demonstrated that the induced current at a given space-time point involved the external current in the vicinity of that point and not the electromagnetic potentials*. This *gauge invariant* result showed that a light wave propagating at remote distances from its source induced no current in the vacuum and was therefore undisturbed in its passage through space. This indicated an *absence of a light quantum self-energy effect*. The *current* induced at a point consisted of two parts, a logarithmically divergent multiple of the *external current* at that point, *which produced an unobservable renormalization of charge*, and a more involved finite contribution, which is the physically significant *induced current. The third section considered the modification of matter field properties arising from the interaction with the vacuum fluctuations of the electromagnetic field*. This analysis was carried out with two alternative formulations, one employing the complete *electromagnetic potential* together with a *supplementary condition*, the other using the *transverse potential* with the variables of the *supplementary condition* eliminated. It was noted that no real processes were produced by the first order coupling between the fields. Alternative *equations of motion* for the *state vector* were constructed from which the first order interaction term had been eliminated and replaced by the *second order coupling* which it generated. The latter included the *self-action* of individual particles and light quanta, the *interaction* of different particles, and a *coupling* between particles and light quanta which produced such effects as Compton scattering and two quantum pair annihilation. It was concluded from a comparison of the alternative procedures that, *for the treatment of virtual light quantum processes, the separate consideration of longitudinal and transverse fields was an inadvisable complication*. The light quantum the *self-energy* term was shown to vanish, while that for a particle had the form for a change in *proper mass* but was logarithmically divergent in agreement with previous calculations. The identification of *self-energy* effect with a change in *proper mass* was confirmed by removing this term from the state vector *equation of motion*, which altered the *matter field equations of motion* in the expected manner. It was verified that *the energy and momentum modifications produced by self-interaction effects were entirely accounted for by the addition of the electromagnetic proper mass to the mechanical proper mass—an unobservable mass renormalization*. An appendix was devoted to the construction of several invariant functions associated with the *electromagnetic* and *matter* fields.

In June 1949, Dyson published a paper [The S Matrix in Quantum Electrodynamics] in which the covariant *quantum electrodynamics* of Tomonaga, Schwinger and Feynman was used as basis for a general treatment of *scattering problems* involving electrons, positrons, and photons. It addressed the relation between Schwinger and Feynman theories when the restriction to one-electron problems was removed. In these more general circumstances the two theories appeared as complementary rather than identical. *The Feynman method was essentially a set of rules for the calculation of the elements of the Heisenberg S matrix* corresponding to any physical process, and could be applied directly to all kinds of scattering problems. *The Schwinger method evaluated radiative corrections by exhibiting them as extra terms appearing in the Schrodinger equation*. The paper showed the practical usefulness of the S matrix as a connecting link between the Feynman technique of calculation and Hamiltonian formulation of quantum electrodynamics. The *Feynman radiation theory* provided a set of rules for the calculation of matrix elements between *states* composed of any number of ingoing and outgoing free particles. It was thus an S matrix theory. The paper showed that *scattering processes*, including the creation and annihilation of particles, were completely described by Heisenberg's S matrix. The elements of this matrix were calculated by a consistent use of perturbation theory to any desired order. Detailed rules were given for carrying out such calculations, and divergences arising from higher order radiative corrections were removed from the S matrix by consistent use of *mass and charge renormalization*. The operators so calculated were divergence-free*, the divergent parts at every stage of the calculation being explicitly dropped after being separated from the finite parts*. This involved *extensive manipulations of infinite quantities* and had to be justified *a posteriori* by the fact that they ultimately lead to a clear separation of finite from infinite expressions. Dyson claimed that such an *a posteriori* justification of dubious manipulations was an inevitable feature of any theory which aims to extract meaningful results from not completely consistent premises. The perturbation theory of this paper was *applicable only to a restricted class of problems* when not only the radiation interaction but also the external potential was small enough to be treated as a perturbation. *It did not give a satisfactory approximation either in problems involving bound states or in scattering problems at low energies*. In other situations the Schwinger theory would have to be used in its original form,. The *problems of extending this treatment to include bound-state phenomena and of proving convergence of the theory as the order of perturbation itself tends to infinity was not addressed*. This analysis suggested that *the divergences of electrodynamics were directly attributable to the fact that* *the Hamiltonian formalism was based upon an idealized conception of measurability*. Now it was no longer a compelling necessity for a future theory to abandon some essential features of the present electrodynamics. The present electrodynamics was certainly incomplete, but was no longer certainly incorrect.

The paper published by Richard Eden in September 1949 [Heisenberg's S Matrix for a System of Many Particles] was included as a vignette. Dr. Eden provided lectures in Nuclear Physics and Quantum Theory to the author in 1963 as part of his M.A. Cantab. degree. Eden's thesis advisors were Dirac and Heisenberg. The author attended a joint lecture by Heisenberg and Dirac on May 23, 1963, (both speaking whilst writing and erasing on dueling pairs of blackboards at the same time) at the old Cavendish Laboratory.

In September 1949, Feynman published a paper [The Theory of Positrons] which was the first of a set of papers dealing with the solution of problems in *quantum electrodynamics* in which the main principle was to deal directly with *solutions* of the Hamiltonian differential equations rather than with equations themselves. This paper *analyzed the motion of electrons and positrons in given external potentials, neglecting interaction*, by dealing directly with the *solutions* of the Hamiltonian time differential equations rather than with the equations. A second paper below considers interactions (*quantum electrodynamics*). The problem of charges in a fixed potential was usually treated by method of second quantization of the electron field using the *theory of holes*. Here Dirac's "hole theory" was replaced by reinterpreting the *solutions* of the Dirac equation. The results were simplified by *following the charge rather than the particles* because *the number of particles is not conserved whereas charge is conserved*. In the approximation of classical *relativistic* theory the creation of an electron pair (electron and positron) was represented by the start of two world lines from the point of creation. *Following the charge rather than the particles corresponds to considering the continuous world line as a whole rather than breaking it up into its pieces*. This over-all *space-time* point of view led to considerable simplification in many problems. Quantum mechanically the direction of the world lines was replaced by the direction of propagation of waves. This was quite different from Hamiltonian method which considered the future as developing continuously from the past. In a scattering problem this over-all view of the complete scattering process was similar to the *S-matrix view-point of Heisenberg*. The temporal order of events during scattering which was analyzed in such detail by the Hamiltonian differential equation was irrelevant. This development stemmed from the idea that in *non-relativistic* quantum mechanics the *amplitude* for a given process could be considered as the sum of the *amplitude* for each *space-time* path available. In the *relativistic* case the restriction that the paths must proceed always in one direction in time was removed. The results were more easily understood from the more familiar physical viewpoint of *scattered waves* used in this paper. After the equations were worked out physically the proof of the equivalence to the *second quantization theory* was found. The solution in terms of boundary conditions on *wave function* contained all possibilities of pair formation and annihilation together with the ordinary scattering processes. Negative energy states appeared in space-time as waves traveling away from the external potential backwards in time (as suggested in Stückelberg

(December 1942)). Such a wave corresponded to a positron approaching the potential and annihilating the electron. A particle moving forward in time (electron) in a potential might be scattered forward in time (ordinary scattering) or backward (pair annihilation). When moving backward (positron) it might be scattered backward in time (positron scattering) or forward (pair production). The *amplitude* for a transition from an initial to a final state could be analyzed to any order in the *potential* by considering it to undergo a sequence of such scatterings. The *amplitude* for a process involving many such particles was the product of the *transition amplitudes* for each particle. Vacuum problems did not arise for charges which did not interact with one another. The equivalence to the *theory of holes* in second quantization was demonstrated in an appendix.

Electrodynamics was viewed as direct *interaction* at a distance between *charges* rather than the behavior of a *field* (Maxwell's equations). The *field point of view*, which separated the production and absorption of light, was most practical for problems involving *real quanta* while the *interaction view* was best for a discussion of *virtual quanta* when dealing with close collisions of particles or their actions on themselves. The Hamiltonian method was not well adapted to represent direct *action at a distance* between *charges* because action was delayed. This forced the use of the field viewpoint rather than the interaction viewpoint. For *collisions* it was much easier to treat the process as a whole. The effects of longitudinal and transverse waves could be combined.

In a continuation of the previous paper in the same issue [Space-Time Approach to Quantum Electrodynamics.] Feynman applied the same technique to include *interactions* and, in that way, to express in simple terms of the solution of problems in *quantum electrodynamics* rather than the differential equations from which they came. (1) It was shown that *a considerable simplification could be attained by writing down matrix elements for complex processes in electrodynamics*. A physical point of view was available which permitted them to be written down directly. The simplification resulted from the fact that previous methods separated into individual terms processes that were closely related physically. (2) *Electrodynamics was modified by altering the interaction of electrons at short distances*. All matrix elements were now finite with the exception of those relating to *vacuum polarization*. The latter were evaluated in a manner suggested by Pauli and Bethe to give finite results. Phenomena directly observable were insensitive to the details of the modification. Feynman began with the solution in *space* and *time* of the Schrodinger equation for particles interacting instantaneously. The solution was expressed in terms of *matrix elements, derived from the Lagrangian form of quantum mechanics*. It was then modified in accordance with the requirements of the Dirac equation and the phenomenon of pair creation. This was made easier by reinterpreting the *theory of holes*. This was generalized to *delayed interactions* of *relativistic* electrons. For practical calculations expressions were developed in a power series of $e^2/\hbar c$. The *derivation would appear in a separate paper*. By forsaking Hamiltonian method, the wedding of *relativity* and *quantum mechanics* was accomplished most naturally. The *relativistic* invariance became self-evident but the *matrix elements for complex processes* and the *self-energy diverged.* Then it was possible to see how the matrix elements could be written down directly. Feynman then introduced a modification in the interaction between charges at short distances - *he assumed that substantial interaction existed as long as the four-dimensional interval was time-like and less than some small length of order of the electron radius* [Feynman (October, 1948). A Relativistic Cut-Off for Classical Electrodynamics.]. *Convergence factors* were then introduced such that the integrals with their convergence factors now converged. The *self-energy* was now convergent, and corresponded to a correction to the electron mass. *All matrix elements were now finite* with exception of those relating to *vacuum polarization*, but *a strict physical basis for rules of convergence was not known.* After *mass and charge renormalization*, the results were equivalent to those of Schwinger in which the terms corresponding to corrections in mass and charge were identified and removed from the expressions for real processes. *Although in the limit the two methods agreed neither method appeared to be completely satisfactory theoretically*. The practical advantage of the new method was that ambiguities could be more easily resolved by direct calculation of otherwise divergent integrals. A complete method was therefore available for calculations of all processes involving electrons and photons. It showed how matrix element for any process could be written down directly in *Feynman diagrams*. This paper included the first published "*Feynman diagram*". *This attempt to find a consistent modification of quantum electrodynamics was incomplete*. It was *not at all clear that the convergence factors did not upset the physical consistency of the theory.*

A covariant form of *quantum electrodynamics* had been developed and applied by Schwinger in the previous articles of this series to two elementary phenomena that are produced by the *vacuum fluctuations of the electromagnetic field*. These applications were the *polarization of the vacuum* expressing the modifications in the properties of an electromagnetic field arising from its interaction with the *matter field* vacuum fluctuations, and the *electromagnetic mass of the electron* embodying the corrections to the mechanical properties of the *matter field* in its single particle aspect. In these problems *the divergences that mar the theory were found to be concealed in unobservable charge and mass renormalization factors*. The previous paper [Quantum Electrodynamics. II. Vacuum Polarization and Self-Energy] was confined to consideration of *vacuum polarization* produced by the field of a prescribed *current* distribution.

In September 1949, Schwinger published a paper [Quantum Electrodynamics. III. The Electromagnetic Properties of the Electron—Radiative Corrections to Scattering.] in which he now *considered how the induction of a current in a vacuum by an electron resulted in an alteration in its electromagnetic properties* revealed by scattering in Coulomb field and energy level displacements. *This paper was concerned with the computation of the second-order corrections to the current operator as modified by the coupling with the vacuum electromagnetic field and its application to electron scattering* by a Coulomb field. It applied canonical transformation to *renormalize the electron mass*. A correction to the *current operator* produced by coupling with electromagnetic field was developed in power series, of which the first- and second-order terms were retained. Second-order modifications in the *current operator* were obtained which were of same general nature as the previously treated *vacuum polarization current*, apart from a contribution in the form of a *dipole current*. The latter implied a fractional increase of $\alpha/2\pi$ in the *spin magnetic moment* of electron. The only flaw in the second-order current correction was a *logarithmic divergence attributable to an infra-red catastrophe*. In the presence of an *external field* the first-order *current* correction introduced a compensating divergence. Thus, the second-order corrections to the electromagnetic properties of a particle could not be completely stated without regard for the manner of exhibiting them by an *external field*. Accordingly, in the second section, the interaction of three systems - the *matter field*, the *electromagnetic field*, and a given *current distribution* - was considered. It was shown that this could be described in terms of an *external potential* coupled to the *current* operator as modified by the interaction with the *vacuum electromagnetic field*. This was applied to *scattering of an electron* by an *external field* which was regarded as a small perturbation. It was convenient to calculate the total rate at which collisions occur and then identify the *cross sections* for individual events. The correction to the *cross section* for radiation-less scattering was determined by the second-order correction to the *current* operator. Scattering that was accompanied by a single quantum emission was a consequence of the first-order *current* correction. The final object of calculation was the *differential cross section* for scattering through a given angle with a prescribed maximum energy loss which was completely free of divergences. An *Appendix* was devoted to an alternative treatment of the *polarization of the vacuum* by an external field. It was noted that *radiative corrections to energy levels would be treated in the next paper of the series* [*but this did not appear*].

In two previous papers Feynman gave rules for the calculation of the matrix element for any process in electrodynamics. No complete proof of the equivalence of these rules to conventional electrodynamics was given, nor was a closed expression given valid to all orders in $e^2/\hbar c$.

In November 1950, Feynman published a paper [Mathematical Formulation of the Quantum Theory of Electromagnetic Interaction.] which addressed these formal omissions, giving the derivations of the formulas of Feynman (September 1949). [Space-Time Approach to Quantum Electrodynamics.] by means of the form of quantum mechanics given in Feynman (April 1948).[Space-Time Approach to Non-Relativistic Quantum Mechanics]. The derivation of the rules used the Lagrangian form of quantum mechanics, which permitted the motion of any part to be solved first and the results to be used in the solution of the motion of other parts. The electromagnetic field is a simple system. The *interaction of matter (electrons and positrons) and the field were analyzed by first solving for the behavior of the field in terms of the coordinates of the matter*. Integration over field oscillator coordinates eliminated field variables from the *equations of motion* of electrons. He then addressed the behavior of electrons. In this way, all of the rules given in the second paper were derived. This was *restricted to cases in which the particle's motion was non-relativistic* but Feynman claimed that the transition of the final formulas to the relativistic case was direct and the proof could have been kept relativistic throughout. The *generalized formulation was unsatisfactory because for situations of importance it gave divergent results*. Problems of divergences were not discussed. The *theory assumed that substantial interaction existed as long as the interval was time-like and less than some small length of the order of the electron radius.*

In October 1951, Feynman published another paper [An Operator Calculus Having Applications in Quantum Electrodynamics] which suggested an alteration in the mathematical notation for handling operators. The new notation permitted a considerable increase in the ease of manipulation of complicated expressions involving operators. *No new results were obtained in this way. The mathematics was not completely satisfactory and no attempt had been made to maintain mathematical rigor*. Feynman believed that to put these methods on a rigorous basis might be quite a difficult task, beyond the abilities of the author.

In October 1962, Dirac published a paper [The Conditions for a Quantum Field Theory to be Relativistic] which noted that a quantum field theory in agreement with *special relativity* could be built up from the infinitesimal operators of translation and rotation. These operators were expressible in terms of a momentum density and an energy density. The momentum density was determined by the geometrical properties of the fields concerned. The energy density had to satisfy commutation relations for which certain conditions hold.

On December 11, 1966, Tomonaga, received the 1965 Nobel Prize in Physics, together with Schwinger and Feynman, "for their fundamental work in *quantum electrodynamics*, with deep-ploughing consequences for the physics of elementary particles". Tomonaga received the prize for his reformulation of the *relativistic* theory for the interaction between charged particles and electromagnetic fields as a consequence of the observation of the Lamb shift in 1947, in which the supposed single energy level within a hydrogen atom was instead proven to be two similar levels. Tomonaga solved this problem in 1948, when he was 42 years old, through a "*renormalization*" and thereby contributed to a new quantum electrodynamics. Tomonaga's Nobel Prize lecture [Tomonaga, S. (May 6, 1966). Nobel Lecture. Development of Quantum Electrodynamics - 1966: Personal recollections.] described the evolution of his work and that of others since 1932 when he started his research career up until 1948.

On December 11, 1966, Feynman received the 1965 Nobel Prize in Physics, together with Tomonaga and Schwinger, "for their fundamental work in *quantum electrodynamics*, with deep-ploughing consequences for the physics of elementary particles". Feynman was awarded the prize for contributing to creating a new quantum electrodynamics by introducing Feynman diagrams in 1948, when he was 30 years old. These were graphic representations of various interactions between different particles which facilitated the calculation of interaction probabilities. Feynman's Nobel Prize lecture [Feynman, R. P. (December 11, 1965). Nobel Lecture. The Development of the Space-Time View of Quantum Electrodynamics.] described the sequence of ideas which occurred and by which he finally came out the other end with an unsolved problem for which he ultimately received the Nobel prize. He represented conventional electrodynamics with *retarded interaction*, not his half-advanced and half-retarded theory. The *action* expression was not used. The idea that *charges* do not act on themselves was abandoned. The *path-integral formulation* of quantum mechanics was useful for guessing at final expressions and at formulating the general theory of electrodynamics in new ways though was not absolutely necessary. The rest of his work was simply to improve the techniques then available for calculations, *making diagrams to help analyze perturbation theory quicker*. He concluded: "*I don't think we have a completely satisfactory relativistic quantum-mechanical model*, … Therefore, I think that the *renormalization theory* was simply a way to sweep the difficulties of the divergences of electrodynamics under the rug".

On December 11, 1966, Schwinger received the 1965 Nobel Prize in Physics, together with Tomonaga and Feynman, "for their fundamental work in *quantum electrodynamics*, with deep-ploughing consequences for the physics of elementary particles". Schwinger received the prize for his reformulation of the *relativistic* theory for the interaction between charged particles and electromagnetic fields as a consequence of the discovery that the electron's *magnetic moment* proved to be somewhat larger than expected. He solved this problem in 1948, when he was 30 years old, through "*renormalization*" and thereby contributed to a new quantum electrodynamics. In his Nobel Prize lecture [Schwinger, J. (December 11, 1965). Nobel Lecture. Relativistic Quantum Field Theory.], Schwinger did not mention any of his 1948 and 1949 papers for which he was awarded the Nobel Prize. Instead, he focused on papers he had written between 1951 and 1965. He started by describing the logical foundations of *quantum field theory* or *relativistic quantum mechanics* which he defined *as the synthesis of quantum mechanics with relativity*. He noted improvements in the formal presentation of quantum mechanical principles by himself, described in a series of six papers on the theory of quantized fields published in *Physical Review* between June 1951 and June 1954, and by Feynman, described in Feynman's April 1948 Space-Time Approach to Non-Relativistic Quantum Mechanics, utilizing the concept of *action*, based on a study of Dirac concerning the correspondence between the quantum *transformation function* and the classical *action*. He identified two distinct formulations of quantum mechanics – his *differential formulation* utilizing a *differential* version of the *action* principle, and the *path integral formulation* of Feynman. He claimed that his *differential* version transcended the *correspondence principle* and incorporated on the same footing the two different kinds of quantum dynamical variable that are demanded empirically by the two known varieties of particles obeying Bose-Einstein or Fermi-Dirac statistics. The *quantum action principle was a differential statement about time transformation functions*; *all quantum-dynamical aspects of the system were derived from a single dynamical principle*. He also claimed that *quantum field theory* had failed no significant test nor could any decisive confrontation be anticipated in the near future, contrary to Feynman's reservations and Schwinger's statements in the preface to his 1958 book. He stated that *classical mechanics was a determinate theory*; knowledge of the *state* at a given time permits precise prediction of the result of measuring any property of the system. *Quantum mechanics was only statistically determinate*, it is the *probability* of attaining a particular result on measuring any property of the system not the outcome of an individual microscopic observation that is predictable from knowledge of the *state*. The *relativistic* structure of the *action principle* was completed by demanding that it present the same form independently of the particular partitioning of *space-time* into space and time. This was facilitated by the appearance of the *action operator* - the time integral of the Lagrangian - as the *space-time* integral of the Lagrange function. He also discussed a further eleven papers that he had published in *Physical Review* between July 1962 and October 1965 which attempted to extend his theory to gravitational fields and develop a *field theory of matter*.

Trevor G. Underwood
18 SE 10th Ave
Fort Lauderdale, FL33301

March 31, 2023

Werner Heisenberg – 1932 Nobel Lecture, December 11, 1933. *The development of quantum mechanics.*

[https://www.nobelprize.org/prizes/physics/1932/heisenberg/lecture/.]

> Heisenberg was awarded the 1932 Nobel Prize "for the creation of quantum mechanics, the application of which has, inter alia, led to the discovery of the allotropic forms of hydrogen".
>
> In Niels Bohr's theory of the atom, electrons absorb and emit radiation of fixed wavelengths when jumping between fixed orbits around a nucleus. The theory provided a good description of the spectrum created by the hydrogen atom, but needed to be developed to suit more complicated atoms and molecules. In 1925, Werner Heisenberg formulated a type of quantum mechanics based on matrices. In 1927 he proposed the "uncertainty relation", setting limits for how precisely the position and velocity of a particle can be simultaneously determined. [Werner Heisenberg – Facts. NobelPrize.org. https://www.nobelprize.org/prizes/physics/1932/heisenberg/facts/.]

In his Nobel Prize lecture, Heisenberg noted that "the impossibility of harmonizing the Maxwellian theory with the pronouncedly visual concepts expressed in the hypothesis of light quanta subsequently compelled research workers to the conclusion that *radiation phenomena can only be understood by largely renouncing their immediate visualization*. ... Classical physics seemed the limiting case of visualization of a fundamentally unvisualizable microphysics, the more accurately realizable the more Planck's constant vanishes relative to the parameters of the system". After describing the development of the current theory of quantum mechanics, he concluded that "a visual description for the atomic events is possible only within certain limits of accuracy - but within these limits the laws of classical physics also still apply. Owing to these limits of accuracy as defined by the uncertainty relations, moreover, a visual picture of the atom free from ambiguity has not been determined. On the contrary the corpuscular and the wave concepts are equally serviceable as a basis for visual interpretation. The laws of quantum mechanics are basically statistical". He noted that "the attention of the research workers was now primarily directed to *the problem of reconciling the claims of the special relativity theory with those of the quantum theory*. ... The attempts made hitherto to achieve a *relativistic* formulation of the quantum theory are all based on visual concepts so close to those of classical physics that it seems impossible to determine the fine-structure constant within this system of concepts".

> Quantum mechanics, on which I am to speak here, arose, in its formal content, from the endeavor to expand Bohr's principle of correspondence to a complete mathematical scheme by refining his assertions. The physically new viewpoints that distinguish quantum mechanics from classical physics were prepared by the researches of various investigators engaged in analyzing the difficulties posed in Bohr's theory of atomic structure and in the radiation theory of light.

In 1900, through studying the law of black-body radiation which he had discovered, Planck had detected in optical phenomena a discontinuous phenomenon totally unknown to classical physics which, a few years later, was most precisely expressed in Einstein's hypothesis of light quanta. The impossibility of harmonizing the Maxwellian theory with the pronouncedly visual concepts expressed in the hypothesis of light quanta subsequently compelled research workers to the conclusion that *radiation phenomena can only be understood by largely renouncing their immediate visualization.* The fact, already found by Planck and used by Einstein, Debye, and others, that the element of discontinuity detected in radiation phenomena also plays an important part in material processes, was expressed systematically in Bohr's basic postulates of the quantum theory which, together with the Bohr-Sommerfeld quantum conditions of atomic structure, led to a qualitative interpretation of the chemical and optical properties of atoms. The acceptance of these basic postulates of the quantum theory contrasted uncompromisingly with the application of classical mechanics to atomic systems, which, however, at least in its qualitative affirmations, appeared indispensable for understanding the properties of atoms. This circumstance was a fresh argument in support of the assumption that the natural phenomena in which Planck's constant plays an important part can be understood only by largely foregoing a visual description of them. Classical physics seemed the limiting case of visualization of a fundamentally unvisualizable microphysics, the more accurately realizable the more Planck's constant vanishes relative to the parameters of the system. This view of classical mechanics as a limiting case of quantum mechanics also gave rise to Bohr's principle of correspondence which, at least in qualitative terms, transferred a number of conclusions formulated in classical mechanics to quantum mechanics. In connection with the principle of correspondence there was also discussion whether the quantum-mechanical laws could in principle be of a statistical nature; the possibility became particularly apparent in Einstein's derivation of Planck's law of radiation. Finally, the analysis of the relation between radiation theory and atomic theory by Bohr, Kramers, and Slater resulted in the following scientific situation:

According to the basic postulates of the quantum theory, an atomic system is capable of assuming discrete, stationary states, and therefore discrete energy values; in terms of the energy of the atom the emission and absorption of light by such a system occurs abruptly, in the form of impulses. On the other hand, the visualizable properties of the emitted radiation are described by a wave field, the frequency of which is associated with the difference in energy between the initial and final states of the atom by the relation

$$E_1 - E_2 = h\nu$$

To each stationary state of an atom corresponds a whole complex of parameters which specify the probability of transition from this state to another. There is no direct relation between the radiation classically emitted by an orbiting electron and those parameters defining the probability of emission; nevertheless, Bohr's principle of correspondence enables a specific term of the Fourier expansion of the classical path to be assigned to each transition of the atom, and the probability for the particular transition follows qualitatively similar laws as the intensity of those Fourier components. Although therefore in the researches carried out by Rutherford, Bohr, Sommerfeld and others, the comparison of the atom with a planetary system of electrons leads to a qualitative interpretation of the optical and chemical properties of atoms, nevertheless the fundamental dissimilarity between the atomic spectrum and the classical spectrum of an electron system imposes the need to relinquish the concept of an electron path and to forego a visual description of the atom.

The experiments necessary to define the electron-path concept also furnish an important aid in revising it. The most obvious answer to the question how the orbit of an electron in its path within the atom could be observed, namely ...?, will perhaps be to use a microscope of extreme resolving power. But since the specimen in this microscope would have to be illuminated with light having an extremely short wavelength, the first light quantum from the light source to reach the electron and pass into the observer's eye would eject the electron completely from its path in accordance with the laws of the Compton effect. Consequently, only one point of the path would be observable experimentally at any one time.

In this situation, therefore, the obvious policy was to relinquish at first the concept of electron paths altogether, despite its substantiation by Wilson's experiments, and, as it were, to attempt subsequently how much of the electron-path concept can be carried over into quantum mechanics.

In the classical theory the specification of frequency, amplitude, and phase of all the light waves emitted by the atom would be fully equivalent to specifying its electron path. Since from the amplitude and phase of an emitted wave the coefficients of the appropriate term in the Fourier expansion of the electron path can be derived without ambiguity, the complete electron path therefore can be derived from a knowledge of all amplitudes and phases. Similarly, in quantum mechanics, too, the whole complex of amplitudes and phases of the radiation emitted by the atom can be regarded as a complete description of the atomic system, although its interpretation in the sense of an electron path inducing the radiation is impossible. In quantum mechanics, therefore, the place of the electron coordinates is taken by a complex of parameters corresponding to the Fourier coefficients of classical motion along a path. These, however, are no longer classified by the energy of state and the number of the corresponding harmonic vibration, but are in each case associated with two stationary states of the atom, and are a measure for the transition probability of the atom from one stationary state to another. A complex of coefficients of this type is comparable with a matrix such as occurs in linear algebra. In exactly the same way each parameter of classical mechanics, e.g. the momentum or the energy of the electrons, can then be assigned a corresponding matrix in quantum mechanics. To proceed from here beyond a mere description of the empirical state of affairs it was necessary to associate systematically the matrices assigned to the various parameters in the same way as the corresponding parameters in classical mechanics are associated by equations of motions. When, in the interest of achieving the closest possible correspondence between classical and quantum mechanics, the addition and multiplication of Fourier series were tentatively taken as the example for the addition and multiplication of the quantum-theory complexes, the product of two parameters represented by matrices appeared to be most naturally represented by the product matrix in the sense of linear algebra - an assumption already suggested by the formalism of the Kramers- Ladenburg dispersion theory.

It thus seemed consistent simply to adopt in quantum mechanics the equations of motion of classical physics, regarding them as a relation between the matrices representing the classical variables. The Bohr-Sommerfeld quantum conditions could also be reinterpreted in a relation between the matrices, and together with the equations of motion they were sufficient to define all matrices and hence the experimentally observable properties of the atom.

Born, Jordan, and Dirac deserve the credit for expanding the mathematical scheme outlined above into a consistent and practically usable theory. These investigators observed in the first place that the quantum conditions can be written as commutation relations between the matrices representing the momenta and the coordinates of the electrons, to yield the equations (p_r, momentum matrices; q_r, coordinate matrices):

$$p_r q_s - q_s p_r = h/2\pi i \, \delta_{rs} \qquad q_r q_s - q_s q_r = 0 \qquad p_r p_s - p_s p_r = 0$$

$$\delta_{rs} = I \text{ for } r = s; = 0 \text{ for } r \neq s.$$

By means of these commutation relations they were able to detect in quantum mechanics as well the laws which were fundamental to classical mechanics: the invariability in time of energy, momentum, and angular momentum.

The mathematical scheme so derived thus ultimately bears an extensive formal similarity to that of the classical theory, from which it differs outwardly by the commutation relations which, moreover, enabled the equations of motion to be derived from the Hamiltonian function.

In the physical consequences, however, there are very profound differences between quantum mechanics and classical mechanics which impose the need for a thorough discussion of the physical interpretation of quantum mechanics. As hitherto defined, quantum mechanics enables the radiation emitted by the atom, the energy values of the stationary states, and other parameters characteristic for the stationary states to be treated. The theory hence complies with the experimental data contained in atomic spectra. In all those cases, however, where a visual description is required of a transient event, e.g. when interpreting Wilson photographs, the formalism of the theory does not seem to allow an adequate representation of the experimental state of affairs. At this point Schrödinger's wave mechanics, meanwhile developed on the basis of de Broglie's theses, came to the assistance of quantum mechanics.

In the course of the studies which Mr. Schrödinger will report here himself he converted the determination of the energy values of an atom into an eigenvalue problem defined by a boundary-value problem in the coordinate space of the particular atomic system. After Schrödinger had shown the mathematical equivalence of wave mechanics, which he had discovered, with quantum mechanics, the fruitful combination of these two different areas of physical ideas resulted in an extraordinary broadening and enrichment of the formalism of the quantum theory. Firstly, it was only wave mechanics which made possible the mathematical treatment of complex atomic systems, secondly analysis of the connection between the two theories led to what is known as the *transformation theory* developed by Dirac and Jordan. As it is impossible within the limits of the present lecture to give a detailed discussion of the mathematical structure of this theory, I should just like to point out its fundamental physical significance. Through the adoption of the physical principles of quantum mechanics into its expanded formalism, the *transformation theory* made it possible in completely general terms to calculate for atomic systems the probability for the occurrence of a particular, experimentally ascertainable, phenomenon under given experimental conditions. The hypothesis conjectured in the studies on the radiation theory and enunciated in precise terms in Born's collision theory, namely that the wave function governs the probability for the presence of a corpuscle, appeared to be a special case of a more general pattern of laws and to be a natural consequence of the fundamental assumptions of quantum mechanics. Schrödinger, and in later studies Jordan, Klein, and Wigner as well, had succeeded in developing as far as permitted by the principles of the quantum theory de Broglie's original concept of visualizable matter waves occurring in space and time, a concept formulated even before the development of quantum mechanics. But for that the connection between Schrödinger's concepts and de Broglie's original thesis would certainly have seemed a looser one by this statistical interpretation of wave mechanics and by the greater emphasis on the fact that Schrödinger's theory is concerned with waves in multidimensional space. Before proceeding to discuss the explicit significance of quantum mechanics it is perhaps right for me to deal briefly with this question as to the existence of matter waves in three-dimensional space, since the solution to this problem was only achieved by combining wave and quantum mechanics.

A long time before quantum mechanics was developed Pauli had inferred from the laws in the Periodic System of the elements the well-known principle that a particular quantum state can at all times be occupied by only a single electron. It proved possible to transfer this principle to quantum mechanics on the basis of what at first sight seemed a surprising result: the entire complex of stationary states which an atomic system is capable of adopting breaks down into definite classes such that an atom in a state belonging to one class can never change into a state belonging to another class under the action of whatever perturbations. As finally clarified beyond question by the studies of Wigner and Hund, such a class of states is characterized by a definite symmetry characteristic of the Schrödinger eigenfunction with respect to the transposition of the coordinates of two electrons. Owing to the fundamental identity of electrons, any external perturbation of the atom remains unchanged when two electrons are exchanged and hence causes no transitions between states of various classes. The Pauli principle and the Fermi-Dirac statistics derived from it are equivalent with the assumption that only that class of stationary states is achieved in nature in which the eigenfunction changes its sign when two electrons are exchanged. According to Dirac, selecting the symmetrical system of terms would lead not to the Pauli principle, but to Bose-Einstein electron statistics.

Between the classes of stationary states belonging to the Pauli principle or to Bose-Einstein statistics, and de Broglie's concept of matter waves there is a peculiar relation. A spatial wave phenomenon can be treated according to the principles of the quantum theory by analyzing it using the Fourier theorem and then applying to the individual Fourier component of the wave motion, as a system having one degree of freedom, the normal laws of quantum mechanics. Applying this procedure for treating wave phenomena by the quantum theory, a procedure that has also proved fruitful in Dirac's studies of the theory of radiation, to de Broglie's matter waves, exactly the same results are obtained as in treating a whole complex of material particles according to quantum mechanics and selecting the symmetrical system of terms. Jordan and Klein hold that the two methods are mathematically equivalent even if allowance is also made for the interaction of the electrons, i.e. if the field energy originating from the continuous space charge is included in the calculation in de Broglie's wave theory. Schrödinger's considerations of the energy-momentum tensor assigned to the matter waves can then also be adopted in this theory as consistent components of the formalism. The studies of Jordan and Wigner show that modifying the commutation relations underlying this quantum theory of waves results in a formalism equivalent to that of quantum mechanics based on the assumption of Pauli's exclusion principle.

These studies have established that the comparison of an atom with a planetary system composed of nucleus and electrons is not the only visual picture of how we can imagine the atom. On the contrary, it is apparently no less correct to compare the atom with a charge cloud and use the correspondence to the formalism of the quantum theory borne by this concept to derive qualitative conclusions about the behavior of the atom. However, it is the concern of wave mechanics to follow these consequences.

Reverting therefore to the formalism of quantum mechanics; its application to physical problems is justified partly by the original basic assumptions of the theory, partly by its expansion in the transformation theory on the basis of wave mechanics, and the question is now to expose the explicit significance of the theory by comparing it with classical physics.

In classical physics the aim of research was to investigate objective processes occurring in space and time, and to discover the laws governing their progress from the initial conditions. In classical physics a problem was considered solved when a particular phenomenon had been proved to occur objectively in space and time, and it had been shown to obey the general rules of classical physics as formulated by differential equations. The manner in which the knowledge of each process had been acquired, what observations may possibly have led to its experimental determination, was completely immaterial, and it was also immaterial for the consequences of the classical theory, which possible observations were to verify the predictions of the theory. In the quantum theory, however, the situation is completely different. The very fact that the formalism of quantum mechanics cannot be interpreted as visual description of a phenomenon occurring in space and time shows that quantum mechanics is in no way concerned with the objective determination of space-time phenomena. On the contrary, the formalism of quantum mechanics should be used in such a way that the probability for the outcome of a further experiment may be concluded from the determination of an experimental situation in an atomic system, providing that the system is subject to no perturbations other than those necessitated by performing the two experiments. The fact that the only definite known result to be ascertained after the fullest possible experimental investigation of the system is the probability for a certain outcome of a second experiment shows, however, that each observation must entail a discontinuous change in the formalism describing the atomic process and therefore also a discontinuous change in the physical phenomenon itself. Whereas in the classical theory the kind of observation has no bearing on the event, in the quantum theory the disturbance associated with each observation of the atomic phenomenon has a decisive role. Since, furthermore, the result of an observation as a rule leads only to assertions about the probability of certain results of subsequent observations, the fundamentally unverifiable part of each perturbation must, as shown by Bohr, be decisive for the non-contradictory operation of quantum mechanics. This difference between classical and atomic physics is understandable, of course, since for heavy bodies such as the planets moving around the sun the pressure of the sunlight which is reflected at their surface and which is necessary for them to be observed is negligible; for the smallest building units of matter, however, owing to their low mass, every observation has a decisive effect on their physical behavior.

The perturbation of the system to be observed caused by the observation is also an important factor in determining the limits within which a visual description of atomic phenomena is possible. If there were experiments which permitted accurate measurement of all the characteristics of an atomic system necessary to calculate classical motion, and which, for example, supplied accurate values for the location and velocity of each electron in the system at a particular time, the result of these experiments could not be utilized at all in the formalism, but rather it would directly contradict the formalism. Again, therefore, it is clearly that fundamentally unverifiable part of the perturbation of the system caused by the measurement itself which hampers accurate ascertainment of the classical characteristics and thus permits quantum mechanics to be applied. Closer examination of the formalism shows that between the accuracy with which the location of a particle can be ascertained and the accuracy with which its momentum can simultaneously be known, there is a relation according to which the product of the probable errors in the measurement of the location and momentum is invariably at least as large as Planck's constant divided by 4π. In a very general form, therefore, we should have

$$\Delta p \, \Delta q \geq h/4\pi$$

where p and q are canonically conjugated variables. These uncertainty relations for the results of the measurement of classical variables form the necessary conditions for enabling the result of a measurement to be expressed in the formalism of the quantum theory. Bohr has shown in a series of examples how the perturbation necessarily associated with each observation indeed ensures that one cannot go below the limit set by the uncertainty relations. He contends that in the final analysis an uncertainty introduced by the concept of measurement itself is responsible for part of that perturbation remaining fundamentally unknown. The experimental determination of whatever space-time events invariably necessitates a fixed frame - say the system of coordinates in which the observer is at rest - to which all measurements are referred. The assumption that this frame is "fixed" implies neglecting its momentum from the outset, since "fixed" implies nothing other, of course, than that any transfer of momentum to it will evoke no perceptible effect. The fundamentally necessary uncertainty at this point is then transmitted via the measuring apparatus into the atomic event.

Since in connection with this situation it is tempting to consider the possibility of eliminating all uncertainties by amalgamating the object, the measuring apparatuses, and the observer into one quantum-mechanical system, it is important to emphasize that the act of measurement is necessarily visualizable, since, of course, physics is ultimately only concerned with the systematic description of space-time processes. The behavior of the observer as well as his measuring apparatus must therefore be discussed according to the laws of classical physics, as otherwise there is no further physical problem whatsoever. Within the measuring apparatus, as emphasized by Bohr, all events in the sense of the classical theory will therefore be regarded as determined, this also being a necessary condition before one can, from a result of measurements, unequivocally conclude what has happened. In quantum theory, too, the scheme of classical physics which objectifies the results of observation by assuming in space and time processes obeying laws is thus carried through up to the point where the fundamental limits are imposed by the unvisualizable character of the atomic events symbolized by Planck's constant. A visual description for the atomic events is possible only within certain limits of accuracy - but within these limits the laws of classical physics also still apply. Owing to these limits of accuracy as defined by the uncertainty relations, moreover, a visual picture of the atom free from ambiguity has not been determined. On the contrary the corpuscular and the wave concepts are equally serviceable as a basis for visual interpretation.

The laws of quantum mechanics are basically statistical. Although the parameters of an atomic system are determined in their entirety by an experiment, the result of a future observation of the system is not generally accurately predictable. But at any later point of time there are observations which yield accurately predictable results. For the other observations only the probability for a particular outcome of the experiment can be given. The degree of certainty which still attaches to the laws of quantum mechanics is, for example, responsible for the fact that the principles of conservation for energy and momentum still hold as strictly as ever. They can be checked with any desired accuracy and will then be valid according to the accuracy with which they are checked. The statistical character of the laws of quantum mechanics, however, becomes apparent in that an accurate study of the energetic conditions renders it impossible to pursue at the same time a particular event in space and time.

For the clearest analysis of the conceptual principles of quantum mechanics we are indebted to Bohr who, in particular, applied the concept of complementarity to interpret the validity of the quantum-mechanical laws. The uncertainty relations alone afford an instance of how in quantum mechanics the exact knowledge of one variable can exclude the exact knowledge of another. This complementary relationship between different aspects of one and the same physical process is indeed characteristic for the whole structure of quantum mechanics. I had just mentioned that, for example, the determination of energetic relations excludes the detailed description of space- time processes. Similarly, the study of the chemical properties of a molecule is complementary to the study of the motions of the individual electrons in the molecule, or the observation of interference phenomena complementary to the observation of individual light quanta. Finally, the areas of validity of classical and quantum mechanics can be marked off one from the other as follows: Classical physics represents that striving to learn about Nature in which essentially we seek to draw conclusions about objective processes from observations and so ignore the consideration of the influences which every observation has on the object to be observed; classical physics, therefore, has its limits at the point from which the influence of the observation on the event can no longer be ignored. Conversely, quantum mechanics makes possible the treatment of atomic processes by partially foregoing their space- time description and objectification.

So as not to dwell on assertions in excessively abstract terms about the interpretation of quantum mechanics, I would like briefly to explain with a well-known example how far it is possible through the atomic theory to achieve an understanding of the visual processes with which we are concerned in daily life. The interest of research workers has frequently been focused on the phenomenon of regularly shaped crystals suddenly forming from a liquid, e.g. a supersaturated salt solution. According to the atomic theory the forming force in this process is to a certain extent the symmetry characteristic of the solution to Schrödinger's wave equation, and to that extent crystallization is explained by the atomic theory. Nevertheless, this process retains a statistical and - one might almost say - historical element which cannot be further reduced: even when the state of the liquid is completely known before crystallization, the shape of the crystal is not determined by the laws of quantum mechanics. The formation of regular shapes is just far more probable than that of a shapeless lump. But the ultimate shape owes its genesis partly to an element of chance which in principle cannot be analyzed further.

Before closing this report on quantum mechanics, I may perhaps be allowed to discuss very briefly the hopes that may be attached to the further development of this branch of research. It would be superfluous to mention that the development must be continued, based equally on the studies of de Broglie, Schrödinger, Born, Jordan, and Dirac. Here *the attention of the research workers is primarily directed to the problem of reconciling the claims of the special relativity theory with those of the quantum theory.* The extraordinary advances made in this field by Dirac about which Mr. Dirac will speak here, meanwhile leave open the question *whether it will be possible to satisfy the claims of the two theories without at the same time determining the Sommerfeld fine-structure constant.* The attempts made hitherto to achieve a *relativistic* formulation of the quantum theory are all based on visual concepts so close to those of classical physics that it seems impossible to determine the fine-structure constant within this system of concepts. The expansion of the conceptual system under discussion here should, furthermore, be closely associated with the further development of the *quantum theory of wave fields*, and it appears to me as if this formalism, notwithstanding its thorough study by a number of workers (Dirac, Pauli, Jordan, Klein, Wigner, Fermi) has still not been completely exhausted. Important pointers for the further development of quantum mechanics also emerge from the experiments involving the structure of the atomic nuclei. From their analysis by means of the Gamow theory, it would appear that between the elementary particles of the atomic nucleus forces are at work which differ somewhat in type from the forces determining the structure of the atomic shell; Stem's experiments seem, furthermore, to indicate that the behavior of the heavy elementary particles cannot be represented by the formalism of Dirac's theory of the electron. Future research will thus have to be prepared for surprises which may otherwise come both from the field of experience of nuclear physics as well as from that of cosmic radiation. But however, the development proceeds in detail, the path so far traced by the quantum theory indicates that an understanding of those still unclarified features of atomic physics can only be acquired by foregoing visualization and objectification to an extent greater than that customary hitherto. We have probably no reason to regret this, because the thought of the great epistemological difficulties with which the visual atom concept of earlier physics had to contend gives us the hope that the abstracter atomic physics developing at present will one day fit more harmoniously into the great edifice of Science.

Erwin Schrödinger – 1933 Nobel Lecture, December 12, 1933. *The fundamental idea of wave mechanics.*

[https://www.nobelprize.org/prizes/physics/1933/schrödinger/lecture/.]

> Schrödinger shared the 1933 Nobel Prize in Physics with Paul Dirac "for the discovery of new productive forms of atomic theory".
>
> In Niels Bohr's theory of the atom, electrons absorb and emit radiation of fixed wavelengths when jumping between fixed orbits around a nucleus. The theory provided a good description of the spectrum created by the hydrogen atom, but needed to be developed to suit more complicated atoms and molecules. Assuming that matter (e.g., electrons) could be regarded as both particles and waves, in 1926 Erwin Schrödinger formulated a wave equation that accurately calculated the energy levels of electrons in atoms. [Erwin Schrödinger – Facts. NobelPrize.org. https://www.nobelprize.org/prizes/physics/1933/schrodinger/facts/.]

Schrödinger describes a fascinating analogy between the path of a ray of light and the path of a mass point. In optics the old system of mechanics corresponds to operating with isolated mutually independent light rays. The new wave mechanics corresponds to the wave theory of light. We are never in a position to say what really is or what really happens, but we can only say what will be observed in any concrete individual case. He concludes that *the ray or the particle path corresponds to a longitudinal relationship of the propagation process (i.e. in the direction of propagation), the wave surface on the other hand to a transversal relationship (i.e. normal to it). Both relationships are without doubt real;* one is proved by photographed particle paths, the other by interference experiments. *To combine both in a uniform system has proved impossible so far. Only in extreme cases does either the transversal, shell-shaped or the radial, longitudinal relationship predominate to such an extent that we think we can make do with the wave theory alone or with the particle theory alone.*

> On passing through an optical instrument, such as a telescope or a camera lens, a ray of light is subjected to a change in direction at each refracting or reflecting surface. The path of the rays can be constructed if we know the two simple laws which govern the changes in direction: the law of refraction which was discovered by Snellius a few hundred years ago, and the law of reflection with which Archimedes was familiar more than 2,000 years ago. As a simple example, Fig. 1 shows a ray A-B which is subjected to refraction at each of the four boundary surfaces of two lenses in accordance with the law of Snellius.

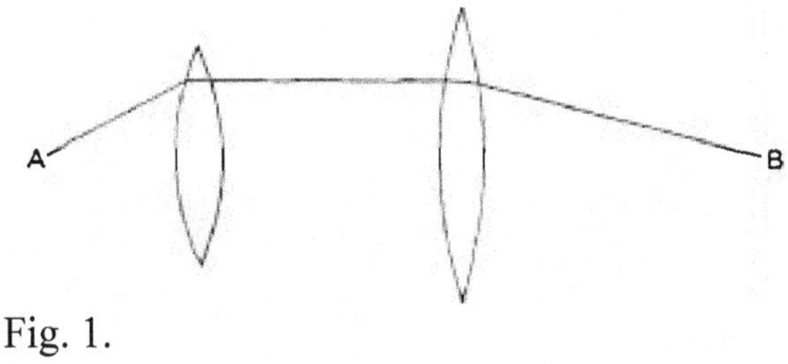

Fig. 1.

Fermat defined the total path of a ray of light from a much more general point of view. In different media, light propagates with different velocities, and the radiation path gives the appearance as if the light must arrive at its destination *as quickly as possible*. (Incidentally, it is permissible here to consider *any two* points along the ray as the starting- and end-points.) The least deviation from the path actually taken would mean a delay. This is the famous Fermat *principle of the shortest light time,* which in a marvelous manner determines the entire fate of a ray of light by a single statement and also includes the more general case, when the nature of the medium varies not suddenly at individual surfaces, but gradually from place to place. The atmosphere of the earth provides an example. The more deeply a ray of light penetrates into it from outside, the more slowly it progresses in an increasingly denser air. Although the differences in the speed of propagation are infinitesimal, Fermat's principle in these circumstances demands that the light ray should curve earthward (see Fig. 2), so that it remains a little longer in the higher "faster" layers and reaches its destination more quickly than by the shorter straight path (broken line in the figure; disregard the square, WWW^1W^1 for the time being).

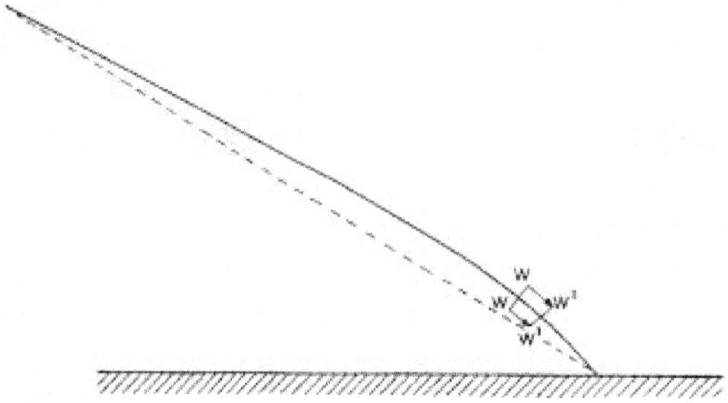

Fig. 2.

I think, hardly any of you will have failed to observe that the sun when it is deep on the horizon appears to be not circular but flattened: its vertical diameter looks to be shortened. This is a result of the curvature of the rays.

According to the wave theory of light, the light rays, strictly speaking, have only fictitious significance. They are not the physical paths of some particles of light, but are a mathematical device, the so-called orthogonal trajectories of wave surfaces, imaginary guide lines as it were, which point in the direction normal to the wave surface in which the latter advances (cf. Fig. 3 which shows the simplest case of concentric spherical wave surfaces and accordingly rectilinear rays, whereas Fig. 4 illustrates the case of curved rays).

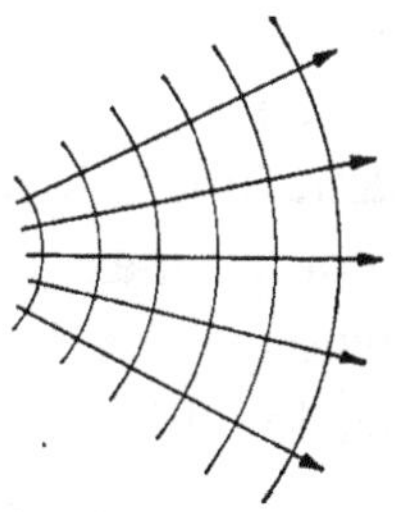

Fig. 3.

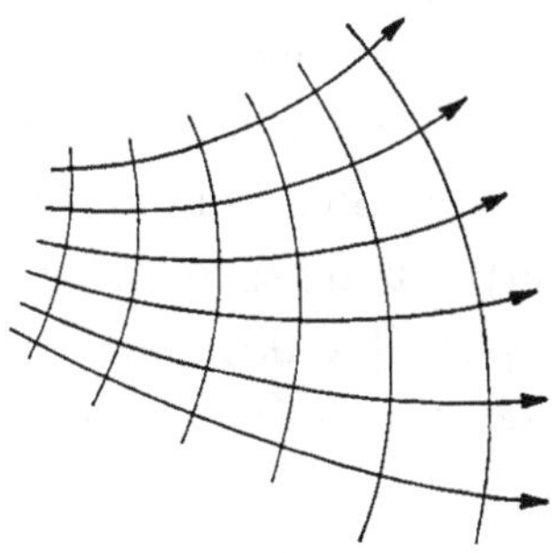

Fig. 4.

It is surprising that a general principle as important as Fermat's relates directly to these mathematical guide lines, and not to the wave surfaces, and one might be inclined for this reason to consider it a mere mathematical curiosity. Far from it. It becomes properly understandable only from the point of view of wave theory and ceases to be a divine miracle. From the wave point of view, the so-called *curvature* of the light ray is far more readily understandable as a *swerving* of the wave surface, which must obviously occur when neighboring parts of a wave surface advance at different speeds; in exactly the same manner as a company of soldiers marching forward will carry out the order "right incline" by the men taking steps of varying lengths, the right-wing man the smallest, and the left-wing man the longest. In atmospheric refraction of radiation for example (Fig. 2) the section of wave surface WW must necessarily swerve to the right towards W^1W^1 because its left half is located in slightly higher, thinner air and thus advances more rapidly than the right part at lower point. (In passing, I wish to refer to one point at which the *Snellius'* view fails. A horizontally emitted light ray should remain horizontal because the refraction index does not vary in the horizontal direction. In truth, a horizontal ray curves more strongly than any other, which is an obvious consequence of the theory of a swerving wave front.) On detailed examination the Fermat principle is found to be completely *tantamount* to the trivial and obvious statement that - given local distribution of light velocities - the wave front must swerve in the manner indicated. I cannot prove this here, but shall attempt to make it plausible. I would again ask you to visualize a rank of soldiers marching forward. To ensure that the line remains dressed, let the men be connected by a long rod which each holds firmly in his hand. No orders as to direction are given; the only order is: let each man march or run as fast as he can. If the nature of the ground varies slowly from place to place, it will be now the right wing, now the left that advances more quickly, and changes in direction will occur spontaneously. After some time has elapsed, it will be seen that the entire path travelled is not rectilinear, but somehow curved. That this curved path is exactly that by which the destination attained at any moment could be attained *most rapidly* according to the nature of the terrain, is at least quite plausible, since each of the men did his best. It will also be seen that the swerving also occurs invariably in the direction in which the terrain is worse, so that it will come to look in the end as if the men had intentionally "by-passed" a place where they would advance slowly.

The Fermat principle thus appears to be the trivial quintessence of the wave theory. It was therefore a memorable occasion when Hamilton made the discovery that the true movement of mass points in a field of forces (e.g. of a planet on its orbit around the sun or of a stone thrown in the gravitational field of the earth) is also governed by a very similar general principle, which carries and has made famous the name of its discoverer since then. Admittedly, the Hamilton principle does not say exactly that the mass point chooses the quickest way, but it does say something so similar - the analogy with the principle of the shortest travelling time of light is so close, that one was faced with a puzzle. It seemed as if Nature had realized one and the same law twice by entirely different means: first in the case of light, by means of a fairly obvious play of rays; and again, in the case of the mass points, which was anything but obvious, unless somehow wave nature were to be attributed to them also. And this, it seemed impossible to do. Because the "mass points" on which the laws of mechanics had really been confirmed experimentally at that time were only the large, visible, sometimes very large bodies, the planets, for which a thing like "wave nature" appeared to be out of the question.

The smallest, elementary components of matter which we today, much more specifically, call "mass points", were purely hypothetical at the time. It was only after the discovery of radioactivity that constant refinements of methods of measurement permitted the properties of these particles to be studied in detail, and now permit the paths of such particles to be photographed and to be measured very exactly (stereo photogrammetrically) by the brilliant method of C. T. R. Wilson. As far as the measurements extend, they confirm that the same mechanical laws are valid for particles as for large bodies, planets, etc. However, it was found that neither the molecule nor the individual atom can be considered as the "ultimate component": but even the atom is a system of highly complex structure. Images are formed in our minds of the structure of atoms *consisting of* particles, images which seem to have a certain similarity with the planetary system. It was only natural that the attempt should at first be made to consider as valid the same laws of motion that had proved themselves so amazingly satisfactory on a large scale. In other words, Hamilton's mechanics, which, as I said above, culminates in the Hamilton principle, were applied also to the "inner life" of the atom. That there is a very close analogy between Hamilton's principle and Fermat's optical principle had meanwhile become all but forgotten. If it was remembered, it was considered to be nothing more than a curious trait of the mathematical theory.

Now, it is very difficult, without further going into details, to convey a proper conception of the success or failure of these classical-mechanical images of the atom. On the one hand, Hamilton's principle in particular proved to be the most faithful and reliable guide, which was simply indispensable; on the other hand, one had to suffer, to do justice to the facts, the rough interference of entirely new incomprehensible postulates, of the so-called quantum conditions and quantum postulates. Strident disharmony in the symphony of classical mechanics - yet strangely familiar - played as it were on the same instrument. In mathematical terms we can formulate this as follows: *whereas the Hamilton principle merely postulates that a given integral must be a minimum, without the numerical value of the minimum being established by this postulate, it is now demanded that the numerical value of the minimum should be restricted to integral multiples of a universal natural constant, Planck's quantum of action*. This incidentally. The situation was fairly desperate. Had the old mechanics failed completely, it would not have been so bad. The way would then have been free to the development of a new system of mechanics. As it was, one was faced with the difficult task of saving the *soul* of the old system, whose inspiration clearly held sway in this microcosm, while at the same time flattering it as it were into accepting the quantum conditions not as gross interference but as issuing from its own innermost essence.

The way out lay just in the possibility, already indicated above, of attributing to the Hamilton principle, also, the operation of a wave mechanism on which the point-mechanical processes are essentially based, just as one had long become accustomed to doing in the case of phenomena relating to light and of the Fermat principle which governs them. Admittedly, *the individual path of a mass point loses its proper physical significance and becomes as fictitious as the individual isolated ray of light*. The essence of the theory, the minimum principle, however, remains not only intact, but reveals its true and simple meaning only under the wave-like aspect, as already explained. Strictly speaking, the new theory is in fact not *new,* it is a completely organic development, one might almost be tempted to say a more elaborate exposition, of the old theory.

How was it then that this new more "elaborate" exposition led to notably different results; what enabled it, when applied to the atom, to obviate difficulties which the old theory could not solve? What enabled it to render gross interference acceptable or even to make it its own?

Again, these matters can best be illustrated by analogy with optics. Quite properly, indeed, I previously called the Fermat principle the quintessence of the wave theory of light: nevertheless, it cannot render dispensable a more exact study of the wave process itself. The so-called refraction and interference phenomena of light can only be understood if we trace the wave process in detail because what matters is not only the eventual destination of the wave, but also whether at a given moment it arrives there with a wave peak or a wave trough. In the older, coarser experimental arrangements, these phenomena occurred as small details only and escaped observation. Once they were noticed and were interpreted correctly, by means of waves, it was easy to devise experiments in which the wave nature of light finds expression not only in small details, but on a very large scale in the entire character of the phenomenon.

Allow me to illustrate this by two examples, first, the example of an optical instrument, such as telescope, microscope, etc. The object is to obtain a sharp image, i.e. it is desired that all rays issuing from a point should be reunited in a point, the so-called focus (cf. Fig. 5 a).

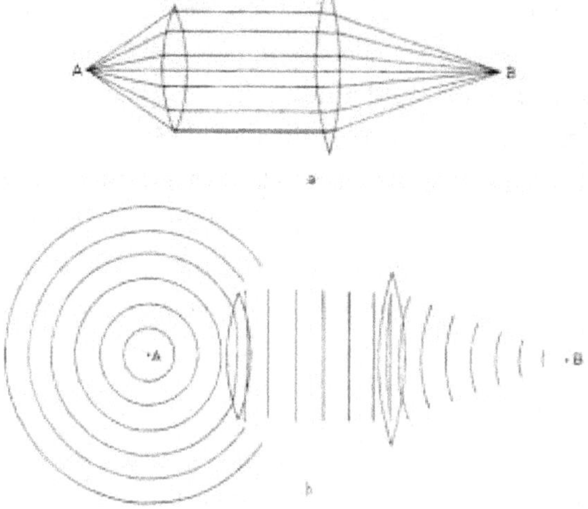

Fig. 5.

It was at first believed that it was only geometrical-optical difficulties which prevented this: they are indeed considerable. Later it was found that even in the best designed instruments focusing of the rays was considerably inferior than would be expected if each ray exactly obeyed the Fermat principle independently of the neighboring rays. The light which issues from a point and is received by the instrument is reunited behind the instrument not in a single point any more, but is distributed over a small circular area, a so-called diffraction disc, which, otherwise, is in most cases a circle only because the apertures and lens contours are generally circular. For, the cause of the phenomenon which we call *diffraction* is that not all the spherical waves issuing from the object point can be accommodated by the instrument. The lens edges and any apertures merely cut out a part of the wave surfaces (cf. Fig. 5b) and - if you will permit me to use a more suggestive expression - the injured margins resist rigid unification in a point and produce the somewhat blurred or vague image. The degree of blurring is closely associated with the wavelength of the light and is completely inevitable because of this deep-seated theoretical relationship. Hardly noticed at first, it governs and restricts the performance of the modern microscope which has mastered all other errors of reproduction. The images obtained of structures not much coarser or even still finer than the wavelengths of light are only remotely or not at all similar to the original.

A second, even simpler example is the shadow of an opaque object cast on a screen by a small point light source. In order to construct the shape of the shadow, each light ray must be traced and it must be established whether or not the opaque object prevents it from reaching the screen. The *margin* of the shadow is formed by those light rays which only just brush past the edge of the body. Experience has shown that the shadow margin is not absolutely sharp even with a point-shaped light source and a sharply defined shadow-casting object. The reason for this is the same as in the first example. The wave front is as it were bisected by the body (cf. Fig. 6) and the traces of this injury result in blurring of the margin of the shadow which would be incomprehensible if the individual light rays were independent entities advancing independently of one another without reference to their neighbors.

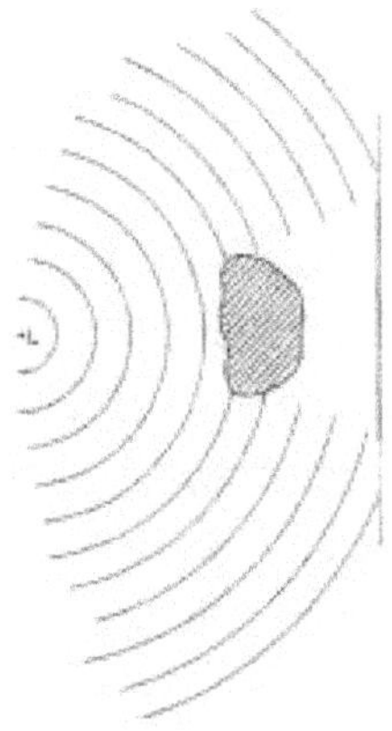

Fig. 6.

This phenomenon - which is also called diffraction - is not as a rule very noticeable with large bodies. But if the shadow-casting body is very small at least in one dimension, diffraction finds expression firstly in that no proper shadow is formed at all, and secondly - much more strikingly - in that the small body itself becomes as it were its own source of light and radiates light in all directions (preferentially to be sure, at small angles relative to the incident light).

All of you are undoubtedly familiar with the so-called "motes of dust" in a light beam falling into a dark room. Fine blades of grass and spiders' webs on the crest of a hill with the sun behind it, or the errant locks of hair of a man standing with the sun behind often light up mysteriously by diffracted light, and the visibility of smoke and mist is based on it. It comes not really from the body itself, but from its immediate surroundings, an area in which it causes considerable interference with the incident wave fronts. It is interesting, and important for what follows, to observe that the area of interference always and in every direction has at least the extent of one or a few wavelengths, no matter how small the disturbing particle may be. Once again, therefore, we observe a close relationship between the phenomenon of diffraction and wavelength. This is perhaps best illustrated by reference to another wave process, i.e. sound. Because of the much greater wavelength, which is of the order of centimeters and meters, shadow formation recedes in the case of sound, and diffraction plays a major, and practically important, part: we can easily *hear* a man calling from behind a high wall or around the corner of a solid house, even if we cannot see him.

Let us return from optics to mechanics and explore the analogy to its fullest extent. *In optics the old system of mechanics corresponds to intellectually operating with isolated mutually independent light rays. The new undulatory mechanics corresponds to the wave theory of light.* What is gained by changing from the old view to the new is that the diffraction phenomena can be accommodated or, better expressed, what is gained is something that is strictly analogous to the diffraction phenomena of light and which on the whole must be very unimportant, otherwise the old view of mechanics would not have given full satisfaction so long. It is, however, easy to surmise that the neglected phenomenon may in some circumstances make itself very much felt, will entirely dominate the mechanical process, and will face the old system with insoluble riddles, *if the entire mechanical system is comparable in extent with the wavelengths of the "waves of matter"* which play the same part in mechanical processes as that played by the light waves in optical processes.

This is the reason why in these minute systems, the atoms, the old view was bound to fail, which though remaining intact as a close approximation for gross mechanical processes, but is no longer adequate for the delicate interplay in areas of the order of magnitude of one or a few wavelengths. It was astounding to observe the manner in which all those strange additional requirements developed spontaneously from the new undulatory view, whereas they had to be forced upon the old view to adapt them to the inner life of the atom and to provide some explanation of the observed facts.

Thus, *the salient point of the whole matter is that the diameters of the atoms and the wavelength of the hypothetical material waves are of approximately the same order of magnitude*. And now you are bound to ask whether it must be considered mere chance that in our continued analysis of the structure of matter we should come upon the order of magnitude of the wavelength at this of all points, or whether this is to some extent comprehensible. Further, you may ask, how we know that this is so, since the material waves are an entirely new requirement of this theory, unknown anywhere else. Or is it simply that this is an *assumption* which had to be made?

The agreement between the orders of magnitude is no mere chance, nor is any special assumption about it necessary; it follows automatically from the theory in the following remarkable manner. That the heavy *nucleus* of the atom is very much smaller than the atom and may therefore be considered as a point center of attraction in the argument which follows may be considered as experimentally established by the experiments on the scattering of alpha rays done by Rutherford and Chadwick. Instead of the *electrons* we introduce hypothetical waves, whose wavelengths are left entirely open, because we know nothing about them yet. This leaves a letter, say *a*, indicating a still unknown figure, in our calculation. We are, however, used to this in such calculations and it does not prevent us from calculating that the nucleus of the atom must produce a kind of diffraction phenomenon in these waves, similarly as a minute dust particle does in light waves. *Analogously, it follows that there is a close relationship between the extent of the area of interference with which the nucleus surrounds itself and the wavelength, and that the two are of the same order of magnitude*. What this is, we have had to leave open; but the most important step now follows: we *identify the area of interference, the diffraction halo, with the atom; we assert that the atom in reality is merely the diffraction phenomenon of an electron wave captured us it were by the nucleus of the atom.* It is no longer a matter of chance that the size of the atom and the wavelength are of the same order of magnitude: it is a matter of course. We know the numerical value of neither, because we still have in our calculation the *one* unknown constant, which we called *a*. There are two possible ways of determining it, which provide a mutual check on one another. First, we can so select it that the manifestations of life of the atom, above all the spectrum lines emitted, come out correctly quantitatively; these can after all be measured very accurately. Secondly, we can select *a* in a manner such that the diffraction halo acquires the size required for the atom. These two determinations of *a* (of which the second is admittedly far more imprecise because "size of the atom" is no clearly defined term) *are in complete agreement with one another.* Thirdly, and lastly, we can remark that the constant remaining unknown, physically speaking, does not in fact have the dimension of a length, but of an action, i.e. energy x time. It is then an obvious step to substitute for it the numerical value of Planck's universal quantum of action, which is accurately known from the laws of heat radiation. It will be seen that *we return,* with the full, now considerable accuracy, *to the first* (most accurate) *determination.*

Quantitatively speaking, the theory therefore manages with a minimum of new assumptions. It contains a single available constant, to which a numerical value familiar from the older quantum theory must be given, first to attribute to the diffraction halos the right size so that they can be reasonably identified with the atoms, and secondly, to evaluate quantitatively and correctly all the manifestations of life of the atom, the light radiated by it, the ionization energy, etc.

I have tried to place before you the fundamental idea of the wave theory of matter in the simplest possible form. I must admit now that in my desire not to tangle the ideas from the very beginning, I have painted the lily. Not as regards the high degree to which all sufficiently, carefully drawn conclusions are confirmed by experience, but with regard to the conceptual ease and simplicity with which the conclusions are reached. I am not speaking here of the mathematical difficulties, which always turn out to be trivial in the end, but of the conceptual difficulties. It is, of course, easy to say that we turn from the concept of a *curved path* to a system of wave surfaces normal to it. The wave surfaces, however, even if we consider only small parts of them (see Fig. 7) include at least a narrow *bundle* of possible curved paths,

Fig. 7.

to all of which they stand in the same relationship. According to the old view, but not according to the new, one of them in each concrete individual case is distinguished from all the others which are "only possible", as that "really travelled". We are faced here with the full force of a logical opposition between an

either – or (point mechanics)

and a

both – and (wave mechanics)

This would not matter much, if the old system were to be dropped entirely and to be *replaced* by the new. Unfortunately, this is not the case. From the point of view of wave mechanics, the infinite array of possible point paths would be merely fictitious, none of them would have the prerogative over the others of being that really travelled in an individual case. I have, however, already mentioned that we have yet really observed such individual particle paths in some cases. The wave theory can represent this, either not at all or only very imperfectly. We find it confoundedly difficult to interpret the traces we see as nothing more than narrow bundles of equally possible paths between which the wave surfaces establish cross-connections. Yet, these cross-connections are necessary for an understanding of the diffraction and interference phenomena which can be demonstrated for the same particle with the same plausibility - and that on a large scale, not just as a consequence of the theoretical ideas about the interior of the atom, which we mentioned earlier. Conditions are admittedly such that we can always manage to make do in each concrete individual case without the two different aspects leading to different expectations as to the result of certain experiments. We cannot, however, manage to make do with such old, familiar, and seemingly indispensable terms as "real" or "only possible"; *we are never in a position to say what really is or what really happens, but we can only say what will be observed in any concrete individual case*. Will we have to be permanently satisfied with this. . . ? On principle, yes. On principle, there is nothing new in the postulate that in the end exact science should aim at nothing more than the description of what can really be observed. The question is only whether from now on we shall have to refrain from tying description to a clear hypothesis about the real nature of the world. There are many who wish to pronounce such abdication even today. But I believe that this means making things a little too easy for oneself.

I would define the present state of our knowledge as follows. *The ray or the particle path corresponds to a longitudinal relationship of the propagation process (i.e. in the direction of propagation), the wave surface on the other hand to a transversal relationship (i.e. normal to it). Both relationships are without doubt real;* one is proved by photographed particle paths, the other by interference experiments. *To combine both in a uniform system has proved impossible so far. Only in extreme cases does either the transversal, shell-shaped or the radial, longitudinal relationship predominate to such an extent that we think we can make do with the wave theory alone or with the particle theory alone.*

Paul Dirac – 1933 Nobel Lecture, December 12, 1933. ***Theory of electrons and positrons.***

[https://www.nobelprize.org/prizes/physics/1933/dirac/lecture/.]

> Dirac shared the 1933 Nobel Prize in Physics with Erwin Schrödinger "for the discovery of new productive forms of atomic theory".
>
> During the intense period of 1925-26 quantum theories were proposed that accurately described the energy levels of electrons in atoms. These equations needed to be adapted to Einstein's theory of relativity, however. In 1928 Paul Dirac formulated a fully relativistic quantum theory. The equation gave solutions that he interpreted as being caused by a particle equivalent to the electron, but with a positive charge. This particle, the positron, was later confirmed through experiments. [Paul A.M. Dirac – Facts. NobelPrize.org. https://www.nobelprize.org/prizes/physics/1933/dirac/facts/.]

In his Nobel Lecture, Dirac described the current state of his theory of electrons and positrons. He focused on electrons and the positrons on the grounds that the theory has been developed further, and hardly anything could be inferred theoretically about the properties of the others. He noted that photons are so simple that they can easily be fitted into any theoretical scheme, and protons and neutrons seem to be too complicated and no reliable basis for a theory of them had yet been discovered. He noted that the question that must first be considered is *how theory can give any information at all about the properties of elementary particles*. There existed at the present time a general quantum mechanics which could be used to describe the motion of any kind of particle, no matter what its properties are. The general quantum mechanics, however, was valid only when the particles have small velocities and failed for velocities comparable with the velocity of light, when the effects of *relativity* come in. There existed no *relativistic* quantum mechanics for particles with large velocities which could be applied to particles with arbitrary properties. But, *subjecting quantum mechanics to relativistic requirements imposes restrictions on the properties of the particle, and in this way information about the particles can be deduced from purely theoretical considerations*. Dirac described how the spin properties of the electron could be deduced, and the existence of positrons with similar spin properties and with the possibility of being annihilated in collisions with electrons could be inferred. He started with the equation connecting the *kinetic energy* W and *momentum* p_r ($r = 1, 2, 3$) of a particle in *relativistic* classical mechanics. From this he obtained a *wave equation* of quantum mechanics, by letting it operate on the wave function ψ, where W and p_r were the operators $ih\partial/\partial t$ and $-ih\partial/\partial x_r$. In order to satisfy the general requirement of quantum mechanics that its *wave equations* be linear in the operator W or $\partial/\partial t$, and the requirement of *relativistic invariance* that they must also be linear in the p's, he replaced this with a *wave equation* linear in kinetic energy W and momentum p_r. This involved four new variables α_r and α_0, which were operators that could operate on ψ. The new variables α, which he introduced to get a *relativistic* wave equation linear in W, gave rise to the spin of the electron. The variables α also gave rise to some unexpected phenomena concerning the motion of the electron. *It was found that an electron which appears to be moving slowly, must actually have a very high frequency oscillatory motion of small amplitude superposed on the regular motion, which resulted in the velocity of the electron at any time being equal to the velocity of light*. This prediction could not be directly verified by experiment, but other consequences of the theory which are inseparably bound up with this one, such as the law of scattering of light by an electron, were confirmed by experiment. A feature of these equations also led to the prediction of the positron. These quantum equations were such that they allowed as the possible results of a measurement of W either something greater than mc^2 or something less than $-mc^2$. An examination of the behavior of these negative-energy states in an electromagnetic field showed that they *corresponded to the motion of an electron with a positive charge instead of the usual negative one*, i.e. a positron. Dirac solved *the difficulty of negative energies* by identifying the positron with a "hole" using his "hole theory". On this view, the positron was just a mirror-image of the electron, having exactly the same mass and opposite charge.

> Matter has been found by experimental physicists to be made up of small particles of various kinds, the particles of each kind being all exactly alike. Some of these kinds have definitely been shown to be composite, that is, to be composed of other particles of a simpler nature. But there are other kinds which have not been shown to be composite and which one expects will never be shown to be composite, so that one considers them as elementary and fundamental.
>
> From general philosophical grounds one would at first sight like to have as few kinds of elementary particles as possible, say only one kind, or at most two, and to have all matter built up of these elementary kinds. It appears from the experimental results, though, that there must be more than this. In fact, the number of kinds of elementary particle has shown a rather alarming tendency to increase during recent years.
>
> The situation is perhaps not so bad, though, because on closer investigation it appears that the distinction between elementary and composite particles cannot be made rigorous. To get an interpretation of some modern experimental results one must suppose that particles can be created and annihilated. Thus, if a particle is observed to come out from another particle, one can no longer be sure that the latter is composite. The former may have been created. The distinction between elementary particles and composite particles now becomes a matter of convenience. This reason alone is sufficient to compel one to give up the attractive philosophical idea that all matter is made up of one kind, or perhaps two kinds of bricks.
>
> I should like here to discuss the simpler kinds of particles and to consider what can be inferred about them from purely theoretical arguments. The simpler kinds of particle are:
>
> (i) the photons or light-quanta, of which light is composed;
> (ii) the electrons, and the recently discovered positrons (which appear to be a sort of mirror image of the electrons, differing from them only in the sign of their electric charge);
> (iii) the heavier particles - protons and neutrons.

Of these, I shall deal almost entirely with the electrons and the positrons - not because they are the most interesting ones, but because in their case the theory has been developed further. There is, in fact, hardly anything that can be inferred theoretically about the properties of the others. The photons, on the one hand, are so simple that they can easily be fitted into any theoretical scheme, and the theory therefore does not put any restrictions on their properties. The protons and neutrons, on the other hand, seem to be too complicated and no reliable basis for a theory of them has yet been discovered.

The question that we must first consider is how theory can give any information at all about the properties of elementary particles. There exists at the present time a general quantum mechanics which can be used to describe the motion of any kind of particle, no matter what its properties are. The general quantum mechanics, however, is valid only when the particles have small velocities and fails for velocities comparable with the velocity of light, when effects of *relativity* come in. There exists no *relativistic* quantum mechanics (that is, one valid for large velocities) which can be applied to particles with arbitrary properties. Thus, when one subjects quantum mechanics to *relativistic* requirements, one imposes restrictions on the properties of the particle. In this way one can deduce information about the particles from purely theoretical considerations, based on general physical principles.

This procedure is successful in the case of electrons and positrons. It is to be hoped that in the future some such procedure will be found for the case of the other particles. I should like here to outline the method for electrons and positrons, showing how one can deduce the spin properties of the electron, and then how one can infer the existence of positrons with similar spin properties and with the possibility of being annihilated in collisions with electrons.

We begin with the equation connecting the kinetic energy W and momentum p_r (r = 1, 2, 3), of a particle in *relativistic* classical mechanics

$$\frac{W^2}{c^2} - p_r^2 - m^2c^2 = 0 \qquad (1)$$

From this we can get a wave equation of quantum mechanics, by letting the left-hand side operate on the wave function ψ and understanding W and p_r to be the operators $ih\partial/\partial t$ and $-ih\partial/\partial x_r$. With this understanding, the wave equation reads

$$\left[\frac{W^2}{c^2} - p_r^2 - m^2c^2\right]\psi = 0 \qquad (2)$$

Now it is a general requirement of quantum mechanics that its wave equations shall be linear in the operator W or $\partial/\partial t$ so this equation will not do. We must replace it by some equation linear in W, and in order that this equation may have *relativistic* invariance it must also be linear in the p's.

We are thus led to consider an equation of the type

$$\left[\frac{W}{c} - \alpha_r p_r - \alpha_0 mc\right] \psi = 0$$

(3)

This involves four new variables α_r and α_0 which are operators that can operate on ψ. We assume they satisfy the following conditions,

$$\alpha_\mu^2 = I \qquad \alpha_\mu \alpha_\nu + \alpha_\nu \alpha_\mu = 0$$

for

$\mu \neq \nu$ and $\mu, \nu = 0, 1, 2, 3$

and also, the α's commute with the p's and W. These special properties for the α's make Eq. (3) to a certain extent equivalent to Eq. (2), since if we then multiply (3) on the left-hand side by $W/c + \alpha_r p_r + \alpha_0 mc$ we get exactly (2).

The new variables α, which we have to introduce to get a *relativistic* wave equation linear in W, give rise to the spin of the electron. From the general principles of quantum mechanics one can easily deduce that these variables a give the electron a spin angular momentum of half a quantum and a magnetic moment of one Bohr magneton in the reverse direction to the angular momentum. These results are in agreement with experiment. They were, in fact, first obtained from the experimental evidence provided by spectroscopy and afterwards confirmed by the theory.

The variables α also give rise to some rather unexpected phenomena concerning the motion of the electron. These have been fully worked out by Schrödinger. *It is found that an electron which seems to us to be moving slowly, must actually have a very high frequency oscillatory motion of small amplitude superposed on the regular motion which appears to us. As a result of this oscillatory motion, the velocity of the electron at any time equals the velocity of light.* This is a prediction which cannot be directly verified by experiment, since the frequency of the oscillatory motion is so high and its amplitude is so small. But one must believe in this consequence of the theory, since other consequences of the theory which are inseparably bound up with this one, such as the law of scattering of light by an electron, are confirmed by experiment.

There is one other feature of these equations which I should now like to discuss, a feature which led to the prediction of the positron. If one looks at Eq. (1), one sees that it allows the kinetic energy W to be either a positive quantity greater than mc^2 or a negative quantity less than $-mc^2$. This result is preserved when one passes over to the quantum equation (2) or (3). These quantum equations are such that, when interpreted according to the general scheme of quantum dynamics, they allow as the possible results of a measurement of W either something greater than mc^2 or something less than $-mc^2$.

Now in practice the kinetic energy of a particle is always positive. We thus see that our equations allow of two kinds of motion for an electron, only one of which corresponds to what we are familiar with. *The other corresponds to electrons with a very peculiar motion such that the faster they move, the less energy they have, and one must put energy into them to bring them to rest.*

One would thus be inclined to introduce, as a new assumption of the theory, that only one of the two kinds of motion occurs in practice. But this gives rise to a difficulty, since we find from the theory that if we disturb the electron, we may cause a transition from a positive-energy state of motion to a negative-energy one, so that, even if we suppose all the electrons in the world to be started off in positive-energy states, after a time some of them would be in negative-energy states.

Thus, in allowing negative-energy states, the theory gives something which appears not to correspond to anything known experimentally, but which we cannot simply reject by a new assumption. We must find some meaning for these states.

An examination of the behavior of these states in an electromagnetic field shows that they *correspond to the motion of an electron with a positive charge instead of the usual negative one* - what the experimenters now call a positron. One might, therefore, be inclined to assume that electrons in negative-energy states are just positrons, but this will not do, because *the observed positrons certainly do not have negative energies*. We can, however, establish a connection between electrons in negative-energy states and positrons, in a rather more indirect way.

We make use of the exclusion principle of Pauli, according to which there can be only one electron in any state of motion. We now make the assumptions that in the world as we know it, nearly all the states of negative energy for the electrons are occupied, with just one electron in each state, and that a uniform filling of all the negative-energy states is completely unobservable to us. Further, any unoccupied negative-energy state, being a departure from uniformity, is observable and is just a positron.

An unoccupied negative-energy state, or hole, as we may call it for brevity, will have a positive energy, since it is a place where there is a shortage of negative energy. *A hole is, in fact, just like an ordinary particle, and its identification with the positron seems the most reasonable way of getting over the difficulty of the appearance of negative energies in our equations*. On this view the positron is just a mirror-image of the electron, having exactly the same mass and opposite charge. This has already been roughly confirmed by experiment. The positron should also have similar spin properties to the electron, but this has not yet been confirmed by experiment.

From our theoretical picture, we should expect an ordinary electron, with positive energy, to be able to drop into a hole and fill up this hole, the energy being liberated in the form of electromagnetic radiation. This would mean a process in which an electron and a positron annihilate one another. The converse process, namely the creation of an electron and a positron from electromagnetic radiation, should also be able to take place. Such processes appear to have been found experimentally, and are at present being more closely investigated by experimenters.

The theory of electrons and positrons which I have just outlined is a self-consistent theory which fits the experimental facts so far as is yet known. One would like to have an equally satisfactory theory for protons. One might perhaps think that the same theory could be applied to protons. This would require the possibility of existence of negatively charged protons forming a mirror-image of the usual positively charged ones. There is, however, some recent experimental evidence obtained by Stern about the spin magnetic moment of the proton, which conflicts with this theory for the proton. As the proton is so much heavier than the electron, it is quite likely that it requires some more complicated theory, though one cannot at the present time say what this theory is.

In any case I think it is probable that negative protons can exist, since as far as the theory is yet definite, there is a complete and perfect symmetry between positive and negative electric charge, and if this symmetry is really fundamental in nature, it must be possible to reverse the charge on any kind of particle. The negative protons would of course be much harder to produce experimentally, since a much larger energy would be required, corresponding to the larger mass.

If we accept the view of complete symmetry between positive and negative electric charge so far as concerns the fundamental laws of Nature, we must regard it rather as an accident that the Earth (and presumably the whole solar system), contains a preponderance of negative electrons and positive protons. It is quite possible that for some of the stars it is the other way about, these stars being built up mainly of positrons and negative protons. In fact, there may be half the stars of each kind. The two kinds of stars would both show exactly the same spectra, and there would be no way of distinguishing them by present astronomical methods.

Dirac, P. A. M. (January, 1930). A theory of electrons and protons.

[*Roy. Soc. Proc., A*, 126, 801, 360-65; https://doi.org/10.1098/rspa.1930.0013.]

Communicated by R. H. Fowler, F.R.S.

Received December 6, 1929.

St. John's College, Cambridge.

In *relativistic* quantum theory in which the *electromagnetic field* is subjected to quantum laws the *wave equation* refers equally well to an electron with charge + e with *negative kinetic energy*, the difficulty is not restricted to the quantum theory of the electron but appears in all relativity theories, it arises because there is an ambiguity in the sign of W, or rather W + eA_0, in the *relativity* Hamiltonian equation of the classical theory, in the quantum theory transitions can take place in which the *energy* of the electron changes from a positive to a negative value, Dirac uses his "*hole theory*" to address the problem by assuming that in a vacuum all negative-energy electron eigenstates are occupied, if negative-energy *eigenstates* are incompletely filled each unoccupied *eigenstate* – called a "hole" – would behave like a positively charged particle, which Dirac initially thought might be a proton, for scattering of radiation by an electron the *exclusion principle* forbids the electron to jump into a state of *negative energy* so Dirac assumes a double transition process in which first one of the negative-energy electrons jumps up into the *final state* for the electron with *absorption* (or *emission*) of a photon, and then the original positive-energy electron drops into hole formed by first transition with *emission* (or *absorption*) of a photon.

§ 1. *Nature of the Negative Energy Difficulty.*

The *relativity* quantum theory of an electron moving in a given electromagnetic field, although successful in predicting the spin properties of the electron, yet *involves one serious difficulty which shows that some fundamental alteration is necessary before we can regard it as an accurate description of nature.* This difficulty is connected with the fact that the *wave equation*, which is of the form

$$\{W/c + e/c \,.\, A_0 + \rho_1 (\boldsymbol{\sigma}, \mathbf{p} + e/c\, \mathbf{A}) + \rho_3 mc\} \psi = 0, \qquad (1)$$

[Dirac, P. A. M. (February, 1928). The Quantum Theory of the Electron:

$$\{p_0 + e/c \,.\, A_0 + \rho_1 (\boldsymbol{\sigma}, \mathbf{p} + e/c\, \mathbf{A}) + \rho_3 mc\} \psi = 0, \qquad (14)$$

where $p_0 = W/c$.]

has, in addition to the wanted solutions for which the kinetic energy of the electron is positive, *an equal number of unwanted solutions with negative kinetic energy for the electron, which appear to have no physical meaning*. Thus, if we take the case of a steady *electromagnetic field*, equation (1) will admit of periodic solutions of the form

$$\psi = u\, e^{-iEt/h}, \qquad (2)$$

where u is independent of t, representing *stationary states*, E being the *total energy* of the *state*, including the *relativity* term mc^2. There will then exist solutions (2) with negative values for E as well as those with positive values; in fact, if we take a matrix representation of the operators $\rho_1\sigma_1$, $\rho_1\sigma_2$, $\rho_1\sigma_3$ with the matrix elements all real, then the conjugate complex of any solution of (1) will be a solution of the *wave equation* obtained from (1) by reversal of the sign of the potentials A, and either the original *wave function* or its *conjugate complex* must refer to a *negative E.*

The difficulty is not a special one connected with the quantum theory of the electron, but is a general one appearing in all relativity theories, also in the classical theory. It arises on account of the fundamental fact that in the relativity Hamiltonian equation of the classical theory, namely,

$$(W/c + e/c\, A_0)^2 - (p + e/c\, \mathbf{A})^2 - m^2c^2 = 0, \qquad (3)$$

there is an ambiguity in the sign of W, or rather $W + eA_0$. Although the operator on the *wave function* in (1) is linear in W, yet it is, roughly speaking, equivalent to the left-hand side of (3) and the ambiguity in sign persists. The difficulty is not important in the classical theory, since here dynamical variables must always vary continuously, so that there will be a sharp distinction between those solutions of the *equations of motion* for which $W + eA_0 > mc^2$ and those for which $W + eA_0 < mc^2$, and we may simply ignore the latter.

We cannot, however, get over the difficulty so easily in the quantum theory. It is true that in the case of a steady *electromagnetic field* we can draw a distinction between those solutions of (1) of the form (2) with E positive and those with E negative and *may assert that only the former have a physical meaning* (as was actually done when the theory was applied to the determination of the *energy* levels of the hydrogen atom), but if a perturbation is applied to the system it may cause transitions from one kind of *state* to the other. In the general case of an arbitrarily varying *electromagnetic field,* we can make no hard-and-fast separation of the solutions of the wave equation into those referring to positive and those to negative *kinetic energy*. Further, in the accurate quantum theory in which the *electromagnetic field* also is subjected to quantum laws, *transitions can take place in which the energy of the electron changes from a positive to a negative value even in the absence of any external field, the surplus energy*, at least $2mc^2$ in amount, *being spontaneously emitted in the form of radiation.* (The laws of conservation of energy and momentum require at least two light-quanta to be formed simultaneously in such a process.) Thus, we cannot ignore the *negative-energy states* without giving rise to ambiguity in the interpretation of the theory.

Let us examine the *wave functions* representing *states* of *negative energy* a little more closely. If we superpose a number of these *wave functions* in such a way as to get a wave packet, the motion of this packet will be along a classical trajectory given by the Hamiltonian (3) *with $W + eA_0$ negative.* Such a trajectory, it is easily seen, is a possible trajectory for an ordinary electron (with positive energy) moving in the *electromagnetic field* with reversed sign, or for an electron of charge e (and positive energy) moving in the original *electromagnetic field.* Thus, *an electron with negative energy moves an external* *field as though it carries a positive charge.* This result has led people to suspect a connection between the *negative-energy electron* and the proton or hydrogen nucleus*.

> * See for example, Weyl, H. (May, 1929). Elektron und Gravitation. (Electron and gravity.) *Zeit. Phys.*, 56: 330-52; also, Weyl, H. (April, 1929). Gravitation and the electron. *PNAS*, 15, 4, 323–34[; attempt to incorporate Dirac theory into the scheme of *general relativity*, introduces *gauge invariance* of *theory of coupled electromagnetic potentials* and Dirac *matter waves*, explains why "anti-symmetric" Pauli-Fermi statistics for electrons lead to "symmetric" Bose-Einstein statistics for photons, barrier which hems progress of quantum theory is quantization of field equations].

One cannot, however, simply assert that a negative-energy electron is a proton, as that would lead to the following paradoxes: —

(i) A transition of an electron from a *state* of positive to one of negative energy would be interpreted as a transition of an electron into a proton, which would violate the *law of conservation of electric charge.*

(ii) Although a *negative-energy electron* moves in an external field as though it has a positive charge, yet, as one can easily see from a consideration of *conservation of momentum*, the field it produces must correspond to it having a negative charge, e.g., the *negative-energy electron* will repel an ordinary *positive-energy electron* although it is itself attracted by the *positive-energy electron.*

(iii) A *negative-energy electron* will have less *energy* the faster it moves and will have to absorb *energy* in order to be brought to rest. No particles of this nature have ever been observed.

A closer consideration of the conditions that we should expect to hold in the actual world suggests that the connection between protons and negative-energy electrons should be on a somewhat different basis and this will be found to remove all the above-mentioned difficulties.

§ 2. *Solution of the Negative Energy Difficulty.*

The most stable states for an electron (the states of lowest energy) are those with negative energy and very high velocity. All the electrons in the world will tend to fall into these *states* with *emission* of radiation. The *Pauli exclusion principle*, however, will come into play and prevent more than one electron going into any one *state*. Let us assume there are so many electrons in the world that all the most stable *states* are occupied, or, more accurately, that all the *states* of *negative energy* are occupied except perhaps a few of small velocity. Any electrons with *positive energy* will now have very little chance of jumping into *negative-energy* states and will therefore behave like electrons are observed to behave in the laboratory. We shall have an infinite number of electrons in *negative-energy states*, and indeed an infinite number per unit volume all over the world, but if their distribution is exactly uniform, we should expect them to be completely unobservable. Only the small departures from exact uniformity, brought about by some of the *negative-energy* states being unoccupied, can we hope to observe.

Let us examine the properties of the vacant *states* or "holes." The problem is analogous to that of the X-ray levels in an atom with many electrons. According to the usual theory of the X-ray levels, the hole that is formed when one of the inner electrons of the atom is removed is describable as an orbit and is pictured as the orbit of the missing electron before it was removed. This description can be justified by quantum mechanics, provided the orbit is regarded, not in Bohr's sense, but as something representable, apart from spin, by a three-dimensional wave function. Thus, the hole or vacancy in a region that is otherwise saturated with electrons is much the same thing as a single electron in a region that is otherwise devoid of them.

In the X-ray case the holes should be counted as things of *negative energy*, since to make one of them disappear (i.e., to fill it up), one must add to it an ordinary electron of *positive energy*. Just the contrary holds, however, for the holes in our distribution of *negative-energy* electrons. These holes will be things of positive *energy* and will therefore be in this respect like ordinary particles. Further, the motion of one of these holes in an external *electromagnetic field* will be the same as that of the *negative-energy* electron that would fill it, and will thus correspond to it possessing a charge + e. *We are therefore led to the assumption that the holes in the distribution of negative-energy electrons are the protons.* When an electron of positive *energy* drops into a hole and fills it up, we have an electron and proton disappearing together with emission of radiation.

A difficulty arises when we consider the field produced by the distribution of negative energy electrons. There is an infinite density of electricity which, according to Maxwell's equation

$$\operatorname{div} \mathbf{E} = -4\pi\rho, \tag{4}$$

should produce an electric field of infinite divergence. It seems natural, however, to interpret the ρ in Maxwell's equation (4) as the departure from the *normal state of electrification of the world, which normal state of electrification, according to the present theory, is the one where every electronic state of negative energy and none of positive energy is occupied.* This ρ will then consist of a charge – e arising from each state of positive energy that is occupied, together with a charge + e arising from each state of negative energy that is unoccupied. Thus, the field produced by a proton will correspond to it having a charge + e.

In this way we can get over the three difficulties mentioned at the end of the preceding section. We require to postulate only one fundamental kind of particle, instead of the two, electron and proton, that were previously necessary. *The mere tendency of all the particles to go into their states of lowest energy results in all the distinctive things in nature having positive energy.*

Can the present theory account for the great dissymmetry between electrons and protons, which manifests itself through their different masses and the power of protons to combine to form heavier atomic nuclei? It is evident that the theory gives, to a large extent, symmetry between electrons and protons. We may interchange their roles and assert that the protons are the real particles and the electrons are merely holes in the distribution of protons of negative energy. The symmetry is not, however, mathematically perfect when one takes interaction between the electrons into account. If one neglects the interaction, the Hamiltonian describing the whole system will be of the form $\Sigma\ H_a$, where H_a is the Hamiltonian or *energy* of an electron in *state a* and the summation is taken over all occupied *states*. This differs only by a constant (i.e., by something independent of which *states* are occupied) from the sum $\Sigma\ (-\ H_a)$ taken over all unoccupied *states*. Thus, we get formally the same dynamical system if we consider the unoccupied *states* or protons each to contribute a term $-\ H_a$ to the Hamiltonian. On the other hand, if we take *interaction* between the electrons into account, we get an extra term of the form $\Sigma\ V_{ab}$ in the Hamiltonian, the summation being taken over all pairs of *occupied states* (*a*, *b*), and this is not equivalent to any sum taken over pairs of *unoccupied states*. The *interaction* would therefore give an essentially different Hamiltonian if we regard the protons as the real particles that occupy *states*.

The consequences of this dissymmetry are not very easy to calculate on relativistic lines, but we may hope it will lead eventually to an explanation of the different masses of proton and electron. Possibly some more perfect theory of the interaction, based perhaps on Eddington's calculation* of the fine structure constant e^2/hc, is necessary before this result can be obtained.

* Eddington, A. S. (1929). The Charge of an Electron. *Roy. Soc. Proc., A*, 122, 358-69.

§ 3. *Application to Scattering.*

As an elementary application of the foregoing ideas, we may consider the problem of the *scattering of radiation by an electron*, free or bound. A scattering process ought, according to theory, to be considered as a *double transition* process, consisting of first an *absorption* of a photon with the electron simultaneously jumping to any *state*, and then an *emission* with the electron jumping into its *final state*, or else of first the *emission* and then the *absorption*.

> [Dirac, P. A. M. (May, 1927). The quantum theory of dispersion. *Roy. Soc. Proc., A*, 114, 769, 710-28; application of Dirac's *non-relativistic quantum electrodynamics* theory to determine the *radiation scattered by the atom*, method used involves finding a solution of Schrodinger equation that satisfies initial conditions corresponding to a given *initial state* for the atom and field, scattered radiation appears as result of two processes, an a*bsorption* and an *emission*, problem of light quanta being emitted not converging at high frequencies arises from approximation of regarding atom as a dipole, but use of exact expression for *interaction energy* too complicated for radiation theory at present, leads to correct

formula for scattering of radiation by a free electron, with neglect of relativity, and thus of the Compton effect, approximation sufficient for *dispersion* and *resonance* but dipole theory inadequate to calculate *breadth of a spectral line*.]

We therefore have to consider altogether three *states* of the whole system, the *initial state* with an incident photon and the electron in its *initial state*, an *intermediate state* with either two or no photons in existence and the electron in any *state*, and the *final state* with the scattered photon and the electron in its *final state*. The *initial* and *final states* of the whole system must have the same *total energy*, but the intermediate *state*, which lasts only a very short time, may have a considerably different *energy*.

The question now arises as to how one is to interpret those scattering processes for which the *intermediate state* is one of *negative energy* for the electron. According to previous ideas these *intermediate states* had no real physical meaning, so it was doubtful whether scattering processes that arise through their agency should be included in the formula for the scattering coefficient. *This gave rise to a serious difficulty, since in some important practical cases nearly all the scattering comes from intermediate states with negative energy for the electron**.

* I am indebted to I. Waller for calling my attention to this difficulty.

In fact, for a free electron and radiation of low *frequency*, where the classical formula holds, *the whole of the scattering comes from such intermediate states*.

According to the theory of the present paper it is absolutely forbidden, by the *exclusion principle*, for the electron to jump into a state of *negative energy*, *so that the double transition processes with intermediate states of negative energy for the electron must be excluded.* We now have, however, another kind of double transition process taking place, namely, that in which first one of the distribution of negative-energy electrons jumps up into the required *final state* for the electron with *absorption* (or *emission*) of a photon, and *then the original positive-energy electron drops into the hole formed by the first transition* with *emission* (or *absorption*) of a photon. Such processes result in a *final state* of the whole system indistinguishable from the *final state* with the more direct processes, in which the *same elec*tron makes two successive jumps. *These new processes just make up for those of the more direct processes that are excluded on account of the intermediate state having negative energy for the electron*, since the matrix elements that determine the transition probabilities are just the same in the two cases, though they come into play in the reverse order. In this way the old scattering formulas, in which no *intermediate states* are excluded, can be justified.

Heisenberg, W. (November, 1930). Die Selbstenergie des Elektrons. (The self-energy of the electron.)

[*Zeit. Phys.*, 65, 1-2, 4-13; https://doi.org/10.1007/BF01397404; (translation in Miller, A. I. (1994). *Early quantum electrodynamics. A sourcebook.* Selected papers, I); https://www.cambridge.org/core/books/abs/early-quantum-electrodynamics/selfenergy-of-the-electron/D1B89DCC1B534F4CA24A23EA2915DCA2).]

Received August 3, 1930.

Leipzig.

Point-like electron results in infinite energy density, but *finite radius for electron is inconsistent with special relativity, chooses to represents electron by point charge*, conditions are investigated under which the self-energy of electron vanishes, shows that one-electron problem could be treated correctly without an infinite *self-energy* if there were solutions of vacuum electrodynamics without a *zero-point energy* but *such solutions do not exist*, difficulties of *field theory* do not come directly from infinite *self-energy* of electron, *a solution of the basic equations has therefore not been found for the time being*, it also is not probable that one will achieve a solution without substantial modification of the quantum theory of wave fields.

Abstract

The behavior of very fast electrons, whose energy is large compared with mc^2 and Mc^2, is investigated. Since the rest mass of the electron can be neglected for such motions, a characteristic electron radius plays no role in the question of self-energy. The conditions are investigated under which the self-energy of the electron vanishes.

1. *Introduction*

In *classical theory*, the *field strengths* $\boldsymbol{E}$ and $\boldsymbol{H}$ become arbitrarily large in the neighborhood of the *point-charge* e, so that the integral over the *energy density* $1/8\pi\ (\boldsymbol{E}^2 + \boldsymbol{H}^2)$ diverges. To overcome this difficulty, one therefore assumes a finite radius r_0 for the electron in classical electron theory. This radius is related to the mass m of the electron in the order-of-magnitude relation $r_0 \sim e^2/mc^2$; the integral over the energy density is then of the order mc^2. *In quantum theory, not only this radius r_0 but possibly also another length $\lambda_0 = h/mc$, which is characteristic of the electron, plays a role in the self-energy*. In a superficial consideration in terms of the *correspondence principle*, one would suspect that the *self-energy* of the point-like electron must also become infinite in quantum theory.

In fact, Oppenheimer[1] and Waller[2] have indeed shown that a perturbation method which proceeds in powers of e does not yield finite values for the *self-energy*.

[1] Oppenheimer, J. R. (March, 1930). Note on the Theory of the Interaction of Field and Matter. *Phys. Rev.*, 35, 5, 461; https://doi.org/10.1103/PhysRev.35.461.

[2] Waller, I. (June, 1930). Bemerkungen über dir Rolle der Eigenenergie des Elektrons in der Quantentheorie der Strahlung. (Remarks on the role of the electron's own energy in the quantum theory of radiation.) *Zeit. Phys*. 62, 673-6; https://doi.org/10.1007/BF01843484.

At first, it appears as if, in *quantum theory* also, only the introduction of a *finite electron radius* can provide a way out of this difficulty. However, *a more detailed discussion shows that introducing such a radius would entail quite radical modifications of our present quantum theoretical ideas, since, according to the present principles, one can always construct arbitrarily small wave packets for an electron*. (This theorem no longer holds generally, if, in the Dirac spin theory, one allows only positive energy states for the construction of wave packets. But I do not believe that this circumstance is essential for the question of self-energy.) If one decides in favor of such a basic modification of quantum theory, it first seems necessary to introduce the radius r_0 more or less in such a fashion that one divides space into cells of the finite size r_0^3, and that one replaces the present differential equations by difference equations. In any case, the self-energy of the electron would be finite in such a lattice world. Although such a lattice world otherwise also has remarkable properties, one must nevertheless consider that it leads to deviations from the present theory which experimentally are not probable. *In particular, the statement that a smallest length exists is no longer relativistically invariant, and no way is presently known to harmonize the requirement of relativistic invariance with the fundamental introduction of a smallest length.*

In the meantime, *it therefore seems more correct not to introduce the length r_0 into the foundations of the theory but to hold fast to relativistic invariance*. If one takes this second standpoint, one obtains a significant simplification of the problem we have raised if one considers only those motions of the electrons and protons where their speed is nearly the speed of light and their energy is very large compared to mc^2 an Mc^2 (M is the proton mass). For such motions, the rest masses of the electron and of the proton can in fact be neglected, and consequently we should always calculate further with $m = M = 0$. In this simplified theory, only the constants h, c, and e appear: incidentally, the equations are now completely symmetric in protons and electrons. *In such a theory, there is no more room to introduce an electron radius, since, purely in terms of dimensions, no length can be formed from the constants h, c, and e.* The *self-energy* of the electron here must remain finite for other reasons. A closer investigation also shows that such profound differences exist between *quantum theory* and *classical* ideas, precisely with respect to the *self-energy*, that considerations in terms of the *correspondence principle* can no longer be meaningful.

We shall briefly enumerate these differences: In *quantum theory*, the *energy* of the electron is not given by $1/8\pi \int(\boldsymbol{E}^2 + \boldsymbol{H}^2)\, dV$, but rather one must add to this term from the *interaction* of the *matter field* and the *Maxwell field*. Further, in *classical theory*, one seeks *stationary* solutions of the equation div $\boldsymbol{E} = 0$ with a singularity at a point. In *quantum theory*, a point-shaped *wave packet* will immediately disperse; therefore, it is not *stationary* solutions of div $\boldsymbol{E} = 0$ with a singularity that are involved but time-variable fields, for which div $\boldsymbol{E}$ behaves exactly like the *wave packet*. In *classical theory, the speed of the electron furthermore is always assumed smaller than c, whereas in the Dirac theory it moves at the speed of light* ($q^{·}_i = \alpha_i \cdot c$, $\alpha_i^2 = 1$). Finally, in *quantum theory*, there exist *commutation relations* between the *field variables* which cause deviations from the *classical theory* in the case of small quantum numbers. Calculation will show that these deviations can be significant because of the empirical numerical value of the electron charge.

2. *Mathematical formulation of the self-energy.*

To investigate the conditions for the appearance of a possible infinite *self-energy*, we first write the basic *equations of quantum electrodynamics*[*]

> [*]Heisenberg, W. & Pauli, W. (July, 1929). Zur Quantendynamik der Wellenfelder. (On the quantum dynamics of wave fields.) *Zeit. Phys.*, 56, 1-61; Heisenberg, W. & Pauli, W. (January, 1930). Zur Quantendynamik der Wellenfelder II. (On the quantum dynamics of wave fields II.) *Zeit. Phys.*, 59, 168-190.

for the special case $m = M = 0$:

$$H^- = \int [\alpha^k_{\rho\sigma}\psi^*_\rho (hc/2\pi i \; \partial\psi_\sigma/\partial x_k + e\; \psi_\sigma\; \Phi_k) + 1/8\pi\; (\boldsymbol{E}^2 + \boldsymbol{H}^2)]\; dV, \qquad (1)$$

> [Heisenberg, W. & Pauli, W. (1930):
> $H^- = \int (H^{(m)} + H^{(s)})\; dV = \int [\alpha^k_{\rho\sigma}\psi^*_\rho (hc/2\pi i\; \partial\psi_\sigma/\partial x_k + e\; \psi_\sigma\; \Phi_k)$
> $+ mc^2\alpha^4_{\rho\sigma}\; \psi^*_\rho\; \psi_\sigma - \frac{1}{2} F_{4k} F_{4k} + \frac{1}{4} F_{ik} F_{ik} + i\; \Phi_4 C]\; dV. \quad (38)$
> $L^{(s)} = -\frac{1}{4} F_{\alpha\beta} F_{\alpha\beta} = \frac{1}{2} (E^2 - H^2)] \qquad\qquad (36b)]$

$$\text{div } \boldsymbol{E} = -4\pi e\; \psi^*_\rho\; \psi_\rho \qquad (1)$$

> [*Ibid.*: $C = \text{div } \mathcal{E} + e \sum_\rho \psi^*_\rho\; \psi_\rho = 0$ (25),
> so div $\mathcal{E} = -e \sum_\rho \psi^*_\rho\; \psi_\rho$]

$$\boldsymbol{H} = \text{rot } U \quad (U = \Phi_1, \Phi_2, \Phi_3), \qquad (1)$$

$$\boldsymbol{E}_i = -4\pi c\; \Pi_i, \qquad (1)$$

> [*Ibid.*: $\Pi_i = -1/4\pi c\; \mathcal{E}_k$ (41)]

$$\psi^*_\rho (P)\; \psi_\sigma (P') + \psi_\sigma (P')\; \psi^*_\rho (P) = \delta(P - P')\; \delta_{\rho\sigma}, \qquad (1)$$

> [*Ibid.*, original page 388: $[\psi^*_\rho\; \psi_\sigma '] = \delta_{\rho\sigma}\; \delta(r, r')$]

$$\Pi_i (P)\; \Phi_k (P') - \Phi_k (P')\; \Pi_i (P) = h/2\pi i\; \delta(P - P')\; \delta_{jk}. \qquad (1)$$

> [*Ibid.*, original page 382: $[\Pi_i, \Phi'_r] = h/2\pi i\; \delta_{ik} \cdot \delta(r, r')$]

Here H^- signifies the *Hamilton function*, ψ_σ the *Dirac function*, α^k the *spin matrices*, Φ_k the components of the *magnetic potential*, δ the Dirac *δ-function* in space, [$\boldsymbol{E}$ and $\boldsymbol{H}$ are the *electric* and *magnetic field strengths*, Π_i are the components of the *electric potential*, and P is the *current vector*]. By slightly changing the variables, one can make the universal constants in the Hamiltonian function appear only in one common factor. We therefore set

$$\begin{array}{ll} U = (2hc)^{1/2}\; \boldsymbol{a}, & \Phi_i = (2hc)^{1/2}\; \varphi_i, \\ \boldsymbol{E}_i = (2hc)^{1/2}\; \boldsymbol{e}_i, & \boldsymbol{H}_i = (2hc)^{1/2}\; \boldsymbol{h}_i, \end{array} \qquad (2)$$

and there follows:

$$H^{-} = \dots , \qquad (3)$$
$$\mathrm{div}\,\mathbf{e} = -\mu\psi_\rho\psi^*_\rho,$$
$$\psi^*_\rho(P)\psi_\sigma(P') + \psi_\sigma(P')\psi^*_\rho(P) = i\delta(P - P')\,\delta_{\rho\sigma}.$$
$$e_k(P)\varphi_l(P') - \varphi_l(P')e_k(P) = i\delta(P - P')\,\delta_{kl}.$$

Here, μ signifies the expression $4\pi e/(2hc)^{1/2}$, that is a pure number whose value is approximately 0.303 ... . Because of the fact that μ is certainly not large compared to unity, it follows that *the deviations which always occur between quantum theory and classical theory in the case of small quantum numbers can be significant for the problem of the self-energy*. The fields **E** and **H** have the dimension of a reciprocal area.

From the matter waves ψ, a simple transformation carries one to the coordinates of the material particles#.

Loc. cit. Heisenberg & Pauli, (1930), §7.

Here, this transformation will be performed under the special assumption that only one electron is present. Furthermore, deviating from the usual notation, we shall designate $2\pi/h$ times the electron momentum by $\mathbf{p}(p_1, p_2, p_3)$. In the new variables one then has:

$$H^{-} = \dots , \qquad (4)$$
$$\mathrm{div}_p\,\mathbf{e} = -\mu\delta(P - P_q),$$
$$p_k q_l - q_l p_k = 1/i\;\delta_{kl},$$
$$e_k(P)\varphi_l(P') - \varphi_l(P')e_k(P) = i\delta(P - P')\,\delta_{kl}.$$

(P_q or q characterizes the position of the electron.) Only the *one-electron problem* will be discussed below since, for the question of the *self-energy*, an extension to several electrons would give only superfluous complications.

For the total *momentum* G of the *wave field*, one calculates (according to *loc. cit.*, Heisenberg & Pauli (1929), eq. (13)):

$$G = \dots . \qquad (5)$$

It is now a special feature of the *one-electron* problem that the electron coordinates can here be eliminated completely from the Hamiltonian function, by using the *total momentum*. Inserting (5) into (4) one obtains:

$$H^{-} = \dots , \qquad (6)$$

This expression can still be transformed somewhat by introducing, following Dirac, the spin matrices $\sigma_1, \sigma_2, \sigma_3$ (written simply '$\boldsymbol{\sigma}$' as a vector) and the matrices ρ_1, ρ_2, ρ_3 which are related to the Dirac α^k as follows: $\alpha^k = \rho_1\sigma^k$; $\alpha^4 = \rho_3$. Equation (6) [the transformed Hamiltonian] then becomes

$$H^{-} = c.\ \alpha^k G_k + hc/2\pi \int dV\ \tfrac{1}{2}\,(\boldsymbol{\sigma}, \rho_2\mathbf{e} - \rho_3\mathbf{h})^2 \qquad (7)$$

[where $\boldsymbol{E}_i = (2\pi c)^{1/2}\,\mathbf{e}_i$, $\qquad \boldsymbol{H}_i = (2\pi c)^{1/2}\,\mathbf{h}_i$, (2) and G_k are the components of the *total momentum*, **G**, of the *wave field*; $\boldsymbol{\sigma} = (\sigma_1, \sigma_2, \sigma_3)$ is Dirac's *spin matrix vector*, and the matrix $\boldsymbol{\rho} = (\rho_1, \rho_2, \rho_3)$.]

The components of the *total momentum* here satisfy the commutation relations [*loc. cit.*, Heisenberg & Pauli (1929), eq. (23)]:

$$G_k q_l - q_l G_k = 1/i\ \delta_{kl}, \quad G_k \varphi_l - \varphi_l G_k = i\ \partial\varphi_l/\partial x_k. \tag{8}$$

Electron coordinates have completely dropped out of (7), and they occur only in the auxiliary condition

$$\mathrm{div}_p\ \mathbf{e} = -\ \mu\delta(P - P_q). \tag{4}$$

For a *force-free* electron, the equation

$$H^- = c.\ \alpha^k\ G_k \tag{9}$$

must hold according to the Dirac theory. *The volume integral which also appears in (7)*

$$[H^- = c.\ \alpha^k\ G_k + hc/2\pi \int dV\ \tfrac{1}{2}\ (\boldsymbol{\sigma}, \rho_2\mathbf{e} - \rho_3\mathbf{h})^2, \qquad (7)]$$

$$\int dV\ \tfrac{1}{2}\ (\sigma, \rho_2\mathbf{e} - \rho_3\mathbf{h})^2$$

can thus be interpreted as the "self-energy" of the electron and, in a correct theory, must vanish for a *force free* electron. This is possible only if, in all space,

$$(\sigma, \rho_2\mathbf{e} - \rho_3\mathbf{h}) = 0. \tag{10}$$

One must therefore seek a Schrodinger functional $\Psi_\rho(\varphi, q)$, ($\rho = 1, 2, 3, 4$), which satisfies the equation

$$(\sigma, \rho_2\mathbf{e} - \rho_3\mathbf{h})\ \Psi_\rho(\varphi, q) = 0 \tag{11}$$

Further condition (11) should persist during the course of time. Consequently, for the special solution (11), expression (10) must commute with H^-, i.e. the following must hold:

$$(\sigma, \rho_2\mathbf{e} - \rho_3\mathbf{h})\ \alpha^k\ G_k\ \Psi_\rho(\varphi, q) = 0. \tag{12}$$

If it is possible to specify solutions of (11) and (12), which furthermore fulfil the condition

$$\mathrm{div}_\rho\ e = -\ \mu\delta(P - P_q), \tag{4}$$

[where $\mu = 4\pi e/(2hc)^{1/2}$, the value of which is approximately 0.303…]

the question of the self-energy will have been satisfactorily solved.

3. *The eigenfield of the electron.*

Before investigating the solutions of (11), (12), and (4), we shall first treat the question of whether such solutions exist for $\mu = 0$ and whether these solutions have a physical meaning.

…

…

We thus see that, for every solution in vacuum electrodynamics, there also exist solutions of (4) for $\mu = 0$. If we choose special solutions of vacuum electrodynamics, for which the energy-momentum vector is a null vector, i.e. solutions which correspond to *one* light quantum (or several light quanta running in the same direction), then, by suitable choice of the solutions of (15), it is always possible simultaneously to satisfy the equation:

$$[H^{-} = c.\ \alpha^k G_k]\ \Psi = 0. \tag{17}$$

Indeed, for its validity, it is only necessary that the *energy* of the system is derived from the absolute magnitude of the *momentum* through multiplication by c.

Equations (11) and (12)

$$[(\sigma, \rho_2\mathbf{e} - \rho_3\mathbf{h})\ \Psi_\rho\ (\varphi, q) = 0, \tag{11}$$

$$(\sigma, \rho_2\mathbf{e} - \rho_3\mathbf{h})\ \alpha^k\ G_k\ \Psi_\rho\ (\varphi, q) = 0, \qquad (12)]$$

are thus also necessarily fulfilled for such a functional Ψ_ρ (q, φ). In particular, if there were solutions of vacuum electrodynamics for completely empty space, in which the *energy* and *momentum* of the *radiation field* would therefore vanish, one could again obtain a solution of (17) by multiplication with any arbitrary solution of (15), and thus obtain a solution of (11) and (12). *As is well known, in the present quantum theory of waves, no such solution exists because of the infinite zero-point energy of the radiation field.*

We now address the actual problem: solution of Equations (11) and (12) for μ # 0. At first, we are not interested in the time behavior and therefore seek to satisfy only the two equations [(10) and (4)]:

$$(\sigma, \rho_2\mathbf{e} - \rho_3\mathbf{h}) = 0, \tag{18}$$

$$\mathrm{div}_\rho\ e = -\ \mu\delta(P - P_q).$$

…

The *one-electron problem* could thus be treated correctly without an infinite *self-energy* if there were solutions of vacuum electrodynamics without a *zero-point energy. Unfortunately, such solutions do not exist.* However, according to Landau and Peierls *, the zero-point energy of radiation can be eliminated by formal artifice.

> * Landau, L & Peierls, R. (March, 1930). Quantenelectrodynamik im Konfigurationsraum. (Quantum electrodynamics in configuration space.) *Zeit. Phys*., 62, 188-200; https://doi.org/10.1007/BF01339793; (Translated by D. H. Delphenich; https://neo-classical-physics.info/electromagnetism. html)[; this paper restates Heisenberg and Pauli's *relativistic* theory in the configuration space of light quanta, analogous to usual quantum mechanics, instead of using the methods of quantized waves. Contrary to the assertion above, this article comes to the same conclusion as Heisenberg. It states "In particular, *the complication that the interaction of a particle with itself becomes infinitely large is not eliminated here. The equations are thus certainly still not the physically correct ones, and we also do not believe that this drawback can be removed by any purely formal alteration*].

But, then the simple form of the Hamiltonian (14) is lost and the application of these artifices to the *one-electron problem* proves impossible. *A solution of the basic equations (9), (11), and (12) has therefore not been found for the time being; it also is not probable that one will achieve a solution without substantial modification of the quantum theory of wave fields*. The purpose of this paper was to show that the difficulties of *field theory* do not come directly from the infinite *self-energy* of the electron but that rather *the foundations of field theory still require modification.*

Dirac, P. A. M. (May, 1932.) Relativistic Quantum Mechanics.

[*Roy. Soc. Proc., A*, 136, 829, 453-64; http://www.jstor.org/stable/95782.]

Received March 24, 1932.

St. John's College, Cambridge.

Alternative to Heisenberg and Pauli's approach to *relativistic* quantum mechanics which regards the field itself as a dynamical system amenable to Hamiltonian treatment and its interaction with the particles as describable by an *interaction energy*, there are serious objections to these views, if we wish to make an observation on a system of interacting particles the only effective method of procedure is to subject them to a field of electromagnetic radiation and see how they react, the role of the *field* is to provide a means for making observations, the nature of an observation requires an interplay between the *field* and the *particles*, we cannot suppose the *field* to be a dynamical system on the same footing as the *particles* and thus something to be observed in the same way as the *particles*, the *field* should appear in the theory as something more elementary and fundamental, in this paper an *interaction representation* is proposed which gives interplay between particles and field, translates *equations of motion* of *relativistic* classical theory directly into equations expressible entirely in terms of *probability amplitudes* referring to one *ingoing field* and one *outgoing field*, assumes passage from the field of ingoing waves to the field of outgoing waves is a quantum jump performed by one field composed of waves passing undisturbed through the electron and satisfying Maxwell's equations, *relativistic* observable quantities always *transition probabilities*, *probability amplitudes* analogous to Heisenberg's matrix elements, quantization assumes *intensities* and *phases* are operators satisfying usual quantum conditions governing the *intensities* and *phases* of Fourier components of electromagnetic field in empty space, determines matrix elements associated with electron jumps, assumes interaction of each electron with the *field* can be described by an *interaction energy* equal to its charge multiplied by the potential at the point where it is situated, wave equation of interactions of two electrons due to motions of each being connected with same field

$\{ih\, \delta/\delta t + h^2/2m_1\, \delta^2/\, \delta x_1^2 + h^2/2m_2\, \delta^2/\, \delta x_2^2 - \varepsilon_1 V(x_1 t) - \varepsilon_2 V(x_2 t)\}\, \psi = 0.$

> [*Quantum gravity in the first half of the twentieth century. A sourcebook.* Blum, A. S. & Rickles, D. (eds.) (2018). Blum, A. S. Chapter 17 Without New Difficulties: Quantum Gravity and the Crisis of the Quantum Field Theory Program, p. 265: "Covariant commutation relations were brought back into the game by Dirac in 1932. In an attempt to relaunch QED, he had effectively reconstructed Heisenberg-Pauli quantum field theory, only now in the *interaction representation* (Dirac 1932). Dirac was much ridiculed by Pauli for having merely produced an equivalent theory, despite his grandiloquent claims to the contrary. In a letter to Dirac of September 11, 1932, Pauli wrote:
>
> > "Your remarks on quantum electrodynamics [...] were—to put it mildly—certainly no masterpiece. After a confused introduction, consisting of sentences that were only halfway understandable because they were only halfway understood, you finally arrive at results for a simplified, one-

dimensional example that are identical with those obtained by applying the formalism of Heisenberg and myself to this example. (The identity is immediately recognizable and was then calculated in too complicated a manner by Rosenfeld.) This conclusion of your work stands in contradiction to the more or less clearly voiced claims in the introduction that you would somehow be able to make a better quantum electrodynamics than Heisenberg and myself." (Hermann, Meyenn, and Weisskopf. (1985). *Wolfgang Pauli: Wissenschaftlicher Briefwechsel mit Bohr, Einstein, Heisenberg u.a.*, p. 115.)].

§ 1. *Introduction.*

The steady development of the quantum theory that has taken place during the present century was made possible only by continual reference to the *Correspondence Principle* of Bohr, according to which, classical theory can give valuable information about quantum phenomena in spite of the essential differences in the fundamental ideas of the two theories.

[The *correspondence principle* states that the behavior of systems described by the theory of quantum mechanics (or by the old quantum theory) reproduces classical physics in the limit of large quantum numbers. In other words, it says that for large orbits and for large energies, quantum calculations must agree with classical calculations. The principle was formulated by Niels Bohr in 1920.]

A masterful advance was made by Heisenberg in 1925*, who showed how equations of classical physics could be taken over in a formal way and made to apply to quantities of importance in quantum theory, thereby establishing the *Correspondence Principle* on a quantitative basis and laying the foundations of the new Quantum Mechanics.

* Heisenberg, W. (July, 1925). Über quantentheoretische Umdeutung kinematischer und mechanischer Beziehungen. (Quantum-theoretical re-interpretation of kinematic and mechanical relations.) *Zeit. Phys.*, 33, 879-93.

Heisenberg's scheme was found to fit wonderfully well with the Hamiltonian theory of classical mechanics and enabled one to apply to quantum theory all the information that classical theory supplies, in so far as this information is consistent with the Hamiltonian form. Thus, one was able to build up a satisfactory quantum mechanics for dealing with any dynamical system composed of interacting particles, *provided the interaction could be expressed by means of an energy term to be added to the Hamiltonian function.* This does not exhaust the sphere of usefulness of the classical theory. *Classical electrodynamics, in its accurate (restricted) relativistic form, teaches us that the idea of an interaction energy between particles is only an approximation and should be replaced by the idea of each particle emitting waves, which travel outward with a finite velocity and influence the other particles in passing over them.* We must find a way of taking over this new information into the quantum theory and must set up a *relativistic* quantum mechanics, before we can dispense with the *Correspondence Principle.*

A preliminary attack on the question of *relativistic* quantum mechanics has been made through the solution of the problem of a *single charged particle moving in a specified classical field.* For the treatment of this problem, it is essential to use Schrodinger's form of quantum mechanics, according to which the motion of the particle is described by a *wave function* involving the space and time co-ordinates in a symmetrical manner. The solution is satisfactory from the point of view of the *Correspondence Principle*, although *it involves a difficulty owing to the appearance of possible negative energy values for the particle*. The difficulty is not due to a misuse of classical information and will not concern us here.

The extension of this *wave-function* method to two or more particles can easily be made so long as we keep to the idea of a given classical field in which the particles are moving. The resulting theory is logically satisfactory, but *is, of course, incomplete, as it gives no interaction between the particles*. It becomes necessary then to abandon the idea of a given classical field and to have instead a field which is of dynamical significance and acts in accordance with quantum laws.

An attempt at a comprehensive theory on these lines has been made by Heisenberg and Pauli*.

> * Heisenberg, W. & Pauli, W. (July, 1929). Zur Quantendynamik der Wellenfelder. (On the quantum dynamics of wave fields.) *Zeit. Phys.*, 56, 1-61[; Heisenberg and Pauli's first attempt to construct their own version of a *relativistically-invariant quantum electrodynamics* to treat interaction between matter and the electromagnetic field and between matter and matter, canonical quantization of both electromagnetic and matter-wave fields, but Lorentz-invariant Lagrangian for interacting *electromagnetic* and *matter-wave fields*, requires working with the *electromagnetic potentials* not just with the *fields*, Lagrangian does not contain a time derivative of the *electric potential* so *there is no corresponding canonical momentum variable,* prevents straightforward implementation of canonical commutation relations, the theory is still afflicted with many defects, *the fundamental difficulties in the relativistic formulation that were emphasized by Dirac remain unchanged,* the formulas of the theory lead to an *infinite zero-point energy* for the radiation and thus include the interaction of an electron with itself as an *infinite* additive constant, however, these difficulties are of a sort that they do not interfere with the application of the theory to many physical problems, used "crude trick" of adding additional terms to the Lagrangian.]; Heisenberg, W. & Pauli, W. (January, 1930). Zur Quantendynamik der Wellenfelder II. (On the quantum dynamics of wave fields II.) *Zeit. Phys.*, 59, 168-190; new approach to Lorentz-invariant Lagrangian problem based on notion of *gauge invariance* of *theory of coupled electromagnetic potentials* and Dirac *matter waves*, integrals of the *equations of motion* derived from *invariance properties* of Hamiltonian function, *invariance properties* of *wave equations* exploited in similar way, an *infinite* interaction of the electron with itself will also result from this approach making application of the theory impossible in many cases, the theory leads to divergent expressions for the *energies* of *stationary states* and the differences between these *energies* (i.e., the actually observed frequencies of spectral lines) came out *infinite*].

These authors regard the field itself as a dynamical system amenable to Hamiltonian treatment and its interaction with the particles as describable by an interaction energy, so that the usual methods of Hamiltonian quantum mechanics may be applied. There are serious objections to these views, apart from the purely mathematical difficulties to which they lead. If we wish to make an observation on a system of interacting particles, the only effective method of procedure is to subject them to a *field* of *electromagnetic radiation* and see how they react. Thus, the role of the *field* is to provide a means for making observations. The very nature of an observation requires an interplay between the *field* and the *particles*. We cannot therefore suppose the *field* to be a dynamical system on the same footing as the *particles* and thus something to be observed in the same way as the *particles*. *The field should appear in the theory as something more elementary and fundamental.*

Again, the *field equations* are always linear and thus of the form typical of the *wave equation* of quantum theory. This suggests deep-lying connections and possibilities for simplification and unification *which are entirely lacking in the Heisenberg-Pauli theory. In the present paper a scheme is proposed which gives the interplay between particles and field apparently correctly and in a surprisingly simple manner*. Full use is made of all the information supplied by the classical theory. The general ideas are applicable with any kind of simple harmonic wave transmitting the interaction between *particles* and providing the means of observation of *particles* (e.g., with longitudinal waves like sound waves) and not merely for the *electromagnetic* case, though presumably only the latter is of interest in atomic theory.

§ 2. *Relativistic Observations*.

A definite advance in the *relativistic* theory of the interaction of two electrons is contained in a recent paper by Moller*, where it is shown that in the calculation of the mutual scattering of two colliding electrons by Born's method of approximation, one may describe the interaction with *retarded potentials* and use *relativistic* ideas throughout, without getting any ambiguity in the scattering coefficient to the first order of approximation.

* Møller, C. (November, 1931). Über den Stoß zweier Teilchen unter Berücksichtigung der Retardation der Kräfte. (About the collision of two particles taking into account the retardation of the forces.) *Zeit. Phys.*, 70, 786-95. https://doi.org/10.1007/BF01340621.

This lack of ambiguity is ground of presumption of the correctness of the result. When, however, one tries to apply similar methods to the higher approximations or to more general problems, *one meets very definitely with ambiguities*.

The method by which Moller obtained his result may be compared with the methods of the *Correspondence Principle* in use before the introduction of Heisenberg's matrix theory, for calculating Einstein's A and B coefficients from classical models. In certain cases, the result obtained was unambiguous (usually those cases for which the result was zero) and was then presumed to be correct. In general, however, there was ambiguity, so that one could get no reliable accurate result.

This analogy suggests that it would be useless to try to extend Moller's method by setting up rules to provide a definite interpretation for ambiguous quantities. Any attempts in this direction would be just as futile as the attempts made in the pre-Heisenberg epoch to calculate Einstein's A's and B's from some sort of mean of classical quantities referring to the *initial* and *final states. One ought to proceed on quite different lines, namely by following the methods introduced by Heisenberg in 1925, which have already met with such great success for non-relativistic quantum mechanics.*

Heisenberg put forward the principle that one should confine one's attention to observable quantities, and set up an algebraic scheme in which only these observable quantities appear. Strictly speaking, it is not the observable quantities themselves (the Einstein A's and B's) that formed the building stones of Heisenberg's algebraic scheme, *but rather certain more elementary quantities, the matrix elements*, having the observable quantities as the squares of their moduli. *The extra phase quantities introduced in this way are essential.*

Let us see what are the corresponding quantities in *relativistic theory*. To make a *relativistic* observation on a system of particles we must, as mentioned in the introduction, *send in some incident electromagnetic radiation and examine the scattered radiation. The numerical quantity that we observe is thus the probability of occurrence of a certain radiative transition process*. This process may be specified by the *intensities* of the various monochromatic components of the ingoing and of the outgoing fields of radiation. (We shall ignore the purely mathematical difficulty that *the total number of these components is an infinity of a high order*.) The *phases* must not be specified together with the *intensities*, as this would violate well-established quantum principles.

In *non-relativistic* quantum mechanics, the probability of occurrence of any *transition process* is always given as the square of the modulus of a certain quantity, of the nature of a matrix element or simply a transformation function, referring to the *initial* and *final states*. It appears reasonable to assume that this will still be the case in *relativistic* quantum mechanics. *Thus, the relativistic observable quantities, which are always transition probabilities, will all appear as the squares of the moduli of certain quantities*. These quantities, which we shall refer to as *probability amplitudes*, will then be the building stones analogous to Heisenberg's matrix elements. We should expect to be able to *set up an algebraic scheme involving only the probability amplitudes and to translate the equations of motion of relativistic classical theory directly into exact equations expressible entirely in terms of these quantities*.

The information that classical theory supplies is thus to be used to give relations between the *probability amplitudes* of different physical processes, rather than to enable one to calculate a particular one of them. *Only in very special cases, of which Moller's paper provides an example, is it possible to evaluate a relativistic transition probability without at the same time evaluating a whole series of them, referring to all the possible ways in which the particles under consideration can react with the radiation field.*

A point of special importance about the building stones of the new theory is that each of them refers to one *field* of ingoing *waves* and one *field* of outgoing *waves*, or to one *initial field* of a transition process and one *final field*. Quantities referring to two initial fields, or to two final fields, are not allowed. *This shows a departure from the theory of Heisenberg and Pauli*, according to which, if one is given any quantity referring to one *initial field* and one *final field*, one can obtain from it a quantity referring to two *initial fields*, or to two *final fields*, by a straightforward application of the transformation theory of quantum mechanics. The Heisenberg-Pauli theory thus involves many quantities which are unconnected with results of observations and which must be removed from consideration if one is to obtain a clear insight into the underlying physical relations.

§ 3. *Equations of Motion.*

We shall now consider in detail the question of how the information contained in classical electrodynamics can be taken over into the quantum theory. We meet at once with the difficulty that the classical theory itself is not free from ambiguity.

To make the discussion precise, *let us suppose we have a single electron interacting with a field of radiation and consider the radiation resolved into ingoing and outgoing waves*. The classical problem is, given the ingoing radiation and suitable initial conditions for the electron, *determine the motion of the electron and the outgoing radiation*. The classical equations which deal with this problem are of two kinds, (i) those that *determine the field produced by the electron* (which field is just the difference of the ingoing and outgoing fields) in terms of the variables describing the *motion of the electron*, and (ii) those that *determine the motion of the electron*. Equations (i) are quite definite and unambiguous, but not so equations (ii). The latter express the acceleration of the electron in terms of field quantities at the point where the electron is situated and *these field quantities in the complete classical picture are infinite and undefined.*

In the usual approximate treatment of the problem one takes for these field quantities just the contributions of the *ingoing waves*. *This treatment is necessarily only approximate, since it does not take into account the reaction on the electron of the waves it emits*. We should expect in an accurate treatment, that the field determining the acceleration of the electron would be in some way associated with *both the ingoing and outgoing waves*. Classical attempts have been made to improve the theory by assuming a definite structure for the electron and calculating the effect on one part of it of the field produced by the rest, but such methods are not permissible in modern physics.

We must recognize at this point that we have reached the limit of classical electromagnetic theory. We have quite definite equations for determining the *motion of the electron* in terms of *field quantities*, but we cannot interpret these field quantities in the usual classical picture and the most we can say about them is that they are related in some non-classical way to two fields, namely, those of the ingoing and of the outgoing waves. Further advance can be made only by introducing quantum ideas.

Let us make the assumption that *the passage from the field of ingoing waves to the field of outgoing waves is just a quantum jump performed by one field*. This assumption is permissible on account of the fact, discussed in the preceding section, that all the quantities in *relativistic quantum mechanics* are of the nature of *probability amplitudes* referring to one *ingoing field* and one *outgoing field*, so that we may associate, say, the right-hand sides of the *probability amplitudes* with *ingoing fields* and the left-hand sides with *outgoing fields. In this way we automatically exclude quantities referring to two ingoing fields or to two outgoing fields* and make a great simplification in the foundations of the theory.

The significance of the new assumption lies in the fact that the classical picture from which we derive our *equations of motion* must contain no reference to quantum jumps. *This classical picture must therefore involve just one field, a field composed of waves passing undisturbed through the electron and satisfying everywhere Maxwell's equations for empty space.* With this picture the *equations of motion* for the electron are perfectly definite and unambiguous. There are no *equations of motion* for the *field*, as the *field* throughout *space-time* is pictured as given. Thus, the interaction between electron and field is introduced into the equations in only one place.

The quantization of the *equations of motion* derived from this picture may conveniently be carried out in two stages. Let us *first quantize only the variables describing the electron.* We then get just the usual quantum theory of the *motion of an electron* in a given classical field, with the difference that in the present case the field must necessarily be resolvable into plane waves and must therefore contain nothing of the nature of a Coulomb force. We have a *Schrodinger equation* of the form

$$F\psi = 0,$$

where the operator [of the *wave equation* of the quantum theory] F is, *neglecting spin*,

$$F = (ih\ \delta/\delta t + eA_0)^2 - (ihc\ \delta/\delta x - eA_x)^2 - \ldots - m^2c^4 \qquad (1)$$

[cf Dirac (February, 1928). The Quantum Theory of the Electron: "§ 1. *Previous Relativity Treatments*. The *relativity* Hamiltonian according to the classical theory for a point electron moving in an arbitrary electro-magnetic field with scalar potential A_0 and vector potential A is

$$F = (W/c + e/c\ A_0)^2 + (\mathbf{p} + e/c\ \mathbf{A})^2 + m^2c^2,$$

where **p** is the momentum vector.]

It should be remembered that *the wave-function ψ involves not only the variables x, y, z, t describing the electron, but also a large number of parameters describing the field, which parameters may conveniently be taken to be the intensities J and phases w of the various Fourier components of the field.* The *potentials* A occurring in F are likewise functions, not only of the variables x, y, z, t describing the momentary position of the electron, but also of the parameters J and *w*.

In the second stage of the quantization, we assume that the J's and *w*'s occurring in ψ and the A's are not numerical, but are operators satisfying the usual quantum conditions governing the *intensities* and *phases* of the Fourier components of the electromagnetic field in empty space. The new *wave equation* obtained in this way is to be treated on the same lines as the previous one. *In particular, it may be used to determine matrix elements associated with electron jumps.* Each such matrix element will now be a function of the non-commuting J's and *w*'s, so that, when we take a representation of the J's and *w*'s, it becomes a set of quantities, each referring to two *states* of the *field* as well as the two electronic *states*, and thus being of the nature of the *probability amplitudes* of § 2.

For the problem of *the interaction of two electrons*, we require a *wave-function* ψ which is a function of the variables x_1, y_1, z_1, t_1 and x_2, y_2, z_2, t_2 describing the two electrons and of one set of J's and *w*'s describing one field. This ψ must satisfy the two *wave equations*

$$F_1\psi = 0, \qquad F_2\psi = 0, \tag{2}$$

where F_1 is the operator obtained from F by substituting $\delta/\delta t_1$, etc., for $\delta/\delta t$, etc., and taking for the A's their values at the point x_1, y_1, z_1, t_1 and similarly for F_2. *These two wave equations describe completely the relations between the two electrons and the field. No terms of the type of a Coulomb interaction energy are required in the operators of the wave equations.* The interaction of the two electrons is due to the motions of both being connected with the same *field*. This interaction manifests itself mathematically through the fact that, if we take a *wave function* ψ_1, a function only of x_1, y_1, z_1, t_1, and the J's and *w*'s, satisfying

$$F_1\psi_1 = 0, \tag{3A}$$

and a second *wave-function* ψ_2, a function only of x_2, y_2, z_2, t_2 and the J's and *w*'s, satisfying

$$F_2\psi_2 = 0, \tag{3B}$$

then neither of the products $\psi_1\psi_2$ and $\psi_2\psi_1$ will satisfy both the *wave equations* (2). The solution of equations (2) is an essentially different and more complicated problem than the solution of (3A) and (3B).

§ 4. *Interaction between Two Particles in One Dimension.*

It may seem rather surprising that a theory in which all the fields are resolvable into plane waves can give anything of the nature of the usual *electrostatic forces* between electrons. We shall therefore illustrate by a simple example the fact that these forces really are contained in our *wave equations*. We shall take the case of two particles moving in a field in one-dimensional space and shall proceed to solve equations (2), making various approximations that are permissible *when we are not interested in relativistic effects*.

Suppose the field to be describable by a *potential function* V satisfying the *wave equation*

$$\delta^2V/\delta x^2 - 1/c^2\, \delta^2V/\delta t^2 = 0,$$

and the classical expression for the *energy* to be

$$H = 1/8\pi \int\{(\delta V/\delta x)^2 + 1/c^2\, (\delta V/\delta t)^2\}dx.$$

If we resolve V into its Fourier components, thus

$$V = \int_{-\infty}^{\infty} \{a_v e^{iv(t+x/c)} + b_v e^{iv(t-x/c)}\}\, dv, \tag{4}$$

the expression for the *energy* will go over into

$$H = 1/c \int_0^\infty v^2\{a_v a_{-v} + b_v b_{-v}\}dv. \qquad (5)$$

Let us now see what are the Poisson bracket relations between the Fourier coefficients *a* and *b*. These relations must be chosen such that the quantities $a_v e^{ivt}$, $b_v e^{ivt}$, considered as dynamical variables, satisfy *equations of motion* of the Hamiltonian form with the Hamiltonian function (5), thus

$$d/dt\,(a_v e^{ivt}) = [a_v e^{ivt}, H]$$

or

$$iva_v = [a_v, H],$$

and similarly, for b_v. It is easily verified that we must have

$$[a_v, a_{v^*}] = [b_v, b_{v^*}] = ic/v\ .\ \delta\,(v + v')$$
$$[a_v, b_{v'}] = 0.$$

In the quantum theory these relations become

$$a_v\, a_{v'} - a_{v'}\, a_v = b_v\, b_{v'} - b_{v'}\, b_v = -\,hc/v\ .\ \delta\,(v + v')$$
$$a_v\, b_{v'} - b_{v'}\, a_v = 0 \qquad (6)$$

We now introduce two particles, of masses m_1 *and* m_2 *and "charges"* ε_1 *and* ε_2, and suppose the interaction of each with the *field* can be described by an *interaction energy* equal to its charge multiplied by the value of V at the point where it is situated. Thus, *if we neglect the relativistic variation of mass with velocity*, we have the two *wave equations*

$$\{ih\ \delta/\delta t_1 + h^2/2m_1\ \delta^2/\ \delta x_1^2 - \varepsilon_1\ V\,(x_1 t_1)\}\ \psi = 0,$$
$$\{ih\ \delta/\delta t_2 + h^2/2m_2\ \delta^2/\ \delta x_2^2 - \varepsilon_2\ V\,(x_2 t_2)\}\ \psi = 0.$$

By putting $t_1 = t_2 = t$, we can reduce these to one *wave equation*

$$\{ih\ \delta/\delta t + h^2/2m_1\ \delta^2/\ \delta x_1^2 + h^2/2m_2\ \delta^2/\ \delta x_2^2 - \varepsilon_1 V(x_1 t) - \varepsilon_2 V(x_2 t)\}\ \psi = 0, \qquad (7)$$

We shall proceed to obtain a solution of this equation in the form of a power series in the ε's. Thus, we put

$$\psi = \psi_0 + \psi_1 + \psi_2 + \ldots$$

where

$$\{ih\ \delta/\delta t + h^2/2m_1\ \delta^2/\ \delta x_1^2 + h^2/2m_2\ \delta^2/\ \delta x_2^2\}\ \psi_0 = 0 \qquad (8)$$
$$\{ih\ \delta/\delta t + h^2/2m_1\ \delta^2/\ \delta x_1^2 + h^2/2m_2\ \delta^2/\ \delta x_2^2\}\ \psi_1 = \{\varepsilon_1 V(x_1 t) + \varepsilon_2 V(x_2 t)\}\ \psi_0 \qquad (9)$$
$$\{ih\ \delta/\delta t + h^2/2m_1\ \delta^2/\ \delta x_1^2 + h^2/2m_2\ \delta^2/\ \delta x_2^2\}\ \psi_2 = \{\varepsilon_1 V(x_1 t) + \varepsilon_2 V(x_2 t)\}\ \psi_1 \qquad (10)$$

We take as the solution of (8)

$$\psi_0 = e^{ip1x1/h}\ e^{ip2x2/h}\ e^{-iWt/h}\ \delta_{j0}$$

where $W = p_1^2/2m_1 + p_2^2/2m_2$, (11)

representing a *state* for which the *particles* have the *momenta* p_1 and p_2, and all the J's, i.e., the *intensities* of the Fourier components of the field, vanish. Now the operator $\varepsilon_1 V(x_1 t) + \varepsilon_2 V(x_2 t)$ occurring on the right-hand sides of (9) and (10), if expressed as a matrix in a representation in which the J's are diagonal, would contain only matrix elements referring to *transitions* in which just one of the J's changes by one quantum. It follows that ψ_1 must consist of a sum of terms each referring to a *state* of the *field* in which just one oscillation is excited by one quantum. Similarly, ψ_2 must consist of a sum of terms each referring either to a two-quantum or to a zero-quantum *state* of the field. The latter are the ones that interest us here, as they may be compared with the terms that would arise from the insertion of an *interaction energy* between the two *particles* into the operator of the *wave equation* (7).

We can obtain the solution of equation (9) by expanding the right-hand side in terms of its Fourier components by means of (4) and dividing each component by the number to which the operator on the left-hand side of (9) is equivalent when it operates on that component. This gives

$$\psi_1 = \varepsilon_1 \int_{-\infty}^{\infty} \{\ldots\}\, dv \,.\, \psi_0 + \varepsilon_2 \int_{-\infty}^{\infty} \{\ldots\}\, dv \,.\, \psi_0.$$

If we use (11) and also neglect terms like p_1/m_1c, hv/m_1c^2 compared with unity, *as is permissible when we are not interested in relativistic effects*, this reduces to

$$\begin{aligned}\psi_1 = &- \varepsilon_1/h \int_{-\infty}^{\infty} \{a_v e^{iv(t+x1/c)} + b_v e^{iv(t-x1/c)}\}\, dv/v \,.\, \psi_0 \\ &- \varepsilon_2/h \int_{-\infty}^{\infty} \{a_v e^{iv(t+x2/c)} + b_v e^{iv(t-x2/c)}\}\, dv/v \,.\, \psi_0 \qquad (12)\end{aligned}$$

When we substitute this value for ψ_1 in the right-hand side of (10) and also substitute for V its expansion given by (4), we obtain an expression consisting of an operator, which is a homogeneous quadratic function of the *a*'s and *b*'s, operating on ψ_0. We must evaluate that particular part of the expression that refers to the *unexcited state* of the *field*. Only those terms of the operator involving products like $a_v a_{-v}$ or $b_v b_{-v}$ will contribute anything to that part. To obtain the contribution of a term involving $a_v a_{-v}$, we observe that, for $v > 0$, a_v and a_{-v} are like the quantities p + iq and p – iq respectively in the problem of the simple harmonic oscillator. Thus $a_v a_{-v}$ with $v > 0$, is proportional to twice the *energy* of the corresponding oscillation (without *zero-point energy*) so that it gives no contribution when multiplied into ψ_0. The first of the quantum conditions (6) now shows that, to get the contribution from a term involving $a_{-v} a_v$ with $v > 0$, we must count $a_{-v} a_v$ as equal to $hc/v \,.\, \delta(v - v')$. In the same way we find that we must count $b_v b_{-v} = 0$ and $b_{-v} b_v = hc/v \,.\, \delta(v - v')$ with $v > 0$.

The term on the right-hand side of (10) arising from the product of $\varepsilon_2 V(x_2 t)$ with the first of the terms for ψ_1 in (12) may be written

$$\begin{aligned}&- \varepsilon_1 \varepsilon_2/h \int_{-\infty}^{\infty} dv' \{a_{-v'} e^{-iv'(t+x2/c)} + b_{-v'} e^{-iv'(t-x2/c)}\} \\ &\quad \times \int_{-\infty}^{\infty} dv/v \{a_v e^{iv(t+x1/c)} + b_v e^{iv(t-x1/c)}\} \,.\, \psi_0\end{aligned}$$

That part of it referring to the unexcited state of the field is, by the forcgoing rules

$$-\varepsilon_1\varepsilon_2/h \int_0^\infty dv' \int_0^\infty dv/v\ hc/v\ .\ \delta(v-v')\{e^{-iv'(t+x2/c)}\ e^{iv(t+x1/c)} + e^{-iv'(t-x2/c)}\ e^{iv(t-x1/c)}\}\ .\ \psi_0$$
$$= -\,2\varepsilon_1\varepsilon_2 c \int_0^\infty dv/v^2 \cos v\,(x_1 - x_2)/c\ .\ \psi_0$$

The coefficient of ψ_0 *here differs only by an infinitely great constant* (independent of x_1 and x_2) from

$$2\varepsilon_1\varepsilon_2 c \int_0^\infty dv/v^2 \{1 - \cos v\,(x_1 - x_2)/c\} = \pi\varepsilon_1\varepsilon_2 |x_1 - x_2|.$$

The other terms on the right-hand side of (10) may be dealt with in the same way and give for the complete part referring to the *unexcited state* of the *field*

$$\{2\pi\varepsilon_1\varepsilon_2 |x_1 - x_2| + K\}\ \psi_0, \qquad (13)$$

where K is an infinite constant.

Equation (10) with expression (13) on its right-hand side is just what we should get if we were solving the *wave equation*

$$\{ih\ \delta/\delta t + h^2/2m_1\ \delta^2/\delta x_1^2 + h^2/2m_2\ \delta^2/\delta x_2^2 - 2\pi\varepsilon_1\varepsilon_2 |x_1 - x_2| - K\}\ \psi = 0,$$

by a method of approximation through expansion in powers of $\varepsilon_1\varepsilon_2$. *Thus, our wave-equation* (7)
$$[\{ih\ \delta/\delta t + h^2/2m_1\ \delta^2/\delta x_1^2 + h^2/2m_2\ \delta^2/\delta x_2^2 - \varepsilon_1 V(x_1 t) - \varepsilon_2 V(x_2 t)\}\ \psi = 0, \quad (7)]$$
contains implicitly an interaction between the particles, expressible approximately by the interaction energy $2\pi\varepsilon_1\varepsilon_2 |x_1 - x_2|$. This *interaction energy* agrees numerically with what we should expect from a one-dimensional *electrostatic* theory. There is, however, a mistake in sign, as it gives an attractive force between like charges.

> [Note added, April 20th. *It has been pointed out to me by Professor Heisenberg that the sign of the interaction energy given by the above calculation is really quite correct,* since with the one-dimensional, longitudinal waves there used the classical theory also requires an attractive force between like charges.]

Summary.

A quantum theory is proposed in which *the interaction between particles takes place by means of vibrations of an intervening medium transmitted with a finite velocity.* The fundamental relations involve only quantities having observational significance, account being taken of the fact that an act of observation necessarily involves an interplay between particles and field. A detailed solution of a one-dimensional problem is given in order to show that forces of *electrostatic* nature are implicitly contained in the theory.

Dirac, P. A. M., Fock, V. A., Podolsky, B. (1932). On quantum electrodynamics.

[*Phys. Zeit. Sowjetunion*, 2, 468; in Schwinger, J. (ed). (1958). *Selected Papers on Quantum electrodynamics*. Dover, New York, pages 29-40; paper unavailable online.]

Cambridge, Leningrad and Kharkov.

Received October 25, 1932.

Relativistic model in which a fixed number of electrons interact through a second-quantized electromagnetic field, applies Dirac's *interaction representation* formulation of quantum field theory to full electrodynamics, formulated with the help of a *multi-time* wave function $\psi(t_1,\mathbf{x}_1,\ldots,t_N,\mathbf{x}_N)\psi(t_1,x_1,\ldots,t_N,x_N)$ that generalizes the Schrödinger's multiparticle wave function to allow for a manifestly *relativistic* formulation of wave mechanics, all non-trivial dynamics due to interaction between charged matter and electromagnetic field relegated to time evolution of state vector, second Maxwell equation - the Ampère-Maxwell law - satisfied only through the action of the field operators on the wave function, implies that electromagnetic field operators still obey the free field equations and consequently the covariant commutation relations of Jordan and Pauli, but method of quantizing by imposing covariant commutation relations still an isolated technique applied only to Maxwell field, shows relationship between *Maxwell's equations* for empty space, the *wave equation* of Heisenberg-Pauli's theory, and the *wave equation of* Dirac's *interaction representation.*

> [*Quantum gravity in the first half of the twentieth century. A sourcebook.* Blum, A. S. & Rickles, D. (eds.) (2018). Blum, A. S. Chapter 17 Without New Difficulties: Quantum Gravity and the Crisis of the Quantum Field Theory Program, p. 266: "Covariant commutation relations were brought back into the game by Dirac in 1932. In an attempt to relaunch QED, he had effectively reconstructed Heisenberg-Pauli quantum field theory, only now in the interaction representation [Dirac. (1932). Relativistic Quantum Mechanics.] Dirac was much ridiculed by Pauli for having merely produced an equivalent theory, despite his grandiloquent claims to the contrary. In a letter to Dirac of September 11, 1932, Pauli wrote:
>
>> "Your remarks on quantum electrodynamics [...] were—to put it mildly—certainly no masterpiece. After a confused introduction, consisting of sentences that were only halfway understandable because they were only halfway understood, you finally arrive at results for a simplified, one-dimensional example that are identical with those obtained by applying the formalism of Heisenberg and myself to this example. (The identity is immediately recognizable and was then calculated in too complicated a manner by Rosenfeld.) This conclusion of your work stands in contradiction to the more or less clearly voiced claims in the introduction that you would somehow be able to make a better quantum electrodynamics than Heisenberg and myself. (Hermann, Meyenn, and Weisskopf 1985, 115)."

But when Dirac, together with Vladimir Fock and Boris Podolsky, actually applied his *interaction representation* formulation of quantum field theory to the full electrodynamics [Dirac, Fock, and Podolsky, (1932)] and not just to the toy example of the first paper, Pauli realized that it had a huge advantage: *All the non-trivial dynamics due to the interaction between charged matter and electromagnetic field were relegated to the time evolution of the state vector*. In other words: Also, the second inhomogeneous Maxwell equation, *the Ampère-Maxwell law, was satisfied only through the action of the field operators on the wave function* and not as an operator identity. This implied that the electromagnetic field operators still obeyed the free field equations and consequently the covariant commutation relations of Jordan and Pauli. On 2 June 1933, Pauli wrote to Heisenberg:

> "As time goes by, I find myself liking the work of Dirac, Fock and Podolsky more and more. It is funny that there the commutation relations of vacuum electrodynamics do not change when particles are present, even for $t \neq t'$. (Hermann, Meyenn, and Weisskopf 1985, 167)."

But the method of quantizing by imposing covariant commutation relations was still an isolated technique, applied only to the Maxwell field, far-removed from the generality of the canonical quantization approach, which could be applied to any classical field theory given in Lagrangian or Hamiltonian form. It was extended in 1938 by Stueckelberg to encompass massive scalar fields, again by explicitly expanding the field operators in terms of solutions of the wave equation (in this case the Klein Gordon equation) [Stückelberg, E. C. G. (1938). Die Wechselwirkungskräfte in der Elektrodynamik und in der Feldtheorie der Kernkräfte: Teil 1. (The Interaction Forces in Electrodynamics and in the Field Theory of Nuclear Forces: Part 1.)], and then also to massive vector fields [Stückelberg, E. C. G. (1938). Die Wechselwirkungskräfte in der Elektrodynamik und in der Feldtheorie der Kernkräfte: Teil 2 und 3. (The interaction forces in electrodynamics and in the field theory of nuclear forces: Parts 2 and 3.)]. In 1939, finally, Markus Fierz gave a general expression for the covariant commutation relations for fields with arbitrary spin [Fierz, M. (January, 1939). Über die relativistische Theorie kräftefreier Teilchen mit beliebigem Spin. (On the relativistic theory of force-free particles with arbitrary spin.)]. The covariant quantization method thereby became a full-fledged quantization method in its own right, applicable to any Lorentz-covariant field theory."]

Abstract

In the first part of this paper the equivalence of the new form of relativistic quantum mechanics[1] to that of Heisenberg and Pauli[2] is proved in a new way which has the advantage of showing their physical relation and serves to suggest further development considered in the second part.

[1] Dirac, P. A. M. (May, 1932). Relativistic Quantum Mechanics. *Roy. Soc. Proc., A*, 136, 829, 453-64.
[2] Heisenberg, W. & Pauli, W. (July, 1929). Zur Quantendynamik der Wellenfelder. (On the quantum dynamics of wave fields.) *Zeit. Phys.*, 56, 1-61; (January, 1930). Zur Quantendynamik der Wellenfelder II. (On the quantum dynamics of wave fields II.) *Ibid.*, 59, 168-190.

Part I. Equivalence of Dirac's and Heisenberg-Pauli's Theories.

§ 1. Recently Rosenfeld showed[3] that the new form of relativistic Quantum Mechanics[1] is equivalent to that of Heisenberg and Pauli[2].

[3] Rosenfeld, L. (November, 1932). Über eine mögliche Fassung des Diracschen Programms zur Quantenelektrodynamik und deren formalen Zusammenhang mit der Heisenberg-Paulischen Theorie. (On a possible version of Dirac's program on quantum electrodynamics and its formal connection with Heisenberg-Paulian theory.) *Zeit. Phys.*, 76, 729-34; https://doi.org/10.1007/BF01341566.

Rosenfeld's proof is, however, obscure and does not bring out some features of the relation between the two theories. To assist in the further development of the theory we give here a simplified proof of the equivalence.

Consider a system, with a Hamiltonian H, consisting of two parts A and B with their respective Hamiltonians H_a and H_b and the interaction V. We have

$$H = H_a + H_b + V, \tag{1}$$

where $H_a = H_a(p_a q_a T)$; $\qquad H_a = H_b(p_b q_b T)$;

$$V = V(p_a q_a p_b q_b T)$$

and T is the time for the entire system. The wave function for the entire system will satisfy the equation[4]

$$(H - i\hbar\, \partial/\partial T)\psi(q_a q_b T) = 0 \tag{2}$$

and will be a function of the variables indicated.

[4] ħ is Planck's constant divided by 2π.

Now, upon performing the canonical transformation

$$\psi^* = e^{i/h\, HbT}\, \psi, \tag{3}$$

by which the dynamical variables, say F, transform as follows

$$F^* = e^{i/h\, HbT}\, F\, e^{-i/h\, HbT}, \tag{4}$$

Eq. (2) takes the form

$$(H_a + V^* - i\hbar\, \partial/\partial T)\psi^* = 0. \tag{5}$$

Since H_a commutes with H_b, $H^*_a = H_a$. On the other hand, since the functional relation between variables is not disturbed by the canonical transformation (3), V* is the same function of the transformed variables p*, q* as V is of p, q. But p_a and q_a commute with H_b so that $p^*_a = p_a$, $q^*_a = q_a$. Therefore

$$V^* = V(p_a q_a p^*_b q^*_b) \tag{6}$$

where

$$q^*_b = e^{i/h\, HbT}\, q_b\, e^{-i/h\, HbT},$$
$$p^*_b = e^{i/h\, HbT}\, p_b\, e^{-i/h\, HbT}, \tag{7}$$

It will be shown in §7, after suitable notation is developed, that Eqs. (7) are equivalent to

$$\partial q^*_b/\partial t = i/\hbar\,(H_b q^*_b - q^*_b H_b),$$
$$\partial p^*_b/\partial t = i/\hbar\,(H_b p^*_b - p^*_b H_b), \tag{8}$$

where t is the separate time of the part B.

These, however, are just the equations of motion for the part B alone, unperturbed by the presence of part A.

§ 2. Now let part B correspond to the field and part A to the particles present.
Eqs. (8)

$$[\partial q^*_b/\partial t = i/\hbar\,(H_b q^*_b - q^*_b H_b),$$
$$\partial p^*_b/\partial t = i/\hbar\,(H_b p^*_b - p^*_b H_b), \qquad (8)]$$

must then be equivalent to *Maxwell's equations for empty space.*
Eq. (2)

$$[(H - i\hbar\,\partial/\partial T)\psi \qquad (q_a q_b T) = 0 \qquad (2)]$$

is then the *wave equation of Heisenberg-Pauli's theory,* while
Eq. (5)

$$(H_a + V^* - i\hbar\,\partial/\partial T)\psi^* = 0 \qquad (5),$$

in which the perturbation is expressed in terms of potentials corresponding to empty space, is the *wave equation of the new theory*. Thus, *this theory corresponds to treating separately a part of the system*, which in some problems is more convenient[1].

[1] This is somewhat analogous to Frenkel's method of treating incomplete systems, see Frenkel, Y. (1932). *Sow. Phys.*, 1, 99.

Now, H_a can be represented as a sum of the Hamiltonians for the separate particles. The interaction between the particles is not included in H_a for this is taken to be the result of interaction between the particles and the field. Similarly, V is the sum of the interactions between the field and the particles. Thus, we may write

$$H_a = \sum_{s=1}^{n} (c\alpha_s . p_s + m_s c^2 \alpha_s^4) = \sum_{s=1}^{n} H_s,$$

and $$V^* = \sum_{s=1}^{n} V^*_s = \sum_{s=1}^{n} \varepsilon_s\, [\Phi(r_s, T) - \alpha_s . A(r_s, T)], \tag{9}$$

where r_s are the coordinates of the s-th particle and n is the number of particles.

Eq. (5) [the *wave equation of the new theory*] takes the form

$$[\sum_{s=1}^{n} (Hs + V^*_s) - i\hbar\, \partial/\partial T)\psi^*(r_s; J; T) = 0, \qquad (10)$$

where J stands for the variables describing the field. Besides the common time T and the *field time* $t_s = t_1, t_2,\ldots,t_n$ is introduced for each particle. Eq. (10) is satisfied by the common solution of the set of equations

$$(Rs - i\hbar\, \partial/\partial T)\psi^* = 0, \qquad (11)$$

where $Rs = c\alpha_s.p_s + m_s c^2 \alpha_s{}^4 + \varepsilon_s\, [\Phi(r_s, T) - \alpha_s\, .\, A(r_s, T)]$
and $\psi^* = \psi^*(r_1, r_2,\ldots,r_n; t_1, t_2,\ldots,t_n; J)$,
when all the t's are put equal to the common time T.

Now, *Eqs. (11) are the equations of Dirac's theory*. They are obviously *relativistically invariant* and form a generalization of Eq. (10). *The obvious relativistic invariance is achieved by the introduction of separate time for each particle*.

§ 3. For further development we shall need some formulas of quantization of electromagnetic fields and shall use for this purpose some formulas obtained by Fock and Podolsky[1]. …

[1] Fock, V. A., & Podolsky, B. (1932). *Sow. Phys.*, 1, 801. For other treatments see Jordan, P. & Pauli, W. (1928). Zur Quantenelektrodynamik ladungsfreier Felder. (On the quantum electrodynamics of charge-free fields.) *Zeit. Phys.*, 47, 73, 151-73, or Fermi, E. (1929). *Rend. Lincei*, 9, 881. The Lagrangian (12) differs from that of Fermi only by a four-dimensional divergence.

…

Part II. The Maxwellian Case.

§ 4. For the Maxwellian case the following additional conditions are necessary. …

…

Dirac, P. A. M. (1933). The Lagrangian in Quantum Mechanics.

[*Phys. Zeit. Sowjetunion*, 3, 1, 64-72; also in Dirac, P. A. M. (1945). *Rev. Mod. Phys.*, 17, 195; also in Schwinger, J. (ed). (1958). *Selected Papers on Quantum electrodynamics.* Dover, New York, pages 312-20, and in *Feynman's Thesis - A New Approach to Quantum Theory.* (2005). Edited by Brown, L. M. World Scientific Publishing Co.; https://doi.org/ 10.1142/ 9789812567635_0003]

Received November 19, 1932.

St John's College, Cambridge.

Alternative formulation of quantum mechanics in terms of Lagrangian in place of Hamiltonian, *coordinates* and *velocities* instead of *coordinates* and *momenta*, allows *equations of motion* to be expressed as stationary property of *action function* which is the time-integral of the Lagrangian and *relativistic invariant*, there is no corresponding *action principle* in terms of *coordinates and momenta* in Hamiltonian theory, Lagrangian closely connected with theory of *contact transformations*, transformation functions have classical analogues expressible in terms of the Lagrangian, function of coordinates at time t and time t + dt rather than of the coordinates and velocities, applies to *field dynamics* using suitable field quantities or potentials as coordinates, "*many-time*" theory, each coordinate a function of four *space-time* variables instead of one time variable in particle theory, *generalized transformation function*.

> [*Quantum gravity in the first half of the twentieth century. A sourcebook.* Blum, A. S. & Rickles, D. (eds.) (2018). Blum, A. S. Chapter 17. Without New Difficulties: Quantum Gravity and the Crisis of the Quantum Field Theory Program, p. 264: "… We begin by discussing his 1933 paper on "The Lagrangian in Quantum Mechanics" (Dirac 1933). Dirac's general idea was to formulate quantum mechanics not in terms of *states* (*wave functions*) propagating in time, but in terms of *transition amplitudes* from the *state* at one time to the *state* at a later time. Dirac called these *transition amplitudes* "*transformation functions*," since he conceptualized them as generating *canonical transformations* from the *canonical coordinates* at one time to those at a later time. In the final section, he also hinted at how to generalize this idea to *quantum field theory*, where the initial and final times would be replaced by an arbitrary (not necessarily space-like) three-dimensional hypersurface of space-time, which formed the integration boundary of the classical action. He dubbed the field theoretic *amplitudes* relating the *canonical* (field) coordinate values on different points on this hypersurface "*generalized transformation functions*"."

Quantum mechanics was built up on a foundation of analogy with the Hamiltonian theory of classical mechanics. This is because the classical notion of *canonical coordinates and momenta* was found to be one with a very simple quantum analogue as a result of which the whole of the classical Hamiltonian theory, which is just a structure built up on this notion, could be taken over in all its details into quantum mechanics.

[*Canonical coordinates* are sets of coordinates on phase space which can be used to describe a physical system at any given point in time. *Phase space* is a space in which all possible states of a system are represented, with each possible state corresponding to one unique point in the phase space. For mechanical systems, the *phase space* usually consists of all possible values of *position* and *momentum* variables.]

Now there is an alternative formulation for classical dynamics, provided by the Lagrangian. This requires one to work in terms of *coordinates* and *velocities* instead of *coordinates* and *momenta*. The two formulations are, of course, closely related, but *there are reasons for believing that the Lagrangian one is the more fundamental*.

In the first place the Lagrangian method allows one to collect together all the *equations of motion* and express them as the stationary property of a certain *action function*. (*This action function is just the time-integral of the Lagrangian.*)

[*Action* is a numerical value describing how a physical system has changed over time. The *equations of motion* of the system can be derived through the *principle of stationary action*. The *stationary-action principle* – also known as the *principle of least action* – is a variational principle that, when applied to the action of a mechanical system, yields the *equations of motion* for that system. The principle states that the trajectories (i.e. the solutions of the *equations of motion*) are stationary points of the system's *action functional*. The term "least action" is a historical misnomer since the principle has no minimality requirement: the value of the action functional need not be minimal (even locally) on the trajectories. *Least action* refers to the absolute value of the action functional being minimized. The *stationary-action principle* can be used to derive Newtonian, Lagrangian and Hamiltonian *equations of motion*, and even *general relativity*. In relativity, a different *action* must be minimized or maximized. In the simple case of a single particle moving with a specified velocity, the *action* is *the momentum of the particle times the distance it moves*, added up along its path, or equivalently, twice its *kinetic energy* times the length of time for which it has that amount of *energy*, added up over the period of time under consideration.]

There is no corresponding *action principle* in terms of the *coordinates and momenta* of the Hamiltonian theory. Secondly *the Lagrangian method can easily be expressed relativistically, on account of the action function being a relativistic invariant*; while the Hamiltonian method is essentially *non-relativistic* in form, since it marks out a particular time variable as the *canonical conjugate* of the Hamiltonian function.

[*Conjugate variables* are pairs of variables mathematically defined in such a way that they become *Fourier transform duals*. The duality relations lead naturally to an *uncertainty relation*—in physics called the *Heisenberg uncertainty principle*—between them. *Conjugate variables* are related by Noether's theorem, which states that if the laws of physics are invariant with respect to a change in one of the *conjugate variables*, then the other *conjugate variable* will not change with time (i.e. it will be *conserved*).

A *Fourier transform* is a mathematical transform that decomposes functions depending on space or time into functions depending on spatial frequency or temporal frequency. A *duality* translates concepts, theorems or mathematical structures into other concepts, theorems or structures, in a one-to-one fashion.]

For these reasons it would seem desirable to take up the question of *what corresponds in the quantum theory to the Lagrangian method of the classical theory.* A little consideration shows, however, that one cannot expect to be able to take over the classical Lagrangian equations in any very direct way. These equations involve partial derivatives of the Lagrangian with respect to the *coordinates* and *velocities* and *no meaning can be given to such derivatives in quantum mechanics.* The only differentiation process that can be carried out with respect to the dynamical variables of quantum mechanics is that of forming Poisson brackets and this process leads to the Hamiltonian theory[#].

[#] Processes for partial differentiation with respect to matrices have been given by Born, Heisenberg and Jordan [Born, M., Heisenberg, W. & Jordan, P. (August, 1926). Zur Quantenmechanik II. (On Quantum Mechanics II.) *Zeit. Phys*., 35, 557-615] but these processes do not give us means of differentiation with respect to dynamical variables, since they are not independent of the representation chosen. As an example of the difficulties involved in differentiation with respect to quantum dynamical variables, consider the three components of an *angular momentum*, satisfying

$$m_x m_y - m_y m_x = \mathrm{i}h m_z.$$

We have here m_z expressed explicitly as a function of m_x and m_y, but we can give no meaning to its partial derivative with respect to m_x or m_y.

We must therefore seek our quantum Lagrangian theory in an indirect way. *We must try to take over the ideas of the classical Lagrangian theory, not the equations of the classical Lagrangian theory.*

Contact Transformations.

Lagrangian theory is closely connected with the theory of *contact transformations.*

[A *contact transformation* is a transformation of curves in the plane or surfaces in space under which tangent curves or surfaces are taken into tangent curves or surfaces. A simple example of a *contact transformation* is the Legendre transformation.]

We shall therefore begin with a discussion of the analogy between classical and quantum contact transformations. Let the two sets of variables be p_r, q_r and P_r, Q_r, ($r = 1, 2 \ldots n$) and suppose the q's and Q's to be all independent, so that any function of the dynamical variables can be expressed in terms of them. It is well known that in the classical theory the *transformation equations* for this case can be put in the form

$$p_r = \partial S/\partial q_r, \qquad P_r = \partial S/\partial Q_r, \qquad (1)$$

where S is some function of the q's and Q's.

In the quantum theory we may take a representation in which the *q's* are diagonal, and a second representation in which the Q's are diagonal. There will be a transformation function (q' | Q') connecting the two representations. We shall now show that this transformation function is the quantum analogue of $e^{iS/h}$.

If α is any function of the dynamical variables in the quantum theory, it will have a "mixed" representative (q' | α | Q'), which may be defined in terms of either of the usual representatives (q' | α | q"), (Q' | α | Q"), by

$$(q' | \alpha | Q') = \int (q' | \alpha | q'')\, dq''\, (q'' | Q') = \int (q' | Q'')\, dQ''\, (Q'' | \alpha | Q').$$

From the first of these definitions, we obtain

$$(q' | q_r | Q') = q'_r\, (q' | Q') \qquad (2)$$
$$(q' | p_r | Q') = -\, ih\, \partial/\partial q'_r\, (q' | Q') \qquad (3)$$

and from the second

$$(q' | Q_r | Q') = q'_r\, (q' | Q') \qquad (4)$$
$$(q' | P_r | Q') = ih\, \partial/\partial Q'_r\, (q' | Q'). \qquad (5)$$

Note the difference in sign in (3) and (5).

Equations (2) and (4) may be generalized as follows. Let f(q) be any function of the q's and g(Q) any function of the Q's. Then

$$(q' | f(q)\, g(Q) | Q') = \int\int (q' | f(q) | q'')\, dq''\, (q'' | Q')\, dQ''\, (Q'' | g(Q) | Q')$$
$$= f(q')\, g(Q')(q' | Q').$$

Further, if $f_k(q)$ and $g_k(Q)$, (k= 1, 2 … , m) denote two sets. of functions of the q's and Q's respectively,

$$(q' | \sum_k f_k(q) g_k(Q) | Q') = \sum_k f_k(q')\, g_k(Q')\, .\, (q' | Q').$$

Thus, if α is any function of the dynamical variables and we suppose it to be expressed as a function α(qQ) of the q's and Q's in a "well-ordered" way, that is, so that it consists of a sum of terms of the form f(q)g(Q), we shall have

$$(q' | \alpha(qQ) | Q') = \alpha(q'\, Q')\, (q' | Q'). \qquad (6)$$

This is a rather remarkable equation, giving us a connection between α(qQ), which is a function of operators, and α(q'Q'), which is a function of numerical variables.

Let us apply this result for $\alpha = p_r$. Putting

$$(q' | Q') = e^{iUh}, \qquad (7)$$

where U is a new function, of the *q''s* and Q''s we get from (3)

$$(q' | p_r | Q') = \partial U(q'\, Q')/\partial q'_r\, (q' | Q').$$

By comparing this with (6) we obtain

$$p_r = \partial U(qQ)/\partial q_r$$

as an equation between operators or dynamical variables, which holds provided $\partial U/\partial q_r$ is well-ordered. Similarly, by applying the result (6) for $\alpha = P_r$ and using (5), we get

$$P_r = - \partial U(qQ)/\partial Q_r,$$

provided $\partial U/\partial Q_r$ is well-ordered. *These equations are of the same form as (1)*

$$[p_r = \partial S/\partial q_r, \quad P_r = \partial S/\partial Q_r, \qquad (1)]$$

and show that the U defined by (7) is the analogue of the classical function S, which is what we had to prove.

Incidentally, we have obtained another theorem at the same time, namely that equations (1) hold also in the quantum theory provided the right-hand sides are suitably interpreted, the variables being treated classically for the purpose of the differentiations and the derivatives being then well-ordered. This theorem has been previously proved by Jordan by a different method#.

Jordan, P. (1926). Über kanonische Transformationen in der Quantenmechanik. II. (On Canonical Transformations in Quantum Mechanics. II.) *Zeit. Phys.*, 38, 513-7 (1926). https://doi.org/10.1007/BF01397170

The Lagrangian and the Action Principle.

The *equations of motion* of the classical theory cause the dynamical variables to vary in such a way that their values q_t, p_t at any time t are connected with their values q_T, p_T at any other time T by a *contact transformation,* which may be put into the form (1) with $q, p = q_t, p_t$; $Q, P = q_T, p_T$ and S equal to the time integral of the Lagrangian over the range T to t. In the quantum theory the q_t, p_t will still be connected with the q_T, p_T by a *contact transformation* and there will be a *transformation function* $(q_t \mid q_T)$ connecting the two representations in which the q_t and the q_T are diagonal respectively. The work of the preceding section now shows that

$$(q_t \mid q_T) \text{ corresponds to } \exp [i \textstyle\int_T^t L dt/h], \qquad (8)$$

where L is the Lagrangian. If we take T to differ only infinitely little from t, we get the result

$$(q_{t+dt} \mid q_t) \text{ corresponds to } \exp [iL\, dt/h]. \qquad (9)$$

The transformation functions in (8) and (9) are very fundamental things in the quantum theory and it is satisfactory to find that they have their classical analogues, expressible simply in terms of the Lagrangian. We have here the natural extension of the well-known result that the *phase* of the *wave function* corresponds to Hamilton's principal function in classical theory. *The analogy (9) suggests that we ought to consider the classical Lagrangian not as a function of the coordinates and velocities, but rather as a function of the coordinates at time t and the coordinates at time t + dt.*

For simplicity in the further discussion in this section we shall take the case of a *single degree of freedom*, although the argument applies also to the general case. We shall use the notation

$$\exp [i \int_T^t Ldt /h] = A(tT),$$

so that A(tT) is the classical analogue of $(q_t \mid q_T)$.

Suppose we divide up the time interval T - t into a large number of small sections $T - t_1$, $t_1 - t_2, \ldots, t_{m-1} - t_m, t_m - t$ by the introduction of a sequence of intermediate times $t_1, t_2, \ldots t_m$. Then

$$A(tT) = A(tt_m)\, A(t_m t_{m-1}) \ldots A(t_2 t_1)\, A(t_1 T). \qquad (10)$$

Now in the quantum theory we have

$$(q_t \mid q_T) = \int (q_t \mid q_m)\, dq_m\, (q_m \mid q_{m-1})\, dq_{m-1} \cdots (q_2 \mid q_1)\, dq_1\, (q_1 \mid q_T), \qquad (11)$$

where q_k denotes q at the intermediate time t_k, (k = 1, 2 … m).

Equation (11) at first sight does not seem to correspond properly to equation (10), since on the right-hand side of (11) we must integrate after doing the multiplication while on the right-hand side of (10) there is no integration.

Let us examine this discrepancy by seeing what becomes of (11) when we regard t as extremely small. From the results (8) and (9) we see that the integrand in (11) must be of the form $e^{iF/h}$ where F is a function of $q_T, q_1, q_2 \ldots q_m, q_t$ which remains finite as h tends to zero. Let us now picture one of the intermediate q's say q_k, as varying continuously while the others are fixed. Owing to the smallness of h, we shall then in general have F/h varying extremely rapidly. This means that $e^{iF/h}$ will vary periodically with a very high frequency about the value zero, as a result of which its integral will be practically zero. The only important part in the domain of integration of q_k is thus that for which a comparatively large variation in q_k produces only a very small variation in *F*. This part is the neighborhood of a point for which F is stationary with respect to small variations in q_k.

We can apply this argument to each of the variables of integration in the right-hand side of (11) and obtain the result that the only important part in the domain of integration is that for which F is stationary for small variations in all the intermediate q's. But, by applying (8) to each of the small time-sections, we see that F has for its classical analogue

$$\int_{tm}^{t} Ldt + \int_{t(m-1)}^{tm} Ldt + \ldots + \int_{t1}^{t2} Ldt + \int_{T}^{t1} j'Ldt = \int_T^t Ldt,$$

which is just the action function which classical mechanics requires to be stationary for small variations in all the intermediate q's. This shows the way in which equation (11) goes over into classical results when h becomes extremely small.

We now return to the general case when h is not small. We see that, for comparison with the quantum theory, equation (10) must be interpreted in the following way. Each of the quantities A must be considered as a function of the q's at the two times to which it refers.

The right-hand side is then a function, not only of q_r and q_t, but also of $q_1, q_2, \ldots q_m$, and in order to get from it a function of q_r and q_t only, which we can equate to the left-hand side, we must substitute for $q_1, q_2, \ldots q_m$ their values given by the *action principle*. This process of substitution for the intermediate q's then corresponds to the process of integration over all values of these q's in (11)

$$[(q_t \mid q_T) = \int (q_t \mid q_m)\, dq_m\, (q_m \mid q_{m-1})\, dq_{m-1} \cdots (q_2 \mid q_1)\, dq_1\, (q_1 \mid q_T), \quad (11).]$$

Equation (11) contains the quantum analogue of the action principle, as may be seen more explicitly from the following argument. From equation (11) we can extract the statement (a rather trivial one) that, if we take specified values for q_r and q_t, then the importance of our considering any set of values for the intermediate q's is determined by the importance of this set of values in the integration on the right hand side of (11). If we now make h tend to zero, this statement goes over into the classical statement that, if we take specified values for q_r and q_t, then the importance of our considering any set of values for the intermediate q's is zero unless these values make the action function stationary. *This statement is one way of formulating the classical action principle.*

Application to Field Dynamics.

We may treat the *problem of a vibrating medium* in the classical theory by Lagrangian methods which form a natural generalization of those for particles. We choose as our coordinates suitable field quantities or potentials. Each coordinate is then a function of the four *space-time* variables x, y, z, t, corresponding to the fact that in particle theory it is a function of just the one variable t. *Thus, the one independent variable t of particle theory is to be generalized to four independent variables x, y, z, t*#.

It is customary in field dynamics to regard the values of a field quantity for two different values of (x, y, z) but the same value of t as two different coordinates, instead of as two values of the same coordinate for two different points in the domain of independent variables, and in this way to keep to the idea of a single independent variable t. This point of view is necessary for the Hamiltonian treatment, but for the Lagrangian treatment the point of view adopted in the text seems preferable on account of its greater space-time symmetry.

We introduce at each point of space-time a Lagrangian density, which must be a function of the coordinates and their first derivatives with respect to x, y, z, and t, corresponding to the Lagrangian in particle theory being a function of coordinates and velocities. The integral of the *Lagrangian density* over any (four-dimensional) region of *space-time* must then be stationary for all small variations of the *coordinates* inside the region, provided the *coordinates* on the boundary remain invariant.

It is now easy to see what the quantum analogue of all this must be. If *S* denotes the integral of the classical *Lagrangian density* over a particular region of *space-time*, we should expect there to be a quantum analogue of $e^{iS/h}$ corresponding to the $(q_t \mid q_T)$ of *particle theory.* This $(q_t \mid q_T)$ is a function of the values of the *coordinates* at the ends of the time interval to which it refers and so we should expect the quantum analogue of $e^{iS/h}$ to be a function (really a *functional*) of the values of the *coordinates* on the boundary of the *space-time* region. This quantum analogue will be a sort of "*generalized transformation function*". It cannot in general be interpreted, like $(q_t \mid q_T)$, as giving a transformation between one set of dynamical variables and another, but it is a four dimensional generalization of $(q_t \mid q_T)$ in the following sense.

Corresponding to the *composition law* for $(q_t \mid q_T)$

$$(q_t \mid q_T) = \int (q_t \mid q_1)\, dq_1\, (q_1 \mid q_T), \tag{12}$$

the *generalized transformation function* (g.t.f.) will have the following *composition law*. Take a given region of space time and divide it up into two parts. Then the g.t.f. for the whole region will equal the product of the g.t.f.'s for the two parts, integrated over all values for the *coordinates* on the common boundary of the two parts.

Repeated application of (12) gives us (11) and repeated application of the corresponding law for g.t.f's will enable us in a similar way to connect the g.t.f. for any region with the g.t.f.'s for the very small sub-regions into which that region may be divided. *This connection will contain the quantum analogue of the action principle applied to fields.*

The square of the modulus of the transformation function $(q_t \mid q_T)$ can be interpreted as the probability of an observation of the coordinates at the later time t giving the result q_t for a *state* for which an observation of the *coordinates* at the earlier time T is certain to give the result q_T. A corresponding meaning for the square of the modulus of the g.t.f. will exist only when the g.t.f. refers to a region of *space-time* bounded by two separate (three-dimensional) surfaces, each extending to infinity in the space directions and lying entirely outside any light-cone having its vertex on the surface. *The square of the modulus of the g.t.f. then gives the probability of the coordinates having specified values at all points on the later surface for a state for which they are given to have definite values at all points on the earlier surface*. The g.t.f. may in this case be considered as a *transformation function* connecting the values of the *coordinates* and *momenta* on one of the surfaces with their value on the other.

We can alternatively consider $| (q_t \mid q_T) |^2$ as giving the relative a priori *probability* of any *state* yielding the results q_T and q_t when observations of the q's are made at time T and at time t (account being taken of the fact that the earlier observation will alter the *state* and affect the later observation). Correspondingly, *we can consider the square of the modulus of the g.t.f. for any space-time region as giving the relative a priori probability of specified results being obtained when observations are made of the coordinates at all points on the boundary*. This interpretation is more general than the preceding one, since it does not require a restriction on the shape of the *space-time* region.

Dirac, P. A. M. (1934). Discussion of the infinite distribution of electrons in the theory of the positron.

[*Proc. Camb. Phil. Soc.*, 30, 2, 150-163; doi:10.1017/S030500410001656X.]

Received February 2, 1934, read March 5, 1934.

St. John's College, Cambridge.

Attempt by Dirac to address problem with his *relativistic* 'hole' theory which implies an infinite number of negative-energy electrons (per unit volume) with energies extending continuously from $-mc^2$ to $-\infty$, when an electromagnetic field is present positive- and negative-energy states cannot be distinguished in *relativistically* invariant way, need to set up assumptions for production of *electromagnetic field* by the electron distribution such that any finite change in distribution produces a change in the field in agreement with Maxwell's equations and such that the infinite field which would be required by Maxwell's equations from an infinite density of electrons is in some way cut out, *assumes each electron has its own individual wave function in space-time* and each electron moves in an *electromagnetic field* which is the same for all electrons part coming from external causes and part from the electron distribution itself, introduces *relativistic density matrix* $\sum_{oc} \psi_{k'}(x'\ t')\ \psi^*_{k''}(x''\ t'')$ referring to two points in space and two times, separates density distribution into two parts where one contains the singularities, and the other describes the *electric* and *current densities* physically present.

1. *Use of the density matrix.*

The quantum theory of the electron allows states of negative *kinetic energy* as well as the usual states of positive *kinetic energy* and also allows transitions from one kind of *state* to the other. Now particles in *states* of negative *kinetic energy* are never observed in practice. *We can get over this discrepancy between theory and observation by assuming that, in the world as we know it, nearly all the states of negative kinetic energy are occupied, with one electron in each state in accordance with Pauli's exclusion principle, and that the distribution of negative-energy electrons is unobservable to us on account of its uniformity.* Any unoccupied negative-energy states would be observable to us, as holes in the distribution of negative-energy electrons, but *these holes would appear as particles with positive kinetic energy* and thus not as things foreign to all our experience. It seems reasonable and in agreement with all the facts known at present to identify these holes with the recently discovered positrons and thus to obtain a theory of the positron*.

* As this theory was first put forward, Dirac (January, 1930). [A theory of electrons and protons. *Roy. Soc. Proc., A*, 126, 801, 360-65] and (October, 1930). [On the Annihilation of Electrons and Protons. *Proc. Camb. Phil. Soc.*, 26, 361-75; https://doi.org/10.1017/S0305004100016091], the holes were assumed to be protons, but this assumption was afterwards seen to be untenable, since it was found that the holes must correspond to particles with the same rest-mass as electrons. See Dirac (September, 1931). [Quantized singularities in the electromagnetic field. *Roy. Soc. Proc., A*, 133, 821, 60-72]; page 61.

We now have a picture of the world *in which there are an infinite number of negative-energy electrons (in fact an infinite number per unit volume) having energies extending continuously from* $-mc^2$ *to* $-\infty$. The problem we have to consider is the way this infinity can be handled mathematically and the physical effects it produces. In particular, we must set up some assumptions for the production of *electromagnetic field* by the electron distribution, which assumptions must be such that any finite change in the distribution produces a change in the field in agreement with Maxwell's equations, *but such that the infinite field which would be required by Maxwell's equations from an infinite density of electrons is in some way cut out.*

These problems are quite simple when we suppose each electron to be moving in a space free of electromagnetic field. *They are not so simple when there is a field present, since the positive- and negative-energy states then get mixed together so intimately that one cannot in general distinguish accurately between them in a relativistically invariant way.* A careful investigation is then necessary, even for such an elementary problem as seeing that a precise meaning can be given to a distribution such as occurs in practice, in which nearly all the negative-energy states are occupied and nearly all the positive-energy ones unoccupied.

To make an exact treatment of the matter would be very complicated and in the present paper only an approximate treatment will be given, on the lines of Hartree's method of the self-consistent field. *We shall suppose that each electron has its own individual wave function in space-time* (instead of there being one wave function in an enormous number of variables to describe the whole distribution), and also, *we shall suppose that each electron moves in a definite electromagnetic field*, which is the same for all the electrons. *This field will consist of a part coming from external causes and a part coming from the electron distribution itself*, the precise way in which the latter part depends on the electron distribution being one of the problems we have to consider.

Let the normalized functions for the electrons at any time be $\psi_a(x)$, where x stands for three positional coordinates x_1, x_2, x_3 of an electron and the suffix a takes on different values for the different electrons. *With electron spin taken into account, each* $\psi_a(x)$ *must have four components*, which may be specified by $\psi_{ak}(x)$ with k = 1, 2, 3, 4. The whole distribution of electrons may now be described by the *density matrix* ρ defined by

$$(x' \mid \rho \mid x'')_{k'k''} = \sum_a \psi_{ak'}(x')\, \psi^*_{ak''}(x''), \qquad (1)$$

in which the sum is taken over all the electrons. This is a matrix in the spin variables k as well as in the positional variables x. It is, of course, a Hermitian matrix. Its properties have been studied previously*,

* Dirac, P. A. M. (January, 1929). The basis of statistical quantum mechanics. *Proc. Camb. Phil. Soc.*, 25, 1, 62-6; https://doi.org/10.1017/S0305004100018570; Dirac, P. A. M. (July, 1930). Note on Exchange Phenomena in the Thomas Atom. *Proc. Camb. Phil. Soc.*, 26, 3, 376-85; https://doi.org/10.1017/s0305004100016108; Dirac, P. A. M. (April, 1931). Note

on the Interpretation of the Density Matrix in the Many-Electron Problem. *Proc. Camb. Phil. Soc.*, 27, 2, 240-3; https://doi.org/10.1017/S0305004100010343.

the chief ones being the equation

$$\rho^2 = \rho, \tag{2}$$

which expresses that the electron distribution satisfies the *exclusion principle*, and the *equation of motion*

$$ih\, d\rho/dt = H\rho - \rho H. \tag{3}$$

Here H is the Hamiltonian for the motion of a single electron in the field, thus

$$H = \alpha_s(p_s + eA_s) - eA_0 + \alpha_4 m, \tag{4}$$

the velocity of light being made equal to unity and a summation being implied over the values s = 1, 2, 3.

An alternative way of regarding the sum on the right-hand side of (1) is as a sum over all the occupied states*.

* The word 'all' used in this connection means each of a set of orthogonal states which is made as large as possible, and does not include states formed by superposition of these orthogonal states.

It may then conveniently be written $\sum_{oc} \psi_{k'}(x')\, \psi^*_{k''}(x'')$. There will be a corresponding sum over the unoccupied states, which may be written $\sum_{un} \psi_{k'}(x')\, \psi^*_{k''}(x'')$. If we add these two sums, we get the sum over all states and this must give us the unit matrix, from the transformation theory of quantum mechanics. Thus

$$\sum_{oc} \psi_{k'}(x')\, \psi^*_{k''}(x'') + \sum_{un} \psi_{k'}(x')\, \psi^*_{k''}(x'') = \delta(x' - x'')\, \delta_{k'\,k''}.$$

Put $$\rho = \tfrac{1}{2}(1 + \rho_1), \tag{5}$$

so that

$$(x' \mid \rho_1 \mid x'')_{k'k''} = \sum_{oc} \psi_{k'}(x')\, \psi^*_{k''}(x'') - \sum_{un} \psi_{k'}(x')\, \psi^*_{k''}(x'').$$

We may now consider the electron distribution as specified by the matrix ρ_1 instead of the matrix ρ. *This has the advantage that it makes a closer symmetry between the electrons and the positrons and leads to neater mathematical expressions*. The *equation of motion* (3) holds unchanged with ρ_1 instead of ρ and equation (2) gets modified to

$$\rho_1^2 = 1. \tag{6}$$

The *density matrices* that we have been discussing up to the present are *non-relativistic* things, since their elements each refer to two points in space x' and x'' but to only one time. *To get a relativistic theory, we must introduce two times, t' and t'', and use instead of ρ the relativistic density matrix R defined by*

$$(x'\,t' \mid R \mid x''\,t'')_{k'k''} = \sum_a \psi_{ak'}(x'\,t')\, \psi^*_{ak''}(x''\,t'')$$

$$= \sum_{oc} \psi_{k'}(x'\ t')\ \psi^*_{k''}(x''\ t''). \qquad (7)$$

Instead of ρ_1 we shall now have R_1, defined by

$$(x'\ t' \mid R_1 \mid x''\ t'')_{k'k''} = \sum_{oc} \psi_{k'}(x'\ t')\ \psi^*_{k''}(x''\ t'') - \sum_{un} \psi_{k'}(x'\ t')\ \psi^*_{k''}(x''\ t''),$$

and instead of equation (5) we shall have

$$R = \tfrac{1}{2}\,(R_F + R_1),$$

where

$$(x'\ t' \mid R_F \mid x''\ t'')_{k'k''} = \sum_{oc} \psi_{k'}(x'\ t')\ \psi^*_{k''}(x''\ t'') + \sum_{un} \psi_{k'}(x'\ t')\ \psi^*_{k''}(x''\ t''),$$

R_F, representing *the full distribution with all possible states occupied*, is no longer simply the unit matrix, but all the same we should expect it to play some fundamental part in the theory.

The new matrices R, R_1, R_F are also Hermitian

[A *Hermitian matrix* (or self-adjoint matrix) is a *complex square matrix* that is equal to its own *conjugate transpose*—that is, the element in the i-th row and j-th column is equal to the *complex conjugate* (two complex numbers having their real parts identical and their imaginary parts of equal magnitude but opposite sign) of the element in the j-th row and i-th column, for all indices i and j:

$\boldsymbol{A}$ Hermitian $\Leftrightarrow$ $a_{ij} = a^*_{ji}$

or in matrix form:

$\boldsymbol{A}$ Hermitian $\Leftrightarrow$ $\boldsymbol{A} = \boldsymbol{A}^{*T}$.

Hermitian matrices can be understood as the complex extension of real symmetric matrices. If the *conjugate transpose* of a matrix $\boldsymbol{A}$ is denoted by $\boldsymbol{A}^H$, then the Hermitian property can be written concisely as

$\boldsymbol{A}$ Hermitian $\Leftrightarrow$ $\boldsymbol{A} = \boldsymbol{A}^H$.

Hermitian matrices are named after Charles Hermite, who demonstrated in 1855 that matrices of this form share a property with real symmetric matrices of always having real eigenvalues.]

and their *equation of motion* is

$$\mathscr{H}R = 0, \qquad \mathscr{H}R_1 = 0, \qquad \mathscr{H}R_F = 0, \qquad (8)$$

where $\mathscr{H}$ is the total operator that operates on the wave function in the *wave equation* for one electron, i.e.

$$\mathscr{H} = W - H$$

W being the operator ih times time-differentiation. Equations (2)

[$\rho^2 = \rho$, (2)

which expresses that the electron distribution satisfies the *exclusion principle*, and the *equation of motion*

$ih\ d\rho/dt = H\rho - \rho H$]

and (6),

[The *equation of motion* (3) holds unchanged with ρ_1 instead of ρ and equation (2) gets modified to

$$\rho_1^2 = 1. \qquad (6)]$$

cannot be concisely expressed in terms of the R's.

To obtain the field produced by the distribution of electrons, we must first get the *electric density* and *current density. For this purpose, we must, according to the usual theory for finite distributions, take a diagonal element of ρ in the positional variables, or a diagonal element of R in the positional and time variables, and form its diagonal sum over the spin variables*. The resulting expression, namely $\sum_k (x \mid \rho \mid x)_{kk}$ or $\sum_k (xt \mid R \mid xt)_{kk}$ would then be the *electric density* (apart from the factor – e). The corresponding *current density* would have for its sth component $\sum_k (x \mid \alpha_s\rho \mid x)_{kk}$ or $\sum_k (xt \mid \alpha_s R \mid xt)_{kk}$.

We can easily verify that this *electric density* and *current density* satisfy the *conservation law of electricity*. In equation (3)

[the *equation of motion*

$$ih\, d\rho/dt = H\rho - \rho H. \qquad (3)]$$

let us take diagonal elements in the positional variables, but keep to the symbolic matrix notation for the spin variables, so that a symbol like $(x \mid \rho \mid x)$ denotes a matrix with four rows and columns, of the same nature as an α. This gives

$$\begin{aligned} ih\, d/dt\, (x \mid \rho \mid x) &= (x \mid H\rho - \rho H \mid x) \\ &= \alpha_s (x \mid p_s\rho - \rho p_s \mid x) + (x \mid \{\alpha_s\rho - \rho\alpha_s\}\{p_s + eA_s\} \mid x) \\ &\qquad + (x \mid \{\alpha_4\rho - \rho\alpha_4 \mid x)\, m. \end{aligned}$$

If we now take the *diagonal sum* with respect to the *spin variables*, the last two terms will contribute nothing, from the rule that the diagonal sum of the product of two matrices is independent of their order, and we shall be left with

$$\begin{aligned} ih\, d/dt \textstyle\sum_k (x \mid \rho \mid x)_{kk} &= \textstyle\sum_{kk'} \alpha_{s\,kk'} (x \mid p_s\rho - \rho p_s \mid x)_{k'k} \\ &= - ih \textstyle\sum_{kk'} \alpha_{s\,kk'}\, \partial/\partial x_s\, (x \mid \rho \mid x)_{k'k}, \end{aligned}$$

i.e. $$d/dt \textstyle\sum_k (x \mid \rho \mid x)_{kk} = - \partial/\partial x_s \textstyle\sum_k (x \mid \alpha_s\rho \mid x)_{kk}, \qquad (9)$$

which is the required conservation law.

In our present theory the electric density and current density given by these formulae would be infinite and some alteration of the assumptions is therefore necessary. *The problem now presents itself of finding some natural way of removing the infinities from $\sum_k (xt \mid R \mid xt)_{kk}$ and $\sum_k (xt \mid \alpha_s R \mid xt)_{kk}$ so as to leave finite remainders*, which we could then assume to be the *electric* and *current densities*. This problem requires us to make a detailed investigation of the singularities in the matrix elements $(x't' \mid R \mid x''t'')_{k'k''}$ near the diagonal $x_s' = x_s''$, $t' = t''$.

2. *Case of no field.*

We shall begin our investigation by taking the case of *no electromagnetic field*, when the Hamiltonian (4)

$$[H = \alpha_s(p_s + eA_s) - eA_0 + \alpha_4 m, \qquad (4)]$$

reduces to

$$H = \alpha_s p_s + \alpha_4 m = (\boldsymbol{\alpha}, \mathbf{p}) + \alpha_4 m. \qquad (10)$$

In this case we can calculate accurately the matrix elements $(x't' \mid R \mid x''t'')_{k'k''}$ *for the distribution of electrons in which all the negative-energy states are occupied and all the-positive-energy ones unoccupied,* and see exactly how these matrix elements behave near the diagonal.

If we try to work directly from the definition (7) we meet with some awkward calculations in taking the *spin variables* into account and summing over the two possible spin orientations. We can avoid these calculations by using *symbolic methods* and first obtaining ρ. *The condition that a wave function ψ contains only Fourier components belonging to negative-energy states may be expressed symbolically by*

$$\{H + \sqrt{(P^2 + m^2)}\}\ \psi = 0,$$

where P denotes the length of the vector **p** and the *positive square root is understood.* Similarly, *the condition that the distribution ρ contains electrons only in negative-energy states may be expressed by*

$$\{H + \sqrt{(P^2 + m^2)}\}\ \rho = 0. \qquad (11)$$

The condition that in the distribution ρ every negative-energy state is occupied, is just the condition that the distribution 1 – ρ contains electrons only in positive-energy states and may thus be expressed by

$$\{H - \sqrt{(P^2 + m^2)}\}\ (1 - \rho) = 0.$$

Adding this equation to (11) we get

$$\{H + \sqrt{(P^2 + m^2)}\}\ (2\rho - 1) = 0$$

or

$$\rho = \tfrac{1}{2}\,[1 - H/\sqrt{(P^2 + m^2)}] = \tfrac{1}{2}\,[1 - \{(\boldsymbol{\alpha}, \mathbf{p}) + \alpha_4 m\}/\sqrt{(P^2 + m^2)}]. \qquad (12)$$

Hence, from the transformation theory,

$$(x' \mid \rho \mid x'') = 1/2h^3 \iint e^{i(x',p')/h}\, dp'\, [1 - \{(\boldsymbol{\alpha}, \mathbf{p}) + \alpha_4 m\}/\sqrt{(P^2 + m^2)}]$$
$$\times\ \delta(p' - p'')\, dp''\, e^{-i(x'',p')/h} \qquad (13)$$

where dp denotes the product $dp_1 dp_2 dp_3$.

…

… Introducing the notation

$$x_s' - x_s'' = x_s, \qquad t' - t'' = t,$$

which we shall keep through the rest of the paper, …

…

It is clear from these equations that there will be singularities, not only at the point $x_s = 0$, $t = 0$, but also everywhere on the light-cone $t^2 - r^2 = 0$. In order to determine these singularities, we may expand the Bessel functions in power series of $\sqrt{(t^2 - r^2)}$ and retain only the first few terms. …

…

The main result of this investigation for the case of no field is that there are two quite distinct kinds of singularity occurring in the matrices R_F [representing the full distribution with all possible states occupied] and R_1 [representing the relativistic density matrix] respectively. The singularities occurring in R_F are all associated with the δ function and those in R_1 with the reciprocal function and logarithm. From the generality of this result, we may expect it to hold also when there is a field present.

2. *Case of an arbitrary field.*

Let us now examine the singularities in $(x't' \mid R_F \mid x''t'')$ and $(x't' \mid R_1 \mid x''t'')$ when there is a general field present. Our method will be to suppose that the singularities are of the same form as in the case of no field, but have unknown coefficients. These coefficients must be functions of x_s', t', x_s'', t'' which are free from singularities and can be expanded as Taylor series for small values of x_s and t. We must try to choose them so that the *equations of motion* (8)

$$[\mathscr{H}R = 0, \qquad \mathscr{H}R_1 = 0, \qquad \mathscr{H}R_F = 0, \qquad (8)]$$

are satisfied.

The application of the method follows a parallel course for R_F and R_1, and we need therefore treat in detail only R_1 which is the *density matrix* we are mainly interested in. …

… In this way all our equations are satisfied and all our unknowns are determined, with the exception of g, which is not itself determined although $\mathscr{H}g$ is. The final result is

$$(x't' \mid R_1 \mid x''t'') = \ldots \qquad (48)$$

…

To do the corresponding work for R_F we put, analogously to (29),

…

… Thus, R_F is completely fixed.

4. *Conclusion.*

From the foregoing work we see that the following results must hold, at least to the accuracy of the Hartree method of approximation:

(i) One can give a precise meaning to the *distribution of electrons* in which every state is occupied. This distribution may be defined as that described by the *density matrix* R_F given by (49), this matrix being completely fixed for any given field.

(ii) One can give a precise meaning to a *distribution of electrons* in which nearly all (i.e. all but a finite number, or all but a finite number per unit volume) of the negative-energy states are occupied and nearly all of the positive-energy ones unoccupied. Such a distribution may be defined as one described by a *density matrix* $R = \frac{1}{2}(R_F + R_1)$, where R_1 is of the form (48). This definition is permissible because *the only possible variations in* R_1, *namely those due to g not being completely defined, are free from singularity and thus correspond to finite changes, or finite changes per unit volume, in the electron distribution.* Our method does not give any precise meaning to which negative-energy states are unoccupied or which positive-energy ones are occupied. It is sufficiently definite, though, to take as the basis of the theory of the position the *assumption that only distributions described by* $R = \frac{1}{2}(R_F + R_1)$ *with R1 of the form (48) occur in nature.*

(iii) A distribution R such as occurs in nature according to the above assumption *can be divided naturally into two parts*

$$R = R_a + R_b,$$

where R_a *contains all the singularities* and is also completely fixed for any given field, so that any alteration one may make in the distribution of electrons and positrons will correspond to an alteration in R_b but to none in R_a. We get this division into two parts by putting the term containing g into R_b and all the other terms into R_a. Thus,

$$R_b = g/4ih.$$

It is easily seen that R_b *is relativistically invariant and gauge invariant*, and it may be verified after some calculation that R_b is Hermitian and that the *electric density* and *current density* corresponding to it satisfy the conservation law (9)

$$[d/dt \sum_k (x \mid \rho \mid x)_{kk} = -\partial/\partial x_s \sum_k (x \mid \alpha_s\rho \mid x)_{kk}. \qquad (9)]$$

It therefore appears reasonable to make the assumption that the electric and current densities corresponding to R_b *are those which are physically present, arising from the distribution of electrons and positrons. In this way we can remove the infinities mentioned at the end of § 1.*

The present paper is incomplete in that the effect of the *exclusion principle*, equation (2) or (6)

$$[\quad \rho^2 = \rho, \qquad (2)$$

$$\rho_1^2 = 1, \qquad (6)]$$

on R_b has not been investigated. Further work that remains to be done is to examine the physical consequences of the foregoing assumption and to see whether it leads to any phenomena of the nature of a *polarization of a vacuum* by an *electromagnetic field.*

Heisenberg, W. (1934). Bemerkungen zur Diracschen Theorie des Positrons. (Remarks on the Dirac theory of positron.)

[*Zeit. Phys*., 90, 3-4, 209-31; https://doi.org/10.1007/BF01333516; (translated by D. H. Delphenich; https://neo-classical-physics.info/ electromagnetism. html).]

Received June 21, 1934.

Leipzig.

Heisenberg's reconstruction of Dirac's *theory of the positron* in the formalism of quantum electrodynamics, demands that symmetry between positive and negative charge should be expressed in the basic equations from the outset, in addition to the well-known difficulties with the divergences no new infinities should appear in the formalism, theory should provide an approximation for the treatment of problems that have been treated by quantum electrodynamics up to now, Dirac [(1934). Discussion of the infinite distribution of electrons in the theory of the positron.] showed that a quantum mechanical system of many electrons that fulfill the Pauli principle and move in a given force field without back-reaction can be characterized by a *density matrix* $(x', t', k' \mid R \mid x'', t'', k'') = \sum n\, \psi^*n\, (x', t', k')\, \psi_n\, (x'', t'', k'')$ when $\psi_n(x', t', k')$ means the normalized eigenfunctions of the states that possess one electron, and x′, t′, k′ (x″, t″, k″, resp.) are position, time, and spin variables, all physically-important properties of quantum-mechanical systems like *charge density, current density*, etc., can be read off from the *density matrix*, the temporal change in the density matrix is determined by the Dirac differential equation, Dirac made different choice of *density matrix* representing external field resulting in a different energy and impulse density, by restricting oneself to an *intuitive analogue theory of matter fields the negative energy levels in the Dirac theory could be avoided by replacing the homogeneous Dirac differential equation with an inhomogeneous equation* where the inhomogeneity is indicative of pair creation, for most practical applications e.g. pair creation, annihilation, Compton scattering, etc. *the theory described here does not yield anything new compared to the formulation of the Dirac theory*, in the Maxwell theory a continuous charge distribution also leads to a finite self-energy; it is the "quantization" that leads to the infinite self-energy, *if one represents the quantization of the electromagnetic field by point-like light quanta then the infinitude of the self-energy also emerges in the intuitive theory of matter waves.*

I. Intuitive theory of matter waves.
1. The inhomogeneous differential equation of the density matrix.
2. The conservation laws.
3. Applications (polarization of the vacuum).

II. Quantum theory of the wave field.
1. Presentation of the field equations.
2. Applications (the self-energy of light quanta).

The purpose of the present paper is to construct the Dirac theory of the positron in the formalism of quantum electrodynamics[1,2].

[1] The paper originated in some discussions that I had with Herren Pauli, Dirac, and Weisskopf, in part written and in part oral, and to them I am deeply grateful.

[2] E.g.: Dirac, P. A. M. (1930). *The Principles of Quantum Mechanics*, Oxford, pp. 255.

Thus, we shall demand that the symmetry in nature between positive and negative charge should be expressed in the basic equations from the outset, and that, in addition to the well-known difficulties with the divergences that quantum electrodynamics leads to, no new infinities should appear in the formalism, moreover; i.e., that the theory should provide an approximation for the treatment of the circle of problems that have been treated by quantum electrodynamics up to now. By the latter postulate, one distinguishes the present effort from the investigations of Fock[3], Oppenheimer and Furry[4], and Peierls[5],

[3] Fock, V. (1933). *C. R. Leningrad (N. S.)*, no. 6, 267-71,

[4] Furry, W. H. & Oppenheimer, J. R. (February, 1934). On the Theory of the Electron and Positive. *Phys. Rev*., 45, 4, 245; https://doi.org/10.1103/PhysRev.45.245,

[5] Peierls, R. to appear,

the last of which is similar to it; he is closely linked with the paper of Dirac[6], moreover.

[6] Dirac, P. A. M. (March, 1934). Discussion of the infinite distribution of electrons in the theory of the positron. *Proc. Camb. Phil. Soc*., 30, 2, 150-63[; attempts to addresses problem with *relativistic* 'hole' theory which implies an infinite number of negative-energy electrons (per unit volume) with energies extending continuously from $-mc^2$ to $-\infty$, when an electromagnetic field is present positive- and negative-energy states cannot be distinguished in *relativistically* invariant way, problem is to set up assumptions for production of *electromagnetic field* by the electron distribution such that any finite change in distribution produces a change in the field in agreement with Maxwell's equations and such that the infinite field which would be required by Maxwell's equations from an infinite density of electrons is in some way cut out, assumes each electron has its own individual wave function in space-time and each electron moves in an electromagnetic field which is the same for all electrons, part coming from external causes and part from the electron distribution itself, substitutes *relativistic* density matrix, separates density distribution into two parts, where one contains the singularities, and the other describes the *electric* and *current densities* physically present] (in what follows, this is always referred to by *loc. cit.*).

In contrast to the Dirac treatment, one has the work on the meaning of the *conservation law* for the total system of *radiation-matter* and the necessity of formulating the basic equations of the theory in a way that grows out of the Hartree approximation.

I. *Intuitive theory of matter waves.*

1. *The inhomogeneous differential equation for the density.* Let the most important result of the aforementioned Dirac paper be briefly summarized as follows: *A quantum-mechanical system of many electrons that fulfill the Pauli principle and move in a given force field without back-reaction can be characterized by a "density matrix* [R]:"

$$(x', t', k' \mid R \mid x'', t'', k'') = \sum_n \psi^*_n (x', t', k')\, \psi_n (x'', t'', k''), \qquad (1)$$

when $\psi_n(x', t', k')$ means the *normalized eigenfunctions* of the *states* that possess one electron, and x′, t′, k′ (x″, t″, k″, resp.) are position, time, and spin variables.

[Dirac (March, 1934). Discussion of the infinite distribution of electrons in the theory of the positron, *loc. cit.*: "The whole distribution of electrons may now be described by the (*non-relativistic*) *density matrix* ρ defined by

$$(x' \mid \rho \mid x'')_{k'k''} = \sum_a \psi_{ak'}(x')\, \psi^*_{ak''}(x''), \qquad (1)$$

in which the sum is taken over all the electrons. This is a matrix in the spin variables k as well as in the positional variables x. … (whose chief properties are)

$$\rho^2 = \rho, \qquad (2)$$

which expresses that the electron distribution satisfies the *exclusion principle*, and the *equation of motion*

$$ih\, d\rho/dt = H\rho - \rho H \qquad (3).$$

… To get a *relativistic* theory, we must introduce two times, t' and t", and use instead of ρ the *relativistic density matrix* R defined by

$$(x't' \mid R \mid x''t'')_{k'k''} = \sum_a \psi_{ak'}(x't')\, \psi^*_{ak''}(x''t'')$$
$$= \sum_{oc} \psi_{k'}(x')\, \psi^*_{k''}(x''). \qquad (7)$$

… The new matrices R, R_1, R_F are also Hermitian and their *equation of motion* is

$$\mathscr{H}R = 0, \qquad \mathscr{H}R_1 = 0, \qquad \mathscr{H}R_F = 0, \qquad (8)$$

where $\mathscr{H}$ is the total operator that operates on the wave function in the *wave equation* for one electron, i.e.

$$\mathscr{H} = W - H$$

W being the operator ih times time-differentiation."]

All physically-important properties of quantum-mechanical systems like *charge density*, *current density*, etc., can be read off from the *density matrix*. In general, this is always true *in the approximation in which the interaction of the electrons can be ignored*; i.e., in which the typical quantum-mechanically intuitive course of events does not enter. The *density matrix* thus mediates an intuitive, corresponding picture of the actual process that is similar to what the classical-mechanical atomic model does. The demand that the ψ_n in (1), which, according to Dirac, can also be expressed in the form (t′ = t):

$$R^2 = R, \qquad (2)$$

should be normalized can be posed in parallel to the quantum conditions of the previous semi-classical theory.

The temporal change in the density matrix will be determined by the Dirac differential equation:

$$\mathscr{H}R = [i\hbar\, \partial/c\partial t' + e/c\, A_0(x') + \alpha_s\{i\hbar\, \partial/\partial x'_s - e/c\, A_s(x')\} + \beta mc]\, R = 0 \qquad (3)$$

[where $[i\hbar\, \partial/c\partial t' + e/c\, A_0(x') + \alpha_s\{i\hbar\, \partial/\partial x'_s - e/c\, A_s(x')\} + \beta mc]$ is the equivalent of the differential operator defined by

$$\mathscr{H} = ih(\partial/\partial t + \alpha_s\partial/\partial x_s) + e(A_0 - \alpha_s A_s) - \alpha_s m, \qquad \text{(30) in } \textit{loc. cit.}]$$

From now on, the following notations shall be applied throughout:

Coordinates:	$ct' = x'_0 = -x^{0'}$,	$x'_i = x^{i'}$,	$x'_\lambda - x_\lambda$, $(x'_\lambda + x''_\lambda)/2 = \xi_\lambda$,	(4)
Potentials:	$A_0 = -A^0$,	$A_i = A^i$,		
Field strengths:	$\partial A^\mu/\partial\xi_\nu - \partial A^\nu/\partial\xi_\mu = F^{\nu\mu}$, $F^{0s} = -F_{0s}$,			
	$(F^{01}, F^{02}, F^{03}) = \mathcal{E}$, $(F^{23}, F^{31}, F^{12}) = \mathcal{H}$			
Spin matrices:	$\alpha^0 = 1$,	$\alpha_0 = 1$,	$\alpha^i = \alpha_i$,	

Greek indices always run from 0 to 3 and Latin ones from 1 to 3. *The raising or lowering of indices shall result from the usual formulas of the theory of relativity.* Doubled indices shall always be summed over. Since the α^ν do not transform simply like a vector, the chosen notation only amounts to a convenient abbreviation for these quantities. Equation (3) now assumes, e.g., the form:

$$[\alpha^\lambda \{i\hbar\, \partial/\partial x'_s - e/c\, A_s(x')\} + \beta mc]\, R = 0.$$

If, as the Dirac *theory of holes* requires, all states of negative energy are occupied, except for finitely many of them, and also only finitely many positive energy states are occupied then the matrix R will be singular on the light-cone that is defined by:

$$x_\rho\, x^\rho = 0. \tag{5}$$

Following Dirac, one then suitably considers the new matrix[1]:

$$R_S = R - \tfrac{1}{2} R_F, \tag{6}$$

in place of the matrix R, in which R_F refers to the value of R for the state of the system in which every electron level is occupied.

[1] The doubled matrix R_S is the matrix that Dirac denoted by R_1.

[Dirac (March, 1934). Discussion of the infinite distribution of electrons in the theory of the positron, *loc. cit.*:

"Put $\rho = \tfrac{1}{2}(1 + \rho_1)$, (5)

so that

$$(x' \mid \rho_1 \mid x'')_{k'k''} = \textstyle\sum_{oc} \psi_{k'}(x')\, \psi^*_{k''}(x'') - \sum_{un} \psi_{k'}(x')\, \psi^*_{k''}(x'').$$

We may now consider the electron distribution as specified by the matrix ρ_1 instead of the matrix ρ. *This has the advantage that it makes a closer symmetry between the electrons and the positrons and leads to neater mathematical expressions.* The *equation of motion* (3) holds unchanged with ρ_1 instead of ρ and equation (2) gets modified to

$$\rho_1^2 = 1. \tag{6}$$

The *density matrices* that we have been discussing up to the present are *non-relativistic* things, since their elements each refer to two points in space x' and x'' but to only one time. *To get a relativistic theory, we must introduce two times, t' and t'', and use instead of ρ the relativistic density matrix R defined by*

$$(x'\, t' \mid R \mid x''\, t'')_{k'k''} = \textstyle\sum_a \psi_{ak'}(x'\, t')\, \psi^*_{ak''}(x''\, t'')$$
$$= \textstyle\sum_{oc} \psi_{k'}(x'\, t')\, \psi^*_{k''}(x''\, t''). \tag{7}$$

Instead of ρ_1 we shall now have R_1, defined by

$(x' t' | R_1 | x'' t'')_{k'k''} = \sum_{oc} \psi_{k'}(x' t') \psi^*_{k''}(x'' t'') - \sum_{un} \psi_{k'}(x' t') \psi^*_{k''}(x'' t'')$,

and instead of equation (5) we shall have

$R = ½ (R_F + R_1)$,

where

$(x' t' | R_F | x'' t'')_{k'k''} = \sum_{oc} \psi_{k'}(x' t') \psi^*_{k''}(x'' t'') + \sum_{un} \psi_{k'}(x' t') \psi^*_{k''}(x'' t'')$,

R_F, representing *the full distribution with all possible states occupied*, is no longer simply the unit matrix, but all the same we should expect it to play some fundamental part in the theory."]

As one easily confirms, for t′ = t, R_F goes to the Dirac δ-function of the variables x′, k′, x″, k″. The matrix R_S already has a symmetry relative to the sign of the charge that will be important in the formalism that follows: Under the addition of ½ R_F, it goes to the corresponding matrix R of *"hole" theory*. Under subtraction of ½ R_F, it goes to the *negative density matrix* of a distribution in which the states of positive energy are occupied and the positive energy states are free. Permuting the points x′, t′, k′ and x″, t″, k″ and switching the sign of R_S *are equivalent to a change of sign in the electric charge*. The singularity of the matrix R_S on the light-cone was investigated by Dirac; one can represent the matrix in the form:

$$(x', k' \mid R_S \mid x'', k'') = \ldots, \qquad (7)$$

in which …

The quantity w is determined uniquely by a differential equation, but v is determined only up to an additive term of the form $x^\lambda x_\lambda \cdot g$. One ordinarily deduces the charge density, current density, etc., from the density matrix R when one makes the Ansatz, e.g., for the charge density:

$$\rho(x) = e \sum_k (x,k \mid R \mid x,k); \qquad (9)$$

corresponding statements are true for the other physical quantities. Now, *due to the singularity of the matrix R, this conclusion is obviously incorrect* – e.g., when no external field is present – *since only the deviation of the density matrix from the matrix of the state in which all of the negative energy levels are filled contributes any charge and current density*. From Dirac, one would then have to subtract from the density matrix, another density matrix that is determined uniquely by the external field in order to obtain the "true" density matrix – we call it (x′, k′ | r | x″, k″) – that is definitive for the charge and current densities, energy densities, etc., corresponding to equation (9). We set:

$$r = R_S - S, \qquad (10)$$

in which S shall be a function of x'_λ, k′ and x''_λ, k″ that is uniquely determined by the potentials A^λ.

In place of the differential equation (3),

$[\mathscr{H} R = [i\hbar\, \partial/c\partial t' + e/c\, A_0(x') + \alpha_s\{i\hbar\, \partial/\partial x'_s - e/c\, A_s(x')\} + \beta mc]\, R = 0 \qquad (3)]$

one now has the equation:

$$\mathscr{H}\mathrm{r} = -\mathscr{H}\mathrm{S}. \tag{11}$$

The right-hand side is a function of the electromagnetic field that must be determined more precisely; the original *homogeneous* Dirac equation (3) will then be replaced with the *inhomogeneous* equation (11). *Such an equation is the natural expression for the fact that matter can be created and destroyed. The type of creation and annihilation will be established by the form of the quantity* HS. If no other external fields are present then S shall be given by the value of R_S for the distribution in which *all negative energy states are occupied.* We then assume that the matrix r vanishes everywhere in field-free space. The set of all matter that is collectively created when an external field is imposed and the again removed can be ascertained without any closer approximation on S by the presence of external fields. Then, when R_S (and therefore r) is known before the imposition of any sort of field the value of R_S can be ascertained from equation (3) after the field is again removed. However, after the field is removed, S again has its original value, so r can also be calculated. Nonetheless, conversely, *the result of the matter created by the imposition and removal of the fields gives the general reference point for the form of the right-hand side of (11) in the presence of fields*. For example, a simple perturbative calculation shows that the total set of matter that is created by the imposition and removal of the field is, in general, already infinite when the temporal differential quotient of the electric or magnetic field strength is sometimes discontinuous in the process of imposing and removing it, and *first becomes correct when the field strengths or potentials are themselves discontinuous*. From this, one concludes that the right-hand side of (11),

$$[\mathscr{H}\mathrm{r} = -\mathscr{H}\mathrm{S}. \qquad (11)]$$

along with the potentials and field strengths, must also contain the first and second derivatives.

Dirac (*loc. cit.*) carried out the determination of S in the presence of external fields in such a way that he described a certain mathematical process that gave the matrix R_S from the sequence of singular parts; Dirac identified the sum of the singular parts thus obtained by S. *However, the mathematical process that was chosen by Dirac did not deliver the aforementioned value of S in the field-free case, but one that differed from it by a matrix that was regular on the light-cone*. Whether or not a unique determination of the inhomogeneity in (11) is therefore hardly possible using formal arguments, by considering the conservation laws for charge, energy, and impulse, one can restrict the possibilities for S in such a way that a definite value can be distinguished as a first hypothesis. …

…

… In regard to the *current* that follows from the *density matrix* r, our assumptions are equivalent to those of Dirac (*loc. cit.*). On the other hand, as Herr Dirac cordially communicated to us, the choice of matrix S that was made here delivers a different *energy* and *impulse density* than Dirac's choice does.

2. *The conservation laws*. The *charge* and *current density* follow from the *density matrix* r in the usual way, and from an investigation by Tetrode[1],

[1] Tetrode, H. (1928). Der Impuls-Energiesatz in der Dirachsen Quantentheorie des Elektrons. (The impulse-energy theorem in Dirac's quantum theory of the electron.) *Zeit. Phys*. 49, 858-64; https://doi.org/10.1007/BF01328632; (translated by D. H. Delphenich; https://neo-classical-physics.info/ electromagnetism. html).

the *energy* and *impulse* tensor of the matter waves can be derived from the following equations:

$$s_\lambda(\xi) = \ldots \qquad (16)$$
$$U_{\nu\mu}(\xi) = \ldots .$$

…

One can also briefly summarize the results up to now in the following way: If one restricts oneself to an intuitive analogue theory of matter fields then the well-known difficulty with *the appearance of negative energy levels in the Dirac theory can be avoided in such a way that one replaces the homogeneous Dirac differential equation (3) with an inhomogeneous equation, where the inhomogeneity is indicative of "pair creation." The usual conservation laws are valid for the matter field that satisfies this equation, as well as the Maxwell field, and at the same time the energies of the matter and the radiation field are always individually positive*.

One can recognize the *invariance of the theory under a change of sign of the elementary charge* most simply in the following way: One replaces + e with – e in equations (11)

$$[\mathscr{H} r = - \mathscr{H} S. \qquad (11)]$$

and (16), as well as (x′, k′ | r | x″, k″) with – (x″, k″ | r⁻ | x′, k′). The original equations (11) and (16) are then valid once more for the matrix r.

3. *Applications*. Two simple examples shall illustrate the application of the methods that were depicted in 1 and 2. We first assume that a scalar *potential* A_0, which is regarded as a small perturbation, is slowly introduced and then kept constant, and then ask what sort of matter is created by it from originally empty space; thus, the *charge density* that gives rise to the *potential* A_0 shall be referred to as the "*external charge density*"[1].

[1] This problem has essentially already been treated by Dirac in his report to the Solvay Conference in 1933.

…

The "*polarization of the vacuum*" first becomes a physical problem for the *temporal variation of external densities*; one imagines, e.g., a charge distribution that moves back and forth periodically. In such a case, one can distribute the *external charge density* in its temporal mean value and a *second density* that oscillates periodically around the null value. The spatial integral of the second part vanishes when the *external charge density* moves back and forth in a finite spatial domain. The considerations up to now are valid for the part, so for them, the "*polarization*" plays no physical role. The total charge of a particle can thus never change by means of the *polarization of the vacuum*. …

… one thus has the extra density:

$$\rho = - 1/15\pi \, e^2/\hbar c \, f^2/(mc)^2 \, . \, \rho_0. \tag{44}$$

… Equation (44) teaches us that the dipole moment that is coupled to an oscillating charge will be reduced by the *polarization of the vacuum*, and indeed even more so as the frequency of oscillation increases. As Dirac has already suggested, this situation necessitates a change in the *scattering formula* of Klein and Nishina, which generally amounts to perhaps one-tenth of a percent in the realm of the Compton wavelength.

If one carries out an analogous calculation, in order to compute, say, the *matter density* that is induced by a *light wave* then this gives the result that the periodically varying field of a monochromatic plane light wave generates either *charge* or *current density*. One can easily see that this result also remains true to an arbitrary approximation: One cannot distinguish any sign for the *charge* by means of an *electromagnetic field* in empty space, so the *induced charge density* must vanish. On the grounds of invariance, the *current density* also vanishes then. Certainly, the vanishing of the *energy density* does not follow even from this, and in fact two plane waves that pass through each other can already give rise to the creation of matter. The *intuitive theory of matter waves* is thus no longer appropriate for the treatment of such problems (pair creation and annihilation), and we thus go on to the *quantum theory of waves*.

II. *Quantum theory of wave fields.*

1. *Presentation of the basic equations*. In the *quantum theory of matter waves*, the *Dirac density matrix* corresponds to the *product of wave functions with their conjugates*; we then set:

$$R = \psi^*(x', k') \, \psi(x'', k''). \tag{45}$$

The commutation relation:

$$\psi^*(x', k') \, \psi(x'', k'') + \psi(x'', k'') \, \psi^*(x', k') = \delta(x'', k'') \, \delta_{k'k''} \tag{46}$$

is true for the *wave function* (for $x'_0 = x''_0$). If one considers the Maxwell field as a given c-field then the Dirac matrix is simply the expectation value for the matrix that is defined by (45). Due to the commutation relation (46), one has, in the quantum theory of waves:

$$R_S = \tfrac{1}{2} \, [\psi^*(x', k') \, \psi(x'', k'') - \psi(x'', k'') \, \psi^*(x', k')]. \tag{47}$$

The equations:

$$\mathcal{H} R_S = 0 \tag{3a}$$

and $R_S = r + S$ remain unchanged and only in the form of the inhomogeneity $\mathcal{H} S$ in:

$$\mathcal{H} r = - \mathcal{H} S \tag{11a}$$

can a change become necessary due to the non-commutation of field strengths with potentials. Now, no non-commuting functions appear in the first term:

$$e^{-ei/\hbar c \int_{p'}^{p''} A_\lambda \, dx_\lambda} \, . \, S_0.$$

Terms enter into S_1 [cf., (13) and (14)] that are quadratic in the field strengths and play a role when one calculates *energy* and *impulse density* from the *density matrix*. As long as one restricts oneself to the calculation of *charge* and *current density* these terms will no longer appear. Now, since the Maxwell equations, together with the inhomogeneous equation (11a), determine the physical evolution completely, the reasoning of the formalism that was depicted in I in the context of quantum theory results from a process that was given for ordinary quantum electrodynamics in a note of the author[1] in connection with previous research of Klein[2].

[1] Heisenberg, W. (1931). Bemerkungen zur Strahlungstheorie. (Remarks on Radiation Theory.) *Ann. Phys*. 401, 3, 338-46; https://doi.org/10.1002/andp.19314010305.

[2] Klein, O. (1927). Elektrodynamik und Wellenmechanik vom Standpunkt des Korrespondenzprinzips. (Electrodynamics and wave mechanics from the point of view of the correspondence principle.) *Zeit. Phys*., 41, 10, 407-22; https://doi.org/10.1007/BF01400205.

This process starts with the *Maxwell equations* and the *wave equation*, which are treated as q-number relations and are integrated according to the usual methods of the intuitive theory. Ordinarily, a perturbation process is applied to the integration of the basic equations, in which one assumes that the interaction between light and matter is small and is developed in powers of the *charge*. The plane *light waves* in empty space and the plane *electron waves* in field-free space then take the form of the unperturbed system. Such a perturbation process is also applicable in the present theory with no further assumptions. *It is then necessary only to also develop the matrix S that is appropriate for the inhomogeneity in the wave equation in powers of charge*, and to consider the individual terms in the development that are produced by the perturbation process in turn at the successive degrees of approximation. In order to deduce R_S and r – and therefore, the *charge* and *current density*, in the zeroth order approximation – one will then only have to subtract the matrix S_0 from R_S. …

…

One finally obtains for the matrix r:

$$r = \sum \ldots - \sum \ldots + \sum \ldots = \ldots . \tag{54}$$

This representation of the *density matrix* agrees with the representation that was chosen by Pauli and Peierls[1], Oppenheimer and Furry, Fock (*loc. cit.*).

[1] I would like to cordially thank Herrn W. Pauli for the written communication of this result.

N_n means the number of electrons, N'_n, that of the positrons, and the symmetry in the theory on the sign of charge is assumed from the outset. This representation is, however, *only correct in the zeroth-order approximation.* If one goes on to the first-order approximation then, on the one hand, the coefficients a_n will also contain, as functions of time, terms that are linear in the field strengths [cf., e.g., *loc. cit., Ann. Phys*. 9, pp. 341, equation (9)], and on the other hand, the terms that are linear in *e* in the definition of r must be subtracted from the matrix S, and thus the terms:

.... . (55)

These terms, together with the terms in the coefficients an that are linear in *e*, then give a contribution to the matrix r that leads to a finite *charge* and *current density* (in the first approximation) and which can therefore assist in the calculation of the *electromagnetic field* in the second approximation, etc. Instead of this process, which is closely connected with the integration methods of the intuitive theory, one can, however, also define a Hamiltonian function in the usual way and then carry out the perturbation theory for the associated Schrödinger equation. ...

...

$H_1 = \ldots$ (58)

The present theory thus agrees with the results of Oppenheimer and Furry in the expressions for H_0 and H_1. We thus obtain terms of higher order that come from the matrix S. ...

In this way, the perturbation process can be, in principle, performed *when no infinite self-energy*, as in quantum electrodynamics up to now, *leads to a divergence in the process*[2].

[2] Cf., on this, Weisskopf, V. (January, 1934). Über die Selbstenergie des Elektrons. (The Self-energy of the Electron.) *Zeit. Phys*., 89, 27-39[; the *self-energy* of the electron is derived in close formal connection to classical radiation theory without direct application of quantum electrodynamics, the radiation field is calculated classically from *current and charge densities* of atom, divides electromagnetic field into rotation-free *electrostatic* part and divergence free *electrodynamic* part, *self-energy* derives from electrostatic part, self-energy of electron at occupied negative energy states diverges logarithmically as in Dirac's "hole theory" in contrast to the linear divergence of the classical theory and the quadratic divergence of the one particle Dirac theory.]; furthermore, on the search for ways to avoid the infinite self-energy of the electron, see Born, M. (January, 1934). On the Quantum Theory of Electromagnetic Field. *Proc. Roy. Soc. (A)*, 143, 849, 410-37; https://doi.org/ 10.1098/ rspa.1934.0010; Born, M. & Infeld, L. (March, 1934). Foundations of the New Field Theory. *Ibid*., 144, 852, 425-51; https://doi.org/ 10.1098/rspa.1934.0059.

The perturbation energy H_2 has the following form:

$H_2 = \ldots$ (59)

Due to the integration over ξ, H_2 gives rise to only matrix elements that correspond to the creation or annihilation of light quanta of the same impulse. For the ordinary processes in which light quanta are emitted or absorbed or scattered, these matrix elements then play no role in the first approximation. ...

2. *Applications*. For the most practical applications – e.g., pair creation, annihilation, Compton scattering, etc. – *the theory described here does not yield anything new compared to the formulation of the Dirac theory all along*. Thus, in all of the cases mentioned, one can break off the perturbation calculation at the second-order approximation and the new terms in H_2, due to their special form, contribute nothing to the transition probabilities that

were sought. Things are different for the aforementioned problem of the scattering of light by light and for the coherent scattering of γ-rays from fixed charge centers that was discussed by Delbrück[3];

[3] Delbrück, M. (1933). Discussion of the experimental results of L. Meitner and her colleagues. *Zeit. Phys.*, 84, 144.

the calculations in these problems are so complicated that they will not be attempted here.

We would therefore like to restrict the applications to an example in which the term H_2 in equation (59) becomes important; we shall treat the *matter density that is linked to a light quantum*, and in particular, the *self-energy of the light quantum* that is given on the basis of this *matter density*. If one first ignores the term H_2 and calculates with the usual methods heretofore then the process can be represented as follows: Since matrix elements appear in H_1 [equation (58)] that correspond to the conversion of a light quanta into a pair, *a light quantum generates a matter field in its neighborhood in a manner that is similar to the way that an electron generates a Maxwell field. The energy of this matter field becomes infinite in complete analogy to the infinite self-energy of electrons*. Now, part of the singular terms in the infinite self-energy of the light quantum vanishes when one considers the perturbation term H_2. They are then arranged such that no infinite self-energy would appear for a classical light wave. Nevertheless, the following calculation shows that *an infinite part of the self-energy that is required by the application of quantum theory remains*. The analogy with the self-energy of the electrons is complete now. In the Maxwell theory, a continuous charge distribution would also lead to a finite self-energy; it is the "quantization" that leads to the infinite self-energy. *If one represents the quantization of the electromagnetic field by point-like light quanta then the infinitude of the self-energy also emerges in the intuitive theory of matter waves*, since the inhomogeneity in equation (11)

$$[\mathscr{H} \mathrm{r} = - \mathscr{H} \mathrm{S}. \qquad (11)]$$

includes the field strengths and their first and second derivatives, which become singular in the context of light quanta.

…

In the *quantum theory of wave field*, the domain of applicability of the Dirac formulation of the theory of positrons is therefore not essentially larger than the domain of applicability for the elementary formulas of Pauli, Peierls, Fock, Oppenheimer, and Furry. Equations (48) to (61) then show how these formulas can be regarded as the first step in a sequence of approximations that satisfy the requirements of *relativistic* and *gauge invariance*. Furthermore, the formalism that is described here also yields finite expectation values for present and *energy densities* in the first approximation where the elementary formulas would give infinite values. The fact that divergences would appear in the second approximation of the quantum theory of wave fields was to be expected from results of quantum electrodynamics up to now. The situation that the application of the quantum theory first leads to divergences that do not appear in the *intuitive theory of wave fields* suggests that *this intuitive theory, in fact, already contains the essence of the correct, corresponding description of how things happen*, so one cannot carry out the transition to quantum theory in the original way that was sought for in the current theory heretofore. In the Dirac theory of positrons, moreover, a pure separation of the fields that are involved into *matter fields* and *electromagnetic fields* is scarcely possible any more. In particular, this comes from the fact that in the quantum theory of waves it is the matrix R_S – not the matrix r – that can be represented simply by the *matter wave functions* ψ. *It is therefore only a unified theory of matter and light fields that gives the Sommerfeld constant $e^2/\hbar c$ a definite value that will make possible a contradiction-free union of the demands of quantum theory with those of a correspondence with intuitive field theory.*

Victor Frederick Weisskopf (September 19, 1908–April 22, 2002).

[Drawn largely from Jackson, J. D. & Gottfried, K. (2004). *Victor Frederick Weisskopf.* National Academy of Sciences. Biographical Memoirs: Volume 84. Washington, DC: The National Academies Press; https://doi.org/10.17226/10992.]

Weisskopf was an Austrian-born American theoretical physicist. He did postdoctoral work with Werner Heisenberg, Erwin Schrödinger, Wolfgang Pauli, and Niels Bohr. He made seminal contributions to the quantum theory of radiative transitions, the self-energy of the electron, the electrodynamic properties of the vacuum, and to the theory of nuclear reactions. During World War II he was Group Leader of the Theoretical Division of the Manhattan Project at Los Alamos, and he later campaigned against the proliferation of nuclear weapons.

Weisskopf was the second of three children born in Vienna, Austria, on September 19, 1908, into a comfortably middle-class Jewish family. His father, Emil, originally from Czechoslovakia, was a successful lawyer; his mother, Martha, was from an upper-middle-class nonobservant Jewish Viennese family. He studied the piano and developed a lifelong love of music. In his teens he attended a gymnasium and for two years the University of Vienna, where he found inspiration in Hans Thirring's lectures in classical theoretical physics. Thirring, sensing Weisskopf's exceptional abilities and knowing that Vienna was not in the forefront in modern physics, recommended that he transfer to Göttingen to continue his studies under Max Born.

After learning quantum mechanics from Gerhard Herzberg, he embarked on research on the interaction of radiation with matter, a broad subject of central importance. He attacked his first unsolved problem: the natural width of spectral lines in emission of radiation by atoms. He was able to make progress on a two-level quantum system but not beyond. He sought help from Eugene Wigner, who was then in Berlin but who came back to Göttingen for regular visits. Wigner became Weisskopf's mentor and the collaboration led to two papers. His first important paper (1930), written with Eugene Wigner, was on the quantum theory of the breadth of spectral lines. It treats the exponential decay of excited atomic states and the natural breadth of the associated spectral lines for all types of transitions. In contrast to the semiclassical result where an intense line was necessarily broad and a weak line narrow, the quantum theory accommodates the occasional puzzling broad but weak line.

After completing his Ph.D. thesis in the Spring of 1931, he went to Leipzig to work under Werner Heisenberg and then in the spring term of 1932 under Erwin Schrödinger in Berlin. For the academic year 1932-33 he received a Rockefeller Fellowship to work in Copenhagen with Niels Bohr and in Cambridge with Paul Dirac.

In the fall of 1933 Weisskopf came to Zürich for two and a half years as Wolfgang Pauli's assistant. While there he published two important papers. The first was on the self-energy of the electron in the framework of Dirac's hole theory, in which he showed that the self-energy diverged only logarithmically with decreasing size of the electron's charge distribution, in contrast to the linear divergence of classical theory and the quadratic divergence of the one particle Dirac theory. [Weisskopf, V. (1934). Uber die Selbstenergie des Elektrons. (The Self-energy of the Electron.) *Zeit. Phys*., 89, 27-39. Received March 13, 1934.] The second paper, coauthored with Pauli, concerned the quantum field theory of charged scalar particles (not the spin 1/2 particles of Dirac). [Pauli, W. & Weisskopf, V. (1934). Uber die Quantisierung der skalaren relativistischen Wellengleichung. (The Quantization of the Scalar Relativistic Wave Equation.) *Helv. Phys. Acta*, 7, 709-731.] They showed that antiparticles were not unique to the Dirac theory but occurred in general and that electrodynamic processes involving the scalar particles were closely similar to those involving spin 1/2 electrons.

In April 1936 Weisskopf accepted a fellowship at Bohr's institute in Copenhagen. While there he completed an impressive analysis of the properties of the vacuum in the presence of electromagnetic fields, clarifying earlier work, giving physical arguments for the removal of certain infinities, and presciently enunciating the concept of *charge renormalization.*

With the increasing persecution of Jews in Nazi Germany and the prospect of war, he and his wife decided to look for ways to escape Western Europe. To enhance his chances Weisskopf began to work in the increasingly important field of nuclear physics, which occupied many at Bohr's institute, and to publish in English. Although he had job offers from the Soviet Union, after a visit in late 1936 he and his wife decided that he would consider them only as a last resort. With Bohr's help he was offered a lectureship at the University of Rochester beginning in the fall of 1937.

He was on the Rochester faculty for five and one-half years. During that time, he continued research in nuclear physics [Weisskopf, V. F. (1938). Excitation of nuclei by bombardment with charged particles. *Phys. R*ev., 53, 1018] but also on the electrodynamics of the electron. He returned to the self-energy problem and in 1939 established a result little appreciated at the time or now: that in the nth order of perturbation theory the self-energy diverges only as the nth power of a logarithm. [Weisskopf, V. F. (July, 1939). On the Self-energy and Electromagnetic Field of the Electron. *Phys. R*ev., 56, 1018.]

In early 1943 Weisskopf was invited to Los Alamos, where he soon became Hans Bethe's deputy in the theoretical physics group. Already famous for his physical intuition, he was much sought after by the experimenters to provide estimates of little known or little understood nuclear processes.

In early 1946 he joined the physics faculty at the Massachusetts Institute of Technology. With his student Bruce French he revisited the electron self-energy problem to explore an earlier suggestion of Hendrik Kramers that one might make sense of the higher-order radiative corrections in electrodynamics in spite of the infinity in the self-energy of the electron. Kramers had pointed out that what is actually observable is the energy difference between free and bound states of electrons. Weisskopf's demonstration that the divergence is only logarithmic made it plausible that differences would be finite and meaningful.

French and Weisskopf had not completed their calculation when, in June 1947, Willis Lamb had announced the results of his microwave experiments on hydrogen, showing a tiny disagreement with existing theory, with a very small energy difference between two levels supposedly degenerate, which became known as the "Lamb shift." Bethe quickly showed *in a nonrelativistic calculation with a cutoff* that Kramer's idea led to a level shift close to that measured by Lamb. His work depended, however, on the plausible but unproven assumption that the logarithmic divergences at high energy exactly canceled.

By early 1948 French and Weisskopf completed the first consistent calculation of the Lamb shift, but Weisskopf would not publish because they had a very small disagreement with the independent calculations of Richard Feynman and Julian Schwinger, who agreed with each other. Weisskopf could not believe that his work with French was correct. The upshot was that French and Weisskopf published their year-old result [French, J. B. & Weisskopf, V. F. (April, 1949). The Electromagnetic Shift of Energy Levels. *Phys. Rev.*, 75, 8, 1240-8] only after a paper by Kroll and Lamb appeared with essentially the same calculation. The Kroll-Lamb theory paper contains a succinct statement about Weisskopf's place in the firmament of theoretical physics: "[Our] calculation," they wrote, "[is] based on the 1927-34 formulation of quantum electrodynamics due to Dirac, Heisenberg, Pauli, and Weisskopf." [Kroll, N. M. & Lamb, Jr., W. E. (1949). On the self-energy of a bound electron. *Phys. Rev.*, 75, 388-98.]

In 1961 at the apex of his career as an academic researcher Weisskopf was invited to be director general of CERN, the European center for high-energy physics near Geneva, Switzerland. For CERN it was an inspired choice. Weisskopf provided intellectual leadership and a vision of the laboratory as an international research center second to none. He successfully promoted construction of the first proton-proton collider, the intersecting storage rings, and saw to the eventual building of a 300-GeV accelerator. He set CERN on its path to be a preeminent, some would say the preeminent, research center in high-energy physics today.

At the end of his five-year term he returned to MIT, where he was named Institute Professor and chaired the Physics Department for six years (1967-73). During these years he pursued occasional research, but devoted increasing fractions of his time to writing, invited lectures, and public service. He published popular expositions of science, collections of essays on science, and his autobiography.

Weisskopf was awarded the Max Planck Medal in 1956 and the Prix mondial Cino Del Duca in 1972, the National Medal of Science (1980), and the Wolf Prize (1981).

His first wife, died in 1989; he was survived by his second wife. Increasingly frail, he died at home in Newton, Massachusetts, on April 22, 2002, at age 93.

Weisskopf, V. (January, 1934). Über die Selbstenergie des Elektrons. (The Self-energy of the Electron.)

[*Zeit. Phys.*, 89, 27-39; https://doi.org/10.1007/BF01333228; (translation in A. I. Miller. (1994). Early Quantum Electrodynamics. A Sourcebook, Cambridge University Press, 157-68; https://doi.org/10.1017/CBO9780511608223.015.]

Received March 13, 1934.

Physikalisches Institut der Eidg. Techn. Hochschule, Zürich.

The *self-energy* of the electron is derived in close formal connection to classical radiation theory without direct application of quantum electrodynamics, the radiation field is calculated classically from *current and charge densities* of atom, divides electromagnetic field into rotation-free *electrostatic* part and divergence free *electrodynamic* part, *self-energy* derives from electrostatic part, self-energy of electron at occupied negative energy states diverges logarithmically as in Dirac's "hole theory" in contrast to the linear divergence of the classical theory and the quadratic divergence of the one particle Dirac theory.

Summary

The *self-energy of the electron* is derived in close formal connection to classical radiation theory, and the *self-energy* of an electron at occupied negative energy states is calculated, which corresponds to the idea of Dirac's "hole theory" of the positive and negative electron. As you might expect, self-energy also diverges in this theory, to the same extent as in ordinary single-electron theory [see correction in footnote 4].

Problem definition

The *self-energy of the electron is the energy of the electromagnetic field which is generated by the electron in addition to the energy of the interaction of the electron with this field.* Waller[1], Oppenheimer[2], and Rosenfeld[3]

[1] Waller, I. (June, 1930). Bemerkungen über dir Rolle der Eigenenergie des Elektrons in der Quantentheorie der Strahlung. (Remarks on the role of the electron's own energy in the quantum theory of radiation.) *Zeit. Phys.* 62, 673-6; https://doi.org/10.1007/ BF01843484.
[2] Oppenheimer, J. R. (March, 1930). Note on the Theory of the Interaction of Field and Matter. *Phys. Rev.*, 35, 5, 461; https://doi.org/10.1103/PhysRev.35.461.
[3] Rosenfeld, L. (July, 1931). Zur Kritik der Diracschen Strahlungstheorie. (On the critique of Dirac's theory of radiation.) *Zeit. Phys.*, 70, 454-62; https://doi.org/10.1007/ BF01330848.

calculated the *self-energy* of the free electron by means of the *Dirac relativistic wave equation* of the electron and the *Dirac theory of the interaction between matter and light*. They here used an approximation method which represents the *self-energy* in powers of the charge e. They found that the first term, which is proportional to e^2, already *becomes infinitely large*. The essential reason for this is that the theory of the interaction of the electron with the electromagnetic field is built on the classical *equations of motion* of a point shaped electron whose *self-energy*, as is well known, also becomes infinite in classical theory[4].

[4] Recently, Wentzel [Wentzel, G. (July, 1933). Über die Eigenkräfte der Elementarteilchen I. (On the intrinsic forces of elementary particles. I.) *Zeit. Phys*., 86, 479-94; https://doi.org/10.1007/BF01341363; (September, 1933). Über die Eigenkräfte der Elementarteilchen II. (On the intrinsic forces of elementary particles. II.) 86, 635-45. https://doi.org/10.1007/BF01338974635] has shown that one can circumvent the divergence of the *self-energy* in classical electron theory by suitable limiting processes. The transfer of these methods to quantum theory has failed, however, since, according to Waller, the degree of infinity in quantum theory is higher than in classical theory. The hope expressed there that the degree of infinity will become smaller in the Dirac formalism of the 'hole' theory, does indeed hold for the electrostatic part but not for the electrodynamic part, so that the Wentzel method must fail here too.

Correction to the paper: The self-energy of the electron Zeitschrift fur Physik, 90: 817-18 (1934). Received 20 July 1934. On [p. 166] of the paper cited above, there is a computational error which has seriously garbled the results of the calculation for the electrodynamic self-energy of the electron according to the Dirac hole theory. I am greatly indebted to Mr. Furry (University of California, Berkeley) for kindly pointing this out to me. The degree of divergence of the self-energy in the hole theory is not, as asserted in [the preceding paper], just as great as in the Dirac one-electron theory, but the divergence is only logarithmic. ... The computational error arose in the transformation of the electrodynamic portion E^D for the case of the hole theory... where $J^k_+(\vec{p})$ is defined on [p. 166] ... As a consequence of the new result, the question raised in note 4 of the paper requires a new examination, whether the Wentzel method, to avoid the infinite self-energy by suitable limiting processes, might not still lead to the objective in the hole theory.

In the present note, the expressions for the self-energy shall be derived *without direct application of quantum electrodynamics*, but by means of the *Heisenberg radiation theory*[5],

[5] Heisenberg, W. (1931). Bemerkungen zur Strahlungstheorie. (Remarks on Radiation Theory.) *Ann. Phys*. 401, 3, 338-46; https://doi.org/10.1002/andp.19314010305; see also W. Pauli's article in Geiger-Scheel, *Handb. d. Phys*., XXIC/1, 2nd edn., pp. 201-10.

which is linked much more closely to classical electrodynamics. The radiation field is calculated classically from the current and charge densities of the atom; however, the *amplitudes* of the *electromagnetic potentials* are regarded as non-commuting in the final result. Just as was shown in a corresponding paper by Casimir[6]

[6] Casimir, H. (July, 1933). Zur korrespondenzmäßigen Theorie der Linienbreite. (On the correspondence theory of line width.) *Zeit. Phys*., 81, 78, 496-506; https://doi.org/ 10.1007/BF01342295.

concerning the natural linewidth, this method yields the same result as explicit quantum electrodynamics and thus leads to the same difficulties. It is equivalent to the latter in every respect. It therefore perhaps is not quite appropriate to designate this method as a correspondence method in contrast to the Dirac radiation theory (see Casimir, *loc. cit.*), if, with Bohr, one sees the essential content of the correspondence argument in the tendency to preserve the consequences of classical theory as long as they do not directly contradict the specific atom-mechanical phenomena.

Furthermore, the calculation shall be performed generally for a multi-electron system; in particular, it will also be evaluated for the case that all negative energy states are occupied. The *self-energy of the electron* in the new *Dirac 'hole' theory* can then be calculated. As is well known, this theory assumes - in order to eliminate the difficulties of negative energy states - that these states are generally occupied. However, only a certain residue of the *density-current vector*, which comes from all the existing electrons, becomes noticeable. This residue is obtained by subtracting that *density-current vector* which is generated by the complete occupation of all negative states of the electron. Consequently, if a negative energy state has remained unoccupied, for example, one obtains a residual amount which corresponds to the *current-density vector* of a positive electron.

In calculating the *self-energy*, we shall limit ourselves to terms proportional to e^2 in the expansion by powers of the *charge*. Consequently, the calculation also does not contain those difficulties and ambiguities which appear in the action of *electromagnetic fields* upon electrons in negative states, since, in the required approximation, these may be assumed to be empty.

2 *The self-energy of the electron.*

The *energy operator* of an electron with a surrounding field is

$$H = c(\vec{\alpha},\vec{p} - e/c\, \vec{A}) + \beta mc^2 + 1/8\pi \int (\vec{E}^2 + \vec{H}^2)\, d\vec{r},$$

whereby $\vec{\alpha}$ and β are the familiar *Dirac spin matrices*, $\vec{A}$ is the vector *potential* of the fields $\vec{E}$, $\vec{H}$. It should be emphasized that the scalar *potential* Φ does not occur at all in this expression[7].

[7] $d\vec{r}$ signifies the volume element dx dy dz.

The *self-energy* is then given by the expectation value of

$$E = -\int (\vec{i}\vec{A})\, d\vec{r} + 1/8\pi \int (\vec{E}^2 + \vec{H}^2)\, d\vec{r}$$

where $\vec{i} = e\vec{\alpha}$ is the *current-density operator* and $\vec{A}$, is, $\vec{E}$, $\vec{H}$ here describe the field which is generated by the *current-density* i and the *charge density* ρ.

We now divide the electromagnetic field into a *rotation-free part* with components $\vec{A}_l$, is, $\vec{E}_l$, ($\vec{H}_l = 0$) and a *divergence-free part*[8] $\vec{A}_{tr}$, $\vec{E}_{tr}$, $\vec{H}_{tr}$.

[8] The indices come from the circumstance that, in a spatial Fourier decomposition, the former are represented by transverse [tr] waves, whereas the latter are represented by longitudinal waves.

The '*electrostatic*' *self-energy* coming from the first part

$$E^S = -\int (\vec{i}\,\vec{A}_1)\, d\vec{r} + 1/8\pi \int \vec{E}_1^2\, d\vec{r}$$

...

3 *Calculation of the self-energy.*

In calculating the energy expressions (3) and (4), we start from the *stationary states* of a free electron, which result from the Dirac equation

$$\hbar c/i\,(\vec{\alpha}, \mathrm{grad})\varphi + \beta mc^2\varphi = E\varphi.$$

The *eigenfunctions* are[9]

[9] $2\pi\hbar = h$ = Planck's constant. Below we shall use both symbols $\hbar$ and h.

...

(a) *The electrostatic self-energy.*

We first calculate the *electrostatic* portion of the *self-energy* E^S, which is associated with a particular occupation of the *states*. ...

... The *electrostatic self-energy* thus diverges logarithmically in the 'hole' theory.

By comparison, let us specify the *self-energy of the electron* according to the Dirac single-electron theory, as calculated according to (14):

$$E^S = \ldots,$$

which diverges linearly.

(b) *The electrodynamic self-energy.*

We now arrive at the calculation of E^D. Here we expand the vector potential $\vec{A}_{tr}$ in powers of e:

$$\vec{A}_{tr} = \vec{A}_{tr}^{\,0} + \vec{A}_{tr}^{\,1} +$$

The portion $\vec{A}_{tr}^{\,0}$ which is independent of e may not be set equal to zero, although it naturally vanishes according to classical electrodynamics. In fact, however, it represents the zero-point fluctuations of the field strengths which exist in empty space. In general, one can take into account the effect of this zero-point field strength, which causes *spontaneous emission*, by writing down the light field existing in space in the following fashion in a Fourier decomposition:

$$\vec{A}_{tr} = \ldots. \qquad (16)$$

The index j designates the two *polarization* directions ($k = |\,k\,|$).

... According to classical electrodynamics, we have:

$$\vec{A}_j^{+}(\vec{k})\,\vec{A}_j^{-}(\vec{k}) + \vec{A}_j^{-}(\vec{k})\,\vec{A}_j^{+}(\vec{k}) = h^2/\pi k^2\, S_j(\vec{k}), \qquad (17)$$

where $S_j(\vec{k})d\vec{k}$ is the *energy density* in the radiation with the *momentum* vector $\vec{k}$ and the *polarization* j.

To satisfy quantum theory, one must at this point take into account the non-commutability of the $\vec{A}_j(k)$. If, in the expression for the *charge density*, one specifies the sequence $e\psi^+\psi$ (instead of $e\psi\psi^+$), one should set

$$2\,\vec{A}_j^+(\vec{k})\,\vec{A}_j^-(\vec{k}) = h^2/\pi k^2\,\{S_j(\vec{k}) + ck/h^3\}, \qquad (18)$$
$$2\,\vec{A}_j^-(\vec{k})\,\vec{A}_j^+(\vec{k}) = h^2/\pi k^2\,S_j(\vec{k}).$$

Whereas the classical expression (17) yields only those transitions of an atomic system that are forced by the external radiation field, *the addition* ck/h^3 *yields the additional spontaneous emission transitions*, since the combination $\vec{A}_j^+\vec{A}_j^-$ occurs only for *emission*, $\vec{A}_j^-\,\vec{A}_j^+$ only for *absorption*[12].

...

[*Original page 166*: see footnote 4:]

... We thus obtain for the *self-energy of an electron* in *state* p_0^k, in the 'hole theory', after subtracting the 'vacuum' energy:

$$E^D = J_+^k(p_0) - J_-^k(p_0).$$

The calculation of the $J^k(p)$ is done with the same calculational methods as the calculation of the *electrostatic energy*. One thus obtains:

$$J_+^k(\vec{p}) = \dots,$$
$$J_-^k(\vec{p}) = \dots, \qquad \text{for } k = 1, 2,$$

...

From this one can also see that *the degree of divergence of the self-energy does not become smaller through the occupation of negative states*[14].

[14] See note 4 [p. 157].

Finally, I wish to express my sincere gratitude to Professor Pauli for stimulating this work as well as for constant help in its performance. I likewise owe special thanks to Professor Niels Bohr for the discussion of the theoretical foundations.

Pauli, W. & Weisskopf, V. (1934). Über die Quantisierung der skalaren relativistischen Wellengleichung. (The Quantization of the Scalar Relativistic Wave Equation.)

[*Helv. Phys. Acta*, 7, 709-31; https://doi.org/10.1007/978-3-322-90270-2_36; (translation in A. I. Miller. (1994). *Early Quantum Electrodynamics. A Sourcebook*, Cambridge University Press, 188-205; https://doi.org/10.1017/CBO9780511608223.017.]

Received July 27, 1934.

Zurich, Physical Institute of the E. T. H.

The *scalar relativistic wave equation* $E^2/c^2 - \sum_{k=1}^{3} p_k^2 - m^2c^2 = 0$ has generally been relinquished in favor of *Dirac's four-component wave equation* $\{p_0 + \rho_1 (\boldsymbol{\sigma}, \mathbf{p}) + \rho_3 mc\}\ \psi = 0$ because the former does not yield the spin of the particles, shows that Dirac's *a priori* arguments based on limitation to a *single-body problem* and *particle density* being a meaningful observable need to be revised, shows that *charge density* is a meaningful observable, no reason for special form of *charge density* $\sum_r \psi^*_r\ \psi_r$, in *scalar relativistic wave equation theory* the energy of material particles is always positive after wave fields have been quantized, no new hypothesis such as the "hole theory" is required, applies Heisenberg-Pauli formalism of quantization of wave fields to *scalar relativistic wave equation* for matter fields for particles without spin and with Bose-Einstein statistics, shows that quantization of the *Klein-Gordon relativistic wave equation* for *scalar particles* gives rise to particles with opposite charge but with same rest mass which can be created or destroyed with absorption or emission of electromagnetic radiation, frequency of pair creation and annihilation processes of same order of magnitude as frequency for particles of same charge and mass which follows from Dirac "hole theory", *also leads to infinite self-energy not only of the particles but also to an infinite polarizability of the vacuum.*

> [Pauli, W., Weisskopf, V.F. (1988). Über die Quantisierung der skalaren relativistischen Wellengleichung. In: Enz, C.P., v. Meyenn, K. (eds). *Wolfgang Pauli.* Vieweg+Teubner Verlag., pp. 407-430 (in German); https://doi.org/ 10.1007/978-3-322-90270-2_36:
>
> > Wentzel (1960), S. 62 f.: "The interest in the *scalar (relativistic) field* subsided until Pauli and Weisskopf (1934) showed that a consistent and physically reasonable theory of the *scalar field* can be constructed just by applying to it the canonical rules of field quantization as set up in 1929 by Heisenberg and Pauli, in their first paper on quantum electrodynamics. [...] Indeed, the topic remained academic until the meson was discovered."
> >
> > Weisskopf (1981), S. 75.: "An interesting episode in the fight for the elimination of vacuum electrons was the quantization of the *Klein-Gordon relativistic wave equation for scalar particles.* [...] The problem attracted the attention of Pauli and myself because we saw that the quantized Klein-Gordon equation gives rise to particles and antiparticles and to pair creation and annihilation processes without introducing a vacuum full of particles.

[...] Pauli called our work the 'anti-Dirac paper', he considered it as a weapon in the fight against the filled vacuum, which he never liked. [...] We had no idea that the world of particles would abound with spin-zero entities a quarter of a century later."]

Summary

The present paper consistently applies the Heisenberg-Pauli formalism of the quantization of wave fields to the scalar relativistic wave equation for matter fields when the particles have Bose-Einstein statistics. Without any further hypothesis, *this yields the existence of particles with opposite charge but with the same rest mass,* which can be created or destroyed with the absorption or respectively emission of electromagnetic radiation. *The frequency of these processes proves to be of the same order of magnitude as the frequency for particles of the same charge and mass, which follows from the Dirac hole theory* (§ 4). Here we investigate *the possibility of oppositely charged particles* without spin, with Bose-Einstein statistics, and conformable to the correspondence principle, which also satisfy *relativity* requirements. *This possibility has the advantage over the hole theory that the energy is by itself always positive. However, just as in the original version of the hole theory, the theory discussed here results not only with the infinitely large self-energies but also with an infinite polarizability of the vacuum.*

§1 *The connection of the scalar relativistic wave equation with the existence of oppositely charged particles.*

By introducing the operators

$$E = ih\, \partial/\partial t, \qquad p_k = -\, ih\, \partial/\partial x^k \qquad (k = 1, 2, 3) \qquad (1)$$

the *scalar relativistic wave equation* in the force-free case can be written

$$E^2/c^2 - \sum_{k=1}^{3} p_k^2 - m^2c^2 = 0. \qquad (2)$$

(Here and below h is always Planck's constant divided by 2π, m is the rest mass of the electron, and c is the speed of light.) As is well known, *this wave equation has generally been relinquished in favor of the Dirac four-component wave equation since the former does not yield the spin of the particles* and thus surely represents for electrons an inadequate approximation to experiment.

[Dirac, P. A. M. (February, 1928). The Quantum Theory of the Electron: "The wave equation (4)

$$[(p_0 + \alpha_1 p_1 + \alpha_2 p_2 + \alpha_3 p_3 + \beta)\, \psi = 0 \qquad (4)]$$

now takes the form [the *relativistic equation of motion* for the *wave function of the electron*, referred to as the *Dirac equation*]

$$\{p_0 + \rho_1 (\boldsymbol{\sigma}, \mathbf{p}) + \rho_3 mc\}\, \psi = 0 \qquad (9)$$

where [**p** is the *momentum* vector, $p_0 = ih/c\, \partial/\partial t$, $p_r = -\, ih\, \partial/\partial x_r$, r = 1, 2, 3; and] **σ** denotes the vector (σ_1, σ_2, σ_3)[, vectors of four complex numbers (known as bispinors)]".

A special justification is therefore required *if below we again resume the discussion of the consequences of the first wave equation.* The empirical discovery of the positron was theoretically interpreted by Dirac in his re-interpretation of the negative energy states which occurred in his original theory. We show that this requires a revision of Dirac's *a priori* arguments, which are based on general quantum-mechanical transformation theory. *These arguments originally substantiated his transition from the scalar relativistic wave equation to his spinor wave equation.* By showing that this revision is necessary, we believe we have given the justification mentioned above. Indeed, *we shall show below that, by applying to the present problem the general methods for quantizing wave fields, which previously were formulated by Heisenberg and Pauli*[1]*, not only can no general objections be maintained against the scalar wave equation from the standpoint of quantum-mechanical transformation theory, but also relativistic and gauge invariance of the theory can be preserved*, and, without any further hypothesis, one arrives at the conclusion that oppositely charged particles exist and that processes occur in which such particle pairs are created and annihilated, and *where furthermore the energy of the matter-wave field automatically always comes out positive*.

[1] Heisenberg, W. & Pauli, W. (July, 1929). Zur Quantendynamik der Wellenfelder. (On the quantum dynamics of wave fields.) *Zeit. Phys.*, 56, 1-61[; Heisenberg and Pauli's first attempt to construct their own version of a *relativistically-invariant quantum electrodynamics* to treat interaction between matter and the electromagnetic field and between matter and matter, canonical quantization of both electromagnetic and matter-wave fields, but Lorentz-invariant Lagrangian for interacting *electromagnetic* and *matter-wave fields*, requires working with the *electromagnetic potentials* not just with the *fields*, Lagrangian does not contain a time derivative of the *electric potential* so *there is no corresponding canonical momentum variable,* prevents straightforward implementation of canonical commutation relations, the theory is still afflicted with many defects, *the fundamental difficulties in the relativistic formulation that were emphasized by Dirac remain unchanged,* the formulas of the theory lead to an *infinite zero-point energy* for the radiation and thus include the interaction of an electron with itself as an *infinite* additive constant, however, these difficulties are of a sort that they do not interfere with the application of the theory to many physical problems, used "crude trick" of adding additional terms to the Lagrangian].

For the particles, one must here assume the statistics of symmetric states (Bose-Einstein statistics), but surely *it is only to be regarded as satisfactory that, without simultaneously introducing the spin, the exclusion principle cannot be introduced while preserving the relativistic invariance of the theory.*

Concerning the above-mentioned *a priori* argument of Dirac against the *scalar relativistic wave equation*[2],

[2] This is presented in most detail in the Leipzig Lectures 1932 (appeared in collective form under the title *Quantum Theory and Chemistry*), p. 85 ff.

this is essentially based on two presuppositions.

1. In *relativistic* quantum theory, it is possible to formulate a single-body problem self-consistently.

2. The spatial particle density $\rho(x)$ (which is to be interpreted statistically) is a meaningful concept. After integration over an arbitrary finite volume, one obtains from it an 'observable' (in the sense of transformation theory) with the eigenvalues 0 and +1.

As soon as the first presupposition applies, it is indeed unnecessary to apply to the problem the formalism of the quantization of wave fields; rather, it is then possible to make do with the ordinary wave field in three-dimensional space. The consequence of the second presupposition is that the *particle density* not only must be the fourth component of a four-vector and must satisfy a continuity equation, but that it must also have the property of never being negative. Furthermore, the *eigenvalues* of the associated *density matrix*, after integration over an infinite volume, become the correct ones, as Dirac had shown, only if the *particle density* has the form:[3]

[3] Only from this form for ρ does one further conclude that the wave equations must be of first order in $\partial/\partial t$.

$$\rho(x) = \sum_r \psi^*_r \psi_r.$$

[Dirac (March, 1934). Discussion of the infinite distribution of electrons in the theory of the positron: "To get a relativistic theory, we must introduce two times, t' and t", and use instead of ρ the relativistic density matrix R defined by

$$(x'\, t' \mid R \mid x''\, t'')_{k'k''} = \sum_a \psi_{ak'}(x'\, t')\, \psi^*_{ak''}(x''\, t'') = \sum_{oc} \psi_{k'}(x'\, t')\, \psi^*_{k''}(x''\, t''). \quad (7)."]$$

On the other hand, the *particle density* which is associated with the *scalar relativistic wave equation* has the form

$$\rho(x) = \psi^* (ih\, \partial\psi/\partial t - e\Phi_0\psi) - (ih\, \partial\psi^*/\partial t + e\Phi_0\psi^*)\, \psi \qquad (3)$$

where e signifies the charge of the particle and Φ_0 is the *external scalar potential. Since this does not have the required form, a contradiction seems to be established.*

As is well known, Dirac - supported by the fact that, on the basis of his wave equation, a wave packet consisting of negative energy states moves in an external field as would correspond to a particle with opposite charge, the same mass, and positive energy - has drawn upon the states of negative energy to interpret the positron in the following fashion. Supposedly only the deviations from the case where all states of negative energy are occupied, the 'holes' in the occupation of the negative energy states, are observable, that is contribute to the 'true' (field producing) charge density and to the actual (then positive) energy.

The difficulties of a self-consistent and relativistically and gauge invariant formulation of this Dirac hole theory of electrons and positrons in the case of external fields have indeed been discussed several times in the literature. Without dealing with these difficulties in more detail, we can make the following observations:

1. Because of the processes of pair creation, and because of the new interpretation of the negative energy states altogether, it is no longer possible to limit oneself to a single-body problem.

2. The *particle density* no longer has a direct physical meaning[4].

[4] If ψ_ρ^+ is the 'positive' (consisting of states of positive energy), and ψ_ρ^- is the 'negative' part of the wave function in the Dirac hole theory, the *charge density* as an operator has the form $\rho(x) = \sum_{\rho=1}^{1} (\psi_\rho^{+*} \psi_\rho^{+} - \psi_\rho^{-*} \psi_\rho^{-} + \psi_\rho^{+*} \psi_\rho^{-*})$. Because of the appearance of the mixed terms, this expression already cannot be divided into two parts in the absence of external forces, such that each part by itself satisfies a continuity equation and forms the four-component of a four-vector.

However, in the force-free case, the number of particles with a given *momentum* (*probability density* in *momentum space*), and consequently also a total number of particles present, are meaningful 'observables'.

3. On the other hand, not only the *total charge* but also the *charge density* p(x) is a meaningful observable. After integration over an arbitrary finite volume, it must have the eigenvalues 0, ±1, ±2, ... , ±N, ... - even when external fields are present - (by applying the formalism of the quantization of waves). These *eigenvalues* now can be positive as well as negative. (In the case of the *exclusion principle*, the number N has an upper limit for a given size of the space region under consideration.) Besides, the *charge density* p(x) and the total number of particles present are not commutable.

These requirements have now been modified with respect to the original requirements for a true *relativistic* single-body problem to such an extent that no reason any longer exists for the special form $\sum_r \psi^*_r \psi_r$ for the *charge density*. Furthermore, *we shall show that the preceding conditions are fulfilled just as well in the scalar relativistic theory for spinless particles with Bose-Einstein statistics as in the Dirac hole theory*. Here, Expression (3)

$$[\rho(x) = \psi^* (ih\, \partial\psi/\partial t - e\Phi_0\psi) - (ih\, \partial\psi^*/\partial t + e\Phi_0\psi^*)\, \psi \qquad (3)]$$

is obviously no longer to be interpreted as a *particle density* but as a *charge density*.

The main interest of the latter theory appears to us to lie in the fact that, in this theory, the energy of the material particles is automatically always positive after the wave fields have been quantized. This is automatic in the sense that no new hypothesis equivalent to the hole idea is required and there are no concepts of passing to the limit and of subtraction, concepts which are foreign to the formalism of quantum theory[5].

[5] Dirac, P. A. M. (March, 1934). Discussion of the infinite distribution of electrons in the theory of the positron. *Proc. Camb. Phil. Soc*., 30, 2, 150-63[; attempt by Dirac to address problem with his *relativistic* 'hole' theory which implies an infinite number of negative-energy electrons (per unit volume) with energies extending continuously from $-mc^2$ to $-\infty$, when an electromagnetic field is present positive- and negative-energy states cannot be distinguished in *relativistically* invariant way, need to set up assumptions for production of *electromagnetic field* by the electron distribution such that any finite change in distribution produces a change in the field in agreement with Maxwell's equations and such

that the infinite field which would be required by Maxwell's equations from an infinite density of electrons is in some way cut out, *assumes each electron has its own individual wave function in space-time* and each electron moves in an *electromagnetic field* which is the same for all electrons part coming from external causes and part from the electron distribution itself, introduces *relativistic density matrix* $\sum_{oc} \psi_{k'}(x' t') \psi^*_{k''}(x'' t'')$ referring to two points in space and two times, separates density distribution into two parts where one contains the singularities, and the other describes the *electric* and *current densities* physically present.]; Peierls, R. (September, 1934). The vacuum in Dirac's theory of the positive electron. *Roy. Soc. Proc., A*, 146, 857, 420-41; http://doi.org/10.1098/rspa.1934.0164; Heisenberg, W. (March, 1934). Bemerkungen zur Diracschen Theorie des Positrons. (Remarks on the Dirac theory of positron.) *Zeit. Phys*., 90, 3-4, 209-31[; Heisenberg's reconstruction of Dirac's *theory of the positron* in the formalism of quantum electrodynamics, demands that symmetry between positive and negative charge should be expressed in the basic equations from the outset, in addition to the well-known difficulties with the divergences no new infinities should appear in the formalism, theory should provide an approximation for the treatment of problems that have been treated by quantum electrodynamics up to now, Dirac [(1934). Discussion of the infinite distribution of electrons in the theory of the positron.] showed that a quantum mechanical system of many electrons that fulfill the Pauli principle and move in a given force field without back-reaction can be characterized by a *density matrix*
$(x', t', k' \mid R \mid x'', t'', k'') = \sum n\ \psi^*n\ (x', t', k')\ \psi_n\ (x'', t'', k'')$ when $\psi_n(x', t', k')$ means the normalized eigenfunctions of the states that possess one electron, and x′, t′, k′ (x″, t″, k″, resp.) are position, time, and spin variables, all physically-important properties of quantum-mechanical systems like *charge density, current density*, etc., can be read off from the *density matrix*, the temporal change in the density matrix is determined by the Dirac differential equation, Dirac made different choice of *density matrix* representing external field resulting in a different energy and impulse density, by restricting oneself to an *intuitive analogue theory of matter fields the negative energy levels in the Dirac theory could be avoided by replacing the homogeneous Dirac differential equation with an inhomogeneous equation* where the inhomogeneity is indicative of pair creation, for most practical applications e.g. pair creation, annihilation, Compton scattering, etc. *the theory described here does not yield anything new compared to the formulation of the Dirac theory*, in the Maxwell theory a continuous charge distribution also leads to a finite self-energy; it is the "quantization" that leads to the infinite self-energy, *if one represents the quantization of the electromagnetic field by point-like light quanta then the infinitude of the self-energy also emerges in the intuitive theory of matter waves*].

The positive value of the energy results from the fact that the Hamiltonian function of the matter-wave field, in the *scalar theory* discussed here, in contrast to the corresponding expression of the Dirac spinor theory, *always assumes the positive definite form*:

$$H = \int dV\ \{\mid ih\ \partial\psi/\partial t - e\Phi_0\psi \mid^2 + \sum_{k=1}{}^3 \mid ihc\ \partial\psi/\partial x^k - e\Phi_k\psi \mid^2 + m^2c^4 \mid \psi \mid^2\}. \quad (4)$$

This *scalar relativistic theory* is free of hypotheses. In the view of this, one might perhaps be surprised at first sight why 'nature has not availed itself' of this possibility of the existence of oppositely charged particles without spin and with Bose statistics, which can arise and vanish by radiation-creating or matter-creating processes[6].

[6] Compare Dirac, P. A. M. (September, 1931). Quantized singularities in the electromagnetic field. *Roy. Soc. Proc., A*, 133, 821, 60-72, especially p. 71 "Under these circumstances one would be surprised if Nature had made no use of it"[; the object of the paper is to show that quantum mechanics does not preclude the existence of *isolated magnetic poles*, addresses *smallest electric charge* e known experimentally to be given by $hc/e^2 = 137$, considers particle whose motion is represented by a wave function, uses *non-relativistic* theory, shows *change in phase* round a closed curve must be same for all *wave functions*, applies to motion of an electron in an electromagnetic field, shows non-integrable derivatives of phase of the wave function represent potentials of the electromagnetic field, connection between *non-integrability of phase* and *electromagnetic field* essentially Weyl's *principle of gauge invariance*, leads to wave equations whose only physical interpretation is in the motion of an electron in the field of a single pole, does not give value for e but shows reciprocity between *electricity* and *magnetism*, strength of pole and electric charge must both be quantized, gives relationship between the strength of quantum of magnetic pole and electronic charge $hc/e\mu_0 = 2$ but *does not explain their magnitudes*, reason that isolated magnetic poles have not been separated probably due to the very large force between two one-quantum poles of opposite sign, $(137/2)^2$ times that of that between electron and proton].

But one must consider that the question of the applicability of the theory of nuclear structure discussed here, for example to α particles, probably lies altogether outside the range of validity of present quantum theory owing to the relevant effects. Also, as will be shown in §4, *the theory discussed here leads to similar infinities in connection with questions about the polarization of the vacuum, as did the original form of the hole theory*[7].

[7] P. A. M. Dirac, *Solvay Conference Report* 1933.

It incidentally also leads to an infinite self-energy, not only of the electric particles but also to an infinite material self-energy of the light quanta[8].

[8] Analogous with W. Heisenberg, *loc. cit.* - One may have doubts concerning the value of the circumstance that, in some formulations of the hole theory, the polarization effects are indeed finite but the self-energies are nevertheless infinite.

Consequently, further progress in answering these questions can probably only be expected by a theoretical understanding of the numerical value of the Sommerfeld fine structure constant.

§2 *Quantization of the wave field*[9] *in the force-free case.*

[9] As regards the expressions for the Lagrange function, energy-momentum tensor, and current vector in the *scalar relativistic theory*, compare e.g. Gordon, W. (January, 1927). Der Comptoneffekt nach der Schrödingerschen Theorie. (The Compton effect according to Schrödinger's theory.) *Zeit. Phys*., 40, 117-33[; Heisenberg and Schrödinger provided alternative methods for determination of quantum *frequencies* and *intensities*, Compton effect already calculated by Dirac (June, 1926) using Heisenberg method, here the same problem treated by Schrödinger method, starts with the same *classic relativistic equation for kinetic energy* in terms of *momentum* and *energy*, which is *Hamiltonian equation* for the system, introduces same imaginary variables for *time* and *energy* to create same space-

time symmetric form, applies in same way to *electron in electromagnetic field* described in terms of *vector potential* and *scalar potential*, and introduces same imaginary variable for scalar potential, adds the same *field energy* to the *kinetic energy* resulting in the same *classical relativistic Hamiltonian equations for a point electron moving in an electromagnetic field*, in accordance with Schrödinger's rules Gordon then substitutes the classical *quantum differential operators* for the momentum vector in the amended *Hamiltonian equation* and applies resulting differential operator to the *wave function* ψ to obtain the *Klein-Gordon equation*, $1/c^2\, \partial^2/\partial t^2\, \psi - \nabla^2 \psi + m^2c^2/h^2\, \psi = 0$, (Dirac [(February, 1928). The Quantum Theory of the Electron.] objected to this substitution on grounds of the interpretation of the wave function, and solutions with negative probabilities, negative energy, and positive charge for the electron); calculates radiation from *current density* and *charge density*, applies to Compton effect].

The *Lagrange function* of the *scalar relativistic theory* reads as follows (with ρ, ν, … –1 to 4 and $x_4 = ict$):

$$L = - h^2c^2 \sum_{\nu=1}^{4} \partial\psi^*/\partial x_\nu\, \partial\psi/\partial x_\nu - m^2c^4\, \psi^*\psi \qquad (5)$$
$$= h^2 \partial\psi^*/\partial t\, \partial\psi/\partial t - h^2c^2 \sum_{k=1}^{3} \partial\psi^*/\partial x^k\, \partial\psi/\partial x^k - m^2c^4\, \psi^*\psi\,.$$

The *relativistic energy-momentum tensor* becomes

$$T_{\mu\nu} = - h^2c^2\, (\partial\psi^*/\partial x^\mu\, \partial\psi/\partial x^\nu + \partial\psi^*/\partial x^\nu\, \partial\psi/\partial x^\mu) - L\delta_{\mu\nu} \qquad (6)$$

and thus, the *energy* (Hamiltonian function)

$$H^- = \int T_{44}\, dV$$
$$= \int \{h^2\partial\psi^*/\partial t\, \partial\psi/\partial t + h^2c^2 \sum_{k=1}^{3} \partial\psi^*/\partial x^k\, \partial\psi/\partial x^k$$
$$+ m^2c^4\, \psi^*\psi\}dV \qquad (7)$$

and the *momentum*

$$G_k = i/c \int T_{4k}\, dV = - \int h^2(\partial\psi^*/\partial t\, \partial\psi/\partial x^k + \partial\psi^*/\partial x^k\, \partial\psi/\partial t)dV. \qquad (8)$$

We must now regard ψ* and ψ as q-numbers (operators acting on the Schrodinger functional), where ψ* is the Hermitian conjugate to ψ. We shall always designate the Hermitian conjugate to a given q-number by means of a superscript*. We then must form the momentum π and π* which are canonically conjugate to ψ* and ψ, in accordance with the rule …

…

As already mentioned in the introduction, *it is not possible to carry through the scalar relativistic wave theory consistently for particles obeying the exclusion principle.* As a more detailed investigation of the Hamiltonian function with the variables *a* and *b* shows, the reason for this is as follows: *When Fermi statistics are valid, the relativistic invariance of the four-current cannot be achieved.*

§3 *The case where external forces are present.*

For a particle with charge e, one goes from the force-free case to the case of the presence of an external electromagnetic field with the four-potential Φ_μ ($\Phi_4 = i\,\Phi_0$), by replacing the operator p_μ by[12]

$$p_\mu \rightarrow p_\mu - e/c\ \Phi_\mu, \qquad (36)$$

[12] In the Dirac theory, the substitution $p_\mu \rightarrow p_\mu + e/c\ \Phi_\mu$ is introduced, since the electron charge is there designated by (– e). Our designations agree with W. Gordon, *loc. cit.*

…

§4 *Pair creation by light quanta and the polarization of the vacuum.*

…

One can then easily see that the additions $H^{-}1$ and $H^{-}2$, which appear in the Hamiltonian function *when external fields are present*, contain terms due to the factors $a^*_k\ b^*_l$ and $b^*_k\ a^*_l$, *which cause pair creation and pair annihilation. These terms do indeed lead to matrix elements between the states which differ precisely by one positive and one negative particle, while the factors* $a^*_k\ a_l$ *and respectively* $b_k\ b^*_l$ *yield only transitions of one positive or respectively one negative charge from one state to another.*

Below we shall now calculate the probability of pair creation through a light quantum of energy $h\nu > 2mc^2$, on the basis of expressions (52) and (53). We shall compare this probability with the corresponding expressions of the hole theory, as calculated by Bethe and Heitler[14].

[14] Bethe, H. & Heitler, W. (August, 1934) On the Stopping of Fast Particles and on the Creation of Positive Electrons. *Roy. Soc. Proc., A*, 146, 83-112; https://doi.org/10.1098/rspa.1934.0140.

Due to the law of conservation of energy and momentum, this probability vanishes in field-free space. We therefore assume that an electric field exists in space, a field which can be represented by a time-independent scalar potential Φ_0 (for example the Coulomb field of a nucleus), which can take up the momentum excess.

We consider the influence of the field only in first approximation, just like Bethe and Heitler, by starting from field-free space and by considering Φ_0 as well as the potential of the light wave as a perturbation.

…

The corresponding expression in hole theory, according to Bethe and Heitler, reads as follows:

… .

This differs from expression (56), which was obtained from the scalar wave equation, only by the third term in curly brackets and in the terms with q^2 that appear there. However, at high energies the latter are negligible, since $hc|\vec{q}| \ll h\nu$ for $h\nu \gg mc^2$.

…

Finally, the polarization of the vacuum through an electrostatic field will be calculated

…

The *induced charge density* has the opposite sign to the *external density* $\rho_0 = -(1/4\pi)\Delta\Phi_0$ and is proportional to it, with the diverging proportionality factor $4\pi K$. This would have as its consequence that every external charge would be completely compensated for by the induced charge. *This result agrees completely with the result calculated by Dirac*[16] *on the basis of his hole theory.*

[16] P. A. M. Dirac, *Solvay Conference Report* 1933.

Even the factor K of the diverging term is the same.

Ernst Carl Gerlach Stueckelberg (February 1, 1905 – September 4, 1984)

Stueckelberg (baptised as Johann Melchior Ernst Karl Gerlach Stückelberg, full name after 1911: Baron Ernst Carl Gerlach Stueckelberg von Breidenbach zu Breidenstein und Melsbach) was a Swiss mathematician and physicist, regarded as one of the most eminent physicists of the 20th century. Despite making key advances in theoretical physics, including the exchange particle model of fundamental forces, causal S-matrix theory, and the renormalization group, his idiosyncratic style and publication in minor journals led to his work not being widely recognized until the mid-1990s.

Born into a semi-aristocratic family in Basel in 1905, Stueckelberg's father Alfred was a lawyer, and his paternal grandfather a distinguished Swiss artist Ernst Stückelberg. A highly gifted school student, Stueckelberg initially began a physics degree at the University of Basel in 1923.

While still a student, Stueckelberg was invited by the distinguished quantum theorist Arnold Sommerfeld, to attend his lectures at the University of Munich. He went on to gain a Ph.D. on cathode physics in 1927. Later that year he went to Princeton University, becoming an assistant professor in 1930. Philip Morse met Stueckelberg at this time.

The Michigan summer physics programme was held every year and, in 1928, Kramers lectured on quantum theory and Ehrenfest on statistical physics. Also lecturing were Sam Goudsmit and George Uhlenbeck. Stueckelberg arrived halfway through the summer and from then on Morse, Kramers and Stueckelberg would go on outings and spend time together drinking beer.

In 1930 Stueckelberg was appointed as a research associate and assistant professor at Princeton. He gave a paper at the New York meeting of the Physical Society held at Columbia University then in the autumn of 1930 then, along with Morse who was collaborating on various research topics, he went to spend time at Sommerfeld's Institute in Munich. There Stueckelberg and Morse worked together on high energy collisions. In the spring Stueckelberg, together with Morse, moved to England to spend time at Cambridge.

However, these were difficult times with signs of the Great Depression evident in Europe during his visit. By the time he returned to Princeton economic conditions were deteriorating and his employment at Princeton came to an end in 1932 as funding became increasingly tight.

He returned to Switzerland in 1932, working first at the University of Basel before switching the following year to the University of Zurich. Gregor Wentzel held the Chair for Theoretical Physics at the University of Zürich at this time and Wolfgang Pauli was professor at the Eidgenössische Technische Hochschule Zürich, so Zürich had a high reputation for theoretical physics. In 1933 Stueckelberg submitted his habilitation thesis to the University of Zürich and gained the right to become a lecturer. Stueckelberg's sojourn in Zurich led him to focus on the emerging theory of elementary particles.

In the winter of 1934, he was called by the University of Geneva to substitute for the deceased Arthur Schidlof, who held the chair of theoretical physics. Stueckelberg was appointed to succeed him in the following year. The University of Geneva, together with the University of Lausanne, became his principal bases for the rest of his career.

In September 1934 Stueckelberg submitted his paper on high energy collision phenomena between electrons and nuclei, to *Annalen der Physik.* [Stueckelberg, E. C. G. (September, 1934). Relativistisch invariante Störungstheorie des Diracschen Elektrons I. Teil: Streustrahlung und Bremsstrahlung. (Relativistically invariant perturbation theory of Dirac's electron Part I: scattered radiation and Bremsstrahlung.) *Ann. Phys*., 413, 4, 367-89]. In this paper he devised a fully Lorentz-covariant perturbation theory for quantum fields. The approach proposed by Stueckelberg was very powerful, but was not adopted by others at the time, and has now been all but forgotten. This paper introduced *Stueckelberg diagrams*, which were later renamed *Feynman diagrams* after he reintroduced them at the Pocono Manor Inn during the spring of 1948, and subsequently published examples in 1949. In his report to the Solvay Conference 1948, J. R. Oppenheimer insisted on preserving covariance in all steps of the calculation if one wants to eliminate the infinities which otherwise occur. He then quoted Stueckelberg's 1934 paper as giving an example of such a covariant theory. [Oppenheimer, J. R. (1948). Electron Theory. Report to the Solvay Conference for Physics at Brussels, Belgium, September 27 to October 2, 1948, pages 6-7. (Reprint in Schwinger, J. (ed). (1958). *Selected Papers on Quantum electrodynamics*, Dover, New York, pages 150-1]. Stueckelberg's paper was also referred to in Dyson, F. J. (February, 1949). The Radiation Theories of Tomonaga, Schwinger, and Feynman.

Pauli had also noticed the 1934 paper. After the publication of Stueckelberg's 1934 paper, Pauli pointed out to Stueckelberg that he first averaged over the photon polarization in the rest system of the electron, and only then Lorentz-transformed it back to the moving frame. Stueckelberg replied to Pauli's objection by remarking that, since unpolarized light is a Lorentz-invariant notion, the averaging is a Lorentz-invariant operation and can therefore be performed in any reference system [Stueckelberg, E. C. G. (1935). Bemerkungen zur Intensitat der Streustrahlung bewegter freier Elektronen. (Remark on the intensity of the radiative scattering of moving free electrons.) *Helv. Phys. Acta*, 8, 197-204].

Pauli wrote to Heisenberg about this paper on February 5, 1937 (over two years after its publication):

> "Concerning the formalization of scattering theory, I wish to draw your attention to a paper of Stueckelberg (1934). This paper is not written very well, but the basic idea (which goes back to Wentzel) seems to me reasonable; it consists of establishing relativistic invariance by the fact that one removes space and time totally from the theory, and directly examines the coefficients of the four-dimensional Fourier expansion of the wave function."

In 1935, Stueckelberg developed the vector boson exchange force model as the theoretical explanation of the strong nuclear force. Discussions with Pauli led Stueckelberg to drop the idea, however. It was rediscovered by Hideki Yukawa, who won a Nobel Prize for his work in 1949, the first of several Nobel Prizes awarded for work which Stueckelberg contributed to, without recognition.

In 1936 he published [Stueckelberg, E. C. G. (1936). Invariante Storungstheorie des Elektron-Neutrino Teilchens unter dem Einfluss von elektromagnetischem Feld und Kernkraftfeld (Feldtheorie der Materie II). (Invariant perturbation theory of the electron-neutrino particle under the influence of electromagnetic field and nuclear force field (Field theory of matter II).) *Helv. Phys. Acta*, 9, 533-54]; and in 1938 he wrote a paper which recognized that massive electrodynamics contains a hidden scalar, and formulated an affine version of what would become known as the Abelian Higgs mechanism [Stueckelberg, E. C. G. (1938). Die Wechselwirkungskrafte in der Elektrodynamik und in der Feldtheorie der Kernkrafte (Teil I, II, III). (The Interaction Forces in Electrodynamics and in the Field Theory of Nuclear Forces (Parts I, II, III).) *Helv. Phys. Acta*, 225-44, 299-328]. This was referenced in Tomonaga, S. (1943). On a Relativistically Invariant Formulation of the Quantum Theory of Wave Fields.

In 1941 he proposed the interpretation of the positron as a positive energy electron traveling backward in time. [Stueckelberg, E. C. G. (1941). The Interpretation of the Positron as a positive Energy Electron traveling backward in time. *Helv. Phys. Acta*, 14, 51-80; (1944). *Idem*, 17, 3; (1945). *Idem*, 18, 195; (1946). *Idem*, 19, 242.] This was referenced together with a series of subsequent papers in Dyson, F. J. (June, 1949). The S Matrix in Quantum Electrodynamics:

> "The idea of using standard electrodynamics as a starting point for an explicit calculation of the S matrix has been previously developed by Stueckelberg, E. C. G. (1941). The Interpretation of the Positron as a positive Energy Electron traveling backward in time. *Helv. Phys. Acta*, 14, 51-80; (1944). *Idem*, 17, 3; (1945). *Idem*, 18, 195; (1946). *Idem*, 19, 242; (January, 1944). An Unambiguous Method of Avoiding Divergence Difficulties in Quantum Theory. *Nature*, 153, 143-4; https://doi.org/10.1038/153143a0; (1947). *Phys. Soc. Cambridge Conference Report*, 199; Rivier, D. & Stueckelberg, E. C. G. (July, 1948). A Convergent Expression for the Magnetic Moment of the Neutron. *Phys. Rev.*, 74, 218; https://doi.org/10.1103/ PhysRev.74.218".

Feynman was also aware of Stueckelberg's papers. Stueckelberg, E. C. G. (December, 1942). La mécanique du point matériel en théorie de relativité et en théorie des quanta. (Mechanics of the material point in relativity theory and quantum theory.) *Helv. Phys.,*15, 23-37, was referred to in Feynman, R. P. (September, 1949) (The Theory of Positrons. *Phys. Rev*., 76, 749-59):

> "In this solution, the "negative energy states" appear in a form which may be pictured (*as by Stückelberg*) in space-time as waves traveling away from the external potential backwards in time. Experimentally, such a wave corresponds to a positron approaching the potential and annihilating the electron. A particle moving forward in time (electron) in a potential may be scattered forward in time (ordinary scattering) or backward (pair annihilation). When moving backward

(positron) it may be scattered backward in time (positron scattering) or forward (pair production). For such a particle the amplitude for transition from an initial to a final state is analyzed to any order in the potential by considering it to undergo a sequence of such scatterings. …

> [7] The idea that positrons can be represented as electrons with proper time reversed relative to true time has been discussed by the author and others, particularly by Stueckelberg (1942). [Stueckelberg, E. C. C. (December, 1942). La mécanique du point matériel en théorie de relativité et en théorie des quanta. (Mechanics of the material point in relativity theory and quantum theory.) *Helv. Phys.,*15, 23-37."

In 1943 Stueckelberg wrote a long paper outlining a complete and correct description of the renormalization procedure for quantum electrodynamics to address the problems of infinities in quantum electrodynamics, but his paper was rejected by the *Physical Review*. As Stueckelberg later recalled: "They said it was not a paper, it was a program, an outline, a proposal ...". He then set about filling in all the details but Schwinger and Feynman published their version first and Stueckelberg received no recognition for his remarkable contributions.

Stueckelberg, E. C. G. (January, 1944). An Unambiguous Method of Avoiding Divergence Difficulties in Quantum Theory. *Nature*, 153, 143-4; https://doi.org/10.1038/ 153143a0, was also referenced in Dyson, F. J. (February, 1949). The Radiation Theories of Tomonaga, Schwinger, and Feynman:

"III. *Introduction of perturbation theory*

Notes added in proof

A covariant perturbation theory similar to that of Section III has previously been developed by Stueckelberg, E. C. G. (September, 1934). Relativistisch invariante Störungstheorie des Diracschen Elektrons. Teil I: Streustrahlung und Bremsstrahlung. (Relativistically invariant perturbation theory of Dirac's electron. Part I: Scattered radiation and Bremsstrahlung.) *Ann. Phys*., 413, 4, 21, 367-89; (January, 1944). An Unambiguous Method of Avoiding Divergence Difficulties in Quantum Theory. *Nature*, 153, 143-4; https://doi.org/10.1038/153143a0; "The classical theory of a point charge and the quantum theory of wave packets contain in their usual form well-known divergences. Dirac (August, 1938) [Classical Theory of Radiating Electrons], has elaborated a classical particle theory which avoids these difficulties, *but his results cannot as yet be applied to quantum theory*."

In 1965 Sin-Itiro Tomonaga, Julian Schwinger and Richard Feynman were jointly awarded the Nobel Prize for Physics; Feynman for "in 1948 … Richard Feynman contributed to creating a new quantum electrodynamics by introducing *Feynman [Stueckelberg] diagrams*: graphic representations of various interactions between different particles. These diagrams facilitate the calculation of interaction probabilities". According to an account by Philip Morse who met Stueckelberg at Princeton, "After receiving the Nobel Prize, Feynman lectured at CERN to an audience which included Stueckelberg. After the lecture, Stueckelberg was making his way out alone ... from the CERN amphitheater, when

Feynman - surrounded by admirers - made the remark: "He [Stueckelberg] did the work and walks alone toward the sunset; and, here I am, covered in all the glory, which rightfully should be his!" Morse also reported that "At the 1948 Solvay Conference, Oppenheimer insisted on preserving covariance in all steps of the calculation if one wants to eliminate the infinities which otherwise occur. He then quoted Stueckelberg's 1934 paper as giving an example of such a covariant theory". [O'Connor, J. J. & Robertson, E. F. (July, 2008). *Ernst Carl Gerlach Stueckelberg*. MacTutor, School of Mathematics and Statistics, University of St Andrews, Scotland.]

In 1951 Stueckelberg and Andre Petermann invented the *renormalization group*. In 1982 Kenneth G Wilson was awarded the Nobel Prize for Physics "for his theory for critical phenomena in connection with phase transitions". The Press Release for this award stated that:

> "Wilson built his theory on an essential modification of a method in theoretical physics called *renormalization group theory*, which was developed already during the fifties and was applied with varying success to different problems."

In his Nobel Lecture, Wilson said:

> "Stueckelberg and Petermann observed that transformation groups could be defined which relate different reparametrizations - they called these groups "groupes de normalization" which is translated "*renormalization group*".

Stueckelberg continued to hold the chair at Geneva until he retired in 1975. He was also appointed professor at the University of Lausanne in 1956, again holding this position until he retired. Although Stueckelberg never received the recognition for his achievement which many felt that he so clearly deserved, he did receive a number of significant honors. In 1976 he was awarded the Max Planck medal. In December 2005 an international symposium celebrating the centenary of his birth was held at Geneva University.

He died in 1984. He is buried in the Cimetière de Plainpalais (Cemetery of Kings), which is considered the Genevan Panthéon (where the reformer John Calvin is also buried).

Stueckelberg, E. C. G. (September, 1934). Relativistisch invariante Störungstheorie des Diracschen Elektrons I. Teil: Streustrahlung und Bremsstrahlung. (Relativistically invariant perturbation theory of Dirac's electron Part I: scattered radiation and Bremsstrahlung.)

Ann. Phys., 413, 4, 367-389; https://doi.org/10.1002/andp.19344130403; also at https://archive-ouverte.unige.ch/unige:161855; translated by T. G. Underwood.

Zurich, Physics. Institute of the University.

Submitted August 31, 1934. Received September 10, 1934.

The main innovation of Stueckelberg's paper is the introduction of a new perturbative scheme yielding relativistic expressions for the *matrix elements* which are manifestly *gauge invariant,* this is achieved by performing a four-dimensional Fourier transformation of the *wave-function* eliminating space and time variables based on the interaction picture of Dirac, Fock and Podolsky, the starting point is the Dirac equation for the *spinor wave-function* $[1/i\ (\gamma,\ \partial/\partial x) + M + eV(x)]\ \Psi(x) = 0$, the contributions of positive and negative energies (corresponding to virtual electrons and positrons) are contained in a single propagation function which corresponds to what later was called the *Feynman propagator*, his method for calculating the cross-section starts with the definition of *on mass-shell wave-functions* and uses integration over the *complex energy-plane*, this paper introduces *Stueckelberg diagrams* later adopted by Feynman and subsequently renamed *Feynman diagrams*, all of Stueckelberg's expressions for matrix elements are identical to those obtained nowadays from Feynman diagrams.

> [Oppenheimer, J. R. (1948). Electron Theory. Report to the Solvay Conference for Physics at Brussels, Belgium, September 27 to October 2, 1948, pages 6-7. (Reprint in Schwinger, J. (ed). (1958). *Selected Papers on Quantum electrodynamics*, Dover, New York, pages 150-1: "Now it is true that the fundamental equations of quantum-electrodynamics are gauge and Lorentz covariant. But they have in a strict sense no solutions expansible in powers of e. If one wishes to explore these solutions, *bearing in mind that certain infinite terms will, in a later theory, no longer be infinite*, one needs a covariant way of identifying these terms, and for that, not merely the field equations themselves, but the whole method of approximation and solution must at all stages preserve covariance. This means that the familiar Hamiltonian methods, which imply a fixed Lorentz frame t = constant, must be renounced; neither Lorentz frame nor gauge can be specified until after, in a given order in e, all terms have been identified, and those bearing on the definition of charge and mass recognized and relegated; then of course, in the actual calculation of transition probabilities and the reactive corrections to them, or in the determination of stationary states in fields which can be treated as static, and in the reactive corrections thereto, the introduction of a definite coordinate system and gauge for these no longer singular and completely well-defined terms can lead to

no difficulty.

It is probable that, at least to order e^2, more than one covariant formalism can be developed. Thus, *Stueckelberg's four-dimensional perturbation theory*[26] would seem to offer a suitable starting point, as also do the related algorithms of Feynman[27].

[26] Stueckelberg, E. C. G. (September, 1934). Relativistisch invariante Störungstheorie des Diracschen Elektrons I. Teil: Streustrahlung und Bremsstrahlung. (Relativistically invariant perturbation theory of Dirac's electron Part I: scattered radiation and Bremsstrahlung.) *Ann. Phys*., 413, 4, 367-89

[27] Feynman, R. P. (November, 1948). Relativistic Cut-Off for Quantum Electrodynamics. *Phys. Rev*., 74, 1430-8.

But a method originally suggested by Tomonaga[28], and independently developed and applied by Schwinger[22], would seem, apart from its practicality, to have the advantage of very great generality and a complete conceptual consistency.

[28] Tomonaga, S. (1943). On a Relativistically Invariant Formulation of the Quantum Theory of Wave Fields. *Bull. I. P. C. R. Riken-iho*, 22, 545 (in Japanese); translation Tomonaga, S. (August, 1946). *Prog. Theor. Phys*. 1, 2, 27-42; Koba, Z., Tati, T. & Tomonaga, S. (1947a). On a Relativistically Invariant Formulation of the Quantum Theory of Wave Fields. II. Case of Interacting Electromagnetic and Electron Fields. § 1-4. *Prog. Theor. Phys*., 2, 101–116; Koba, Z., Tati, T. & Tomonaga, S. (1947b). On a Relativistically Invariant Formulation of the Quantum Theory of Wave Fields. III. Case of Interacting Electromagnetic and Electron Fields. § 5-7. *Prog. Theor. Phys*., 2, 198–208.

[22] Schwinger, J. (November, 1948). Quantum Electrodynamics. I. A Covariant Formulation. *Phys. Rev*., 74, 10, 1439-61, and in press.

It has also been shown by Dyson[29] how Feynman's algorithms can be derived from the Tomonaga equations.

[29] Dyson, F. J., *Phys. Rev*., in press."]

[Lacki, J. Ruegg H. & Telegdi, V. L. (March 15, 1999). *The Road to Stueckelberg's Covariant Perturbation Theory as Illustrated by Successive Treatments of Compton Scattering*; https://doi.org/10.48550/arXiv.physics/ 9903023: "As we show in the present paper … *Stueckelberg in 1934 used the interaction picture of Dirac, Fock and Podolsky* and again in later publications [Stueckelberg, E. C. G. (1936). Invariante Storungstheorie des Elektron-Neutrino Teilchens unter dem Einfluss von elektromagnetischem Feld und Kernkraftfeld (Feldtheorie der Materie II). (Invariant perturbation theory of the electron-neutrino particle under the influence of electromagnetic field and nuclear force field (Field theory of matter II).) *Helv. Phys. Acta*, 9, 533-54; Stueckelberg, E. C. G. (1938). Die Wechselwirkungskrafte in der Elektrodynamik und in der Feldtheorie der Kernkrafte (Teil I, II, III). (The Interaction Forces in Electrodynamics and in the Field Theory of Nuclear Forces (Parts I, II, III).) *Helv. Phys. Acta*, 225-44, 299-328)]. He discussed the *multi-time formalism* in 1935 and 1938. *The main innovation of the 1934 paper is the introduction of a new perturbative scheme yielding manifestly relativistic expressions for the matrix elements.* This is achieved by performing a four-dimensional Fourier transformation of the wave-function, thus eliminating space and time variables. ... Stueckelberg's procedure is thus the first departure from the "older (Dirac) form of the perturbation theory". The approach proposed by Stueckelberg is far more powerful, but was not adopted by others at the time. This lack of interest appears retrospectively as very unfortunate, as was recognized by V. Weisskopf [The Development of Field Theory in the last Fifty Years, *Physics Today*, Nov. 1981, pp. 69-85], ..., in his recollections of that period. Weisskopf reminds us of the difficulties encountered in the higher order corrections to Q.E.D., and remarks [p. 78]: "Already in 1934 [...] it seemed that a systematic theory could be developed in which these infinities [divergent radiative corrections] are circumvented. At that time nobody attempted to formulate such a theory [...]. There was one tragic exception [...], and that was Ernst C. G. Stueckelberg. He wrote several important papers in 1934-38 putting forward a manifestly invariant formulation of field theory. This could have been a perfect basis for developing the ideas of *renormalization*. Later on, he actually carried out a complete renormalization procedure in papers with D. Rivier, independently of the efforts of other authors. Unfortunately, his writings and his talks were rather obscure, and it was very difficult to understand them or to make use of his methods ...

In Stueckelberg's 1934 paper, thanks to the use of the four-dimensional Fourier transform, manifest covariance is kept throughout. The factor $(E - E')^{-1}$ is now replaced by the covariant expression $(p^2 + M^2)^{-1}$ where *p* is the *four-momentum* of the intermediate state, $M = mc/\hbar$, and m is the *electron mass*, corresponding to a virtual particle off mass-shell $(p^2 + M^2 \neq 0)$. *Energy* and *three-momentum* are now conserved at the vertices. The contributions of positive and negative energies (corresponding to virtual electrons and positrons) are contained in the single propagation function $(p^2 + M^2)^{-1}$, which corresponds to what later was called *Feynman propagator*, and by Rivier-Stueckelberg 1948, causal function. *All of Stueckelberg's expressions for matrix elements are identical to those obtained nowadays from Feynman diagrams.*

Let us mention two other important features. First, because he doesn't commit himself to any specific gauge, Stueckelberg's *matrix elements* are manifestly *gauge invariant*, ...

Next, in evaluating his expressions, Stueckelberg uses integration over the *complex energy-plane*, making explicit use of the singularity structure as a device for putting external particle states on *mass-shell*".]

(With 2 figures)

Introduction and Executive Summary

The usual perturbation theory of quantum mechanics develops the solution of the tested problem according to eigenfunctions of an unperturbed system and determines the dependence of the variable expansion coefficients on time by a proximity method.

The elastic impact of an electron at a nucleus (electromagnetic field) is then given for high velocities by the first proximity, and its radiation in an electromagnetic field by the second proximity, if the zero proximity is considered to be the free electron. For example, if the perturbing field is that of a light wave, the Klein-Nishina formula (K.-N.-F.) is obtained. If it is the field of a nucleus, the formula for the bremsstrahlung is obtained.

For many cases it is now advantageous to write invariant the expressions obtained by perturbation methods for the movement and radiation of an electron. This is achieved by a Fourier expansion of the variable coefficients according to time. The solution is then written as a four-dimensional Fourier series with *constant expansion coefficients,* which are determined by proximity processes. Since time and space are included in the calculation in the same way, the results are relativistically invariant and one avoids, for example, the problem of the Lorentz transformation in the calculation of the K.-N.-F. for moving electrons from that for resting electrons (§ 5)[1].

[1] Pauli, W. (1933). *Helv. Phys. Acta*, 6, 279.

The bremsstrahlung of an electron in any electromagnetic field can be derived by a gauge and Lorentz transformation from the generalized K.-N.-F. (§ 3) if the light properties (speed of light and transverse wave) of the primary waves are not used in their derivation. The result is consistent with that of Bethe and Heitler and Sauter[2].

[2] Bethe, H. & Heitler, W. (August, 1934) On the Stopping of Fast Particles and on the Creation of Positive Electrons. Roy. Soc. Proc., A, 146, 83-112; https://doi.org/ 10.1098/ rspa.1934.0140; Sauter, F. (1934). *Ann. Phys*., 5, 20, 404.

For high velocities of the electron, the non-transverse and sub-light velocity can be reduced[3] (§ 6 and § 7).

[3] Weizsäcker, C. F. v. (September, 1934). Ausstrahlung bei Stößen sehr schneller Elektronen. (Radiation at collisions of very fast electrons.) *Zeit. Phys*., 88, 612-25; https://doi.org/10.1007/BF01333110.

The application of the method to the problem of pair generation will be carried out in two parts.

§ 1 *The wave equation*

We write the wave equation in the form*:

(1,1) $[1/i\,(\gamma, \partial/\partial x) + C + V(x)]\,\psi(x) = 0.$

* See, for example, Pauli, *Handb. d. Physik*, XXIV Formula (II) p. 220. The γ_μ of the present calculations are the γ_μ of Pauli multiplied by i.

[Lacki, J. Ruegg H. & Telegdi, V. L. (March 15, 1999), pages 57-58:
"6.3 *The spinor case*
We are now in a position to review Stueckelberg's 1934 paper for spin 1/2 electrons. The essential complication with respect to the scalar case above is the necessity to take into account the spin degrees of freedom. Furthermore, Stueckelberg generalizes slightly the formalism in order to derive an expression which is then suitable to be applied to Compton scattering and also to Bremsstrahlung and pair production. The starting point is now the Dirac equation for the spinor wave-function $\Psi(x)$

$$[1/i\,(\gamma, \partial/\partial x) + M + eV(x)]\,\Psi(x) = 0 \qquad (98)$$

(where M is the reciprocal Compton wavelength, $C = \mu c/\hbar$ and $e^2/\hbar cM = r_0$ is the classical radius of the electron).]

Here and below, small letters *a*, *b*, ... world vectors [e.g. $x = (x_1, x_2, x_3, x_4)$; $x_4 = i\,ct$] with imaginary time. (*a*, *b*) is their scalar product. $a = (a_1, a_2, a_3)$ represents the spatial fraction of *a* in a given Lorentz system. $(\vec{a}, \vec{b})$ is the spatial scalar product. The real magnitude *a* is the time fraction of *a* divided by the imaginary unit i. γ is the world vector formed from the Hermitian Dirac operators:

$$\gamma = (\beta\vec{\alpha};\, i\beta).$$

Its components correspond to the relations:

(1,2) $\gamma_\mu\gamma_\nu + \gamma_\nu\gamma_\mu = -\,2\delta_{\mu\nu},$

$\gamma_\mu^{+} = -\,\gamma_\mu$

A '+' means "Hermitian conjugate".

V is the interaction energy of electron and field (divided by ħc*)* if the field energy is eliminated[4] and $C = \mu c/\hbar$ is the reciprocal Compton wavelength. (Where ħ is Planck's constant divided by 2π, μ the electron mass and c the speed of light.)

[4] See, for example, Dirac, P.A.M., Fock, V.A., & Podolsky, B. (1932). On quantum electrodynamics. *Phys. Time. Soviet Union*, 2, 468[; *relativistic* model in which a fixed number of electrons interact through a second-quantized electromagnetic field, applies Dirac's interaction *representation* formulation of quantum field theory to full electrodynamics, all non-trivial dynamics due to interaction between charged matter and electromagnetic field relegated to time evolution of state vector, second Maxwell equation, the Ampère-Maxwell law, satisfied only through the action of the field operators on the

wave function, but method of quantizing by imposing covariant commutation relations still an isolated technique, applied only to Maxwell field].

If we describe in the usual way only the transverse field part by quantum electrodynamics, then the operators Γ_k and Γ_k^+ occur there. If one thinks of ψ according to eigenfunctions of the radiation field $u(N^j)$ develops, where

$$N^j = (N_1^j\{ \ldots N[_k^j \ldots)$$

is the totality of the photon numbers in the light wave numbered by *k*, and the index *j* passes through all possible distributions, then:

(1,3) $\psi = \sum_j \varphi^j(x)\, u(N^j)$.

The operators are defined by:

$$\Gamma_k .\, u(N_l \ldots ,N_k,\ldots) = \sqrt{N_k}.\, u(N_l \ldots ,N_k - 1,\ldots),$$
$$\Gamma^+_{-k} .\, u(N_1 \ldots ,N_k \ldots) = \sqrt{(N_k + 1)}.\, u(N_l \ldots ,N_k + 1,\ldots).$$

[In Lacki, J. Ruegg H. & Telegdi, V. L. (March 15, 1999), page 52, this is written as:
"Applied to an eigenstate $|N_1, \ldots, N_k, \ldots >$ of the photon number operator N, with eigenvalues N_i, they yield the well-known result

$$a_k\, |N_l \ldots ,N_k,\ldots > = \sqrt{N_k}\, |N_l \ldots ,N_k - 1,\ldots >, \qquad (73)$$
$$a^\dagger_k\, |N_1 \ldots ,N_k \ldots > = \sqrt{(N_k + 1)}\, |N_l \ldots ,N_k + 1,\ldots >.$$

(where $\Gamma_k = a_k$, $\Gamma^+_k = a^\dagger_k$, and $u(N_l \ldots ,N_k,\ldots) = |N_l \ldots ,N_k - 1,\ldots >$).]

The Fourier expansion of the transverse part of V is then

(1,4) $V^t + V^{+t} = \sum_{k\rightarrow} T_k (\sigma^k, \gamma)[\Gamma_k e^{i(k,x)} + \Gamma^+_k e^{-i(k,x)}]$.

[In Lacki, J. Ruegg H. & Telegdi, V. L. (March 15, 1999), page 51, this is written as:
"In the interaction picture … , A_μ is a free field with the Fourier expansion

$$A^\mu = \sum_k e^\mu_k V_k\, [a_k e^{i(k.x)} + a^\dagger_k e^{-i(k.x)}],$$

(where $\Gamma_k = a_k$, $\Gamma^+_k = a^\dagger_k$, $\sigma^k = e_k$, and $T_k = eV_k$".]

The summation can only be extended over a three-dimensional area, since

(1,5) $(k, k) = 0;\ k_0 > 0$

always applies, and over two mutually and perpendicular polarization directions σ^k. So, it is

(1,6) $(k, \sigma^k) = 0$.

Let us choose

$$(\sigma^k, \sigma^k) = 1$$

and if we think of the Fourier expansion in a certain Lorentz system, then if G there is anormal periodicity and *e* is the elementary electric charge:

(1,7) $T_k^2 = 2\pi e^2/Gk_0c\hbar$.

[In Lacki, J. Ruegg H. & Telegdi, V. L. (March 15, 1999), page 52, this is written as:
"The sum over k refers to photons in a box G with
$V_k^2 = 2\pi e^2/Gk_0c\hbar$, (70)
(where $T_k = eV_k$)".]

The part dominated by charges can be represented by

(1,8) $V^L = \sum_{k\rightarrow} M_k (\sigma^k, \gamma)\, e^{i(k,x)}$

If the summation is performed in the Lorentz system where this fraction is a spherically symmetric field with a scalar potential

$Ze/r\ f(r)$,

then $\sigma^k = 0$, $\sigma_4{}^k = i$ and

(1,9) $M_k = 4\pi Ze^2/G|k^{\rightarrow}|^2\hbar c\ g(|k^{\rightarrow}|)$,
$g|k^{\rightarrow}| = \lim_{a=0} \int_0^\infty e^{-ar} f(r)\, |k^{\rightarrow}| \sin |k^{\rightarrow}|\, r\, dr$.

For a Coulomb field, $f = g = 1$.

§ 2 *The perturbation theory*

We summarize the Fourier developments of the fields V^i, V^{i+} and V^L as V_k, the operator Γ_k, Γ^+_{-k} and 1 as P_k, and define the matrix elements as:

(2,1) $P_k\, u(N^j) = \sum_i u(N^i)\, P_{kij}$

(e.g. in the case of V^t: $P_k = \Gamma_k$; $P_{kij} = \sqrt{N^j{}_k}$, if the distribution N^i differs from N^j by the increase of the light quantum number by 1 in *k*. In all other cases, $P_{kij} = 0$; $V_k = T_k$).

The *wave function* ψ, *written as a four-dimensional Fourier series*, is:

(2,2) $\psi(x,N) = \sum_j \int dl^4\, e^{i(l,x)} (N^i)\, A^j(l)\, u(N^j) = \sum_j \varphi^j(x)\, u(N^j)$

$\int dl^4$ means the quadruple integral extended by l_1, l_2, l_3 and l_4. $\int dl^3$ is the symbol for integration about l_1, l_2, and l_3.

[Lacki, J. Ruegg H. & Telegdi, V. L. (March 15, 1999), page 52-4:
"Following Stueckelberg the interacting *wave-function* is expanded on the photon number *eigenstates* |Nj ≻ with coefficients φ(x), according to Dirac [Dirac, P. A. M. (March, 1927). The Quantum Theory of Emission and Absorption of Radiation. *Proc. Roy. Soc. A*, 114, 767, 243-65], section 4.1
$\varphi^j(x) = \sum_j \varphi^j(x)\, |N^j \succ$ (74)

Here N^j denotes all possible *photon configurations* $\{N_1, N_2, ..., Nk, ...\}$. Now comes the departure of Stueckelberg's method from those of his predecessors, Waller, Tamm, Heitler, etc. He introduces the *four-dimensional Fourier transformation* for the "electron" *wave function*

$$\varphi^j(x) = \int d^4 p e^{i(p.x)} \chi^j(p) \qquad (75)$$

The result will be the elimination in the perturbation expansion of time together with space. …

The general approach of Stueckelberg is to use a perturbation expansion in powers of the charge e. The perturbation expansion applies to the time-independent functions $\chi^j(p)$. Thus, the zeroth order is given by the free state equation

$$\sum_j \int d^4 p e^{i(p.x)} [p^2 + M^2] \chi^{j(0)}(p) = 0 \qquad (77)$$

…

For the *second approximation*, $\chi^{j(1)}$ is introduced into the second term and $\chi^{j(0)}$ into the third one [of (76)]

$$\sum_j \int d^4 p' e^{i(p'.x)} [p'^2 + M^2] \chi^{j(2)}(p') |N^j > \qquad (79)$$

$$= \ldots = 0$$

This expression gives a matrix element symmetric in k and k'. For Compton scattering, we take p,k for the 4-momenta of the initial "electron" and photon. $a^\dagger_{-k'} a_k$ annihilates the photon k and creates a photon with momentum $-k'$. The value of the matrix element of $a^\dagger_{-k'} a_k$ is one. In order to describe an outgoing photon, we change the sign of k'. With p' the final "electron" momentum the conservation law reads

$$p + k = p' + k',$$

(where $l^0 = p$, $l = p'$, $-p$ (or m) $= k'$)

so that finally the relation between the initial wave-function $\chi^{j(0)}(p)$ and the final $\chi^{j(2)}(p')$ is given by

$$\chi^{j(2)}(p') = e^2/(p'^2 + M^2)\, V_k V_{k'}\, \Omega \chi^{j(0)}(p) \qquad (80)$$

$$\Omega = (2p' + k') \cdot e_{k'} (2p + k) \cdot e_k/\{(p + k)^2 + M^2\} \qquad (81)$$
$$+ (2p' - k) \cdot e_k (2p + k') \cdot e_{k'}/\{(p - k')^2 + M^2\} - 2e_k \cdot e_{k'}.$$

We see that Stueckelberg's scheme is entirely within *momentum space p*. The matrix element $\Omega(p)$ is obviously Lorentz invariant. It is also invariant under the gauge transformations $e_k \to e_k + \lambda k$, $e_{k'} \to e_{k'} + \lambda' k'$ if the external electrons are on *mass-shell. The first two terms with denominators contain the contribution of intermediate states whereas the last term corresponds to the direct interaction.* It is important to realize that Stueckelberg didn't need to modify the overall scheme of iteration to achieve this result; such is not the case with the previous (Dirac theory) where a refinement of the perturbative method had to be made to obtain a contribution through intermediate states …

Equation (81) is a witness to the modernity of Stueckelberg's approach, since one had to wait until 1948 to find similar expressions. Ω is actually identical (except for normalizations) to the corresponding factor given by Bjorken and Drell 1964, eq. (9.30), using Feynman techniques.
(81) greatly simplifies in the rest system of the initial charged particle ($\mathbf{p} = 0$) and in the Coulomb or radiation gauge where $e_0 = e'_0 = 0$ and hence $p \cdot e_k = p \cdot e_{k'} = 0$. In this case $\Omega_{lab.} = -2e_k \cdot e_{k'}$.
The steps above illustrate the iteration method of Stueckelberg."]

Now let's define the operators:

$$H(l) = (l, \gamma) + C,$$
(2,3) $$K(l) = -(l, \gamma) + C,$$
$$K(l)H(l) = H(l)K(l) = R(l) = (l, l) + C^2$$

and put (2,2)

[(2,2) $\psi(x,N) = \sum_j \int dl^4\, e^{i(l,x)}\, (N^i)\, A^j(l)\, u(N^j) = \sum_j \varphi^j(x)\, u(N^j)$]

in the *wave equation* (1,1),

[(1,1) $[1/i\, (\gamma, \partial/\partial x) + C + V(x)]\, \psi(x) = 0$,]

then

(2,4) $$\sum_j \int dl^4\, e^{i(l,x)}\, u(N^j)\, [H(l)\, A^j(l) + \sum_{k,i} V_k\, P_{kji}(\sigma^k, \gamma)\, A^i(l-k)] = 0.$$

As a zeroth approximation we choose the unperturbed eigenfunction (V = 0), whose expansion coefficients must be sufficient for each j

(2,5) $$\int dl^4\, e^{i(l,x)}\, H(l)\, A^j(l) = 0$$

The $A^j(l)$ are spinors. The (2,5) fulfilling condition

$$H(l)A^j(l) = 0$$

corresponds to four homogeneous equations with four unknowns.

[Lacki, J. Ruegg H. & Telegdi, V. L. (March 15, 1999), pages 57-58:
"6.3 *The spinor case*
We are now in a position to review Stueckelberg's 1934 paper for spin 1/2 electrons. The essential complication with respect to the *scalar case* above is the necessity to take into account the *spin degrees of freedom.* Furthermore, Stueckelberg generalizes slightly the formalism in order to derive an expression which is then suitable to be applied to Compton scattering and also to Bremsstrahlung and pair production. The starting point is now the Dirac equation for the *spinor wave-function* $\Psi(x)$

$$[1/i\, (\gamma, \partial/\partial x) + M + eV(x)]\, \Psi(x) = 0 \qquad (98)$$

(where M is the reciprocal Compton wavelength, $C = \mu c/\hbar$ and $e^2/\hbar c M = r_0$ is the classical radius of the electron)

where the coupling of the *matter current* to the electromagnetic field is

$$V(x)\ [= A^{\mu}/e^{\mu}_{k}] = \sum_k V_k\,(e_k \cdot \gamma)\,[a_k\, e^{i(k.x)} + a^{\dagger}_{k}\, e^{-i(k.x)}]\ (99)$$

V_k is given by (70),

$$V_k^2 = 2\pi e^2/Gk_0c\hbar, \qquad (70)$$

$(e_k \cdot \gamma)$ denotes the scalar product of the *photon polarization 4-vector* e^{μ}_{k}, and the Dirac matrices γ_μ, $\mu = 0, 1, 2, 3$ (we now use the following convention for γ matrices: $\gamma_\mu\gamma_\nu + \gamma_\nu\gamma_\mu = -2g_{\mu\nu}$). The analogue of the expansion (74)

$$[\varphi^j(x) = \sum_j \varphi^j(x)\,|N^j \succ \qquad (74)]$$

involves now spinor functions $\varphi^j(x)$:

$$\Psi(x,N) = \sum_j \varphi^j(x)\,|N^j \succ$$

and as previously Stueckelberg Fourier expands

$$\Psi(x,N) = \sum_j \int d^4\, pe^{i(p.x)}\, u^j(p)\,|\,N^j\rangle \qquad (101)$$

with $u^j(p)$ the spinor in *momentum space*, [and where $A^j(l) = u^j(p)$.]

The Fourier-transform of the full Dirac equation (98)

$$[[1/i\,(\gamma, \partial/\partial x) + M + eV(x)]\,\Psi(x) = 0 \qquad (98)]$$

can be written

$$\sum_j \int d^4\, pe^{i(p.x)}\,\{[(\gamma \cdot p) + M]\,u^j(p) + e\sum_{k,i} V_k\,P_{kji}(e_k \cdot \gamma)\,u^i\,(p-k)\}\,|N^j \succ = 0.\ (102)$$

Here P_{kji} is a matrix element of P_k the latter standing for one of the operators a_k, a^{+}_{-k}, or 1. This notation allows a unified treatment of the above-mentioned processes

$$P_k\,|N^j \succ = \sum_i |N^i \succ P_{kij} \qquad (103)$$

and V_k is one term in the sum of eq. (99)

$$[V(x)\ [= A^{\mu}/e^{\mu}_{k}] = \sum_k V_k\,(e_k \cdot \gamma)\,[a_k\, e^{i(k.x)} + a^{\dagger}_{k}\, e^{-i(k.x)}]\ (99).]$$

In order for their determinant to disappear, the determinant defined in (2,3) must be

(2,6) $R(l) = 0$.

In real l space, these points lie on a double-shell hyper-rotational hyperboloid, corresponding to the possible states of positive and negative energy of the Dirac electron. The three-dimensional manifold of the (2,6) fulfilling vectors of the one shell is denoted by *l*. Then suffices:

(2,7) $A^j(l) = 1/\pi\ B^j(l)/R(l)$

which is singular on the hyperboloid, the condition (2,5), if $Bj(l)$ is a continuous function for real $l^{\rightarrow}$ and l_0, which satisfies the condition on the one hyperboloid shell $l_4 = l^{-}_{4}(l^{\rightarrow})$ and vanishes sufficiently strongly at a great distance from it*

(2,8) $H\,(l^{-})\,B^j\,(l^{-}) = 0$.

* l^{-} is the four-vector with the components l_1, l_2, l_3 and $l^{-}_{4} = l^{-}_{0}(l^{\rightarrow}) = i\,l^{-}_{0}(l^{\rightarrow}) = \pm i\sqrt{\{C^2 + (l^{\rightarrow}, l^{\rightarrow})\}}$. Depending on the choice of sign, *l* passes through the totality of points

(i.e. possible states of the Dirac electron) of positive or negative energy.

[Lacki, J. Ruegg H. & Telegdi, V. L. (March 15, 1999), pages 54-5:
"We now pass to the discussion of Stueckelberg's method for calculating the cross-section. It starts with the definition of "on mass-shell wave-functions". We first introduce the on mass-shell *4-momentum*, denoted by

$p^{\bar{}} = (p, p^{\bar{}}_0); p^{\bar{}2}_0 = p^2 + M^2$ (82)

To solve the free Klein-Gordon equation (77),

$[\sum_j \int d^4 pe^{i(p.x)} [p^2 + M^2] \chi^{j(0)}(p) = 0$ (77)]

Stueckelberg makes the Ansatz

$\chi^{(0)}(p) = 1/i\pi\, \omega^{(0)}(p)/(p^2 + M^2)$ (83)

where $\omega^{(0)}(p)$ is a continuous function which vanishes fast enough at infinity.

[Nowadays, instead of (83) one writes $\chi^{(0)}(p) = \delta(p^2 + M^2)\omega^{(0)}(p)$.]

Then $\chi^{(0)}(p)$ satisfies (77) if $\omega^{(0)}(p^{\bar{}})$ obeys

$(p^{\bar{}2} + M^2)\, \omega^{(0)}(p^{\bar{}}) = 0$ (84)

Notice that $p^2 + M^2 = p^{\bar{}}_0{}^2 - p_0{}^2$ which makes explicit the singularity of (83) in the energy variable p_0. The reason of this special way in which Stueckelberg expresses the solution of (77) is the following. It is different from zero only for those values of *p* where the $p^{\bar{}}_0{}^2 = \mathbf{p}^2 + M^2$. Elsewhere, it vanishes. The form of (83) with its explicit singularity makes it possible to write both cases in a single closed form, which is then suitable for being reinserted into iteration steps. Each order of iteration will add a new singularity the impact of which being crucial during contour-integration. The information about the values of *p* where the non-trivial solution is valid is then retrieved by complex contour-integration around the singularity. (This feature of Stueckelberg's method is quite original and contributes to the full automatism of his perturbative calculus.)"]

Then the integration via dl_4 can be executed by switching to the complex l_4 level**:

(2.7) $\int dl_4\, e^{il_4x_4} A^j(l) = B^j(l)/l_0\, e^{il^{\bar{}}_4x_4}$,

thus fulfilling (2.5).

** In the case of integration for positive times, in the complex l_0 domain, the point $l_0 = l^{\bar{}}_0$ must be bypassed in a positive sense if $l^{\bar{}}$ is positive, and in the negative sense if $l^{\bar{}}_0$ is negative. In the case of the bypass in the opposite sense, the integration gives the value zero.

The spatial Fourier expansion is

(2,9) $\varphi^j = \int dl^{\to 3}\, e^{i(l^{\bar{}},x)}\, B^j(l^{\bar{}})/l^{\bar{}}_0$.

The number of particles is:

(2.10) $n^j = \int \rho^j dx^{\to 3} = \ldots = \ldots$

and is (always in zeroth proximity) of course an invariant.

For our problems we choose in particular such a solution of zero proximity, for which the $A^j(l)$ for all distributions j are zero, except for a certain light quantum distribution, which is characterized by the index $j = 0$. (2,4) is then in zeroth proximity (2,5) with $j = 0$.

The largest part of the *first proximity* are then the first member of (2,4) summed over $j \neq 0$, and in the second member those terms of the sum for which $i = 0$. The A^j ($j \neq 0$) are then determined in first proximity by the known A^0, if the bracket of (2,4) is set equal to zero. If the operators from (2,3) are used, then in the *first approximation*:

(2,11) $A^j(l) = -\, 1/R(l).\ K(l) \sum_k V_k P_{kj0}(\sigma^k, \gamma)\, A^0(l-k)$.

If we write this in the form (2,7), then the $B^j(l)$ first proximity now has a singularity on the hyperboloid $R(l-k) = 0$ associated with a fixed k. The coefficients of the three-dimensional Fourier expansion φ^j therefore become time-dependent. The meaning of the formula (2,11) is given by Fig. 1:

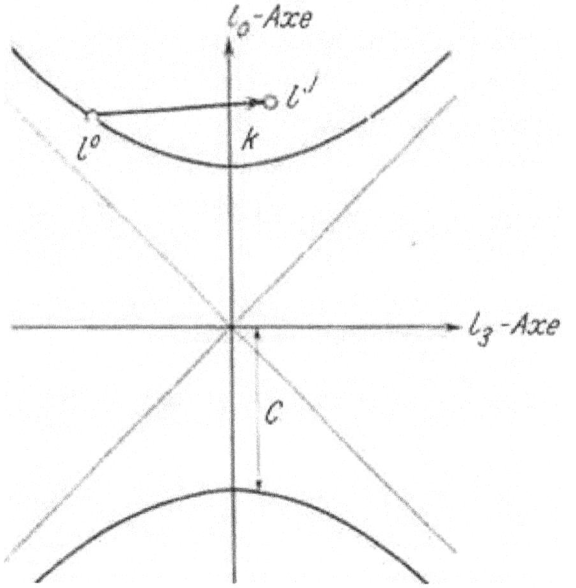

Fig. 1

$l_0\, l_3$ -section through the (real) energy momentum space. The perturbation caused by the *kth* Fourier component forms in the first proximity the area l^0 on the region $l_j = l^0 + k$ ab. Only if l_j lies on the hyperbola $l_0^2 - l_3^2 = C^2$, a perturbation of the first proximity occurs.

The real l, l_0 -space is drawn on the average l_3, l_0. We consider only the perturbation caused by the kth wave of the field. The region in which the $A^0(l)$ are substantially different from zero is on the hyperboloid because of $R(l)$ in (2,7). If the zeroth proximity represents a wave packet with mean impulse l^0 and mean energy l_0^0, so A^0 is different from zero only in the region l_0. Then (2,11) says that $A^j(l)$ can only take different values in the region $l = l^j = l^0 + k$ from zero. Because of $R(l)$ in (2.11), however, this value is infinitely larger if l^j is also on the hyperboloid than if this is not the case. If the perturbation field in the drawn space-time system is a temporally constant field, then this area l^j, generated by the totality of all k from l^0, lies because of $k_4 = 0$, on the hypercircle generated by the hyperplane parallel to l by l^0 with the hyperhyperboloid. This corresponds to the elastic scattering of the electron ($l_4^j = l_4^0$). If the k-light vectors that are parallel to the surface lines of the asymptote hypercone of the hyperhyperboloid because of $(k, k) = 0$, then none of the regions $l^0 + k$ lies on the hyperhyperboloid. An interaction between light and electron does not occur in the first proximity.

The *second approximation* we obtain from the first in the same way as the first was developed from the second:

(2,12) $A^j(l) = 1/R(l) \sum_{pk} V_p V_k P_{pji} P_{ki0}\, \Omega(l^0) A^0(l^0)$,

(2,13) $\Omega(l^0) = K(l)\, [\{(\sigma^p, \gamma)\, K(l-p)(\sigma^k, \gamma)\}/R(l-p)$
$+ \{(\sigma^k, \gamma) K\{(l-p)(\sigma^p, \gamma)\}/ R(l-k)]$

with $l^0 = l - k - p$.

> [Lacki, J. Ruegg H. & Telegdi, V. L. (March 15, 1999), page 59:
> "Similarly, for the second order, the result is (compare (80) and (81)
>
> $[\chi^{j(2)}(p') = e^2/(p'^2 + M^2)\, V_k V_{k'}\, \Omega \chi^{j(0)}(p)$ (80)
> $\Omega = (2p' + k') \cdot e_{k'}\, (2p + k) \cdot e_k/\{(p + k)^2 + M^2\}$ (81)
> $+ (2p' - k) \cdot e_k\, (2p + k') \cdot e_{k'}/\{(p - k')^2 + M^2\} - 2e_k \cdot e_{k'}]$:
>
> $u^{(2)}(p') = -\, 1/(p'^2 + M^2) \sum_{k',k} V_{k'} V_k (P_{k'})_{ji} (P_k)_{i0}\, \Omega(p) u^{(0)}(p)$ (105)
> $\Omega(p) = ((\gamma \cdot p') - M)\, [\{(e_{k'} \cdot \gamma)((\gamma \cdot p' - k') - M)(ek \cdot \gamma)\}/\{(p' - k')^2 + M^2\}$
> $+ \{(e_k \cdot \gamma)((\gamma \cdot p' - k) - M)(e_{k'} \cdot \gamma)\}/\{(p' - k)^2 + M^2\}$ (106)
>
> with $p = p' - k - k'$ and $(P_k)_{ij}$ are the matrix elements of the operators a_k, $a^\dagger_k$ and 1, that is $\sqrt{N_k}$ and $\sqrt{(N_k + 1)}$ for the first two. Here, one gets only the contributions corresponding to intermediate states as there is no direct transition, contrary to the scalar case. The expressions (105) and (106) are obviously Lorentz-invariant and correspond exactly to the expressions obtained by "modern" Feynman rules."]

Fig. 2 represents the relationships of the formula (2,12).

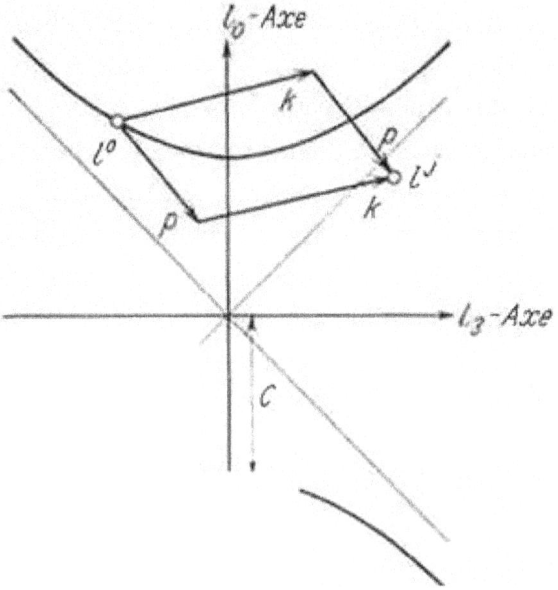

Fig. 2

As Fig. 1. The perturbation caused by the *k-th* and *p*-th Fourier components forms the area l^0 on the area

$$l^j = l^0 + k + p$$

ab. Only if l^j also lies on the hyperbola, a perturbation of second proximity occurs. The intermediate states $l^{i1} = l^0 + k$ and $l^{i2} = l + p$ are then *not* on the hyperbola. (The symbols l^i of the intermediate state are not entered in the figure).

Again, $A^j(l)$ on the hyperhyperboloid is infinitely larger than in other areas. The two paths in Fig. 2 correspond to the sum of two members in the operator Ω. Extracting the matrices of the P-operators from the sum is allowed because $P_{pji1} P_{ki10} = P_{kji2} P_{pi20}$ is always allowed. The summation over i can be omitted in (2,12), since for each k there exists only one matrix element P_{kji}.

The time-dependent coefficients of the three-dimensional Fourier expansion are derived from the relation

$$\frac{B^j(\bar{l};x_4)}{l_0} = \int dl_4\, e^{i(l_4-\bar{l}_4)x_4} A^j(l) \qquad (2.14)$$

by moving the integration path into the complex*.

* See second note on p. 371.

(The relation (2.14) applied to A^0 trivially gives $B^0(l; x_4) = B^0(l)$.)

From (2,12) follows:

$$(2,15)\ \frac{B^j(\bar{l};x_4)}{\bar{l}_0} = \sum_{kp} V_k V_p P_{kji} P_{pio} \frac{1-e^{is_4x_4}}{2s_4} \frac{\Omega(\bar{l}^0)}{l_1} \frac{B^0(\bar{l}^0)}{\bar{l}^0_0}$$

mit

$$(2,16)\qquad s_4 = \bar{l}_4 - (\bar{l}^0_4 + k_4 + p_4)\,.$$

The time-dependent number of particles in state *j* is

(2,17) $n^j(x_4) = 1/i \int dx^3 \varphi^j + \gamma^4 \varphi^j = \dots$.

§ 3 *Light scattering at the free particle (generalized K.-N.-F.)*

The distribution function of zero proximity is chosen

$N^0 = (000 \dots 0, N_k, 0 \dots)$

and this determines the number of particles.

$N^{jm} = (000 \dots\ 0, 1_m, 0 \dots\ 0, N_k - 1, 0 \dots)$

These correspond to the intermediate states

$N^i(000.0, N_k - 1.0 \dots)$ and $(000 \dots 0, 1_m, 0 \dots 0, N_k, 0 \dots)$.

In (2,15) *k* and – *p* = *m* become light vectors. This is the only possible perturbation if transitions to states on the other hyperhyperboloid shell are not taken into account. [States of negative energy are achieved from such positive energy if – *k* and – *p* are light vectors. In Dirac's theory of the positive electron [positron], the process corresponds to the recombination radiation (emission of two light quanta) of positive and negative electrons if an unoccupied state of negative energy is considered a positive electron.] The matrices $P_{kji} P_{pi0}$ are equal to $\sqrt{N_k}$.

Instead of *p* we write – *m* so that *k* and *m* are light vectors: $(m, m) = (k, k) = 0$; m_0 and $k_0 > 0$. The "initial state" then corresponds to the presence of N_k quanta in the *k*th light wave, which produce a current density in the direction *k* of N_k c/G light quanta of the frequency $\nu_k = k_0 c$. The "final state" corresponds to the presence of $N_k - 1$ quanta in *k* and one quantum in *m*. The summation over *k* and *p* or *m* in (2.15) is therefore omitted. If we ask about the number of particles corresponding to the occurrence of a light quantum of certain polarization σ^m in solid angle $d\omega_m$, then $n^{jm}(x_4)$ is obtained from (2.17) by summing over all *m* lying in $d\omega_m$. This summation corresponds to the integration over $(2\pi)^{-3}\, G m_0^2\, dm_0\, d\omega_m$. ...

...

§ 4. *The expected value of* $\gamma_4\, \Omega + \Omega$

...

§ 5 *The Klein-Nishina-Formal for moving electrons*

Following Pauli, we use the abbreviations

$D_k = 1 - 1/c\ (k/k_0, \upsilon)$

and according to D_m, one obtains the K.-N.-F. for an electron moving at velocity υ, when the polarization direction σ^k is averaged and summed over σ^m, the simplest way is as follows:

...

§ 6 *The derivation of the braking formula from the generalized Klein-Nishina formula*

> [*Bremsstrahlung*, from bremsen "to brake" and Strahlung "radiation"; i.e., "braking radiation" or "deceleration radiation", *is electromagnetic radiation produced by the deceleration of a charged particle when deflected by another charged particle*, typically an electron by an atomic nucleus. The moving particle loses kinetic energy, which is converted into radiation (i.e., photons), thus satisfying the law of conservation of energy. The term is also used to refer to the process of producing the radiation. Bremsstrahlung has a continuous spectrum, which becomes more intense and whose peak intensity shifts toward higher frequencies as the change of the energy of the decelerated particles increases.]

The formula for bremsstrahlung in the electrostatic field was derived from Bethe and Heitler and from Sauter[2] in the second proximity. Through a qualitative consideration, v. Wiezsacker[3] obtained an approximate formula which agrees with the formula of Bethe and Heitler for large initial energy of the electron and large energy of the emitted light quanta compared to μc^2. His reasoning can be presented in our formalism as follows:

In the space-time system, in which the field appears static (i.e. where, for example, the nucleus is at rest), the Fourier decomposition of the field (1,8) into stationary partial waves applies. This system is called the *core system*. The electron moves in the *core system* with the velocity υ. The splitting of a world vector according to this system is characterized by unprimed components. *k* in (18) thus has spatial components in the *core system* only from zero.

In the space-time system, where the electron is initially at rest (briefly referred to as electron system and defined by deleted components of world vectors), the partial waves then move at a speed $\upsilon' = -\upsilon$. If the initial energy is » μc^2, then $|\upsilon'|$ is almost equal to c. A recalibration of the polarization vector σ^k, which according to § 2 has only one non-zero time-component in the core system, into a world vector σ^{k0}, which appears purely spatial in the electron system, results in *almost* transverse waves for the most important partial waves with small ^{k}.

v. Weizsacker simply applies the K.-N.-F. to these quasi-light waves in the electron system for stationary electrons [(5,1) with $D_m = D_k = 1$].

...

§ 7 *The strict derivation of the braking formula*

In the derivation of the braking formula given in the previous paragraph from the generalized K.-N.-F. two points need to be justified:

Firstly, the identification of the normalization factor G' with the volume of the wave packet $L'_1L'_2L'_3$ in the amplitudes of the individual partial waves and secondly, the incoherent superposition of their effects. ...

...

Since there is no specialization in positive energies anywhere in the calculations, especially in formula (4,6), the following three problems can be calculated on the basis of Dirac's interpretation of the states of negative energy:

1. The generation of *an electron pair from two light quanta*. This case is to be treated by analogy with the derivation of the K.-N.-F. in § 3. l^0 then represents a state of *negative energy* and, instead of m, $p = -m$ is a light vector.

2. The *generation of an electron pair by the interaction of a light quantum and the nuclear field*. The inverse process, as Bethe and Heitler[2] have shown, is a "bremsstrahlung" in which the final state l is a state of negative energy.

3. The *generation of an electron pair by the collision of a fast particle at the nuclear field*. The field of the impacting particle is decomposed by a Fourier expansion, and then the pair generation by each individual partial wave is calculated according to 2. The incoherent superposition of the contributions of the individual waves gives the desired result, as long as one allows only collisions in which the momentum of the impacting particle changes only by a small amount compared to its initial momentum.

These calculations are to be carried out in a *second part*.

I owe the inspiration for the elaboration of the present *invariant perturbation theory* to Prof. G. Wentzel, to whom I would like to express my thanks at this point.

Einstein, A., Podolsky, B. & Rosen, N. (May, 1935). Can Quantum-Mechanical Description of Physical Reality Be Considered Complete?

[*Phys. Rev.*, 47, 10, 777-80; https://doi.org/10.1103/PhysRev.47.777.]

Received 25 March 1935.

Institute for Advanced Study, Princeton, New Jersey.

The description of reality as given by a wave function in quantum mechanics is not complete.

Abstract

In a complete theory there is an element corresponding to each element of reality. A sufficient condition for the reality of a physical quantity is the possibility of predicting it with certainty, without disturbing the system. In quantum mechanics in the case of two physical quantities described by non-commuting operators, the knowledge of one precludes the knowledge of the other. Then either (1) the description of reality given by the wave function in quantum mechanics is not complete or (2) these two quantities cannot have simultaneous reality. Consideration of the problem of making predictions concerning a system on the basis of measurements made on another system that had previously interacted with it leads to the result that if (1) is false then (2) is also false. One is thus led to conclude that the description of reality as given by a wave function is not complete.

Bloch, F. & Nordsieck, A*. (July, 1937). Note on the Radiation Field of the Electron.

[*Phys. Rev.*, 52, 54; https://doi.org/10.1103/PhysRev.52.54; also in Schwinger, J. (ed). (1958). *Selected Papers on Quantum electrodynamics*. Dover, New York, pages 129-34.]

Received May 14, 1937.

Stanford University, California.

*National Research Fellow.

Abstract

Previous methods of treating radiative corrections in non-stationary processes such as the scattering of an electron in an atomic field or the emission of a β-ray, by an expansion in powers of $e^2/\hbar c$, are defective in that they predict infinite low frequency corrections to the transition probabilities. This difficulty can be avoided by a method developed here which is based on the alternative assumption that $e^2\omega/mc^3$, $\hbar\omega/mc^2$ and $\hbar\omega/c\Delta p$ (ω = angular frequency of radiation, Δp = change in momentum of electron) are small compared to unity. In contrast to the expansion in powers of $e^2/\hbar c$, this permits the transition to the classical limit $\hbar$=0. External perturbations on the electron are treated in the Born approximation. It is shown that for frequencies such that the above three parameters are negligible the quantum mechanical calculation yields just the directly reinterpreted results of the classical formulae, namely that the total probability of a given change in the motion of the electron is unaffected by the interaction with radiation, and that the mean number of emitted quanta is infinite in such a way that the mean radiated energy is equal to the energy radiated classically in the corresponding trajectory.

John Archibald Wheeler (July 9, 1911 – April 13, 2008)

Wheeler was an American theoretical physicist. He was largely responsible for reviving interest in general relativity in the United States after World War II. Wheeler also worked with Niels Bohr in explaining the basic principles behind nuclear fission. He is best known for popularizing the term "black hole," as to objects with gravitational collapse already predicted during the early 20th century, for inventing the terms "quantum foam", "neutron moderator", "wormhole" and "it from bit", and for hypothesizing the "one-electron universe".

Wheeler was born in Jacksonville, Florida on July 9, 1911, to librarians Joseph Lewis Wheeler and Mabel Archibald (Archie) Wheeler. He was the oldest of four children, having two younger brothers, Joseph and Robert, and a younger sister, Mary. They grew up in Youngstown, Ohio, but spent a year in 1921 to 1922 on a farm in Benson, Vermont, where Wheeler attended a one-room school.

After graduating from the Baltimore City College high school in 1926, Wheeler entered Johns Hopkins University with a scholarship from the state of Maryland. He published his first scientific paper in 1930, as part of a summer job at the National Bureau of Standards. He earned his doctorate in 1933. His dissertation research work, carried out under the supervision of Karl Herzfeld, was on the "*Theory of the Dispersion and Absorption of Helium*". He received a National Research Council fellowship, which he used to study under Gregory Breit at New York University in 1933 and 1934, and then in Copenhagen under Niels Bohr in 1934 and 1935. In a 1934 paper, Breit and Wheeler introduced the Breit–Wheeler process, a mechanism by which photons can be potentially transformed into matter in the form of electron-positron pairs.

For most of his career, Wheeler was a professor of physics at Princeton University, which he joined in 1938, remaining until his retirement in 1976. At Princeton he supervised 46 PhD students, more than any other professor in the Princeton physics department.

In a 1937 paper "*On the Mathematical Description of Light Nuclei by the Method of Resonating Group Structure*", Wheeler introduced the S-matrix – short for scattering matrix – "a unitary matrix of coefficients connecting the asymptotic behavior of an arbitrary particular solution [of the integral equations] with that of solutions of a standard form". Heisenberg subsequently developed the idea of the S-matrix in the 1940s. Due to the problematic divergences present in quantum field theory at that time, Heisenberg was motivated to isolate the essential features of the theory that would not be affected by future changes as the theory developed. In doing so he was led to introduce a unitary "characteristic" S-matrix, which became an important tool in particle physics.

Considering the notion that positrons were electrons that were traveling backwards in time, he came up in 1940 with his one-electron universe postulate: that there was in fact only one electron, bouncing back and forth in time. Richard Feynman, his graduate student, found this hard to believe, but the idea that positrons were electrons traveling backwards in time intrigued him and Feynman incorporated the notion of the reversibility of time into his Feynman diagrams. *Feynman received a Ph.D. from Princeton* in 1942; Wheeler was his *thesis advisor*.

Soon after the Japanese bombing of Pearl Harbor brought the United States into World War II, Wheeler accepted a request from Arthur Compton to join the Manhattan Project's Metallurgical Laboratory at the University of Chicago. He moved there in January 1942, joining Eugene Wigner's group, which was studying nuclear reactor design. He co-wrote a paper with Robert F. Christy on "Chain Reaction of Pure Fissionable Materials in Solution", which was important in the plutonium purification process. It would not be declassified until December 1955.

After the United States Army Corps of Engineers took over the Manhattan Project, it gave responsibility for the detailed design and construction of the reactors to DuPont. Wheeler became part of the DuPont design staff. He worked closely with its engineers, commuting between Chicago and Wilmington, Delaware, where DuPont had its headquarters. He moved his family to Wilmington in March 1943. DuPont's task was not just to build nuclear reactors, but an entire plutonium production complex at the Hanford Site in Washington. As work progressed, Wheeler relocated his family again in July 1944, this time to Richland, Washington, where he worked in the scientific buildings known as the 300 area.

In August 1945 Wheeler and his family returned to Princeton, where he resumed his academic career. Working with Feynman, he explored the possibility of physics with particles, but not fields, and they published a paper on the Wheeler–Feynman absorber theory, an interpretation of electrodynamics derived from the assumption that the solutions of the electromagnetic field equations must be invariant under time-reversal transformation, as are the field equations themselves. [Wheeler, J. A. & Feynman R. P. (April, 1945). Interaction with the Absorber as the Mechanism of Radiation. *Rev. Mod. Phys*., 17, 157-81.]

Wheeler also carried out theoretical studies of the muon with Jayme Tiomno, resulting in a series of papers on the topic, including a 1949 paper in which Tiomno and Wheeler introduced the "Tiomno Triangle", which related different forms of radioactive decay. He also suggested the use of muons as a nuclear probe. This paper, written and privately circulated in 1949 but not published until 1953, resulted in a series of measurements of the Chang radiation emitted by muons. Muons are a component of cosmic rays, and Wheeler became the founder and first director of Princeton's Cosmic Rays Laboratory, which received a substantial grant of $375,000 from the Office of Naval Research in 1948. He received a Guggenheim Fellowship in 1946, which allowed him to spend the 1949–50 academic year in Paris.

The 1949 detonation of Joe-1 by the Soviet Union prompted an all-out effort by the United States, led by Teller, to develop the more powerful hydrogen bomb in response. Henry D. Smyth, Wheeler's department head at Princeton, asked him to join the effort. At Los Alamos, Wheeler and his family moved into the house on "Bathtub Row" that had been occupied by Robert Oppenheimer and his family during the war. In 1950 there was no practical design for a hydrogen bomb. Calculations by Stanisław Ulam and others showed that Teller's "Classical Super" would not work. Teller and Wheeler created a new design known as "Alarm Clock", but it was not a true thermonuclear weapon. Not until January 1951 did Ulam come up with a workable design.

General relativity had been considered a less respectable field of physics, being detached from experiment. Wheeler was a key figure in the revival of the subject, leading the school at Princeton University, while Dennis William Sciama and Yakov Borisovich Zel'dovich developed the subject at Cambridge University and the University of Moscow, respectively. Wheeler and his students made substantial contributions to the field during the Golden Age of General Relativity.

While working on mathematical extensions to Einstein's general relativity in 1957, Wheeler introduced the concept and word wormhole to describe hypothetical "tunnels" in space-time. Bohr asked if they were stable and further research by Wheeler determined that they are not. His work in general relativity included the theory of gravitational collapse. He used the term black hole in 1967 during a talk he gave at the NASA Goddard Institute of Space Studies (GISS), although the term had been used earlier in the decade. Wheeler said the term was suggested to him during a lecture when a member of the audience was tired of hearing Wheeler say "gravitationally completely collapsed object". Wheeler was also a pioneer in the field of quantum gravity due to his development, with Bryce DeWitt, of the Wheeler–DeWitt equation in 1967. Stephen Hawking later described Wheeler and DeWitt's work as the equation governing the "wave function of the Universe".

Wheeler left Princeton University in 1976 at the age of 65. He was appointed as the director of the Center for Theoretical Physics at the University of Texas at Austin in 1976 and remained in the position until 1986, when he retired and became a professor emeritus.

With Kent Harrison, Kip Thorne and Masami Wakano, Wheeler wrote *Gravitation Theory and Gravitational Collapse* (1965). This led to the voluminous general relativity textbook *Gravitation* (1973), co-written with Misner and Thorne. Its timely appearance during the golden age of general relativity and its comprehensiveness made it an influential relativity textbook for a generation. Wheeler teamed up with Edwin F. Taylor to write *Spacetime Physics* (1966) and *Scouting Black Holes* (1996).

Wheeler won numerous prizes and awards, including the Golden Plate Award of the American Academy of Achievement in 1966, the Enrico Fermi Award in 1968, the Franklin Medal in 1969, the Einstein Prize in 1969, the National Medal of Science in 1971, the Niels Bohr International Gold Medal in 1982, the Oersted Medal in 1983, the J. Robert Oppenheimer Memorial Prize in 1984 and the Wolf Foundation Prize in 1997.

On April 13, 2008, Wheeler died of pneumonia at the age of 96 in Hightstown, New Jersey.

Wheeler, J. A. (December, 1937). On the Mathematical Description of Light Nuclei by the Method of Resonating Group Structure.

[*Phys. Rev.*, 52 (11): 1107–1122. doi:10.1103/physrev.52.1107.]

Received August 17, 1937.

University of North Carolina, Chapel Hill, North Carolina.

Introduction of S-matrix.

Abstract

The wave function for the composite nucleus is written as a properly, antisymmetrized combination of partial wave functions, corresponding to various possible ways of distributing the neutrons and protons into various groups, such as alpha-particles, di-neutrons, etc. The dependence of the total wave function on the intergroup separations is determined by the variation principle. The analysis is carried out in detail for the case that the configurations considered contain only two groups. Integral equations are derived for the functions of separation. The associated Fredholm determinant completely determines the stable energy values of the system, connects the asymptotic behavior of an arbitrary particular solution with that of solutions possessing a standard asymptotic form. With its help, the Fredholm determinant also determines all scattering and disintegration cross sections, without the necessity of actually obtaining the intergroup wave functions. The expressions obtained for the cross sections, taking account of spin effects, have general validity. Details of the application of the method of resonating group structure to actual problems are discussed.

Pauli, W. & Fierz, M. (March, 1938). Zur Theorie der Emission langwelliger Lichtquanten. (On the theory of the emission of long-wave light quanta.)

[*Nuovo Cimento*, 15, 3, 167-88; https://doi.org/10.1007/BF02958939.]

Zürich.

Summary

As is well known, the usual radiation theory provides an infinitely large value for the effective cross-section dq of a charged particle when passing through a force field when deflected to a given angular range ("*ultrared catastrophe*"). If one prescribes that the energy loss of the particle should lie between E and E + dE, then for small E according to this theory dq = const. dE/E, which diverges logarithmically at the point E = 0 when integrated. *The present study investigates in more detail what quantum electrodynamics yields for this cross-section when a finite expansion is attributed to the charged body*. It turns out that then infinity is always eliminated and that the distractions considered radiationless in the ordinary theory appear here as those with finite, albeit very small energy loss. On the other hand, according to the exact theory, the closer course of dq for very small energy losses E depends so much on the expansion of the charged body that a direct application of the result to real electrons is possible. *It must therefore be concluded that the problem in question is essentially linked to the fundamental difficulties of quantum electrodynamics which have not yet been resolved.*

Dirac, P. A. M. (August, 1938). Classical Theory of Radiating Electrons.

[*Roy. Soc. Proc., A*, 167, 929, 148-169; https://royalsocietypublishing.org/doi/pdf/10.1098/rspa.1938.0124.]

Received March 15, 1938.

St. John's College, Cambridge.

Dirac referenced Frenkel (1925) in describing *relativistic* form of classical theory of radiating electrons assuming electron to be a *point charge* with no volume, *an extended electron inconceivable in special theory of relativity due to intrinsic connection between space and time*, Lorentz *model of electron as possessing mass on account of the electromagnetic field around it fails* without further assumptions if the electromagnetic field varies too rapidly or if the acceleration of the electrons is too great, no natural way of introducing further assumptions has been discovered, discovery of neutron introduced a form of mass that is difficult to believe is of electromagnetic nature, theory of the positron in which positive and negative values of an electron play symmetrical roles cannot be fitted into idea of electromagnetic idea of mass which requires all mass to be positive, *departure from electromagnetic theory of mass removes main reason for believing in finite size of the electron*, results in difficulty that the *field* in the immediate neighborhood of the electron has an infinite mass, in quantum mechanics results in divergence in the solution of the equations that describe the interaction of an electron with an electromagnetic field and prevents its application to high-energy radiation processes, *Dirac chose to represent electron by point charge and to avoid the difficulties with the infinite energy by direct omission or subtraction of unwanted terms* as had been used in the *theory of the positron. new theory must be in agreement with special relativity* and the conservation of energy and momentum, first addressed problem of a *single electron moving in an electromagnetic field*, used equations for electromagnetic *potentials* in terms of the *charge-current density* vector, *assumed that the charge-current density vector vanishes everywhere except on the world-line of the electron where it is infinitely great*, derived *field* quantities from electromagnetic *potentials*, obtained solutions in terms of *retarded* and *advanced* potentials and potentials representing the incoming and outgoing radiation, calculated field of radiation produced by the electron, in agreement with classical theory but provided a value for the field of radiation throughout space-time, obtained exact *equations of motion* of the electron in a specified incident *field* from the laws of conservation of energy and momentum within limits of classical theory, imposed condition that solutions that occur in Nature when there is no incident field are those for which velocity in terms of proper time is constant, requires solutions of equations of motion *for which the final acceleration as well as the initial position and velocity are prescribed*, electron responds to pulse of electromagnetic radiation before it reaches the center of the electron, behaves as if it has a radius of order $1/a$ where $a = 3m/e^2$, implies that signal can be transmitted *faster than the speed of light* through the interior of an electron, *interior of electron is a region of failure of some of the elementary properties of space-time*, showed how theory can be extended to any number of electrons interacting with each other and with a field of radiation.

Introduction

The Lorentz model of the electron as a small sphere charged with electricity, possessing mass on account of the energy of the electric field around it, has proved very valuable in accounting for the motion and radiation of electrons in a certain domain of problems, in which the electromagnetic field does not vary too rapidly and the accelerations of the electrons are not too great. Beyond this domain it will not go unless supplemented by further assumptions about the forces that hold the charge on an electron together. No natural way of introducing such further assumptions has been discovered, and it seems that the Lorentz model has reached the limit of its usefulness and must be abandoned before we can make further progress.

One of the most attractive ideas in the Lorentz model of the electron, *the idea that all mass is of electromagnetic origin, appears at the present time to be wrong*, for two separate reasons. First, *the discovery of the neutron has provided us with a form of mass which it is very hard to believe could be of electromagnetic nature*. Secondly, we have the theory of the positron— a theory in agreement with experiment so far as is known—in which positive and negative values for the *mas*s of an electron play symmetrical roles. *This cannot be fitted in with the electromagnetic idea of mass, which insists on all mass being positive, even in abstract theory.*

The departure from the electromagnetic theory of the nature of mass removes the main reason we have for believing in the finite size of the electron. It seems now an unnecessary complication not to have the field equations holding all the way up to the electron's center, which would then appear as a point of singularity. *In this way we are led to consider a point model for the electron.*

[This appears to be rather extreme.]

Further reasons for preferring the *point-electron* have been given by Frenkel (1925)*.

> * [Frenkel, J. (December, 1925). Zur elektrodynamik punktförmiger Elektronen. (On the electrodynamics of point-like electrons.) *Zeit. Phys*., 32, 518-34; argues that electrons and protons should be treated as *point charges* and their *masses* considered to be a primary property independent of *charge* in place of *electromagnetic theory of mass*, based on erroneous assumption of *mass defect* of helium relative to hydrogen at time when neutron had not been discovered, but claims that *an extended electron is inconceivable in the special theory of relativity since due to the intrinsic connection between space and time an invariant definition of a geometrically invariable (i.e., rigid) body is impossible for arbitrary motions*, notes that electrons are not only physically but geometrically indivisible and have no extension in space, no internal forces between the elements of an electron because such elements do not exist, the electromagnetic explanation for *mass* then goes away]. See pp. 526 and 527 of Frenkel's paper. The reason on p. 518 is not valid according to present-day knowledge.

> [The latter was based on an erroneous assumption of *mass defect* of helium relative to hydrogen at time when neutron had not been discovered. The neutron was

> discovered in 1932 by the English physicist James Chadwick. The real issue (on pages 526 and 527) was that *an extended electron is inconceivable in special theory of relativity due to intrinsic connection between space and time*, an invariant definition of a geometrically invariable body is impossible for arbitrary motions. This, on its face, raises serious questions about the applicability of the *special theory of relativity* to the constitution of matter, but the finite size of the electron reappears.]

We are now faced with the difficulty that, if we accept Maxwell's theory, the field in the immediate neighborhood of the electron has an *infinite mass*. This difficulty has recently received much prominence in quantum mechanics (which uses a *point model* of the electron), *where it appears as a divergence in the solution of the equations that describe the interaction of an electron with an electromagnetic field and prevents one from applying quantum mechanics to high-energy radiative processes*. One may think that this difficulty will be solved only by a better understanding of the structure of the electron according to quantum laws. However, it seems more reasonable to suppose that the electron is too simple a thing for the question of the laws governing its structure to arise, and thus quantum mechanics should not be needed for the solution of the difficulty. *Some new physical idea is now required*, an idea which should be intelligible both in the classical theory and in the quantum theory, and our easiest path of approach to it is to keep within the confines of the classical theory.

A possible line of attack is to modify Maxwell's theory so as to make the energy of the field around the singularity that represents an electron finite. This method has been tried by Born, but it leads to great complexity, and its present outlook is not very hopeful.

> [Dirac's solution was, as before, to stick with *special relativity*, give up on obtaining a mathematical model of the electron, and omit or subtract unwanted terms from the equations used to calculate results that can be obtained from experiment.]

We shall here proceed from the opposite point of view. *We shall retain Maxwell's theory to describe the field right up to the point-singularity which represents our electron and shall try to get over the difficulties associated with the infinite energy by a process of direct omission or subtraction of unwanted terms, somewhat similar to what has been used in the theory of the positron*. Our aim will be *not so much to get a model of the electron* as to get a simple scheme of equations which can be used to calculate all the results that can be obtained from experiment. *The scheme must be mathematically well-defined and self-consistent, and in agreement with well-established principles, such as the principle of relativity and the conservation of energy and momentum*. Provided these conditions are satisfied, it should not be considered an objection to the theory that it is not based on a model conforming to current physical ideas.

We shall be working all the time with the ordinary Maxwell theory, and thus we shall not be led to any essentially new equations. We shall be concerned rather with getting a satisfactory physical interpretation for equations which are already well known. *It may be wondered why this problem was not solved long ago*. A great deal of work has been done in the past in examining the general implications of Maxwell's theory, but it was nearly all done before the discovery of quantum mechanics in 1925, when people gave all their attention to the question of *how an electron could remain in an atomic orbit without radiating*—a question we now know can be answered only by going outside classical theory—and were thus not interested in simply looking for the most natural interpretation their equations will allow.

The fields associated with an electron

We shall deal first with the problem of a single electron moving in an electromagnetic field. We shall use *relativistic* notation throughout, taking the velocity of light to be unity. When we want to pass from the contravariant form K_μ to the covariant form K^μ of a vector, we shall use the fundamental tensor $g^{\mu\nu}$ appropriate for flat space-time, with components $g^{00} = 1$, $g^{11} = g^{22} = g^{33} = -1$, the other components vanishing. We shall also sometimes use the scalar product notation

$$(K, L) = K_\mu L^\mu = K_0L_0 - K_1L_1 - K_2L_2 - K_3L_3 \tag{1}$$

Let us suppose the world-line of our electron in space-time to be known and to be described by the equations

$$z_\mu = z_\mu(s), \tag{2}$$

where the $z_\mu(s)$ are given functions of the proper-time s. We have, of course, $dz_0/ds > 0$. The electromagnetic potentials A_μ at the point x_μ satisfy the equations

$$\partial A_\mu/\partial x_\mu = 0, \tag{3}$$

$$\Box A_\mu = 4\pi\, j_\mu, \tag{4}$$

where j_μ is the *charge-current density vector*. With our present model of the electron, j_μ vanishes everywhere except on the world-line of the electron, where it is infinitely great. The singularity in j_μ may conveniently be expressed in terms of δ-functions, and is easily seen to be

$$j_\mu = e \int dz_\mu/ds\ \delta(x_0 - z_0)\ \delta(x_1 - z_1)\ \delta(x_2 - z_2)\ \delta(x_3 - z_3)\ ds \tag{5}$$

for an electron of charge e. From the potentials A_μ the field quantities $F^{\mu\nu}$ can be derived according to

$$F^{\mu\nu} = \partial A^\nu/\partial x_\mu - \partial A^\mu/\partial x_\nu. \tag{6}$$

Equations (3) and (4) have many solutions and thus do not suffice to fix the field. One solution is provided by the well-known *retarded potentials* of Lienard and Wiechert. We shall call the field derived from these potentials $F^{\mu\nu}{}_{ret}$. We can obtain other solutions by adding to this one any solution of (3) and

$$\Box A_\mu = 0, \tag{7}$$

representing a *field of radiation.* The particular solution that represents the actual conditions in our one-electron problem will be the sum of the retarded potentials and the potentials satisfying (7) that represent the *incoming electromagnetic waves* incident on our electron. Thus, calling the actual field $F^{\mu\nu}{}_{act}$ and the incident field $F^{\mu\nu}{}_{in}$ we have

$$F^{\mu\nu}{}_{act} = F^{\mu\nu}{}_{ret} + F^{\mu\nu}{}_{in}. \tag{8}$$

Another important solution of (3) and (4) is that provided by the *advanced potentials.* We call the field derived from it $F^{\mu\nu}{}_{adv}$. We should expect $F^{\mu\nu}{}_{adv}$ to play a symmetrical role to $F^{\mu\nu}{}_{ret}$ in all questions of general theory. Let us therefore put, corresponding to (8),

$$F^{\mu\nu}{}_{act} = F^{\mu\nu}{}_{adv} + F^{\mu\nu}{}_{out}, \tag{9}$$

defining in this way a new field $F^{\mu\nu}{}_{out}$. This field is derivable from potentials satisfying (7) and thus represents some field of radiation, and we should expect it to play a symmetrical role in general theory to $F^{\mu\nu}{}_{in}$ so that it should be interpretable as the field of outgoing radiation leaving the neighborhood of the electron. The difference

$$F^{\mu\nu}{}_{rad} = F^{\mu\nu}{}_{out} - F^{\mu\nu}{}_{in} \tag{10}$$

would then be *the field of radiation produced by the electron.* This difference may, from (8) and (9), be expressed as

$$F^{\mu\nu}{}_{rad} = F^{\mu\nu}{}_{ret} - F^{\mu\nu}{}_{adv}, \tag{11}$$

which shows that it is completely determined by the world-line of the electron. In the appendix it is calculated near the world-line and found to be free from singularity and to have the value on the world-line

$$F_{\mu\nu\ rad} = 4e/3\ (d^3z_\mu/ds^3\ dz_\nu/ds - d^3z_\nu/ds^3\ dz_\mu/ds). \tag{12}$$

Let us compare our result (11) with the usual one for the radiation produced by an accelerating electron. For definiteness, suppose we have an electron initially moving with constant velocity, then undergoing an acceleration and finally getting again into a state of constant velocity. According to the usual theory, the radiation emitted by the electron while it is accelerating will be given by the value of $F^{\mu\nu}{}_{ret}$ at great distances from the electron and at correspondingly great times after the time of the acceleration. This will be equal to the value of the right-hand side of (11)

$$[F^{\mu\nu}{}_{rad} = F^{\mu\nu}{}_{ret} - F^{\mu\nu}{}_{adv}, \qquad (11)]$$

at these distances and times, since $F^{\mu\nu}{}_{adv}$ will be zero in this part of space-time. *Thus, our result (11) for the radiation produced by an accelerating electron is in agreement with the usual one in that region of space-time where the usual one is defined.*

Our result goes beyond the usual one in providing a definite value for the field of radiation throughout space-time, which is an advantage in giving us a meaning for the radiation field close to the electron. (The usual theory gives this field inextricably mixed up with the Coulomb field.) *Our result also gives a meaning for the radiation field before its time of emission, when it can have no physical significance.* This, however, is unavoidable if we are to have the radiation field well defined near the electron.

Our theory has attained symmetry between the use of retarded and advanced potentials. This symmetry will be maintained in all general theoretical discussions, but *it will not apply to practical applications*, since we then have $F^{\mu\nu}{}_{in}$ given (very often it is zero), while its counterpart $F^{\mu\nu}{}_{out}$ is unknown.

A field that we shall need in the future is $f^{\mu\nu}$, defined as the actual field minus the mean of the advanced and retarded fields,

$$f^{\mu\nu} = F^{\mu\nu}{}_{act} - \tfrac{1}{2}\,(F^{\mu\nu}{}_{ret} + F^{\mu\nu}{}_{adv}). \qquad (13)$$

It is derivable from potentials satisfying (7) and is free from singularity on the world-fine of the electron. It is, in fact, from (8) and (9), just the mean of the ingoing and outgoing fields of radiation,

$$f^{\mu\nu} = \tfrac{1}{2}\,(F^{\mu\nu}{}_{in} + F^{\mu\nu}{}_{out}).$$

The equations of motion of an electron

To complete our theory of the interaction between an electron and the electromagnetic field we require *equations of motion* for the electron— equations to determine the *world-line* which we assumed known in the preceding section. We can get information on this question from the *laws of conservation of energy and momentum*. We surround the singular *world-line* in space-time by a thin tube, whose radius is much smaller than any length of physical importance in our one-electron problem, and we calculate the flow of *energy* and *momentum* across the (three-dimensional) surface of this tube, using the *stress tensor* $T_{\mu\rho}$ of Maxwell's theory, calculated from the actual field $F^{\mu\nu}{}_{act}$ according to the formula*

$$4\pi\, T_{\mu\rho} = F_{\mu\nu}F_{\rho}{}^{\nu} + \tfrac{1}{4}\, g_{\mu\rho}\, F_{\alpha\beta}F^{\alpha\beta}. \qquad (14)$$

* The usual derivation of the *stress-tensor* is valid only for continuous charge distributions and we are here using it for point charges. This involves adopting as a fundamental assumption the point of view that *energy and momentum are localized in the field* in accordance with Maxwell's and Poynting's ideas.

We then have the information that the total flow of *energy* (or *momentum*) out from the surface of any finite length of tube must equal the difference in the *energy* (or *momentum*) residing within the tube at the two ends of this length, and must thus depend only on conditions at the two ends of this length. In mathematical language, the rate of flow of *energy* (or *momentum*) out from the surface of the tube must be a perfect differential.

It is easily seen that the information obtained in this way is independent of what shape and size we give to our tube (provided it is sufficiently small for the Taylor expansions used in the calculations to be valid). If we take two tubes surrounding the singular *world-line*, the divergence of the *stress tensor* $\partial T_{\mu\rho}/\partial x_{\rho}$ will vanish everywhere in the region of *space-time* between them, since there are no singularities in this region and (7) is satisfied throughout it. Expressing the integral

$$\int\int\int\int \partial T_{\mu\rho}/\partial x_\rho \, . \, dx_0 dx_1 dx_2 dx_3$$

over the region of *space-time* between a certain length of the two tubes as a surface integral over the (three-dimensional) surface of this region, we obtain immediately that the difference in the flows of *energy* (or *momentum*) across the surfaces of the two tubes depends only on conditions at the two ends of the length considered. Thus, the information provided by the conservation laws is well defined.

The calculations involved in getting this information are rather long and are given in the appendix. We choose the simplest shape of tube, a tube which is spherical, of constant radius ε, for each value of the *proper-time*, in that *Lorentz frame of reference* in which the electron is then at rest. To save writing we put υ_μ for dz_μ/ds and use dots to denote differentiations with respect to s. We note for future use the elementary equations

$$v^2 = 1 \tag{15}$$
$$(vv^{\cdot}) = 0 \tag{16}$$
$$(vv^{\cdot\cdot}) + v^{\cdot 2} = 0 \tag{17}$$

in the vector notation (1).

The result of the appendix is now that, with $f^{\mu\nu}$ defined by (13)

[$f^{\mu\nu} = F^{\mu\nu}{}_{act} - \frac{1}{2}(F^{\mu\nu}{}_{ret} + F^{\mu\nu}{}_{adv})$, (13)]

the flow of *energy* and *momentum* out from the surface of any length of tube is given by the vector integrated over the length of tube, where terms that vanish with ε are neglected. This must depend only on conditions at the two ends of the length of tube, so that the integrand must be a perfect differential, i.e.

$$\tfrac{1}{2} e^2\varepsilon^{-1}\upsilon_\mu^{\cdot} - e\upsilon_\nu f_\mu{}^\nu = B^{\cdot}_\mu \tag{18}$$

This is as far as we can get from the laws of conservation of energy and momentum. To fix our *equations of motion* for the electron, we must make some further assumption to fix the vector B_μ. From (18) we see that

$$(v\mathbf{B}) = \tfrac{1}{2} e^2\varepsilon^{-1}(\mathbf{vv^{\cdot}}) = 0, \tag{19}$$

with the help of (16)

[$(vv^{\cdot}) = 0$, (16)]

which restricts our choice of B_μ. Apart from this, B_μ may be any vector function of υ_μ and its derivatives. The simplest B_μ satisfying the restriction (19) is

$$B_\mu = k\upsilon_\mu, \tag{20}$$

where k is a constant. There are other possible expressions for B_μ satisfying (19), for example,

$$B_\mu = k' \, [v^{\cdot 4}\, \upsilon_\mu + 4(v^{\cdot}v^{\cdot\cdot})\, \upsilon_\mu^{\cdot}],$$

k' being another constant, but they are all much more complicated than (20), so that one would hardly expect them to apply to a simple thing like an electron. *We therefore assume (20).*

Substituting (20) into the right-hand side of (18), we see that the constant k must be of the form

$$k = \tfrac{1}{2} e^2\varepsilon^{-1} - m, \qquad (21)$$

where m is another constant independent of ε, *in order that our equations may have a definite limiting form when ε tends to zero.* We then get

$$m\dot{\upsilon}_\mu = e\upsilon_\nu f_\mu{}^\nu, \qquad (22)$$

as our *equations of motion for the electron*. They are of the usual form of the *equations of motion* of an electron in an external electromagnetic field, with m playing the part of the *rest-mass* of the electron and $f_\mu{}^\nu$, the actual field minus the mean of the advanced and retarded fields, playing the part of the *external field.*

Equations (22) are not in a form suitable for application to practical problems where we are given, not $f_\mu{}^\nu$, but the incident field $F_\mu{}^\nu{}_{in}$. The connection between these two fields is, from (13), (8) and (11),

$$\begin{aligned} f_\mu{}^\nu &= F_\mu{}^\nu{}_{in} + \tfrac{1}{2} F_\mu{}^\nu{}_{rad} \\ &= F_\mu{}^\nu{}_{in} + 2/3\, e(\upsilon_\mu{}^{\cdot\cdot}\upsilon^\nu - \upsilon^{\nu\cdot\cdot}\upsilon_\mu) \end{aligned} \qquad (23)$$

with the help of (12)

$$[F_{\mu\nu\ rad} = 4e/3\ (d^3z_\mu/ds^3\ dz_\nu/ds - d^3z_\nu/ds^3\ dz_\mu/ds). \qquad (12)]$$

Substituting this into (22) and using (15) and (17), we get

$$m\dot{\upsilon}_\mu - 2/3\, e^2\ddot{\upsilon}_\mu - 2/3\, e^2\dot{v}^2\upsilon_\mu = e\upsilon_\nu F_\mu{}^\nu{}_{in}. \qquad (24)$$

These are the same as the equations of motion obtained from the Lorentz theory of the extended electron by equating the total force on the electron to zero, if one neglects terms involving higher derivatives of v than the second. But whereas these equations, as derived from the Lorentz theory, are only approximate, we now see *that there is good reason for believing them to be exact*, within the limits of the classical theory.

To discuss the significance of equations (24), let us take the equation with $\mu = 0$, describing the *energy balance*. The right-hand side, which gives the rate at which the incident field does work on the electron, is equated to the sum of the three terms $m\dot{\upsilon}_0$, $- 2/3\, e^2\ddot{\upsilon}_0$ and $- 2/3\, e^2\dot{v}^2\upsilon_0$. The first two of these are perfect differentials and the things they are the differentials of, namely $m\upsilon_0$ and $- 2/3\, e^2\dot{\upsilon}_0$, may be considered as *intrinsic energies of the electron*. The former is just the usual expression for the kinetic energy of a particle of rest-mass m, while the latter is what is called the "*acceleration energy*" of the electron [Schott, G. A. (1915). On the Motion of the Lorentz Electron. *Phil. Mag.*, 29, 49-62; https://doi.org/10.1080/14786440108635280]. Changes in the *acceleration energy* correspond to *a reversible form of emission or absorption of field energy*, which never gets very far from the electron. The third term $-2/3\, e^2 \dot{v}^2 \upsilon_0$ corresponds to *irreversible emission of radiation* and gives the effect of *radiation damping* on the motion of the electron. It is necessarily positive, since, according to (16), $\dot{\upsilon}_\mu$ is orthogonal to the time-like vector υ_μ and is thus a space-like vector, and hence its square is negative.

Let us see how the kinetic energy term arises. The B_μ introduced in (18) can be interpreted as *minus the vector of energy and momentum* residing within the tube at any value of the proper-time. Thus, from (20) and (21), *the energy within the tube must be negative and must tend to* $-\infty$ *as* ε *tends to zero*. This negative *energy* is needed to compensate for the large positive *energy* of the *Coulomb field* just outside the tube, to keep the *total energy* down to the value appropriate to the *rest-mass* m. *If we want a model of the electron, we must suppose that there is an infinite negative mass at its center such that, when subtracted from the infinite positive mass of the surrounding Coulomb field, the difference is well defined and is just equal to m.* Such a model is hardly a plausible one according to current physical ideas but, as discussed in the Introduction, this is not an objection to the theory provided we have a reasonable mathematical scheme.

Application of the equations of motion

Let us now accept equations (24) as giving an exact description of the motion of an electron in a specified incident field and investigate some of their consequences. *The equations involve* $\ddot{\upsilon}_\mu$ *which means that the motion will not be determined if we are given only the position and velocity of the electron at one instant of time. We must also be given its acceleration.* The equations can then be used to determine the rate of change of the acceleration and the whole motion will then be fixed.

As an interesting special case let us suppose there is no incident field, so that we have the *equations of motion*

$$a\dot{\upsilon}_\mu - \ddot{\upsilon}_\mu - \dot{v}^2\upsilon_\mu = 0, \qquad (25)$$

where $a = 3m/2e^2$. In general, *the electron will not now be moving with constant velocity, as it would according to ordinary ideas, since we may suppose it to be started off with a non-zero acceleration and it cannot then suddenly lose its acceleration.* Let us pass over to the notation x, y, z, t instead of z_1, z_2, z_3, z_0 and choose a system of co-ordinates so that the initial velocity and acceleration are in the direction of the x-axis, or, in *relativistic* language, *the initial velocity and acceleration four-vectors lie in the (xt) plane*. Then from symmetry considerations the motion must lie entirely in this plane.

The equations (25) now become

$$a\ddot{x} - \dddot{x} - \dot{x}(\ddot{t}^2 - \ddot{x}^2) = 0, \qquad (26)$$

$$a\ddot{t} - \dddot{t} - \dot{t}(\ddot{t}^2 - \ddot{x}^2) = 0. \qquad (27)$$

Equations (15), (16) and (17) give

$$\dot{x}^2 - \dot{t}^2 = 1, \qquad (28)$$

$$\dot{t}\ddot{t} - \dot{x}\ddot{x} = 0, \qquad (29)$$

$$(\dot{t}\dddot{t} - \dot{x}\dddot{x}) + (\ddot{t}^2 - \ddot{x}^2) = 0,$$

from which we see that $\dot{x}$ times equation (26) minus $\dot{t}$ times equation (27) vanishes identically, so that either of these equations is equivalent to the other. Eliminating $\dot{t}$ from (28) and (29), we get

$$\dot{t} = \ddot{x}\,\dot{x}/(1 + \dot{x}^2)^{1/2},$$

which gives, on being substituted into (26),

$$a\ddot{x} - \dddot{x}/\ddot{x} + \dot{x}\ddot{x}/(1 + \dot{x}^2) = 0.$$

If x does not vanish, we may write this as

$$a - \dddot{x} - \dot{x}\ddot{x}^2/(1 + \dot{x}^2) = 0,$$

and integrate immediately, to get

$$as - \log \ddot{x} + \tfrac{1}{2} \log (1 + \dot{x}^2) = c,$$

where c is a constant of integration. It is convenient to choose the origin from which s is measured to make $c = -\log a$. We then have

$$\ddot{x}/(1 + \dot{x}^2)^{1/2} = a e^{as},$$

which can be integrated again, to give

$$\log [\dot{x} + (1 + \dot{x}^2)^{1/2}] = e^{as} + b,$$

where b is another constant of integration. We may write this result as

$$\left.\begin{aligned} x &= \sinh (e^{as} + b), \\ t &= \cosh (e^{as} + b). \end{aligned}\right) \qquad (30)$$

We now have that, as s tends to $-\infty$, the velocity tends to a constant value, given by $x = \sinh b$ or $dx/dt = \tanh b$. As s increases from $-\infty$ the motion departs from the asymptotic motion of constant velocity, and the velocity steadily increases. For large values of s the velocity tends to the velocity of light according to an extremely rapid law.

At this stage one would be inclined to say that there is a mistake in sign in our equations and that we ought to have e^{-as} instead of e^{as} in (30). With this alteration we should have a theory in which, if an electron is disturbed in any way and then left alone, it would rapidly settle down into a state of constant velocity, with emission of radiation while it is settling down. This would be a reasonable behavior for an electron according to our present-day physical ideas. However, *it is not possible to tamper with the signs in our theory in any relativistic way to obtain this result, without getting equations of motion which would make the electron in the hydrogen atom spiral outwards, instead of spiraling inwards and ultimately falling into the nucleus, as it should in the classical theory*. We are therefore forced to keep the signs in (30) as they are and to see what interpretation we can give to the equations as they stand. This will lead us to the most beautiful feature of the theory.

The motion (30) is certainly never observed for an electron in the absence of incident radiation. This does not give a contradiction between our theory and observation, however, since our equations of motion (25)

$$[a\upsilon^{\cdot}_{\mu} - \upsilon_{\mu}^{\cdot\cdot} - v^{\cdot 2}\upsilon_{\mu} = 0, \qquad (25)]$$

admit of alternative solutions, namely υ_{μ} = constant, which are in agreement with observation. *We must merely impose the condition that these solutions are the ones that occur in Nature*. Let us see how this condition works in a practical problem.

Suppose we have an electron which is disturbed by some electromagnetic radiation incident on it and is then left alone. We may use our general equations of motion (24)

$$[m\upsilon^{\cdot}_{\mu} - 2/3\ e^2\upsilon_{\mu}^{\cdot\cdot} - 2/3\ e^2 v^{\cdot 2}\upsilon_{\mu} = e\upsilon_{\nu}F_{\mu}{}^{\nu}{}_{in}. \qquad (24)]$$

and must *restrict ourselves to those solutions for which the velocity is constant during the final period when the electron is left alone*. This will in general prevent us from having the velocity exactly constant during the initial period before the disturbance has arrived and will compel us then to have a motion of the form (30), with the electron gradually building up a small acceleration. The difference of such a motion from one of constant velocity is too small to show itself as a contradiction with observation, so that we can get agreement between the solutions of our equations and observed motions on these lines.

We now have a striking departure from the usual ideas of mechanics. We must obtain solutions of our equations of motion for which the initial position and velocity of the electron are prescribed, together with its final acceleration, instead of solutions with all the initial conditions prescribed.

Motion of an electron disturbed by a pulse

To study the rather unexpected results of the preceding section more closely, let us take a typical special case, which is sufficiently simple to be worked out completely. Suppose we have an electron initially at rest, disturbed by a pulse of *electromagnetic radiation* passing over it. To simplify the mathematics as much as possible, let us suppose the duration of the pulse to be infinitely short, so that the *electric force* is represented by a δ-function, say

$$E_x = k\ \delta(t - y) \qquad E_y = E_z = 0, \qquad (31)$$

in the *xyzt* notation, where k is a constant, giving us a pulse polarized in the direction of the x-axis and moving in the direction of the y-axis. We shall consider a motion for which the electron remains always in the plane y = 0, z = 0 in space-time (neglecting the magnetic effects of the pulse).

Substituting (31) into our equations of motion (24)

$$[m\upsilon^{\cdot}_{\mu} - 2/3\ e^2\upsilon_{\mu}^{\cdot\cdot} - 2/3\ e^2 v^{\cdot 2}\upsilon_{\mu} = e\upsilon_{\nu}F_{\mu}{}^{\nu}{}_{in}, \qquad (24)]$$

we get

$$a x^{\cdot\cdot} - x^{\cdot\cdot\cdot} - x^{\cdot}(t^{\cdot\cdot 2} - x^{\cdot\cdot 2}) = \kappa\ \delta(t)\ t^{\cdot}, \qquad (32)$$

where κ = 3k/2e. The equation involving $t^{\cdot\cdot\cdot}$ is not independent of this one and may be ignored. Let us now suppose k to be sufficiently small for the velocity acquired by the electron to be small compared with the velocity of light, so that we may neglect *relativistic* effects. Equation (32) then reduces to

$$a\ddot{x} - \dddot{x} = \kappa\, \delta(t), \qquad (33)$$

in which we may take the dot to denote differentiation with respect to t. Equation (33) shows that, at the time t = 0, $\ddot{x}$ increases discontinuously by an amount $-\kappa$, and before and after this time we have

$$a\ddot{x} - \dddot{x} = 0. \qquad (34)$$

According to the conclusions of the preceding section, we must take a motion for which, after t = 0, $\dot{x}$ is a constant, q say. We now have $\ddot{x}$ zero just after t = 0, so it must have the value κ just before. The general solution of (34) is

$$\dot{x} = c_1\, e^{at} + c_2,$$

where c_1 and c_2 are constants of integration. To obtain the motion of our electron before t = 0, we must choose these constants of integration so that $\ddot{x} = 0$ for $t = -\infty$ and $\ddot{x} = \kappa$ for t = 0, the former condition taking into account that the electron is initially at rest. This fixes $c_2 = 0$ and $c_1 = \kappa/a$. Finally, we have the condition that $\dot{x}$ must be continuous at t = 0 (since there is no δ-function in x), which gives us $q = c_1$. Thus, the solution of our *equations of motion* is

$$\left.\begin{aligned} \dot{x} &= \kappa/a \,.\, e^{at}, \quad t < 0, \\ &= \kappa/a, \quad\quad\;\; t > 0. \end{aligned}\right) \qquad (35)$$

We can describe the motion by saying that the electron is, to a high approximation, at rest for large negative values of t, but as t approaches zero it acquires a velocity and acceleration, in accordance with equations (30), of such amounts that just before t = 0 the acceleration has the right value to be exactly cancelled by the effect of the pulse, so that after t = 0 the electron is left moving with constant velocity. It would appear here that we have a contradiction with elementary ideas of causality. The electron seems to know about the pulse before it arrives and to get up an acceleration (as the *equations of motion* allow it to do), just sufficient to balance the effect of the pulse when it does arrive. The electron will, of course, radiate all the time it is accelerating and will thus be radiating before t = 0.

The behavior of our electron can be interpreted in a natural way, however, if we suppose the electron to have a finite size. There is then no need for the pulse to reach the center of the electron before it starts to accelerate. *It starts to accelerate and radiate as soon as the pulse meets its outside.* Mathematically, the electron has no sharp boundary and must be considered as extending to infinity, but *for practical purposes it may be considered to have a radius of order a^{-1} this being the distance within which the pulse must arrive before the acceleration and radiation are appreciable.*

The above interpretation leaves one further point to be examined. Suppose we have a pulse sent out from a place A and a receiving apparatus for *electromagnetic waves* set up at a place B, and suppose there is an electron on the straight-line joining A to B. Then the electron will be radiating appreciably at a time a^{-1} before the pulse has reached its center and this emitted radiation will be detectable at a time a^{-1} earlier than when the pulse, which travels from A to B with the velocity of light, arrives. In this way a signal can be sent from A to faster than light. *This is a fundamental departure from the ordinary ideas of relativity* and is to be interpreted by saying that *it is possible for a signal to be transmitted faster than light through the interior of an electron. The finite size of the electron now reappears in a new sense*, the interior of the electron being a region of failure, not of the field equations of electromagnetic theory, but of some of the elementary properties of space-time. In spite of this departure from ordinary *relativistic* ideas, our whole theory is the Lorentz invariant.

Some interesting points about the solution (35) of our problem may be noted. In the first place, the total momentum acquired by the electron is

$$m\kappa/a = ek.$$

This is precisely the same as the momentum the electron would acquire on an elementary theory which ignores the radiation damping completely.

Secondly, let us determine the spectral distribution of the radiation emitted by the electron. The rate of emission of energy is $2/3\ e^2\ddot{x}^2$ and its spectral distribution is given by the Fourier resolution of $\ddot{x}$. Thus, the energy emitted per unit frequency range will be

$$E_\nu = 2/3\ e^2\ |\int_{-\infty}^{\infty} \ddot{x}\, e^{2\pi i\nu t}\, dt\ |^2 = 2/3\ e^2\kappa^2|\int_{-\infty}^{\infty} e^{(a+2\pi i\nu)t}\, dt\ |^2$$

$$= 2e^2\kappa^2/3(a^2 + 4\pi^2\nu^2) = 2e^4\kappa^2/3m^2(1 + 4\pi^2\nu^2/a^2). \qquad (36)$$

If we resolve the incident pulse into its Fourier components and suppose each component to be scattered in accordance with Thomson's formula, we should obtain for the energy scattered per unit frequency range the value $2e^4\kappa^2/3m^2$. This is the same as (36) for values of ν small compared with a.

It has been suggested by Oppenheimer (1935) [Oppenheimer, J. R. (January, 1935). Are the Formulae for the Absorption of High Energy Radiations Valid? *Phys. Rev.*, 47, 44; https://doi.org/10.1103/PhysRev.47.44], as a possible escape from the discrepancy between theory and cosmic ray observations for the penetrating power of fast electrons, that it may not be permissible to calculate the scattering of the various Fourier components of the radiation incident on an electron independently, but that the presence of the high-frequency components may reduce the scattering of the low-frequency components. *We now see that this suggestion is not supported by our theory* and some other explanation of the discrepancy, such as the recently proposed heavy electron, is needed.

Extension to several electrons

The foregoing theory may easily be extended to apply to any number of electrons interacting with each other and with a field of radiation. Let us label the electrons by a suffix n or m. The nth electron will have its own retarded field $F^{\mu\nu}{}_{n,ret}$ and advanced field $F^{\mu\nu}{}_{n,adv}$, determined by its world-line, and the difference

$$F^{\mu\nu}{}_{n,rad} = F^{\mu\nu}{}_{n,ret} - F^{\mu\nu}{}_{n,adv} \qquad (37)$$

may be considered as the radiation field produced by the nth electron. Equation (12)

$$[F_{\mu\nu\ rad} = 4e/3\ (d^3z_\mu/ds^3\ dz_\nu/ds - d^3z_\nu/ds^3\ dz_\mu/ds). \qquad (12)]$$

will still apply for the value of this field on the world-line of the electron producing it.

Generalizing (8),

$$[F^{\mu\nu}{}_{act} = F^{\mu\nu}{}_{ret} + F^{\mu\nu}{}_{in}. \qquad (8)]$$

we have that the actual field is equal to the incident field plus the sum of the retarded fields of all the electrons,

$$F^{\mu\nu}{}_{act} = F^{\mu\nu}{}_{in} + \textstyle\sum_n F^{\mu\nu}{}_{n,ret} \qquad (38)$$

Similarly, generalizing (9),

$$[F^{\mu\nu}{}_{act} = F^{\mu\nu}{}_{adv} + F^{\mu\nu}{}_{out}, \qquad (9)]$$

we have

$$F^{\mu\nu}{}_{act} = F^{\mu\nu}{}_{out} + \textstyle\sum_n F^{\mu\nu}{}_{n,adv} \qquad (39)$$

Subtracting, we find that the total radiation field produced by the assembly of electrons is

$$F^{\mu\nu}{}_{rad} = F^{\mu\nu}{}_{out} - F^{\mu\nu}{}_{in} = \textstyle\sum_n (F^{\mu\nu}{}_{n,ret} - F^{\mu\nu}{}_{n,adv}),$$

which is thus the sum of all the above-defined radiation fields produced by the electrons individually. Of course, only the total radiation field is observable, that for an individual electron being merely a mathematical concept.

To get the field $f^{\mu\nu}{}_n$ which must be inserted into the *equations of motion* (22)

$$[m\upsilon^{\cdot}{}_\mu = e\upsilon_\nu f_\mu{}^\nu, \qquad (22)]$$

for the nth electron we must, generalizing (13),

$$[f^{\mu\nu} = F^{\mu\nu}{}_{act} - \tfrac{1}{2}\,(F^{\mu\nu}{}_{ret} + F^{\mu\nu}{}_{adv}). \qquad (13)]$$

subtract the mean of the retarded and advanced fields of this electron from the actual field, thus

$$f^{\mu\nu}{}_n = F^{\mu\nu}{}_{act} - \tfrac{1}{2}\,(F^{\mu\nu}{}_{n,ret} + F^{\mu\nu}{}_{n,adv}). \qquad (40)$$

…

Suppose now we are concerned with a problem in which

$$F^{\mu\nu}{}_{out} - F^{\mu\nu}{}_{in} = 0. \qquad (43)$$

Then the *field* $f^{\mu\nu}{}_n$ to be used in the *equations of motion* (22)

$$[m\upsilon^{\cdot}{}_\mu = e\upsilon_\nu f_\mu{}^\nu, \qquad (22)]$$

for the nth electron is just the sum of the mean of the *retarded* and *advanced fields* of all the other electrons. The interaction of the electrons for this case has been studied by Fokker [Fokker, A. D. (May, 1929). Ein invarianter Variationssatz für die Bewegung mehrerer elektrischer Massenteilchen. (An invariant set of variations for the movement of several electrical mass particles.) *Zeit. Phys.*, 58, 386-93; https://doi.org/10.1007/ BF01340389], who finds that the motion of all the electrons may now be described by a *variation principle*

$$\delta I = 0. \tag{44}$$

In our present notation I would have the form

$$I = \sum_n \int m \, ds_n + \tfrac{1}{2} \sum_n \sum_{m \neq n} \int \ldots \, ds_n + \tfrac{1}{2} \sum_n \sum_{m \neq n} \int \ldots \, ds_n, \tag{45}$$

where the two double sums are to be taken over all pairs of electrons, and in the first we must take for z_m and v_m their *retarded values* (with respect to the point on s_n considered), and in the second we must take their *advanced values*.

When the condition (43) does not apply, Fokker's *variation principle* (44) will still hold provided we add on to I the terms

$$\sum_n e \int \tfrac{1}{2} (A^{\mu}_{in} + A^{\mu}_{out}) \upsilon_{\mu n} \, ds_n, \tag{46}$$

involving the *potentials* of the mean of the *ingoing* and *outgoing* fields. The *variation principle* will not be of much use in practical problems, where this mean field is unknown, but it may be of value in suggesting how the quantum generalization of our theory is to be made.

Appendix

…

Summary

The object of the paper is to set up in the classical theory a self-consistent scheme of equations which may be used to calculate all the results that can be obtained from experiment about the interaction of electrons and radiation. *The electron is treated as a point charge and the difficulties of the infinite Coulomb energy are avoided by a procedure of direct omission or subtraction of unwanted terms*, somewhat similar to what has been used in the theory of the positron. The equations obtained are of the same form as those already in current use, but in their physical interpretation *the finite size of the electron reappears in a new sense, the interior of the electron being a region of space through which signals can be transmitted faster than light.*

Dancoff, S. M. (May, 1939). On Radiative Corrections for Electron Scattering.

[*Phys. Rev.*, 55, 959; https://doi.org/10.1103/PhysRev.55.959.]

Received 27 March 1939.

University of California, Berkeley, California.

Abstract

A *relativistic* treatment of the radiative correction of order $e^2/\hbar c$ to the elastic scattering cross section leads to the following results: (a) For the scattering in an electrostatic field of a particle described by the Pauli-Weisskopf theory, the correction is finite … . (b) For the scattering of a Dirac electron in an electrostatic field, the correction diverges logarithmically and is positive. (c) The convergence or divergence of the correction depends critically on the type of scattering potential considered.

Dirac, P. A. M. (July, 1939). A new notation for quantum mechanics.

[*Proc. Camb. Phil. Soc.*, 35, 416-8; https://doi.org/10.1017/S0305004100021162.]

Received April 29, 1939.

St John's College Cambridge.

Introduction of *bra ket* notation, quantum mechanics deals with *vectors in Hilbert space*, representing the *states* of a dynamical system, and with *linear operators*, representing dynamical *variables*, sometimes one makes calculations using the *vectors and linear operators* directly treating them as *abstract quantities*, at other times one works with *coordinates* (or *representatives*) of these *quantities*, the *bra ket* notation provides concise way of writing the *abstract quantities* themselves and their *coordinates* in a single scheme, leads to a unification of ideas.

In mathematical theories the question of notation, while not of primary importance, is yet worthy of careful consideration, since a good notation can be of great value in helping the development of a theory, by making it easy to write down those quantities or combinations of quantities that are important, and difficult or impossible to write down those that are unimportant. The summation convention in tensor analysis is an example, illustrating how especially appropriate a notation can be.

The notation in current use in quantum mechanics is fairly well suited to its purposes, but has some drawbacks. One has to deal with *vectors in Hilbert space*, representing the states of a dynamical system, and with *linear operators*, representing dynamical variables, and one sometimes makes calculations using the *vectors and linear operators* directly, treating them as *abstract quantities* which can be combined together algebraically according to certain rules, while at other times one works with *coordinates* (or *representatives*, as they are called) *of these quantities*.

[*Hilbert spaces* (named after David Hilbert) allow generalizing the methods of linear algebra and calculus from (finite-dimensional) Euclidean vector spaces to spaces that may be infinite-dimensional. Hilbert spaces arise naturally and frequently in mathematics and physics, typically as function spaces. Formally, a Hilbert space is a vector space equipped with an *inner product* that defines a distance function for which the space is a complete metric space.

The *inner product* of two vectors in the space is a scalar, often denoted with angle brackets such as in $< a, b >$. Inner products allow formal definitions of intuitive geometric notions, such as lengths, angles, and orthogonality (zero inner product) of vectors.]

For the two styles of calculation two distinct notations are used, which do not fit together very naturally and which give rise to an awkward jump in the flow of one's thoughts when one changes from one to the other. In the present note *a new notation is set up, which provides a neat and concise way of writing, in a single scheme, both the abstract quantities themselves and their coordinates*, and thus leads to a unification of ideas.

A *Hilbert-space vector*, which was denoted in the old notation by the letter ψ, will now be denoted by a special new symbol $>$. If we are concerned with a particular vector, specified by a label, *a* say, which would be used as a suffix to the ψ in the old notation, we write it $|\,a>$. It may be that the label is very complicated, consisting of many letters, but we can always write down the vector conveniently in the new notation, simply by enclosing the label between $|$ on the left and $>$ on the right. We have also to deal with another kind of *Hilbert-space vector*, the *conjugate imaginary* of the first kind. This was denoted in the old notation by ϕ or ψ^-, and will now be denoted by $<$. If one of them is specified by a label *a*, we write it $<a\,|$.

A pair of vectors, one of each kind, have a symbolic product, which is a number. In the old notation such a product was denoted by $\phi\psi_a$, $\phi_a\psi$, or $\phi_a\psi_b$ according as the vectors have labels or not. In the new notation these products will be denoted by $<\,>$, $<\,|\,a>$, $<a\,|\,>$ and $<a\,|\,b>$ respectively. Note that in the last of these it is not necessary to have the $|$ occurring twice, as a simple juxtaposition of $<a\,|$ and $|\,b>$ would give.

Let us now introduce a representation, say the one in which each of a certain complete set of commuting observables *q* is diagonal. This gives us a set of basic vectors in the Hilbert space, one for each set of eigenvalues q' for the *q*'s. These basic vectors, which were denoted in the old notation by $\psi(q')$, will now be denoted by $|\,q'>$, the q' being treated as an ordinary label. Similarly, the *conjugate imaginary* basic vectors, which were denoted in the old notation by $\phi(q')$, will now be denoted by $<q'\,|$. To get the representative of any vector $|\,a>$, we must form its product with the basic vector $<q'\,|$, which gives us $<q'\,|\,a>$. This is similar to the bracket expression $(q'\,|\,a)$ for the representative of ψ_a in the old notation. *But while the bracket expression $(q'\,|\,a)$ of the old notation has to be introduced as an extraneous symbol, independent of what has gone before, the $<q'\,|\,a>$ of the new notation appears naturally as a symbolic product.*

This illustrates the main feature of the new notation. With the old notation there are often two quite different ways of writing a quantity, one as a product of abstract symbols and the other by means of the bracket notation - for example, the above bracket expression $(q'\,|\,a)$ could also be written as the product $\phi(q')\,\psi_a$ - but in the new notation there is always only one, and complete continuity is preserved when one passes from expressions involving abstract symbols to representatives.

This feature is illustrated also by transformation functions. If we take a second representation in which, say, each of the complete set of commuting observables ξ is diagonal, the transformation function would be written in the old notation either as the bracket expression $(q'\,|\,\xi')$ or as the symbolic product of two basic vectors $\phi(q')\,\psi(\xi')$. In the new notation these two ways of writing it coalesce into $<q'\,|\,\xi'>$.

The development of the new notation to include linear operators and observables can be effected without difficulty. Below is a list of the various types of quantity involving a linear operator α, written on the left in the old notation and on the right in the new,

$\alpha\psi$	$\alpha >$
$\alpha\psi_a$	$\alpha \vert a >$
$\phi\alpha$	$< \alpha$
$\phi_a\alpha$	$< a \vert \alpha$
$\phi_a\alpha\psi$	$< a \vert \alpha >$ or $< a \vert \alpha \vert >$
$\phi\alpha\psi_a$	$< \alpha \vert a >$ or $< \vert \alpha \vert a >$
$\phi_a\alpha\psi_b$	$< a \vert \alpha \vert b >$
$\phi(q')\alpha\psi(q'')$ or $\phi(q')\alpha(q'')$	$< q' \vert \alpha \vert q'' >$

The last of these is the representative of a linear operator and provides a further example of a quantity that can be written in two quite different ways with the old notation, but only in one with the new. Where two forms are given for writing an expression in the new notation, the former may be used for brevity when there is no danger of α being mistaken for the label of a vector in Hilbert space, otherwise the latter must be used.

Two general rules in connection with the new notation may be noted, namely, *any quantity in brackets < > is a number, and any expression containing an unclosed bracket symbol < or > is a vector in Hilbert space, of the nature of a* ϕ *or* ψ *respectively*. As names for the new symbols < and > to be used in speech, I suggest the words *bra* and *ket* respectively.

Weisskopf, V. F. (July, 1939). On the Self-energy and Electromagnetic Field of the Electron.

[*Phys.* Rev., 56, 1, 72; https://doi.org/10.1103/PhysRev.56.72; also in Schwinger, J. (ed). (1958). *Selected Papers on Quantum electrodynamics*. Dover, New York, pages 68-81.]

Received April 12, 1939.

University of Rochester, Rochester, New York.

The main purpose of this paper is to show the physical significance of the logarithmic divergence of the self-energy of the electron and to demonstrate the reason for its occurrence, the *self-energy of the electron is its total energy in free space when isolated from other particles or light quanta*, $W = T + (1/8\pi) \int (H^2 + E^2)\, dr$ where T is the *kinetic energy* of the electron and H and E are the *magnetic* and *electric field strengths* at point r, identifies three reasons why quantum theory of the electron results in infinite *self-energy* of the electron, claims that *quantum kinematics shows that the radius of the electron must be assumed to be zero*, resulting in infinite energy of the *electrostatic field*, the contributions of the electric and magnetic fields of the spin to the *self-energy* of the electron cancel one another, *quantum theory of the electromagnetic field postulates the existence of field strength fluctuations in empty space*, gives rise to an additional energy which diverges more strongly that the electrostatic self-energy, induces the electron to perform vibrations with energy that diverges quadratically *for an infinitely small radius*, Dirac's positron theory implies that the charge and magnetic dipole of the electron are extended over a finite region, explains why the *self-energy* is only *logarithmically infinite*, *divergences are consequence of assumption of point electron*.

Abstract

The *charge distribution*, the *electromagnetic field* and the *self-energy* of an electron are investigated. It is found that, as a result of Dirac's positron theory, the charge and the magnetic dipole of the electron are extended over a finite region; the contributions of the spin and of the fluctuations of the radiation field to the *self-energy* are analyzed, and the reasons that the *self-energy* is only *logarithmically infinite* in positron theory are given. It is proved that the latter result holds to every approximation in an expansion of the *self-energy* in powers of e^2/hc. The *self-energy* of charged particles obeying Bose statistics is found to be quadratically divergent. Some evidence is given that the "*critical length*" of positron theory is as small as $h/(mc)\cdot\exp(-hc/e^2)$.

1. *Introduction and Discussion of Results*

The *self-energy of the electron is its total energy in free space when isolated from other particles or light quanta*. It is given by the expression

$$W = T + (1/8\pi) \int (H^2 + E^2)\, dr \qquad (1)$$

Here T is the *kinetic energy* of the electron; H and E are the *magnetic* and *electric field strengths* [at point r]. In classical electrodynamics the *self-energy of an electron* of radius a at rest and without spin is given by

$$W \sim mc^2 + e^2/a$$

and consists solely of the energy of the rest mass and of its electrostatic field. This expression diverges linearly *for an infinitely small radius*. If the electron is in motion, other terms appear representing the energy produced by the magnetic field of the moving electron. These terms, of course, can be obtained by a Lorentz transformation of the former expression.

> [*Lorentz transformations* are a six-parameter family of linear transformations from a coordinate frame in spacetime to another frame that moves at a constant velocity relative to the former. The most common form of the transformation, parametrized by the real constant υ, representing a velocity confined to the x-direction, is expressed as
>
> $$t' = \gamma\left(t - \frac{vx}{c^2}\right)$$
> $$x' = \gamma(x - vt)$$
> $$y' = y$$
> $$z' = z$$
>
> where (t, x, y, z) and (t′, x′, y′, z′) are the coordinates of an event in two frames with the origins coinciding at t = t′ = 0, where the primed frame is seen from the unprimed frame as moving with speed υ along the x-axis, where c is the speed of light, and light, and
>
> $$\gamma = \left(\sqrt{1 - \frac{v^2}{c^2}}\right)^{-1}$$
>
> is the Lorentz factor. When speed υ is much smaller than c, the Lorentz factor is negligibly different from 1, but as υ approaches c, γ grows without bound. The value of υ must be smaller than c for the transformation to make sense.]

The quantum theory of the electron has put the problem of *self-energy* into a critical state. There are three reasons for this:

(a) *Quantum kinematics shows that the radius of the electron must be assumed to be zero.* [???] It is easily proved that the product of the charge densities at two different points, $\rho(r - \xi/2) \times \rho(r + \xi/2)$, is a delta function $e^2\delta(\xi)$. In other words: if one electron alone is present, the probability of finding a charge density simultaneously at two different points is zero for every finite distance between the points. Thus, *the energy of the electrostatic field is infinite* as

$W_{st} = \lim_{(a=0)} e^2/a$.

(b) The quantum theory of the *relativistic electron* attributes a *magnetic moment* to the electron, so that the electron is surrounded by a magnetic field. The energy … of this field … corresponds to the *field energy* of a magnetic dipole of the moment eh/2mc which is spread over a volume of dimension *a*. The spin, however, does not only produce a magnetic field, it also gives rise to an alternating electric field. The closer analysis of the Dirac wave equation has shown[1] that the *magnetic moment* of the spin is produced by an irregular circular fluctuation movement (Zitterbewegung) of the electron which is superimposed to the translatory motion.

[1] Schroedinger, E. (1930). Uber die kraftfreie Bewegung in der relativistischen Quantenmechanik. (About forceless motion in relativistic quantum mechanics.) *Sitzungsber. Preuss. Akad. Wiss. Phys.*, 24, 418-28.

The instantaneous value of the velocity is always found to be zero. It must be expected that this motion will also create an alternating current. … The fact that the above expression does not vanish for an electron at rest proves the existence of a *solenoidal field*[2] apart from the irrotational electric field of the charge.

[2] A *solenoidal electric field* is necessarily an alternating field for its time average vanishes in a stationary state, whereas the time average of a magnetic field does not vanish if stationary currents are present.

The energies of the electric and magnetic fields of the spin are found to be equal. … If the *self-energy* is expressed in terms of the field energies, the electric and magnetic parts have opposite signs[3], so that *the contribution of the electric and magnetic fields of the spin cancel one another*.

[3] This at first sight unfamiliar result is connected with the well-known fact that a system of steady currents *increases* its *magnetic energy* if it performs mechanical work, whereas a system of charges *decreases* its *field energy* by performing mechanical work.

(c) *The quantum theory of the electromagnetic field postulates the existence of field strength fluctuations in empty space*. These give rise to an additional energy, which diverges more strongly that the electrostatic self-energy. … consider an electron with radius *a*. The field fluctuations in a volume a^3 are of the order $E^2 \sim hc/a^4$.[4]

[4] The fluctuations are of the order of magnitude of the field-strength of one light quantum with wavelength .. [???].

The mean frequency of the fluctuation is $\nu = c/a$. This field induces the electron to perform vibrations with amplitude $x \sim e^2E^2/mv^2 \sim$ e2h/mca2. This energy diverges quadratically *for an infinitely small radius*. …

A new situation is created by Dirac's *theory of the positron*: the *self-energy* diverges only logarithmically with infinitely small radius. This fact has been proved only for the first approximation of the *self-energy* expanded in powers of e^2/hc.5

[5] Weisskopf, V. (January, 1934). Über die Selbstenergie des Elektrons. (The Self-energy of the Electron.) *Zeit. Phys.*, 89, 27-39.

However, it will be shown in Section VI that the divergence is logarithmic in every approximation. *The main purpose of this paper is to show the physical significance of the logarithmic divergence and to demonstrate the reason for its occurrence.*

…

II. *The Charge Distribution of the Electron*

…

III. *The Electromagnetic Field of the Electron*

…

IV. *The Self-energy of the Electron*

…

V. *The Self-energy of a particle obeying the Bose Statistics*

…

IV. *The higher approximations of the Self-energy in the Theory of the Positron*

…

Hideki Yukawa (January 23, 1907 – September 8, 1981)

Yukawa was a Japanese theoretical physicist and the first Japanese Nobel laureate in 1949 "for his prediction of the existence of mesons on the basis of theoretical work on nuclear forces."

Atomic nuclei consist of protons and neutrons held together by a strong force. Yukawa assumed that this force is borne by particles and that there is a relationship between the range of the force and the mass of the force-bearing particle. In 1934, Yukawa predicted that this particle should have a mass about 200 times that of an electron. He called this particle a "meson". Mesons' existence was verified in later experiments. [Hideki Yukawa – Facts. NobelPrize.org. https://www.nobelprize.org/prizes/physics/1949/yukawa/ facts/.]

Hideki Yukawa was born in Tokyo, Japan, on 23rd January, 1907, the third son of Takuji Ogawa, who later became Professor of Geology at Kyoto University. His father, for a time, considered sending him to technical college rather than university since he was "not as outstanding a student as his older brothers". However, when his father broached the idea with his middle school principal, the principal praised his "high potential" in mathematics and offered to adopt Yukawa himself in order to keep him on a scholarly career. At that, his father relented.

Yukawa decided against becoming a mathematician when in high school; his teacher marked his exam answer as incorrect when Yukawa proved a theorem but in a different manner than the teacher expected. He decided against a career in experimental physics in college when he demonstrated clumsiness in glassblowing, a requirement for experiments in spectroscopy.

In 1929, after receiving his degree from Kyoto Imperial University, he stayed on as a lecturer for four years. After graduation, he was interested in theoretical physics, particularly in the theory of elementary particles.

In 1932, he married Sumi Yukawa. In accordance with Japanese customs of the time, since he came from a family with many sons but his father-in-law Genyo had none, he was adopted by Genyo and changed his family name from Ogawa to Yukawa. The couple had two sons, Harumi and Takaaki.

Between 1932 and 1939 he was a lecturer at the Kyoto University and lecturer and Assistant Professor at the Osaka University. While at Osaka University, in 1935, he published a paper entitled "On the Interaction of Elementary Particles. I." [Yukawa, H. (1935). *Proc. Phys.-Math. Soc. Japan*, 17, 48], in which he proposed a new field theory of nuclear forces and *predicted the existence of the meson.* Encouraged by the discovery by American physicists of one type of meson in cosmic rays, in 1937, he devoted himself to the development of the meson theory, on the basis of his original idea.

Yukawa gained the D.Sc. degree in 1938 and in 1940 was appointed Professor of Theoretical Physics at Kyoto University. In 1940 he won the Imperial Prize of the Japan Academy, in 1943 the Decoration of Cultural Merit from the Japanese government.

Yukawa was invited as Visiting Professor to the Institute for Advanced Study at Princeton, U.S.A., in 1948, and in 1949 he became a visiting professor at Columbia University, the same year he received the Nobel Prize in Physics, after the discovery by Cecil Frank Powell, Giuseppe Occhialini and César Lattes of Yukawa's predicted pi meson in 1947.

Yukawa became the first chairman of Yukawa Institute for Theoretical Physics in 1953. He was an editor of Progress of Theoretical Physics, and published the books *Introduction to Quantum Mechanics* (1946) and *Introduction to the Theory of Elementary Particles* (1948).

Yukawa retired from Kyoto University in 1970 as a Professor Emeritus. Owing to increasing infirmity, in his final years he appeared in public in a wheelchair. He died at his home in Sakyo-ku, Kyoto, on September 8, 1981 from pneumonia and heart failure, aged 74. His tomb is in Higashiyama-ku, Kyoto.

Yukawa, H. (1942). *Kagaku* (Science), 7, 249; 8, 282; 9, 322 (in Japanese).

[unavailable online.]

Attempt at extended space-time description of quantum field theory, non-separability of cause and effect, density matrix, field equation with additional conditions, Lagrangian and Hamiltonian, numerable elementary particles, point-like space-time.

MEISO-KAI (Wondering Meeting) on 24 April 1942. Yukawa reported his attempt on an extended space-time description in the quantum field theory in this meeting. He discussed i) non-separability of cause and effect, ii) the density matrix, iii) the field equation with additional conditions, iv) Lagrangian and Hamiltonian, and v) elementary particles – numerable; space-time – point-like. He recorded names of 21 participants and 9 questioners including Tomonaga, who got a clue to his *super many-time theory* from Yukawa's talk.

Dirac, P. A. M. (March, 1942.) Bakerian Lecture - The physical interpretation of quantum mechanics.

[*Roy. Soc. Proc., A*, 180, 980, 1-40, https://doi.org/10.1098/ rspa.1942.0023.]

Delivered June 19, 1941, at Burlington House, whilst London was being bombed.

Received September 23, 1941.

Describes how a satisfactory *non-relativistic* quantum mechanics had been established, *Heisenberg method* focuses on quantities which enter into experimental results, connects together in one calculation probability coefficients from all initial states to all final states, *Schrodinger method* connects together in one calculation probability coefficients for transitions from one particular initial state to any final state, both methods rested on same mathematical formalism, generalization of Hamilton form of classical dynamics involving linear operators instead of ordinary algebraic variables, methods of physical interpretation differ, probably are still not finally settled, *the theory is not in agreement with the theory of special relativity* as is evident by the special role played by the time t, works very well in the *non-relativistic* region of low velocities where it appears to be in complete agreement with experiment, can be considered only as an approximation, setting up the mathematical formalism for a *relativistic* theory is fairly straightforward, the classical mechanics needs to be put into a *relativistic* Hamiltonian form taking into account that that the various particles comprising the dynamical system interact through the medium of the electromagnetic field, use Lorentz's *equations of motion* including the dampening terms which express the reaction of radiation, this Hamitonian formulation can then be made into a quantum theory following the procedure from *non-relativistic* quantum mechanics, appears satisfactory mathematically but meets serious difficulties in its physical interpretation, *it results in states of negative energy and negative probability*, it appears that whether one is dealing with particles of integral spin or of half-odd integral spin the mathematical methods at present in use in quantum mechanics *are capable of direct interpretation only in terms of a hypothetical world differing very markedly from the actual one*, all that *relativistic* theory does is provide a consistent means of calculating experimental results.

Modern developments of atomic theory have required alterations in some of the most fundamental physical ideas. This has resulted in its being usually easier to discover the equations that describe some particular phenomenon than just how the equations are to be interpreted. The quantum mechanics of Heisenberg and Schrodinger was first worked out for a number of simple examples, from which a general mathematical scheme was constructed, and afterwards people were led to the general physical principles governing the interpretation, such as the *superposition of states* and the *indeterminacy principle*. In this way *a satisfactory non-relativistic quantum mechanics was established.*

In extending the theory to make it *relativistic*, the developments needed in the mathematical scheme are easily worked out, but difficulties arise in the interpretation. If one keeps to the same basis of interpretation as in the *non-relativistic* theory, one finds that particles have states of *negative kinetic energy* as well as their usual states of positive energy, and, further, for particles whose spin is an integral number of quanta, there is the added difficulty that *states of negative energy occur with a negative probability*.

With electrons the *negative-probability* difficulty does not arise, and one can get a sensible interpretation of the *negative-energy states* by assuming them to be nearly all occupied and an unoccupied one to be a positron. This model, however, is excessively complicated to work with and one cannot get any results from it without making very crude approximations. The simple accurate calculations that one can make would apply to a world which is almost saturated with positrons, and it appears to be a better method of interpretation to make the general assumption that *transition probabilities* obtained from these calculations for this hypothetical world are the same as those for the actual world.

With photons one can get over the negative-energy difficulty by considering the states of positive and negative energy to be associated with the emission and absorption of a photon respectively, instead of, as previously, with the existence of a photon. The simplest way of developing the theory would make it apply to a hypothetical world in which *the initial probability of certain states is negative, but transition probabilities calculated for this hypothetical world are found to be always positive*, and it is again reasonable to assume that these *transition probabilities* are the same as those for the actual world.

Non-relativistic theory

Heisenberg, with his discovery of quantum mechanics, introduced a new outlook on the nature of physical theory. Previously, it was always considered essential that there should be a detailed description of what is taking place in natural phenomena, and one used this description to calculate results comparable with experiment. *Heisenberg put forward the view that it is sufficient to have a mathematical scheme from which one can calculate in a consistent manner the results of all experiments*—a detailed description in the traditional sense is unnecessary and may very well be impossible to establish.

Heisenberg's method, focuses attention on to the quantities which enter into experimental results. It was first applied to the *theory of spectra*, for which these quantities are the *energy levels* of the atomic system and certain *probability coefficients* which determine the probability of a radiative transition taking place from one level to another. The method sets up equations connecting these quantities and allows one to calculate them, but does not go beyond this. *It does not provide any description of radiative transition processes*. It does not even allow one to deduce how the results of a calculation are to be used, but requires one to assume Einstein's *laws of radiation* (the laws which tell how the *probability of a radiative transition process depends on the intensity of the incident radiation*), and to assume that certain quantities determined by the calculation are the coefficients appearing in the laws.

Shortly after Heisenberg's discovery, Schrodinger set up independently another form of quantum mechanics, which also enables one to calculate *energy levels* and *probability coefficients* and gives results agreeing with those of Heisenberg, but which introduces an important new feature. *It connects together, in one calculation, a set of probability coefficients which act together under certain conditions in Nature*; for example, the set of *probability coefficients* referring to *transitions from one particular initial state to any final state*. In this respect it is to be contrasted with Heisenberg's method, which connects together in one calculation all the *probability coefficients* for a dynamical system, i.e. the *probability coefficients* from all *initial states* to all *final states*.

This feature of Schrodinger's method gives it two important advantages. First, as a consequence of its enabling one to obtain fewer results at a time, it makes the computation much simpler. Secondly, it supplies, in a certain sense, a description of what is taking place in Nature, since a calculation leading to results which come into play together under certain conditions in Nature will be in close correspondence with the physical process that is taking place under those conditions, various points in the calculation having their counterparts in the physical process. *A description in this limited sense seems to be the most that is possible for atomic processes*. It implies a much less complete connection between the mathematics and the physics than one has in classical mechanics, and one might be disinclined to call it a description at all, but one may at least consider it as an appropriate generalization of what one usually means by a description. On account of Schrodinger's method allowing a description in this new sense while Heisenberg's allows none, Schrodinger's method introduces an outlook on the nature of physical theory intermediate between Heisenberg's and the old classical one.

When Heisenberg's and Schrodinger's theories were developed it was soon found that they both rested on the *same mathematical formalism* and differed only with regard to the method of physical interpretation. The formalism is *a generalization of the Hamiltonian form of classical dynamics, involving linear operators instead of ordinary algebraic variables*, and is so natural and beautiful as to make one feel sure of its correctness as the foundation of the theory. The question of its interpretation, however, which involved unifying Heisenberg's and Schrodinger's ideas into a satisfactory comprehensive scheme, was not so easily settled, and is probably still not finally settled.

The situation of a formalism becoming established before one is clear about its interpretation should not be considered as surprising, but as a natural consequence of the drastic alterations which the development of physics has required in some of the basic physical concepts. This makes it an easier matter to discover the mathematical formalism needed for a fundamental physical theory than its interpretation, since the number of things one has to choose between in discovering the formalism is very limited, the number of fundamental ideas in pure mathematics being not very great, while with the interpretation most unexpected things may turn up.

The best way of seeking the interpretation in such cases is probably from a discussion of simple examples. This way was used for the theory of quantum mechanics and *led eventually to a satisfactory interpretation applicable to all phenomena for which relativistic effects are negligible*. This interpretation is more closely connected with Schrodinger's method than Heisenberg's, as one would expect on account of the former affording in some sense a description of Nature, and *is centered round a Schrodinger's wave function, which is one of the things that can be operated on by the linear operators which the dynamical variables have become*. The correspondence which the existence of a description implies between the mathematics and the physics *makes a wave function correspond to a state of motion of the atomic system*, in such a way that, for example, a calculation which gives the *transition probabilities* from a particular initial *state* to any final *state* would be based on that *wave function* which represents the motion ensuing from this initial *state*. *A wave function is a complex function $\psi(q_1,q_2, ...,q_n, t)$ of all of the co-ordinates $q_1,q_2, ...,q_n$ of the system and of the time t*, and it receives the interpretation that the square of its modulus, $|\psi(q_1,q_2, ...,q_n, t)|^2$ is the *probability*, for the *state of motion* it corresponds to, of the co-ordinates having values in the neighborhood of $q_1, q_2, ...,q_n$, per unit volume of co-ordinate space, at the time t.

A *wave function* can be transformed so as to refer to other dynamical variables, for example, the *momenta* $p_1, p_2, ...,p_n$, when it is said to be in another *representation*. The square of its modulus $|\psi(p_1, p_2, ...,p_n, t)|^2$ is then the *probability*, per unit volume of *momentum space*, of the *momenta* having values in the neighborhood of $p_1, p_2, ...,p_n$ at the time t. *A wave function itself never has an interpretation, but only the square of its modulus*, and the need for distinguishing between two *wave functions* having the same squares of their moduli arises only because, if they are transformed to a different representation, the squares of their moduli will in general become different. *This brings out the incompleteness of the kind of description which is possible with quantum mechanics.*

One may make a slight modification in the *wave functions* in any *representation* by introducing a weight factor λ and arranging for the *probability* to be $\lambda\ |\psi|^2$ instead of $|\psi|^2$. The weight factor may be any positive function of the variables occurring in the *wave function.*

Wave functions have to satisfy a certain equation, namely, the equation

$$ih\ d\psi/dt = H\psi, \qquad (1)$$

where H a *Hermitian (or self-adjoint) linear operator*, and is the Hamiltonian of the system expressed in the representation concerned. *The wave equation (1) is a generalization of the Hamilton-Jacobi equation of classical mechanics.*

> [The *Hamilton–Jacobi equation* is an alternative formulation of classical mechanics, equivalent to other formulations such as Newton's laws of motion, Lagrangian mechanics and Hamiltonian mechanics. The Hamilton–Jacobi equation is particularly useful in identifying conserved quantities for mechanical systems, which may be possible even when the mechanical problem itself cannot be solved completely. It is also the only formulation of mechanics in which the motion of a particle can be represented as a wave. In this sense, it fulfilled a long-held goal of theoretical physics (dating at least to Johann Bernoulli in the eighteenth century) of finding an analogy between the propagation of light and the motion of a particle.]

If S is a solution of the latter equation [Hamilton–Jacobi equation], then

$$\psi = e^{iS/h} \qquad (2)$$

will give a first approximation to a solution of the former [the wave equation].

An important property of the *wave equation* (1) is that it yields the *conservation law* - the *total probability of the variables occurring in the wave function having any value is constant.* The *wave function* should be normalized so as to make this probability initially unity and then it always remains unity. *This conservation law is a mathematical consequence of the wave equation being linear in the operator d/dt and of H being a self-adjoint operator.*

The wave equation is linear and homogeneous in the wave function ψ, and so are the transformation equations. In consequence, one can add together two ψ's and get a third. The correspondence between ψ's and *states of motion* now allows one to infer that there is a relationship between the *states of motion*, such that one can add or superpose two *states* to get a third. This relationship constitutes the *Principle of Superposition of States*, one of the general principles governing the interpretation of quantum mechanics that we are here considering.

Another of these principles is *Heisenberg's Principle of Indeterminacy.* This is a consequence of the transformation laws connecting ψ(q) and ψ(p), which show that each of these functions is the Fourier transform of the other, apart from numerical coefficients, so that one meets the same limitations in giving values to a q and p as in giving values to the position and frequency of a train of waves.

These general principles serve to bring out the departures needed from ordinary classical ideas. They are of so drastic and unexpected a nature that it is not to be wondered at that they were discovered only indirectly, as consequences of a previously established mathematical scheme, instead of being built up directly from experimental facts.

Difficulties of the relativistic theory

The theory outlined above is not in agreement with the restricted principle of relativity, as is at once evident from the special role played by the time t. Thus, *while it works very well in the non-relativistic region of low velocities, where it appears to be in complete agreement with experiment, it can be considered only as an approximation*, and one must face the task of extending it to make it conform to restricted *relativity.* (*General relativity need not be considered, since gravitational effects are negligible in atomic theory.*) One should be prepared for possible further alterations being needed in basic physical concepts, and hence one should follow the route of first setting up the mathematical formalism and then seeking its physical interpretation.

Setting up the mathematical formalism is a fairly straightforward matter. One must first put classical mechanics into *relativistic* Hamiltonian form. One must take into account that the various particles comprising the dynamical system interact through the medium of the *electromagnetic field*, and one must use *Lorentz's equations of motion* for them, including the damping terms which express the reaction of radiation. This is done in Appendix I, where, with the help of an *action principle*, the *equations of motion* are obtained in the Hamiltonian form (I-40) with the Hamiltonians F_i one for each particle, given by (I-39)

[$F_i = \{p_i - e_i A(z'_i)\}^2 - m_i^2 = 0.$ (I-39)

There is one of these equations for each particle. The expressions F_i may be used as Hamiltonians to determine how any dynamical variable ξ varies with the proper-times s'_i, in accordance with the equations

$\kappa\, d\xi/ds'_i = [\xi, F_i].$ (I-40)]

Conversion of relativistic classical theory into relativistic quantum theory results in particles with states with negative energies and negative probabilities. Negative energies occur also in *relativistic* classical theory but do cause trouble because a particle in a positive-energy state can never make a transition to a negative-energy one.

This Hamiltonian formulation may now be made into a quantum theory by following rules which have become standardized from the *non-relativistic* quantum mechanics. *The resulting formalism appears to be quite satisfactory mathematically, but when one proceeds to consider its physical interpretation, one meets with serious difficulties.*

Take an elementary example, that of a *free particle without spin*, moving in the absence of any field. The classical Hamiltonian for this system is the left-hand side of the equation

$$p_0^2 - p_1^2 - p_2^2 - p_3^2 - m^2 = 0, \qquad (3)$$

where p_0 is the *energy* and p_1, p_2, p_3 the *momentum* of the particle, the velocity of light being taken as unity. Passing over to quantum theory by the standard rules, one gets from this Hamiltonian the *wave equation*

$$(h^2\Box + m^2)\,\psi = 0, \qquad (4)$$

where $\Box = \delta^2/\delta x_0^2 - \delta^2/\delta x_1^2 - \delta^2/\delta x_2^2 - \delta^2/\delta x_3^2$.

The *wave function* ψ here is a *scalar*, involving the co-ordinates x_1, x_2, x_3 and the time x_0 on the same footing, and so it is suitable for a *relativistic* theory.

If one now tries to use the old interpretation that $|\psi|^2$ is the *probability* per unit volume of the particle being in the neighborhood of the point x_1, x_2, x_3 at the time x_0, *one immediately gets into conflict with relativity, since this probability ought to transform under Lorentz transformations like the time component of a 4-vector, while* $|\psi|^2$ *is a scalar*. Also, *the conservation law for total probability would no longer hold*, the usual proof of it failing on account of the *wave equation* (4) not being linear in $\delta/\delta x_0$.

An important step forward was taken by Gordan and Klein

[Gordon, W. (January, 1927). Der Comptoneffekt nach der Schrödingerschen Theorie. (The Compton effect according to Schrödinger's theory.); Heisenberg and Schrödinger provided alternative methods for determination of quantum *frequencies* and *intensities*, Compton effect already calculated by Dirac (June, 1926) using Heisenberg method, here the same problem treated by Schrödinger method, starts with the same *classic relativistic equation for kinetic energy* in terms of *momentum* and *energy*, which is *Hamiltonian equation* for the system, introduces same imaginary variables for *time* and *energy* to create same space-time symmetric form, applies in same way to *electron in electromagnetic field* described in terms of *vector potential* and *scalar potential*, and introduces same imaginary variable for scalar potential, adds the same *field energy* to the *kinetic energy* resulting in the same *classical relativistic Hamiltonian equations for a point electron moving in an electromagnetic field*, in accordance with Schrödinger's rules Gordon then substitutes the classical *quantum differential operators* for the momentum vector in the amended *Hamiltonian equation* and applies resulting differential operator to the *wave function* ψ to obtain the *Klein-Gordon equation*, $1/c^2\, \partial^2/\partial t^2\, \psi - \nabla^2\, \psi + m^2c^2/h^2\, \psi = 0$, (Dirac [(February, 1928). The Quantum Theory of the Electron.] objected to this substitution on grounds of the interpretation of the wave function, and solutions with negative probabilities, negative energy, and positive charge for the electron); calculates radiation from *current density* and *charge density*, applies to Compton effect.

Klein, O. (October, 1927). Elektrodynamik und Wellenmechanik vom Standpunkt des Korrespondenzprinzips. (Electrodynamics and wave mechanics from the standpoint of the correspondence principle.); alternative calculation of Compton effect restricted to the *one-electron problem*, starts from Maxwell-Lorentz field equations, describes motion of an electron in an electromagnetic field by *four-potential* and *scalar potential*, regards *Hamilton-Jacobi differential equation* for the action function (Klein–Gordon equation) as expression for motion of the electron, following de Broglie and Schrodinger replaces this first order equation with a second-order linear equation representing *relativistic* generalization of Schrödinger's wave equation for one-electron problem, evaluates equations determining the electromagnetic field with the help of wave mechanics using the correspondence principle to determine wave-mechanical expressions for *electric density* and *current vector*, after neglecting relativity results in the same expressions as those obtained by Schrodinger, applies to a "bound" electron moving in an axially symmetric electrostatic field over which a weak homogeneous magnetic field is superimposed to derive normal Zeeman effect, applies to scattered radiation from a light wave on a "force-free" electron to obtain the Compton effect, five-dimensional wave mechanics.]

who proposed that instead of $|\psi|^2$ *one should use the expression*

$$1/4\pi i\ \{\delta\psi^*/\delta x_0\ \psi - \psi^*\delta\psi/\delta x_0\}. \qquad (5)$$

[for the *probability* per unit volume of the particle being in the neighborhood of the point x_1, x_2, x_3 at the time x_0]

This expression is the time component of a 4-vector. Further, it is easily verified that *the divergence of this 4-vector vanishes, which gives the conservation law in relativistic form*. Thus (5) is evidently the correct mathematical form to use. *This form leads to trouble on the physical side*, however, since, although it is real, it is not positive definite like $|\psi|^2$. *Its employment would result in one having at times a negative probability for the particle being in a certain place*. This is not the only difficulty. Let us consider the *energy* and *momentum* of the particle, and take for simplicity a state for which these variables have definite values. The corresponding *wave function* will be of the form of plane waves,

$$\psi = e^{-i(p0x0 - p1x1 - p2x2 - p3x3)/h}$$

In order that the *wave equation* (4)

$$[(h^2\Box + m^2)\,\psi = 0, \qquad (4)$$
$$\text{where } \Box = \delta^2/\delta x_0{}^2 - \delta^2/\delta x_1{}^2 - \delta^2/\delta x_2{}^2 - \delta^2/\delta x_3{}^2]$$

may be satisfied, the *energy* and *momentum* values p_0, p_1, p_2, p_3 here must satisfy the classical equation (3)

$$[p_0{}^2 - p_1{}^2 - p_2{}^2 - p_3{}^2 - m^2 = 0, \qquad (3)]$$

This equation allows of negative values for the energy as well as positive ones and is, in fact, symmetrical between positive and negative energies. The negative energies occur also in the classical theory, but do not then cause trouble, since a particle started off in a positive-energy state can never make a transition to a negative-energy one. In the quantum theory, however, such transitions are possible and do in general take place under the action of perturbing forces.

The *wave function* may be transformed to the *momentum* and *energy* variables. The Gordon-Klein expression (5)

$$[1/4\pi i\ \{\delta\psi^*/\delta x_0\ \psi - \psi^*\delta\psi/\delta x_0\}. \qquad (5)]$$

then goes over into

$$|\ \psi(p_0p_1p_2p_3)\ |^2\ p_0{}^{-1}dp_1dp_2dp_3, \qquad (6)$$

as the probability of the *momentum* having a value within the small domain $dp_1dp_2dp_3$ about the value p_1, p_2, p_3, with the *energy* having the value p_0, which must be connected with p_1, p_2, p_3 by (3). The weight factor $p_0{}^{-1}$ appears in (6) and makes it *Lorentz invariant*, since $\psi(p)$ is a *scalar*—it is defined in terms of $\psi(x)$ to make it so—and the differential element $p_0{}^{-1}dp_1dp_2dp_3$ is *Lorentz invariant. This weight factor may be positive or negative, and makes the probability positive or negative accordingly. Thus, the two undesirable things, negative energy and negative probability, always occur together.*

Let us pass on to another simple example, that of *a free particle with spin half a quantum*. The *wave equation* is of the same form (4)

$$[(h^2\Box + m^2)\,\psi = 0, \qquad (4)$$
$$\text{where } \Box = \delta^2/\delta x_0{}^2 - \delta^2/\delta x_1{}^2 - \delta^2/\delta x_2{}^2 - \delta^2/\delta x_3{}^2]$$

as before, *but the wave function ψ is no longer a scalar. It must have two components, or four if there is a field present,* and the way they transform under Lorentz transformations is given by the general connection between the *theory of angular momentum* in quantum mechanics and *group theory*. The expression $\Sigma \mid \psi(x) \mid^2$, summed for the components of ψ, turns out to be the time component of a 4-vector, and further the divergence of this 4-vector vanishes. Thus, it is satisfactory to use this expression as the *probability* per unit volume of the particle being at any place at any time. *One does not now have any negative probabilities in the theory. However, the negative energies remain, as in the case of no spin.*

We may go on and consider particles of higher spin. *The general result is that there are always states of negative energy as well as those of positive energy. For particles whose spin is an integral number of quanta, the negative-energy states occur with a negative probability and the positive-energy ones with a positive probability, while for particles whose spin is a half-odd integral number of quanta, all states occur with a positive probability.*

Negative energies and probabilities should not be considered as nonsense. They are well-defined concepts mathematically, like a negative sum of money, since the equations which express the important properties of *energies* and *probabilities* can still be used when they are negative. Thus, negative *energies* and *probabilities* should be considered simply as things which do not appear in experimental results. *The physical interpretation of relativistic quantum mechanics that one gets by a natural development of the non-relativistic theory involves these things and is thus in contradiction with experiment.* We therefore have to consider *ways of modifying or supplementing this interpretation.*

Particles of half-odd integral spin

> [The *negative-energy* difficulty for electrons can be addressed using the 'hole theory', by assuming that nearly all *negative-energy states* are occupied. An unoccupied *negative-energy state* now appears as a 'hole' in the distribution of occupied *negative-energy states* and thus has a deficiency of *negative energy*, i.e. a *positive energy*. An electron jumping from a positive- to a negative-energy state in the theory is now interpreted as an *annihilation of an electron and a positron*, and one jumping from a negative- to a positive-energy state as a *creation of an electron and a positron. This hypothetical world having nearly every quantum state for a positron occupied differs very much from the actual world*, but one can now calculate the *probability* of any kind of collision process occurring in this hypothetical world. *It does not provide a complete physical theory, since it enables one to calculate only those experimental results that are reducible to collision probabilities*, and some branches of physics, e.g. the structure of solids, do not seem to be so reducible. However, *collision probabilities* are the things for which a *relativistic* theory is at present most needed.]

Let us first consider particles with a *half-odd integral spin*, for which there is only the *negative-energy* difficulty to be removed. The chief particle of this kind for which a *relativistic* theory is needed is the electron, with spin half a quantum. Now electrons, and also, it is believed, all particles with a half-odd integral spin, satisfy the *exclusion principle* of Pauli, according to which not more than one of them can be in any quantum *state*. (This principle is obtained in quantum mechanics from the requirement that *wave functions shall be antisymmetric in all the particles*.) With this principle there are only two alternatives for a *state*, either it is unoccupied or it is occupied by one particle, and a symmetry appears with respect to these two alternatives.

Some time ago I proposed a way of dealing with the negative-energy difficulty for electrons, based on a theory in which nearly all their negative-energy states are occupied. *An unoccupied negative-energy state now appears as a 'hole' in the distribution of occupied negative-energy states and thus has a deficiency of negative energy, i.e. a positive energy*. From the wave equation one finds that a hole moves in the way one would expect a positively charged electron to move. *It becomes reasonable to identify the holes with the recently discovered positrons*, and thus to get an interpretation of the theory involving positrons together with electrons. *An electron jumping from a positive- to a negative-energy state in the theory is now interpreted as an annihilation of an electron and a positron, and one jumping from a negative- to a positive-energy state as a creation of an electron and a positron.*

The theory involves an *infinite density of electrons* everywhere. It becomes necessary to assume that the distribution of electrons for which all positive-energy states are unoccupied and all negative-energy states occupied—what one may call the *vacuum distribution*, as it corresponds to the absence of all electrons and positrons in the interpretation—*is completely unobservable. Only departures from this distribution are observable and contribute to the electric density and current which give rise to electro-magnetic field in accordance with Maxwell's equations. The above theory does provide a way out from the negative-energy difficulty, but it is not altogether satisfactory*. The infinite number of electrons that it involves requires one to deal with wave functions of very great complexity and leads to such complicated mathematics that one cannot solve even the simplest problems accurately, but must resort to crude and unreliable approximations. Such a theory is a most inconvenient one to have to work with, and on general philosophical grounds *one feels that it must be wrong.*

Let us see whether one can modify the theory so as to make it possible to work out simple examples accurately, while retaining the basic idea of identifying unoccupied negative-energy *states* with positrons. The simple calculations that one can make involve simple *wave functions*, referring to only one or two electrons, and thus referring to nearly all the negative-energy *states* being unoccupied. The calculations therefore apply to a world almost saturated with positrons, i.e. having nearly every quantum state for a positron occupied. *Such a world, of course, differs very much from the actual world*. One can now calculate the *probability* of any kind of collision process occurring *in this hypothetical world* (in so far as electrons and positrons are concerned). One can deduce the *probability coefficient* for the process, i.e. the probability per unit number of incident particles or per unit intensity of the beam of incident particles, for each of the various kinds of incident particle taking part in the process. For this purpose, one must use the *laws of statistical mechanics*, which tell how the *probability* of a collision process depends on the number of incident particles, paying due attention to the modified form of these laws arising from the *exclusion principle* of Pauli.

Let us now assume that probability coefficients so calculated for the hypothetical world are the same as those of the actual world. This single assumption provides a general physical interpretation for the formalism, enabling one to calculate *collision probabilities* in the actual world. *It does not provide a complete physical theory, since it enables one to calculate only those experimental results that are reducible to collision probabilities*, and some branches of physics, e.g. the structure of solids, do not seem to be so reducible. However, *collision probabilities* are the things for which a *relativistic* theory is at present most needed, and one may hope in the future to find ways of extending the scope of the theory to make it include the whole of physics.

Comparing the new theory with the old, one may say that *the new assumption, identifying collision probability coefficients in the actual world with those in a certain hypothetical world, replaces the old assumption about the non-observability of the vacuum distribution of negative-energy electrons*. The approximations needed for working out simple examples in the old theory are equivalent in their mathematical effect to making the new assumption; e.g. *these approximations include the neglect of the Coulomb interaction between electron and positron in the calculation of the probability of pair creation and annihilation*, and this interaction cannot appear in the new theory, since the calculation there is concerned with a one-electron system. Thus, the new theory may be considered as a precise formulation of the old theory together with some general approximations needed for applying it.

The new theory for dealing with the *negative-energy states* of the electron may be applied to any kind of elementary particle with spin half a quantum, and probably also to particles with other half-odd integral spin values, provided, of course, they satisfy *Pauli's exclusion principle*. It may thus be applied to protons and neutrons. *It requires for each particle the possibility of existence of an antiparticle of the opposite charge, if the original particle is charged. If the original particle is uncharged, one can arrange for the antiparticle to be identical with the original.*

Particles of integral spin

Most of the elementary particles of physics have half-odd integral spin, but there is the important exception of the light-quantum or *photon*, with spin one quantum, and there is the new cosmic-ray particle, the *meson*, also probably with spin one quantum, and other such particles may be discovered in the future. All these kinds of particle, it is believed, satisfy the Bose statistics, a statistics which allows any number of particles to be in the same quantum state with the same *a priori* probability. (This statistics is obtained in quantum mechanics from the requirement that wave functions shall be symmetric in all the particles.) For these kinds of particles, the previous method of dealing with the negative-energy states is therefore no longer applicable, and *there is the further difficulty of the negative probabilities*.

When dealing with particles satisfying the Bose statistics, it is useful to consider the operators corresponding to the *absorption* of a particle from a given *state* or the *emission* into a given *state*. These operators can be treated as *dynamical variables*, although they do not have any analogues in classical mechanics. If one works out their *equations of motion* and *transformation equations*, one finds a remarkable correspondence. The *absorption* operators from a set of independent *states* have the same *equations of motion* and *transformation equations* as the *wave function* ψ representing a single particle, and similarly for the *emission* operators and the *conjugate complex wave function* $\psi^{\sim}$. Thus, one can pass from a one-particle theory to a many-particle theory by making the ψ and $\psi^{\sim}$ describing the one particle into *absorption* and *emission operators*, which must satisfy the appropriate *commutation relations*. Such a passage is called *second quantization*.

One can get over the difficulties of negative energies and negative probabilities for Bose particles by abandoning the attempt to get a satisfactory theory of a single particle and passing on to consider the problem of many particles, using a method given by Pauli & Weisskopf for *electrons* having no spin and satisfying the Bose statistics.

> [Pauli, W. & Weisskopf, V. (1934). Über die Quantisierung der skalaren relativistischen Wellengleichung. (The Quantization of the Scalar Relativistic Wave Equation.); the *scalar relativistic wave equation* $E^2/c^2 - \sum_{k=1}^{3} p_k^2 - m^2c^2 = 0$ has generally been relinquished in favor of *Dirac's four-component wave equation* $\{p_0 + \rho_1 (\boldsymbol{\sigma}, \mathbf{p}) + \rho_3 mc\}\ \psi = 0$ because the former does not yield the spin of the particles, shows that Dirac's *a priori* arguments based on limitation to a *single-body problem* and *particle density* being a meaningful observable need to be revised, shows that *charge density* is a meaningful observable, no reason for special form of *charge density* $\sum_r \psi^*_r\, \psi_r$, in *scalar relativistic wave equation theory* the energy of material particles is always positive after wave fields have been quantized, no new hypothesis such as the "hole theory" is required, applies Heisenberg-Pauli formalism of quantization of wave fields to *scalar relativistic wave equation* for matter fields for particles without spin and with Bose-Einstein statistics, shows that quantization of the *Klein-Gordon relativistic wave equation* for *scalar particles* gives rise to particles with opposite charge but with same rest mass which can be created or destroyed with absorption or emission of electromagnetic radiation, frequency of pair creation and annihilation processes of same order of magnitude as frequency for particles of same charge and mass which follows from Dirac "hole theory", *also leads to infinite self-energy not only of the particles but also to an infinite polarizability of the vacuum.*]

(Such electrons are not known experimentally, but there is no known theoretical reason why they should not exist.) *The method of Pauli & Wiesskopf is to work entirely with positive-energy states. The operators of absorption from and emission into negative-energy states, arising in the application of second quantization to the one-electron theory, are replaced by the operators of emission into and absorption from positive-energy states of electrons with the opposite charge, respectively*. This replacement does not disturb the *laws of conservation of charge, energy and momentum*. The resulting theory involves spinless electrons of both kinds of *charge* together, and leads to pair creation and annihilation, as with ordinary electrons and positrons.

The method of Pauli & Wiesskopf may be applied in a degenerate form to photons and leads to the quantum electrodynamics of Heisenberg & Pauli

> [Heisenberg, W. & Pauli, W. (July, 1929). Zur Quantendynamik der Wellenfelder. (On the quantum dynamics of wave fields.) *Zeit. Phys*., 56, 1-61[; Heisenberg and Pauli's first attempt to construct their own version of a *relativistically-invariant quantum electrodynamics* to treat interaction between matter and the electromagnetic field and between matter and matter, canonical quantization of both electromagnetic and matter-wave fields, but Lorentz-invariant Lagrangian for interacting *electromagnetic* and *matter-wave fields*, requires working with the *electromagnetic potentials* not just with the *fields*, Lagrangian does not contain a time derivative of the *electric potential* so *there is no corresponding canonical momentum variable,* prevents straightforward implementation of canonical commutation relations, the theory is still afflicted with many defects, *the fundamental difficulties in the relativistic formulation that were emphasized by Dirac remain unchanged,* the formulas of the theory lead to an *infinite zero-point energy* for the radiation and thus include the interaction of an electron with itself as an *infinite* additive constant, however, these difficulties are of a sort that they do not interfere with the application of the theory to many physical problems, used "crude trick" of adding additional terms to the Lagrangian.]; Heisenberg, W. & Pauli, W. (January, 1930). Zur Quantendynamik der Wellenfelder II. (On the quantum dynamics of wave fields II.) *Zeit. Phys*., 59, 168-190; new approach to Lorentz-invariant Lagrangian problem based on notion of *gauge invariance* of *theory of coupled electromagnetic potentials* and Dirac *matter waves*, integrals of the *equations of motion* derived from *invariance properties* of Hamiltonian function, *invariance properties* of *wave equations* exploited in similar way, an *infinite* interaction of the electron with itself will also result from this approach making application of the theory impossible in many cases, the theory leads to divergent expressions for the *energies* of *stationary states* and the differences between these *energies* (i.e., the actually observed frequencies of spectral lines) came out *infinite*].

To take into account that photons have no charge, one must start with a one-particle theory in which the *wave functions* are real, so that $\psi^* = \psi$. The part of ψ referring to positive-energy *states* is then made into the *absorption operators* from positive-energy *states*, and the part referring to negative-energy states into the *emission operators* into positive-energy *states*. The resulting scheme of operators, involving only positive-energy photon *states*, may then be put into correspondence with classical electrodynamics, according to the usual laws governing the correspondence between quantum and classical theory.

It would seem that in this way the difficulties of negative energies and probabilities for Bose particles can be overcome, but *a new difficulty appears. When one tries to solve the wave equation* (or the *wave equations* if there are several particles with their respective Hamiltonians) *one gets divergent integrals* in the solution, of the form, in the case of photons,

$$\int_0^\infty f(\nu)\, d\nu, \qquad f(\nu) \sim \nu^n \text{ for large } \nu, \qquad (7)$$

ν being the *frequency* of a photon. The values 1, 0 and — 1 for *n* are the chief ones occurring in simple examples. *Thus, the wave equation really has no solutions and the method fails.*

I have made a detailed study of the *divergent integrals occurring in quantum electrodynamics* and have shown (Dirac, P. A. M. (1939). *Ann. Inst. Poincare*; this paper as not yet appeared in England) [and is unavailable online] that all those with even values of n can be eliminated by introducing into the equations a certain limiting process, which one can justify by showing that a corresponding limiting process is needed in classical electrodynamics to get the equations of motion into Hamiltonian form. (It appears accordingly in the *action principle* of Appendix I.) The divergent integrals with odd values of n remain, however, and indicate *something more fundamentally wrong with the theory.*

> [*Appendix I. The action principle of classical electrodynamics*
>
> There are various forms which the *action principle* of classical electro-dynamics may take, but most of them involve awkward conditions concerning the singularities of the field where the charged particles are situated and are not suitable for a subsequent passage to quantum mechanics.
>
> > [The *stationary-action principle* – also known as the *principle of least action* – is a variational principle that, when applied to the action of a mechanical system, yields the *equations of motion* for that system. The principle states that the trajectories (i.e. the solutions of the *equations of motion*) are stationary points of the system's *action functional.*]
>
> Fokker (1929) set up a form of *action principle* which does not refer to the singularities of the *field* and which appears to be the best starting point for getting a quantum theory. [Fokker, A. D. (May, 1929). Ein invarianter Variationssatz für die Bewegung mehrerer elektrischer Massenteilchen. (An invariant set of variations for the movement of several electrical mass particles.) *Zeit. Phys.*, 58, 386-93; https://doi.org/10.1007/BF01340389.] Fokker's *action integral* may conveniently be written with the help of the δ function and is then
>
> $$S = S_1 + S_2,$$
>
> where $S_1 = \sum_i m_i \int ds_i$ (I-1)
>
> and $S_2 = \sum_i \sum_{j \neq i} e_i e_j \iint \delta(z_i - z_j)^2 (v_i, v_j)\, ds_i\, ds_j.$ (I-2)
>
> …

The *action integral* as it stands is not a general one covering all possible *states* of motion. To make it general one must, as has been pointed out by the author (1938), add to it a term of the form

$$S_3 = \sum_i e_i \int (M_\mu(z_i) v^\mu_i \, ds_i. \qquad \text{(I-5)}$$

[Dirac, P. A. M. (August, 1938). Classical Theory of Radiating Electrons: "... Suppose now we are concerned with a problem in which

$$F^{\mu\nu}_{out} - F^{\mu\nu}_{in} = 0. \qquad (43)$$

Then the *field* $f^{\mu\nu}_n$ to be used in the *equations of motion* (22)

$$[m\upsilon'_\mu = e\upsilon_\nu f_\mu{}^\nu, \qquad (22)]$$

for the nth electron is just the sum of the mean of the *retarded* and *advanced fields* of all the other electrons. The interaction of the electrons for this case has been studied by Fokker (1929) who finds that the motion of all the electrons may now be described by a *variation principle*

$$\delta I = 0. \qquad (44)$$

In our present notation I would have the form

$$I = \sum_n \int m \, ds_n + \tfrac{1}{2} \sum_n \sum_{m \neq n} \int \ldots ds_n$$
$$+ \tfrac{1}{2} \sum_n \sum_{m \neq n} \int \ldots ds_n, \qquad (45)$$

where the two double sums are to be taken over all pairs of electrons, and in the first we must take for z_m and v_m their *retarded values* (with respect to the point on s_n considered), and in the second we must take their *advanced values*.

When the condition (43) does not apply, Fokker's *variation principle* (44) will still hold provided we add on to I the terms

$$\sum_n e \int \tfrac{1}{2} (A^\mu_{in} + A^\mu_{out}) \upsilon_{\mu n} \, ds_n, \qquad (46)$$

involving the *potentials* of the mean of the *ingoing* and *outgoing* fields. The *variation principle* will not be of much use in practical problems, where this mean field is unknown, but it may be of value in suggesting how the quantum generalization of our theory is to be made.]

The 4-vector potential $M_\mu(x)$ may be left for the present an arbitrary function of the *field point* x.

For the purpose of deducing the *equations of motion* one may take the limits of integration in the various integrals to be $-\infty$ and ∞, as was done by Fokker, *but in order to introduce momenta and get the equations into Hamiltonian form one must take finite limits*. Let us therefore suppose each s_i goes from s^0_i to s'_i, and let the corresponding z_i and v_i be z^0_i, z'_i and v^0_i, v'_i. It is desirable to restrict the initial values s^0_i so that the points z^0_i all lie outside each other's light-cones, and similarly with the final values s'_i. Thus,

$$(z^0_i - z^0_j)^2 < 0, \qquad (z'_i - z'_j)^2 < 0, \quad (i \neq j). \qquad \text{(I-6)}$$

...

A difficulty appears in the work at this stage. In varying the *action integral*, it was implicitly assumed, by the use of (I-11), that $M^{\nu}(x)$ is a *continuous field function* in the neighborhood of the *world-lines* of the particles, but the expression subsequently obtained for it, (I-18), has discontinuities. One can easily see that those discontinuities on the *world-line* of any particle due to the terms in the sum in (I-18) referring to the other particles are not serious and do not invalidate the *equations of motion* (they produce the same sort of effects as a pulse in the incident radiation would). But *those at the end-points of the world-line of any particle due to the term in the sum in (I-18) referring to that particle are serious, as they prevent the passage to the Hamiltonian formulation.*

To get over the difficulty one must introduce a certain limiting process. Let λ be a small 4-vector whose direction is within the future light-cone, so that

$$\lambda^2 > 0, \qquad \lambda_0 > 0, \qquad \text{(I-22)}$$

and let the $\delta(z_i - z_j)^2$ in (I-2) be replaced by

$$\begin{aligned}\delta^*(z_i, z_j)^2 &= \delta(z_i - z_j + \lambda)^2 \text{ for } (z_i - z_j)_0 > 0 \quad)\\ &= \delta(z_i - z_j - \lambda)^2 \text{ for } (z_i - z_j)_0 < 0 \quad)\end{aligned} \qquad \text{(I-23)}$$

This will cause interaction to occur between pairs of particles at points on their world-lines that lie one just outside the light-cone from the other, instead of on it as previously. (*It is assumed that two particles never approach to a distance of the order λ from one another.*) …

… the restrictions (I-6)

$$[(z^0_i - z^0_j)^2 < 0, \qquad (z'_i - z'_j)^2 < 0, \;\; (i \neq j). \qquad \text{(I-6)}]$$

having to be sharpened to

$$(z^0_i - z^0_j \pm \lambda)^2 < 0, \qquad (z'_i - z'_j \pm \lambda)^2 < 0, \qquad (i \neq j). \qquad \text{(I-25)}$$

The singularities in $M^{\nu}(x)$ are now displaced a little from the end-points of the world-lines and no longer come into play, even if the initial points z^0_i are also varied, provided the variations in the world-lines are smaller than λ. The *equations of motion* resulting from the variation principle are no longer exact, but are correct only in the limit $\lambda \to 0$. *The limiting process here introduced appears to be unavoidable for setting up a satisfactory action theory and plays an important role in the later development, leading to the elimination of those divergent integrals* (7)

$$[\textstyle\int_0^\infty f(\nu)\, d\nu, \qquad f(\nu) \sim \nu^n \text{ for large } \nu, \qquad (7)$$

ν being the frequency of a photon]

in quantum electrodynamics for which n is even.

One can now pass to the Hamiltonian formulation of the *equations of motion. …*

…

The above Hamiltonian formulation of the equations of classical electro-dynamics may be taken over into the quantum theory in the usual way, by making the *momenta* into operators satisfying *commutation relations* corresponding to the *Poisson Bracket* relations (I-29), (I-30). Equation (I-38) in the limit $\lambda \rightarrow 0$ goes over into the quantum equation (9) [below,

$$A_\mu(x'), A_\nu(x'')] = g_{\mu\nu} \Delta(x'-x''). \qquad (9)]$$

The Hamiltonians (I-39) provide the *wave equations*

$$F_i\psi = 0, \qquad \text{(I-43)}$$

in which the *wave function* ψ is a function of the co-ordinates z_i' of all the particles, subject to the inequalities (I-25), and of the field variables $M_\mu(x)$. One can apply the theory to spinning electrons instead of spinless particles, by modifying the Hamiltonians F_i in the appropriate way.

The present form of quantum electrodynamics differs from that of Heisenberg & Pauli in two respects. First, *it involves the λ limiting process.* This limiting process may be introduced into the theory of Heisenberg & Pauli, as shown by the author [Dirac, P. A. M. (1939). *Ann. Inst. Poincare*], and *leads to the elimination of the divergent integrals with even n-values.* Secondly, it is expressed in terms of a *different representation, with the wave functions involving the $M_\mu(x)$ as co-ordinates.* This change in the representation *eliminates the divergent integrals with odd n-values. The new representation cannot be interpreted in terms of only positive-energy photons, as is the case with the representation of Heisenberg & Pauli, but it can be interpreted in terms of both positive- and negative-energy photons*, as will be shown in Appendix II.]

Divergent integrals are a general feature of [relativistic] quantum field theories, and it has usually been supposed that they should be avoided by altering the forces or the laws of interaction between the elementary particles at small distances, so as to get the integrals cut off for some high value of ν. *However, one can easily see that this is wrong, in the case of electrodynamics at any rate*, by referring to the corresponding classical theory. The *wave function* should have its analogue in the solution of the *Hamilton-Jacobi equation*, in accordance with equation (2)

[The *wave equation* (1)

$$ih\, d\psi/dt = H\psi, \qquad (1)$$

is a generalization of the Hamilton-Jacobi equation of classical mechanics. If S is a solution of the latter equation, then

$$\psi = e^{iS/h} \qquad (2)$$

will give a first approximation to a solution of the former,]

but already *when one tries to solve the Hamilton-Jacobi equation of classical electrodynamics corresponding to the wave equation of Heisenberg & Pauli's quantum electrodynamics, one meets with divergent integrals.* Now the classical *equations of motion* concerned, namely, *Lorentz's equations* including radiation damping, have definite solutions when treated by straightforward methods and *if, on trying to get these solutions by a Hamilton-Jacobi method, one meets with divergent integrals, it means simply that the Hamilton-Jacobi method is an unsuitable one, and not that one should try to alter the physical laws of interaction to get the integrals to converge*. The correspondence between the quantum and classical theories is so close that one can infer that the corresponding *divergent integrals* in the quantum theory must also be due to an unsuitable mathematical method.

The appearance of divergent integrals with odd n-values in Heisenberg & Pauli's form of quantum electrodynamics may be ascribed to the unsymmetrical treatment of positive- and negative-energy photon states. If instead of using Pauli & Weisskopf's method one keeps to plain *second quantization*, one can build up a form of quantum electrodynamics symmetrical between *positive- and negative-energy photon states*, as is done in Appendix II. The new theory leads to similar equations as the old one, but with integrals of the type

$$\int_{-\infty}^{\infty} f(v)\, dv, \tag{8}$$

instead of (7)

$$[\int_{0}^{\infty} f(v)\, dv, \qquad f(v) \sim v^n \text{ for large } v, \qquad (7)]$$

and since f(v) is always a rational algebraic function, and it is reasonable on physical grounds to approach the upper and lower limits of integration in (8) at the same rate, *the divergencies with odd n-values all cancel out.*

[*Appendix II. Relativistic second quantization*

The author's original method of *second quantization* for an assembly of similar Bose particles, [Dirac, P. A. M. (1939). *Ann. Inst. Poincare*; (unavailable online); detailed study of the divergent integrals occurring in quantum electrodynamics, shows that all those with even values of n can be eliminated by introducing into the equations a certain limiting process, justified by showing that a corresponding limiting process is needed in classical electrodynamics to get the equations of motion into Hamiltonian form], consisting in the direct introduction as *dynamical variables* of the numbers of particles in the various quantum *states* for a particle, *is not very suitable for the purpose of relativistic generalization*. Fock (1934, §1) gave an improved method, which presents the basic ideas in a more direct manner *and which allows itself to be easily developed into a relativistic theory*. [Fock, V. (1934). Zur Quantenelektrodynamik. (To quantum electrodynamics.) *Phys. Zeit. Sowjet.*, 6, 425-69 (unavailable online).] The main features of Fock's method will first be mentioned here, to provide a starting point. The author's bracket notation will be used, as it seems to be the neatest and most concise one for this kind of work. [Dirac, P. A. M. (1939). A new notation for quantum mechanics. *Proc. Camb. Phil. Soc.*, 35, 416-8.]

A *state* of an assembly of u particles satisfying the Bose statistics is represented by a symmetrical *wave function* where each value for a variable q represents a *state* for one particle. Let us suppose for definiteness that the values a q can take on are discrete. Then ...

...

Hence … so that

$$\xi^*_x \xi_{x0} - \xi_{x0} \xi^*_x = i\, \Delta(x - x^0) \qquad \text{(II-28)}$$

The above *relativistic* theory for spinless particles of zero rest-mass may easily be extended to apply to photons. The *wave function* for a single particle must be made into a 4-vector with components corresponding to the four components of the *electromagnetic potential*. The 4-vector *wave function* may be considered as a function of the four coordinates x and of another variable μ taking on the four values 0,1, 2, 3, and may be written $< x_\mu | >$. …

…

The further development of these extensions is obvious and need not be mentioned in detail. The result is that the *emission operator* becomes a 4-vector $\xi_{x\mu}$, and satisfies with its adjoint $\xi^*_{x\mu}$ in the x_μ representation the *commutation relation*

$$\xi^*_{x\mu} \xi_{x'\nu} - \xi_{x'\nu} \xi^*_{x\mu} = i g_{\mu\nu}\, \Delta(x - x') \qquad \text{(II-33)}$$

corresponding to (II-28).

To establish a connection between the present theory of photons and the quantum electrodynamics of the end of Appendix I, one must put

$$M_\mu(x) = (\tfrac{1}{2} h)^{1/2} \xi_{x\mu}, \quad N_\mu(x) = (\tfrac{1}{2} h)^{1/2} \xi^*_{x\mu}, \qquad \text{(II-34)}$$

so as to make the commutation relation (I-33)

$$[[N_\mu(x), M_\nu(x')] = \tfrac{1}{2} g_{\mu\nu}\, \Delta(x - x') \qquad \text{(I-33)]}$$

go over into (II-33). The wave function ψ of Appendix I, which is of the form of a power series in the variables $M_\mu(x)$ may now be identified with $< \xi | >$. *The interpretation of ψ that one would expect from the rules of non-relativistic quantum mechanics, namely, that $|\psi|^2$ gives the probability of the variables $M_\mu(x)$ having specified values, is no longer applicable*. Instead, one must work from (II-4) suitably modified to fit in with the Gordon-Klein expression for the *probability* for one particle. One may make a Fourier resolution of the $\xi_{x\mu}$ and replace the continuous range of Fourier components by a discrete set. The *wave function* then becomes a power series in ξ variables referring to the Fourier components and its coefficients will give, according to the formula (III-24) of Appendix III, the *probabilities* of there being various numbers of photons in various *momentum states, some with positive and some with negative energy.* These *probabilities* are then to be re-interpreted, in accordance with one of the fundamental assumptions of the text, as the *probabilities* of various photons having been *emitted* or *absorbed.*

…

This result is sufficient to ensure the *conservation law*—that if the *wave function* is initially normalized it always remains so—and makes the existence of other real dynamical variables not corresponding to self-adjoint operators unimportant.

Appendix III. The transformation connecting the two forms of quantum electrodynamics

The new form of quantum electrodynamics set up in Appendices I and II and the old form of Heisenberg & Pauli (*amended by the introduction of the λ limiting process*) involve *two different representations of the same set of operators describing the electromagnetic field*, and one can discuss the transformation from one of these representations to the other on the lines of the general *transformation theory* of quantum mechanics. One cannot, as before mentioned, carry out the transformation for solutions of the *wave equations*, since such solutions in the old representation contain *divergent integrals*, but one can carry it out for other representatives of *states* of the *field* of a sufficiently general kind to make the methods of the *transformation theory* applicable. …]

The work of Appendices I and II shows that the new form of quantum electrodynamics also corresponds to classical electrodynamics in accordance with the usual laws, with the exception that operators corresponding to real dynamical variables in the classical theory are no longer always self-adjoint. This exception is not important, as it rather stands apart from the general mathematical connection between quantum and classical theory. The *Hamilton-Jacobi equation* corresponding to the *wave equation* of the new quantum electrodynamics differs from that of the old one only through being expressed in terms of a different set of coordinates, but the new *Hamilton-Jacobi equation* can be solved *without divergent integrals* and is connected with a satisfactory *action principle*, namely, that of Appendix I. Thus, the correspondence with classical theory of the new form of quantum electrodynamics is more far-reaching than that of the old form, which provides a strong reason for preferring the new form. *It now becomes necessary to find some new physical interpretation to avoid the difficulties of negative energies and probabilities.*

Let us consider in more detail the relation between the two forms of quantum electrodynamics. In either form the electromagnetic potentials A at two points x' and x" must satisfy the *commutation relations*

$$[A_\mu(x'), A_\nu(x'')] = g_{\mu\nu}\,\Delta(x'-x''), \qquad (9)$$

obtained from analogy with the classical theory, Δ being the *four-dimensional Lorentz-invariant function* introduced by Jordan & Pauli (1927), which has a singularity on the light-cone and vanishes everywhere else. [Jordan, P. & Pauli, W. (1927). Zur Quantenelektrodynamik ladungsfreier Felder. (On the quantum electrodynamics of charge-free fields.) *Zeit. Phys.*, 47, 151-73.] In the quantum electrodynamics of Heisenberg and Pauli the A's are operators referring to the *absorption* and *emission* of photons into *positive-energy states*. Let us call such operators $\mathbf{A}^1$. One could introduce a similar set of operators referring to the *absorption* and *emission* of photons into *negative-energy states*. Let us call these operators A^2. They satisfy the same *commutation relations* (9) and commute with the $\mathbf{A}^1$'s. One can now introduce a third set of operators

$$A^3 = 2^{-1/2}\,(A^1 + A^2),$$

which operate on *wave functions* referring to photons in both *positive- and negative-energy states*, and which satisfy the same *commutation relations* (9). The use of this third set gives the new form of quantum electrodynamics arising from plain *second quantization.*

The three sets of **A**'s may be expressed in terms of their Fourier components thus:

$$A^1(x) = \iiint \{R_k e^{i(k,x)} + R_k{}^* e^{-i(k,x)}\}\, k_0{}^{-1} dk_1 dk_2 dk_3, \qquad (10)$$

where $k_0 = \sqrt{(k_1{}^2 + k_2{}^2 + k_3{}^2)}$,

$\mathbf{R}_k$ being the *emission operator* and $\mathbf{R}^*{}_k$ the *absorption operator,*

$$A^2(x) = \iiint \ldots \qquad (11)$$

where $k_0 = -\sqrt{(k_1{}^2 + k_2{}^2 + k_3{}^2)}$,

$$A^3(x) = 2^{-1/2}\, \Sigma\, k_0 \iiint \{R_k e^{i(k,x)} + R_k{}^* e^{-i(k,x)}\}\, k_0{}^{-1} dk_1 dk_2 dk_3. \qquad (12)$$

where $k_0 = +/- \sqrt{(k_1{}^2 + k_2{}^2 + k_3{}^2)}$.

Since the three sets of **A**'s all satisfy the same *commutation relations*, they must correspond merely to three different representations of the same dynamical variables, and the passage from one to another must be a transformation of the linear type usual in quantum mechanics. Thus, after obtaining the divergency-free solution of the wave equation in the representation corresponding to $\mathbf{A}^3$, one could apply a transformation to get the solution in the $\mathbf{A}^1$ representation. *However, the transformation would then introduce the same divergent integrals as appear with the direct solution of the wave equation in the* $\mathbf{A}^1$ *representation, so one would not get any further this way.*

In working with the $\mathbf{A}^3$ representation one has redundant *dynamical variables*. It is as though, in dealing with a system of one degree of freedom with the variables q, p, one decided to treat it as a system of two degrees of freedom by putting

$$q = 2^{-1}(q_1 + q_2), \qquad p = 2^{-1}(p_1 + p_2).$$

This would be quite a correct procedure, but would introduce an unnecessary complication. In the case of quantum electrodynamics, the complication is a necessary one, to avoid the *divergent integrals*. Let us put

$$\boldsymbol{B}(x) = 2^{-1}\{A_1(x) - A_2(x)\}. \qquad (13)$$

Then the ***B***'s commute with the $\mathbf{A}^3$'s, and thus with all the *dynamical variables* appearing in the Hamiltonian, so they are the redundant variables.

To determine the significance of redundant variables in quantum mechanics one may consider a general case and work in a representation which separates the redundant variables from the non-redundant ones. One then sees immediately that a solution of the *wave equation* corresponds in general, *not to a single state, but to a set of states with a certain probability for each—what in the classical theory is called a Gibbs ensemble*. The probabilities of the various states depend on the weights attached to the various *eigenvalues* of the redundant variables in the *wave function*, these weights being arbitrary, depending on the weight factor in the representation used. If one works in a representation which does not separate the redundant and non-redundant variables, as is the case in quantum electrodynamics with the representation corresponding to the use of $\mathbf{A}^3$, the general result that *wave functions* represent *Gibbs ensembles* and not single *states* must still be valid. Thus, one can conclude that *there are no solutions of the wave equation of quantum electrodynamics representing single states, but only solutions representing Gibbs ensembles. The problem remains of interpreting the negative energies and probabilities occurring with these Gibbs ensembles*.

For any x, $\boldsymbol{B}$(x) commutes with the Hamiltonian and is a constant of the motion. We may give it any value we like, subject to not contradicting the *commutation relations*. Instead of $\boldsymbol{B}$(x) it is more convenient to work with the *potential field*, **B**(x) say, obtained from $\boldsymbol{B}$(x) by changing the sign of all the Fourier components containing e^{ik0x0} with negative values of k_0. From (13), (10) and (11),

$$\mathbf{B}(x) = 2^{-1}\,\Sigma \iiint \{R_k e^{i(k,x)} - R_k{}^* e^{-i(k,x)}\}\, k_0{}^{-1} dk_1 dk_2 dk_3 \qquad (14)$$

the Σ meaning a summation over both values of k_0, as in equation (12).

Let us now take **B** equal to the initial value of $\mathbf{A}^3$, a proceeding which does not contradict the *commutation relations* since its consequences are self-consistent. Then for the initial *wave function* ψ

$$\{\mathbf{B}(x) - \mathbf{A}^3(x)\}\,\psi = 0,$$

or, from (12) and (14),

$$\mathbf{R}^*{}_k\,\psi = 0, \qquad (15)$$

with k_0 either positive or negative. Thus, any *absorption operator* applied to the initial *wave function* gives the result zero, which means that the corresponding *state* is one with no photons present.

The following natural interpretation for the *wave function* at some later time now appears. That part of it corresponding to *no photons present* may be supposed to give (through the square of its modulus) the probability of *no change having taken place in the field of photons*; that part corresponding to *one positive-energy photon present* may be supposed to give the probability of *a photon having been emitted*; that corresponding to *one negative-energy photon present* may be supposed to give the *probability of a photon having been absorbed*; and so on for the parts corresponding to two or more photons present. *The various parts of the wave function which referred to the existence of positive- and negative-energy photons in the old interpretation now refer to the emissions and absorptions of photons. This disposes of the negative-energy difficulty in a satisfactory way, conforming to the laws of conservation of energy and momentum*. It is possible only because of the redundant variables enabling one to arrange that the initial *wave function* shall correspond in its entirety to no *emissions* or *absorptions* having taken place.

The interpretation is not yet complete, because *the theory at present would give a negative probability for a process involving the absorption of a photon, or the absorption of any odd number of photons*. To find the origin of these negative probabilities, one must study the probability distribution of the photons initially present in the Gibbs ensemble, which one can do by transforming to the representation corresponding to the $\mathbf{A}^1$ *potentials*. It is true that one cannot apply this transformation to a solution of the *wave equation* without getting *divergent integrals*, as has already been mentioned, *but one can apply it to the initial wave function*, which is of an especially simple form in the photon variables. This is done in Appendix III, and it is found that the probability of there being n photons initially in any photon state is $P_n = +/- 2$, according to whether n is even or odd. Strictly, to make $\Sigma_{n=0}^{\infty} P_n$ converge to the limit unity, one must consider P_n as a limit,

$$P_n = 2(\varepsilon - 1)^n, \qquad (16)$$

with ε a small positive quantity tending to zero.

Probabilities 2 and – 2 are, of course, not physically understandable, but one can use them mathematically in accordance with the rules for working with a Gibbs ensemble. *One can suppose a hypothetical mathematical world with the initial probability distribution (16) for the photons, and one can work out the probabilities of radiative transition processes occurring in this world.* One can deduce the corresponding probability coefficients, i.e. the probabilities per unit intensity of each beam of incident radiation concerned, by using *Einstein's laws of radiation*. For example, for a process involving the *absorption of a photon*, if the probability coefficient is B, the probability of the process is

$$\Sigma_{n=0}^{\infty} nP_nB = -\tfrac{1}{2} B, \qquad (17)$$

and for a process involving the *emission of a photon*, if the probability coefficient is A, the probability of the process is

$$\Sigma_{n=0}^{\infty} (n+1)\, P_nA = -\tfrac{1}{2} A. \qquad (18)$$

Now the *probability* of an *absorption process*, as calculated from the theory, is negative, and that for an *emission process* is positive, so that, equating these calculated probabilities to (17) and (18) respectively, one obtains positive values for both B and A. Generally, it is easily verified that any *radiative transition probability coefficient* obtained by this method is positive.

It now becomes reasonable to assume that these probability coefficients obtained for a hypothetical world are the same as those of the actual world. One gets in this way a general physical interpretation for the quantum theory of photons. When applied to elementary examples, it gives the same results as Heisenberg & Pauli's quantum electrodynamics with neglect of the divergent integrals, since the extra factor $2^{-\frac{1}{2}}$ occurring in the matrix elements of the present theory owing to the $2^{-\frac{1}{2}}$ in the right-hand side of (12) compensates the factor

½ in the right-hand side of (17) or (18). The present general method of physical interpretation is probably applicable to any kind of particle with an integral spin.

Conclusion

It appears that, whether one is dealing with particles of integral spin or of half-odd integral spin, one is led to a similar conclusion, namely, that the mathematical methods at present in use in quantum mechanics *are capable of direct interpretation only in terms of a hypothetical world differing very markedly from the actual one*. These mathematical methods can be made into a physical theory by the assumption that results about collision processes are the same for the hypothetical world as the actual one. *One thus gets back to Heisenberg's view about physical theory—that all it does is to provide a consistent means of calculating experimental results. The limited kind of description of Nature which Schrodinger's method provides in the non-relativistic case is possible relativistically only for the hypothetical world*, and even then is rather more indefinite (e.g. the *principle of superposition of states* no longer applies), because of the need to use a Gibbs ensemble for describing the photon distribution.

To have a description of Nature is philosophically satisfying, though not logically necessary, and *it is somewhat strange that the attempt to get such a description should meet with a partial success, namely, in the non-relativistic domain, but yet should fail completely in the later development. It seems to suggest that the present mathematical methods are not final*. Any improvement in them would have to be of a very drastic character, because *the source of all the trouble, the symmetry between positive and negative energies arising from the association of energies with the Fourier components of functions of the time*, is a fundamental feature of them.

References

Dirac, P. A. M. (August, 1938). Classical Theory of Radiating Electrons. *Roy. Soc. Proc., A*, 167, 929, 148-69.
Dirac, P. A. M. (1939a). A new notation for quantum mechanics. *Proc. Camb. Phil. Soc.*, 35, 416-18.
Dirac, P. A. M. (1939b). *Ann. Inst. Poincare*. This paper has not yet appeared in England.
Fock, V. (1934). Zur Quantenelektrodynamik. (To quantum electrodynamics.) *Phys. Zeit. Sowjet.*, 6, 425-69.
Fokker, A. D. (1929). Ein invarianter Variationssatz für die Bewegung mehrerer elektrischer Massenteilchen. (An invariant set of variations for the movement of several electrical mass particles.) *Zeit. Phys.*, 58, 386-93.
Gordon, W. (January, 1927). Der Comptoneffekt nach der Schrödingerschen Theorie. (The Compton effect according to Schrodinger's theory.) *Zeit. Phys.*, 40, 117-33.
Heisenberg, W. & Pauli, W. (1929a). Zur Quantendynamik der Wellenfelder. (On the quantum dynamics of wave fields.) *Zeit. Phys.*, 56, 1-61.
Heisenberg, W. & Pauli, W. (1929b). Zur Quantendynamik der Wellenfelder. II. (On the quantum theory of wave fields. II.) *Zeit. Phys.* 59, 168-90.

Jordan, P. & Pauli, W. (1927). Zur Quantenelektrodynamik ladungsfreier Felder. (On the quantum electrodynamics of charge-free fields.) *Zeit. Phys.*, 47, 151-73.
Klein, O. (1927). Elektrodynamik und Wellenmechanik vom Standpunkt des Korrespondenzprinzips. (Electrodynamics and wave mechanics from the point of view of the correspondence principle.) *Zeit. Phys.*, 41, 10, 407-42
Pauli, W. & Weisskopf, V. (1934). Uber die Quantisierung der skalaren relativistischen Wellengleichung. (The Quantization of the Scalar Relativistic Wave Equation.) *Helv. Phys.*, 7, 709-31.
Weiss, P. (1936). On the Quantization of a Theory Arising from a Variational Principle for Multiple Integrals with Application to Born's Electrodynamics. *Roy. Soc. Proc., A,* 156, 887, 192-220.

Heisenberg, W. (July, 1943). Die beobachtbaren Größen in der Theorie der Elementarteilchen. (The "observable quantities" in the theory of elementary particles.)

[*Zeit. Phys.*, 120, 7-10, 513-38; https://doi.org/10.1007/bf01329800.]

Received September 8, 1942.

Describes S-matrix theory as a principle of particle interactions, the present work attempts to extract from the conceptual structure of the quantum theory of wave fields those terms that are unlikely to be affected by the future change and which will therefore also form a part of the future theory, it avoids the notion of space and time by replacing it with abstract mathematical properties of the S-matrix, the S-matrix relates the infinite past to the infinite future in one step, without being decomposable into intermediate steps corresponding to time-slices.

[S-matrix theory was proposed as a principle of particle interactions by Heisenberg in 1943, following John Archibald Wheeler's 1937 introduction of the S-matrix. It was a proposal for replacing local quantum field theory as the basic principle of elementary particle physics. It avoided the notion of space and time by replacing it with abstract mathematical properties of the S-matrix. In S-matrix theory, the S-matrix relates the infinite past to the infinite future in one step, without being decomposable into intermediate steps corresponding to time-slices.

This program was very influential in the 1960s, because it was a plausible substitute for quantum field theory, which was plagued with the zero-interaction phenomenon at strong coupling. Applied to the strong interaction, it led to the development of string theory.

S-matrix theory was largely abandoned by physicists in the 1970s, as quantum chromodynamics was recognized to solve the problems of strong interactions within the framework of field theory. But in the guise of string theory, S-matrix theory is still a popular approach to the problem of quantum gravity.]

[Feynman (September, 1949). Space-Time Approach to Quantum Electrodynamics: "In many problems, for example, the close collisions of particles, we are not interested in the precise temporal sequence of events. It is not of interest to be able to say how the situation would look at each instant of time during a collision and how it progresses from instant to instant. Such ideas are only useful for events taking a long time and for which we can readily obtain information during the intervening period. *For collisions it is much easier to treat the process as a whole*[6].

[6] This is the viewpoint of the theory of the S matrix of Heisenberg."]

Abstract

The well-known divergence difficulties in the theory of elementary particles show that the future theory will contain in its foundations a universal constant of the dimension of a length, which apparently cannot be integrated into the previous form of the theory without contradiction. With regard to this later modification of the theory, *the present work attempts to extract from the conceptual structure of the quantum theory of wave fields those terms that are unlikely to be affected by the future change and which will therefore also form a part of the future theory*. The work is divided into the following sections:

I. The observable quantities and their mathematical representation. a) Formulation of the basic assumptions. b) The mathematical representation of the observable quantities.
c) The singularities in the impulse space. d) The characteristic matrix. (e) The emission of several particles. —

II. The properties of the *matrixS.* a) The general quantum mechanical properties of *S.*
b) The singular parts of the *matrixS.* c) Consideration of spin and statistics of the various elementary particles. (d) The relativistic behavior of *MatrixS.* —

III. Relationships between observable quantities. a) The eigenvalue problem given by *matrixS*. (b) relationships between scattering coefficients and emission coefficients.
c) relationships between the observable quantities that do not follow from the general properties of *matrixS*.

Yoshio Nishina (December 6, 1890 – January 10, 1951)

Nishina was a Japanese physicist who was called "the founding father of modern physics research in Japan". He led the efforts of Japan to develop an atomic bomb during World War II.

Nishina was born in Satoshō, a village near Okayama, Japan. After high school in Okayama, he enrolled in Tokyo Imperial University, graduating with a degree in Electrical Engineering in 1919. He received a silver watch from the emperor as he graduated at the top of his class. He turned down an engineering job offer and enrolled in graduate school, first studying electrochemistry but soon switching to physics. In addition to his graduate studies, he became a research assistant at the Institute of Physical and Chemical Research (known by its Japanese acronym, *RIKEN*).

In 1921, *RIKEN* sent Nishina to study in Europe. He studied at the Cavendish Laboratory at Cambridge, under Ernest Rutherford. In 1923, he moved to Copenhagen to work with Niels Bohr and they became good friends. He also visited Gottingen and Hamburg. Before he left Copenhagen in 1928, he published a paper on Compton scattering, a collaboration with Oscar Klein, from which the Klein–Nishina formula derives. After he returned home, he convinced Werner Heisenberg to visit Japan and give a number of lectures. Later, in 1937, Bohr and his wife also visited Nishina in Tokyo.

In 1929, he returned to Japan, where he endeavored to foster an environment for the study of quantum mechanics. In 1931, he established his own laboratory at RIKEN, and invited some Western scholars to Japan including Heisenberg, Dirac and Bohr to stimulate Japanese physicists. It was also in 1931 that he lectured about the Dirac theory in Kyoto, which was attended by Hideki Yukawa and Sin-Itiro Tomonaga. His lab became focused on quantum mechanics, nuclear physics, cosmic rays, and high-energy proton beams.

Nishina completed a "small" cyclotron in 1937, the first cyclotron constructed outside the United States (and the second in the world). While he was constructing a larger cyclotron, he consulted Ernest Lawrence for help. Lawrence assisted him in buying an electromagnet from an American company, and hosted some of Nishina's assistants in 1940. Communication between the two physicists was cut off by the start of the war. The "large" cyclotron would not be complete until 1944.

In 1940, Nishina was asked by Lieutenant General Takeo Yasuda to research nuclear fission. In April 1941, the Imperial Japanese Army (IJA) officially authorized *Ni-Go*, the project to research an atomic bomb. *Nishina's team's initial conclusion was that an atomic bomb was theoretically, but not technically, feasible,* but they continued to pursue uranium enrichment via gaseous thermal diffusion (a process that had been abandoned by the Allies in 1941). Nishina also participated in a committee convened by the Technical Research Institute of the Imperial Japanese Navy (IJN) between 1942 and 1943. After several meetings, the committee disbanded with pessimistic conclusions. Meanwhile, Nishina continued his research for the IJA. In June 1945, a few months after *RIKEN* facilities were bombed, Nishina declared the project over, and resources were handed over to the IJN's *F-Go* project.

The day after the US dropped an atomic bomb on Hiroshima on August 9, 1945, Nishina received a copy of a restricted press release about the bomb that had come from President Truman. Nishina confirmed to the government the veracity of the claims about atomic weaponry. In October 1945, after surrender, Nishina asked the occupying forces for permission to continue using his remaining cyclotrons for biological and medical research. Permission was initially granted, then rescinded under orders from the Secretary of War. Every cyclotron in Japan was destroyed. The ones from *RIKEN* were disassembled and thrown into the Gulf of Tokyo.

> *In 1945 the US detonated two nuclear weapons over the Japanese cities of Hiroshima and Nagasaki. On 6 August, "Little Boy", an enriched uranium gun-type fission weapon was dropped on Hiroshima; on August 9, 1945, "Fat Man", a plutonium implosion-type nuclear weapon was dropped on Nagasaki. Japan surrendered to the Allies on August 15.*

After the war, Nishina became president of *RIKEN* and helped salvage the institute from the ashes of war. He died on January 10, 1951, age 60. *RIKEN* later named its Center for Accelerator-Based Science after him, and the Nishina Memorial Foundation gives out awards in his name on an annual basis.

Shinichiro Tomonaga (March 31, 1906 – July 8, 1979)

[Includes material from Nobel Lectures, Physics 1963-1970, Elsevier Publishing Company, Amsterdam, 1972.]

Shinichiro Tomonaga, usually cited as Sin-Itiro Tomonaga in English, was a Japanese physicist, influential in the development of quantum electrodynamics, work for which he was jointly awarded the Nobel Prize in Physics in 1965 with Julian Schwinger and Richard P. Feynman, "for their fundamental work in quantum electrodynamics, with deep-ploughing consequences for the physics of elementary particles". "Following the establishment of the theory of relativity and quantum mechanics, an initial *relativistic* theory was formulated for the interaction between charged particles and electromagnetic fields. The theory had to be reformulated, however, partly due to the observation of the Lamb shift in 1947, in which the supposed single energy level within a hydrogen atom was instead proven to be two similar levels. Tomonga solved this problem in 1948 through a "*renormalization*" and thereby contributed to a new quantum electrodynamics." [Sin-Itiro Tomonaga – Facts. NobelPrize.org. <https://www.nobelprize.org/prizes/ physics/1965/ tomonaga/facts/].

Tomonaga was born in Tokyo in 1906, the eldest son of Sanjuro and Hide Tomonaga. In 1913 his family moved to Kyoto when his father was appointed a professor of philosophy at Kyoto Imperial University. He is a graduate of the Third Higher School, Kyoto, a renowned senior high school which educated a number of leading personalities in prewar Japan.

He entered the Kyoto Imperial University in 1926. Hideki Yukawa, also a Nobel laureate, was one of his classmates during undergraduate school. Tomonaga completed work for Rigakushi (bachelor's degree) in physics at Kyoto Imperial University in 1929. He was engaged in graduate work for three years at the same university and in 1932 was then appointed a research associate by Yoshio Nishina at the Institute of Physical and Chemical Research, Tokyo, where he started to work in a newly developed frontier of theoretical physics quantum electrodynamics under the guidance of Nishina.

Dirac's 1932 paper [Relativistic Quantum Mechanics] attracted Tomonaga's interest, and Nishina also showed a great interest in this paper and suggested that Tomonaga investigate the possibility of predicting some new phenomena by this theory.

Tomonaga found that Dirac's *many-time* formalism turned out to be equivalent to the Heisenberg-Pauli theory, but recognized that it had the advantage that it gave the possibility of generalizing the former interpretation of the *probability amplitude*.

Tomonaga stayed in Leipzig, Germany, from 1937 to 1939, to study *nuclear physics* and the quantum field theory in collaboration with the theoretical group of Heisenberg, where he published a paper on nuclear materials, "*Innere Reibung und Wärmeleitfähigkeit der Kernmaterie*" ("Internal friction and thermal conductivity of the nuclear matter"), which was chosen as his thesis for Rigakuhakushi (Doctor of Science) at Tokyo Imperial

University in December, 1939, after he returned to Japan due to the outbreak of the Second World War.

> [At the end of 1938 Otto Hahn and Fritz Strassmann in Berlin showed that nuclear fission not only released a lot of energy, but that it also released additional neutrons which could cause fission in other uranium nuclei and possibly a self-sustaining chain reaction leading to an enormous release of energy. Lise Meitner and her nephew Otto Frisch, working under Niels Bohr, then explained this by suggesting that the neutron was captured by the nucleus, causing severe vibration leading to the nucleus splitting into two not quite equal parts. They calculated the energy release from this fission as about 200 million electron volts. Frisch then confirmed this figure experimentally in January 1939. Bohr soon proposed that fission was much more likely to occur in the uranium-235 isotope than in U-238 and that fission would occur more effectively with slow-moving neutrons than with fast neutrons. Niels Bohr and John Wheeler extended these ideas into what became the classical analysis of the fission process. Their paper on the mechanism of nuclear fission appeared in the *Physical Review* of September 1, 1939, the same day that World War II began. [Bohr, N. & Wheeler, J. A. (1939). The Mechanism of Nuclear Fission. *Phys. Rev*., 56, 426; Published September 1, 1939. (Received June 28, 1939)]

In 1940, Tomonaga was married to Ryoko Sekiguchi, daughter of Dr. K. Sekiguchi, the former Director of the Tokyo Metropolitan Observatory. They had two sons, Atsushi and Makoto and one daughter. Their daughter was married in 1965 to Dr. Y. Nagashima, research associate of the Physics Department, University of Rochester.

> [*Japan entered World War II on 22 September 1940 when it invaded French Indochina. During World War II, Heisenberg was the head of the German nuclear weapon program.*]

In 1941, Tomonaga joined the faculty of Tokyo Bunrika University (which was absorbed into the Tokyo University of Education in 1949, and subsequently became Tsukuba University) as Professor of Physics.

> [*On Sunday, December 7, 1941, the Imperial Japanese Navy Air Service launched a surprise military strike on the United States naval base at Pearl Harbor in Honolulu, Hawaii, which led to the formal entry of the US into World War II the next day*.]

Stimulated by comments made by Yukawa in 1942*, Tomanaga developed the covariant formulation of the quantum field theory, *covariant field theory*, in which the concept of the *quantum state* was generalized so as to be *relativistically* covariant.

> * At the MEISO-KAI (Wondering Meeting) on 24 April 1942, Yukawa reported his attempt on an extended *space-time* description in the quantum field theory in this meeting. He discussed i) non-separability of cause and effect, ii) the density matrix,

iii) the field equation with additional conditions, iv) Lagrangian and Hamiltonian, and v) elementary particles – numerable; space-time – point-like. He recorded the names of 21 participants and 9 questioners including Tomonaga, who got a clue to his *super many-time theory* from Yukawa's talk. [*The Legacy of Hideki Yukawa, Sin-itiro Tomonaga, and Shoichi Sakata: Some Aspects from their Archives*. Proceedings of the 12th Asia Pacific Physics Conference held at Makuhari, Chiba, Japan on July 15, 2013.]

[*In September 1942, Robert Oppenheimer was appointed to head up the nuclear weapons laboratory of the newly formed US "Manhattan Project"*].

Tomonaga was first to summarize and extend the *intermediate coupling theory*, in order to clarify the structure of the meson cloud around the nucleon. [Tomonaga, S. (June, 1943). *Interactions between nuclear particles and mesons, Part 1 A study based on the cut-off hypothesis*. Advance lecture note for the Informal Discussion Meeting on Mesons held in Sept. 1943.]

The largest informal discussion meeting called the Meson Symposium was held at RIKEN on 26 and 27 September 1943. Around 50 participants attended from all over Japan in spite of severe wartime. This was the last gathering in this series during the World War II. The number of speakers was 5 for 2 days including Tomonaga and Shoichi Sakata. All of them prepared detailed texts, which were circulated to expected nation-wide participants in advance. Tomonaga discussed methods of approximation on nucleon-meson interactions. [*The Legacy of Hideki Yukawa, Sin-itiro Tomonaga, and Shoichi Sakata: Some Aspects from their Archives*. Proceedings of the 12th Asia Pacific Physics Conference held at Makuhari, Chiba, Japan on July 15, 2013.]

In 1943, he published the "*super-many-time theory*", which reconciled quantum mechanics with the *theory of relativity*. [Tomonaga, S. (Aug-Sep, 1946). On a Relativistically Invariant Formulation of the Quantum Theory of Wave Fields." *Prog. Theor. Phys*., 1, 27–42. Institute of Physics, Tokyo Bunrika University. Received May 17, 1946. Translated from the paper (1943). *Bull. I. P. C. R. Riken-iho*, 22, 545 (appeared originally in Japanese).]

[*In 1945 the US detonated two nuclear weapons over the Japanese cities of Hiroshima and Nagasaki. On 6 August, "Little Boy", an enriched uranium gun-type fission weapon was dropped on Hiroshima; on August 9, 1945, "Fat Man", a plutonium implosion-type nuclear weapon was dropped on Nagasaki. Japan surrendered to the Allies on August 15.*]

In the summer of 1946, as noted in part V, together with members of the seminary of the theoretical physics at the Institute of Physics, Tokyo Bunrika University, he published parts II, III, IV, and V of his 1943 paper, applying *super-many-time theory* to the cases of interacting electromagnetic and electron and meson fields*.

* Koba, Z., Tati, T. & Tomonaga, S. (1947a). On a Relativistically Invariant Formulation of the Quantum Theory of Wave Fields. II. Case of Interacting Electromagnetic and Electron Fields. § 1-4. *Prog. Theor. Phys*., 2, 101–116. Received November 6, 1946.

Koba, Z., Tati, T. & Tomonaga, S. (1947b). On a Relativistically Invariant Formulation of the Quantum Theory of Wave Fields. III. Case of Interacting Electromagnetic and Electron Fields. § 5-7. *Prog. Theor. Phys.*, 2, 198–208. Received November 6, 1946.

Kanesawa, S. & Tomonaga, S. (1948). On a Relativistically Invariant Formulation of the Quantum Theory of Wave Fields. IV. Case of Interacting Electromagnetic and Meson Fields. § 1-4. *Prog. Theor. Phys.* 3, 1–13. Received November 6, 1946.

Kanesawa, S. & Tomonaga, S. (1948). On a Relativistically Invariant Formulation of the Quantum Theory of Wave Fields. V. Case of Interacting Electromagnetic and Meson Fields. § 5-7. *Prog. Theor. Phys.* 3, 101–113. Received November 6, 1946.

Shoichi Sakata's work on his subsequently abandoned *C-meson theory* [Sakata, S. & Hara, O. (1947). The Self-Energy of the Electron and the Mass Difference of Nucleons. *Progr. Theoret. Phys. (Kyoto)*, 2, 30-1; published January 1, 1947], and a 1939 paper by Sidney Dancoff [Dancoff, S. M. (May, 1939). On Radiative Corrections for Electron Scattering. *Phys. Rev.*, 55, 959] that had attempted, but failed, to show that the infinite quantities that arise in quantum electro-dynamics can be canceled with each other, helped Tomonaga's group to reorganize the properties of various divergences appearing in quantum electro-dynamics. He applied his *super-many-time theory* and a *relativistic* method based on the *non-relativistic* method of Pauli and Fierz to greatly speed up and clarify the calculations. He and his students found that Dancoff had overlooked one term in the perturbation series. With this term, the theory gave finite results.

In a lecture at the symposium on the theory of elementary particles in Kyoto, Nov. 1947, Tomanaga proposed in connection with it a "self-consistent" subtraction method*.

* Tomonaga, S. Lecture at the symposium on the theory of elementary particles in Kyoto, November, 1947; subsequently published as Tati, T. & Tomonaga, S. (December, 1948). A Self-Consistent Subtraction Method in the Quantum Field Theor. I. *Prog. Theor. Phys.*, 3, 4, 391-406.

They eventually reached the *renormalization* idea, in a series of letters published in the same issue of Progress of Theoretical Physics [Ito, D., Koba, Z. & Tomonaga, S. (1947). Correction due to the Reaction of "Cohesive Force Field" for the Elastic Scattering of an Electron. *Prog. Theor. Phys.*, 2, 216 (containing "serious oversight"); Ito, D., Koba, Z. & Tomonaga, S. (1947). Errata to the Above Letter. *Prog. Theor. Phys.*, 2, 217 (correction); Koba, Z. & Tomonaga, S. (1947). Application of the "Self-Consistent" Subtraction Method to the Elastic Scattering of an Electron. *Prog. Theor. Phys.*, 2, 218]. A full account was subsequently published in a two-part paper in 1948 [Ito, D., Koba, Z. & Tomonaga, S. (1948). Corrections due to the Reaction of "Cohesive Force Field" to the Elastic Scattering of an Electron. I. *Prog. Theor. Phys.*, 3, 3, 276-89; 3, 4, 325-37 [*abandoned C-meson field theory*].

They concluded that the ordinary formalism of the *quantum electrodynamics* (i.e. the formalism without C-meson field) gives rise to at least two types of diverging corrections to the *scattering cross-section*, the one related to the *self-energy*, and the other to the *vacuum polarization* effect, which are to be distinguished by their origins as well as by their forms and which cannot be eliminated simultaneously by one single modification of prescription introduced in the theory.

> [The *self-energy* represents the *potential* felt by the electron due to the surrounding medium's interactions with it. *Vacuum polarization* describes a process in which a background *electromagnetic field* produces *virtual electron - positron pairs* that change the distribution of *charges* and *currents* that generated the original *electromagnetic field.*]

Koba & Tomonaga (1947) found that the diverging radiative correction in the scattering process is eliminated *if one re-interprets the mass of the electron on one hand and the scattering potential on the other in such a manner that they already include the electromagnetic mass correction and the vacuum polarization effect in themselves*. Thus, Tomonaga discovered the *renormalization* method independently of Julian Schwinger and calculated physical quantities such as the Lamb shift at the same time.

A few weeks after Feynman and Schwinger reported their own efforts in this area at the Pocono conference in March 1948, Oppenheimer received a letter from Tomonaga describing his work and immediately recognized the overlap in their efforts.

In 1948, Oppenheimer published a joint paper with Tomonaga [Tomonaga, S. & Oppenheimer, J. R. (1948). On Infinite Field Reactions in Quantum Field Theory." *Phys. Rev.* 74, 224-5] and in 1949 invited him to work at the Institute for Advanced Study at Princeton, where he was engaged in the investigation of a *many-body problem* on the collective oscillations of a quantum-mechanical system, a one-dimensional fermion system.

His book, *Quantum Mechanics*, was published in 1949 and translated into English in 1963. Tomonaga's honors and awards include the Japan Academy Prize (1948); the Order of Culture (1952); the Lomonosov Medal, U.S.S.R. (1964). In 1965, he was awarded the Nobel Prize in Physics, with Julian Schwinger and Richard P. Feynman.

He died of throat cancer in Tokyo in 1979, at age 73.

Tomonaga, S. (1943). On a Relativistically Invariant Formulation of the Quantum Theory of Wave Fields.

[*Bull. I. P. C. R. Riken-iho*, 22, 545 (in Japanese); translation Tomonaga, S. (August, 1946). *Prog. Theor. Phys*. 1, 2, 27–42; also in Schwinger, J. (ed). (1958). *Selected Papers on Quantum electrodynamics*. Dover, New York, pages 156-68.]

Translation received May 17, 1946.

Institute of Physics, Tokyo Bunrika University.

Existing formalism of quantum field theory not perfectly relativistic, commutation relations refer to points in space at different times, Schrodinger equation for the vector representing the state of the system is a function of time, time variable plays a different role than space variables, *probability amplitude* not *relativistically invariant* in the space-time world, follows Dirac (1932) [Relativistic Quantum Mechanics] in *generalizing the notion of probability amplitude as far as is required by the theory of special relativity*, substitutes four-dimensional form of the commutation relations, *generalizes the Schrodinger equation* following the Dirac (1933) [The Lagrangian in Quantum Mechanics] *many-time* formalism, then introduces his *super many-time theory* [$\{H_{12}(P) + h/i\ \partial/\partial C_P\}\ \Psi[C] = 0$ at point P on surface C with infinitely many time variables representing the local time for each position in the space, results in *relativistic interaction representation*, three-dimensional manifold (space-like "surface") in four-dimensional space-time world, not necessary to also assume time-like surfaces for the variable surface as was required by Dirac, previous formalism built up in way too analogous to ordinary *non-relativistic* mechanics, theory was divided into one section giving the kinematical relations between various quantities at the same instant of time and another section determining the causal relations between quantities at different instants of time with the *commutation relations* belong to the first section and the *Schrodinger equation* to the second, this way of separating the theory into two sections very *unrelativistic*, "same instant of time" plays a distinct role, *new formalism consists of one section giving laws of behavior of the fields when they are left alone and the other giving the laws determining the deviation from this behavior due to interactions*, can be carried out *relativistically*, although theory in more satisfactory form no new contents added, *divergence difficulties inherited*, fundamental equations admit only catastrophic solutions *due to non-vanishing zero-point amplitudes of the fields which inheres in the operator $H_{12}(P)$, a more profound modification of the theory is required in order to remove this fundamental difficulty*.

> [MEISO-KAI (Wondering Meeting) on 24 April 1942. Yukawa reported his attempt on an extended space-time description in the quantum field theory in this meeting. He discussed i) non-separability of cause and effect, ii) the density matrix, iii) the field equation with additional conditions, iv) Lagrangian and Hamiltonian, and v) elementary particles – numerable; space-time – point-like. He recorded names of 21 participants and 9 questioners including Tomonaga, who got a clue to his *super many-time theory* from Yukawa's talk.]

> [Oppenheimer, J. R. (1948). Electron Theory. *Report to the Solvay Conference for*

Physics at Brussels, Belgium, September 27 to October 2, 1948, pages 6-7. (Reprint in Schwinger, J. (ed). (1958). *Selected Papers on Quantum electrodynamics*, Dover, New York, pages 150-1: "Now it is true that the fundamental equations of quantum-electrodynamics are gauge and Lorentz covariant. But they have in a strict sense no solutions expansible in powers of e. If one wishes to explore these solutions, bearing in mind that certain infinite terms will, in a later theory, no longer be infinite, one needs a covariant way of identifying these terms, and for that, not merely the field equations themselves, but *the whole method of approximation and solution must at all stages preserve covariance. This means that the familiar Hamiltonian methods, which imply a fixed Lorentz frame t = constant, must be renounced*; neither Lorentz frame nor gauge can be specified until after, in a given order in e, all terms have been identified, and those bearing on the definition of charge and mass recognized and relegated; then of course, in the actual calculation of transition probabilities and the reactive corrections to them, or in the determination of stationary states in fields which can be treated as static, and in the reactive corrections thereto, the introduction of a definite coordinate system and gauge for these no longer singular and completely well-defined terms can lead to no difficulty.

It is probable that, at least to order e^2, more than one covariant formalism can be developed. Thus, *Stueckelberg's four-dimensional perturbation theory*[26] would seem to offer a suitable starting point, as also do the related algorithms of Feynman[27].

[26] Stueckelberg, E. C. G. (September, 1934). Relativistisch invariante Störungstheorie des Diracschen Elektrons I. Teil: Streustrahlung und Bremsstrahlung. (Relativistically invariant perturbation theory of Dirac's electron Part I: scattered radiation and Bremsstrahlung.) *Ann. Phys*., 413, 4, 367-389
[27] Feynman, R. P. (November, 1948). Relativistic Cut-Off for Quantum Electrodynamics. *Phys. Rev*., 74, 1430-8.

But *a method originally suggested by Tomonaga*[28], and independently developed and applied by Schwinger[22], would seem, apart from its practicality, to have the advantage of very great generality and a complete conceptual consistency.

[28] Tomonaga, S. (1943). On a Relativistically Invariant Formulation of the Quantum Theory of Wave Fields. *Bull. I. P. C. R. Riken-iho*, 22, 545 (in Japanese); translation Tomonaga, S. (August, 1946). *Prog. Theor. Phys*. 1, 2, 27-42; Koba, Z., Tati, T. & Tomonaga, S. (1947a). On a Relativistically Invariant Formulation of the Quantum Theory of Wave Fields. II. Case of Interacting Electromagnetic and Electron Fields. § 1-4. *Prog. Theor. Phys*., 2, 101–116; Koba, Z., Tati, T. & Tomonaga, S. (1947b). On a Relativistically Invariant Formulation of the Quantum Theory of Wave Fields. III. Case of Interacting Electromagnetic and Electron Fields. § 5-7. *Prog. Theor. Phys*., 2, 198–208.
[22] Schwinger, J. (November, 1948). Quantum Electrodynamics. I. A Covariant Formulation. *Phys. Rev*., 74, 10, 1439-61, and in press.

It has also been shown by Dyson[29] how Feynman's algorithms can be derived from the Tomonaga equations.

[29] Dyson, F. J., *Phys. Rev.*, in press."]

§ 1. *The formalism of the ordinary quantum theory of wave fields.*

Recently Yukawa[1] has made a comprehensive consideration about the basis of the *quantum theory of wave fields.*

[1] Yukawa, H. (1942). *Kagaku* (Science), 7, 249; 8, 282; 9, 322 (in Japanese); unavailable online; attempt at extended space-time description of quantum field theory, non-separability of cause and effect, density matrix, field equation with additional conditions, Lagrangian and Hamiltonian, numerable elementary particles, point-like space-time.

In his article he has pointed out the fact that *the existing formalism of the quantum field theory is not yet perfectly relativistic.*

Let υ(xyz) be the quantity specifying the *field*, and λ(xyz) denote its *canonical conjugate*. Then the quantum theory requires the *commutation relations* of the form:

$$[\upsilon(xyzt), \upsilon(x'y'z't)] = [\lambda(xyzt), \lambda(x'y'z't)] = 0$$
$$[\upsilon(xyzt), \lambda(x'y'z't)] = ih\delta(x - x')\delta(y - y')\delta(z - z'), \qquad (1)^{\#}$$

but these have quite *non-relativistic* forms.

[#] [A, B] = AB – BA. We assume that the field obeys the Bose statistics. Our considerations apply also to case of Fermi statistics.

The equations (1) give namely the *commutation relations* between the quantities at different points (xyz) and (x'y'z') at the same instant of time t. *The concept "same instant of time at different points" has, however, a definite meaning only when one specifies some definite Lorentz frame of reference*. Thus, this is not a *relativistically invariant* concept.

Further, the *Schrodinger equation* for the ψ-vector representing the *state* of the system has the form;

$$(H^{-} + h/i\ \delta/\delta t)\psi = 0, \qquad (2)$$

where H^{-} is the operator representing the *total energy* of the *field* which is given by the space integral of a function of υ and λ. *As we adopt here the Schrodinger picture, υ and λ are operators independent of time*. The vector representing the *state* is in this picture a function of the time, and its dependence on t is determined by (2).

Also, the differential equation (2) is no less *non-relativistic*. *In this equation the time variable t plays a role quite distinguished from the space coordinates x, y and z*. This situation is closely connected with the fact that *the notion of probability amplitude does not fit with the relativity theory.*

As is well known, the vector ψ has, as the *probability amplitude*, the following physical meaning: Suppose the representation which makes the field quantity υ(xyz) diagonal. Let ψ[υ'(xyz)] denote the representative of ψ in this representation[#].

> [#] We use the square brackets to indicate a functional. Thus ψ[υ'(xyz)] means that ψ is a functional of the variable function υ'(xyz). When we use ordinary brackets ( ), as ψ(υ'(xyz)), we consider ψ as an ordinary function of the function υ'(xyz). For example: the *energy density* is written as H(υ(xyz), λ(xyz)) and this is also a function of x, y and z, whereas the *total energy* H = ∫ H(υ(xyz), λ(xyz))dυ is a functional of υ(xyz) and λ(xyz) and is written as H[υ(xyz), λ(xyz)].

Then the representative ψ[υ'(xyz)] is called *probability amplitude*, and its absolute square

$$W[\upsilon'(xyz)] = |\,\psi\,[\upsilon'(xyz)]\,|^2 \qquad (3)$$

gives the *relative probability* of υ(xyz) having the specified functional form υ'(xyz) at the instant of time t. In other words: Suppose a plane[$] which is parallel to the xyz-plane and intercepts the time axis at t.

> [$] We call a three-dimensional manifold in the four-dimensional space-time world simply "surface".

Then the *probability* that the *field* has the specified functional υ'(xyz) on this plane is given by (3).

As one sees, a plane parallel to the xyz-plane plays here a significant role. But such a plane is only defined by referring to a certain frame of reference. *Thus, the probability amplitude is not a relativistically invariant concept in the space-time world.*

§ 2. *Four-dimensional form of the commutation relations.*

As stated above, the laws of the *quantum theory of wave fields* are usually expressed as mathematical relations between quantities having their meanings only in some specified Lorentz frame of reference. *But since it is proved that the whole contents of the theory are of course relativistically invariant*, it must be certainly possible to build up the theory on the basis of concepts having *relativistic space-time* meanings. Thus, in his consideration, Yukawa has required with Dirac[2] to generalize the notion of *probability amplitude* so that it fits with the *relativity theory.*

> [the Dirac (1932) and Dirac (1933) references were incorrectly switched over]

> [2] Dirac, P. A. M. (May, 1932). Relativistic Quantum Mechanics. *Roy. Soc. Proc., A*, 136, 829, 453-64[; alternative to Heisenberg and Pauli's approach to *relativistic* quantum mechanics which regards the field itself as a dynamical system amenable to Hamiltonian treatment and its interaction with the particles as describable by an *interaction energy*, there are serious objections to these views, if we wish to make an observation on a system of interacting particles the only effective method of procedure is to subject them to a field of electromagnetic radiation and see how they react, the role of the *field* is to provide a means for making observations, the nature of an observation requires an interplay between the

field and the *particles*, we cannot suppose the *field* to be a dynamical system on the same footing as the *particles* and thus something to be observed in the same way as the *particles*, the *field* should appear in the theory as something more elementary and fundamental, in this paper an *interaction representation* is proposed which gives interplay between particles and field, translates *equations of motion* of *relativistic* classical theory directly into equations expressible entirely in terms of *probability amplitudes* referring to one *ingoing field* and one *outgoing field*, assumes passage from the field of ingoing waves to the field of outgoing waves is a quantum jump performed by one field composed of waves passing undisturbed through the electron and satisfying Maxwell's equations, *relativistic* observable quantities always *transition probabilities*, *probability amplitudes* analogous to Heisenberg's matrix elements, quantization assumes *intensities* and *phases* are operators satisfying usual quantum conditions governing the *intensities* and *phases* of Fourier components of electromagnetic field in empty space, determines matrix elements associated with electron jumps, assumes interaction of each electron with the *field* can be described by an *interaction energy* equal to its charge multiplied by the *potential* at the point where it is situated, *wave equation* of interactions of two electrons due to motions of each being connected with same field
$\{ih\, \delta/\delta t + h^2/2m_1\, \delta^2/\delta x_1^2 + h^2/2m_2\, \delta^2/\delta x_2^2 - \varepsilon_1 V(x_1t) - \varepsilon_2 V(x_2t)\}\ \psi = 0$].

We shall now show below that the generalization of the theory on these lines is in fact possible *to the relativistically necessary and sufficient extent.* Our results are, however, not so general as expected by Dirac and by Yukawa, but *are already sufficiently general in so far as it is required by the relativity theory.*

Let us suppose for simplicity that there are only two *fields* interacting with each other. The case of a greater number of *fields* can also be treated in the same way. Let υ_1 and υ_2 denote the quantities specifying the *fields*. The *canonically conjugate* quantities be λ_1 and λ_2 respectively. Then between these quantities the *commutation relations*

$$[\upsilon_r(xyzt), \upsilon_s(x'y'z't)] = 0$$
$$[\lambda_r(xyzt), \lambda_s(x'y'z't)] = 0 \qquad r, s = 1, 2$$
$$[\upsilon_r(xyzt), \lambda_s(x'y'z't)] = ih\delta(x - x')\delta(y - y')\delta(z - z')\delta_{rs}, \qquad (4)$$

must hold. The ψ-vector satisfies the Schrodinger equation

$$(H^{-}_1 + H^{-}_2 + H^{-}_{12} + h/i\ \delta/\delta t)\psi = 0. \qquad (5)$$

In this equation H^{-}_1 and H^{-}_2 mean respectively the *energy* of the first and the second *field.* H^{-}_1 is given by the space integral of a function of υ_1 and λ_1, H^{-}_1 by the space integral of a function of υ_2 and λ_2. Further, H^{-}_{12} is the *interaction energy* of the *fields* and is given by the space integral of a function of both υ_1, λ_1 and υ_2, λ_2. We assume (i) that the integrand of H^{-}_{12}, i.e. the *interaction-energy density*, is a scalar quantity, and (ii) that the *energy densities* at two different points (but at the same instant of time) commute with each other. In general, these two facts follow from the *single assumption: the interaction term in the Lagrangean does not contain the time derivatives of* υ_1 *and* υ_2.

If this *energy density* is denoted by H_{12} then we have

$$H^{\cdot}_{12} = \int H_{12}\, dx\, dy\, dz. \tag{6}$$

As we adopt here the Schrodinger picture, the quantifies υ and λ in H_1, H_2 and H_{12} are all operators independent of time.

Thus far we have merely summarized the well-known facts. Now, as the first stage of making the theory *relativistic*, we suppose the unitary operator

$$U = \exp\{i/h\,(H^{\cdot}_1 + H^{\cdot}_2)\, t\} \tag{7}$$

and introduce the following *unitary transformations* of υ and λ, and the corresponding transformation of ϕ:

$$\{V_r = U\upsilon_r U^{-1}, \quad \Lambda_r = U\lambda_r U^{-1} \qquad r = 1, 2$$
$$\boldsymbol{\Psi} = U\psi. \tag{8}$$

As stated above, υ and λ in (5) are quantities independent of time. But V and Λ obtained from them by means of (8) contain t through U. Thus, they depend on t by

$$ih\, V^{\cdot}_r = V_r\, H^{\cdot}_r - H^{\cdot}_r\, V_r \qquad r = 1, 2 \tag{9}$$
$$ih\, \Lambda^{\cdot}_r = \Lambda_r\, H^{\cdot}_r - H^{\cdot}_r\, \Lambda_r.$$

These equations must necessarily have covariant forms against Lorentz transformations, because they are just the *field equations* for the *fields* when they are left alone without interacting with each other.

Now, the solutions of these "*vacuum equations*", the equations which the *fields* must satisfy when they are left alone, together with the *commutation relations* (4)

$$[[\upsilon_r(xyzt), \upsilon_s(x'y'z't)] = 0$$
$$[\lambda_r(xyzt), \lambda_s(x'y'z't)] = 0 \qquad r, s = 1, 2$$
$$[\upsilon_r(xyzt), \lambda_s(x'y'z't)] = ih\delta(x - x')\delta(y - y')\delta(z - z')\delta_{rs}, \qquad (4)]$$

give rise to the relations of the following forms:

$$[V_r(xyzt), V_s(x'y'z't)] = A_{rs}\,(x - x', y - y', z - z', t - t') = 0$$
$$[\Lambda_r(xyzt), \Lambda_s(x'y'z't)] = B_{rs}\,(x - x', y - y', z - z', t - t') = 0$$
$$[V_r(xyzt), \Lambda_s(x'y'z't)] = C_{rs}\,(x - x', y - y', z - z', t - t') = 0 \tag{10}$$

where A_{rs}, B_{rs} and C_{rs} are functions which are combinations of the so-called four-dimensional *δ-functions* and their derivatives[3].

[3] Pauli, W. (1939). *Solvey Berichte.*

One denotes usually these four-dimensional *δ-functions* by $D_r(xyzt)$, r =1, 2. They are defined by

$$D_r(xyzt) = 1/16\pi^2 \iiint \{e^{i(kxx + kyy + kzz + ckrt)}/ik_r$$
$$- e^{i(kxx + kyy + kzz - ckrt)}/ik_r\}\, dk_x\, dk_y\, dk_z \tag{11}$$

with

$$kr = \sqrt{(k_x^2 + k_y^2 + k_z^2 + k_r^2)}, \qquad (12)$$

x_r being the constant characteristic to the field r. It can be easily proved that these functions are *relativistically invariant**.

* Suppose that a surface in the k_x k_y k_z k-space is defined by means of the equation $k^2 = k^2_x + k^2_y + k^2_z + \kappa^2$. Then this surface has the invariant meaning in this space, since $k^2_x + k^2_y + k^2_z - k^2$ is invariant against Lorentz transformations. The area of the surface element of this surface is given by $dS = \sqrt{\{(\partial k/\partial k_x)^2 + (\partial k/\partial k_y)^2 + (\partial k/\partial k_z)^2 - 1\}}\ dk_x\ dk_y\ dk_z = \kappa\ (dk_x\ dk_y\ dk_z)/k$. Now, since dS has the invariant meaning, we can thus conclude that $(dk_x\ dk_y\ dk_z)/k$ is an invariant, and this results that the function defined by (11) is invariant.

Since (10) gives, in contrast with (4), the *commutation relations* between the *fields* at two different *world points* (xyzt) and (x'y'z't'), it contains no more the notion of same instant of time. Therefore, (10) *is sufficiently relativistic presupposing no special frame of reference*. We call (10) the four-dimensional form of the commutation relations.

One property of D(xyzt) will be mentioned here: When the *world point* (xyzt) lies outside the light cone whose vertex is at the origin, then D(xyzt) vanishes identically:

$$D(xyzt) = 0 \text{ for } x^2 + y^2 + z^2 - c^2t^2 > 0. \qquad (13)$$

It follows directly from (13) that, if the *world point* (x'y'z't') lies outside the light cone whose vertex is at the *world point* (xyzt), the right-hand sides of (10) always vanish. In words: Suppose two world points P and P'. When these points lie outside each other's light cones, the field quantities at P and field quantities at P' commute with each other.

§ 3. *Generalization of the Schrodinger equation.*

Next, we observe the vector $\boldsymbol{\Psi}$ obtained from ψ by means of the *unitary transformation* U. We see from (5), (7) and (8)

$$[(H^{\cdot}_1 + H^{\cdot}_2 + H^{\cdot}_{12} + h/i\ \delta/\delta t)\psi = 0, \qquad (5)$$

$$U = \exp\{i/h\ (H^*_1 + H^*_2)\ t\}, \qquad (7)$$

$$\{V_r = U\upsilon_r U^{-1}, \quad \Lambda_r = U\lambda_r U^{-1} \qquad r = 1, 2$$

$$\boldsymbol{\Psi} = U\psi. \qquad (8)]$$

that this $\boldsymbol{\Psi}$, considered as a function of t, satisfies

$$\{\int H_{12}(V_1(xyzt), \Lambda_1(xyzt), V_2(xyzt), \Lambda_2(xyzt)\ dx\ dy\ dz + h/i\ \partial/\partial t\}\ \boldsymbol{\Psi} = 0. \qquad (14)$$

One sees that t plays also here a role distinguished from x, y and z: also, here a plane parallel to the xyz-plane has a special significance. So, we must in some way remove this unsatisfactory feature of the theory.

This improvement can be attained in the way similar to that in which Dirac[4] has built up the so-called *many-time formalism* of the quantum mechanics.

[4] Dirac, P. A. M. (1933). The Lagrangian in Quantum Mechanics. *Phys. Zeit. Sowjetunion*, 3, 1, 64-72[; alternative formulation of quantum mechanics in terms of Lagrangian in place

of Hamiltonian, *coordinates* and *velocities* instead of *coordinates* and *momenta*, allows *equations of motion* to be expressed as stationary property of *action function* which is the time-integral of the Lagrangian and *relativistic invariant*, there is no corresponding *action principle* in terms of *coordinates and momenta* in Hamiltonian theory, Lagrangian closely connected with theory of *contact transformations*, transformation functions have classical analogues expressible in terms of the Lagrangian, function of coordinates at time t and time t + dt rather than of the coordinates and velocities, applies to *field dynamics* using suitable field quantities or potentials as coordinates, "*many-time*" theory, each coordinate a function of four *space-time* variables instead of one time variable in particle theory, *generalized transformation function*].

We will now recall this theory.

The Schrodinger equation (or the system containing N charged *particles* interacting with the *electromagnetic field* is given by

$$(H^{-}_{el} + \sum_{n=1}^{N} H_n\{q_n, p_n, a(q_n)\} + h/i\ \delta/\delta t)\psi = 0. \qquad (15)$$

Here H^{-}_{el} means the *energy* of the *electromagnetic field*, H_n the *energy* of the n-th *particle*. H_n contains, besides the *kinetic energy* of the n-th particle, the *interaction energy* between this particle and the field through $a(q_n)$, q_n being the *coordinates* of the particle and a the *potential* of the field. p_n in (15) means as usual the *momentum* of the n-th particle.

We consider now the *unitary operator*

$$u = \exp\{i/h\ H^{-}_{el}\} \qquad (16)$$

and introduce the *unitary transformation* [frame change] *a*:

$$\boldsymbol{U} = u a u^{-1} \qquad (17)$$

and the corresponding transformation of ψ:

$$\boldsymbol{\Phi} = u\psi. \qquad (18)$$

Then we see that **Φ** satisfies the equation

$$\{\sum_n H_n\{q_n, p_n, \boldsymbol{U}(q_n, t)\} + h/i\ \delta/\delta t\}\ \boldsymbol{\Phi} = 0. \qquad (19)$$

In contrast with *a*, which was independent of times (Schrodinger picture), ***U*** contains t through u. To emphasize this, we have written t explicitly as argument of ***U***. We can prove that ***U*** satisfies the Maxwell equations in vacuum (accurately speaking, we need special considerations for the equation div ***E*** = 0).

The equation (19) is the starting point of the many-time theory. In this theory one introduces then the function $\boldsymbol{\Phi}(q_1t_1, q_2t_2, \ldots, q_Nt_N)$ containing as many time variables $t_1, t_2, \ldots, t_N$ as the number of the *particles* in place of the function $\boldsymbol{\Phi}(q_1, q_2, \ldots, q_N, t)$ containing only one time variable*,

* Here we suppose the representation which makes the coordinates $q_1, q_2, \ldots, q_N$ diagonal. Thus, the vector **Φ** is represented by a function of these coordinates.

and suppose that this $\mathbf{\Phi}(q_1t_1, q_2t_2, \ldots, q_Nt_N)$ satisfies simultaneously the following N equations;

$$\{H_n\{q_n, p_n, \boldsymbol{U}(q_n, t)\} + h/i\ \delta/\delta t_n\}\ \mathbf{\Phi}(q_1t_1, q_2t_2, \ldots, q_Nt_N) = 0 \qquad n = 1, 2 \ldots, N. \qquad (20)$$

This $\mathbf{\Phi}(t_1, t_2, \ldots, t_N)$, which is a fundamental quantity in the *many-time theory* is related to the ordinary *probability amplitude* $\mathbf{\Phi}(t)$ by

$$\mathbf{\Phi}(t) = \mathbf{\Phi}(t_1, t_2, \ldots, t_N). \qquad (21)$$

Now, the simultaneous equations (20) can be solved when and only when the N^2 conditions

$$(H_nH_{n'} - H_{n'}H_n)\ \mathbf{\Phi}(q_1t_1, q_2t_2, \ldots, q_Nt_N) = 0 \qquad (22)$$

are satisfied for all pairs of n and n'. If the *world point* (q_nt_n) lies outside the light cone whose vertex is at the point $(q_{n'}t_{n'})$, we can prove $H_nH_{n'} - H_{n'}H_n = 0$. As the result, the function satisfying (20) can exist in the region where

$$(q_n - q_{n'})^2 - c^2(t_n - t_{n'})^2 \geq 0 \qquad (23)$$

is satisfied simultaneously for all values of n and n'.

According to Bloch[5]

[5] Bloch, F. (1943). Die physikalische Bedeutung mehrerer Zeiten in der Quantenelektrodynamik. (The physical significance of multiple times in quantum electrodynamics.) *Phys. Zeit. Sowjet.*, 5, 301-15.

we can give $\mathbf{\Phi}(q_1t_1, q_2t_2, \ldots, q_Nt_N)$ a physical meaning when its arguments lie in the region given by (23). Namely

$$W(q_1t_1, q_2t_2, \ldots, q_Nt_N) = |\ \mathbf{\Phi}(q_1t_1, q_2t_2, \ldots, q_Nt_N)\ |^2 \qquad (24)$$

gives the *relative probability* that one finds the value q_1 in the measurement of the position of the first particle at the instant of time t_1, the value q_2 in the measurement of the position of the second particle at the instant of time t_2, … and the value q_N in the measurement of the position of the Nth particle at the instant of time t_N.

This is the outline of the many-time formalism of the quantum mechanics.

We will now return to our main subject. If we compare our equation (14)
$[\{\int H_{12}(V_1(xyzt), \Lambda_1(xyzt), V_2(xyzt), \Lambda_2(xyzt)\ dx\ dy\ dz + h/i\ \partial/\partial t\}\ \mathbf{\Psi} = 0. \quad (14)]$
with the equation (19)

$$[\{\textstyle\sum_n H_n\{q_n, p_n, \boldsymbol{U}(q_n, t)\} + h/i\ \delta/\delta t\}\ \mathbf{\Phi} = 0. \qquad (19)]$$

of the *many-time theory*, we notice a marked similarity between these two equations. In (19) stands the suffix n, which designates the *particle*, while in (14) stand the variables x, y and z, which designate the *position in space*. Further, **Φ** is a function of the N independent variables $q_1, q_2, \ldots, q_N$, q_n giving the *position* of the n-th particle, while $\boldsymbol{\Psi}$ is a functional of the infinitely many "independent variables" $\upsilon_1(xyz)$ and $\upsilon_2(xyz)$, $\upsilon_1(xyz)$ and $\upsilon_2(xyz)$ giving the *fields* at the *position* (xyz). Corresponding to the sum $\sum_n H_n$ in (19) the integral $\int H_{12}dxdydz$ stands in (14). In this way, to the suffix n in (19) which takes the values 1, 2, 3, ... , N correspond the variables x, y and z which take continuously all values from $-\infty$ to $+\infty$.

Such a similarity suggests us to introduce infinitely many time variables t_{xyz}, which we may call *local time** each for one *position* (xyz) in space as we have introduced N time variables, *particle times*, $t_1, t_2, \ldots, t_N$, each for one *particle*.

* The notion of *local time* of this kind has been occasionally introduced by Stueckelberg[6]

[6] Stueckelberg, E. (1938). Die Wechselwirkungskräfte in der Elektrodynamik und in der Feldtheorie der Kernkräfte (Teil I). (Forces of interaction in electrodynamics and in the field theory of nuclear forces. (Part I)). *Helv. Phys.*, 11, 225-44, §5; (translation by D. H. Delphenich; https://neo-classical-physics.info/electromagnetism. html)[; the quantum theory of wave fields leads to the same expressions for the interaction of charges as the classical treatment of retarded potentials. The interaction operator has the following form: Retarded (or advanced) potential of one charge at the position of the second one, times the second charge. If one of those two charges can emit radiation in the first approximation then one must choose the retarded potential of that charge or the advanced potential of the other].

The only difference consists in that we use in our case infinitely many time variables whereas we have used N time variables in the ordinary many-time theory.

Corresponding to the transition from the use of the function with *one time variable* to the use of the function of *N time variables*, we must now consider the transition from the use of $\boldsymbol{\Psi}(t)$ to the use of a *functional* $\boldsymbol{\Psi}[t_{xyz}]$ *of infinitely many time variables* t_{xyz}.

We regard now t_{xyz} as a function of (xyz) and consider its variation ε_{xyz} which differs from zero only in a small domain V_0 in the neighborhood of the point $(x_0y_0z_0)$. We will define the partial differential coefficient of the functional $\boldsymbol{\Psi}[t_{xyz}]$ with respect to the variable t_{x0y0z0} in the following manner:

$$\delta\boldsymbol{\Psi}/\delta t_{x0y0z0} = \lim_{\varepsilon\to 0,\ V0\to 0} \{ \boldsymbol{\Psi}[t_{xyz} + \varepsilon_{xyz}] - \boldsymbol{\Psi}[t_{xyz}]\}/ \iiint \varepsilon_{xyz}\, dx\, dy\, dz \qquad (25)$$

We then generalize (14), and regard

$$\{H_{12}(x, y, z, t) + h/i\ \partial/\partial t_{xyz}\}\ \boldsymbol{\Psi} = 0 \qquad (26)$$

the infinitely many simultaneous equations corresponding to the N equations (20)

$$[\{H_n\{q_n, p_n, \boldsymbol{U}(q_n, t)\} + h/i\ \delta/\delta t_n\}\ \boldsymbol{\Phi}(q_1t_1, q_2t_2, \ldots, q_Nt_N) = 0 \qquad n = 1, 2 \ldots, N, \qquad (20)]$$

as the *fundamental equations* of our theory. In (26) we have written, for simplicity, $H_{12}(x, y, z, t)$ in place of $H_{12}(V_1(xyz, t), V_2(xyz, t), \ldots)$. In general, when we have a function $F(V, A)$ of V and A, we will write simply $F(x, y, z, t)$ for $F(V(xyz, t_{xyz}), A(xyz, t_{xyz}))$, or still simpler F(*P*), *P* denoting the world point with the coordinates (xyz, t_{xyz}). Thus F(*P*') means F(x', y', z', t') or, more precisely, F(V(x'y'z', $t_{x'y'z'}$), (x'y'z', $_{x'y'z'}$)).

We will now adopt the equation (26) as the basis of our theory. For V_1(P), V_2(P), Λ_1(P), and Λ_2(P) in H_{12} the commutation relations (10)

$$[[V_r(xyzt), V_s(x'y'z't)] = A_{rs}(x - x', y - y', z - z', t - t') = 0$$
$$[\Lambda_r(xyzt), \Lambda_s(x'y'z't)] = B_{rs}(x - x', y - y', z - z', t - t') = 0$$
$$[V_r(xyzt), \Lambda_s(x'y'z't)] = C_{rs}(x - x', y - y', z - z', t - t') = 0 \qquad (10)]$$

hold, where D(xyzt) has the property (13)

$$[D(xyzt) = 0 \text{ for } x^2 + y^2 + z^2 - c^2t^2 > 0. \qquad (13)]$$

As the consequence, we have

$$H_{12}(P)H_{12}(P') - H_{12}(P')H_{12}(P) = 0 \qquad (27)$$

when the point P lies a finite distance apart from P' and outside the light cone whose vertex is at P. Further, from our assumption (ii) the relation (27) holds also when P and P' are two adjacent points approaching in a space-like direction. Thus, our system of equations (26) is integrable when the surface defined by the equations $t = t_{xyz}$, considering t_{xyz} as a function of x, y and z, is space-like.

In this way, a *functional* of the *variable surface* in the *space-time world* is determined by the functional partial differential equations (26). Corresponding to the relation (21)

$$[\mathbf{\Phi}(t) = \mathbf{\Phi}(t_1, t_2, \ldots, t_N). \qquad (21)]$$

in case of *many-time theory*, $\mathbf{\Psi}[t_{xyz}]$ reduces to the ordinary $\mathbf{\Psi}[t]$ when the surface reduces to a plane parallel to the xyz-plane.

The dependent variable surface $t = t_{xyz}$ can be of any (space-like) form in the space-time world, and we need not presuppose any Lorenz frame of reference to define such a surface. *Therefore, this* $\mathbf{\Psi}[t_{xyz}]$ *is a relativistically invariant concept.* The restriction that the surface must be space-like makes no harm since the property that a surface is space-like or time-like does not depend on a special choice of the reference system. *It is not necessary, from the stand-point of the relativity theory, to admit also time-like surfaces for the variable surface, as was required by Dirac and by Yukawa.* Thus, we consider that $\mathbf{\Psi}[t_{xyz}]$ introduced above is already the sufficient generalization of the ordinary ψ-vector, and assume that the quantum-theoretical *state** of the *fields* is represented by this functional vector.

* The word *state* is here used in the *relativistic space-time* meaning. Cf. Dirac's book [Dirac, P.A.M. (1935). *The Principles of Quantum Mechanics*, (second edition). Oxford University Press], §6.

Let C denote the surface defined by the equation $t = t_{xyz}$. Then $\mathbf{\Psi}$ is a functional of the surface C. We write this as $\mathbf{\Psi}$[C]. On C we take a point P, whose coordinates are (xyz, t_{xyz}), and suppose a surface C' which overlap C except in a small domain about P. We denote the volume of the small world lying between Cand C' with $d\omega_P$. Then we may write (25)

$$[\delta\mathbf{\Psi}/\delta t_{x0y0z0} = \lim_{\varepsilon\to 0,\, V0\to 0} \{ \mathbf{\Psi}[t_{xyz} + \varepsilon_{xyz}] - \mathbf{\Psi}[t_{xyz}]\}/ \iiint \varepsilon_{xyz}\, dx\, dy\, dz \qquad (25)]$$

also in the form:

$$\delta\boldsymbol{\Psi}[C]/\delta C_P = \lim_{C'\to 0}\{\boldsymbol{\Psi}[C'] - \boldsymbol{\Psi}[C]\}/d\omega_P \quad (28)$$

Then (26)

$$[\{H_{12}(x, y, z, t) + h/i\ \partial/\partial t_{xyz}\}\ \boldsymbol{\Psi} = 0 \quad (26)]$$

can be written in the form:

$$\{H_{12}(P) + h/i\ \partial/\partial C_P\}\ \boldsymbol{\Psi}[C] = 0. \quad (29)$$

This equation (29) has now a perfect space-time form. In the first place, H_{12} is a scalar according to our assumption (i); in the second place; the *commutation relations* between $V(p)$ and $\Lambda(P)$ contained in H_{12} has the four-dimensional form as (10)

$$[[V_r(xyzt), V_s(x'y'z't)] = A_{rs}\,(x - x', y - y', z - z', t - t') = 0$$
$$[\Lambda_r(xyzt), \Lambda_s(x'y'z't)] = B_{rs}\,(x - x', y - y', z - z', t - t') = 0$$
$$[V_r(xyzt), \Lambda_s(x'y'z't)] = C_{rs}\,(x - x', y - y', z - z', t - t') = 0, \quad (10)]$$

and finally, the differentiation $\delta/\delta C_P$ is defined by (28) quite independently of any frame of reference.

A direct conclusion obtained from (29) is that $\boldsymbol{\Psi}[C']$ is obtained from $\boldsymbol{\Psi}[C]$ by the following infinitesimal transformation:

$$\delta\boldsymbol{\Psi}[C] = \{1 - 1/h\ H_{12}(P)d\omega_P\}\ \boldsymbol{\Psi}[C] = 0. \quad (30)$$

When there exist in the *space-time world* two surfaces C_1 and C_2 a finite distance apart, we need only to repeat the infinitesimal transformations in order to obtain $\boldsymbol{\Psi}[C_2]$ from $\boldsymbol{\Psi}[C_1]$. Thus

$$\delta\boldsymbol{\Psi}[C_2] = \prod_{C1}^{C2}\{1 - 1/h\ H_{12}(P)d\omega_P\}\ \boldsymbol{\Psi}[C_1] = 0. \quad (31)$$

The meaning of this equation is as follows: We divide the world region lying between C_1 and C_2 in small elements $d\omega_P$ (it is necessary that each *world element* is surrounded by two space-like surfaces). We consider for each *world element* the infinitesimal transformation $1 - i/h\ H_{12}(P)d\omega_P$. Then we take the product of these transformations, the order of the factor being taken from C_1 to C_2. This product transforms then $\boldsymbol{\Psi}[C_1]$ into $\boldsymbol{\Psi}[C_2]$.

The surfaces C_1 and C_2 must. be here both space-like, but otherwise they may have any form and any configuration. Thus, C_2 does not necessarily lie afterward against C_1; C_1 and C_2 may even cross with each other.

The relation of the form (31)

$$[\delta\boldsymbol{\Psi}[C_2] = \prod_{C1}^{C2}\{1 - 1/h\ H_{12}(P)d\omega_P\}\ \boldsymbol{\Psi}[C_1] = 0. \quad (31)]$$

has been already introduced by Heisenberg[7].

[7] Heisenberg, W. (1938). Die Grenzen der Anwendbarkeit der bisherigen Quantentheorie. (The limits of the applicability of the previous quantum theory.) *Zeit. Phys.*, 110, 251-66; reprinted in *Collected Works* AII, 315–330.

It can be regarded as the integral form of our generalized Schrodinger equation (29)

$$[\{H_{12}(P) + h/i\ \partial/\partial C_P\}\ \boldsymbol{\Psi}[C] = 0. \quad (29)]$$

§ 4. *Generalized probability amplitude.*

We must now find the physical meaning of the functional **Ψ**[C]. As regards this we can make a similar consideration as Bloch has done for the case of ordinary *many-time theory*. Besides the fact that in our case there appear infinitely many time variables, one point differs from Bloch's case that in (16) the *unitary operator* u is commutable with the coordinates q_1, q_2, ... q_N, our U is not commutable with the field quantities $\upsilon_1(xyz)$) and $\upsilon_2(xyz)$. Noting this difference *and treating the continuum infinity as the limit of an enumerable infinity by some artifice*, for instance, by the procedure of Heisenberg and Pauli[8], Bloch's consideration can be applied also here almost without any alteration.

[8] Heisenberg, W. & Pauli, W. (July, 1929). Zur Quantendynamik der Wellenfelder. (On the quantum dynamics of wave fields.) *Zeit. Phys.*, 56, 1-61[; Heisenberg and Pauli's first attempt to construct their own version of a *relativistically-invariant quantum electrodynamics* to treat interaction between matter and the electromagnetic field and between matter and matter, canonical quantization of both electromagnetic and matter-wave fields, but Lorentz-invariant Lagrangian for interacting *electromagnetic* and *matter-wave fields*, requires working with the *electromagnetic potentials* not just with the *fields*, Lagrangian does not contain a time derivative of the *electric potential* so *there is no corresponding canonical momentum variable,* prevents straightforward implementation of canonical commutation relations, the theory is still afflicted with many defects, *the fundamental difficulties in the relativistic formulation that were emphasized by Dirac remain unchanged,* the formulas of the theory lead to an *infinite zero-point energy* for the radiation and thus include the interaction of an electron with itself as an *infinite* additive constant, however, these difficulties are of a sort that they do not interfere with the application of the theory to many physical problems, used "crude trick" of adding additional terms to the Lagrangian.]

We shall give here only the results.

Let us suppose that the *fields* are in the *state* represented by a vector **Ψ**[C]. We suppose that we make measurements of a function $f(\upsilon_1, \upsilon_2, \lambda_1, \lambda_2)$ at every point on a surface C_1 in the *space-time world.* Let P_1 denote the variable point on C_1, then, if $f(P_1)$ at any two "values" of P_1 commute with each other, the measurements of f at each of these two points do not interfere with each other. Our first conclusion says that in this case the *expectation value* of $f(P_1)$ is given by

$$f(P_1)^{-} = ((\boldsymbol{\Psi}[C_1], f(P_1)\, \boldsymbol{\Psi}[C_1])) \qquad (32)$$

where $f(P_1)$ means $f(V_1(P_1), \ldots)$ according to our convention on [original] page 35, and the symbol ((A, B)) with double brackets is the scalar product of two vectors A and B. *It is impossible in case of continuously many degrees of freedom to represent this scalar product by an integral of the product of two functions*. For this purpose, we must replace the continuum infinity by an at least enumerable infinity.

More generally, we suppose a functional $F[f(P_1)]$ of the independent variable function $f(P_1)$, regarding $f(P_1)$ as a function of P_1. Then the *expectation value* of this F is given by

$$F[f(P_1)]^{-} = ((\boldsymbol{\Psi}[C_1], F[f(P_1)\, \boldsymbol{\Psi}[C_1])) \qquad (33)$$

A physically interesting F is the *projective operator* $M[\upsilon_1'(P_1), \upsilon_2'(P_1), V_1(P_1), V_2(P_1)]$ belonging to the "*eigen-value*" $\upsilon_1'(P_1), \upsilon_2'(P_1)$ of $V_1(P_1), V_2(P_1)$. Then its expectation value

$$M[\upsilon_1'(P_1), \upsilon_2'(P_1), V_1(P_1), V_2(P_1)]^{-} = ((\boldsymbol{\Psi}[C_1], M[\upsilon_1'(P_1), \upsilon_2'(P_1); V_1(P_1), V_2(P_1)]\, \boldsymbol{\Psi}[C_1])) \qquad (34)$$

gives the *probability* that the *field* 1 and the *field* 2 have respectively the functional form $\upsilon_1'(P_1)$ and $\upsilon_2'(P_1)$ on the surface C_1. As C_1 is assumed to be *space-like*, the measurement of the functional M is possible (the measurements of $V_1(P_1)$ and $V_2(P_1)$ at all points on C_1 means just the measurement of M).

Thus far we have made no mention of the *representation* of $\boldsymbol{\Psi}[C]$. We use now the special representation in which $V_1(P_1)$ at all points on C_1 are simultaneously diagonal. It is always possible to make all $V_1(P_1)$ and $V_2(P_1)$ diagonal when the surface C_1 is space-like. In this representation $\boldsymbol{\Psi}[C_1]$ is represented by a functional $\boldsymbol{\Psi}\,[\upsilon_1'(P_1), \upsilon_2'(P_1); C_1]$ of the eigenvalues $\upsilon_1'(P_1)$ and $\upsilon_2'(P_1)$ of $V_1(P_1)$ and $V_2(P_1)$. The projection operator M has in this representation such diagonal form that (34) is simplified as follows

$$W[\upsilon_1'(P_1), \upsilon_2'(P_1)] = M[\upsilon_1'(P_1), \upsilon_2'(P_1), V_1(P_1), V_2(P_1)]^{-} = |\, \boldsymbol{\Psi}\,[\upsilon_1'(P_1), \upsilon_2'(P_1); C_1]\,|^2. \qquad (35)$$

In this sense we can call $\boldsymbol{\Psi}\,[\upsilon_1'(P_1), \upsilon_2'(P_1); C_1]$ *"generalized probability amplitude".*

§5. *Generalized transformation functional.*

We have stated above that between $\boldsymbol{\Psi}[C_1]$ and $\boldsymbol{\Psi}[C_2]$ the relation (31)

$$[\delta\boldsymbol{\Psi}[C_2] = \prod_{C_1}^{C_2}\{1 - 1/h\, H_{12}(P)d\omega_P\}\, \boldsymbol{\Psi}[C_1] = 0. \qquad (31)]$$

holds, where C_1 and C_2 are two specie-like surfaces in the *space-time* world. We see thus that the *transformation operator*

$$T[C_2; C_1] = \prod_{C_1}^{C_2}(1 - 1/h\, H_{12}d\omega) \qquad (36)$$

plays an important role. It is evident that also this operator has a *space-time* meaning.

Similarly as the special representative of the ϕ-vector, the *probability amplitude*, has a distinct physical meaning, there is a special representation in which the *representative* of the *transformation operator* $T[C_2; C_1]$ has a distinct physical meaning.

We introduce namely the mixed *representative* of $T[C_2; C_1]$ whose rows refer to the representation in which $V_1(P_1)$ and $V_2(P_1)$ at all points on C_1 become diagonal and whose column refer to the representation in which $V_1(P_2)$ and $V_2(P_2)$ at all points on C_2 become diagonal. We denote this *representation* by

$$[\upsilon_1''(P_2), \upsilon_2''(P_2)\,|\, T[C_2; C_1]\,|\, \upsilon_1'(P_1), \upsilon_2'(P_1)], \qquad (37)^*$$

or simpler:

$$[\upsilon_1''(P_2), \upsilon_2''(P_2)\,|\, \upsilon_1'(P_1), \upsilon_2'(P_1)]. \qquad (38)^*$$

* As the matrix elements are functionals of $\upsilon(P)$, we use here the square brackets.

If we note here the relation (35)

$$[W[\upsilon_1'(P_1), \upsilon_2'(P_1)] = M^*[\upsilon_1'(P_1), \upsilon_2'(P_1), V_1(P_1), V_2(P_1)] = |\,\mathbf{\Psi}\,[\upsilon_1'(P_1), \upsilon_2'(P_1); C_1]\,|^2, \qquad (35)]$$

we see that we can give the matrix elements of this *representation* the following meaning: One measures the *field* quantities V_1 and V_2 at all points on C_2 when the *fields* are prepared in such a way that they have certainly the values $\upsilon_1'(P_1)$ and $\upsilon_2'(P_1)$ at all points on C_1. Then

$$W[\upsilon_1''(P_2), \upsilon_2''(P_2); \upsilon_1'(P_1), \upsilon_2'(P_1)) = |\,[\upsilon_1''(P_2), \upsilon_2''(P_2)\,|\,\upsilon_1'(P_1), \upsilon_2'(P_1)]\,|^2 \qquad (39)$$

gives the *probability* that one obtains the result $\upsilon_1''(P_1)$ and $\upsilon_2''(P_2)$ in this measurement. In this proposition we have assumed that C_2 lies afterward against C_1.

From this physical interpretation we may regard the matrix element (37) or (38), considered as a *functional* of $\upsilon_1''(P_2)$, $\upsilon_2''(P_2)$ and $\upsilon_1'(P_1)$, $\upsilon_2'(P_1)$, as the generalization of the ordinary *transformation function* $(q_{t2}''\,|\,q_{t1}')$.

As a special case it may happen that C_2 lies apart from C_1 only in a portion S_2 and a portion S_1 of C_2 and C_1 respectively, the other parts of C1 and C2 overlapping with each other (see Fig. 1 [on original page 40]).

In this case the matrix elements of $T[C_2; C_1]$ depend only on the values of the *fields* on the portions S_1 and S_2 of the surfaces C_1 and C_2. In this case we need for calculating $T[C_2; C_1]$ to take the product in (36)

$$[T[C_2; C_1] = \prod_{C1}^{C2}(1 - 1/h\, H_{12} d\omega) \qquad (36)]$$

only in the closed domain surrounded by S_1 and S_2, thus

$$T[S_2; S_1] = \prod_{C1}^{C2}(1 - 1/h\, H_{12} d\omega) \qquad (40)$$

The matrix elements of the *mixed representation* of this T is a *functional* of $\upsilon_1'(p_1)$, $\upsilon_2'(p_1)$ and $\upsilon_1''(p_2)$, $\upsilon_2''(p_2)$ where p_1 denotes the moving point on the portion S_1, and p_2 the moving point on the portion S_2. This matrix is independent on the *field* quantities on the other portions of the surfaces C_1 and C_2.

The matrix element of $T[S_2; S_1]$ regarded as a functional of $\upsilon_1'(p_1)$, $\upsilon_2'(p_1)$ and $\upsilon_1''(p_2)$, $\upsilon_2''(p_2)$ has *the properties of g.t.f. (generalized transformation functional) of Dirac. But in defining our g.t.f. we had to restrict the surfaces S_1 and S_2 to be space-like*, while Dirac required his g.t.f. to be defined also referring to the time-like surfaces. As mentioned above, however, such a generalization as required by Dirac is superfluous so far as the *relativity theory* is concerned.

It is to be noted that for the physical interpretation of $[\upsilon_1''(P_2), \upsilon_2''(P_2)\,|\,\upsilon_1'(P_1), \upsilon_2'(P_1)]$ it is not necessary to assume C_2 to lie afterward against C_1. Also, when the inverse is the case, we can as well give the physical meaning for *W* of (39): One measures. the *field* quantities V_1 and V at all points on C_2 when the *fields* are prepared in such a way that they would have certainly the values $\upsilon_1'(P_1)$ and $\upsilon_2'(P_1)$ at all points on C_1 if the *fields* were left alone until C_l without being measured before on C_2. Then W gives the *probability* that one finds the results $\upsilon_1''(P_2)$ and $\upsilon_2''(P_2)$ in this measurement on C_2.

§ 6. *Concluding remark.*

We have thus shown that *the quantum theory of wave fields can be really brought into a form which reveals directly the invariance of the theory against Lorentz transformations. The reason why the ordinary formalism of the quantum field theory is so unsatisfactory lies in the fact that one has built up this theory in the way which is too much analogous to the ordinary non-relativistic mechanics.* In this ordinary formalism of the quantum theory of fields the theory is divided into two distinct sections: the section giving the *kinematical relations* between various quantities at the same instant of time, and the section determining the *causal relations* between quantities at different instants of time. Thus, the *commutation relations* (1) belong to the first section and the *Schrodinger equation* (2) to the second. As stated before, this way of separating the theory into two sections is very *unrelativistic*, since here the concept "*same instant of time*" plays a distinct role. Also, in our formalism the theory is divided into two sections. But now the separation is introduced in another place: *In our formalism the theory consists of two sections, one of which gives the laws of behavior of the fields when they are left alone, and the other of which gives the laws determining the deviation from this behavior due to interactions.* This way of separating the theory can be carried out *relativistically*. Although in this way the theory can be brought into more satisfactory form, *no new contents are added thereby*. So, *the well-known divergence difficulties of the theory are inherited also by our theory.* Indeed, *our fundamental equations (29) admit only catastrophic solutions as can be seen directly in the fact that the unavoidable infinity due to non-vanishing zero-point amplitudes of the fields inheres in the operator $H_{12}(P)$.* Thus, *a more profound modification of the theory is required in order to remove this fundamental difficulty.*

It is expected that such a modification of the theory would possibly be introduced *by some revision of the concept of interaction*, because we meet no such difficulty when we deal with the non-interacting fields. This revision would then result that in the separability of the theory into two sections, *one for free fields and one for interactions*, some uncertainty would be introduced. This seems to be implied by the very fact that, when we formulate the quantum field theory in a *relativistically* satisfactory manner, this way of separation has revealed itself as the fundamental element of the theory.

Dirac, P. A. M. (April, 1945). On the Analogy Between Classical and Quantum Mechanics.

[*Rev. Mod. Phys*., 17, 2-3, 195-9; https://doi.org/10.1103/ RevModPhys.17.195.]

St. John's College, Cambridge, England.

Mathematical methods available for working with *non-commuting quantities* much weaker than those available for *commuting quantities* owing to the fact that the only functions of *non-commuting variables* that one has been able to define are those expressible algebraically, shows how this difficulty can be avoided in the case when the *non-commuting quantities* are *observables*, can set up a theory of functions of them (a generalization of the concept known as *well-ordered functions*) of almost the same degree of generality as the usual functions of *commuting variables*, can use this theory to make closer analogy between classical and quantum mechanics, enables discussion of trajectories for motion of a particle in quantum mechanics, method is given for defining *general functions of non-commuting observables*, method is developed to provide formal *probability* for *non-commuting observables* to have numerical values (in general a complex number), also enables the analogy between classical and quantum *contact transformations* to be set up on a more general basis.

There are two forms in which quantum mechanics may be expressed, based on Heisenberg's matrices and Schrodinger's wave functions respectively. The second of these is not connected very directly with classical mechanics. The first is in close analogy with classical mechanics, as it may be obtained from classical mechanics simply by making the variables of classical mechanics into non-commuting quantities satisfying the correct commutation relations. *The development of this analogy has been greatly hampered by the mathematical methods available for working with non-commuting quantities being much weaker than those available for commuting quantities*, owing to the fact that the only functions of non-commuting variables that one has been able to define are those expressible algebraically. The present paper will show how this difficulty can be avoided. In the case *when the non-commuting quantities are observables, one can set up a theory of functions of them of almost the same degree of generality as the usual functions of commuting variables and one can use this theory to make closer the analogy between classical and quantum mechanics*.

The non-commuting quantities of quantum mechanics are called *observables* when they are real (Hermitian or self-adjoint), and satisfy the condition of having sufficient eigenvectors to form a complete set, i.e., any of the vectors on which the non-commuting quantities operate can be expanded in terms of the eigenvectors. The latter condition enables one to define a general function of a single observable[1].

[1] See P. A. M. Dirac, *Principles of Quantum Mechanics*, second edition, §11.

The method of this definition can be extended to provide functions of two non-commuting observables in the following way. Consider two observables α and β. Let f(*ab*) be any function of two real variables *a* and *b* which is defined whenever *a* is an eigenvalue of α and b is an eigenvalue of β. We can then give a meaning to f(αβ). We first define the function f(αb) of the single observable α, involving b, an eigenvalue of β, as a parameter, by means of the equation

$$f(\alpha b) \mid \alpha'\rangle = f(\alpha' b) \mid \alpha'\rangle, \qquad (1)$$

where | α'> denotes an eigenvector of α belonging to the eigenvalue α'. We require Eq. (1) to hold for every eigenvector of α. It then fixes the result of f(αb) applied to a general vector | X>, since this general vector can be expanded in terms the eigenvectors, and so it fixes the linear operator f(αb). We now define f(αβ)by the equation

$$f(\alpha\beta) \mid \beta'\rangle = f(\alpha\beta') \mid \beta'\rangle, \qquad (2)$$

where | β'> is any eigenvector of β belonging to the eigenvalue β', the right-hand side of (2) having a meaning from the fore-going definition of f(αb). Equation (2) holding for every eigenvector of β fixes the result of f(αβ) applied to a general vector | X> and so fixes the linear operator f(αβ).

In this way we can define a general function of two non-commuting observables α and β. However, *this definition is associated with an order for α and β*—we first define f(αb) and then use it to define f(αβ). We could equally well use the reverse order, first defining f(aβ) and then using it to define f(αβ). The resulting f(αβ) would, in general, differ from the previous one. If α and β commute, the two f(αβ)'s are equal.

We may take a function f(ab) consisting of a sum of terms, each of which is a function of a multiplied by a function of b, i.e.,

$$f(ab) = \sum_n u_n(a)v_n(b). \qquad (3)$$

It is then easily seen that f(αβ) defined by (1) and (2) is equal to the expression obtained by substituting α for a and β for b in the right-hand side of (3), the α-factor being always put to the left of the β factor, i.e.,

$$f(\alpha\beta) = \sum_n u_n(\alpha)v_n(\beta). \qquad (4)$$

Functions of *non-commuting observables* of this kind have been used by Jordan[2] for discussing *contact transformations* in quantum mechanics.

[2] Jordan, P. (June, 1926). Über kanonische Transformationen in der Quantenmechanik. II. (On Canonical Transformations in Quantum Mechanics. II.) *Zeit. Phys.*, 38, 513-17; https://doi.org/10.1007/BF01397170.

They are called *well-ordered functions*. The method of Eqs. (1) and (2) thus provides a generalization of the concept of *well-ordered functions*, which does not require the use of the special algebraic form (4).

The method may readily be extended to provide a, definition for a function f(αβγδ…ζ) of any number of non-commuting observables α, β, γ, δ, …ζ, where f($abcd…z$) is a function of the real variables $a, b, c, d, \ldots z$, which is defined for $a, b, c, d, \ldots z$ eigenvalues of α, β, γ, δ, ...ζ respectively. We first define $f(\alpha bcd\ldots z)$ by

$$f(\alpha bcd\ldots z) \mid \alpha'\rangle = f(\alpha' bcd\ldots z) \mid \alpha'\rangle, \tag{5}$$

then use this to define $f(\alpha\beta cd\ldots z)$ by

$$f(\alpha\beta cd\ldots z) \mid \beta'\rangle = f(\alpha\beta' cd\ldots z) \mid \beta'\rangle, \tag{6}$$

and then use this to define $f(\alpha\beta\gamma d\ldots z)$ by

$$f(\alpha\beta\gamma d\ldots z) \mid \gamma'\rangle = f(\alpha\beta\gamma' d\ldots z) \mid \gamma'\rangle, \tag{7}$$

and so on. We finish with a definition of $f(\alpha\beta\gamma\delta\ldots\zeta)$ associated with a certain order α, β, γ, δ, ...ζ of the observables concerned. It is easily seen that if some of the observables occupying consecutive positions in the order commute, one can change the order of these observables among themselves without affecting the function.

The observables that one uses in practice in Heisenberg's form of quantum mechanics are the values of dynamical variables at particular times. They fall into a natural linear order, namely the order of the times to which they refer. Our theory now enables us to set up general functions of them, based on this order. The functions must not involve two observables referring to exactly the same time, unless they commute. Apart from this limitation, the power of forming functions that we now have is just as general as in the classical theory.

If we are working with a *relativistic* theory, our observables will each refer to a certain point in space-time and they will no longer have a unique time order. However, observables referring to points in space-time lying outside each other's light-cones always commute with one another, so that the order of such observables in constructing functions does not matter. We may thus order our observables according to their times in some particular Lorentz frame, and functions defined relative to this order will be the same as those defined relative to the order of their times in a different Lorentz frame. The only limitation we now have is that the functions must not involve two non-commuting observables referring to exactly the same point in space-time.

In working with functions of non-commuting observables, care must be taken not to use ordinary algebraic procedures when they are not permissible for these functions. For example, if one is given an equation between observables, A = B, this does not in general allow one to substitute B for A in functions of A and other observables, since, if A and B contain observables referring to different times, such a substitution may spoil the order of the observables in the function.

Application to probabilities

Take the function $f(abc\ldots)$ which is equal to unity when $a = \alpha'$, $b = \beta'$, $c = \gamma'$, ... , (α', β', γ', ... being eigenvalues of the observables α, β, γ, ...,) and which vanishes otherwise, and form the function $f(\alpha\beta\gamma\ldots)$. Let us evaluate the average of this function for the dynamical system in a certain state. If the state corresponds to the normalized vector | X> (a fixed vector in the Heisenberg scheme of quantum mechanics), and <X | is the conjugate imaginary vector, then this average may be written as a scalar product

$$\langle X \mid f(\alpha\beta\gamma\ldots) \mid X\rangle. \tag{8}$$

According to ordinary ideas of probability, expression (8) would be just the probability of α having the value α', β having the value β', γ having the value γ', and so on, for the state concerned. However, the number (8) is in general complex. Thus, the theory allows one formally to give a value for the probability of non-commuting observables having specified numerical values, but this probability is in general a complex number, so it does not have an immediate physical application. All the same, if the probability is close to zero it can be interpreted as meaning that the observables α, β, γ, ... are unlikely to have the values α', β', γ', ..., so there is a limited application for it. If some of the observables α, β, γ, ... have continuous ranges of eigenvalues, we should take f(*abc*...) to equal unity when the corresponding variables lie in small ranges, instead of when they have precise values, and so get the probability of those observables lying in small ranges.

We are now able to set up a formal probability for non-commuting observables to have specified values, subject to the limitation that there must not be two non-commuting observables referring to exactly the same time. This probability would give correctly the average value of any function of the observables by means of the formula

$$\text{average } f(\alpha\beta\gamma\ldots) = \text{sum or integral of } f(abc\ldots)P(abc\ldots), \tag{9}$$

P being the probability, provided the function is defined in accordance with our general definition, Eqs. (5) - (7). It would not, in general, give the correct average value for a function defined differently, for example, for a function defined by an algebraic formula in which the non-commuting observables are not properly ordered.

The possibility of setting up in quantum mechanics a probability for non-commuting observables to have specified values has been previously considered by J. E. Moyal[3].

[3] This work is not yet published. I am indebted to J. E. Moyal for letting me see the manuscript.

Moyal obtained a probability for a *coordinate q* and a *momentum p* to have specified values at any time, which probability would give by Eq. (9) the correct average value for any quantity of the form $e^{i(aq+bp)}$, where a and b are real numbers. Moyal's probability is always real, though not always positive, and is thus one step more physical than the probability of the present paper, but its region of applicability is rather restricted, as it does not seem to be connected with a general theory of functions like the present one.

We can use the formal probability to set up a quantum picture rather close to the classical picture in which the coordinates q of a dynamical system have definite values at any time. We take a number of times t_1, t_2, t_3, ... following closely one after another and set up the formal probability for the q's at each of these times lying within specified small ranges, this being permissible since the q's at any time all commute. We then get a formal probability for the trajectory of the system in quantum mechanics lying within certain limits. This enables us to speak of some trajectories being improbable and others being likely.

An analytical expression for this formal probability in terms of the transformation functions may be obtained in the following way. Denote by q_1 the q's at time t_1, by q_2 the q's at time t_2, and so on, and assume that the q's at any time form a complete set of commuting observables. Then there exists a representation whose basic vectors are $|q_1'\rangle$, $\langle q_1'|$, labelled by values for the q_1's, a second representation whose basic vectors are are $|q_2'\rangle$, $\langle q_2'|$, labelled by values for the q_2's, and so on. Let us restrict ourselves to three times t_1, t_2, t_3 to save writing, so that we have only three sets of q's, namely q_1, q_2 and q_3. A general function of these q's is then defined, corresponding to (5)-(7)

$$[f(\alpha bcd\ldots z)\,|\alpha'\rangle = f(\alpha' bcd\ldots z)\,|\alpha'\rangle, \qquad (5)$$
$$f(\alpha\beta cd\ldots z)\,|\beta'\rangle = f(\alpha\beta' cd\ldots z)\,|\beta'\rangle, \qquad (6)$$
$$f(\alpha\beta\gamma d\ldots z)\,|\gamma'\rangle = f(\alpha\beta\gamma' d\ldots z)\,|\gamma'\rangle, \qquad (7)]$$

by

$$f(q_1q_2'q_3')\,|q_1'\rangle = f(q_1'q_2'q_3')\,|q_1'\rangle,$$
$$f(q_1q_2q_3')\,|q_2'\rangle = f(q_1q_2'q_3')\,|q_2'\rangle,$$
$$f(q_1q_2q_3)\,|q_3'\rangle = f(q_1q_2q_3')\,|q_3'\rangle,$$

Thus, for any vector $|X\rangle$,

$$f(q_1q_2q_3)\,|X\rangle$$
$$= \int f(q_1q_2q_3)\,|q_3'\rangle dq_3'\langle q_3'|X\rangle$$
$$= \ldots$$
$$\ldots$$
$$= \iiint f(q_1'q_2'q_3')\,|q_1'\rangle dq_1'\langle q_1'|q_2'\rangle\, dq_2'\langle q_2'|q_3'\rangle\, dq_3'\langle q_3'|X\rangle. \qquad (10)$$

If $|X\rangle$ is normalized, the average value of $f(q_1q_2q_3)$ for the state corresponding to it is

$$\langle X|f(q_1q_2q_3)|X\rangle = \iiint f(q_1'q_2'q_3')\,\langle X|q_1'\rangle$$
$$\times\, dq_1'\langle q_1'|q_2'\rangle\, dq_2'\langle q_2'|q_3'\rangle\, dq_3'\langle q_3'|X\rangle.$$

Taking f equal unity for the q_1's in the ranges q_1' to $q_1' + dq_1'$, the q_2's in the ranges q_2' to $q_2' + dq_2'$, and the q_3's in the ranges q_3' to $q_3' + dq_3'$, and zero otherwise, we get

$$\langle X|f|X\rangle = \langle X|q_1'\rangle dq_1'\langle q_1'|q_2'\rangle\, dq_2'\langle q_2'|q_3'\rangle\, dq_3'\langle q_3'|X\rangle. \qquad (11)$$

This is the probability of the q_1's, q_2's, q_3's lying in the ranges q_1' to $q_1' + dq_1'$, q_2' to $q_2' + dq_2'$, and q_3' to $q_3' + dq_3'$, respectively. The extension of this result to the case of more than three times is obvious. The extension to the case when the q's do not form a complete set of commuting observables (owing, say, to the presence of spin variables), is also obvious, as one has only to insert the necessary extra variables in the transformation functions and sum or integrate over them.

The theory provides us with a rather more definite picture of the motion of a particle in quantum mechanics than we had previously. For example in a *relativistic* theory, if the time intervals are very short, the *transformation functions* $<q_r' \mid q_{3+1}'>$ are very small unless the difference between q_r' and q_{3+1}' corresponds to the particle having moved with the velocity of light during the time interval t_r to t_{3+1}. Thus we see that the probability is very small for the particle to move in any way except along a sequence of straight paths with the velocity of light all the time. A closer investigation shows that the duration of these straight paths is likely to be of the order of magnitude of the period of oscillation of the de Broglie waves associated with the particle. These results hold both for a particle with a spin like an electron and for a non-spinning Klein-Gordon particle.

Application to contact transformations

The similarity between classical and quantum *contact transformations* has already been shown by Jordan and by the author[4].

[4] P. Jordan, see reference 2 [Jordan, P. (June, 1926). Über kanonische Transformationen in der Quantenmechanik. II. (On Canonical Transformations in Quantum Mechanics. II.)]; P. A. M. Dirac, reference 1 [P. A. M. Dirac, *Principles of Quantum Mechanics*, second edition], §30.

[The roots of contact geometry appear in work of Christiaan Huygens, Isaac Barrow and Isaac Newton. The theory of *contact transformations* (i.e. transformations preserving a contact structure) was developed by Sophus Lie, with the dual aims of studying differential equations (e.g. the Legendre transformation or canonical transformation) and describing the 'change of space element'.]

The present methods enable one to establish this similarity on a more general basis.

Consider two sets of coordinates, q's and Q's say, for a system in quantum mechanics describable in terms of coordinates and conjugate momenta (i.e., no spins), and suppose the q's refer to one time and the Q's to a later time. We can now set up general functions of the q's and Q's. The question arises whether every dynamical variable can be expressed as such a function or not.

From the method that led to Eq. (10) we find that any function $f(qQ)$ of the q's and Q's is represented in the mixed q-Q representation by

$$<q' \mid f(qQ) \mid Q'> = f(q'Q') <q' \mid Q'>. \qquad (12)$$

If a dynamical variable α can be expressed as a function of the q's and Q's, its representative $<q' \mid \alpha \mid Q'>$ must be of the form of the right-hand side of (12). Now for a general dynamical variable $<q' \mid \alpha \mid Q'>$ is an arbitrary function of the q''s and Q''s, and the condition that it shall always be expressible in the form of the righthand side of (12) is that $<q' \mid Q'>$ shall not vanish for any values of the q''s and Q''s in the domains of these variables. This is then the condition that every dynamical variable can be expressed as a function of the q's and Q's.

Assuming this condition is fulfilled, define the function $S(q'Q')$ by

$$<q' \mid Q'> = \exp iS(q'Q')/h).$$

Then

$$<q' \mid p_r \mid Q'> = \ldots = <q' \mid \partial S(qQ)/\partial q_r \mid Q'>$$

from (12). Thus

$$p_r = \partial S(qQ)/\partial q_r. \quad (13)$$

Similarly

$$<q' \mid P_r \mid Q'> = \ldots = - <q' \mid \partial S(qQ)/\partial Q_r \mid Q'>,$$

so that

$$P_r = - \partial S(qQ)/\partial Q_r. \quad (14)$$

Equations (13) and (14) are of the same form as the classical equations of a *contact transformation* in the case when every dynamical variable can be expressed as a function of the q's and Q's. *However, $S(q'Q')$ is not in general a real function of the q''s and Q''s, so the analogy is not perfect.*

In the case when $<q' \mid Q'>$ vanishes for some values of the q''s and Q''s, we can choose a function $F(q'Q')$ which vanishes everywhere except where $<q' \mid Q'>$ vanishes, and then

$$F(q'Q')<q' \mid Q'> =0$$

everywhere. Equation (12) now shows that

$$F(qQ) = 0$$

so that we have a connection between the q's and the Q's. There may be several independent connections like this, say

…

…

Summary

A method is given for defining *general functions of non-commuting observables* in quantum mechanics, with a certain limitation. *The method is developed to provide a formal probability for non-commuting observables to have numerical values.* This probability turns out to be in general a complex number, but all the same it has some physical meaning, since when it is close to zero one can say that the numerical values are unlikely. The method enables one to discuss trajectories for the motion of a particle in quantum mechanics and thus makes quantum mechanics more closely resemble classical mechanics. The method also enables one to set up the analogy between classical and quantum *contact transformations* on a more general basis.

Richard Phillips Feynman (May 11, 1918 – February 15, 1988)

[This includes material from Feynman's own account of his work given in his Nobel Lecture; https://www.nobelprize.org/ prizes/physics/1965/feynman/lecture/.]

Feynman was an American theoretical physicist, known for his work on the *path integral formulation* of quantum mechanics, the theory of quantum electrodynamics, the physics of the superfluidity of supercooled liquid helium, as well as his work in particle physics for which he proposed the parton model. Feynman received the Nobel Prize in Physics in 1965, jointly with Shin'ichirō Tomonaga and Julian Schwinger, "for their fundamental work in quantum electrodynamics, with deep-ploughing consequences for the physics of elementary particles". "Following the establishment of the theory of relativity and quantum mechanics, an initial relativistic theory was formulated for the interaction between charged particles and electromagnetic fields. This needed to be reformulated, however. In 1948 in particular, Richard Feynman contributed to creating a new quantum electrodynamics *by introducing Feynman diagrams: graphic representations of various interactions between different particles. These diagrams facilitate the calculation of interaction probabilities*." [Richard P. Feynman – Facts. NobelPrize.org. Nobel Prize https://www.nobelprize.org/ prizes/physics/1965/feynman/facts/]

Feynman developed a widely used pictorial representation scheme for the mathematical expressions describing the behavior of subatomic particles, which later became known as Feynman diagrams. These were introduced at a conference in Ponoco in the spring of 1948, and first published in his September 1949 paper, "*Space-Time Approach to Quantum Electrodynamics*". Along with his work in theoretical physics, Feynman has been credited with pioneering the field of quantum computing and introducing the concept of nanotechnology.

Feynman was born on May 11, 1918, in Queens, New York City, to Lucille née Phillips, a homemaker, and Melville Arthur Feynman, a sales manager originally from Minsk in Belarus (then part of the Russian Empire). Feynman's parents were both from Jewish families but not religious, and by his youth, Feynman described himself as an "avowed atheist". Feynman was a late talker, and did not speak until after his third birthday. As an adult he spoke with a New York accent strong enough to be perceived as an affectation or exaggeration, so much so that his friends Wolfgang Pauli and Hans Bethe once commented that Feynman spoke like a "bum".

The young Feynman was heavily influenced by his father, who encouraged him to ask questions to challenge orthodox thinking, and who was always ready to teach Feynman something new. From his mother, he gained the sense of humor that he had throughout his life. As a child, he had a talent for engineering, maintained an experimental laboratory in his home, and delighted in repairing radios. This radio repairing was probably the first job Feynman had, and during this time he showed early signs of an aptitude for his later career in theoretical physics, when he would analyze the issues theoretically and arrive at the solutions.

When Feynman was five, his mother gave birth to a younger brother, Henry Phillips, who died at age four weeks. Four years later, his sister Joan was born and the family moved to Far Rockaway, Queens. Though separated by nine years, Joan and Richard were close, and they both shared a curiosity about the world. Though their mother thought women lacked the capacity to understand such things, Richard encouraged Joan's interest in astronomy, and Joan eventually became an astrophysicist.

Feynman attended Far Rockaway High School, which was also attended by fellow Nobel laureates Burton Richter and Baruch Samuel Blumberg. Upon starting high school, Feynman was quickly promoted to a higher math class. An IQ test administered in high school estimated his IQ at 125—high but "merely respectable", according to biographer James Gleick. His sister Joan, who scored one point higher, later jokingly claimed to an interviewer that she was smarter.

When Feynman was 15, he taught himself trigonometry, advanced algebra, infinite series, analytic geometry, and both differential and integral calculus. Before entering college, he was experimenting with and deriving mathematical topics such as the half-derivative using his own notation.

Feynman applied to Columbia University but was not accepted because of their quota for the number of Jews admitted. Instead, he attended the Massachusetts Institute of Technology, where he joined the Pi Lambda Phi fraternity. Although he originally majored in mathematics, he later switched to electrical engineering, as he considered mathematics to be too abstract. Noticing that he "had gone too far", he then switched to physics, which he claimed was "somewhere in between".

As an undergraduate, he published two papers in the *Physical Review*. One of these, which was co-written with Manuel Vallarta, was entitled "*The Scattering of Cosmic Rays by the Stars of a Galaxy*". The other was his senior thesis, on "Forces in Molecules", based on an idea by John C. Slater, who was sufficiently impressed by the paper to have it published. Today, it is known as the Hellmann–Feynman theorem.

In 1939, Feynman received a bachelor's degree and was named a Putnam Fellow. He attained a perfect score on the graduate school entrance exams to Princeton University in physics—an unprecedented feat—and an outstanding score in mathematics, but did poorly on the history and English portions.

He was Research Assistant at Princeton from 1940 to 1941. Attendees at Feynman's first seminar, which was on the classical version of the Wheeler–Feynman absorber theory, included Albert Einstein, Wolfgang Pauli, and John von Neumann. Pauli made the prescient comment that the theory would be extremely difficult to quantize, and Einstein said that one might try to apply this method to gravity in general relativity, which Sir Fred Hoyle and Jayant Narlikar did much later as the Hoyle–Narlikar theory of gravity.

According to Feynman's account in his Nobel Lecture, in 1941, Feynman went to a beer party at the Nassau Tavern. He sat with a physicist lately arrived from Europe, Herbert

Jehle, who asked Feynman what he was working on. Feynman explained and asked in turn whether Jehle knew of any application of the *least-action principle* in quantum mechanics. Jehle pointed out that Feynman's own hero, Dirac, had published a paper on just that subject eight years before. The next day Jehle and Feynman looked at it together in the library. It was short. They found it in the bound volumes of *Physikalische Zeitschrift der Sowjetunion*, not the best-read of journals. [Dirac, P. A. M. (1933). The Lagrangian in Quantum Mechanics. *Phys. Zeit. Sowjetunion*, 3, 1, 64–72.] *Dirac had worked out the beginnings of a least-action approach in just the style Feynman was seeking, a way of treating the probability of a particle's entire path over time*. Dirac considered only one detail, a piece of mathematics for carrying the *wave function* forward in time by an infinitesimal amount.

In the past eight years neither Dirac nor any other physicist had been able to follow up on the notion of a Lagrangian in quantum mechanics-- a way of expressing a particle's history in terms of quantity of *action*. Where Dirac had pointed the way to calculating how the wave function would evolve in an infinitesimal slice of time, Feynman extended this through finite time. Making use of Dirac's infinitesimal slice required a piling up an infinite number of steps. Each step required an integration, a summing of algebraic quantities, resulting in a sequence of multiplications and compounded integrals. Feynman realized, he had to make a complex integral encompassing every possible coordinate through which a particle could move, creating a sum of the probabilities of every path from the starting position to the final position. The quantity that emerged was, once again, a form of the *action*, and a new formulation of quantum mechanics.

In 1941, with World War II raging in Europe but the United States not yet at war, Feynman spent the summer working on ballistics problems at the Frankford Arsenal in Pennsylvania. After the attack on Pearl Harbor brought the United States into the war, Feynman was recruited by Robert R. Wilson, who was working on means to produce enriched uranium for use in an atomic bomb, as part of what would become the Manhattan Project. At the time, Feynman had not earned a graduate degree. Wilson's team at Princeton was working on a device called an isotron, intended to electromagnetically separate uranium-235 from uranium-238. This was done in a quite different manner from that used by the calutron that was under development by a team under Wilson's former mentor, Ernest O. Lawrence, at the Radiation Laboratory of the University of California. On paper, the isotron was many times more efficient than the calutron, but Feynman and Paul Olum struggled to determine whether or not it was practical. Ultimately, on Lawrence's recommendation, *the isotron project was abandoned.*

Feynman received a Ph.D. from Princeton in 1942; his thesis advisor was John Archibald Wheeler. In his doctoral thesis Feynman applied the *principle of stationary action* to problems of quantum mechanics, and laid the groundwork for the *path integral* formulation and Feynman diagrams. [Feynman, R. P. (1942). *The Principle of Least Action in Quantum Mechanics*. Ph.D. thesis, Princeton University (in Brown, L. M. (ed.) (2005). *Feynman's Thesis. A New Approach to Quantum Theory*. World Scientific).]

At this juncture, in early 1943, Robert Oppenheimer was establishing the Los Alamos Laboratory, a secret laboratory on a mesa in New Mexico where atomic bombs would be designed and built. An offer was made to the Princeton team to be redeployed there. "Like a bunch of professional soldiers," Wilson later recalled, "we signed up, *en masse*, to go to Los Alamos." They were among the first to depart for New Mexico, leaving on a train on March 28, 1943.

At Los Alamos, Feynman was assigned to Hans Bethe's Theoretical (T) Division, and impressed Bethe enough to be made a group leader. He and Bethe developed the Bethe–Feynman formula for calculating the yield of a fission bomb, which built upon previous work by Robert Serber. As a junior physicist, he was not central to the project. He administered the computation group of human computers in the theoretical division. With Stanley Frankel and Nicholas Metropolis, he assisted in establishing a system for using IBM punched cards for computation. He invented a new method of computing logarithms that he later used on the Connection Machine. Other work at Los Alamos included calculating neutron equations for the Los Alamos "Water Boiler", a small nuclear reactor, to measure how close an assembly of fissile material was to criticality.

On completing this work, Feynman was sent to the Clinton Engineer Works in Oak Ridge, Tennessee, where the Manhattan Project had its uranium enrichment facilities. He aided the engineers there in devising safety procedures for material storage so that criticality accidents could be avoided, especially when enriched uranium came into contact with water, which acted as a neutron moderator. He insisted on giving the rank and file a lecture on nuclear physics so that they would realize the dangers. He explained that while any amount of unenriched uranium could be safely stored, the enriched uranium had to be carefully handled. He developed a series of safety recommendations for the various grades of enrichments. He was told that if the people at Oak Ridge gave him any difficulty with his proposals, he was to inform them that Los Alamos "could not be responsible for their safety otherwise".

Returning to Los Alamos, Feynman was put in charge of the group responsible for the theoretical work and calculations on the proposed uranium hydride bomb, which ultimately proved to be infeasible. He was sought out by physicist Niels Bohr for one-on-one discussions. He later discovered the reason: most of the other physicists were too much in awe of Bohr to argue with him. Feynman had no such inhibitions, vigorously pointing out anything he considered to be flawed in Bohr's thinking. He said he felt as much respect for Bohr as anyone else, but once anyone got him talking about physics, he would become so focused he forgot about social niceties. Perhaps because of this, Bohr never warmed to Feynman.

As early as October 30, 1943, Bethe had written to the chairman of the physics department of his university, Cornell, to recommend that Feynman be hired. On February 28, 1944, this was endorsed by Robert Bacher, also from Cornell, and one of the most senior scientists at Los Alamos. This led to an offer being made in August 1944, which Feynman accepted. Oppenheimer had also hoped to recruit Feynman to the University of California, but the head of the physics department, Raymond T. Birge, was reluctant. He made Feynman an offer in May 1945, but Feynman turned it down. Cornell matched its salary offer of $3,900 per annum. Feynman became one of the first of the Los Alamos Laboratory's group leaders to depart. He was Professor of Theoretical Physics at Cornell from 1945 to 1950.

In 1945, Wheeler and Feynman published a paper on the *Wheeler–Feynman absorber theory*, an interpretation of electrodynamics derived from the assumption that the solutions of the electromagnetic field equations must be invariant under time-reversal transformation, as are the field equations themselves. [Wheeler, J. A. & Feynman R. P. (April, 1945). Interaction with the Absorber as the Mechanism of Radiation. *Rev. Mod. Phys*. 17, 157.]

Because Feynman was no longer working at the Los Alamos Laboratory, he was no longer exempt from the draft. At his induction physical, Army psychiatrists diagnosed Feynman as suffering from a mental illness and the Army gave him a 4-F exemption on mental grounds. His father died suddenly on October 8, 1946, and Feynman suffered from depression. Unable to focus on research problems, Feynman began tackling physics problems, not for utility, but for self-satisfaction. He read the work of Sir William Rowan Hamilton on quaternions, and tried *unsuccessfully* to use them to formulate a *relativistic* theory of electrons.

Feynman was not the only frustrated theoretical physicist in the early post-war years. *Quantum electrodynamics suffered from infinite integrals in perturbation theory.* These were clear mathematical flaws in the theory, which Feynman and Wheeler had tried, unsuccessfully, to work around. "Theoreticians", noted Murray Gell-Mann, "were in disgrace." In June 1947, leading American physicists met at the Shelter Island Conference. For Feynman, it was his "first big conference with big men ... I had never gone to one like this one in peacetime." The problems plaguing quantum electrodynamics were discussed, but the theoreticians were completely overshadowed by the achievements of the experimentalists, who reported the discovery of the Lamb shift, the measurement of the magnetic moment of the electron, and Robert Marshak's two-meson hypothesis.

Bethe took the lead from the work of Hans Kramers and derived a renormalized *non-relativistic* quantum equation for the Lamb shift. The next step was to create a *relativistic* version. Feynman thought that he could do this, but when he went back to Bethe with his solution, it did not converge. *Feynman carefully worked through the problem again, applying the path integral formulation that he had used in his thesis. Like Bethe, he made the integral finite by applying a cut-off term.* The result corresponded to Bethe's version. Feynman presented his work to his peers at the Pocono Conference in March 1948. It did not go well. *Julian Schwinger gave a long presentation of his work in quantum electrodynamics*, and Feynman then offered his version, entitled "*Alternative Formulation of Quantum Electrodynamics*". The unfamiliar Feynman diagrams, used for the first time, puzzled the audience. Feynman failed to get his point across, and Paul Dirac, Edward Teller and Niels Bohr all raised objections.

Kaiser, D. March-April, 2005. *American Scientist*, 93, 156-165: "By using the diagrams to organize the calculation problem, Feynman had solved a long-standing puzzle that had stymied the world's best theoretical physicists for years. Looking back, we might expect the reception from his colleagues at the Pocono Manor Inn to have been appreciative, at the very least. Yet things did not go well at the meeting. For one thing, the odds were stacked against Feynman: His presentation followed a marathon daylong lecture by Harvard's *Wunderkind*, Julian Schwinger. Schwinger had arrived at a different method (independent of any diagrams) to remove the infinities from QED calculations, and the audience sat glued to their seats— pausing only briefly for lunch—as Schwinger unveiled his derivation. Coming late in the day, Feynman's blackboard presentation was rushed and confused. No one seemed able to follow what he was doing. He suffered frequent interruptions from the likes of Niels Bohr, Paul Dirac and Edward Teller, each of whom pressed Feynman on how his new doodles fit in with the established principles of quantum physics. Others asked more generally, in exasperation, what rules governed the diagrams' use. By all accounts, Feynman left the meeting disappointed, even depressed."

Murray Gell-Mann always referred to *Feynman diagrams* as *Stueckelberg diagrams*, after the Swiss physicist, Ernst Stueckelberg, who devised a similar notation many years earlier. In September 1934, Stueckelberg submitted a paper to *Annal. Physik* which looked at high energy collision phenomena between electrons and nuclei. [Stueckelberg, E.C.G. (1934), Relativistisch invariante Störungstheorie des Diracschen Elektrons I. Teil: Streustrahlung und Bremsstrahlung. (Relativistically invariant perturbation theory of Dirac's electron Part I: scattered radiation and Bremsstrahlung.) *Ann. Phys.*, 413, 4, 367-389; https://doi.org/10.1002/andp.19344130403; also in Lacki, J., Ruegg, H., Wanders, G. (eds) (2009). *E.C.G. Stueckelberg, An Unconventional Figure of Twentieth Century Physics*. Birkhäuser, Basel. https://doi.org/10.1007/978-3-7643-8878-2_10.].

Stueckelberg was motivated by the need for a manifestly covariant formalism for quantum field theory, but did not provide as automated a way to handle symmetry factors and loops, although he was first to find the correct physical interpretation in terms of forward and backward in time particle paths, all without the path-integral.

To Freeman Dyson, one thing at least was clear: Tomonaga, Schwinger and Feynman understood what they were talking about even if no one else did, but had not published anything. He was convinced that Feynman's formulation was easier to understand, and ultimately managed to convince Oppenheimer that this was the case. *Dyson published a paper in 1949, which added new rules to Feynman's that told how to implement renormalization.* Feynman was prompted to publish his ideas in a series of papers over three years. His April 1948 paper "Space-Time Approach to Non-Relativistic Quantum Mechanics", setting out his formulation of *non-relativistic* quantum theory. His October 1948 paper on "*A Relativistic Cut-Off for Classical Electrodynamics*" attempted to explain what he had been unable to get across at Pocono in March 1948 where he first introduced his diagrams. His September 1949 paper on "*The Theory of Positrons*" addressed the Schrödinger equation and Dirac equation, and introduced what is now called the Feynman propagator; and the second part published on September 15, 1949, "*Space-Time Approach to Quantum Electrodynamics*", included the first published Feynman diagram. Finally, in papers on the "*Mathematical Formulation of the Quantum Theory of Electromagnetic Interaction*" in 1950 and "*An Operator Calculus Having Applications in Quantum Electrodynamics*" in 1951, he developed the mathematical basis of his ideas, derived familiar formulae and advanced new ones.

While papers by others initially cited Schwinger, papers citing Feynman and employing Feynman diagrams appeared in 1950, and soon became prevalent. Students learned and used the powerful new tool that Feynman had created. Computer programs were later written to compute Feynman diagrams, providing a tool of unprecedented power. It is possible to write such programs because the Feynman diagrams constitute a formal language with a formal grammar. To Schwinger, however, the Feynman diagram was "pedagogy, not physics".

By 1949, Feynman was becoming restless at Cornell. He never settled into a particular house or apartment, living in guest houses or student residences, or with married friends "until these arrangements became sexually volatile". He liked to date undergraduates, hire prostitutes, and sleep with the wives of friends. He was not fond of Ithaca's cold winter weather, and pined for a warmer climate. Above all, at Cornell, he was always in the shadow of Hans Bethe. Despite all of this, Feynman looked back favorably on the Telluride House, where he resided for a large period of his Cornell career. In an interview, he described the House as "a group of boys that have been specially selected because of their scholarship, because of their cleverness or whatever it is, to be given free board and lodging and so on, because of their brains". He enjoyed the house's convenience and said that "it's there that I did the fundamental work" for which he won the Nobel Prize.

Feynman spent several weeks in Rio de Janeiro in July 1949. That year, the Soviet Union detonated its first atomic bomb, generating concerns about espionage. Fuchs was arrested as a Soviet spy in 1950 and the FBI questioned Bethe about Feynman's loyalty. Physicist David Bohm was arrested on December 4, 1950 and emigrated to Brazil in October 1951. Because of the fears of a nuclear war, a girlfriend told Feynman that he should also consider moving to South America. He had a sabbatical coming for 1951–52, and elected to spend it in Brazil, where he gave courses at the Centro Brasileiro de Pesquisas Físicas. In Brazil, Feynman was impressed with samba music, and learned to play the frigideira, a metal percussion instrument based on a frying pan. He was an enthusiastic amateur player of bongo and conga drums and often played them in the pit orchestra in musicals. He spent time in Rio with his friend Bohm, but Bohm could not convince Feynman to investigate Bohm's ideas on physics.

Feynman did not return to Cornell. Bacher, who had been instrumental in bringing Feynman to Cornell, had lured him to the California Institute of Technology (Caltech), where he was first a Visiting Professor and thereafter appointed Professor of Theoretical Physics at the California Institute of Technology from 1950.

Part of the deal was that he could spend his first year on sabbatical in Brazil. He had become smitten by Mary Louise Bell from Neodesha, Kansas. They had met in a cafeteria in Cornell, where she had studied the history of Mexican art and textiles. While he was in Brazil, she taught classes on the history of furniture and interiors at Michigan State University. He proposed to her by mail from Rio de Janeiro, and they married in Boise, Idaho, on June 28, 1952, shortly after he returned. They frequently quarreled and she was frightened by his violent temper. Their politics were different; although he registered and voted as a Republican, she was more conservative, and her opinion on the 1954 Oppenheimer security hearing ("Where there's smoke there's fire") offended him. They separated on May 20, 1956. An interlocutory decree of divorce was entered on June 19, 1956, on the grounds of "extreme cruelty". The divorce became final on May 5, 1958.

At Caltech, Feynman investigated the physics of the superfluidity of supercooled liquid helium, where helium seems to display a complete lack of viscosity when flowing. Feynman provided a quantum-mechanical explanation for the Soviet physicist Lev Landau's theory of superfluidity. Applying the Schrödinger equation to the question showed that the superfluid was displaying quantum mechanical behavior observable on a macroscopic scale. This helped with the problem of superconductivity, but the solution eluded Feynman. It was solved with the BCS theory of superconductivity, proposed by John Bardeen, Leon Neil Cooper, and John Robert Schrieffer in 1957.

With Murray Gell-Mann, Feynman developed a model of weak decay, which showed that the current coupling in the process is a combination of vector and axial currents (an example of weak decay is the decay of a neutron into an electron, a proton, and an antineutrino). Although E. C. George Sudarshan and Robert Marshak developed the theory nearly simultaneously, Feynman's collaboration with Murray Gell-Mann was seen as seminal because the weak interaction was neatly described by the vector and axial currents. It thus combined the 1933 beta decay theory of Enrico Fermi with an explanation of parity violation.

Feynman attempted an explanation, called the parton model, of the strong interactions governing nucleon scattering. The parton model emerged as a complement to the quark model developed by Gell-Mann. The relationship between the two models was murky; Gell-Mann referred to Feynman's partons derisively as "put-ons". In the mid-1960s, physicists believed that quarks were just a bookkeeping device for symmetry numbers, not real particles; the statistics of the omega-minus particle, if it were interpreted as three identical strange quarks bound together, seemed impossible if quarks were real.

The SLAC National Accelerator Laboratory deep inelastic scattering experiments of the late 1960s showed that nucleons (protons and neutrons) contained point-like particles that scattered electrons. It was natural to identify these with quarks, but Feynman's parton model attempted to interpret the experimental data in a way that did not introduce additional hypotheses. For example, the data showed that some 45% of the energy momentum was carried by electrically neutral particles in the nucleon. These electrically neutral particles are now seen to be the gluons that carry the forces between the quarks, and their three-valued color quantum number solves the omega-minus problem. Feynman did not dispute the quark model; for example, when the fifth quark was discovered in 1977, Feynman immediately pointed out to his students that the discovery implied the existence of a sixth quark, which was discovered in the decade after his death.

After the success of quantum electrodynamics, Feynman turned to quantum gravity. By analogy with the photon, which has spin 1, he investigated the consequences of a free massless spin 2 field and derived the Einstein field equation of general relativity, but little more. The computational device that Feynman discovered then for gravity, "ghosts", which are "particles" in the interior of his diagrams that have the "wrong" connection between spin and statistics, have proved invaluable in explaining the quantum particle behavior of the Yang–Mills theories, for example, quantum chromodynamics and the electro-weak theory.

In the early 1960s, Feynman acceded to a request to "spruce up" the teaching of undergraduates at Caltech. After three years devoted to the task, he produced a series of lectures that later became *The Feynman Lectures on Physics*. This occupied two physicists, Robert B. Leighton and Matthew Sands, as part-time co-authors for several years. Even though the books were not adopted by universities as textbooks, they continue to sell well because they provide a deep understanding of physics. Many of his lectures and miscellaneous talks were turned into other books, including *The Character of Physical Law*, *QED: The Strange Theory of Light and Matter*, *Statistical Mechanics*, *Lectures on Gravitation*, and the *Feynman Lectures on Computation*.

In the wake of the 1957 Sputnik crisis, the U.S. government's interest in science rose for a time. Feynman was considered for a seat on the President's Science Advisory Committee, but was not appointed. At this time, the FBI interviewed a woman close to Feynman, possibly his ex-wife Bell, who sent a written statement to J. Edgar Hoover on August 8, 1958: “I do not know—but I believe that Richard Feynman is either a Communist or very strongly pro-Communist—and as such is a very definite security risk. This man is, in my opinion, an extremely complex and dangerous person, a very dangerous person to have in a position of public trust ... In matters of intrigue Richard Feynman is, I believe immensely clever—indeed a genius—and he is, I further believe, completely ruthless, unhampered by morals, ethics, or religion—and will stop at absolutely nothing to achieve his ends.”

The U.S. government nevertheless sent Feynman to Geneva for the September 1958 Atoms for Peace Conference. On the beach at Lake Geneva, he met Gweneth Howarth, who was from Ripponden, Yorkshire, and working in Switzerland as an au pair. Feynman's love life had been turbulent since his divorce; his previous girlfriend had walked off with his Albert Einstein Award medal and, on the advice of an earlier girlfriend, had feigned pregnancy and extorted him into paying for an abortion, then used the money to buy furniture. When Feynman found that Howarth was being paid only $25 a month, he offered her $20 a week to be his live-in maid. Feynman knew that this sort of behavior was illegal under the Mann Act, so he had a friend, Matthew Sands, act as her sponsor. Howarth pointed out that she already had two boyfriends, but decided to take Feynman up on his offer, and arrived in Altadena, California, in June 1959. She made a point of dating other men, but Feynman proposed in early 1960. They were married on September 24, 1960, at the Huntington Hotel in Pasadena. They had a son, Carl, in 1962, and adopted a daughter, Michelle, in 1968. Besides their home in Altadena, they had a beach house in Baja California, purchased with the money from Feynman's Nobel Prize.

Feynman tried marijuana and ketamine at John Lilly's sensory deprivation tanks, as a way of studying consciousness. He gave up alcohol when he began to show vague, early signs of alcoholism, as he did not want to do anything that could damage his brain. Despite his curiosity about hallucinations, he was reluctant to experiment with LSD.

Feynman was also interested in the relationship between physics and computation. He was one of the first scientists to conceive the possibility of quantum computers. In the 1980s he began to spend his summers working at Thinking Machines Corporation, helping to build some of the first parallel supercomputers and considering the construction of quantum computers. In 1984–1986, he developed a variational method for the approximate calculation of path integrals, which has led to a powerful method of converting divergent perturbation expansions into convergent strong-coupling expansions (variational perturbation theory) and, as a consequence, to the most accurate determination of critical exponents measured in satellite experiments.

Feynman was a keen popularizer of physics through both books and lectures, including a 1959 talk on top-down nanotechnology called *There's Plenty of Room at the Bottom*. He also became known through his autobiographical books. In the 1960s, Feynman began thinking of writing an autobiography, and he began granting interviews to historians. In the 1980s, working with Ralph Leighton (Robert Leighton's son), he recorded chapters on audio tape that Ralph transcribed. The book was published in 1985 as *Surely You're Joking, Mr. Feynman!* and became a best-seller.

Gell-Mann was upset by Feynman's account in the book of the weak interaction work, and threatened to sue, resulting in a correction being inserted in later editions. This incident was just the latest provocation in decades of bad feeling between the two scientists. Gell-Mann often expressed frustration at the attention Feynman received; he remarked: "[Feynman] was a great scientist, but he spent a great deal of his effort generating anecdotes about himself."

Feynman was elected a Foreign Member of the Royal Society in 1965, received the Oersted Medal in 1972, and the National Medal of Science in 1979. He was elected a Member of the National Academy of Sciences, but ultimately resigned. In 1965 he was elected a foreign member of the Royal Society, London (Great Britain). He held the following awards in addition to the 1965 Nobel Prize in Physics: Albert Einstein Award (1954,

Princeton); Einstein Award (Albert Einstein Award College of Medicine); Lawrence Award (1962).

In 1978, Feynman sought medical treatment for abdominal pains and was diagnosed with liposarcoma, a rare form of cancer. Surgeons removed a tumor the size of a football that had crushed one kidney and his spleen. Further operations were performed in October 1986 and October 1987. He was again hospitalized at the UCLA Medical Center on February 3, 1988. A ruptured duodenal ulcer caused kidney failure, and he declined to undergo the dialysis that might have prolonged his life for a few months. Watched over by his wife Gweneth Howarth, sister Joan, and cousin Frances Lewine, he died on February 15, 1988, at age 69. His burial was at Mountain View Cemetery and Mausoleum in Altadena, California.

Wheeler, J. A.[**] & Feynman R. P.[***] (April, 1945). Interaction with the Absorber as the Mechanism of Radiation[†*].

[*Rev. Mod. Phys.*, 17, 157-81; https://doi.org/10.1103/ RevModPhys.17.157.]

Palmer Physical Laboratory, Princeton University, Princeton, New Jersey

* A preliminary account of the considerations which appear in this paper was presented by us at the Cambridge meeting of the American Physical Society, February 21, 1941, Phys. Rev. 59, 683 (1941).

** On leave of absence from Princeton University.

*** Now a member of the faculty of Cornell University, but on leave of absence from that institution.

The motive of the analysis was to clear the present quantum theory of interacting particles of those of its difficulties which have a purely classical origin, method was to define as closely as one can within the bounds of classical theory the proper use of the *field* concept in the description of nature, this paper represents the third part of the survey, an analysis of the *mechanism of radiation* believed to complete the last tie between *action at a distance* and *field theory*, only section now finished, difficulties to obtain a satisfactory account of the *field* generated by an accelerated charge at a remote point and to understand the source of the force experienced by the charge itself as a result of its motion, takes up suggestion by Tetrode (1922) that *the act of radiation should have some connection with the presence of an absorber*, develops this idea into the thesis that the force of radiative reaction arises from the action on the source owing to the half-advanced fields of the particles of the absorber, absorber response as the mechanism of radiative reaction, expresses in terms of *action at a distance*, *Wheeler–Feynman absorber theory, assumes solutions of the electromagnetic field equations* must be invariant under time-reversal transformation as are the field equations themselves, considers an accelerated charge located in the absorbing system as the source of radiation, *elementary particles not self-interacting*, a complete correspondence is established between *action at a distance* and the usual formulation of *field theory* in the case of a completely absorbing system.

[Sakata, S. (October, 1947). The Theory of the Interaction of Elementary Particles. I. The Method of the Theory of Elementary Particles. *Progr. Theor. Phys*., 2, 3, 145-50; https://doi.org/10.1143/ptp/2.3.145: “Interesting in this connection is Wheeler's recent work. Previous theories of elementary particles not only considered all forces to be conveyed by *fields*, but also treated matter itself as a *field*. In short, they were "*theories of fields*". But he attempts to replace *fields* by interactions between *matter*, from the standpoint of *action at a distance*. This is a "*theory of matter*", in contrast to the previous "*theories of fields*". Further, he holds the view that the action of all *matter* existing in nature must be taken into account even when discussing the motion of a single electron, and thus makes clear the intrinsic nature of the reaction of *fields* in previous theories. This coincides with the dialectic view that the world must be interacted in correlation as a whole.]

[Feynman, R. P. (September, 1949). Space-Time Approach to Quantum Electrodynamics. *Phys. Rev.* 76, 6, 769-89: "Having a term representing the mutual interaction of a pair of charges, we must include similar terms to represent the interaction of a charge with itself. For under some circumstances what appears to be two distinct electrons may, according to I, be viewed also as a single electron (namely in case one electron was created in a pair with a positron destined to annihilate the other electron). *Thus, to the interaction between such electrons must correspond the possibility of the action of an electron on itself*[9].

> [9] These considerations *make it appear unlikely that the contention of J. A. Wheeler and R. P. Feynman* [Wheeler, J. A. & Feynman, R. P. (1945). Interaction with the Absorber as the Mechanism of Radiation. *Rev. Mod. Phys.*, 17, 157-81], *that electrons do not act on themselves, will be a successful concept in quantum electrodynamics*.]

"*We must, therefore, be prepared to find that further advance into this region will require a still more extensive renunciation of features which we are accustomed to demand of the space time mode of description.*" - *Niels Bohr*[1]

> [1] Bohr, N. (1934). Atomic Theory and the Description of Nature. (Cambridge University Press, Teddington, England.

---

† *Introductory Note* - In commemoration of the sixtieth birthday of Niels Bohr it had been hoped to present a critique of classical field theory which has been in preparation since before the war by the writer and his former student, R. P. Feynman. The accompanying joint article, *representing the third part of the survey*, is however the only section now finished. The war has postponed completion of the other parts. As reference to them is made in the present section, it may be useful to outline the plan of the survey.

The motive of the analysis is to clear the present quantum theory of interacting particles of those of its difficulties which have a purely classical origin. The method of approach is to define as closely as one can within the bounds of classical theory the proper use of the field concept in the description of nature. *Division I* is intended first to recall the possibility of idealizing to the case of arbitrarily small quantum effects, a possibility which is offered by the freedom of choice in the present quantum theory for the dimensionless ratio (quantum of angular momentum) (velocity of light)/ (electronic charge $)^2$; then however to recognize the possible limitations placed on this analysis by the relatively large value, 137, of the ratio in question in nature; and finally to present a general summary of the conclusions drawn from the more technical parts of the survey. The plan of the *second article* is a derivation and resume of the theory of *action at a distance* of Schwarzschild and Fokker, to prepare this theory as a tool to analyze the field concept. From the correlation of the two points of view, one comes to Frenkel's solution of the problem of self-energy in the classical field theory and to new expressions for the energy of electromagnetic interaction in the theory of action at a distance. *The third division, which is published herewith, is an analysis of the mechanism of radiation believed to complete the last tie between action at a distance and field theory and to remove the obstacle which has so far prevented the use of both points of view as complementary tools in the description of nature*. It is the plan of a *subsequent division* to discuss the problems which arise when the fields are regarded as subordinate entities with no degrees of freedom of their own. An infinite number of degrees of freedom are found to be attributed to the particles themselves by the theory of propagated action at a distance. However, it appears that the additional modes of motion are divergent and have on this account to be excluded by a general principle of selection. Acceptance of this principle leads to the conclusion that the union of action at a distance and field theory constitutes the natural and self-consistent generalization of Newtonian mechanics to the four-dimensional space of Lorentz and Einstein. - J. A. W.

Past failure of action at a distance to account for the mechanism of radiation

It was the 19th of March in 1845 when Gauss described the conception of an *action at a distance* propagated with a finite velocity, the natural generalization to electrodynamics of the view of force so fruitfully applied by Newton and his followers. In the century between then and now what obstacle has discouraged the general use of this conception in the study of nature?

The difficulty has not been that of giving to the idea of propagated *action at a distance* a suitable embodiment of electromagnetic equations. This problem, to be true, remained unsolved to Gauss and his successors for three quarters of the century. But the formulation then developed by Schwarzschild and Fokker, described and amplified in another article[2], demonstrated that the conception of Gauss is at the same time mathematically self-consistent, in agreement with experience on static and current electricity, and in complete harmony with Maxwell's equations.

[2] Unpublished, see *Introductory Note*.

To find the real obstacle to acceptance of the tool of Newton and Gauss for the analysis of forces, we have to go beyond the bounds of steady-state electromagnetism to the phenomena of emission and propagation of energy. No branch of science has done more than radiation physics to favor the evolution of present concepts of field or more to pose difficulties for the idea of action at a distance. *The difficulties have been twofold - to obtain a satisfactory account of the field generated by an accelerated charge at a remote point and to understand the source of the force experienced by the charge itself as a result of its motion*:

(a) An accelerated charge generates a field given, according to the formulation of Schwarzschild and Fokker, by half the usual retarded solution of Maxwell's equations, plus half the advanced solution. From the presence of the advanced field in the expression for the electric vector, it follows that a distant test body will experience a premonitory force well before the source itself has commenced to move. To avoid a conclusion so opposed to experience *Ritz*[3] *and Tetrode*[4] *proposed to abandon the symmetry in time of the elementary law of force.*

[3] Ritz, W. (1908). Recherches critiques sur l'Électrodynamique générale. (Critical Research on General Electrodynamics.) *Ann. Chem. et Phys.*, 13, 145-275; summarized in (August, 1908). Recherches critiques sur les théories électrodynamiques de Cl. Maxwell et de H.-A. Lorentz, in *Archives des sciences physiques et narurelles*, 26, 209-36.

[4] Tetrode, H. (December, 1922). Über den Wirkungszusammenhang der Welt. Eine Erweiterung der klassischen Dynamik. (On the interdependence of the world. An extension of classical dynamics.) *Zeit. Phys.*, 10, 317-28; http://dx.doi.org/10.1007/ BF01332574.

However, *it was then necessary to give up the possibility to derive the equations of motion and all the electromagnetic forces consistently from a single unified principle of least action* like that of Fokker. More important, the sacrifice made to alleviate one difficulty of the theory of *action at a distance* did not help to solve the other, *the problem of the origin of the force of radiative reaction.*

(b) Experience indicates that an accelerated charge suffers a force of damping which is simultaneous with the moment of acceleration. However, the theory of *action at a distance* predicts that an accelerated charge in otherwise charge-free space will experience no electric force. To exclude the acceleration and thus to avoid the issue does not appear reasonable. Uncharged particles can be present and can accelerate the charge via gravitational forces. It seems just as difficult to explain the reactive force when other charged particles are present. They will indeed be set into motion and will act back on the source. However, if these elementary interactions have the purely retarded character assumed by Ritz, and also by Frenkel[5], the reaction will arrive at the accelerated particle too late and will have the wrong magnitude[6] to produce the damping phenomenon.

[5] Frenkel, J. (December, 1925). Zur elektrodynamik punktförmiger Elektronen. (On the electrodynamics of point-like electrons.) *Zeit. Phys.*, 32, 518-34[; argues that electrons and protons should be treated as *point charges* and their *masses* considered to be a primary property independent of *charge* in place of *electromagnetic theory of mass*, based on erroneous assumption of *mass defect* of helium relative to hydrogen at time when neutron had not been discovered, but claims that *an extended electron is inconceivable in the special theory of relativity since due to the intrinsic connection between space and time an invariant definition of a geometrically invariable (i.e., rigid) body is impossible for arbitrary motions*, notes that electrons are not only physically but geometrically indivisible and have no extension in space, no internal forces between the elements of an electron because such elements do not exist, the electromagnetic explanation for *mass* then goes away].

[6] Synge, J. L. (December, 1940). On the electromagnetic two-body problem. *Roy. Soc. Proc., A*, 177, 968, 118-39; http://doi.org/10.1098/rspa.1940.0114.

On the other hand, interactions symmetrical between past and future - *the half-retarded, half-advanced fields of the unified theory of action at a distance* - have so far appeared to be equally incapable of accounting for the observed force of radiative reaction, with its definitely irreversible character.

It is clear why the viewpoint of Newton and Gauss has not been generally applied in recent times; it has so far failed to give a satisfactory account of the mechanism of radiation.

The failure of *action at a distance* cannot pass unnoticed by *field theory*. The two points of view, according to the thesis of the present critique, are not independent, but mutually complementary. Consequently, *field theory*, too, faces in the radiation problem a significant issue: does this theory give an explanation for the observed force of *radiative reaction* which can be translated into the particle mechanics of Schwarzschild and Fokker, or does it likewise fail to provide a complete picture of the mechanism of radiation?

In attacking the radiation problem our first move, following the above reasoning, is to review the status of the reaction force in existing classical field theory. No more intelligible clue is found to the physical origin of the force in this theory than in the theory of *action at a distance*. Stopped on this approach, *we take up a suggestion made long ago by Tetrode that the act of radiation should have some connection with the presence of an absorber. We develop this idea into the thesis that the force of radiative reaction arises from the action on the source owing to the half-advanced fields of the particles of the absorber*; or, more briefly, that radiation is a matter as much of statistical mechanics as of pure electrodynamics. *We find that this thesis leads to a quantitative solution of the radiation problem*. Finally, we examine some of the implications of this thesis for the conception of causality.

The status of radiative reaction in field theory

A charged particle on being accelerated sends out electromagnetic energy and itself loses energy. This loss is interpreted as caused by a force acting on the particle given in magnitude and direction by the expression

2 (charge)^2 (time rate of change of acceleration)/3 (velocity of light)^3

when the particle is moving slowly, and by a more complicated expression when its speed is appreciable relative to the velocity of light. The existence of this force of radiative reaction is well attested: (a) by the electrical potential required to drive a wireless antenna; (b) by the loss of energy experienced by a charged particle which has been deflected, and therefore accelerated, in its passage near an atomic nucleus; and (c) by the cooling suffered by a glowing body.

The origin of the force of *radiative reaction* has not been nearly so clear as its existence.

Lorentz[7] considers the charged particle to have a finite size and attributes the force in question to the retarded action of one part of the particle on another.

[7] Lorentz H. A. (1892). La Théorie electromagnétique de Maxwell et son application aux corps mouvants. (Maxwell's Electromagnetic Theory and its Application to Moving Bodies.) *Arch. Nèerl. Sci*., 25, 363-552, republished in his *Collected Papers*, Vol. II, pp. 281 and 343. See also his treatise *The Theory of Electrons* (Leipzig, 1909), pp. 49 and 253.

His expression for the force appears as a series in powers of the radius of the particle. The first term in the series gives the expression already mentioned. Otherwise, the derivation leads to difficulties:

(a) All higher terms depend explicitly upon the structure assumed for the entity. These dubious terms enter in a more and more important way into the calculated law of radiative reaction as the frequency of oscillation of the particle is raised, and the period approaches the time required for light to cross the system.

(b) Non-electric forces are required to hold together the charge distribution, according to Poincare[8], for to neglect such forces is to violate the relativistic relation between mass and energy.

[8] Poincare, H. (December, 1906). Sur la dynamique de l'électron. (On the dynamics of the electron.) *Rend. Palermo*, 21, 129-175; https://doi.org/10.1007/BF03013466.

A composite system of this kind would possess an infinite number of internal degrees of freedom of oscillation. No consistent model has been found for the Lorentz electron in either classical or quantum mechanics.

Briefly, Lorentz attempts to propose a physical mechanism behind the radiative reaction, but arrives at a mathematically incomplete expression for this force.

Dirac[9], in contrast, advances no explanation for the origin of the radiative damping, but supplies a well-defined and relativistically invariant prescription to calculate its magnitude:

[9] Dirac, P. A. M. (August, 1938). Classical Theory of Radiating Electrons. *Roy. Soc. Proc., A*, 167, 929, 148-69[; Dirac referenced Frenkel (1925) in describing *relativistic* form of classical theory of radiating electrons assuming electron to be a *point charge* with no volume, *an extended electron inconceivable in special theory of relativity due to intrinsic connection between space and time*, Lorentz *model of electron as possessing mass on account of the electromagnetic field around it fails* without further assumptions if the electromagnetic field varies too rapidly or if the acceleration of the electrons is too great, no natural way of introducing further assumptions has been discovered, discovery of neutron introduced a form of mass that is difficult to believe is of electromagnetic nature, theory of the positron in which positive and negative values of an electron play symmetrical roles cannot be fitted into idea of electromagnetic idea of mass which requires all mass to be positive, *departure from electromagnetic theory of mass removes main reason for believing in finite size of the electron*, results in difficulty that the *field* in the immediate neighborhood of the electron has an infinite mass, in quantum mechanics results in divergence in the solution of the equations that describe the interaction of an electron with an electromagnetic field and prevents its application to high-energy radiation processes, *Dirac chose to represent electron by point charge and to avoid the difficulties with the infinite energy by direct omission or subtraction of unwanted terms* as had been used in the *theory of the positron. new theory must be in agreement with special relativity* and the conservation of energy and momentum, first addressed problem of a *single electron moving in an electromagnetic field*, used equations for electromagnetic *potentials* in terms of the *charge-current density* vector, *assumed that the charge-current density vector vanishes everywhere except on the world-line of the electron where it is infinitely great*, derived *field* quantities from electromagnetic *potentials*, obtained solutions in terms of *retarded* and *advanced* potentials and potentials representing the incoming and outgoing radiation, calculated field of radiation produced by the electron, in agreement with classical theory but provided a value for the field of radiation throughout space-time, obtained exact *equations of motion* of the electron in a specified incident *field* from the laws of

conservation of energy and momentum within limits of classical theory, imposed condition that solutions that occur in Nature when there is no incident field are those for which velocity in terms of proper time is constant, requires solutions of equations of motion *for which the final acceleration as well as the initial position and velocity are prescribed*, electron responds to pulse of electromagnetic radiation before it reaches the center of the electron, behaves as if it has a radius of order $1/a$ where $a = 3m/e^2$, implies that signal can be transmitted *faster than the speed of light* through the interior of an electron, *interior of electron is a region of failure of some of the elementary properties of space-time*, showed how theory can be extended to any number of electrons interacting with each other and with a field of radiation].

Let the motion of the particle be given. Calculate the field produced by the particle from Maxwell's equations, with the boundary condition that at large distances from the particle this field shall contain only outgoing waves. In addition to the so-defined *retarded field* of the particle, calculate its *advanced field*, the sole change being the existence of only convergent waves at large distances. Define half the difference between *retarded* and *advanced fields* as the *radiation field* (half the quantity denoted as radiation field by Dirac). This field is everywhere finite. Evaluate it at the position of the particle and multiply by the magnitude of the charge to obtain the *force of radiative reaction.*

Dirac's prescription is appealing. (a) It is well-defined. (b) The calculated force reduces for slowly moving particles to the simple expression which was given above and which has been well-tested at non-relativistic velocities. (c) The calculation treats the elementary charge as being localized at a mathematical point, a picture which is not only physically reasonable but also translatable into quantum mechanics. (d) The elements of the prescription involve no more than standard electromagnetic theory plus the assumption that the *radiation field*, as above defined, is the source of the force.

The physical origin of Dirac's radiation field is nevertheless not clear. (a) This field is defined for times before as well as after the moment of acceleration of the particle. (b) The field has no singularity at the position of the particle and by Maxwell's equations must, therefore, be attributed either to sources other than the charge itself or to radiation coming in from an infinite distance.

We accept as reasonable Dirac's results. His concept of *radiation field*, however, we cannot adopt as an assumption subject to no further analysis. To do so would be to add to *field theory* a principle incapable of translation into the language of *action at a distance*.

To carry the analysis further requires us to find a new idea. We go back to a suggestion once made by Tetrode[10].

[10] Tetrode, H. (December, 1922). Über den Wirkungszusammenhang der Welt. Eine Erweiterung der klassischen Dynamik. (On the interdependence of the world. An extension of classical dynamics.) *Zeit. Phys.*, 10, 317-28; http://dx.doi.org/10.1007/ BF01332574. When we gave a preliminary account of the considerations which appear in this paper (Cambridge meeting of the American Physical Society, February 21, 1941, [(1941). *Phys. Rev.* 59, 683]) we had not seen Tetrode's paper. We are indebted to Professor Einstein for

> bringing to our attention the ideas of Tetrode and also of Ritz, who is cited in this article. An idea similar to that of Tetrode was subsequently proposed by G. N. Lewis, [(1926). *Nat. Acad. Sci. Proc.*, 12, 22]: "I am going to make the ... assumption that an atom never emits light except to another atom; and to claim that it is as absurd to think of light emitted by one atom regardless of the existence of a receiving atom as it would be to think of an atom absorbing light without the existence of light to be absorbed. I propose to eliminate the idea of mere emission of light and substitute the idea of transmission, or a process of exchange of energy between two definite atoms or molecules." Lewis went nearly as far as it is possible to go without explicitly recognizing the importance of other absorbing matter in the system, a point touched upon by Tetrode, and shown below to be essential for the existence of the normal radiative mechanism.

He proposed to abandon the conception of electromagnetic radiation as an elementary process and to interpret it as a consequence of an interaction between a source and an absorber. In his words:

> "The sun would not radiate if it were alone in space and no other bodies could absorb its radiation ... If for example I observed through my telescope yesterday evening that star which let us say is 100 light years away, then not only did I know that the light which it allowed to reach my eye was emitted 100 years ago, but also the star or individual atoms of it knew already 100 years ago that I, who then did not even exist, would view it yesterday evening at such and such a time.... One might accordingly adopt the opinion that the amount of material in the universe determines the rate of emission. Still, this is not necessarily so, for two competing absorption centers will not collaborate but will presumably interfere with each other. If only the amount of matter is great enough and is distributed to some extent in all directions, further additions to it may well be without influence." [Nonsense.]
>
> Tetrode's idea that the absorber may be an essential element in the mechanism of radiation has been neglected, perhaps partly because it appears to conflict with customary notions of causality, and partly also because of his mistaken belief that the new point of view could by itself explain quantum phenomena. In this connection he assumed that the interaction between charged particles should be described by forces more complicated than those given by electromagnetic theory. Finally, as Tetrode remarks, "on the last pages we have let our conjectures go rather far beyond what has mathematically been proven."

Absorber response as the mechanism of radiative reaction

We take up the proposal of Tetrode that the absorber may be an essential element in the mechanism of radiation. Using the language of the theory of *action at a distance*, we give the idea the following definite formulation:

(1) An accelerated point charge in otherwise charge-free space does not radiate electromagnetic energy.

(2) The fields which act on a given particle arise only from other particles.

(3) These fields are represented by one-half the retarded plus one-half the advanced Lienard-Wiechert solutions of Maxwell's equations. This law of force is symmetric with respect to past and future. In connection with this assump- tion we may recall an inconclusive but illuminating discussion carried on by Ritz and Einstein in 1909, in which

"Ritz treats the limitation to retarded potentials as one of the foundations of the second law of thermodynamics, while Einstein believes that the irreversibility of radiation depends exclusively on considerations of probability.[11]"

> [11] Ritz, W. & Einstein, A. *Phys. Zeitschr.* 10, 323-4 (1909) [Walther Ritz's joint communique with Einstein on their differing viewpoints of the advanced and retarded solutions of Maxwell's equations. Einstein argues that the physical restriction to retarded solutions is not a law, but probabilistic; Ritz states that the same restriction is the basis of the 2nd law of thermodynamics]; see also Ritz, W. (1908). Recherches critiques sur l'Électrodynamique générale. (Critical Research on General Electrodynamics.) *Ann. Chem. et Phys*., 13, 145-275, loc. *cit.*

Tetrode, himself, like Ritz, was willing to assume elementary interactions which were not symmetric in time. However, *complete reversibility is assumed here because it is an essential element in a unified theory of action at a distance*. In proceeding on the basis of this symmetrical law of interaction, we shall be testing not only Tetrode's idea of absorber reaction, but also Einstein's view that the one-sidedness of the force of radiative reaction is a purely statistical phenomenon. This point leads to our final assumption:

(4) Sufficiently many particles are present to absorb completely the radiation given off by the source.

On the basis of these assumptions, we shall consider as the source of radiation an accelerated charge located in the absorbing system. A disturbance travels outward from the source. By it each particle of the absorber is set in motion and caused to generate a field, half-advanced and half-retarded. The sum of the advanced effects of all particles of the absorber, evaluated in the neighborhood of the source, gives a field which we find to have the following properties:

(1) It is independent of the properties of the absorbing medium.

(2) It is completely determined by the motion of the source.

(3) It exerts on the source a force which is finite, is simultaneous with the moment of acceleration, and is just sufficient in magnitude and direction to take away from the source the energy which later shows up in the surrounding particles.

(4) It is equal in magnitude to one-half the retarded field minus one-half the advanced field generated by the accelerated charge. In other words, the absorber is the physical origin of Dirac's radiation field.

(5) This field combines with the half-retarded, half-advanced field of the source to give for the total disturbance the full retarded field which accords with experience.

It will be sufficient to establish these results in order to have both in *field theory* and in the theory of *action at a distance* a solution of the problem of radiation, including an explanation of the force of radiative damping.

We shall present four derivations of the reaction of radiation on the source of successively increasing generality. In the *first* we consider an absorber in which the particles are far from one another. We assume without proof that the disturbance which passes through the medium is the full retarded field of experience. In the second derivation we examine the field of the absorber in the neighborhood of the source and find it just such as to compensate the advanced field of the accelerated charge and to give a retarded field of the previously assumed magnitude. In this case we have allowed the medium to have arbitrary density. The *third* derivation - in contrast to the first two, where the source was taken to be at rest or moving only slowly - considers the case of motion with arbitrary velocity and leads to the same relativistic expression which Dirac has given for the force of radiative reaction. All three treatments proceed by adding up the fields owing to the individual particles of the absorber. A *fourth* derivation uses a much more general approach, assuming only that the medium is a complete absorber.

The radiative reaction: derivation I

For a first analysis of the mechanism of radiative reaction, we shall simplify as much as possible the properties of the absorber:

(a) it is taken to be composed of free-charged particles;

(b) these corpuscles are at rest or are moving only slowly with respect to the particle which we treat as the source;

(c) the charged entities are well separated from one another;

(d) the particles occupy space to distances sufficiently great to bring about essentially complete absorption of radiation from the source.

We begin by considering the reaction set up between the source and a typical charge in the absorber when the particle of the source receives an acceleration by collision with a third particle or otherwise. The source has a charge +e and, therefore, sends out an electromagnetic disturbance. ...

...

... so long as the velocities of all particles remain non-relativistic. In this respect we have a quite general derivation of the law of radiative reaction generally accepted as correct for a slowly moving particle subjected to an arbitrary acceleration.

We conclude that *the force of radiative reaction arises, not from the direct action of a particle upon itself, but from the advanced action upon this charge caused by the future motion of the particles of the absorber*.

Radiative reaction: derivation II

...

Radiative reaction: derivation III

…

… This result establishes the connection between two different methods of evaluating the force of *radiative reaction, one based on the properties of the retarded field of the source at great distances, the other containing half the difference of retarded and advanced fields at the location of the source itself.*

Radiative reaction: derivation IV

From the preceding applications of the *absorber theory* of radiation, it has become clear that such properties of the absorber as refractive index and density have no bearing on the magnitude of the force of *radiative reaction.* The only essential point is that the medium should be a complete absorber. We therefore expect that there should somehow be a means to take this point into account in a very general way.

…

… The special property of a completely absorbing medium is expressed by the equation

$$\sum_k F^{(k)}{}_{ret} - F^{(k)}{}_{adv} = 0 \text{ (everywhere).} \qquad (37)$$

The consequences of Eq. (37) for the force on a typical particle are easily deduced. On the ath charge the entire field acting is given, according to the theory of *action at a distance*, by the sum

$$\sum_{k \neq a} (\tfrac{1}{2} F^{(k)}{}_{ret} + \tfrac{1}{2} F^{(k)}{}_{adv}) \qquad (38)$$

This expression can be broken down into three parts:

$$\sum_{k \neq a} F^{(k)}{}_{ret} + (\tfrac{1}{2} F^{(k)}{}_{ret} - \tfrac{1}{2} F^{(k)}{}_{adv}) - \tfrac{1}{2} \sum_{\text{all k}} (F^{(k)}{}_{ret} - F^{(k)}{}_{adv}) \qquad (39)$$

…

… In arriving at this equation, we have shown that the half-advanced, half-retarded fields of the theory of action at a distance lead to a satisfactory account of the mechanism of *radiative reaction* and to a description of the action of one particle on another in which no evidence of the advanced fields is apparent. *We find in the case of an absorbing universe a complete equivalence between the theory of Schwarzschild and Fokker on the one hand and the usual formalism of electrodynamics on the other*. This is what was to be proved.

The irreversibility of radiation

An oscillating charge surrounded by an absorbing medium loses energy. Why does radiation have this irreversible character even in a formulation of electrodynamics which is from the beginning symmetrical with respect to the interchange of past and future?

It might at first sight appear that the irreversibility is connected with the property of complete absorption. This is not the case. …

…

In this equation, however, the *force of radiative reaction* appears with a sign just opposite to its usual one. Evidently the explanation of the one-sidedness of radiation is not purely a matter of electrodynamics.

We have to conclude with Einstein[11] that the irreversibility of the emission process is a phenomenon of statistical mechanics connected with the asymmetry of the initial conditions with respect to time. In our example the particles of the absorber were either at rest or in random motion before the time at which the impulse was given to the source. … …

That it is solely the nature of the initial conditions which governs the direction of the radiation process can be seen by imagining a reversal of the direction of time in the preceding example. We have then a solution of the equations of motion just as consistent as the original solution. However, our interpretation of the solution is different. As the result of chaotic motion going on in the absorber, we see each one of the particles receiving at the proper moment just the right impulse to generate a disturbance which converges upon the source at the precise instant when it is accelerated. The source receives energy and the particles of the absorber are left with diminished velocity. No electrodynamic objection can be raised against this solution of the equations of motion. Small a priori probability of the given initial conditions provides our only basis on which to exclude such phenomena.

A comparison of radiation with heat conduction is illuminating. Both processes convert ordered into disordered motion although every elementary interaction involved is microscopically reversible.

Consider for the moment the question of the irreversibility of heat conduction, later to be put into relation with the problem of the one-sidedness of radiation. A portion of matter observed at the present moment to be warmer than its surroundings will cool off in the future with a probability overwhelmingly greater than the chance for it to grow hotter. About the past of the same portion of matter Boltzmann's H theorem however also predicts an enormously greater likelihood that the body warmed up to its present state rather than cooled clown to it. In other words, we are asked to understand the present temperature of the body as the result of a simple statistical fluctuation in the distribution of energy through the entire system. This deduction is based on the premise that the system was isolated before observation. However, common experience tells us that the given portion of matter probably acquired its abnormal temperature, not via an internal statistical fluctuation, but because it had earlier not been isolated from the outside.

For the radiative analogy of this example of heat conduction, conceive a charged particle bound to a position of equilibrium by a quasi-elastic force. Furthermore, suppose its energy at the moment of observation is large in comparison with the agitation of the surrounding absorber particles. There is then an overwhelming probability that the oscillator will lose energy to the absorber at a rate in close accord with the law of radiative damping. What can be said of the particle prior to the moment of acceleration? In an ideal absorbing system completely free of special disturbances, there is an equally over-whelming chance that the energy of the charge was then increasing at a rate given approximately by the inverse of the law of radiative damping. In this case as in heat conduction the abnormally high energy of the object is to be interpreted as the result of a statistical fluctuation. However, that the sun at some past age acquired its energy by such a fluctuation no one now would seriously propose. Obviously, the universe is a special system with respect to the origin of which probability considerations cannot freely be applied.

We conclude that radiation and radiative damping come under the head, not of pure electrodynamics, but of statistical mechanics. The conventional expression for the force of radiative reaction, like those for frictional resistance and viscous drag, represents a statistical average only. Application of this concept is not required in such an instance as the case of complete thermodynamical equilibrium, where the relative fluctuations of the actual forces about the conventional values are substantial. The concept of radiative damping is of real value only when we deal with the conversion of organized into disorganized energy, as in wireless transmission or light production.

Complete and incomplete absorption

In the picture of radiation which we have built on the foundation of Tetrode's suggestion, the absorber plays a role of hitherto unsuspected importance. On this account we should investigate not only how much the mechanism depends upon the completeness of the interception, but also the question what should be said of the absorption in the case of the actual universe.

In discussing the case of incomplete interception, we require a convenient means to take into account the initial conditions which so clearly control the irreversibility of the force of radiative reaction. For this reason, we shall break down the half-retarded, half-advanced fields of the theory of action at a distance into three parts … . With this decomposition of the field, we arrive at a description of the behavior of a system of particles which is entirely equivalent to the theory of action at a distance but which in the equation of motion,

…

conceals from view the existence of the advanced part of the fields of Schwarzschild and Fokker. …

The field which enters the third term in the equation of motion … vanished in the case of a completely absorbing system. Its appearance in the present case has led us to give it the name of "incident field," which Dirac applied to a quantity having an identical role in the *equation of motion*. However, on the origin of this field we go beyond Dirac's treatment in giving a prescription for its unique determination in terms of the movements of all the particles of the system. This prescription reveals that the field in question contains the *advanced* effects of the theory of *action at a distance*.

…

Generally, we may say that the explicit appearance of advanced effects is unavoidable in the case of a system which is an incomplete absorber. However, *in neither of the examples so far examined do advanced fields of the source produce explicit advanced effects on any other particle than a test charge*: in the first example, because there is no other charged particle; and in the second case, because the incident field is restricted to a region of space where there are no particles to be disturbed, except a possible test particle.

Advanced effects associated with incomplete absorption

Recognizable *advanced effects* appear for the first time in a third example, a source at the center of a cavity completely absorbing except for a single passage to charge-free space. ...

In this third example an equally consistent solution of the problem of advanced effects may be briefly outlined ... . The fields are in this case such that a test charge placed within the cavity experiences only the full retarded field of experience. The particles on the wall of the cavity are set into motion only after the. time when the source was struck and caused to radiate. These particles, by the now familiar mechanism of absorber response, generate fields, the advanced parts of which in the cavity cancel the advanced field of the source and bring its retarded field up to full strength. In particular, the advanced field of the source in the direction of the passage is compensated, so that an external test charge <?n that side of the absorber will experience no advanced disturbance. It will, however, on our picture undergo a full-strength retarded disturbance. How the half-retarded disturbance of the source in this direction is built up to full strength, and how the half-advanced field in the direction of the anti-passage is cancelled, is a question still to be cleared up. ...

...

Pre-acceleration

...

Pre-acceleration as witness to the interaction of past and future

...

Summary

Use of *action at a distance* with *field theory* as equivalent and complementary tools for the description of nature has so far been prevented by inability of the first point of view fully to account for the mechanism of radiation. Elucidation of this process in both theories comes from a 23-year-old suggestion of Tetrode, that the absorber may be an essential element of the act of emission. A quantitative formulation of this idea is given here on the basis of the following postulates: (1) An accelerated charge in otherwise charge-free space does not radiate energy. (2) The fields which act on a given particle arise only from other particles. (3) These fields are represented by one-half the retarded plus one-half the advanced Lienard-Wiechert solutions of Maxwell's equations.

In a system containing particles sufficient in number ultimately to absorb all radiation, the absorber reacts upon an accelerated charge with a field, the advanced part of which, evaluated in the neighborhood of the source on the basis of these postulates, is found to have the following properties: (1) It is independent of the properties of the absorbing medium. (2) It is completely determined by the motion of the source. (3) It exerts on the source a force which is finite, is simultaneous with the moment of acceleration, and is just sufficient in magnitude and direction to take away from the source the energy which the act of radiation imparts to the surrounding particles. (4) It is equal in magnitude to one-half the retarded field minus one-half the advanced field of the accelerated charge itself, just the field postulated by Dirac as the source of the force of radiative reaction. (5) This field compensates the half-advanced field of the source and combines with its half-retarded field to produce the full retarded disturbance which is required by experience. Radiation is concluded to be a phenomenon as much of statistical mechanics as of pure electrodynamics. A complete correspondence is established between *action at a distance* and the usual formulation of *field theory in the case of a completely absorbing system*. In such a system the phenomenon of pre-acceleration appears as the sole evidence of the *advanced effects* of the theory of *action at a distance*. Other *advanced effects* appear in the case of an incompletely absorbing system and are also discussed.

Lamb, Jr., W. E. & Retherford, R. C. (August, 1947). Fine Structure of the Hydrogen Atom by a Microwave Method.

[*Phys. Rev.*, 72, 3, 241-3* **; https://doi.org/ 10.1103/ PhysRev.72.241; also in Schwinger, J. (ed). (1958). *Selected Papers on Quantum electrodynamics*. Dover, New York, pages 136-8.]

Received June 18, 1947.

Columbia Radiation Laboratory, Department of Physics, Columbia University, New York, New York.

* Publication assisted by the Ernest Kempton Adams Fund for Physical Research of Columbia University, New York.

** Work supported by the Signal Corps under contract number W 36-039 sc-32003.

The spectrum of hydrogen has a fine structure of the energy levels which according to the *Dirac wave equation* for an electron moving in a Coulomb field is due to the combined effects of *relativistic* variation of mass with velocity and spin-orbit coupling, according to this theory the $2^2S_{1/2}$ state should exactly coincide in energy with the $2^2P_{1/2}$ state which is the lower of the two P states, previous attempts at measurement have alternated between finding confirmation and discrepancies of as much as eight percent, using microwave method depending on a novel property of the $2^2S_{1/2}$ level results indicate that contrary to *Dirac wave equation* the $2^2S_{1/2}$ state is higher than the $2^2P_{1/2}$ by about 1000 Mc/sec.

The spectrum of the simplest atom, hydrogen, has a fine structure[1] which according to the Dirac *wave equation* for an electron moving in a Coulomb field is due to the combined effects of *relativistic* variation of mass with velocity and spin-orbit coupling.

[1] For a convenient account, see White, H. E. (1934). *Introduction to Atomic Spectra*. McGraw-Hill Book Company, New York, Chap. 8.

It has been considered one of the great triumphs of Dirac's theory that it gave the "right" fine structure of the energy levels. However, the experimental attempts to obtain a really detailed confirmation through a study of the Balmer lines have been frustrated by the large Doppler effect of the lines in comparison to the small splitting of the lower or n =2 states. The various spectroscopic workers have alternated between finding confirmation[2] of the theory and discrepancies[3] of as much as eight percent.

[2] Drinkwater, J. W., Richardson, O. & Williams, W. E. (February, 1940). Determinations of the Rydberg constants, e/m, and the fine structures of H_α and D_α by means of a reflexion echelon. *Proc. Roy. Soc.*, 174, 957, 164-88; https://doi.org/10.1098/ rspa.1940.0013.

[3] Houston, W. V. (March, 1937). A New Method of Analysis of the Structure of H_α and D_α. *Phys. Rev.*, 51, 446; https://doi.org/10.1103/PhysRev.51.446; Williams, R. C. (December, 1938). The Fine Structures of H_α and D_α Under Varying Discharge Conditions. *Phys. Rev.* 54, 558; https://doi.org/10.1103/PhysRev.54.558. Pasternack, S. (December, 1938). Note on the Fine Structure of H_α and D_α. *Phys. Rev.*, 54, 12, 1113; https://doi.org/

10.1103/PhysRev.54.1113, has analyzed these results in terms of an upward shift of the S level by about 0.03 cm^{-1}.

More accurate information would clearly provide a delicate test of the form of the correct *relativistic* wave equation, as well as information on the possibility of line shifts due to coupling of the atom with the radiation field and clues to the nature of any non-Coulombic interaction between the elementary particles: electron and proton.

The calculated separation between the levels $2^2P_{1/2}$ and $2^2P_{3/2}$ is 0.365 cm^{-1} and corresponds to a wave-length of 2.74 cm. The great wartime advances in microwave techniques in the vicinity of three centimeters wave-length make possible the use of new physical tools for a study of the n = 2 fine structure states of the hydrogen atom. A little consideration shows that it would be exceedingly difficult to detect the direct absorption of radiofrequency radiation by excited H atoms in a gas discharge because of their small population and the high background absorption due to electrons. *Instead, we have found a method depending on a novel property of the $2^2S_{1/2}$ level. According to the Dirac theory, this state exactly coincides in energy with the $2^2P_{1/2}$ state which is the lower of the two P states.* The S state in the absence of external electric fields is metastable. The radiative transition to the ground state $1^2S_{1/2}$ is forbidden by the selection rule $\Delta L = \pm 1$. Calculations of Breit and Teller[4] have shown that the most probable decay mechanism is *double quantum emission* with a lifetime of 1/7 second.

[4] Breit, G. & Teller, E. (March, 1940). Metastability of Hydrogen and Helium Levels. *Astrophys.*, 91, 215; https://doi.org/10.1086/144158.

This is to be contrasted with a lifetime of only 1.6×10^{-9} second for the non-metastable 2^2P states. The metastability is very much reduced in the presence of external electric fields[5] owing to Stark effect mixing of the S and E levels with resultant rapid decay of the combined state.

[5] H. A. Bethe in *Handbuch der Physik*, 24/1, §43.

If for any reason, the $2^2S_{1/2}$ level does not exactly coincide with the $2^2P_{1/2}$ level, the vulnerability of the state to external fields will be reduced. Such a removal of the accidental degeneracy may arise from any defect in the theory or may be brought about by the Zeeman splitting of the levels in an external magnetic field.

In brief, the experimental arrangement used is the following: Molecular hydrogen is thermally dissociated in a tungsten oven, and a jet of atoms emerges from a slit to be cross-bombarded by an electron stream. About one part in a hundred million of the atoms is thereby excited to the metastable $2^2S_{1/2}$ state. The metastable atoms (with a small recoil deflection) move on out of the bombardment region and are detected by the process of electron ejection from a metal target. The electron current is measured with an FP-54 electrometer tube and a sensitive galvanometer.

If the beam of metastable atoms is subjected to any perturbing fields which cause a transition to any of the 2^2P states, the atoms will decay while moving through a very small distance. As a result, the beam current will decrease, since the detector does not respond to atoms in the ground state. Such a transition may be induced by the application to the beam of a static electric field somewhere between source and detector. Transitions may also be induced by radiofrequency radiation for which *hv* corresponds to the energy difference between one of the Zeeman components of $2^2S_{1/2}$ and any component of either $2^2P_{1/2}$ or $2^2P_{3/2}$. Such measurements provide a precise method for the location of the $2^2S_{1/2}$ state relative to the P states, as well as the distance between the latter states.

We have observed an electrometer current of the order of 10^{-14} ampere which must be ascribed to metastable hydrogen atoms. The strong quenching effect of static electric fields has been observed, and the voltage gradient necessary for this has a reasonable dependence on magnetic field strength.

We have also observed the decrease in the beam of metastable atoms caused by microwaves in the wave-length range 2.4 to 18.5 cm in various magnetic fields. In the measurements, the frequency of the r-f is fixed, and the change in the galvanometer current due to interruption of the r-f is determined as a function of magnetic field strength. A typical curve of quenching versus magnetic field is shown in Fig. 1.

See Fig 1 on page 242 of the article.

Fig. 1. A typical plot of galvanometer deflection due to interruption of the microwave radiation as a function of magnetic field. The magnetic field was calibrated with a flip coil and may be subject to some error which can be largely eliminated in a more refined apparatus. The width of the curves is probably due to the following causes. (1) the radiative line width of about 100 Mc/sec. of the ^2P states, (2) hyperfine splitting of the ^2S state which amounts to about 88 Mc/sec., (3) the use of an excessive intensity of radiation which gives increased absorption in the wings of the lines, and (4) inhomogeneity of the magnetic field. No transitions from the state $2^2S_{1/2}$ (m = – ½) have been observed, but atoms in this state may be quenched by stray electric fields because of the more nearly exact degeneracy with the Zeeman pattern of the ^2P states.

We have plotted in Fig. 2 the resonance magnetic fields for various frequencies in the vicinity of 10,000 Mc/sec.

See Fig 2 on page 242 of the article.

Fig. 2. Experimental values for resonance magnetic fields for various frequencies are shown by circles. The solid curves show three of the theoretically expected variations, and the broken curves are obtained by shifting these down by 1000 Mc/sec. This is done merely for the sake of comparison, and it is not implied that this would represent a "best fit". The plot covers only a small range of the frequency and magnetic field scale covered by our data, but a complete plot would not show up clearly on a small scale, and the shift indicated by the remainder of the data is quite compatible with a shift of 1000 Mc.

The theoretically calculated curves for the Zeeman effect are drawn as solid curves, while for comparison with the observed points, the calculated curves have been shifted downward by 1000 Mc/sec. (broken curves). *The results indicate clearly that, contrary to theory but in essential agreement with Pasternack's hypothesis*[3]*, the $2^2S_{1/2}$ state is higher than the $2^2P_{1/2}$ by about 1000 Mc/sec.* (0.033 cm^{-1} or about 9 percent of the *spin relativity doublet separation.*)

[[3] Pasternack, S. (December, 1938). Note on the Fine Structure of H_α and D_α.. *Loc. cit.* has analyzed these results in terms of an upward shift of the S level by about 0.03 cm^{-1}.]

The lower frequency transitions $2^2S_{1/2}$ (m = ½)→$2^2P_{1/2}$ (m = ± ½) have also been observed and agree well with such a shift of the $^2S_{1/2}$ level. With the present precision, we have not yet detected any discrepancy between the Dirac theory and the doublet separation of the P levels. (According to most of the imaginable theoretical explanations of the shift, the doublet separation would not be affected as much as the relative location of the S and P states.) With proposed refinements in sensitivity, magnetic field homogeneity, and calibration, it is hoped to locate the S level with respect to each P level to an accuracy of at least ten Mc/sec. By addition of these frequencies and assumption of the theoretical formula $\Delta v = 1/16\ \alpha^2 R$ for the doublet separation, it should be possible to measure the square of the fine structure constant times the Rydberg frequency to an accuracy of 0.1 percent.

By a slight extension of the method, it is hoped to determine the hyperfine structure of the $2^2S_{1/2}$ state. All of these measurements will be repeated for deuterium and other hydrogen-like atoms.

A paper giving a fuller account of the experimental and theoretical details of the method is being prepared, and this will contain later and more accurate data.

The experiments described here were discussed at the Conference on the Foundations of Quantum Mechanics held at Shelter Island on June 1-3, 1947 which was sponsored by the National Academy of Sciences.

Hans Albrecht Bethe (July 2, 1906 – March 6, 2005)

Bethe was a German-American nuclear physicist who made important contributions to astrophysics, quantum electrodynamics, and solid-state physics, and who won the 1967 Nobel Prize in Physics for his work on the theory of stellar nucleosynthesis. For most of his career, Bethe was a professor at Cornell University.

During World War II, he was head of the Theoretical Division at the secret Los Alamos laboratory that developed the first atomic bombs. There he played a key role in calculating the critical mass of the weapons and developing the theory behind the implosion method used in both the Trinity test and the "Fat Man" weapon dropped on Nagasaki in August 1945.

After the war, Bethe also played an important role in the development of the hydrogen bomb, although he had originally joined the project with the hope of proving it could not be made. Bethe later campaigned with Albert Einstein and the Emergency Committee of Atomic Scientists against nuclear testing and the nuclear arms race. He helped persuade the Kennedy and Nixon administrations to sign, respectively, the 1963 Partial Nuclear Test Ban Treaty and 1972 Anti-Ballistic Missile Treaty (SALT I).

His scientific research never ceased and he was publishing papers well into his nineties, making him one of the few scientists to have published at least one major paper in his field during every decade of his career, which in Bethe's case spanned nearly seventy years. Freeman Dyson, once his doctoral student, called him the "supreme problem-solver of the 20th century".

Bethe was born in Strasbourg, which was then part of the Reichsland Elsaß-Lothringen, Germany, on July 2, 1906, the only child of Anna and Albrecht Bethe, a privatdozent of physiology at the University of Strasbourg. Although his mother, the daughter of a professor at the University of Strasbourg, had a Jewish background, Bethe was raised Protestant, like his father; and he became an atheist later in life.

His father accepted a position as professor and director of the Institute of Physiology at the University of Kiel in 1912, and the family moved into the director's apartment at the Institute. Initially, he was schooled privately by a professional teacher as part of a group of eight girls and boys. The family moved again in 1915 when his father became the head of the new Institute of Physiology at the University of Frankfurt am Main.

Bethe attended the Goethe-Gymnasium in Frankfurt, Germany. His education was interrupted in 1916, when he contracted tuberculosis, and he was sent to Bad Kreuznach to recuperate. By 1917, he had recovered sufficiently to attend the local realschule and the following year, he was sent to the Odenwaldschule, a private, coeducational boarding school. He attended the Goethe-Gymnasium again for his final three years of secondary schooling, from 1922 to 1924.

Having passed his abitur, Bethe entered the University of Frankfurt in 1924. He decided to major in chemistry. The instruction in physics was poor, and while there were distinguished mathematicians in Frankfurt such as Carl Ludwig Siegel and Otto Szász, Bethe disliked their approaches, which presented mathematics without reference to the other sciences. Bethe found that he was a poor experimentalist who destroyed his lab coat by spilling sulfuric acid on it, but he found the advanced physics taught by the associate professor, Walter Gerlach, more interesting. Gerlach left in 1925 and was replaced by Karl Meissner, who advised Bethe that he should go to a university with a better school of theoretical physics, specifically the University of Munich, where he could study under Arnold Sommerfeld.

Bethe entered the University of Munich in April 1926, where Sommerfeld took him on as a student on Meissner's recommendation. Sommerfeld taught an advanced course on differential equations in physics, which Bethe enjoyed. Because he was such a renowned scholar, Sommerfeld frequently received advance copies of scientific papers, which he put up for discussion at weekly evening seminars. When Bethe arrived, Sommerfeld had just received Erwin Schrödinger's papers on wave mechanics.

For his PhD thesis, Sommerfeld suggested that Bethe examine electron diffraction in crystals. As a starting point, Sommerfeld suggested Paul Ewald's 1914 paper on X-ray diffraction in crystals. Bethe later recalled that he became too ambitious, and, in pursuit of greater accuracy, his calculations became unnecessarily complicated. When he met Wolfgang Pauli for the first time, Pauli told him: "After Sommerfeld's tales about you, I had expected much better from you than your thesis." "I guess from Pauli," Bethe later recalled, "that was a compliment."

After Bethe received his doctorate, Erwin Madelung offered him an assistantship in Frankfurt, and in September 1928 Bethe moved in with his father, who had recently divorced his mother. His father had met Vera Congehl earlier that year and married her in 1929. They had two children, Doris, born in 1933, and Klaus, born in 1934.

Bethe did not find the work in Frankfurt very stimulating, and in 1929 he accepted an offer from Ewald at the Technische Hochschule in Stuttgart. While there, he wrote what he considered to be his greatest paper, *Zur Theorie des Durchgangs schneller Korpuskularstrahlen durch Materie* (*"The Theory of the Passage of Fast Corpuscular Rays Through Matter"*). Starting from Max Born's interpretation of the Schrödinger equation, Bethe produced a simplified formula for collision problems using a Fourier transform, which is known today as the *Bethe formula*. He submitted this paper for his habilitation in 1930.

Sommerfeld recommended Bethe for a Rockefeller Foundation Travelling Scholarship in 1929. This provided $150 a month (about $2,000 in 2021 dollars) to study abroad. In 1930, Bethe chose to do postdoctoral work at the Cavendish Laboratory at the University of Cambridge in England, *where he worked under the supervision of Ralph Fowler*. At the request of Patrick Blackett, who was working with cloud chambers, Bethe created a *relativistic* version of the *Bethe formula*.

Bethe was known for his sense of humor, and with Guido Beck and Wolfgang Riezler, two other postdoctoral research fellows, created a hoax paper *On the Quantum Theory of the Temperature of Absolute Zero* where he calculated the fine structure constant from the absolute zero temperature in Celsius units. The paper poked fun at a certain class of papers in theoretical physics of the day, which were purely speculative and based on spurious numerical arguments, such as Arthur Eddington's attempts to explain the value of the fine structure constant from fundamental quantities in an earlier paper. They were forced to issue an apology.

For the second half of his scholarship, Bethe chose to go to Enrico Fermi's laboratory in Rome in February 1931. He was greatly impressed by Fermi and regretted that he had not gone to Rome first. Bethe developed the *Bethe ansatz*, a method for finding the exact solutions for the eigenvalues and eigenvectors of certain one-dimensional quantum many-body models. He was influenced by Fermi's simplicity and Sommerfeld's rigor in approaching problems and these qualities influenced his own later research.

The Rockefeller Foundation offered an extension of Bethe's fellowship, allowing him to return to Italy in 1932. In the meantime, Bethe worked for Sommerfeld in Munich as a privatdozent. Since Bethe was fluent in English, Sommerfeld had Bethe supervise all his English-speaking postdoctoral fellows. Bethe accepted a request from Karl Scheel to write an article for the *Handbuch der Physik* on the quantum mechanics of hydrogen and helium. Reviewing the article decades later, Robert Bacher and Victor Weisskopf noted that it was unusual in the depth and breadth of its treatment of the subject that required very little updating for the 1959 edition. Bethe was then asked by Sommerfeld to help him with the *Handbuch* article on electrons in metals. The article covered the basis of what is now called solid state physics. Bethe took a very new field and provided a clear, coherent, and complete coverage of it. His work on the *Handbuch* articles occupied most of his time in Rome, but *he also co-wrote a paper with Fermi on another new field, quantum electrodynamics*, describing the relativistic interactions of charged particles.

In 1932, Bethe accepted an appointment as an assistant professor at the University of Tübingen, where Hans Geiger was the professor of experimental physics. One of the first laws passed by the new National Socialist government was the Law for the Restoration of the Professional Civil Service. Due to his Jewish background, Bethe was dismissed from his job at the University, which was a government post. Geiger refused to help, but Sommerfeld immediately gave Bethe back his fellowship at Munich. Sommerfeld spent much of the summer term of 1933 finding places for Jewish students and colleagues.

Bethe left Germany in 1933, moving to England after receiving an offer for a position as lecturer at the University of Manchester for a year through Sommerfeld's connection to William Lawrence Bragg. He moved in with his friend Rudolf Peierls and Peierls' wife Genia. Peierls was a fellow German physicist who had also been barred from academic positions in Germany because he was Jewish. This meant that Bethe had someone to speak to in German and he did not have to eat English food. Their relationship was professional as well as personal. Peierls aroused Bethe's interest in nuclear physics. After James

Chadwick and Maurice Goldhaber discovered the photodisintegration of deuterium, Chadwick challenged Bethe and Peierls to come up with a theoretical explanation of this phenomenon. This they did on the four-hour train ride from Cambridge back to Manchester. Bethe would investigate further in the years ahead.

In 1933, the physics department at Cornell was looking for a new theoretical physicist, and Lloyd Smith strongly recommended Bethe. This was supported by Bragg, who was visiting Cornell at the time. In August 1934, Cornell offered Bethe a position as an acting assistant professor. Bethe had already accepted a fellowship for a year to work with Nevill Mott at the University of Bristol for a semester, but Cornell agreed to let him start in the spring of 1935. Before leaving for the United States, he visited the Niels Bohr Institute in Copenhagen in September 1934, where he proposed to Hilde Levi, who accepted. The match was opposed by Bethe's mother, who despite having a Jewish background, did not want him to marry a Jewish woman. A few days before their wedding date in December, Bethe broke off their engagement. Niels Bohr and James Franck were so shocked by this action by Bethe that he was not invited to the Institute again until after World War II.

Bethe arrived in the United States in February 1935, and joined the faculty at Cornell University on a salary of $3,000. Bethe's appointment was part of a deliberate effort on the part of the new head of its physics department, Roswell Clifton Gibbs, to move into nuclear physics. Gibbs had hired Stanley Livingston, who had worked with Ernest Lawrence, to build a cyclotron at Cornell. To complete the team, Cornell needed an experimentalist, and, on the advice of Bethe and Livingston, recruited Robert Bacher. Bethe received requests to visit Columbia University, Princeton University, University of Rochester, Purdue University, the University of Illinois at Urbana–Champaign, and Harvard University. Gibbs moved to prevent Bethe from being poached by having him appointed as a regular assistant professor in 1936, with an assurance that promotion to professor would soon follow.

Together with Bacher and Livingston, Bethe published a series of three articles, which summarized most of what was known on the subject of nuclear physics until that time, an account that became known informally as "Bethe's Bible".

> [Bethe, H. & Bacher, R (April, 1936). Nuclear Physics. A: Stationary States of Nuclei. *Rev. Mod. Phys.*, 8, 2, 82–229; https://doi.org/10.1103/RevModPhys.8.82; Bethe, H. (1937). Nuclear Physics. B: Nuclear Dynamics, Theoretical. *Rev. Mod. Phys.*, 9, 2, 69–244; https://doi.org/10.1103/ RevModPhys.9.69; Bethe, H.; Livingston, M. S. (1937). Nuclear Physics. C: Nuclear Dynamics, Experimental. *Rev. Mod. Phys.*, 9, 2, 245–390; https://doi.org/10.1103/ RevModPhys.9.245.]

It remained the standard work on the subject for many years. In this account, he also continued where others left off, filling in gaps in the older literature. Loomis offered Bethe a full professorship at the University of Illinois at Urbana–Champaign, but Cornell matched the position offered, and the salary of $6,000. He wrote to his mother:

> "I am about the leading theoretician in America. That does not mean the best. Wigner is certainly better and Oppenheimer and Teller probably just as good. But I do more and talk more and that counts too."

On March 17, 1938, Bethe attended the Carnegie Institute and George Washington University's fourth annual Washington Conference of Theoretical Physics. Bethe initially declined the invitation to attend, because the conference's topic, stellar energy generation, did not interest him, but Edward Teller persuaded him to go. At the conference, Bengt Strömgren detailed what was known about the temperature, density, and chemical composition of the Sun, and challenged the physicists to come up with an explanation. George Gamow and Carl Friedrich von Weizsäcker had proposed in a 1937 paper that the Sun's energy was the result of a proton–proton chain reaction, but this did not account for the observation of elements heavier than helium. By the end of the conference, Bethe, working in collaboration with Charles Critchfield, had come up with a series of subsequent nuclear reactions that explained how the Sun shines. That this did not explain the processes in heavier stars was not overlooked. At the time there were doubts about whether the proton–proton cycle described the processes in the Sun, but more recent measurements of the Sun's core temperature and luminosity show that it does. When he returned to Cornell, Bethe studied the relevant nuclear reactions and reaction cross sections, leading to his discovery of the carbon-nitrogen-oxygen cycle (CNO cycle). The two papers, one on the proton–proton cycle, co-authored with Critchfield, and the other on the carbon-oxygen-nitrogen (CNO) cycle, were sent to the *Physical Review* for publication.

After Kristallnacht, Bethe's mother had become afraid to remain in Germany. Taking advantage of her Strasbourg origin, she was able to emigrate to the United States in June 1939 on the French quota, rather than the German one, which was full. Bethe's graduate student Robert Marshak noted that the New York Academy of Sciences was offering a $500 prize for the best unpublished paper on the topic of solar and stellar energy. So, Bethe, in need of $250 to release his mother's furniture, withdrew the CNO cycle paper and sent it in to the New York Academy of Sciences. It won the prize, and Bethe gave Marshak $50 finder's fee and used $250 to release his mother's furniture. The paper was subsequently published in the *Physical Review* in March. It was a breakthrough in the understanding of the stars, and would win Bethe the Nobel Prize in Physics in 1967.

Bethe married Rose Ewald, the daughter of Paul Ewald, on September 13, 1939, in a simple civil ceremony. She had emigrated to the United States and was a student at Duke University and they met while Bethe was lecturing there in 1937. They had two children, Henry and Monica.

Bethe became a naturalized citizen of the United States in March 1941. Writing to Sommerfeld in 1947, Bethe confided that "I am much more at home in America than I ever was in Germany. As if I was born in Germany only by mistake, and only came to my true homeland at 28."

When the Second World War began, Bethe wanted to contribute to the war effort, but was unable to work on classified projects until he became a citizen. Following the advice of the Caltech aerodynamicist Theodore von Kármán, Bethe collaborated with his friend Edward Teller on a theory of shock waves that are generated by the passage of a projectile through a gas. Bethe considered it one of their most influential papers. He also worked on a theory of armor penetration, which was immediately classified by the army, thus making it impossible for Bethe (who was not an American citizen at the time) to access further research on the theory.

After receiving security clearance in December 1941, Bethe joined the MIT Radiation Laboratory, where he invented the Bethe-hole directional coupler, which is used in microwave waveguides such as those used in radar sets. In Chicago in June 1942, and then in July at the University of California, Berkeley, he participated in a series of meetings at the invitation of Robert Oppenheimer, which discussed the first designs for the atomic bomb. They went over the preliminary calculations by Robert Serber, Stan Frankel, and others, and discussed the possibilities of using uranium-235 and plutonium. Teller then raised the prospect of a thermonuclear device, Teller's "Super" bomb. At one point Teller asked if the nitrogen in the atmosphere could be set alight. It fell to Bethe and Emil Konopinski to perform the calculations demonstrating the virtual impossibility of such an occurrence. "The fission bomb had to be done," he later recalled, "because the Germans were presumably doing it."

When Oppenheimer was put in charge of forming a secret weapons design laboratory, Los Alamos, he appointed Bethe director of the T (Theoretical) Division, the laboratory's smallest, but most prestigious division. This move irked the equally qualified, but more difficult to manage Teller and Felix Bloch, who had coveted the job. A series of disagreements between Bethe and Teller between February and June 1944 over the relative priority of Super research led to Teller's group being removed from T Division and placed directly under Oppenheimer. In September it became part of Fermi's new F Division.

Bethe's work at Los Alamos included calculating the critical mass and efficiency of uranium-235 and the multiplication of nuclear fission in an exploding atomic bomb. Along with Richard Feynman, he developed a formula for calculating the bomb's explosive yield. After August 1944, when the laboratory was reorganized and reoriented to solve the problem of the implosion of the plutonium bomb, Bethe spent much of his time studying the hydrodynamic aspects of implosion, a job that he continued into 1944. In 1945, he worked on the neutron initiator, and later, on radiation propagation from an exploding atomic bomb. The Trinity nuclear test validated the accuracy of T Division's results. When it was detonated in the New Mexico desert on July 16, 1945, Bethe's immediate concern was for its efficient operation, and not its moral implications. He is reported to have commented: "I am not a philosopher."

After the war, Bethe argued that a crash project for the hydrogen bomb should not be attempted, although after President Harry Truman announced the beginning of such a project and the outbreak of the Korean War, Bethe signed up and played a key role in the weapon's development. Although he saw the project through to its end, Bethe had hoped that it would be impossible to create the hydrogen bomb. He later remarked in 1968 on the apparent contradiction in his stance, having first opposed the development of the weapon and later helping to create it:

> "Just a few months before, the Korean war had broken out, and for the first time I saw direct confrontation with the communists. It was too disturbing. The cold war looked as if it were about to get hot. I knew then I had to reverse my earlier position. If I did not work on the bomb, somebody else would—and I had thought if I were around Los Alamos, I might still be a force for disarmament. So, I agreed to join in developing the H-bomb. It seemed quite logical. But sometimes I wish I were a more consistent idealist."

As for his own role in the project and its relation to the dispute over who was responsible for the design, Bethe later said that:

> "After the H-bomb was made, reporters started to call Teller the father of the H-bomb. For the sake of history, I think it is more precise to say that Ulam is the father, because he provided the seed, and Teller is the mother, because he remained with the child. As for me, I guess I am the midwife."

In 1954, Bethe testified on behalf of J. Robert Oppenheimer during the Oppenheimer security hearing. Specifically, Bethe argued that Oppenheimer's stances against developing the hydrogen bomb in the late 1940s had not hindered its development, a topic which was seen as a key motivating factor behind the hearing. Bethe contended that the developments that led to the successful Teller–Ulam design were a matter of serendipity and not a question of manpower or logical development of previously existing ideas. During the hearing, Bethe and his wife also tried hard to persuade Edward Teller against testifying. However, Teller did not agree, and his testimony played a major role in the revocation of Oppenheimer's security clearance. While Bethe and Teller had been on very good terms during the prewar years, the conflict between them during the Manhattan Project, and especially during the Oppenheimer episode, permanently marred their relationship.

After the war ended, Bethe returned to Cornell. In June 1947, he participated in the Shelter Island Conference. Sponsored by the National Academy of Sciences and held at the Ram's Head Inn on Shelter Island, New York, the conference on the "Foundations of Quantum Mechanics" was the first major physics conference held after the war. It was a chance for American physicists to come together, pick up where they had left off before the war, and establish the direction of post-war research.

A major talking point at the conference was the discovery by Willis Lamb and his graduate student, Robert Retherford, shortly before the conference began, that one of the two possible quantum states of hydrogen atoms had slightly more energy than that predicted by the theory of Paul Dirac; this became known as the *Lamb shift*. Oppenheimer and Weisskopf suggested that this was a result of quantum fluctuations of the electromagnetic field, which gave the electron more energy. *According to pre-war quantum electrodynamics, the energy of the electron consisted of the bare energy it had when uncoupled from an electromagnetic field, and the self-energy resulting from the electromagnetic coupling*, but both were unobservable, since the electromagnetic field cannot be switched off. *Quantum electrodynamics gave infinite values for the self-energies*; but the Lamb shift showed that they were both real and finite. Hans Kramers proposed *renormalization* as a solution, but no one knew how to do the calculation.

Bethe managed to perform the calculation on the train from New York to Schenectady, where he was working for General Electric. *He did so by realizing that it was a non-relativistic process, which greatly simplified the calculation. The bare energy was easily removed as it was already included in the observed mass of the electron. The self-energy term now increased logarithmically instead of linearly, making it mathematically convergent.* Bethe arrived at a value for the Lamb shift of 1040 MHz, extremely close to that obtained experimentally by Lamb and Retherford. His paper, published in the *Physical Review* in August 1947, was only three pages long and contained just twelve mathematical equations, but was enormously influential. [Bethe, H. A. (August, 1947). The Electromagnetic Shift of Energy Levels. *Phys. Rev*., 72, 339-41.] It had been presumed that the infinities indicated that *quantum electrodynamics* was fundamentally flawed, and that a new, radical theory was required; Bethe demonstrated that this was not necessary.

Bethe believed that the atomic nucleus was like a quantum liquid drop. He investigated the nuclear matter problem by considering the work conducted by Keith Brueckner on perturbation theory. Working with Baird Brandow and Albert Petschek, he came up with an approximation that converted the scattering equation into an easily solved differential equation. He then used these techniques to examine the neutron stars, which have densities similar to those of nuclei. Bethe continued to do research on supernovae, neutron stars, black holes, and other problems in theoretical astrophysics into his late nineties. At age 85, Bethe wrote an important article about the solar neutrino problem, in which he helped establish the conversion mechanism for electron neutrinos into muon neutrinos.

Bethe received numerous honors and awards in his lifetime and afterward. He was awarded the Max Planck Medal in 1955, the Franklin Medal in 1959, the Royal Astronomical Society Eddington Medal and the United States Atomic Energy Commission Enrico Fermi Award in 1961, the Rumford Prize in 1963, and the Nobel Prize in Physics in 1967. Bethe was elected Foreign Member of the Royal Society in 1957, and he gave the 1993 Bakerian Lecture at the Royal Society on the Mechanism of Supernovae.

Bethe loved the outdoors and was an enthusiastic hiker all his life, exploring the Alps and the Rockies. He died in his home in Ithaca, New York, on March 6, 2005, at age 98, of congestive heart failure. He was survived by his wife, and their two children.

Bethe, H. A. (August, 1947). The Electromagnetic Shift of Energy Levels.

[*Phys. Rev.*, 72, 339-41; https://doi.org/10.1103/PhysRev.72.339; also in Schwinger, J. (ed). (1958). *Selected Papers on Quantum electrodynamics*. Dover, New York, pages 139-41.]

Received 27 June 1947.

Cornell University, Ithaca, New York.

Lamb and Retherford results show that fine structure of second quantum state of hydrogen does not agree with *Dirac wave equation*, Schwinger, Weisskopf, and Oppenheimer suggest might be due to *shift of energy levels by interaction of electron with the radiation field*, this shift comes out infinite in all existing theories and has therefore always been ignored, possible to identify the most strongly (linearly) divergent term in the level shift with an *electromagnetic mass effect* which must exist for a bound as well as for a free electron, already included in the *observed mass* of the electron so should be subtracted, assumes *relativistic cut-off* in quantum energies (frequencies) of included atomic states, then calculation of Lamb shift for hydrogen atom using *non-relativistic* ordinary radiation theory gives shift of the levels due to *radiation interaction* in close agreement with observed value, removes discrepancy with Dirac theory, did not carried out *relativistic* calculations.

By very beautiful experiments, Lamb and Retherford[1] have shown that *the fine structure of the second quantum state of hydrogen does not agree with the prediction of the Dirac theory.*

[1] Lamb, Jr., W. E. & Retherford, R. C. (August, 1947). Fine Structure of the Hydrogen Atom by a Microwave Method. *Phys. Rev.*, 72, 3, 241-3.

The 2s level, which according to Dirac's theory should coincide with the $2p_{1/2}$ level, is actually higher than the latter by an amount of about 0.033 cm^{-1} or 1,000 megacycles. This discrepancy had long been suspected from spectroscopic measurements[2,3].

[2] Houston, W. V. (March, 1937). A New Method of Analysis of the Structure of H_α and D_α. *Phys. Rev.* 51, 446; https://doi.org/10.1103/PhysRev.51.446.
[3] Williams, R. C. (October, 1938). The Fine Structures of H_α and D_α Under Varying Discharge Conditions. *Phys. Rev.* 54, 558; https://doi.org/10.1103/PhysRev.54.558.

However, so far, no satisfactory theoretical explanation has been given. Kemble and Present, and Pasternack[4] have shown that the shift of the 2s level cannot be explained by a nuclear interaction of reasonable magnitude, and Uehling[5] has investigated the effect of the "*polarization of the vacuum*" in the Dirac hole theory, and has found that *this effect also is much too small and has, in addition, the wrong sign.*

[4] Kemble. E. C. & Present, R. D. (December, 1932). On the Breakdown of the Coulomb Law for the Hydrogen Atom. *Phys. Rev.*, 44, 1031; https://doi.org/10.1103/PhysRev.44.1031.2; Pasternack, S. (December, 1938). Note on the Fine Structure of H_α and D_α. *Phys. Rev.*, 54, 12, 1113; https://doi.org/ 10.1103/PhysRev.54.1113.

[5] Uehling, E. A. (July, 1935). Polarization Effects in the Positron Theory. *Phys. Rev.*, 48, 55; https://doi.org/10.1103/PhysRev.48.55.

Schwinger and Weisskopf, and Oppenheimer have suggested that a *possible explanation might be the shift of energy levels by the interaction of the electron with the radiation field.* This shift comes out infinite in all existing theories, and has therefore always been ignored. However, it is possible to identify the most strongly (linearly) divergent term in the level shift with an *electromagnetic mass effect* which must exist for a bound as well as for a free electron.

[In Dirac's August 1938 paper, Classical Theory of Radiating Electrons, describing a relativistic form of the classical theory of radiating electrons assuming electron to be a *point charge* with no volume, Dirac noted that an extended electron was inconceivable in the *special theory of relativity* due to the intrinsic connection between space and time, and that the Lorentz *model of electron as possessing mass on account of the electromagnetic field around it fails* without further assumptions if the electromagnetic field varies too rapidly or if the acceleration of the electrons is too great. The discovery of the neutron introduced a form of mass that was difficult to believe to be of electromagnetic nature. Also, theory of the positron in which positive and negative values of an electron play symmetrical roles cannot be fitted into idea of electromagnetic idea of mass, which requires all mass to be positive. He concluded that this required a *departure from the electromagnetic theory of mass.*]

This effect should properly be regarded as already included in the observed mass of the electron, and we must therefore subtract from the theoretical expression, the corresponding expression for a free electron of the same average kinetic energy. *The result then diverges only logarithmically (instead of linearly) in non-relativistic theory*: Accordingly, it may be expected that in the *hole theory*, in which the main term (*self-energy* of the electron) diverges only logarithmically, *the result will be convergent after subtraction of the free electron expression*[6].

[6] It was first suggested by Schwinger and Weisskopf that hole theory must be used to obtain convergence in this problem.

This would set an effective upper limit of the order of mc^2 *to the frequencies of light which effectively contribute to the shift of the level of a bound electron. I have not carried out the relativistic calculations*, but I shall assume that such an effective *relativistic* limit exists.

The ordinary radiation theory gives the following result for the self-energy of an electron in a quantum state m, due to its interaction with transverse electromagnetic waves:

$$W = -(2e^2/3\pi hc^3) \times \int_0^K k\, dk \sum_n |v_{mn}|^2/(E_n - E_m + k), \qquad (1)$$

where $k = h\omega$ is the energy of the quantum and **v** is the velocity of the electron which, in *non-relativistic theory*, is given by

$$\mathbf{v} = \mathbf{p}/m = (h/im)\nabla \qquad (2)$$

[where ∇ is the *nabla operator* = $(\partial/\partial x_1, \partial/\partial x_2, \ldots \partial/\partial x_n)$ or

$p_r = - ih\, \partial/\partial x_r, \qquad r = 1, 2, 3.$]

*Relativistically, **v** should be replaced by c**α** where **α** is the Dirac operator. Retardation* has been neglected and can actually be shown to make no substantial difference. The sum in (1) goes over all atomic states n, the integral over all quantum energies k up to some maximum K to be discussed later.

For a *free electron*, **v** has only diagonal elements and (1) is replaced by

$$W_0 = - (2e^2/3\pi hc^3) \times \int k\, dk\, \mathbf{v}^2/k. \qquad (3)$$

This expression represents the change of the kinetic energy of the electron for fixed momentum, due to the fact that electromagnetic mass is added to the mass of the electron. This *electromagnetic mass* is already contained in the experimental *electron mass*; the contribution (3) to the energy should therefore be disregarded. For a *bound electron*, $\mathbf{v}^2$ should be replaced by its expectation value, $(\mathbf{v}^2)_{mm}$. But the matrix elements of **v** satisfy the sum rule

$$\sum_n |\, v_{mn} \,|^2 = (\mathbf{v}^2)_{mm}. \qquad (4)$$

Therefore, the relevant part of the *self-energy* becomes

$$W' = W - W_0 = + (2e^2/3\pi hc^3) \times \int_0^K dk \sum_n |\, v_{mn} \,|^2 (E_n - E_m)/(E_n - E_m + k). \qquad (5)$$

This we shall consider as a true shift of the levels due to *radiation interaction.*

It is convenient to integrate (5) first over k. Assuming X to be large compared with all energy differences $E_n - E_m$ in the atom,

$$W' = (2e^2/3\pi hc^3) \sum_n |\, v_{mn} \,|^2 (E_n - E_m) \ln K/|\, E_n - E_m \,|. \qquad (6)$$

(If $E_n - E_m$ is negative, it is easily seen that the principal value of the integral must be taken, as was done in (6).) *Since we expect that relativity theory will provide a natural cut-off for the frequency k, we shall assume that in (6)*

$$K = mc^2. \qquad (7)$$

(This does not imply the same limit in Eqs. (2) and (3).) The argument in the logarithm in (6) is therefore very large; accordingly, it seems permissible to consider the logarithm as constant (independent of n) in first approximation. We therefore should calculate

$$A = \sum_n A_{nm} = \sum_n |\, \mathbf{p}_{nm} \,|^2 (E_n - E_m). \qquad (8)$$

This sum is well known; it is

$$A = \sum_n |\, \mathbf{p}_{nm} \,|^2 (E_n - E_m) = - h^2 \int \psi_m^* \nabla V.\, \nabla \psi_m\, d\tau$$
$$= \tfrac{1}{2} h^2 \int \nabla^2 V \psi_m^2\, d\tau = 2\pi h^2 e^2 Z \psi_m^2(0), \qquad (9)$$

for a nuclear charge Z. For any electron with angular momentum $l \neq 0$, the *wave function* vanishes at the nucleus; therefore, the sum A = 0. For example, for the 2p level the negative contribution $A_{1S,2P}$ balances the positive contributions from all other transitions. For a state with $l = 0$, however,

$$\psi_m^2(0) = (Z/na)^3/\pi, \qquad (10)$$

where n is the *principal quantum number* and a is the *Bohr radius.*

Inserting (10) and (9) into (6) and using relations between atomic constants, we get for an S state

$$W_{ns}' = 8/3\pi\ (e^2/hc)^3\ Ry\ Z^4/n_3 \ln K/[E_n - E_m]_{Av}, \qquad (11)$$

where Ry is the *ionization energy* of the *ground state* of hydrogen. The shift for the 2p state is negligible; the logarithm in (11) is replaced by a value of about – 0.04. The *average excitation energy* $[E_n - E_m]_{Av}$ for the 2s state of hydrogen has been calculated numerically[7] and found to be 17.8 Ry, *an amazingly high value.*

[7] I am indebted to Dr. Stehn and Miss Steward for the numerical calculations.

Using this figure and $K = mc^2$, the logarithm has the value 7.63, and we find

$$W_{ns}' = 136 \ln \{K/[E_n - E_m]\} = 1040 \text{ megacycles.} \qquad (12)$$

This is in excellent agreement with the observed value of 1000 megacycles.

A relativistic calculation to establish the limit K is in progress. Even without exact knowledge of E, however, the agreement is sufficiently good to give confidence in the basic theory. This shows

(1) that the level shift due to *interaction with radiation* is a real effect and is of finite magnitude,

(2) that the effect of the *infinite electromagnetic mass of a point electron* can be eliminated by proper identification of terms in the Dirac radiation theory,

(3) *that an accurate experimental and theoretical investigation of the level shift may establish relativistic effects (e.g., Dirac hole theory).* These effects will be of the order of unity in comparison with the logarithm in Eq. (11).

If the present theory is correct, the level shift should increase roughly as Z^4 but not quite so rapidly, because of the variation of $[E_n - E_m]_{Av}$ in the logarithm. For example, for He^+, the shift of the 2s level should be about 13 times its value for hydrogen, giving 0.43 cm^{-1}, and that of the 3.s level about 0.1.3 cm^{-1}. For the x-ray levels L*I* and L*II*, this effect should be superposed upon the effect of screening which it partly compensates. An accurate theoretical calculation of the screening is being undertaken to establish this point.

This paper grew out of extensive discussions at the Theoretical Physics Conference on Shelter Island, June 2 to 4, 1947. The author wishes to express his appreciation to the National Academy of Science which sponsored this stimulating conference.

Tati, T. & Tomonaga, S. (December, 1948). A Self-Consistent Subtraction Method in the Quantum Field Theory, I.

[*Prog. Theor. Phys.*, 3, 4, 391-406; https://doi.org/10.1143/ptp/3.4.391; based on November 1947 lecture at the symposium on the theory of elementary particles in Kyoto.]

Received May 23, 1948.

Notes procedure previously used by Bloch and Nordsieck and then by Pauli and Fierz in the treatment of problems with the self-field of an electron by first separating the radiation field into part bound to the electron and part of unbound photons by means of a canonical transformation, they then obtained a term in the Hamiltonian that could be interpreted as the *interaction energy* of electron with the radiation field bound to it, this term though infinite gave rise to a modification of mass of electron so that it could be amalgamated into the mass term in the Hamiltonian for the free electron, the term was dropped off to obtain *observed mass* when electron mass reinterpreted as already including it, same result as Bethe for level-shift of bound electron in *non-relativistic* treatment, Tati and Tomonaga considered it desirable to obtain *relativistic* generalization of this treatment using a canonical transformation, paper addresses *field reaction problem when no external field* to e^2 approximation, does not address infinity related to *vacuum polarization* effect that occurs in *relativistic* treatment, *omitted as cannot be amalgamated into equation for free radiation*, starts from *relativistic* formalism of quantum field theory proposed by Tomanaga [August, 1946. On a Relativistically Invariant Formulation of the Quantum Theory of Wave Fields] in which Schrodinger equation with Hamiltonian for interaction density between radiation and electron fields is $\{H(P) - i\, \partial/\partial C_P\}\, \Psi[C] = 0$ where H(P) is the *interaction energy density* between radiation and electron fields at the world point P, Ψ[C] is the generalized Schrodinger functional which is a functional of the space-like variable surface C in the four-dimensional world and $\partial/\partial C_P$ its partial functional differentiation at the point P, the point P being considered as lying on C, decomposes fields into parts oscillating with positive and negative frequencies corresponding to the electron and positron, transforms Schrodinger functional retaining terms up to the order e^2, expands integrand into Fourier integral and defines 4-vector to obtain Hamiltonian for transformed equation that enables use of first order calculation, allows processes through intermediate states such as emission and reabsorption of virtual particles (self-energy of an electron) to be treated as direct processes, calculates the commutator and rearranges factors in each term into the correct order, identifies (1) terms to represent *interactions between electrons* caused by exchange of virtual photons including ordinary M¢ller interaction between two electrons (2) terms connected with *self-energy* of an electron (3) terms responsible for the *scattering of a photon* by a free electron, terms for *creation of a pair* by two photons (4) variables describing the radiation field only representing modification of the radiation field in vacuo due to the *polarization of the vacuum* that cause an infinite energy level shift of the vacuum itself and an infinite *self-energy* of a photon, and (5) a *mass-modifying term* with a logarithmic diverging quantity representing the electromagnetic mass of an electron, assumes *mass-modifying term* already included in free field equation so no level-shift caused by the interaction between the electron and radiation, subtraction method "*self-consistent*" in this sense, gives *relativistic* generalization of transformation that separates *radiation field* into field of "unbound" photons and field of photons bound to the electron, procedure similar to Hartree method of the *self-consistent* field in which *interaction* between electrons is considered as a perturbation but some part of its effect is already included.

§ 1. *Introduction*

As is well known the present formalism of the *quantum field theory* contains a fundamental defect which reveals itself most directly in the infinite *self-energy* of elementary particles. The infinities of the same nature appear also when one deals with collision problems involving *field* reactions, e. g., the *radiative correction to the cross-section for the elastic scattering of an electron in an external field of force*. As was discussed by Pauli and Fierz[1] in the *non-relativistic* case and by Dancoff [2] in the *relativistic*, the correction due to the *radiation reaction* in the latter problem turned out infinite.

[1] Pauli, W. & Fierz, M. (March, 1938). Zur Theorie der Emission langwelliger Lichtquanten. (On the theory of the emission of long-wave light quanta.) *Nuovo Cimento*, 15, 3, 167-88; the usual radiation theory provides an infinitely large value for the effective cross-section dq of a charged particle when passing through a force field when deflected to a given angular range ("*ultrared catastrophe*"). If one prescribes that the energy loss of the particle should lie between E and E + dE, then for small E according to this theory dq = const. dE/E, which diverges logarithmically at the point E = 0 when integrated. The present study investigates in more detail what quantum electrodynamics yields for this cross-section when a finite expansion is attributed to the charged body. It turns out that then infinity is always eliminated and that the distractions considered radiationless in the ordinary theory appear here as those with finite, albeit very small energy loss. On the other hand, according to the exact theory, the closer course of dq for very small energy losses E depends so much on the expansion of the charged body that a direct application of the result to real electrons is possible. *It must therefore be concluded that the problem in question is essentially linked to the fundamental difficulties of quantum electrodynamics which have not yet been resolved*].

[2] Dancoff, S. M. (May, 1939). On Radiative Corrections for Electron Scattering. *Phys. Rev*., 55, 959[; a *relativistic* treatment of the radiative correction of order $e^2/\hbar c$ to the elastic scattering cross section leads to the following results: (a) for the scattering in an electrostatic field of a particle described by the Pauli-Weisskopf theory, the correction is finite; (b) for the scattering of a Dirac electron in an electrostatic field, the correction diverges logarithmically and is positive; (c) the convergence or divergence of the correction depends critically on the type of scattering potential considered].

Such a circumstance implies that the satisfactory solution of this problem would be only reached when the fundamental difficulty of the current *quantum field theory* has found its ultimate solution.

In such a situation one used to resort to some procedure to get rid of the difficulty, such as *cutting off of high frequency effects* or *subtracting infinite terms* by a suitable prescription. But it goes without saying that such procedures are only makeshifts far from true solution having neither theoretical basis nor any connection with experimental facts.

A remarkable progress was brought about, however, when Lamb and Retherford[3] confirmed the level-shift of the hydrogen atom by their ingenious experiment

[3] Lamb, Jr., W. E. & Retherford, R. C. (August, 1947). Fine Structure of the Hydrogen Atom by a Microwave Method. *Phys. Rev.*, 72, 3, 241-3[; the spectrum of hydrogen has a fine structure of the energy levels which according to the *Dirac wave equation* for an electron moving in a Coulomb field is due to the combined effects of *relativistic* variation of mass with velocity and spin-orbit coupling, according to this theory the $2^2S_{1/2}$ state should exactly coincide in energy with the $2^2P_{1/2}$ state which is the lower of the two P states, previous attempts at measurement have alternated between finding confirmation and discrepancies of as much as eight percent, using microwave method depending on a novel property of the $2^2S_{1/2}$ level results indicate that contrary to the *Dirac wave equation* the $2^2S_{1/2}$ state is higher than the $2^2P_{1/2}$ by about 1000 Mc/sec].

and Schwinger, Weisskopf, Oppenheimer and Bethe gave its theoretical interpretation in terms of the *radiative reaction*. Especially Bethe[4] proposed a method how to manage the infinity in this problem and enabled us to treat the *field-reaction problem* for the first time in close connection with reliable experimental data.

[4] Bethe, H. A. (August, 1947). The Electromagnetic Shift of Energy Levels. *Phys. Rev.*, 72, 339-41[; Lamb and Retherford results show that fine structure of second quantum state of hydrogen does not agree with *Dirac wave equation*, Schwinger, Weisskopf, and Oppenheimer suggest might be due to *shift of energy levels by interaction of electron with the radiation field*, this shift comes out infinite in all existing theories and has therefore always been ignored, possible to identify the most strongly (linearly) divergent term in the level shift with an *electromagnetic mass effect* which must exist for a bound as well as for a free electron, already included in the *observed mass* of the electron so should be subtracted, assumes *relativistic cut-off* in quantum energies (frequencies) of included atomic states, then calculation of Lamb shift for hydrogen atom using *non-relativistic* ordinary radiation theory gives shift of the levels due to *radiation interaction* in close agreement with observed value, removes discrepancy with Dirac theory, did not carried out *relativistic* calculations].

The fundamental assumption of Bethe's theory lies in the following hypothesis: *the self-energy of a free electron, though infinite in the current formalism of the quantum electrodynamics, is already included in its mass which we really observe*; when, however, the electron is in a bound state, a finite deviation of the self-energy from that in the free state will appear. This deviation is then observed just as the level-shift of the bound electron in question.

Although this hypothesis was in this manner for the first time formulated explicitly and clearly in the light of the experimental evidence, implicitly such an idea had been used in the earlier considerations of Bloch and Nordsieck[5] as well as Pauli and Fierz [*loc. cit.*] about the *self-field* of an electron.

[5] Bloch, F. & Nordsieck, A. (July, 1937). Note on the Radiation Field of the Electron. *Phys. Rev.*, 52, 54[; previous methods of treating radiative corrections in non-stationary processes such as the scattering of an electron in an atomic field or the emission of a β-ray, by an expansion in powers of $e^2/\hbar c$, are defective in that they predict infinite low frequency corrections to the transition probabilities, difficulty can be avoided by a method developed here which is based on the alternative assumption that $e^2\omega/mc^3$, $\hbar\omega/mc^2$ and

$\hbar\omega/c\Delta p$ (ω = angular frequency of radiation, Δp = change in momentum of electron) are small compared to unity, in contrast to the expansion in powers of $e^2/\hbar c$ this permits the transition to the classical limit $\hbar$=0, external perturbations on the electron are treated in the Born approximation, it is shown that for frequencies such that the above three parameters are negligible the quantum mechanical calculation yields just the directly reinterpreted results of the classical formulae, namely that the total probability of a given change in the motion of the electron is unaffected by the interaction with radiation, and that the mean number of emitted quanta is infinite in such a way that the mean radiated energy is equal to the energy radiated classically in the corresponding trajectory].

These authors separate first the radiation field into the part bound to the electron and the part of unbound photons by means of a canonical transformation. Then they obtain a term in the Hamiltonian which can be interpreted as the *interaction energy* of the electron with the *radiation field* bound to it. This term, though infinite, has such form that its effect is simply giving rise to a modification of the *mass* of the electron under consideration so that it can be amalgamated into the *mass* term in the Hamiltonian for the free electron. One may then drop off this *mass*-modifying term from the Hamiltonian *when one reinterprets the electron mass such that the effect of this term if already included in it* and this *compound mass* has, though infinite theoretically, just the value we observe. Then the remaining terms in the Hamiltonian will give the deviation of the *energy* of our system from the free state as considered by Bethe. It is, in fact, not difficult to prove that by this method *the same result as Bethe can be obtained for the level-shift of a bound electron in the non-relativistic treatment.* This method of separating the term to be dropped off by means of a canonical transformation seems preferable to Bethe's because of its mathematical closedness.

It becomes thus desirable to obtain a relativistic generalization of the [unrelativistic] canonical transformation of Pauli and Fierz and find a general prescription to separate the infinite terms which can be amalgamated into the mass of the electron. In this and the following papers it will be shown that such a generalization is in fact possible, and the *field reaction* problems can be treated in a consistent way without touching the inherent infinities of the current *quantum field theory.*

The effectiveness of *our procedure of dropping off the mass-modifying term* was recently demonstrated by some examples lying close at hand by Koba and one of us[6] and Koba and Takeda[7].

[6] Koba, Z. & Tomonaga, S. (December, 1947). Application of the "Self-Consistent" Subtraction Method to the Elastic Scattering of an Electron. *Prog. Theor. Phys.*, 2, 4, 218[below; Bethe's theory assumes *self-energy* of free electron already included in observed mass of electron, applies Tati & Tomonaga "self-consistent" subtraction method to *elastic scattering of electron by a fixed electrostatic potential* to adjust mass for perturbation energy, "counter self-energy" term gives rise to additional correction that cancels divergence in previous result]; (1948). *Ibid.* 3, 4, 208.

[7] Koba, Z. & Takeda, G. (March, 1948). Radiative Corrections for Compton Scattering. *Prog. Theor. Phys.*, 3, 1, 98-99 [below]; (June, 1948). Radiative Corrections in e^2 for an

Arbitrary Process Involving Electrons. Positrons, and Light. *Ibid.*, 3, 2, 203-5 [below]; *Ibid.*, 3, 1, 387.

Similar considerations were made also by two American authors, Lewis[8] and Epstein[9],

[8] Lewis, H. W. (January, 1948). On the Reactive Terms in Quantum Electrodynamics. *Phys. Rev.*, 73, 173; https://doi.org/10.1103/PhysRev.73.173.

[9] Epstein, S. T. (January, 1948). Remarks on H. W. Lewis' Paper "On the Reactive Terms in Quantum Electrodynamics". *Phys. Rev.*, 73, 177; https://doi.org/10.1103/PhysRev. 73.177. According to a letter to the editor of the Physical Review, Schwinger is working on a similar line; Schwinger, J. (February, 1948). On quantum-electrodynamics and the magnetic moment of the electron. *Phys. Rev.,* 73, 4, 416-7 [below].

arriving at the same conclusion: *one part of the infinities occurring in the collision problems has the same origin as that appearing in the mass-modifying term so that it disappears when we reinterpret the electron mass in the way mentioned above and use there the observed finite value in place of the infinite theoretical value*. We shall discuss in this paper the *field reaction problem* in the case where no external *field* is present by means of our *relativistic* canonical transformation. We further confine ourselves to e^2-approximation throughout this paper. The case where an external *electromagnetic field* is present and the case where higher approximations become necessary will be treated in later papers.

As was alluded to in the paper of Koba and one of us

[Koba, Z. & Tomonaga, S. (December, 1947). Application of the "Self-Consistent" Subtraction Method to the Elastic Scattering of an Electron.]

there occurs in the relativistic treatment an infinity of another kind besides that which can be amalgamated into the electron mass. That is the infinity related to the *vacuum polarization* effect which causes an infinite *energy level shift* of the vacuum itself even when no charges nor any external *fields* are present and also gives rise to an infinite *self-energy* of a photon. *Because of its more complicated structure this term cannot be amalgamated into the equation for the free radiation with a reasoning so simple and plausible as that used in the case of the mass-modifying term.* A mathematically very pathological character of this term makes it difficult to "evaluate" it explicitly; it is even possible to find an argument which seems to prove in a somewhat quibbling way that this term vanishes. In such a situation *we must, for the present, content ourselves with simply dropping off this term,* leaving the detailed discussions in a further paper.

§ 2. *Relativistic canonical transformation*

It seems convenient to start from the *relativistic* formalism of the *quantum field theory* proposed by one of us[10].

[10] Tomonaga, S. (1943). On a Relativistically Invariant Formulation of the Quantum Theory of Wave Fields. *Bull. I. P. C. R. Riken-iho*, 22, 545 (in Japanese); translation Tomonaga, S. (August, 1946). *Prog. Theor. Phys.* 1, 2, 27-42[; *existing formalism of quantum field theory not perfectly relativistic*, commutation relations refer to points in

space at different times, Schrodinger equation for the vector representing the state of the system is a function of time, time variable plays a different role than space variables, *probability amplitude* not *relativistically invariant* in the space-time world, follows Dirac (1932) [Relativistic Quantum Mechanics] in *generalizing the notion of probability amplitude as far as is required by the theory of special relativity*, substitutes four-dimensional form of the commutation relations, *generalizes the Schrodinger equation* following the Dirac (1933) [The Lagrangian in Quantum Mechanics] *many-time* formalism, then introduces his *super many-time theory* [$\{H_{12}(P) + h/i\ \partial/\partial C_P\}\ \Psi[C] = 0$ at point P on surface C with infinitely many time variables representing the local time for each position in the space, results in *relativistic interaction representation*, three-dimensional manifold (space-like "surface") in four-dimensional space-time world, not necessary to also assume time-like surfaces for the variable surface as was required by Dirac, previous formalism built up in way too analogous to ordinary *non-relativistic* mechanics, theory was divided into one section giving the kinematical relations between various quantities at the same instant of time and another section determining the causal relations between quantities at different instants of time with the *commutation relations* belong to the first section and the *Schrodinger equation* to the second, this way of separating the theory into two sections very *unrelativistic*, "same instant of time" plays a distinct role, *new formalism consists of one section giving laws of behavior of the fields when they are left alone and the other giving the laws determining the deviation from this behavior due to interactions*, can be carried out *relativistically*, although theory in more satisfactory form no new contents added, *divergence difficulties inherited*, fundamental equations admit only catastrophic solutions *due to non-vanishing zero-point amplitudes of the fields which inheres in the operator $H_{12}(P)$, a more profound modification of the theory is required in order to remove this fundamental difficulty*.]; Koba, Z., Tati, T. & Tomonaga, S. (October, 1947). On a Relativistically Invariant Formulation of the Quantum Theory of Wave Fields. II. Case of Interacting Electromagnetic and Electron Fields. § 1-4. *Prog. Theor. Phys*., 2, 3, 101-116; (December, 1947) III. § 5-7. *Ibid*., 2, 4, 198-208.

In this formalism the Schrodinger equation has the form

$$\{H(P) - i\ \partial/\partial C_P\}\ \Psi[C] = 0 \qquad (1)$$

where H(P) is the *interaction energy density* between radiation and electron fields at the world point P, $\Psi[C]$ the generalized Schrodinger functional which is a functional of the space-like variable surface C in the four-dimensional world and $\partial/\partial C_P$ its partial functional differentiation at the point P, the point P being considered as lying on C. The interaction energy density H(P) has the form

$$H(P) = -\ e\ S_\mu(P)A_\mu(P) \qquad (2)$$

with

$$S_\mu(P) = i\psi^\dagger(P)\gamma^\mu\psi(P) \qquad (2')$$

where ψ is the quantized spinor describing the electron field and $\psi^\dagger$ its adjoint (i.e. $\psi^*\gamma^4$), γ^μ being the Dirac matrices*.

* Strictly speaking, we must already here carry out the decomposition of $\psi^\dagger$ and ψ according to (12) and (12') and rearrange the factors in the correct order and drop off the infinite terms appearing in S_μ by this process. Our further calculations are carefully performed in such a way that we always drop the terms arising from this infinite term in S_μ.

We denote the four-vector describing the radiation field by A_μ and the coordinate of a world point by X_μ, the fourth component being always measured in the imaginary unit. In the following the Greek indices such as λ, μ, ν ... are employed to specify the component of world vectors. Further the natural unit system $\hbar = c = 1$ is used throughout.

We now transform the Schrodinger functional Ψ into Ψ_1 by means of

$$\Psi_1 = U_1[C]\,\Psi[C] \tag{3}$$

with

$$U_1[C] = \exp\{i \int^C H(X)\, dV_4 \tag{3'}$$

dV_4, being the real volume element of the four-dimensional world. It can immediately be shown that $\Psi_1[C]$ satisfies then

$$\{-i/2\,[H(P), \int^C H(X)\, dV_4] - i\,\partial/\partial C_P\}\,\Psi_1[C] = 0 \tag{4}$$

if we retain only terms up to the order of e^2. In (3') the notation of the form $\int^C F(X)\, dV_4$ is employed to represent a quantity which gives $F(P)$ when the differentiation $\partial/\partial C_P$ is performed:

$$\partial/\partial C_P \int^C F(X)\, dV_4 = F(P) \tag{5}$$

Of course, there exist infinitely many solutions of this differential equation, because the solution can contain an arbitrary operator independent of C, but it is convenient for our purpose to fix this constant in the following way: we first expand the integrand into a Fourier integral:

$$F(X) = \int F(M_1M_2M_3M_4)\, e^{iM_\lambda X_\lambda}\, dM_1dM_2dM_3dM_4. \tag{6}$$

Next, we define a four vector $G_\mu(X)$ by means of

$$G_\mu(X) = 1/i \int F(M_1M_2M_3M_4)\, M_\mu\, e^{iM_\lambda X_\lambda} /|\, M_\lambda M_\lambda \,|\, dM_1dM_2dM_3dM_4. \tag{6'}$$

Then our $\int^C F(X)\, dV_4$ is defined by the surface integral

$$\int^C F(X)\, dV_4 = -\int_C N_\mu(X)\, G_\mu(X)\, dF_X \tag{7}$$

where the integral on the right-hand side is extended over the surface C, $N_\mu(X)$ being the unit normal of this surface at the point X with the direction from past to future. It can immediately be shown that this $\int^C F(X)\, dV_4$ satisfies (5) by the help of the Gauss theorem. In the following consideration there occur repeatedly integrals of this form; then the integration over V_4 has always to be interpreted in this manner.

It can be proved that the quantity

$- i/2\ [H(P), \int^C H(X)\ dV_4]$ (8)

depends only on P but not on the form of C. This is because the commutator [H(P), H(X)] vanishes always when the point X is space like with respect to the point P. This fact allows us to regard this quantity as the *Hamiltonian density* for the transformed equation (4).

$[\{- i/2\ [H(P), \int^C H(X)\ dV_4] - i\ \partial/\partial C_P\}\ \Psi_1[C] = 0$ (4)]

It is convenient for the treatment of processes that are second order in e to use this transformed equation, because, instead of carrying out a second order calculation using the original equation, we can treat the same problem more simply by means of a first order calculation if we use the transformed equation which is by itself already second order in e.

Processes through intermediate states can thus be treated as direct processes, especially the second order phenomena involving the *emission* and *reabsorption* of virtual particles, e. g. the *self-energy* of an electron, or the retarded *interaction* between two electrons (the M¢ller interaction) can be treated in the following way: We rearrange the variables describing the field in $[H(P), \int^C H(X)\ dV_4]$ in such a manner that the variables with asterisk (or daggar) stand everywhere on the right-hand side of the variables without asterisk (or dagger). (As to the more precise meaning of this statement see below). Then we obtain, for instance, *terms containing only* $\psi^\dagger$ *and* ψ*, i.e. the variables describing the electron field only without containing the variables* A_μ *that describe the radiation.* Such terms are responsible for the processes in which no ultimate *emission* nor *absorption* nor *scattering* of photons is taking parts. Terms of this form can be further classified into two types, one containing $\psi^\dagger$ and ψ bilinearly, and the other containing four ψ's. *The term of the first type is related closely to the self-energy of an electron because it has the same structure as the energy of free electron field.* The terms of the second type, on the other hand, *represent the interactions of two electrons* such as of M¢ller type, but positron-theoretically interpreted. In the same way we obtain, by rearranging the variables, a term that is quadratic in A_μ but contains no ψ's. Such a term describes a phenomenon in which no ultimate *emission* nor *absorption* nor *scattering* of electrons is taking parts, so that it describes the behavior of the radiation field in vacuo. *The effect of this term corresponds to the polarization of the vacuum due to the emission and reabsorption of virtual pair out of vacuum.*

Following this program, we now calculate the commutator in (8)

$[- i/2\ [H(P), \int^C H(X)\ dV_4]$ (8)]

and carry through the rearrangement of factors in each term. For this purpose, it is necessary to know the commutation relations between various field variables. Let the Latin indices r, s, *l* ..... denote the components of the spinors $\psi^\dagger$ and ψ. Then we have following commutation relations for the *electron field*;

$[\psi_r, \psi_s^{\dagger\prime}] = ...$ $= [\psi_r, \psi_s^{\dagger\prime}]_+ = 0$ (9)

κ being the mass of the electron. The commutation relations for the *radiation fields* are

... (10)

Here the four-dimensional delta-functions $D_0(X)$ and $D_n(X)$ are defined by

$$D_0(X) = -\tfrac{1}{2}(2\pi)^3 \int (e^{iK\lambda X\lambda} - e^{-iK\lambda X\lambda})\, dk/K_4 \qquad (10')$$
$$D_n(X) = -\tfrac{1}{2}(2\pi)^3 \int (e^{iL\lambda X\lambda} - e^{-iL\lambda X\lambda})\, dl/\{L_4\} \qquad (9')$$

where K_μ is the four-dimensional propagation vector for the *light waves*, k being its space component. The fourth component of K_μ is always assumed to be i times the positive value:

$K_\mu = i \mid k \mid = ik.$

In the same way L_μ in (9') is the four-dimensional propagation vector for *electron waves*, the space component being *l* and the fourth component

$\{L_4\} = i\sqrt{(l^2 + \kappa^2)} = i\sqrt{(l^2 + \kappa^2)}.$ [???]

The bracket { } is used here in order to emphasize that *this fourth component belongs to electron waves.*

It is necessary to decompose the fields into two parts, one oscillating with positive frequencies and the other with negative ones. For the *radiation field* this decomposition is defined by

$A_\mu(X) = \ldots$ (11)

with

… (11')

and for the *electron field*

$\psi(X) = \ldots$ (12)

with

… (12')

Physically the decomposition of ψ into ψ^+ and ψ^- corresponds, roughly speaking, to the decomposition of the field into the parts belonging to the negatron [electron] and the positron respectively.

The *commutation relations* for these decomposed fields can be found easily. They are

… (13)

with

… (13')

and

… (14)

with

… (14')

We proceed now to calculate the *commutator* in (8).

[$- i/2\, [H(P), \int^C H(X)\, dV_4]$ (8)]

We obtain

$$[H(X), H(X')] = e^2(I + II + III), \qquad (15)$$

with

$I = \ldots$

II = …
III = … (15')

We must here decompose $\psi^\dagger$ and ψ in (15') according to (11) and (12) and rearrange the factors in each term into the correct order. Then we obtain for I

$$I = I' - [\dots]\, D_0(X - X'), \qquad (16)$$

where each term in I' contains four ψ's (two of them without prime and the remaining two with prime) out of $\psi^{+\dagger}$, ψ^+, $\psi^{-\dagger}$ and ψ^- (and their primed) arranged in the correct order. These terms, whose explicit forms are not necessary for our present consideration, *represent the interactions between electrons caused by an exchange of a virtual photon*, among which the term containing $\psi^{+\dagger}$, ψ^+, $\psi^{+\dagger\prime}$, $\psi^{+\prime}$ corresponds to the ordinary M¢ller interaction between two electrons. The terms in [ ] on the right-hand side of (16), which have appeared by the rearrangement of factors, contain $\psi^\dagger$ and ψ bilinearly. *These terms are, therefore, closely connected to the self-energy of an electron*. It will be shown below that, combined with the terms in III which are of the same structure, *they give rise to the mass-modifying term mentioned in §1*. By a similar procedure we obtain for II

$$I = I'' - [\dots]\, (a_\nu' a_\mu + a_\nu{}^{*\prime} a_\mu + a_\mu{}^{*} a_\nu' + a_\nu{}^{*\prime} a_\mu{}^{*}), \qquad (17)$$

where terms in II' contain two ψ's (one with prime and the other without prime) out of $\psi^{+\dagger}$, ψ^+, $\psi^{-\dagger}$ and ψ^- (and their primed) and two *a*'s (one primed and the other unprimed) out of a^*, a (and their primed) arranged in the correct order. Among various such terms *the term containing, for instance, $\psi^{+\dagger}$, $\psi^{+\prime}$, a^* and a' is responsible for the scattering of a photon by a free electron* and *the term containing $\psi^{+\dagger}$, $\psi^{-\dagger\prime}$, a and a' is responsible for the creation of a pair by two photons*. The explicit form of these terms will not be given here because it is not necessary for our present consideration. *The terms in [ ] on the right-hand side of (17)* contain only variables describing the *radiation field* but no variables describing the electron. Therefore, these terms *represent the modification of the radiation field in vacuo due to the polarization of the vacuum. These are the terms mentioned in § 1 that cause an infinite energy level shift of the vacuum itself and an infinite self-energy of a photon.*

The terms in [ ] of (16) combined with the terms of III gives

$$III + (I - I') = \dots . \qquad (18)$$

It will be shown in the next section that, integrated over V_4' in the sense of (7)

$$[\textstyle\int^C F(X)\, dV_4 = - \int_C N_\mu(X)\, G_\mu(X)\, dF_X \qquad (7)]$$

and multiplied by $-ie^2/2$, this will yield the term of the form

$$\delta H_{mass} = - ie^2/2 \textstyle\int^C \{III + (I - I')\}\, dV' = \delta\kappa\, \psi^\dagger(P)\, \psi(P) \qquad (19)$$

with a logarithmic diverging quantity:

$$\delta\kappa = e^2\kappa/4\pi^2 \textstyle\int [2/\sqrt{(k^2 + \kappa^2)} - 1/\{k + \sqrt{(k^2 + \kappa^2)}\}]\, dk. \qquad (19')$$

the quantity which is familiar to us as *the electromagnetic mass of an electron*[11].

[11] Weisskopf, V. F. (July, 1939). On the Self-energy and Electromagnetic Field of the Electron. *Phys. Rev.*, 56, 1, 72-85[; *the main purpose of this paper is to show the physical significance of the logarithmic divergence of the self-energy of the electron and to demonstrate the reason for its occurrence*, the *self-energy of the electron is its total energy in free space when isolated from other particles or light quanta*,
$W = T + (1/8\pi) \int (H^2 + E^2)\, dr$ where T is the *kinetic energy* of the electron and H and E are the *magnetic* and *electric field strengths* at point r, identifies three reasons why quantum theory of the electron results in infinite *self-energy* of the electron, claims that *quantum kinematics shows that the radius of the electron must be assumed to be zero*, resulting in infinite energy of the *electrostatic field*, the contributions of the electric and magnetic fields of the spin to the *self-energy* of the electron cancel one another, *quantum theory of the electromagnetic field postulates the existence of field strength fluctuations in empty space*, gives rise to an additional energy which diverges more strongly that the electrostatic self-energy, induces the electron to perform vibrations with energy that diverges quadratically *for an infinitely small radius*, Dirac's positron theory implies that the charge and magnetic dipole of the electron are extended over a finite region, explains why the *self-energy* is only *logarithmically infinite*, *divergences are consequence of assumption of point electron*].

Therefore, this term gives just the mass-modifying term mentioned in § 1.

§ 3. *Infinite term of mass-modifying type*

In this section we shall calculate the contribution to (8)

$$[- i/2\, [H(P), \int^C H(X)\, dV_4] \qquad (8)]$$

from the terms of the mass-modifying type. Observing (18) we see that these terms contain each two D-functions or their derivatives. This fact results in that, integrated over V_4', they give rise to infinities. Indeed, by tracing the calculations in this section, one can ascertain that *the infinity occurring in the self-energy is closely connected to the D-function type singularities in the commutation relations between field variables*.

We shall now prove (19) and (19')

$$\delta H_{mass} = - ie^2/2 \int^C \{III + (I - I')\}\, dV' = \delta\kappa\, \psi^\dagger(P)\, \psi(P) \qquad (19)$$

$$\delta\kappa = e^2\kappa/4\pi^2 \int[2/\sqrt{(k^2 + \kappa^2)} - 1/\{k + \sqrt{(k^2 + \kappa^2)}\}]\, dk. \qquad (19')$$

The integral $\int^C \{III + (I - I')\}\, dV_4'$ is computed according to (7)

$$[\int^C F(X)\, dV_4 = - \int_C N_\mu(X)\, G_\mu(X)\, dF_X \qquad (7)]$$

by decomposing the integrand into a Fourier integral of the form (6)

$$[F(X) = \int F(M_1M_2M_3M_4)\, e^{iM\lambda X\lambda}\, dM_1dM_2dM_3dM_4. \qquad (6)]$$

Such a Fourier integral can be obtained immediately as the integral representation of D-functions as well as those for $\psi^{\dagger\prime}$ and ψ' are known as (9'), (10'), (11'), (12'), (13') and (14'). Using (12') we first write this integral in the form:

$$\int^C \{III + (I - I')\}\, dV_4' = \ldots \qquad (20)$$

with

$$A_\lambda = \int^C \ldots\, dV_4' \qquad (20')$$

$$B = \int^C \ldots\, dV_4'.$$

We notice here that because of the factors of the form $\{D_0^+ D_\kappa^+ - D_0^- D_\kappa^-\}$

$\{D_0^+ \, \partial D_\kappa^+/\partial X_\lambda - D_0^- \, \partial D_\kappa^-/\partial X_\lambda\}$ which vanish when X' is space-like with respect to X, the integrals occurring in A_μ and B are independent of the form of the surface C so far as the latter is space-like and passes through X. *The integration is, therefore, carried out assuming that C is a plane whose normal has the same direction with the propagation vector L_μ. Such a prescription for the integration is necessary because otherwise we can ascribe no definite value to the integration owing to the pathological nature of the D-functions.* Then the integration in (20') can be performed according to (7) most simply by referring to a Lorentz frame whose time axis coincides with L_μ. Physically this procedure corresponds to calculating A_μ and B in the rest system for the electron whose propagation vector is L_μ. We find then, using the Fourier representation for D-functions in (13') and (14') and carrying through the integration over V_4' according to (7), the following expressions for A_μ and B in this reference system

$$A_{1,2,3} = 0, \qquad A_4 = -\, i\kappa/8\pi^2 \int[1/\{k + \sqrt{(k^2 + \kappa^2)}\}]\, dk,$$
$$B = -\, i/8\pi^2 \int[1/\sqrt{(k^2 + \kappa^2)}]\, dk,$$

which in the original reference system

$$A_\mu = L_\lambda/\kappa\, (-\, i\kappa/8\pi^2) \int[1/\{k + \sqrt{(k^2 + \kappa^2)}\}]\, dk,$$
$$B = -\, i/8\pi^2 \int[1/\sqrt{(k^2 + \kappa^2)}]\, dk. \qquad (20'')$$

If we substitute (20") into (20) and noticing that

... ...

we find that the required integral has the form

$$\int^C \{III + (I - I')\}\, dV_4' = i\chi/2\pi^2 \int[2/\sqrt{(k^2 + \kappa^2)} - 1/\{k + \sqrt{(k^2 + \kappa^2)}\}]\, dk\psi^\dagger(X)\, \psi(X) \qquad (20''')$$

which gives (19) with (19') of the preceding section.

We have thus proved that an infinite term having the same structure as the mass term $\psi^\dagger\kappa\psi$ *appears in the transformed Schrodinger equation* (4) by the rearrangement of factors. *Because of this structure this term can be amalgamated into the mass term in the Hamiltonian for the free electron.* It is evident that *this procedure is equivalent to the procedure that one modifies first the mass of the electron by the amount of* $\delta\kappa$*, but, on the other hand, supplements the interaction energy with the counter term* $\psi^\dagger\kappa\psi$ *and applies this modified interaction to the problem to be solved.* This is the method employed previously by Koba and one of us [Koba, Z. & Tomonaga, S. (December, 1947), Application of the "Self-Consistent" Subtraction Method to the Elastic Scattering of an Electron, *loc. cit.*].

The procedure in which one includes some part of the effect which would be caused by the perturbation beforehand in the Hamiltonian for the unperturbed system is not new. A well-known example of such a procedure is the Hartree method of the self-consistent field in which one considers the interaction between electrons as perturbation but some part of its effect is taken into account already in the zero-th approximation, and in such a manner that the perturbation energy has no diagonal elements and, consequently, no level-shift is caused in the first order approximation. *The inclusion of the mass-modifying term in the free field equation in our procedure corresponds just to this procedure*. We have here considered that the *mass*-modifying term is already included in the *free field* equation and, consequently, no level-shift is caused by the interaction between electron and *radiation*. *Our subtraction method is thus "self-consistent" in this sense*.

§ 4. *Separation of the fields into" Unbound" and "Bound" parts*

As mentioned in § 1 Pauli and Fierz have found a transformation which separates the *radiation field* into the *field* of "unbound" photons and that of bound to the electron. We shall now give the *relativistic generalization* of this procedure that will be necessary for the later considerations. The treatment in the *non-relativistic* case suggests that this separation would be performed by defining the new fields A^-, $\psi^{\dagger -}$ and ψ^- by means of the canonical transformation

$$A(X) = \dots , \qquad \partial A(X)/\partial X = \dots$$
$$\psi(X) = \dots , \qquad \psi^{\dagger}(X) = \dots , \qquad (21)$$

with

$$U_1^-(X) = \dots \qquad (21')$$

C(X) being a space-like surface passing through the point X, and regard these barred fields as representing the "unbound parts". ...

...

which shows that the total Hamiltonian contains in fact no *interaction* term when expressed in terms of the barred variables.

Note added in proof. During the eight months since we prepared this MS, the problem of photon *self-energy* has been investigated by many authors. As was kindly remarked by Prof. Oppenheimer in his note to the letter of one of us

> [Tomonaga, S. & Oppenheimer, J. R. (July, 1948). On Infinite Field Reactions in Quantum Field Theory. *Phys. Rev.*, 74, 224; https://doi.org/10.1103/PhysRev.74.224.]

Prof. Schwinger has obtained a vanishing value for the photon self-energy by his elaborate calculation. Prof. Wentzel, on the other hand, has presented another result:

> [Wentzel, G. (November, 1948). New Aspects of the Photon Self-Energy Problem. *Phys. Rev.*, 74, 1070; https://doi.org/ 10.1103/PhysRev.74.1070; "*A finite but non-vanishing value for the self-energy of the photon, corresponding to a finite rest-mass, can be deduced from the new invariant formulation of quantum electrodynamics* developed by Tomonaga and Schwinger, in the e^2 order approximation."]

As for our Tokyo group, we are not yet quite sure, how to manage the ambiguity concerning this point, nor does it seem to us quite evident that the problem has been entirely settled by Prof. Schwinger's formalism. *We should like to express our sincere thanks to Prof. Oppenheimer for his kind information about Prof. Schwinger's work*, and to Prof. Rabi who was kind enough to show us the manuscript of this work before publication.

Koba, Z. & Tomonaga, S. (September, 1948). On Radiation Reactions in Collision Processes. I: Application of the "Self-Consistent" Subtraction Method to the Elastic Scattering of an Electron.

[*Progr. Theoret. Phys. (Kyoto)*, 3, 3, 290-303; https://doi.org /10.1143/ptp/3.3.290.]

Received March 1, 1948.

Physics Institutes, Tokyo University, and Tokyo Bunrika Daigaku.

* A preliminary report of the main part of this work has been published in the letter to the editor column of this journal: Koba, Z. & Tomonaga, S. (December, 1947). Application of the "Self-Consistent" Subtraction Method to the Elastic Scattering of an Electron. *Prog. Theor. Phys.*, 2, 4, 218, letter to the editor; https://doi.org/10.1143/ptp/2.4.218.

Applies Tati and Tomonaga's "self-consistent" subtraction method to the elastic scattering of an electron by a fixed electrostatic potential, *formal infinity associated with mass-modifying term attributed to defect of current theory so empirical value substituted for theoretical value* (assuming it was already included in the free field equation), maintains total Hamilton describing the interaction of the electron and electromagnetic fields unaltered, requires interaction-energy part of the Hamiltonian to undergo corresponding change by inclusion of "counter-self-energy" term ("mass-type" correction to scattering cross-section), results in finite value for self-energy of electron in the e^2 approximation, when new formalism applied to elastic scattering of electron *effective cross-section* for scattering by a fixed potential in zeroth approximation has value in good agreement with experiment but as soon as reaction of electromagnetic field with electron is taken into account the correction becomes infinite, first infinite term in modified Hamiltonian eliminated using *subtraction hypothesis of positron theory*, second difficulty disappears by applying *"self-consistent" subtraction method* using modified Hamiltonian, first term can also be eliminated if one interprets the sum of the external potential and its infinite correction due the *vacuum polarization* effect as the physically observable potential and *substitute the empirical value for it* whilst interaction part of the Hamiltonian supplemented by an additional "counter-vacuum polarization" term, *method by no means give the real solution of the fundamental difficulty of the quantum electrodynamics* but reveals nature of various diverging terms and reduces them to two quantities - the *self-energy* and the *vacuum polarization*, in this way *it becomes possible in an unambiguous and consistent manner to treat the field reaction problem without touching the fundamental difficulty by employing the finite empirical values instead of the infinite "theoretical" values for these two quantities.*

§ 1. *Introduction.*

Directly after the striking confirmation of Lamb and Retherford[1]

[1] Lamb, Jr., W. E. & Retherford, R. C. (August, 1947). Fine Structure of the Hydrogen Atom by a Microwave Method. *Phys. Rev.*, 72, 3, 241-3[; the spectrum of hydrogen has a fine structure of the energy levels which according to the *Dirac wave equation* for an electron moving in a Coulomb field is due to the combined effects of *relativistic* variation of mass with velocity and spin-orbit coupling, according to this theory the $2^2S_{1/2}$ state

should exactly coincide in energy with the $2^2P_{1/2}$ state which is the lower of the two P states, previous attempts at measurement have alternated between finding confirmation and discrepancies of as much as eight percent, using microwave method depending on a novel property of the $2^2S_{1/2}$ level results indicate that contrary to the *Dirac wave equation* the $2^2S_{1/2}$ state is higher than the $2^2P_{1/2}$ by about 1000 Mc/sec].

that the energy levels of the hydrogen atom deviate from the prediction obtained by Dirac's theory, Bethe[2]

[2] Bethe, H. A. (August, 1947). The Electromagnetic Shift of Energy Levels. *Phys. Rev.*, 72, 339-41[; Lamb and Retherford results show that fine structure of second quantum state of hydrogen does not agree with *Dirac wave equation*, Schwinger, Weisskopf, and Oppenheimer suggest might be due to *shift of energy levels by interaction of electron with the radiation field*, this shift comes out infinite in all existing theories and has therefore always been ignored, possible to identify the most strongly (linearly) divergent term in the level shift with an *electromagnetic mass effect* which must exist for a bound as well as for a free electron, already included in the *observed mass* of the electron so should be subtracted, assumes *relativistic cut-off* in quantum energies (frequencies) of included atomic states, then calculation of Lamb shift for hydrogen atom using *non-relativistic* ordinary radiation theory gives shift of the levels due to *radiation interaction* in close agreement with observed value, removes discrepancy with Dirac theory, did not carried out *relativistic* calculations].

published a note on his theory that explained this experimental fact in terms of the *reaction of the radiation field.* This attempt draws our attention much, because it may indicate a possible path to overcome the fundamental difficulty of the quantum field theory, and that for the first time in close connection with reliable experimental data. The essential point of Bethe's theory lies in the following hypothesis: *the self-energy of a free electron, i.e. the reaction of the field accompanied by the electron on itself when no external force is acting, is, though divergent in the present formalism of the theory, already included in the mass of the electron which we really observe*; when, however, the electron is put in an external field, the finite deviation of the *self-energy* of this electron from that in the standard *state*, which is of a free electron with the same *kinetic energy* as the average value of the bound one under consideration, is to be taken into account as causing e.g. the level-shift of the hydrogen atom in question.

In order to formulate this idea of Bethe in a more general and closed form, one of us proposed a "self-consistent" subtraction method[3]

[3] Tati, T. & Tomonaga, S. (December, 1948). A Self-Consistent Subtraction Method in the Quantum Field Theory, I. *Prog. Theor. Phys.*, 3, 4, 391-406[; notes procedure previously used by Bloch and Nordsieck and then by Pauli and Fierz in the treatment of problems with the self-field of an electron by first separating the radiation field into part bound to the electron and part of unbound photons by means of a canonical transformation, they then obtained a term in the Hamiltonian that could be interpreted as the *interaction energy* of electron with the radiation field bound to it, this term though infinite gave rise to a modification of mass of electron so that it could be amalgamated into the mass term in the

Hamiltonian for the free electron, the term was dropped off to obtain *observed mass* when electron mass reinterpreted as already including it, same result as Bethe for level-shift of bound electron in *non-relativistic* treatment, Tati and Tomonaga considered it desirable to obtain *relativistic* generalization of this treatment using a canonical transformation, paper addresses *field reaction problem when no external field* to e^2 approximation, does not address infinity related to *vacuum polarization* effect that occurs in *relativistic* treatment, *omitted as cannot be amalgamated into equation for free radiation*, starts from *relativistic* formalism of quantum field theory proposed by Tomanaga [August, 1946. On a Relativistically Invariant Formulation of the Quantum Theory of Wave Fields] in which Schrodinger equation with Hamiltonian for interaction density between radiation and electron fields is $\{H(P) - i\, \partial/\partial C_P\}\ \Psi[C] = 0$ where H(P) is the *interaction energy density* between radiation and electron fields at the world point P, Ψ[C] is the generalized Schrodinger functional which is a functional of the space-like variable surface C in the four-dimensional world and $\partial/\partial C_P$ its partial functional differentiation at the point P, the point P being considered as lying on C, decomposes fields into parts oscillating with positive and negative corresponding to electron and positron, transforms Schrodinger functional retaining terms up to the order e^2, expands integrand into Fourier integral and defines 4-vector to obtain Hamiltonian for transformed equation that enables use of first order calculation, allows processes through intermediate states such as emission and reabsorption of virtual particles (self-energy of an electron) to be treated as direct processes, calculates the commutator and rearranges factors in each term into the correct order, identifies (1) terms to represent *interactions between electrons* caused by exchange of virtual photons including ordinary M¢ller interaction between two electrons (2) terms connected with *self-energy* of an electron (3) terms responsible for the *scattering of a photon* by a free electron, terms for *creation of a pair* by two photons (4) variables describing the radiation field only representing modification of the radiation field in vacuo due to the *polarization of the vacuum* that cause an infinite energy level shift of the vacuum itself and an infinite *self-energy* of a photon, and (5) a *mass-modifying term* with a logarithmic diverging quantity representing the electromagnetic mass of an electron, assumes *mass-modifying term* already included in free field equation so no level-shift caused by the interaction between the electron and radiation, subtraction method "*self-consistent*" in this sense, gives *relativistic* generalization of transformation that separates *radiation field* into field of "unbound" photons and field of photons bound to the electron, procedure similar to Hartree method of the *self-consistent* field in which *interaction* between electrons is considered as a perturbation but some part of its effect is already included].

Let one calculate the *self-energy* of an electron with mechanical mass m and obeying Dirac equation in a many-body treatment, then one obtains an expression which can be written in the form δm, i.e. correction to the original mass m. This value has been calculated by Weisskopf [4] in e^2-approximation:

$$\delta m = 3e^2 m/2\pi \int_{k=0}^{\infty} dk/k + \text{finite term.} \quad (1.1)$$

[4] Weisskopf, V. F. (July, 1939). On the Self-energy and Electromagnetic Field of the Electron. *Phys. Rev*., 56, 1, 72-85[; *the main purpose of this paper is to show the physical significance of the logarithmic divergence of the self-energy of the electron and to*

demonstrate the reason for its occurrence, the *self-energy of the electron is its total energy in free space when isolated from other particles or light quanta*, $W = T + (1/8\pi)\int (H^2 + E^2)\, dr$ where T is the *kinetic energy* of the electron and H and E are the *magnetic* and *electric field strengths* at point r, identifies three reasons why quantum theory of the electron results in infinite *self-energy* of the electron, claims that *quantum kinematics shows that the radius of the electron must be assumed to be zero*, resulting in infinite energy of the *electrostatic field*, the contributions of the electric and magnetic fields of the spin to the *self-energy* of the electron cancel one another, *quantum theory of the electromagnetic field postulates the existence of field strength fluctuations in empty space*, gives rise to an additional energy which diverges more strongly that the electrostatic self-energy, induces the electron to perform vibrations with energy that diverges quadratically *for an infinitely small radius*, Dirac's positron theory implies that the charge and magnetic dipole of the electron are extended over a finite region, explains why the *self-energy* is only *logarithmically infinite, divergences are consequence of assumption of point electron*].

Here and in the following we employ the natural unit system $\hbar = c = 1$. In accordance with Bethe's assumption, one has to interpret $m + \delta m \equiv m^o$ as the electron *mass* actually observed; *the formal infinity in δm is thereby to be attributed to a defect of the current theory, the remedy of which is yet unknown, and so this gap in the theoretical value must be supplemented by the empirical one.*

Now we write down the Hamilton function, which describes our system of interacting electronic and electromagnetic fields, substituting the *rest mass* of the free electron by the above defined m^o and still *maintaining the total Hamiltonian unaltered.* Consequently, the *interaction-energy* part of the Hamiltonian necessarily undergoes a corresponding change so as to compensate the shift of the unperturbed part:

$$\begin{aligned} H_{total} &= \int \phi^*\{(a, p) + m\beta\}\phi dx + H_{radiation} + H_{interaction} - \delta m \int \phi^*\beta\phi\, dx \\ &= \int \phi^*\{(a, p) + (m + \delta m)\beta\}\phi dx + H_{rad} + H_{int} - \delta m \int \phi^*\beta\phi\, dx \\ &= \int \phi^*\{(a, p) + m^o\beta\}\phi dx + H_{rad} + H^o_{int}, \qquad (1.2) \end{aligned}$$

where

$$H^o_{int} = H_{int} - \delta m^o \int \phi^*\beta\phi\, dx. \qquad (1.3)$$

We use here δm^o instead of δm, because, since δm is by itself proportional to e^2, the difference of these two quantities

$$\delta m^o - \delta m = 0(e^4) \qquad (1.4)$$

can be safely neglected in the e^2-approximation. H^o_{int} of (1.3) is our modified *interaction* Hamiltonian. It is from the derivation evident, that with this new *interaction energy* one obtains for the *self-energy* of an electron with *rest mass* m^o a finite value in the approximation considered. *The divergence difficulty of the quantum electrodynamics is, as it were, imprisoned in a single quantity* m^o *and the latter quantity is substituted by the empirical value*. Strictly speaking, also the Hamiltonian of the free *electromagnetic field* H_{rad} should be modified in order to strike out the vacuum effect, which Weisskopf subtracted ad hoc in his calculation of the *self-energy* of an electron (*loc. cit.*). This point will be discussed further in a later paper.

The above mentioned H^{o}_{int} of (1.3) is, to be sure, sufficient but by no means necessary in order to make the *self-energy* finite, because only the diagonal part of the additional term $\{- \delta m^{o} \int \phi^{*}\beta\phi \, dx \}$ - we shall call this term "*counter-self-energy term*" - is responsible for the *self-energy*, while its non-diagonal part with the common infinite factor δm^{o} could be altered at one's discretion without touching the value of the *self-energy*. *Nevertheless, we have settled the modified Hamiltonian in the form of (1.3) as the most plausible choice because in this way the total Hamiltonian remains unaltered.* It is the main purpose of the present note to investigate the effect of this non-diagonal part and to demonstrate the usefulness of our rather conservative choice; conservative because this choice is introduced to keep the total Hamiltonian unmodified.

A problem in which the non-diagonal part of the *counter-self-energy term* plays a decisive role can be found in *transition processes*, and so we shall apply our new formalism to the *elastic scattering of an electron*, a simple but interesting problem, where many a new proposal failed to overcome the divergence except the theory of cohesive-force field[5]

> [5] Ito, D., Koba, Z. & Tomonaga, S. (December, 1947). Correction due to the Reaction of "Cohesive Force Field" for the Elastic Scattering of an Electron. *Progr. Theoret. Phys. (Kyoto)*, 2, 4, 216–217; and *errata*: Ito, D., Koba, Z. & Tomonaga, S. (December, 1947). Errata to the Above Letter. *Prog. Theor. Phys*., 2, 4, 217; (1948). *ibid*., 3, 276, and in press. [Abandoned *C-meson field theory*, in which Sakata and Hara introduced a hypothetical *scalar particle* (called *C-meson*) in order to solve the divergence problem.]

As is well known, the *effective cross-section* for the process that an electron is scattered by a *fixed electrostatic potential* has in the zeroth approximation a reasonable value which is in good agreement with experiment. *But as soon as one will improve the calculation taking into account the reaction of the electromagnetic field with which the electron is always interacting, the correction turns out infinity according to the current formalism of the quantum electrodynamics*. The explicit expression for this correction was first calculated by Dancoff[6]

> [5] Dancoff, S. M. (May, 1939). On Radiative Corrections for Electron Scattering. *Phys. Rev*., 55, 959[; *relativistic* treatment of the radiative correction of order $e^2/\hbar c$ to the elastic scattering cross section leads to the following results: (a) for the scattering in an electrostatic field of a particle described by the Pauli-Weisskopf theory, the correction is finite; (b) for the scattering of a Dirac electron in an electrostatic field, the correction diverges logarithmically and is positive; (c) the convergence or divergence of the correction depends critically on the type of scattering potential considered],

who took all the *relativistic* effects into consideration and carried out a perturbation calculation up to e^2-approximation; his result has been then rectified into the following by our more detailed investigation[7]

> [7] Ito, D., Koba, Z. & Tomonaga, S. (December, 1947), *loc. cit.* [; abandoned *C-meson field theory*, in which Sakata and Hara introduced a hypothetical *scalar particle* (called *C-meson*) in order to solve the divergence problem]:

$$\delta\sigma/\sigma_0 = -\,3e^2/4\pi \int_{l=0}^{\infty} dl/l + 3e^2/2\pi\, m^2(p-q)^2/E^2\{E^2 + (pq) + m^2\} \int_{k=0}^{\infty} dk/k + \text{finite terms.} \qquad (1.5)$$

Here **p** and **q** denote the *momentum* of the incident electron and that of the scattered one respectively; E-means the total energy $E = \sqrt{(p^2 + m^2)} = \sqrt{(q^2 + m^2)}$, σ_0 is the cross-section in the zeroth approximation, $\delta\sigma$ its e^2-correction. *The first term on the right-hand side of the expression (1.5) is infinite but can be eliminated if one resorts to the subtraction hypothesis of the positron theory, while the second remains as the fatal divergence resulting from the reaction of the electromagnetic field, and this fact leads one to discredit of the current quantum field theory.*

It will be seen, however, in the next section that this difficulty disappears as soon as one applies our *"self-consistent " subtraction method*, making use of the Hamiltonian (1. 3)

$$[H^o_{int} = H_{int} - \delta m^o \int \phi^*\beta\phi\, dx, \qquad (1.3)]$$

and one can expect a reasonable result. Further we see in §3, that one can eliminate the first diverging term, too, by a method quite analogous to the above procedure concerning the self-energy: one interprets the sum of the external potential and its infinite correction due to the vacuum polarization effect as a physically observable potential and substitutes it by the actual value of the given potential while, on the other hand, the interaction part of the Hamiltonian is supplemented by an additional "counter-vacuum-polarization" term.

The whole argument in this note is confined to e^2-approximation with regard to the reaction of the self-field, and to the first approximation in the scattering potential V. As an example of the problem involving higher approximations, we have also investigated the radiative corrections to the Klein-Nishina formula and arrived at a similar conclusion. This result will be published soon[8].

[8] Koba, Z. & Takeda, G. (June, 1948). Radiative Corrections in e^2 for an Arbitrary Process Involving Electrons. Positrons, and Light Quanta. *Prog. Theor. Phys.*, 3, 2, 203-5, letter to the editor; cf *ibid.* (March, 1948). Radiative Corrections for Compton Scattering. *Prog. Theor. Phys.*, 3, 1, 98-9, letter to the editor.

§ 2. *The "Mass-Type" Correction to the Scattering Cross-Section.*

Our modified Hamilton function (1.2), (1.3)

$$\begin{aligned}[H_{total} &= \int \phi^*\{(a, p) + m\beta\}\phi dx + H_{radiation} + H_{interaction} - \delta m \int \phi^*\beta\phi\, dx \\ &= \int \phi^*\{(a, p) + (m + \delta m)\beta\}\phi dx + H_{rad} + H_{int} - \delta m \int \phi^*\beta\phi\, dx \\ &= \int \phi^*\{(a, p) + m^o\beta\}\phi dx + H_{rad} + H^o_{int}, \qquad (1.2)\end{aligned}$$

where

$$H^o_{int} = H_{int} - \delta m^o \int \phi^*\beta\phi\, dx, \qquad (1.3)]$$

when applied to our scattering problem, will 'yield in e^2-approximation a result that is different from the usual one in two points:

1) The rest mass m is substituted everywhere in (1.5) by $m^o = m + \delta m$, which, however, does not lead to any significant effect, now that m^o is to be identified with the empirical value.

2) A new correction term in the second order presents itself because of the counter-self-energy term. The perturbed initial wave function contains two new terms; its unperturbed state **[p]** is connected by the counter-self-energy term in the interaction Hamiltonian to two virtual states, in which the total momentum is conserved. ...

...

The above analysis reveals an essential feature of our method, clarifying the ground of its effectiveness and also indicating the limitation of its applicability. We have a simple criterion with regard to this point: one could expect a convergent result only when diverging parts of the correction can be reduced to the form $(\partial\sigma_0/\partial m)\delta m$, which is not always the case. An artificial but simple example where our method fails is afforded when one deals with the scattering of an electron by a world scalar potential. We hope, however, that suitable generalizations of our method might overcome divergence difficulties also in such cases; even when this attempt should fail, our consideration would be, at least, useful in classifying various types of divergent terms.

After we have analyzed in this way the feature of the divergences appearing in our scattering problem, it may be no longer surprising that the theory of cohesive-force field, too, has succeeded in eliminating this type of divergence. It is because the correction to the scattering cross-section due to the *C-meson field* is also related to the self-energy in the same way as in the case of the electromagnetic field so that our choice $2e^2 = f^2$, which has been introduced in order to eliminate the diverging self-energy, can at the same time just eliminate the diverging part of the corrections caused by these two fields. One has here to remember that *this theory aims at a so to speak substantialization of the subtraction operation to get a finite self-energy and a finite scattering cross-section. Of course, the introduction of an independent cohesive-force field means something more than a mere subtraction procedure or use of our modified interaction energy.* Such circumstances will be discussed further in later papers.

§ 3. *The "Polarization-Type" Correction to the Scattering Cross-Section.*

Let us consider the first term on the right-hand side of (1.5)

$$[\delta\sigma/\sigma_0 = -\,3e^2/4\pi \int_{l=0}^{\infty} dl/l +$$
$$3e^2/2\pi\, m^2(p-q)^2/E^2\{E^2 + (pq) + m^2\} \int_{k=0}^{\infty} dk/k + \text{finite terms.} \quad (1.5)]$$

which has been so far ad hoc struck off as the *vacuum polarization effect* caused by the *scattering potential*. We now try to apply our self-consistent subtraction method also to this type of divergence, then we find that by this new method the same result - at least in the approximation under consideration - is obtained, but with a more plausible reasoning for the subtraction procedure.

...

It is to be noticed here that *when a more general type of scattering potential is in question, for example the case of magnetic scatterings, the treatment becomes more complicated than in the present problem.* The case where the scattering potential is electrostatic is a particularly simple and favorable one.

§ 4. *Conclusion.*

Now our prescription introduced in this work can be summarized in the following way. The Hamilton function that describes interacting electron and electromagnetic fields under the action of an external field

$$H_{total} = \ldots \qquad (4.1)$$

is first re-arranged in e^2-approximation into

$$H_{total} = \ldots \qquad (4.2)$$

with

$$m^o = m + \delta m, \quad \delta m = 3e^2 m/2\pi \int_{k=0}^{\infty} dk/k + \ldots, \quad \delta m^o = \delta m + 0(e^4) \qquad (4.3)$$
$$V^o = V + \delta V, \quad \delta V = -2e^2 V/3\pi \int_{l=0}^{\infty} dl/l, \quad \delta V^o = \delta V + 0(e^4).$$

Then we avoid the infinity by *re-interpreting m^o and V^o* in the re-arranged Hamiltonian (4.2) respectively as *the rest mass of the electron and the given external potential which we really observe. In this way the correction due to the electromagnetic field reaction to the elastic scattering cross-section of an electron ($\delta\sigma/\sigma_0$) can be made finite by virtue of the counter terms introduced in (4.2).*

It is true that our method does by no means give the real solution of the fundamental difficulty of the quantum electrodynamics but it has, at least, revealed the nature of various diverging terms appearing in our scattering problem and reduced them into two quantities: the self-energy and the vacuum polarization. In this way it becomes possible in an unambiguous and consistent manner to treat the field reaction problem without touching the fundamental difficulty by employing the finite empirical values instead of the infinite "theoretical" values for these two quantities.

The subtraction of the vacuum effect which appears also in the absence of any charge or any external field and causes an infinite energy-level shift of the vacuum itself could and should be treated in the same manner, reinterpreting the Hamiltonian of the free electromagnetic field. But the discussions about this point shall be left to a later paper.

Finally, we may add a few lines on the so-called *infra-red catastrophe* of the *Bremsstrahlung* of an electron in an electrostatic field,

> [*Bremsstrahlung* is electromagnetic radiation produced by the deceleration of a charged particle when deflected by another charged particle.]

which was the very origin of the discussions about the field reactions in the scattering problem. Bloch and Nordsieck[9]

> [9] Bloch, F. & Nordsieck, A. (July, 1937). Note on the Radiation Field of the Electron. *Phys. Rev.*, 52, 54; https://doi.org/10.1103/PhysRev.52.54.

were the first to obtain in a *non-relativistic* treatment a very natural and plausible result; but afterwards it was pointed out by Pauli and Fierz[10]

[10] Pauli, W. & Fierz, M. (March, 1938). Zur Theorie der Emission langwelliger Lichtquanten. (On the theory of the emission of long-wave light quanta.) *Nuovo Cimento*, 15, 3, 167-88; cf. also Bethe, H. A. & Oppenheimer, J. R. (October, 1946). Reaction of Radiation on Electron Scattering and Heitler's Theory of Radiation Damping. *Phys. Rev.*, 70, 451; https://doi.org/10.1103/ PhysRev.70.451.

that the former authors unreasonably disregarded the *conservation of energy*. On the other hand, since Bloch and Nordsieck had omitted the electromagnetic mass correction of the electron, they could have with equal right struck off the reaction of high frequency components of the radiation field by some cut-off procedure. This simplification was actually done by Pauli and Fierz, who could thus treat the problem more rigorously taking account of the *conservation of energy*. But their result depended, to their regret, critically on the cut-off frequency, so that they made in the conclusion a pregnant prediction: *the satisfactory solution of this problem would not be reached without inquiring into the fundamental difficulty of the quantum electrodynamics, i.e. of the infinite self-energy of the electron*. Now we believe that this is indeed the case, but we think to have indicated an unambiguous way to attain a consistent and reasonable result, which will perhaps turn out very close to the quite natural conclusion of Bloch and Nordsieck. This point will be clarified by the computation of the exact value for our corrected cross-section, which is now being calculated by our collaborator[11].

[11] Endo, S., Kinoshita, T. & Koba, Z. (September, 1948). Reactive Corrections for the Elastic Scattering of an Electron. *Prog. Theor. Phys.*, 3, 3, 320-1; https://doi.org/10.1143/ ptp/3.3.320, in press.

Acknowledgement. The authors wish to express their hearty thanks to Dr. M. Taketani for his valuable discussions.

Addendum. Just after we had finished this paper the January 15 issue of the *Physical Review* arrived, in which papers of Lewis and Epstein dealing with the same problem were published.

[Lewis, H. W. (January, 1948). On the Reactive Terms in Quantum Electrodynamics. *Phys. Rev.*, 73, 173; https://doi.org/10.1103/PhysRev.73.173; Epstein, S. T. (January, 1948). Remarks on H. W. Lewis' Paper "On the Reactive Terms in Quantum Electrodynamics". *Phys. Rev.*, 73, 177; https://doi.org/ 10.1103/PhysRev.73.177.]

Although our conclusions are perfectly the same as these authors', it may be allowed to us to publish our results too, not only because our work has been carried out independently of them and we have promised its publication in a preliminary report (the footnote on page 290), but also because this paper shall play the role of an introduction to a series of papers dealing with the field-reaction problems which have been and are being investigated by our group. We hope that the readers will be good enough to acknowledge also our paper with Lewis and Epstein's.

Koba, Z. & Takeda, G. (December, 1948). Radiation Reaction in Collision Process, II*: Radiative Corrections for Compton Scattering**.

[*Prog. Theor. Phys.*, 3, 4, 407-21; https://doi.org/10.1143/ptp/3.4.407.]

Received May 23, 1948.

Institute of Physics, Tokyo University.

* The part I of this work has appeared in this journal; Koba, Z. & Tomonaga, S. (September, 1948). On Radiation Reactions in Collision Processes. I: Application of the "Self-Consistent" Subtraction Method to the Elastic Scattering of an Electron. *Progr. Theor. Phys.*, 3, 3, 290-303 [; applies Tati and Tomonaga's *"self-consistent" subtraction method* to *elastic scattering of electron by fixed electrostatic potential* to adjust mass for perturbation energy, "counter self-energy" term gives rise to additional correction that cancels divergence in previous result, defect of current theory, gap in theoretical value must be supplemented by empirical one; settles on modification leaving total Hamiltonian unaltered, as soon as reaction of the electromagnetic field is taken into account correction turns out to be infinite, reduces diverging terms in scattering problem to *self-energy* and *vacuum polarization*, treats field reaction problem *by employing the finite empirical values instead of infinite "theoretical" values for these two quantities*]. In the following it will be cited as I.

** A preliminary report of the main part of this work has been published in the letter to the editor column of this journal. Koba, Z. & Takeda, G. (March, 1948). Radiative Corrections for Compton Scattering. *Prog. Theor. Phys.*, 3, 1, 98-9, letter to the editor; https://doi.org/10.1143/ptp/3.1.98. Here we give only an outline of our calculation on account of limited space; as for its details the readers are referred to the article by the present authors to be published elsewhere in Japanese.

Previous paper applied Tati & Tomonaga's *"self-consistent" subtraction method* to *elastic scattering of an electron*, all divergences were eliminated by counter-terms in modified form of the Hamilton function, but this was a simple and particularly favorable case, need to examine whether method is still effective in fundamental processes between elementary particles, applies method to calculate e^2-corrections to *Klein-Nishina formula for Compton scattering* using *perturbation method,* two types of diverging terms, one related to *polarization of vacuum* and other to *self-energy* of electron, sufficient to add two new terms to Hamiltonian function to cancel out these infinite terms, *self-energy* logarithmic divergence eliminated by introducing new term into interaction Hamiltonian derived in a plausible way as a "counter-term" which compensates change in free field Hamiltonian and conserves total Hamiltonian unaltered, one of *vacuum polarization* terms could be treated in similar way, while other term could not be foisted into the theory without radically changing the Maxwell equation for the free electromagnetic field, *far from final settlement of the fundamental problem in theory of elementary particles*.

§ 1. *Introduction*

A reasonable method is now being developed-as has been circumstantially described by Tati and Tomonaga[1] in the introductory section of the foregoing paper-in order to avoid the divergence difficulty in the quantum field theory: the *self-consistent subtraction method.*

[1] Tati, T. & Tomonaga, S. (December, 1948). A Self-Consistent Subtraction Method in the Quantum Field Theory, I. *Prog. Theor. Phys.*, 3, 4, 391-406[; notes procedure previously used by Bloch and Nordsieck and then by Pauli and Fierz in the treatment of problems with the self-field of an electron by first separating the radiation field into part bound to the electron and part of unbound photons by means of a canonical transformation, they then obtained a term in the Hamiltonian that could be interpreted as the *interaction energy* of electron with the radiation field bound to it, this term though infinite gave rise to a modification of mass of electron so that it could be amalgamated into the mass term in the Hamiltonian for the free electron, the term was dropped off to obtain *observed mass* when electron mass reinterpreted as already including it, same result as Bethe for level-shift of bound electron in *non-relativistic* treatment, Tati and Tomonaga considered it desirable to obtain *relativistic* generalization of this treatment using a canonical transformation, paper addresses *field reaction problem when no external field* to e^2 approximation, does not address infinity related to *vacuum polarization* effect that occurs in *relativistic* treatment, *omitted as cannot be amalgamated into equation for free radiation*, starts from *relativistic* formalism of quantum field theory proposed by Tomanaga [August, 1946. On a Relativistically Invariant Formulation of the Quantum Theory of Wave Fields] in which Schrodinger equation with Hamiltonian for interaction density between radiation and electron fields is $\{H(P) - i\, \partial/\partial C_P\}\ \Psi[C] = 0$ where H(P) is the *interaction energy density* between radiation and electron fields at the world point P, Ψ[C] is the generalized Schrodinger functional which is a functional of the space-like variable surface C in the four-dimensional world and $\partial/\partial C_P$ its partial functional differentiation at the point P, the point P being considered as lying on C, decomposes fields into parts oscillating with positive and negative corresponding to electron and positron, transforms Schrodinger functional retaining terms up to the order e^2, expands integrand into Fourier integral and defines 4-vector to obtain Hamiltonian for transformed equation that enables use of first order calculation, allows processes through intermediate states such as emission and reabsorption of virtual particles (self-energy of an electron) to be treated as direct processes, calculates the commutator and rearranges factors in each term into the correct order, identifies (1) terms to represent *interactions between electrons* caused by exchange of virtual photons including ordinary M¢ller interaction between two electrons (2) terms connected with *self-energy* of an electron (3) terms responsible for the *scattering of a photon* by a free electron, terms for *creation of a pair* by two photons (4) variables describing the radiation field only representing modification of the radiation field in vacuo due to the *polarization of the vacuum* that cause an infinite energy level shift of the vacuum itself and an infinite *self-energy* of a photon, and (5) a *mass-modifying term* with a logarithmic diverging quantity representing the electromagnetic mass of an electron, assumes *mass-modifying term* already included in free field equation so no level-shift caused by the interaction between the electron and radiation, subtraction method "*self-consistent*" in this sense, gives *relativistic* generalization of transformation that separates *radiation field* into field of "unbound" photons and field of photons bound to the electron, procedure similar to Hartree method of the *self-consistent* field in which *interaction* between electrons is considered as a perturbation but some part of its effect is already included].

Though the latter may yet be far from the final settlement of the fundamental problem in the theory of elementary particles, it has at least offered a plausible procedure by which one can, both in stationery and transition problems, take into account the radiative reactions without being frustrated by those cumbersome infinite terms.

Indeed, we have in I

> [Koba & Tomonaga (September, 1948). On Radiation Reactions in Collision Processes. I: Application of the “Self-Consistent” Subtraction Method to the Elastic Scattering of an Electron]

applied the method to the *elastic scattering of an electron* and have seen that all the divergences are really eliminated by virtue of the counter-terms in the modified form of the Hamilton function. This case was, however, very simple and perhaps particularly favorable one and we could not, in view of the subtle features of the reaction problems, yet insist upon the general expedience of our formalism, so that it has been desired to examine whether our method is still effective in fundamental processes between elementary particles[2].

> [2] The necessity of investigating the field reaction in the collision process of the elementary particles was first pointed out by Prof. S. Sakata at the meeting of National Research Council in July, 1947.

The later inquiry of our seniors[3]

> [3] Tati, T. & Tomonaga, S., *loc. cit.*

into the *subtraction method* by means of canonical transformation has in fact indicated the existence of those *vacuum polarization* (or *self-energy* of the photon) infinities, which the case treated in I could not reveal exhaustively, and *has brought forward an urgent problem as to the nature of these terms and the way to manage them.* It will be, therefore, of interest to attack the matter on a certain collision problem involving light quanta, which shall afford a foothold for the general argument of the above-mentioned authors.

Thus, we take up the Compton scattering in this paper and shall compute the e^2-corrections to the *Klein-Nishina formula* by a straightforward application of the *perturbation method.*

> [The *Klein–Nishina formu*la gives the differential cross section (i.e. the "likelihood" and angular distribution) of photons scattered from a single free electron, calculated in the lowest order of quantum electrodynamics. It was first derived in 1928 by Oskar Klein and Yoshio Nishina, constituting one of the first successful applications of the Dirac equation.]

First, we formulate the problem and classify the radiative corrections into two types (§ 2), then examine them separately and investigate what modifications one has to introduce into the Hamiltonian in order to cancel out infinities (§ 3-4), and finally discuss how to interpret these modifications from the view point of the *self-consistent subtraction procedure*, special emphasis being laid upon the treatment of quadratically divergent integrals in *vacuum polarization effect* (§5).

§ 2. *Formulation, classification of corrections*

We consider the process of Compton scattering in the center of gravity system: an incident photon with *momentum* – **p** runs against an electron with *momentum* **p** and after collision a photon with *momentum* **q** and an electron with *momentum* – **q** are produced. In this reference system | **p** | = | **q** | as a consequence of the *energy conservation law*.

The Hamiltonian of our system, which consists of electronic and electromagnetic fields in interaction, is given by

$$H = \int u^*(X) \{(\alpha p) + m\beta\}u(X)\, dV + 1/8\pi \int\{\mathbf{E}^2(X) + \mathbf{H}^2(X)\}\, dV$$
$$+ e \int u^*(X)(\alpha.\mathbf{A}(X)u(X)\, dV$$
$$+ e^2/2 \iint u^*(X)u(X)u^*(X')u(X')/|\, \mathbf{X} - \mathbf{X}'\,|\, dVdV' \qquad (2.1)$$

where u(X) is the *quantized amplitude* of electronic field and **A**(X) is the transverse part of the *vector potential* of electromagnetic field. We put …

…

Treating the last two terms of (2.1) as perturbation, we solve the problem in the form of a power series in e. For the *Compton scattering process* the first non-vanishing matrix elements are those in e^2 and there are four of them corresponding to the possible connections through four different *states*. These well-known transitions are described in our reference system as follows, + and ~ indicating a *positron* and a *photon* respectively:

$$\mathbf{p}, -\mathbf{p}^{\sim} \rightarrow \mathbf{o} \text{ ----------------------} \rightarrow \mathbf{q}, -\mathbf{q}^{\sim}, \qquad (2.3)$$
$$\mathbf{p}, -\mathbf{p}^{\sim} \rightarrow \mathbf{p}, -\mathbf{p}^{\sim}, \mathrm{q}, \mathbf{o}^{+}, -\mathbf{q}^{\sim} \quad \rightarrow \mathbf{q}, -\mathbf{q}^{\sim}, \qquad (2.4)$$
$$\mathbf{p}, -\mathbf{p}^{\sim} \rightarrow (\mathbf{p} + \mathbf{q}), -\mathbf{p}^{\sim}, -\mathbf{q}^{\sim} \quad \rightarrow \mathbf{q}, -\mathbf{q}^{\sim}, \qquad (2.5)$$
$$\mathbf{p}, -\mathbf{p}^{\sim} \rightarrow \mathbf{p}, \mathbf{q}, (-\mathbf{p} - \mathbf{q})^{+} \quad \rightarrow \mathbf{q}, -\mathbf{q}^{\sim}. \qquad (2.6)$$

With these matrix elements in e^2 *the Klein-Nishina formula is obtained.* But we proceed further to the fourth order perturbation, *aiming at e^2-correction to this formula*. For this purpose, it is necessary to generalize to some extent the ordinary formulas of non-stationary perturbation calculus, since one has to take into account also those transition processes in which an *intermediate state* is no other but the initial one or the final state itself, and in such cases the conventional expressions fail because of vanishing denominator. So, we give here the formula for the e^2-correction $\delta\sigma(\sim e^6)$ to the zeroth cross section $\sigma_0(\sim e^4)$,

$$\delta\sigma = \ldots - \ldots - \ldots . \qquad (2.7)$$

The derivation of this formula and some other remarks will be mentioned elsewhere[5].

[5] Z. Koba and G. Takeda, unpublished.

In the fourth order we find 1080 kinds of possible paths connecting the initial and the final *states*[6].

[6] We take into account two polarization directions of the intermediate photon and also count a case of momentum transfer through Coulomb interaction twofold, since, for

example, Coulomb interaction between electrons **p** and **q**: {**p**, **q**→**p**–**k**, **q**–**k**} can be described both as transfer of *momentum* **k** from **p** to **q** and as that of –**k** from **q** to **p**. The formula for the number of possible connections and their classification in general cases will be given in our next work.

It is expedient to classify them as follows:

Possible connections	*Mass type [M] 720*	...
1080 in all	*Vacuum polarization type [P]360*	...

Here [M]'s represent those connections which involve *emission* and *reabsorption* of a *virtual photon* with arbitrary *momentum*, while [P]'s involve creation and annihilation of a *virtual electron-positron pair* with arbitrary *momentum*; the latter are subdivided into two kinds, of which [P, I] is characterized by the *emission* and *reabsorption* of a *virtual photon* **p**˜, –**p**˜, **q**˜, or –**q**˜ and [P, II] by that of a *virtual photon* (**p**–**q**)˜ or (**q**–**p**)˜. [M] and [P, I] consist of four groups corresponding to the second order connections and as they can be treated quite in parallel, we shall give the examples corresponding to (2.3) only[7].

[7] The exact definition of "correspondence" will be given in our next work.

Summing up contributions of all the matrix elements in the fourth order, we have to integrate with respect to the above-mentioned arbitrary *momentum* of a *virtual photon* (in case of [M]'s) or of a virtual pair (in case of [P]'s). *Most of these integrals, however, do not converge as a consequence of the defective current formalism of the field theory*, and we shall pertain in the following mainly to these formally infinite corrections.

...

§ 3. *Mass type divergence [M]*

Now we give here the schemes expressing the first group of [M] connection, i.e. those which can be regarded as the modification of (2.3). Roughly speaking they may be characterized by the appearance of an intermediate electron **o**, except some connections which do not contribute to diverging corrections and can be disregarded here. For convenience's sake we assort them in some subgroups. (The titles should not be taken too literally.)

(I) *Intermediate electron emits and reabsorbs virtual photon (–j˜).*

p, –**p**˜ →**o** -------→**j**, –**j**˜ -----→**o** -----------→ **q**, –**q**˜ (3.1)

-------→ **o**, **o**, **j**⁺, –**j**˜ ----

... (3.2)

... (3.3)

... (3.4)

(II) *Final electron **q** emits and reabsorbs virtual photon (–j˜).*

...

(III) *Initial electron **p** emits and reabsorbs virtual photon (–j˜).*

…

(IV) *Intermediate electron **o** and final electron **q** exchange virtual photon* $\tilde{j}$ *or* $-\tilde{j}$.

…

(V) *Intermediate electron **o** and initial electron **p** exchange virtual photon* $\tilde{j}$ *or* $-\tilde{j}$.

…

…

Now that we have investigated the diverging terms thoroughly and have arranged them in the forms of (3.18) - (3.25), *it is not difficult to seek for a new interaction term to be introduced in order to eliminate infinite terms*. If we assume, namely, that

$$\delta m \int u^*(X)\, \beta u(X)\, dV \tag{3.26}$$

is added to the Hamiltonian function, being *the self-energy of an electron*

$$\delta m = (3me^2/2\pi) \int^{\infty} dj/j + \text{finite terms}, \tag{3.27}$$

this term will give rise to the following new connections

…

…

…

…

These will yield diverging corrections to the cross section, just opposite to the above gained ones, so that we shall be left with finite corrections only.

…

§ 4. *Vacuum polarization type divergence [P]*

As examples of [P, I] connections we give here the schemes for the first group corresponding to (2.3).

…

§ 5. *Interpretation of the results*

In the preceding two sections we have found the expressions for the infinite terms, (3.26) with (3.27) and (4.9), (4.12) with (4.13), which, when introduced into the interaction Hamiltonian, are sufficient to cancel out all divergences appearing in the e^2*-correction to the Klein-Nishina formula.*

Now we shall discuss whether these modifications can be derived in a plausible way by generalizing the self-consistent subtraction applied in I.

Then the following prescription will be suggested:

The Hamilton function describing the total system is given by (2.1)

$$[H = \int u^*(X) \{(\alpha p) + m\beta\}u(X)\, dV + 1/8\pi \int\{\mathbf{E}^2(X) + \mathbf{H}^2(X)\}\, dV$$
$$+ e \int u^*(X)(\alpha.\mathbf{A}(X)u(X)\, dV$$
$$+ e^2/2 \iint u^*(X)u(X)u^*(X')u(X')/| \mathbf{X} - \mathbf{X}' |\, dVdV' \qquad (2.1)]$$

or

$$H = \int u^*\{(\boldsymbol{\alpha}.\mathbf{p}) + m\beta\}u\, dV + \ldots + \ldots + \ldots \qquad (5.1)$$

with

$$H_{el.mag.fld.} = 1/8\pi \int (\mathbf{E}^2 + \mathbf{H}^2)\, dV. \qquad i = u^*\boldsymbol{\alpha}u, \qquad \rho = u^*u. \qquad (5.2)$$

…

In any of the above alternatives our self-consistent argument turns out not so natural and plausible in dealing with the quadratically diverging terms as in the cases of other types of infinities. Such a situation indicates, as we believe, that more fundamental considerations are required in order to settle this difficulty. Tati and Tomonaga, on the other hand, have also found a corresponding term in their *relativistic* formulation of the self-consistent subtraction and have tried to investigate its Lorentz and gauge invariance but *there still remain some unclarified points.* In our perturbational treatment the problem of the invariance does not become explicit, for we are necessarily referred to special frames with regard to both kinds of transformation.

By the way we should like to touch the effects of the neutral scalar fields. As is well known, if we put $2e^2 = f^2$ where f is the interaction constant of the electron with this hypothetical field, the self-energy of an electron becomes finite. So, we can expect this field can also eliminate the mass type divergences. In fact, we obtain similar processes to those in § 3. by replacing a transverse photon $-\mathbf{j}^{\sim}$ by a quantum $-\mathbf{j}$ of the new field (i.e. *C-meson*). And the same condition $2e^2 = f^2$ really compensates the self-energy type divergences. *But those of the vacuum polarization type cannot be eliminated by introducing this new field alone, because the latter does not give rise to divergences of this type in the case of e^2-corrections to the Klein-Nishina formula.* This is the same result as in the case of elastic scattering of an electron[13].

[13] Ito, D., Koba, Z. & Tomonaga, S. (September, 1948). Corrections due to the Reaction of "Cohesive Force Field" to the Elastic Scattering of an Electron. I. *Prog. Theor. Phys.*, 3, 3, 276-89; https://doi.org/10.1143/ptp/3.3.276; (October, 1948). Corrections due to the Reaction of " Cohesive Force Field" to the Elastic Scattering of an Electron. II. *Idib.*, 3, 4, 325-37; https://doi.org/10.1143/ptp/3.4.325 [*abandoned C-meson field theory*].

§ 6. *Concluding remarks*

We have calculated the e^2-correction to the Klein-Nishina formula and have discussed to what extent the infinities can be disposed of by generalizing the *self-consistent subtraction method* applied in I. We believe to have determined in this way the modified form of the Hamiltonian describing the electronic and electromagnetic field in interaction, which should give reasonable result for any physical quantity, up to the order of e^3. Moreover, in deriving the above counter-terms in e^2 and e^3 by modification of m, e and $H_{el.mag.fld.}$ some of the e^4-terms are also suggested. For example, as the counter-term to modified Coulomb interaction, we introduce

...

into the interaction Hamiltonian. *This term does not contribute to the e^2-correction to the Klein-Nishina formula, to be sure, but it turns out effective to cancel out the divergent corrections in the case of electron-electron collision.* To complete our modified Hamiltonian up to e^4, however, it is necessary to investigate more complicated cases such as the fourth-order self-energy of the electron and the photon, etc.

Since we have made here only a first step of the inquiry into the problem of the field reaction, we are not of course able to insist upon the general validity of our prescription in approximations of still higher order. Nevertheless, it might not be quite unreasonable to expect that similar procedure could overcome the divergence difficulty in every approximation and, in most favorable case, we might arrive at a closed form of these modifications. It the next paper we shall treat the reaction problem in a more general manner.

Acknowledgement We wish to express our heartiest thanks to Prof. S. Tomonaga, without whose far-sighted suggestion, invaluable guidance and continual encouragement this work could not have been carried through at all. We are also very much obliged to Dr. M. Taketani for his kind interest taken in this work and for his pregnant discussions, and further to Mr. H. Umezawa and Mr. J. Yukawa of the Nagoya University, who visited our institute and discussed the problem with us (cf. footnote 12).

[12] Private conversation at Tokyo University, March 1948.

Note added in proof. With regard to the problematic quadratically divergent integral, refer to the remarks of Tati and Tomonaga added in proof to the foregoing paper.

Koba, Z. & Takeda, G. (June, 1948). Radiative Corrections in e^2 for an Arbitrary Process Involving Electrons. Positrons, and Light Quanta.

[*Prog. Theor. Phys*., 3, 2, 203-5, letter to the editor; https://doi.org/10.1143/ptp/3.2.203.]

Physics Institute, Tokyo University.

Letter to the editor dated May 22, 1948.

[Referenced in Dyson (February, 1949). The Radiation Theories of Tomonaga, Schwinger, and Feynman, but "full account in later issue of this journal" not found online, nor were any references found.]

Investigated further the *radiation correction in collision processes between electrons, positrons and photons* in a general manner and confirmed modified Hamiltonian sufficient to eliminate divergence difficulty *as long as we are concerned with the first radiative correction*, *introduces "transition diagram method"* [similar to the Stueckelberg (1934) diagrams, later introduced by Feynman at Ponoco in spring 1948 and published in September 1949, and subsequently renamed *Feynman diagrams*] *to analyze complicated connection between initial and final states through a number of intermediate ones*, expresses *electrons* and *positrons* as *world lines* in *momentum space*, while *emission* and *absorption* of *photons* are "*leaps*" of these *world lines*, diagram acquires two additional *leaps* (the *emission* and *reabsorption* of a *virtual photon*) when *radiative corrections* are taken into account, two distinct ways of attaching them, existing *world lines* gain new *leaps* or additional closed *world line* is introduced and coupled with existing ones through *virtual photon*, first case implies "*mass-type*" or "*self-energy-type*" divergences when *emission* and *reabsorption* of the *virtual photon* take place in succession which can be eliminated by *counter-self-energy term*, second case represents *vacuum polarization* corrections which can be subdivided according to number of *leaps* in additional closed *world line*, which afford quadratic and logarithmic divergences that are cancelled by the "*counter-terms*" and found not to contribute to divergence.

> [No reference to Stueckelberg's (1934) paper was included in this letter, despite Koba and Takeda being aware of it. Tomanaga, who discussed this paper with them (see acknowledgments), had referred to it in his paper Tomonaga (1943). On a Relativistically Invariant Formulation of the Quantum Theory of Wave Fields. Feynman referred to it in Feynman (April, 1948). Space-Time Approach to Non-Relativistic Quantum Mechanics.]

In a previous letter[1]

[1] Koba, Z. & Takeda, G. (March, 1948). Radiative Corrections for Compton Scattering. *Prog. Theor. Phys*., 3, 1, 98-9, letter to the editor; https://doi.org/10.1143/ptp/3.1.98 [; Bethe's theory assumes *self-energy* of free electron already included in observed mass of electron, applies Tati & Tomonaga "self-consistent" subtraction method to calculate e^2-corrections to the Klein-Nishina formula for the Compton scattering using *perturbation method,* two types of diverging terms, one related to *polarization of vacuum* and other to *self-energy* of electron; sufficient to add two new terms to Hamiltonian function to cancel out these infinite terms, *self-energy* logarithmic divergence eliminated by introducing new

term into interaction Hamiltonian derived in a plausible way as a "counter-term" which compensates change in free field Hamiltonian and conserves total Hamiltonian unaltered, one of *vacuum polarization* terms could be treated in similar way, while other term could not be foisted into the theory without radically changing the Maxwell equation for the free electromagnetic field].

we reported the results of our calculation concerning the e^2-correction to the Klein-Nishina formula due to the reaction of *radiation field. We have found that in order to obtain a convergent correction one has to modify the Hamilton function describing electronic and electromagnetic fields in interaction into the following*:

$$\begin{aligned} H = &\int u^*(X)\,\{(\alpha p) + m\beta\}u(X)\,dv + 1/8\pi \int\{\mathbf{E}^2(X) + \mathbf{H}^2(X)\}\,dv \\ &+ e\int\{u^*(X)\alpha u(X),\mathbf{A}(X)\}\,dv \\ &+ e^2/2 \iint u^*(X)u(X)u^*(X')u(X')/|\,\mathbf{X} - \mathbf{X}'\,|\,dvdv' \\ &- \delta m\int u^*(X)\beta u(X)\,dv - \delta e\int\{u^*(X)\alpha u(X),\mathbf{A}(X)\}\,dv \\ &- e\delta e \iint u^*(X)u(X)u^*(X')u(X')/|\,\mathbf{X} - \mathbf{X}'\,|\,dvdv' \\ &+ (e^2/3\pi)\int^\infty r\,dr \int \mathbf{A}^2\,dv \qquad (1) \end{aligned}$$

with

$$\delta m = (3me^2/2\pi)\int^\infty k\,dk$$
$$\delta e = -(e^3/3\pi)\int^\infty k\,dk$$

Here u, u* are quantized *amplitudes* of *electron fields,* **A** the transverse part of the *electromagnetic potential.* The derivation and interpretation of these modification are discussed in our recent work[2].

[2] Koba, Z. & Takeda, G. *Prog. Theor. Phys*., in press [Koba, Z. & Takeda, G. (December, 1948). Radiation Reaction in Collision Process, II: Radiative Corrections for Compton Scattering. *Prog. Theor. Phys*., 3, 4, 407-21; previous paper applied Tati & Tomonaga's *"self-consistent" subtraction method* to *elastic scattering of an electron,* all divergences were eliminated by counter-terms in modified form of the Hamilton function, but this was a simple and particularly favorable case, need to examine whether method is still effective in fundamental processes between elementary particles, applies method to calculate e^2-corrections to *Klein-Nishina formula for Compton scattering* using *perturbation method,* two types of diverging terms, one related to *polarization of vacuum* and other to *self-energy* of electron, sufficient to add two new terms to Hamiltonian function to cancel out these infinite terms, *self-energy* logarithmic divergence eliminated by introducing new term into interaction Hamiltonian derived in a plausible way as a "counter-term" which compensates change in free field Hamiltonian and conserves total Hamiltonian unaltered, one of *vacuum polarization* terms could be treated in similar way, while other term could not be foisted into the theory without radically changing the Maxwell equation for the free electromagnetic field, *far from final settlement of the fundamental problem in theory of elementary particles*]. Cf. also Tati, T. & Tomonaga, S., *ibid*., in press. [Tati, T. & Tomonaga, S. (December, 1948). A Self-Consistent Subtraction Method in the Quantum Field Theory, I. *Prog. Theor. Phys*., 3, 4, 391-406[; notes procedure previously used by Bloch and Nordsieck and then by Pauli and Fierz in the treatment of problems with the self-field of an electron by first separating the radiation field into part bound to the electron and part of unbound photons by means of a canonical transformation, they then obtained a

term in the Hamiltonian that could be interpreted as the *interaction energy* of electron with the radiation field bound to it, this term though infinite gave rise to a modification of mass of electron so that it could be amalgamated into the mass term in the Hamiltonian for the free electron, the term was dropped off to obtain *observed mass* when electron mass reinterpreted as already including it, same result as Bethe for level-shift of bound electron in *non-relativistic* treatment, Tati and Tomonaga considered it desirable to obtain *relativistic* generalization of this treatment using a canonical transformation, paper addresses *field reaction problem when no external field* to e^2 approximation, does not address infinity related to *vacuum polarization* effect that occurs in *relativistic* treatment, *omitted as cannot be amalgamated into equation for free radiation*, starts from *relativistic* formalism of quantum field theory proposed by Tomanaga [August, 1946. On a Relativistically Invariant Formulation of the Quantum Theory of Wave Fields] in which Schrodinger equation with Hamiltonian for interaction density between radiation and electron fields is $\{H(P) - i\, \partial/\partial C_P\}\, \Psi[C] = 0$ where H(P) is the *interaction energy density* between radiation and electron fields at the world point P, Ψ[C] is the generalized Schrodinger functional which is a functional of the space-like variable surface C in the four-dimensional world and $\partial/\partial C_P$ its partial functional differentiation at the point P, the point P being considered as lying on C, decomposes fields into parts oscillating with positive and negative corresponding to electron and positron, transforms Schrodinger functional retaining terms up to the order e^2, expands integrand into Fourier integral and defines 4-vector to obtain Hamiltonian for transformed equation that enables use of first order calculation, allows processes through intermediate states such as emission and reabsorption of virtual particles (self-energy of an electron) to be treated as direct processes, calculates the commutator and rearranges factors in each term into the correct order, identifies (1) terms to represent *interactions between electrons* caused by exchange of virtual photons including ordinary M¢ller interaction between two electrons (2) terms connected with *self-energy* of an electron (3) terms responsible for the *scattering of a photon* by a free electron, terms for *creation of a pair* by two photons (4) variables describing the radiation field only representing modification of the radiation field in vacuo due to the *polarization of the vacuum* that cause an infinite energy level shift of the vacuum itself and an infinite *self-energy* of a photon, and (5) a *mass-modifying term* with a logarithmic diverging quantity representing the electromagnetic mass of an electron, assumes *mass-modifying term* already included in free field equation so no level-shift caused by the interaction between the electron and radiation, subtraction method "*self-consistent*" in this sense, gives *relativistic* generalization of transformation that separates *radiation field* into field of "unbound" photons and field of photons bound to the electron, procedure similar to Hartree method of the *self-consistent* field in which *interaction* between electrons is considered as a perturbation but some part of its effect is already included].

Now we have investigated further the radiation correction in collision processes between electrons, positrons and photons in a general manner and confirmed that the above-mentioned Hamiltonian (1) is sufficient to eliminate the divergence difficulty as long as we are concerned with the first radiative correction, i.e. correction $\delta\sigma \sim e^{2n+2}$ to the zeroth cross section $\sigma_0 \sim e^{2n}$ (n being the order of perturbation required to obtain a non-vanishing matrix element for the process considered).

The detailed proof of this assertion is given in our full account which will appear in a later issue of this journal [not found].

The outline of our reasoning is as follows.
1) *As an effective tool to analyze complicated connection between initial and final states through a number of intermediate ones, a "transition diagram method" is introduced.*

It expresses *electrons* and *positrons* as *world lines* in the *momentum space*, while *emission* and *absorption* of *photons* are described as "*leaps*" of these *world lines*. Any process can be characterized by the number of *world lines* N and that of their *leaps* L. L is equal to Q+2(N − 1) in case of *no radiative corrections*, Q being the number of real *photons* involved.
2) When we take *radiative corrections* into account, our diagram acquires two additional *leaps* (the *emission* and *reabsorption* of a *virtual photon*). There are two distinct ways of attaching them: i) existing *world lines* gain new *leaps*, or ii) an additional closed *world line* is introduced and coupled with existing ones through the *virtual photon*.
3) The first case implies what we have called "*mass-type*" or "*self-energy-type*" divergences, i.e. when the *emission* and *reabsorption* of the *virtual photon* take place in succession (including Coulomb interaction). But such an effect can be eliminated by virtue of our *counter-self-energy term*, either its diagonal or non-diagonal part being efficacious.
4) The second case represents *vacuum polarization* corrections which can be subdivided into Q groups, [P, I]. [P, II]. ... [P, Q] according to the number of *leaps* in the additional closed *world line*. When one examines [P, I]'s in detail it is confirmed that they afford quadratic and logarithmic divergences which are just cancelled by the "*counter-terms*" in (1) [P, II]'s ... [P, Q]'s are found not to contribute to divergence.

We are very much obliged to Prof. S. Tomonaga and Dr. T. Miyazima for their kind interest taken in our work and for their valuable discussions.

Julian Seymour Schwinger (February 12, 1918 – July 16, 1994)

[Includes material from Nobel Lectures, Physics 1963-1970, Elsevier Publishing Company, Amsterdam, 1972.]

Schwinger is best known for his work on *quantum electrodynamics* (QED), in particular for developing a *relativistically* invariant *perturbation theory*, and for *renormalizing* QED to one loop order. He was jointly awarded the Nobel Prize in Physics in 1965, along with Shin'ichirō Tomonaga and Richard Feynman, "for their fundamental work in quantum electrodynamics, with deep-ploughing consequences for the physics of elementary particles". "Following the establishment of the theory of relativity and quantum mechanics, an initial *relativistic theory* was formulated for the interaction between charged particles and electromagnetic fields. However, partly because the electron's magnetic moment proved to be somewhat larger than expected, the theory had to be reformulated. Julian Schwinger solved this problem in 1948 through "renormalization" and thereby contributed to a new quantum electrodynamics." [Julian Schwinger – Facts. NobelPrize.org. https://www.nobelprize.org/prizes/physics/1965/ schwinger/facts/.]

Schwinger was born in New York City, to Ashkenazi Jewish parents, Belle (née Rosenfeld) and Benjamin Schwinger, a garment manufacturer, who had emigrated from Poland to the United States. Both his father and his mother's parents were prosperous clothing manufacturers, although the family business declined after the Wall Street Crash of 1929. The family followed the Orthodox Jewish tradition. Julian's older brother Harold Schwinger was born in 1911, seven years before Julian who was born in 1918.

Schwinger was a precocious student. He attended the Townsend Harris High School from 1932 to 1934, a highly regarded high school for gifted students at the time. During high school, he had already started reading *Physical Review* papers by authors such as Paul Dirac in the library of the City College of New York, in whose campus Townsend Harris was then located.

In the fall of 1934, Schwinger entered the City College of New York as an undergraduate. CCNY automatically accepted all Townsend Harris graduates at the time, and both institutions offered free tuition. Due to his intense interest in physics and mathematics, he performed very well in those subjects despite often skipping classes and learning directly from books. On the other hand, his lack of interest for other topics such as English led to academic conflicts with teachers of those subjects.

After he had joined CCNY, his brother Harold, who had previously graduated from CCNY, asked his ex-classmate Lloyd Motz to "get to know [Julian]". Lloyd was a CCNY physics instructor and Ph.D. candidate at Columbia University at the time. Lloyd made the acquaintance, and soon recognized Schwinger's talent. Noticing that he had academic problems, Lloyd decided to ask Isidor Isaac Rabi who he knew at Columbia for help. Rabi also immediately recognized Schwinger's capabilities on their first meeting, and then made arrangements to award him with a scholarship to study at Columbia. At first his bad grades in some subjects at CCNY prevented the scholarship award. But Rabi persisted and showed an unpublished paper on quantum electrodynamics written by Schwinger to Hans Bethe, who happened to be passing by New York. Bethe's approval of the paper and his reputation in that domain were then enough to secure the scholarship for Schwinger, who then transferred to Columbia. His academic situation at Columbia was much better than at CCNY. He accepted into the Phi Beta Kappa society and received his B.A. in 1936.

During his graduate studies, Rabi felt that it would be good for Schwinger to visit other institutions around the country, and he was awarded a travelling fellowship for the year 1937/38 which he spent at the University of Wisconsin in Madison working with Gregory Breit and Eugene Wigner. During this time, Schwinger, who previously had already had the habit of working until late at night, went further and made the day/night switch more complete, working at night and sleeping during the day, a habit he would carry throughout his career. Schwinger later commented that this switch was in part a way to retain greater intellectual independence and avoid being "dominated" by Breit and Wigner by simply reducing the duration of contact with them by working different hours.

Schwinger obtained his PhD overseen by Rabi in 1939 at the age of 21. During the fall of 1939 Schwinger started working at the University of California, Berkeley under J. Robert Oppenheimer, where he stayed for two years as an NRC fellow.

After having worked with Oppenheimer, Schwinger's first regular academic appointment was at Purdue University in 1941. While on leave from Purdue, he worked at the MIT Radiation Laboratory instead of at the Los Alamos National Laboratory during World War II. He provided theoretical support for the development of radar.

He first approached electromagnetic radar problems as a nuclear physicist, but soon began to think of nuclear physics in the language of electrical engineering. That would eventually emerge as the effective range formulation of nuclear scattering. Then, being conscious of the large microwave powers available, Schwinger began to think about electron accelerators, which led to the question of radiation by electrons in magnetic fields. *In studying the latter problem, he was reminded, at the classical level, that the reaction of the electron's field alters the properties of the particle, including its mass*. This would be significant in the intensive developments of quantum electrodynamics, which were soon to follow.

After the war, Schwinger left Purdue for Harvard University, where he taught from 1945 to 1974; accepting an appointment as associate professor and two years later he becoming a full professor. That was also the year of his marriage to Clarice Carrol of Boston. At first he continued research in nuclear physics and in classical diffraction. The Shelter Island conference of 1947 changed all that. He and Weisskopf suggested to Bethe that electrodynamic processes were responsible for the Lamb shift, which had been known for some time as the Pasternack effect. Immediately, however, Schwinger saw that the most direct consequence of quantum electrodynamics lay in the hyperfine anomaly reported for the first time at Shelter Island. He anticipated that the effect was due to an induced *anomalous magnetic moment* of the electron. The actual calculation had to wait three months, while Schwinger took an extended honeymoon, but by December 1947 Schwinger had his famous result for the gyromagnetic ratio. *In the process he invented renormalization of mass and charge*, only dimly prefigured by Kramers.

This first formulation of quantum electrodynamics was rather crude, being *noncovariant*; to obtain the correct Lamb shift, a *relativistic* formulation was required, which followed the next year. A comedy of errors ensued: both Feynman and Schwinger made an incorrect patch between hard and soft photon processes, and so obtained identical, but incorrect, predictions for the Lamb shift, and the weight of their reputations delayed the publication of the correct, if pedestrian, calculation by French and Weisskopf in 1948.

Schwinger later claimed that his first noncovariant approach had yielded the correct result, except that he had not trusted it. Schwinger developed an affinity for Green's functions from his radar work, and he used these methods to formulate quantum field theory in terms of local Green's functions in a *relativistically* invariant way. This allowed him to calculate unambiguously the first corrections to the electron *magnetic moment* in quantum electrodynamics. Earlier non-covariant work had arrived at infinite answers, but the extra symmetry in his methods allowed Schwinger to isolate the correct finite corrections. Schwinger developed *renormalization*, formulating *quantum electrodynamics* unambiguously to one-loop order.

Between 1946 and 1948, Japanese physicist Sin-Itiro Tomonaga published a series of papers on a formulation of quantum electrodynamics, which was *relativistically* invariant and eliminated ultraviolet divergences via *renormalization.* A few weeks after Feynman and Schwinger reported their own efforts in this area at the Pocono conference in March 1948, Robert Oppenheimer received a letter from Tomonaga describing his work and immediately recognized the overlap in their efforts.

In the same era, he introduced non-perturbative methods into quantum field theory, by calculating the rate at which electron–positron pairs are created by tunneling in an electric field, a process now known as the "Schwinger effect." This effect could not be seen in any finite order in *perturbation theory.*

Schwinger's foundational work on quantum field theory constructed the modern framework of *field correlation functions* and their *equations of motion.* His approach started with a quantum *action* and allowed bosons and fermions to be treated equally for the first time, using a differential form of Grassman integration. He gave elegant proofs for the *spin-statistics theorem* and the CPT theorem, and noted that the field algebra led to anomalous Schwinger terms in various classical identities, because of short distance singularities. These were foundational results in *field theory*, instrumental for the proper understanding of anomalies.

In other notable early work, Rarita and Schwinger formulated the abstract Pauli and Fierz theory of the spin-3/2 field in a concrete form, as a vector of Dirac spinors, Rarita–Schwinger equation. In order for the spin-3/2 field to interact consistently, some form of *supersymmetry* is required, and Schwinger later regretted that he had not followed up on this work far enough to discover *supersymmetry*.

Schwinger discovered that neutrinos come in multiple varieties, one for the electron and one for the muon. Nowadays there are known to be three light neutrinos; the third is the partner of the tau lepton.

In the 1960s, Schwinger formulated and analyzed what is now known as the Schwinger model, quantum electrodynamics in one space and one time dimension, the first example of a confining theory. He was also the first to suggest an *electroweak gauge theory*, an SU(2) gauge group spontaneously broken to electromagnetic U(1) at long distances. This was extended by his student Sheldon Glashow into the accepted pattern of *electroweak unification*. He attempted to formulate a theory of quantum electrodynamics with point *magnetic monopoles*, a program which met with limited success because monopoles are strongly interacting when the quantum of charge is small.

Having supervised 73 doctoral dissertations, Schwinger is known as one of the most prolific graduate advisors in physics. Four of his students won Nobel prizes: Roy Glauber, Benjamin Roy Mottelson, Sheldon Glashow and Walter Kohn (in chemistry).

Schwinger had a mixed relationship with his colleagues, because he always pursued independent research, different from mainstream fashion. In particular, Schwinger developed the *source theory*, a phenomenological theory for the physics of *elementary particles*, which is a predecessor of the modern *effective field theory*. It treats quantum fields as long-distance phenomena and uses auxiliary 'sources' that resemble currents in classical field theories. The *source theory* is a mathematically consistent field theory with clearly derived phenomenological results. The criticisms by his Harvard colleagues led Schwinger to leave the faculty in 1972 for UCLA. It is a story widely told that Steven Weinberg, who inherited Schwinger's paneled office in Lyman Laboratory, there found a pair of old shoes, with the implied message, "think you can fill these?" At UCLA, and for the rest of his career, Schwinger continued to develop the *source theory* and its various applications.

After 1989 Schwinger took a keen interest in the non-mainstream research of *cold fusion*. He wrote eight theory papers about it. He resigned from the American Physical Society after their refusal to publish his papers. He felt that *cold fusion* research was being suppressed and academic freedom violated. He wrote, "The pressure for conformity is enormous. I have experienced it in editors' rejection of submitted papers, based on venomous criticism of anonymous referees. The replacement of impartial reviewing by censorship will be the death of science."

In his last publications, Schwinger proposed a theory of *sonoluminescence* as a long-distance quantum radiative phenomenon associated not with atoms, but with fast-moving surfaces in the collapsing bubble, where there are discontinuities in the dielectric constant. The mechanism of *sonoluminescence* now supported by experiments focuses on superheated gas inside the bubble as the source of the light.

Schwinger's awards and honors were numerous even before his 1965 Nobel win. They include the first Albert Einstein Award (1951), the U.S. National Medal of Science (1964), honorary D.Sc. degrees from Purdue University (1961) and Harvard University (1962), and the Nature of Light Award of the U.S. National Academy of Sciences (1949). In 1987, Schwinger received the Golden Plate Award of the American Academy of Achievement.

As a famous physicist, Schwinger was often compared to another legendary physicist of his generation, Richard Feynman. Schwinger was more formally inclined and favored symbolic manipulations in quantum field theory. He worked with *local field operators*, and found relations between them, and he felt that physicists should understand the algebra of *local fields*, no matter how paradoxical it was. By contrast, Feynman was more intuitive, believing that the physics could be extracted entirely from the Feynman diagrams, which gave a *particle picture*. Schwinger commented on Feynman diagrams in the following way,

> "Like the silicon chips of more recent years, the Feynman diagram was bringing computation to the masses."

Schwinger disliked Feynman diagrams because he felt that they made the student focus on the *particles* and forget about local *fields*, which in his view inhibited understanding. He went so far as to ban them altogether from his class, although he understood them perfectly well. The true difference is however deeper, and it was expressed by Schwinger in the following passage,

> "Eventually, these ideas led to Lagrangian or *action* formulations of quantum mechanics, appearing in two distinct but related forms, which I distinguish as differential and integral. The latter, spearheaded by Feynman has had all the press coverage, but I continue to believe that the differential viewpoint is more general, more elegant, more useful."

Despite sharing the Nobel Prize, Schwinger and Feynman had a different approach to *quantum electrodynamics* and to *quantum field theory* in general. Feynman used a *regulator*, while Schwinger was able to formally renormalize to one loop without an explicit *regulator*. Schwinger believed in the formalism of *local fields*, while Feynman had faith in the *particle paths*. They followed each other's work closely, and each respected the other.

Schwinger died of pancreatic cancer on 16 July 1994, in Los Angeles, CA. He is buried at Mount Auburn Cemetery; $\alpha/2\pi$, where α is the fine structure constant, is engraved above his name on his tombstone. These symbols refer to his calculation of the correction ("anomalous") to the *magnetic moment* of the electron.

Schwinger, J. (February, 1948). On quantum-electrodynamics and the magnetic moment of the electron.

[*Phys. Rev.*, 73, 4, 416–417, letter to the editor; https://doi.org/10.1103/ PhysRev.73.416; also in Schwinger, J. (ed). (1958). *Selected Papers on Quantum electrodynamics*. Dover, New York, page 142.]

Sent December 30, 1947.

Harvard University, Cambridge, Massachusetts.

Electrodynamics unquestionably requires revision at ultra-relativistic energies, desirable to isolate those aspects of the current theory that essentially involve high energies and are subject to modification by a more satisfactory theory, this goal has been achieved by transforming the Hamiltonian of current *hole theory* electrodynamics to exhibit explicitly the logarithmically divergent self-energy of a free electron which arises from the virtual emission and absorption of light quanta, the electromagnetic *self-energy* of a free electron can be ascribed to an *electromagnetic mass* which must be added to the mechanical mass of the electron, new Hamiltonian involves experimental *electron mass* rather than unobservable *mechanical mass*, electron now interacts with radiation field only in presence of external field such that only an accelerated electron can emit or absorb a light quantum, *interaction energy* of electron with external field subject to *finite* radiative correction, *polarization of the vacuum* still produces logarithmically divergent term proportional to *interaction energy* of electron in an external field, such term equivalent to altering value of *electron charge* by constant factor with only final value being identified with experimental charge, interaction between matter and radiation produces *renormalization* of electron charge and mass, all divergences contained in *renormalization* factors, radiative correction for energy of electron in external magnetic field corresponds to *additional magnetic moment associated with electron spin* of magnitude $\delta\mu/\mu = (\frac{1}{2}\pi)e^2/hc = 0.001162$, experimental measurements on hyperfine splitting of ground states of atomic hydrogen and deuterium larger than expected from directly measured nuclear moments, finds additional *electron spin magnetic moment* to account for measured hydrogen and deuterium *hyperfine structures* to be $\delta\mu/\mu = 0.00126$ and $\delta\mu/\mu = 0.00131$ respectively, these discrepancies accounted for by additional spin magnetic moment to the electron of $\delta\mu/\mu = 0.0018 \pm 0.00003$, *values yielded by relativistic calculation of Lamb shift differ only slightly from those conjectured by Bethe on basis of non-relativistic calculation and are in good accord with experiment.*

> [The first calculation of the Lamb shift by Bethe was rather crude, being *noncovariant*; to obtain the correct Lamb shift, a *relativistic* formulation was required, which followed the next year. A comedy of errors ensued: both Feynman and Schwinger made an incorrect patch between hard and soft photon processes, and so obtained identical, but incorrect, predictions for the Lamb shift, and the weight of their reputations delayed the publication of the correct, if pedestrian, calculation by French and Weisskopf in 1948. Schwinger later claimed that his first noncovariant approach had yielded the correct result, except that he had not trusted it.]

Attempts to evaluate *radiative corrections* to electron phenomena have heretofore been beset by divergence difficulties, attributable to *self-energy* and *vacuum polarization* effects. *Electrodynamics unquestionably requires revision at ultra-relativistic energies*, but is presumably accurate at moderate relativistic energies. It would be desirable, therefore, to isolate those aspects of the current theory that essentially involve high energies, and are subject to modification by a more satisfactory theory, from aspects that involve only moderate energies and are thus relatively trustworthy. *This goal has been achieved by transforming the Hamiltonian of current hole theory electrodynamics to exhibit explicitly the logarithmically divergent self-energy of a free electron, which arises from the virtual emission and absorption of light quanta. The electromagnetic self-energy of a free electron can be ascribed to an electromagnetic mass, which must be added to the mechanical mass of the electron.* Indeed, the only meaningful statements of the theory involve this combination of masses, which is the *experimental mass* of a free electron. It might appear, from this point of view, that the divergence of the *electromagnetic mass* is unobjectionable, since the individual contributions to the *experimental mass* are unobservable. However, *the transformation of the Hamiltonian is based on the assumption of a weak interaction between matter and radiation, which requires that the electromagnetic mass be a small correction* ($\approx (e^2/hc)\, m_0$) *to the mechanical mass* m_0.

The new Hamiltonian is superior to the original one in essentially three ways:
[1] it involves the experimental *electron mass*, rather than the unobservable *mechanical mass*;
[2] an electron now interacts with the radiation field only in the presence of an external field, that is, *only an accelerated electron can emit or absorb a light quantum**;

*A classical *non-relativistic theory* of this type was discussed by H. A. Kramers at the Shelter Island Conference, held in June 1947 under the auspices of the National Academy of Sciences.

[3] the *interaction energy* of an electron with an external field is now subject to a *finite* radiative correction.

In connection with the last point, it is important to note that the inclusion of the *electromagnetic mass* with the *mechanical mass* does not avoid all divergences; the *polarization of the vacuum* produces a logarithmically divergent term proportional to the *interaction energy* of the electron in an external field. However, it has long been recognized that such a term is equivalent to altering the value of the *electron charge* by a constant factor, only the final value being properly identified with the *experimental charge*. Thus, *the interaction between matter and radiation produces a renormalization of the electron charge and mass, all divergences being contained in the renormalization factors.*

The simplest example of a radiative correction is that for the energy of an electron in an external magnetic field. The detailed application of the theory shows that *the radiative correction to the magnetic interaction energy corresponds to an additional magnetic moment associated with the electron spin, of magnitude* $\delta\mu/\mu = (\frac{1}{2}\pi)e^2/hc = 0.001162$. It is indeed gratifying that recently acquired experimental data confirm this prediction.

Measurements on the hyperfine splitting of the *ground states* of atomic hydrogen and deuterium[1] have yielded values that are definitely larger than those to be expected from the directly measured *nuclear moments* and an *electron moment* of one Bohr magneton.

[1] Nafe, J. E., Nelson, E. B. & Rabi, I. I. (June, 1947). The Hyperfine Structure of Atomic Hydrogen and Deuterium. *Phys, Rev.*, 71, 12, 914; https://doi.org/10.1103/PhysRev.71.914; Nagel, D. E., Julian, R. S. & Zacharias, J. R. (November, 1947). Hyperfine Structure of Atomic Hydrogen and Deuterium. *Phys. Rev.*, 72, 10, 971.

These discrepancies can be accounted for by a small additional electron spin magnetic moment[2].

[2] Breit, G. (1947). *Phys. Rev.*, 71, 984. However, Breit has not correctly drawn the consequences of his empirical hypothesis. The effects of a nuclear magnetic field and a constant magnetic field do not involve different combinations of μ and $\delta\mu$.

Recalling that the *nuclear moments* have been calibrated in terms of the *electron moment*, we find the additional *moment* necessary to account for the measured hydrogen and deuterium *hyperfine structures* to be $\delta\mu/\mu = 0.00126 \pm 0.00019$ and $\delta\mu/\mu = 0.00131 \pm 0.00025$, respectively. These values are not in disagreement with the theoretical prediction. More precise conformation is provided by measurement of the g values for the $^2S_{1/2}$, $^2P_{1/2}$, and $^2P_{3/2}$ states of sodium and gallium[3].

[3] Kusch, P. & Foley, H. M. (December, 1947). Precision Measurement of the Ratio of the Atomic `g values' in the $^2P_{3/2}$ and $^2P_{1/2}$ States of Gallium. *Phys. Rev.*, 72, 12, 1256; https://doi.org/10.1103/PhysRev.72.1256.2, and further unpublished work.

To account for these results, it is necessary to ascribe the following additional spin magnetic moment to the electron, $\delta\mu/\mu = 0.00118 \pm 0.00003$.

The radiative correction to the energy of an electron in a Coulomb field will produce a shift in the *energy levels* of hydrogen-like atoms, and *modify the scattering of electrons in a Coulomb field.* Such energy level displacements have recently been observed in the fine structures of hydrogen[4], deuterium, and ionized helium[5].

[4] Lamb, Jr., W. E. & Retherford, R. C. (August, 1947). Fine Structure of the Hydrogen Atom by a Microwave Method. *Phys. Rev.*, 72, 3, 241-3[; the spectrum of hydrogen has a fine structure of the energy levels which according to the *Dirac wave equation* for an electron moving in a Coulomb field is due to the combined effects of *relativistic* variation of mass with velocity and spin-orbit coupling, according to this theory the $2^2S_{1/2}$ state should exactly coincide in energy with the $2^2P_{1/2}$ state which is the lower of the two P states, previous attempts at measurement have alternated between finding confirmation and discrepancies of as much as eight percent, using microwave method depending on a novel property of the $2^2S_{1/2}$ level results indicate that contrary to the *Dirac wave equation* the $2^2S_{1/2}$ state is higher than the $2^2P_{1/2}$ by about 1000 Mc/sec].

[5] Mack, J. E. & Austern, N. (1947). *Phys. Rev.*, 72, 972.

The values yielded by our theory differ only slightly from those conjectured by Bethe[6] *on the basis of a non-relativistic calculation*, and are, thus, in good accord with experiment.

[6] Bethe, H. A. (August, 1947). The Electromagnetic Shift of Energy Levels. *Phys. Rev.*, 72, 339-41[; Lamb and Retherford results show that fine structure of second quantum state of hydrogen does not agree with *Dirac wave equation*, Schwinger, Weisskopf, and Oppenheimer suggest might be due to *shift of energy levels by interaction of electron with the radiation field*, this shift comes out infinite in all existing theories and has therefore always been ignored, possible to identify the most strongly (linearly) divergent term in the level shift with an *electromagnetic mass effect* which must exist for a bound as well as for a free electron, already included in the *observed mass* of the electron so should be subtracted, assumes *relativistic cut-off* in quantum energies (frequencies) of included atomic states, then calculation of Lamb shift for hydrogen atom using *non-relativistic* ordinary radiation theory gives shift of the levels due to *radiation interaction* in close agreement with observed value, removes discrepancy with Dirac theory, did not carried out *relativistic* calculations].

Finally, *the finite radiative correction to the elastic scattering of electrons by a Coulomb field provides a satisfactory termination to a subject that has been beset with much confusion.* A paper dealing with the details of this theory and its applications is in course of preparation.

Feynman, R. P. (April, 1948). Space-Time Approach to Non-Relativistic Quantum Mechanics.

[*Rev. Mod. Phys.*, 20, 2, 367-87; https://doi.org/10.1103/ RevModPhys.20.367; also in Schwinger, J. (ed). (1958). *Selected Papers on Quantum electrodynamics*. Dover, New York, pages 321-41.]

Cornell University, Ithaca, New York.

Third formulation of *non-relativistic* quantum theory in addition to *differential equation of Schroedinger* and *matrix algebra of Heisenberg*, *path integral formulation* utilizing the *action* principle as suggested in Dirac (1933) [The Lagrangian in Quantum Mechanics] and Dirac (1945) [On the Analogy Between Classical and Quantum Mechanics], in quantum mechanics the probability of an event which can happen in several different ways is the absolute square of a sum of complex contributions one from each alternative way $\varphi_{ac} = \sum_b \varphi_{ab}\varphi_{bc}$ where φ_{ab}, φ_{bc}, φ_{ac} are complex numbers such that $P_{ab} = |\varphi_{ab}|^2$, $P_{bc} = |\varphi_{bc}|^2$, and $P^q_{ac} = |\varphi_{ac}|^2$, and P_{ab} as the probability that if measurement A gave the result *a*, then measurement B will give the result *b*, and P^q_{ac} is the quantum mechanical probability that a measurement of C results in c when it follows a measurement of A giving a, the probability that a particle will be found to have a path lying somewhere within a region of space time is the absolute square of a sum of contributions - one from each path in the region, the contribution from a single path is postulated to be an exponential whose (imaginary) phase is the classical *action* for the path in question where *action* refers to time integral of Lagrangian along a path, restriction to finite time interval, the total contribution from all paths reaching x, t from the past is the *wave function* $\psi(x, t)$, shown to satisfy Schroedinger's equation, *probability amplitude for a space-time path associated with entire motion of particle as function of time rather than with position of particle at particular time*, establishes postulates that describe *non-relativistic quantum mechanics neglecting spin*, mathematically equivalent to Heisenberg and Schroedinger formulations, no fundamentally new results, *suffers serious drawbacks*, requires unnatural and cumbersome division of the time interval, not formulated so that it is physically obvious that it is invariant under unitary transformations, improvements could be made through use of notation and concepts of mathematics of functionals.

> [Feynman, R. P. – Nobel Lecture, December 11, 1965. *The Development of the Space-Time View of Quantum Electrodynamics*: "I worked on this problem about eight years until the final publication in 1947."]
>
> [As noted by Feynman, "This article contains most of what was in the thesis. [Feynman, R. P. (1942). *The Principle of Least Action in Quantum Mechanics*. Ph.D. thesis, Princeton University (in Brown, L. M. (ed.) (2005). *Feynman's Thesis. A New Approach to Quantum Theory*. World Scientific); *non-relativistic* quantum mechanics, *space-time* viewpoint, principle of *least action*, functional, Lagrangian, *action functions*.] The thesis contained in addition a discussion of the relations between constants of motion such as energy and momentum and invariance properties of an *action* functional. Further there is a much more thorough discussion of the possible generalization of quantum mechanics to apply to more

general functionals than appears in the *Review* article. Finally, the properties of a system interacting through intermediate harmonic oscillator is discussed in more detail."]

[Oppenheimer, J. R. (1948). Electron Theory. *Report to the Solvay Conference for Physics at Brussels*, Belgium, September 27 to October 2, 1948, pages 6-7. (Reprint in Schwinger, J. (ed). (1958). *Selected Papers on Quantum electrodynamics*, Dover, New York, pages 150-1: "Now it is true that the fundamental equations of quantum-electrodynamics are gauge and Lorentz covariant. But they have in a strict sense no solutions expansible in powers of e. If one wishes to explore these solutions, bearing in mind that certain infinite terms will, in a later theory, no longer be infinite, one needs a covariant way of identifying these terms, and for that, not merely the field equations themselves, but *the whole method of approximation and solution must at all stages preserve covariance. This means that the familiar Hamiltonian methods, which imply a fixed Lorentz frame t = constant, must be renounced*; neither Lorentz frame nor gauge can be specified until after, in a given order in e, all terms have been identified, and those bearing on the definition of charge and mass recognized and relegated; then of course, in the actual calculation of transition probabilities and the reactive corrections to them, or in the determination of stationary states in fields which can be treated as static, and in the reactive corrections thereto, the introduction of a definite coordinate system and gauge for these no longer singular and completely well-defined terms can lead to no difficulty.

It is probable that, at least to order e^2, more than one covariant formalism can be developed. Thus, *Stueckelberg's four-dimensional perturbation theory*[26] would seem to offer a suitable starting point, as also do the related algorithms of Feynman[27].

[26] Stueckelberg, E. C. G. (September, 1934). Relativistisch invariante Störungstheorie des Diracschen Elektrons I. Teil: Streustrahlung und Bremsstrahlung. (Relativistically invariant perturbation theory of Dirac's electron Part I: scattered radiation and Bremsstrahlung.) *Ann. Phys.*, 413, 4, 367-89.

[27] Feynman, R. P. (November, 1948). Relativistic Cut-Off for Quantum Electrodynamics. *Phys. Rev.*, 74, 1430-8.]

Abstract

Non-relativistic quantum mechanics is formulated here in a different way. It is, however, mathematically equivalent to the familiar formulation. In quantum mechanics the probability of an event which can happen in several different ways is the absolute square of a sum of complex contributions, one from each alternative way. The probability that a particle will be found to have a path x(t) lying somewhere within a region of space time is the square of a sum of contributions, one from each path in the region. The contribution from a single path is postulated to be an exponential whose (imaginary) phase is the classical *action* (in units of $\hbar$) for the path in question. The total contribution from all paths reaching x, t from the past is the wave function $\psi(x, t)$. This is shown to satisfy Schroedinger's equation. The relation to matrix and operator algebra is discussed. Applications are indicated, in particular to eliminate the coordinates of the field oscillators from the equations of quantum electrodynamics.

1. *Introduction*

It is a curious historical fact that modern quantum mechanics began with two quite different mathematical formulations: the *differential equation of Schroedinger*, and the *matrix algebra of Heisenberg*. The two, apparently dissimilar approaches, were proved to be mathematically equivalent. These two points of view were destined to complement one another and to be ultimately synthesized in Dirac's *transformation theory.*

This paper will describe what is essentially a *third formulation of non-relativistic quantum theory. This formulation was suggested by some of Dirac's*[1,2] *remarks concerning the relation of classical action*[3] *to quantum mechanics.*

[1] Dirac, P. A. M. (1935). *The Principles of Quantum Mechanics*. The Clarendon Press, Oxford, second edition, Section 33; also, Dirac, P. A. M. (1933). The Lagrangian in Quantum Mechanics. *Phys. Zeit. Sowjet.*, 3, 1, 64-72[; alternative formulation of quantum mechanics in terms of Lagrangian in place of Hamiltonian, *coordinates* and *velocities* instead of *coordinates* and *momenta*, allows *equations of motion* to be expressed as stationary property of *action function* which is the time-integral of the Lagrangian and *relativistic invariant*, there is no corresponding *action principle* in terms of *coordinates and momenta* in Hamiltonian theory, Lagrangian closely connected with theory of *contact transformations*, transformation functions have classical analogues expressible in terms of the Lagrangian, function of coordinates at time t and time t + dt rather than of the coordinates and velocities, applies to *field dynamics* using suitable field quantities or potentials as coordinates, "*many-time*" theory, each coordinate a function of four *space-time* variables instead of one time variable in particle theory, *generalized transformation function*].

[2] Dirac, P. A. M. (April, 1945). On the Analogy Between Classical and Quantum Mechanics. *Rev. Mod. Phys*., 17, 195[; mathematical methods available for working with *non-commuting quantities* much weaker than those available for *commuting quantities* owing to the fact that the only functions of *non-commuting variables* that one has been able to define are those expressible algebraically, shows how this difficulty can be avoided in the case when the *non-commuting quantities* are *observables*, can set up a theory of functions of them (a generalization of the concept known as *well-ordered functions*) of almost the same degree of generality as the usual functions of *commuting variables*, can use this theory to make closer analogy between classical and quantum mechanics, enables discussion of trajectories for motion of a particle in quantum mechanics, method is given for defining *general functions of non-commuting observables*, method is developed to provide formal *probability* for *non-commuting observables* to have numerical values (in general a complex number), also enables the analogy between classical and quantum *contact transformations* to be set up on a more general basis].

[3] Throughout this paper *the term "action" will be used for the time integral of the Lagrangian along a path.* When this path is the one actually taken by a particle, moving classically, the integral should more properly be called *Hamilton's first principle function*.

A probability amplitude is associated with an entire motion of a particle as a function of time, rather than simply with a position of the particle at a particular time.

[The *probability amplitude* was first mentioned in Dirac (May, 1932), Relativistic Quantum Mechanics, page 456.]

The formulation is mathematically equivalent to the more usual formulations. There are, therefore, no fundamentally new results. However, there is a pleasure in recognizing old things from a new point of view. Also, there are problems for which the new point of view offers a distinct advantage. For example, if two systems A and B, *interact*, the coordinates of one of the systems, say B, may be eliminated from the equations describing the motion of A. The *interaction* with B is represented by a change in the formula for the *probability amplitude* associated with a motion of A. It is analogous to the classical situation in which the effect of B, can be represented by a change in the *equations of motion* of A (by the introduction of terms representing forces acting on A). In this way the coordinates of the transverse, as well as of the longitudinal field oscillators, may be eliminated from the equations of quantum electrodynamics.

In addition, there is always the hope that the new point of view will inspire an idea for the modification of present theories, a modification necessary to encompass present experiments.

We first discuss the general concept of the *superposition of probability amplitudes* in quantum mechanics. We then show how this concept can be directly extended to define a *probability amplitude* for any motion or path (position vs. time) in space-time. The ordinary quantum mechanics is shown to result from the postulate that *this probability amplitude has a phase proportional to the action, computed classically, for this path.* This is true when the *action* is the time integral of a quadratic function of velocity. The relation to matrix and operator algebra is discussed in a way that stays as close to the language of the new formulation as possible. There is no practical advantage to this, but the formulae are very suggestive if a generalization to a wider class of *action functionals* is contemplated. Finally, we discuss applications of the formulation. As a particular illustration, we show how the *coordinates* of a harmonic oscillator may be eliminated from the *equations of motion* of a system with which it *interacts*. This can be extended directly for application to *quantum electrodynamics. A formal extension which includes the effects of spin and relativity is described.*

2. *The Superposition of Probability Amplitudes*

The formulation to be presented contains as its essential idea the concept of a *probability amplitude associated with a completely specified motion as a function of time*. It is, therefore, worthwhile to review in detail the quantum-mechanical concept of the *superposition of probability amplitudes*. We shall examine the essential changes in physical outlook required by the transition from classical to quantum physics.

For this purpose, consider an imaginary experiment in which we can make three measurements successive in time: first of a quantity A, then of B, and then of C. There is really no need for these to be of different quantities, and it will do just as well if the example of three successive position measurements is kept in mind. Suppose that a is one of a number of possible results which could come from measurement A, b is a result that could arise from B, and c is a result possible from the third measurement C[4].

[4] For our discussion it is not important that certain values of a, b, or c might be excluded by quantum mechanics but not by classical mechanics. For simplicity, assume the values are the same for both but that the probability of certain values may be zero.

We shall assume that the measurements A, B, and C are the type of measurements that completely specify a *state* in the quantum-mechanical case. That is, for example, the state for which B has the value b is not degenerate.

It is well known that quantum mechanics deals with *probabilities*, but naturally this is not the whole picture. In order to exhibit, even more clearly, the relationship between classical and quantum theory, we could suppose that classically we are also dealing with probabilities but that all probabilities either are zero or one. A better alternative is to imagine in the classical case that the probabilities are in the sense of classical statistical mechanics (where, possibly, internal coordinates are not completely specified).

We define P_{ab} as the probability that if measurement A gave the result a, then measurement B will give the result b. Similarly, P_{bc} is the probability that if measurement B gives the result b, then measurement C gives c. Further, let P_{ac} be the chance that if A gives a, then C gives c. Finally, denote by P_{abc} the probability of all three, i.e., if A gives a, then B gives b, and C gives c. If the events between a and b are independent of those between b and c, then

$$P_{abc} = P_{ab}P_{bc}. \qquad (1)$$

This is true according to quantum mechanics when the statement that B is b is a complete specification of the state.

In any event, we expect the relation

$$P_{ac} = \sum_b P_{abc}. \qquad (2)$$

This is because, if initially measurement A gives a and the system is later found to give the result c to measurement C quantity B must have had some value at the time intermediate to A and C. The probability that it was b is P_{abc}. We sum, or integrate, over all the mutually exclusive alternatives for b (symbolized by $\sum_b$).

Now, *the essential difference between classical and quantum physics lies in Eq. (2). In classical mechanics it is always true. In quantum mechanics it is often false*. We shall denote the quantum-mechanical probability that a measurement of C results in c when it follows a measurement of A giving a by P^q_{ac}. Equation (2) is replaced in quantum mechanics by this remarkable law[5].

[5] We have assumed b is a non-degenerate state, and that therefore (1) is true. Presumably, if in some generalization of quantum mechanics (1) were not true, even for pure states b, (2) could be expected to be replaced by: There are complex numbers φ_{abc} such that $P_{abc} = |\varphi_{abc}|^2$. The analog of (5) is then $\varphi_{ac} = \sum_b \varphi_{abc}$.

There exist complex numbers φ_{ab}, φ_{bc}, φ_{ac} such that

$$P_{ab} = |\varphi_{ab}|^2, \qquad P_{bc} = |\varphi_{bc}|^2, \qquad \text{and} \qquad P^q_{ac} = |\varphi_{ac}|^2. \tag{3}$$

The classical law, obtained by combining (1) and (2)

$$P_{ac} = \sum_b P_{ab} P_{bc} \tag{4}$$

is replaced by

$$\varphi_{ac} = \sum_b \varphi_{ab}\varphi_{bc}. \tag{5}$$

If (5) is correct, ordinarily (4) is incorrect. The logical error made in deducing (4) consisted, of course, in assuming that to get from a to c the system had to go through a condition such that B had to have some definite value, b.

If an attempt is made to verify this, i.e., if B is measured between the experiments A and C, then formula (4) is, in fact, correct. More precisely, if the apparatus to measure B is set up and used, but no attempt is made to utilize the results of the B measurement in the sense that only the A to C correlation is recorded and studied, then (4) is correct. This is because the B measuring machine has done its job; if we wish, we could read the meters at any time without disturbing the situation any further. The experiments which gave a and c can, therefore, be separated into groups depending on the value of b. Looking at probability from a frequency point of view (4) simply results from the statement that in each experiment giving a and c, B had some value. The only way (4) could be wrong is the statement, "B had some value," must sometimes be meaningless. Noting that (5) replaces (4) only under the circumstance that we make no attempt to measure B, we are led to say that the statement, "B had some value," may be meaningless whenever we make no attempt to measure B[6].

[6] It does not help to point out that we could have measured B had we wished. The fact is that we did not.

Hence, we have different results for the correlation of a and c, namely, Eq. (4) or Eq. (5), depending upon whether we do or do not attempt to measure B. No matter how subtly one tries, the attempt to measure B must disturb the system, at least enough to change the results from those given by (5) to those of (4)[7].

[7] How (4) actually results from (5) when measurements disturb the system has been studied particularly by J. von Neumann [(1943). *Mathematische Grundlagen der Quantenmahanik.*

Dover Publications, New York]. The effect of perturbation of the measuring equipment is effectively to change the phase of the interfering components, by θ_b, say, so that (5) becomes $\varphi_{ac} = \sum_b e^{i\theta b}\varphi_{ab}\varphi_{bc}$. However, as von Neumann shows, the phase shifts must remain unknown if B is measured so that the resulting probability P_{ac} is the square of φ_{ac}, averaged over all phases, θ_b. This results in (4).

That measurements do, in fact, cause the necessary disturbances, and that, essentially, (4) could be false was first clearly enunciated by Heisenberg in his uncertainty principle. *The law (5)*

$$[\varphi_{ac} = \sum_b \varphi_{ab}\varphi_{bc}. \qquad (5)]$$

is a result of the work of Schroedinger, the statistical interpretation of Born and Jordan, and the transformation theory of Dirac[8].

[8] If A and B are the operators corresponding to measurements A and B, and if ψa, and ψb are solutions of $A\psi_a = a\psi_a$, and $B\chi_b = b\chi_b$, then $\varphi_{ab} = \int \chi^*_b\psi_a dx = (\chi^*_b, \psi_a)$. Thus, ψ_{ab} is an element $(a|b)$ of the transformation matrix for the transformation from a representation in which A is diagonal to one in which B is diagonal.

Equation (5) is a typical representation of the wave nature of matter. Here, the chance of finding a particle going from *a* to *c* through several different routes (values of *b*) may, if no attempt is made to determine the route, be represented as the square of a sum of several complex quantities - one for each available route.

Probability can show the typical phenomena of interference, usually associated with waves, whose intensity is given by the square of the sum of contributions from different sources. *The electron acts as a wave, (5), so to speak, as long as no attempt is made to verify that it is a particle*; yet one can determine, if one wishes, by what route it travels just as though it were a particle; but when one does that, (4) applies and it does act like a particle.

These things are, of course, well known. They have already been explained many times[9].

[9] See, for example, Heisenberg, W. (1930). *The Physical Principles of the Quantum Theory*. University of Chicago Press, Chicago), particularly Chapter IV.

However, it seems worthwhile to emphasize the fact that they are all simply direct consequences of Eq. (5), for *it is essentially Eq. (5) that is fundamental in my formulation of quantum mechanics*. The generalization of Eqs. (4) and (5) to a large number of measurements, say A, B, C, D, … , K, is, of course, that the probability of the sequence *a, b, c, d … , k*, is

$$P_{abcd...k} = |\varphi_{abcd...k}|^2.$$

The probability of the result *a, c, k*, for example, if *b, d*, . . . are measured, is the classical formula:

$$P_{ack} = \sum_b \sum_d \ldots P_{abcd...k}, \qquad (6)$$

while the probability of the same sequence *a, c, k* if no measurements are made between A and C and between C and K is

$$P_{ack} = |\sum_b \sum_d \ldots \varphi_{abcd\ldots k}|^2. \qquad (7)$$

The quantity $\phi_{abcd\ldots k}$ we can call the *probability amplitude* for the condition A = *a*, B = *b*, C = *c*, D = *d*, … , K = *k*. (It is, of course, expressible as a product $\varphi_{ab}\varphi_{bc}\varphi_{cd} \ldots \varphi_{jk}$.)

3. *The Probability Amplitude for a Space-Time Path*

The physical ideas of the last section may be readily extended to define a *probability amplitude* for a particular completely specified *space-time path*. To explain how this may be done, we shall limit ourselves to a one-dimensional problem, as the generalization to several dimensions is obvious.

…

… Suppose a measurement is made which is capable only of determining that the path lies somewhere within R.

The measurement is to be what we might call an "ideal measurement." We suppose that no further details could be obtained from the same measurement without further disturbance to the system. I have not been able to find a precise definition. We are trying to avoid the extra uncertainties that must be averaged over if, for example, more information were measured but not utilized. We wish to use Eq. (5) or (7) for all x_i and have no residual part to sum over in the manner of Eq. (4).

We expect that the probability that the particle is found by our "ideal measurement" to be, indeed, in the region R is the square of a complex number $|\varphi(R)|^2$. The number $\varphi(R)$, which we may call the *probability amplitude* for region R is given by Eq. (7) with *a*, *b*, … replaced by x_i, x_{i+1}, … and summation replaced by integration:

$$\varphi(R) = \mathrm{Lim}_{\varepsilon\to 0} \int R \times \Phi(\ldots x_i, x_{i+1} \ldots) \ldots dx_i dx_{i+1} \ldots. \qquad (9)$$

The complex number $\Phi(\ldots x_i, x_{i+1} \ldots)$ is a function of the variables x_i, defining the path. Actually, we imagine that the time spacing ε approaches zero so that Φ essentially depends on the entire path x(t) rather than only on just the values of x_i, at the particular times t_i, $x_i = x(t_i)$. We might call Φ the *probability amplitude functional* of paths x(t).

We may summarize these ideas in our first postulate:
I. *If an ideal measurement is performed, to determine whether a particle has a path lying in a region of space-time, then the probability that the result will be affirmative is the absolute square of a sum of complex contributions, one from each path in the region.*

The statement of the postulate is incomplete. The meaning of a sum of terms one for "each" path is ambiguous. The precise meaning given in Eq. (9) is this: A path is first defined only by the positions x_i; through which it goes at a sequence of equally spaced times[10], $t_i = t_{i-1} + \varepsilon$.

[10] There are very interesting mathematical problems involved in the attempt to avoid the subdivision and limiting processes. Some sort of complex measure is being associated with the space of functions x(t). Finite results can be obtained under unexpected circumstances

> because the measure is not positive everywhere, but the contributions from most of the paths largely cancel out. These curious mathematical problems are sidestepped by the subdivision process. However, one feels as Cavalieri must have felt calculating the volume of a pyramid before the invention of calculus.

Then all values of the coordinates within R have an equal weight. The actual magnitude of the weight depends upon ε and can be so chosen that the probability of an event which is certain shall be normalized to unity. It may not be best to do so, but we have left this weight factor in a proportionality constant in the second postulate. The limit $\varepsilon \to 0$ must be taken at the end of a calculation.

When the system has several degrees of freedom the coordinate space x has several dimensions so that the symbol x will represent a set of coordinates $(x^{(1)}, x^{(2)}, \ldots x^{(k)})$ for a system with k degrees of freedom. A path is a sequence of configurations for successive times and is described by giving the configuration x_i, or $(x^{(1)}_i, x^{(2)}_i, \ldots, x^{(k)}_i)$, i.e., the value of each of the k coordinates for each time t_i. The symbol dx_i, will be understood to mean the volume element in k dimensional configuration space (at time t_i). The statement of the postulates is independent of the coordinate system which is used.

The postulate is limited to defining the results of position measurements. It does not say what must be done to define the result of a momentum measurement, for example. This is not a real limitation, however, because in principle the measurement of momentum of one particle can be performed in terms of position measurements of other particles, e.g., meter indicators. Thus, an analysis of such an experiment will determine what it is about the first particle which determines its momentum.

4. *The Calculation of the Probability Amplitude for a Path*

The first postulate prescribes the type of mathematical framework required by quantum mechanics for the calculation of probabilities. The second postulate gives a particular content to this framework by prescribing how to compute the important quantity Φ [the *probability amplitude*] for each path:

II. *The paths contribute equally in magnitude but the phase of their contribution is the classical action (in units of h); i.e., the time integral of the Lagrangian taken along the path.*

That is to say, the contribution $\Phi[x(t)]$ from a given path x(t) is proportional to $\exp\{i/h\, S[x(t)]\}$, *where the action* $S[x(t)] = \int L\{\dot{x}(t), x(t)\}\, dt$ *is the time integral of the classical Lagrangian* $L(\dot{x}, x)$ *taken along the path in question.* The Lagrangian, which may be an explicit function of the time, is a function of *position* and *velocity*. If we suppose it to be a quadratic function of the velocities, we can show the mathematical equivalence of the postulates here and the more usual formulation of quantum mechanics.

To interpret the first postulate, it was necessary to define a path by giving only the succession of points x_i, through which the path passes at successive times t_i. To compute $S = \int L(\dot{x}, x)\, dt$ we need to know the path at all points, not just at x_i. We shall assume that the function $x(t)$ in the interval between t_i and t_{i+1} is the path followed by a classical particle, with the Lagrangian L, which starting from x_i, at t_i reaches x_{i+1} at t_{i+1}. This assumption is required to interpret the second postulate for discontinuous paths. The quantity $\Phi(\ldots x_i, x_{i+1}, \ldots)$ can be normalized (for various ε) if desired, so that the probability of an event which is certain is normalized to unity as $\varepsilon \rightarrow 0$.

There is no difficulty in carrying out the *action* integral because of the sudden changes of velocity encountered at the times t_i as long as L does not depend upon any higher time derivatives of the position than the first. Furthermore, unless L is restricted in this way the end points are not sufficient to define the classical path. Since *the classical path is the one which makes the action a minimum*, we can write

$$S = \sum_i S(x_{i+1}, x_i), \qquad (10)$$

where

$$S(x_{i+1}, x_i) = \text{Min.} \int_{ti}^{ti+1} L\{x^{\cdot}(t), x(t)\}\, dt. \qquad (11)$$

Written in this way, *the only appeal to classical mechanics is to supply us with a Lagrangian function.* Indeed, one could consider postulate two as simply saying, "Φ is the exponential of i times the integral of a real function of x(t) and its first time-derivative". Then the classical *equations of motion* might be derived later as the limit for large dimensions. The function of x and $x^{\cdot}$ then could be shown to be the classical Lagrangian within a constant factor.

Actually, the sum in (10), even for finite ε is infinite and hence meaningless (because of the infinite extent of time). *This reflects a further incompleteness of the postulates. We shall have to restrict ourselves to a finite, but arbitrarily long, time interval.*

Combining the two postulates and using Eq. (10). we find

$$\varphi(R) = \text{Lim}_{\varepsilon \rightarrow 0} \int_R \times \exp\,[i/h \sum_i S(x_{i+1}, x_i) \ldots dx_{i+1}/A\; dx_i/A \ldots , \qquad (12)$$

where we have let the *normalization* factor be split into a factor 1/A (whose exact value we shall presently determine) for each instant of time. The integration is just over those values $x_i, x_{i+1}, \ldots$ which lie in the region R. *This equation, the definition (11) of $S(x_{i+1}, x_i)$, and the physical interpretation of $|\varphi\ (R)|^2$ as the probability that the particle will be found in R, complete our formulation of quantum mechanics.*

5. *Definition of the Wave Function*

We now proceed to show the equivalence of these postulates to the ordinary formulation of quantum mechanics. This we do in two steps. *We show in this section how the wave function may be defined from the new point of view. In the next section we shall show that this function satisfies Schroedinger's differential wave equation.*

We shall see that it is the possibility, (10)

$$[S = \sum_i S(x_{i+1}, x_i), \qquad (10)]$$

of expressing S as a sum, and hence Φ as a product, of contributions from successive sections of the path, which leads to the possibility of defining a quantity having the properties of a *wave function*.

To make this clear, let us imagine that we choose a particular time t and divide the *region* R in Eq. (12) into pieces, future and past relative to t. We imagine that R can be split into: (a) a *region* R', restricted in any way in space, but lying entirely earlier in time than some t', such that t' < t; (b) a *region* R" arbitrarily restricted in space but lying entirely later in time than t", such that t" > t; (c) the region between t' and t" in which all the values of x coordinates are unrestricted, i.e., all of space-time between t' and t". The region (c) is not absolutely necessary. It can be taken as narrow in time as desired. However, it is convenient in letting us consider varying t a little without having to redefine R' and R". Then $|\varphi(R', R'')|^2$ is the probability that the path occupies R' and R". Because R' is entirely previous to R", considering the time t as the present, we can express this as the probability that the path had been in region R' and will be in region R". If we divide by a factor, the probability that the path is in R', to *renormalize* the probability we find: $|\varphi(R', R'')|^2$ is the (relative) probability that if the system were in region R' it will be found later in R".

This is, of course, the important quantity in predicting the results of many experiments. We prepare the system in a certain way (e.g., it was in region R') and then measure some other property (e.g., will it he found in region R"?). What does (12)

$$[\varphi(R) = \mathrm{Lim}_{\varepsilon\to 0} \int_R \times \exp\,[i/h \textstyle\sum_i S(x_{i+1}, x_i) \ldots dx_{i+1}/A\; dx_i/A\ldots, \qquad (12)]$$

say about computing this quantity, or rather the quantity φ(R', R") of which it is the square?

Let us suppose in Eq. (12) that the time t corresponds to one particular point k of the subdivision of time into steps ε, i.e., assume $t = t_k$, the index k, of course, depending upon the subdivision ε. Then, the exponential being the exponential of a sum may be split into a product of two factors

$$\exp\,[i/h \textstyle\sum_{i=k}^{\infty} S(x_{i+1}, x_i)] \cdot \exp\,[i/h \textstyle\sum_{i=-\infty}^{k-1} S(x_{i+1}, x_i)]. \qquad (13)$$

The first factor contains only coordinates with index k or higher, while the second contains only coordinates with index k or lower. This split is possible because of Eq. (10), *which results essentially from the fact that the Lagrangian is a function only of positions and velocities*. First, the integration on all variables x_i for i > k can be performed on the first factor resulting in a function of x_k (times the second factor). Next, the integration on all variables x_i, for i < k can be performed on the second factor also, giving a function of x_k. Finally, the integration on x_k can be performed. That is, φ(R', R") can be written as the integral over x_k of the product of two factors. We will call these $\chi^*(x_k, t)$ and $\psi(x_k, t)$:

$$\varphi(R', R'') = \int \chi^*(x, t)\psi(x, t)\, dx, \qquad (14)$$

where

$$\psi(x_k, t) = \mathrm{Lim}_{\varepsilon\to 0} \int_{R'} \exp\,[i/h \textstyle\sum_{i=-\infty}^{k-1} S(x_{i+1}, x_i)]dx_{k-1}/A\; dx_{k-2}/A\ldots, \qquad (15)$$

and

$$\chi^*(x_k, t) = \mathrm{Lim}_{\varepsilon\to 0} \int_{R''} \exp\,[i/h \textstyle\sum_{i=k}^{\infty} S(x_{i+1}, x_i)]\;.\; 1/A\; dx_{k+1}/A\; dx_{k+2}/A\ldots, \qquad (16)$$

The symbol R' is placed on the integral for ψ to indicate that the coordinates are integrated over the region R', and, for t_i between t' and t, over all space. In like manner, the integral for $\chi*$ is over R" and over all space for those coordinates corresponding to times between t and t". The asterisk on $\chi*$ denotes *complex conjugate*, as it will be found more convenient to define (16) as the *complex conjugate* of some quantity, χ.

The quantity ψ depends only upon the region R' previous to t, and is completely denned if that region is known. It does not depend, in any way, upon what will be done to the system after time t. This latter information is contained in χ. *Thus, with ψ and χ we have separated the past history from the future experiences of the system.* This permits us to speak of the relation of past and future in the conventional manner. Thus, if a particle has been in a region of space-time R' it may at time t be said to be in a certain condition, or state, determined only by its past and described by the so-called wave function $\psi(x, t)$. This function contains all that is needed to predict future probabilities. For, suppose, in another situation, the region R' were different, say r', and possibly the Lagrangian for times before t were also altered. But, nevertheless, suppose the quantity from Eq. (15)

$$[\psi(x_k, t) = \mathrm{Lim}_{\varepsilon \to 0} \int_{R'} \exp [i/h \sum_{i=-\infty}^{k-1} S(x_{i+1}, x_i)] dx_{k-1}/A\ dx_{k-2}/A \ldots , \quad (15)]$$

turned out to be the same. Then, according to (14)

$$[\varphi(R', R") = \int \chi*(x, t)\psi(x, t)\ dx, \quad (14)]$$

the probability of ending in any region R" is the same for R' as for r'. Therefore, future measurements will not distinguish whether the system had occupied R' or r'. Thus, the *wave function* $\psi(x, t)$ is sufficient to define those attributes which are left from past history which determine future behavior.

Likewise, the function $\chi(x, t)$ characterizes the experience, or, let us say, experiment to which the system is to be subjected. If a different region, r" and different Lagrangian after t, were to give the same $\chi*(x, t)$ via Eq. (16)

$$[\chi*(x_k, t) = \mathrm{Lim}_{\varepsilon \to 0} \int_{R"} \exp [i/h \sum_{i=k}^{\infty} S(x_{i+1}, x_i)]\ .\ 1/A\ dx_{k+1}/A\ dx_{k+2}/A \ldots , (16)]$$

as does region R", then no matter what the preparation, ψ, Eq. (14) says that the chance of finding the system in R" is always the same as finding it in r". The two "experiments" R" and r" are equivalent, as they yield the same results. We shall say loosely that these experiments are to determine with what probability the system is in state χ. Actually, this terminology is poor. The system is really in *state* ψ. *The reason we can associate a state with an experiment is, of course, that for an ideal experiment there turns out to be a unique state (whose wave function is $\chi(x, t)$) for which the experiment succeeds with certainty.*

Thus, we can say: *the probability that a system in state ψ will be found by an experiment whose characteristic state is χ* (or, more loosely, the chance that a system in *state* ψ will appear to be in χ) is

$$| \int \chi*(x, t)\psi(x, t)\ dx |^2. \quad (17)$$

These results agree, of course, with the principles of ordinary quantum mechanics. *They are a consequence of the fact that the Lagrangian is a function of position, velocity, and time only.*

6. *The Wave Equation*

To complete the proof of the equivalence with the ordinary formulation we shall have to show that the *wave function* defined in the previous section by Eq. (15) actually satisfies the Schroedinger *wave equation.* Actually, we shall only succeed in doing this *when the Lagrangian L in (11) is a quadratic, but perhaps inhomogeneous, form in the velocities $x^{\cdot}(t)$.* This is not a limitation, however, as *it includes all the cases for which the Schroedinger equation has been verified by experiment.*

The *wave equation* describes the development of the *wave function* with time. We may expect to approach it by noting that, for finite ε, Eq. (15)

$$[\psi(x_k, t) = \mathrm{Lim}_{\varepsilon\to 0} \int_{R'} \exp\,[i/h \textstyle\sum_{i=-\infty}^{k-1} S(x_{i+1}, x_i)]dx_{k-1}/A\ dx_{k-2}/A\ldots, \qquad (15)]$$

permits a simple recursive relation to be developed. Consider the appearance of Eq. (15) if we were to compute ψ at the next instant of time:

$$\psi(x_{k+1}, t + \varepsilon) = \int_{R'} \exp\,[i/h \textstyle\sum_{i=-\infty}^{k} S(x_{i+1}, x_i)]dx_k/A\ dx_{k-1}/A\ldots, \qquad (15')$$

This is similar to (15) except for the integration over the additional variable x_k, and the extra term in the sum in the exponent. This term means that the integral of (15') is the same as the integral of (15) except for the factor $(1/A)\exp(i/h)S(x_{k+1}, x_k)$. Since this does not contain any of the variables x_i, for i less than k, all of the integrations on dx, up to dx_{k-1} can be performed with this factor left out. However, the result of these integrations is by (15) simply $\psi(x_k, t)$. Hence, we find from (15') the relation

$$\psi(x_{k+1}, t + \varepsilon) = \int \exp\,[i/h\ S(x_{k+1}, x_k)]\ \psi(x_k, t)dx_k/A. \qquad (18)$$

This relation giving the development of ψ with time will be shown, for simple examples, with suitable choice of A, to be equivalent to Schroedinger's equation. …

…

We shall illustrate the relation of (18) to Schroedinger's equation by applying it to the simple case of a particle moving in one dimension in a potential V(x). Before we do this, however, we would like to discuss some approximations to the value $S(x_{i+1}, x_i)$ given in (11)

$$[S(x_{i+1}, x_i) = \mathrm{Min.} \int_{ti}^{ti+1} L\{x^{\cdot}(t), x(t)\}\ dt. \qquad (11)]$$

which will be sufficient for expression (18).

…

Canceling ψ(x, t) from both sides, and comparing terms to first order in ε and multiplying by − h/i one obtains

$$-\,h/i\ \partial\psi/\partial t = 1/2m^{\cdot}\ (h/i\ \partial/\partial x)^2\ \psi + V\,(x)\psi, \qquad (30)$$

which is *Schroedinger's equation* for the problem in question.

The equation for χ* can be developed in the same way, but adding a factor *decreases* the time by one step, i.e., χ* satisfies an equation like (30) but with the sign of the time reversed. By taking *complex conjugates* we can conclude that χ satisfies the same equation as ψ, i.e., an experiment can be defined by the particular state χ to which it corresponds.

…

It is clear that the statement (11) is independent of the coordinate system. Therefore, to find the *differential wave equation* it gives in any coordinate system, the easiest procedure is first to find the equations in Cartesian coordinates and then to transform the coordinate system to the one desired. It suffices, therefore, to show the relation of the postulates and Schroedinger's equation in rectangular coordinates. *The derivation given here for one dimension can be extended directly to the case of three-dimensional Cartesian coordinates for any number, K, of particles interacting through potentials with one another, and in a magnetic field, described by a vector potential. …*

…

… The Lagrangian is the classical Lagrangian for the same problem, and the Schroedinger equation resulting will be that which corresponds to the classical Hamiltonian, derived from this Lagrangian. The equations in any other coordinate system may be obtained by transformation. *Since this includes all cases for which Schroedinger's equation has been checked with experiment, we may say our postulates are able to describe what can be described by non-relativistic quantum mechanics, neglecting spin.*

7. *Discussion of the Wave Equation*

The Classical Limit

This completes the demonstration of the equivalence of the new and old formulations. We should like to include in this section a few remarks about the important equation (18).

[$\psi(x_{k+1}, t + \varepsilon) = \int \exp [i/h\ S(x_{k+1}, x_k)]\ \psi(x_k, t)dx_k/A.$ (18)]

This equation gives the development of the *wave function* during a small time-interval. It is easily interpreted physically as the expression of Huygens' principle for *matter waves.* In geometrical optics the rays in an inhomogeneous medium satisfy Fermat's principle of least time. We may state Huygens' principle in wave optics in this way: *If the amplitude of the wave is known on a given surface, the amplitude at a nearby point can be considered as a sum of contributions from all points of the surface. Each contribution is delayed in phase by an amount proportional to the time it would take the light to get from the surface to the point along the ray of least time of geometrical optics.* We can consider (22) in an analogous manner starting with Hamilton's first principle of least action for classical or "geometrical" mechanics. If the amplitude of the wave ψ is known on a given "surface," in particular the "surface" consisting of all x at time t, its value at a particular nearby point at time $t + \varepsilon$, is a sum of contributions from all points of the surface at t. Each contribution is delayed in phase by an amount proportional to the *action* it would require to get from the surface to the point along the path of least action of classical mechanics[16].

[16] See in this connection the very interesting remarks of Schroedinger (1926). [Schrodinger, E. (1926). Quantisierung als Eigenwertproblem. II. (Quantization as an eigenvalue problem II.) *Ann. Phys.*, 79, 6, 489-527; https://doi.org/10.1002/andp.19263840602.]

Actually, Huygens' principle is not correct in optics. It is replaced by Kirchoff's modification which requires that both the amplitude and its derivative must be known on the adjacent surface. This is a consequence of the fact that the wave equation in optics is second order in the time. *The wave equation of quantum mechanics is first order in the time*; therefore, *Huygens' principle is correct for matter waves, action replacing time*.

The equation can also be compared mathematically to quantities appearing in the usual formulations. In Schroedinger's method the development of the *wave function* with time is given by

$$-h/i\ \partial\psi/\partial t = \mathbf{H}\psi, \tag{31}$$

which has the solution (for any ε if **H** is time independent)

$$\psi(x, t + \varepsilon) = \exp(-\,i\varepsilon\mathbf{H}/h)\psi(x, t). \tag{32}$$

Therefore, Eq. (18)

[$\psi(x_{k+1}, t + \varepsilon) = \int \exp\,[i/h\ S(x_{k+1}, x_k)]\ \psi(x_k, t)dx_k/A.$ (18)]

expresses the operator exp (− iε**H**/h) by an approximate integral operator for small ε.

From the point of view of Heisenberg, one considers the position at time t, for example, as an operator x. The position x' at a later time t + ε can be expressed in terms of that at time t by the operator equation

$$x' = \exp\,(i\varepsilon\mathbf{H}/h)\ x\ \exp^{-(i\varepsilon\mathbf{H}/h)}. \tag{33}$$

The *transformation theory* of Dirac allows us to consider the wave function at time t + ε, ψ(x', t + ε), as representing a state in a representation in which x' is diagonal, while ψ(x, t) represents the same state in a representation in which x is diagonal. They are, therefore, related through the transformation function $(x'|x)_\varepsilon$, which relates these representations:

$$\psi(x', t + \varepsilon) = \int (x'|x)_\varepsilon\ \psi(x, t)\ dx.$$

Therefore, the content of Eq. (18) is to show that for small ε we can set

$$(x'|x) = (1/A)\ \exp\,(iS(x', x)/h) \tag{34}$$

with S(x', x) defined as in (11)

[$S(x_{i+1}, x_i) = \text{Min.} \int_{ti}^{ti+1} L\{x^{\cdot}(t), x(t)\}\ dt.$ (11)]

> [Dirac (1933). The Lagrangian in Quantum Mechanics: "Lagrangian theory is closely connected with the theory of *contact transformations. We shall therefore begin with a discussion of the analogy between classical and quantum contact transformations*. Let the two sets of variables be p_r, q_r and P_r, Q_r, *(*r = 1, 2 ... n) and suppose the q's and Q's to be all independent, so that any function of the dynamical variables can be expressed in terms of them. It is well known that in the classical theory the *transformation equations* for this case can be put in the form
>
> $$p_r = \partial S/\partial q_r, \quad P_r = \partial S/\partial Q_r, \tag{1}$$
>
> where S is some function of the q's and Q's.
>
> In the quantum theory we may take a representation in which the *q's* are diagonal, and a second representation in which the Q's are diagonal. There will be a

transformation function (q' | Q') connecting the two representations. We shall now show that this *transformation function* is the quantum analogue of $e^{iS/h}$.

...

The *equations of motion* of the classical theory cause the dynamical variables to vary in such a way that their values q_t, p_t at any time t are connected with their values q_T, p_T at any other time T by a *contact transformation*, which may be put into the form (1) with q, p = q_t, p_t, Q, P = q_T, p_T and S equal to the time integral of the Lagrangian over the range T to t. In the quantum theory the q_t, p_t will still be connected with the q_T, p_T by a *contact transformation* and there will be a *transformation function* (q_t | q_T) connecting the two representations in which the q_t and the q_T are diagonal respectively. The work of the preceding section now shows that

(q_t | q_T) corresponds to exp [i $\int_T^t$ Ldt/h], (8)

where L is the Lagrangian. If we take T to differ only infinitely little from t, we get the result

(q_{t+dt} | q_t) corresponds to exp [iL dt/h]. (9)

The transformation functions in (8) and (9) are very fundamental things in the quantum theory and it is satisfactory to find that they have their classical analogues, expressible simply in terms of the Lagrangian. We have here the natural extension of the well-known result that the *phase* of the *wave function* corresponds to Hamilton's principal function in classical theory. The analogy (9) suggests that we ought to consider the classical Lagrangian not as a function of the coordinates and velocities, but rather as a function of the coordinates at time t and the coordinates at time t + dt.]

The close analogy between $(x'|x)_\varepsilon$ and the quantity exp (iS(x', x)/h) has been pointed out on several occasions by Dirac.[17]

[17] Dirac, P. A. M. (1935). *The Principles of Quantum Mechanics*. (The Clarendon Press, Oxford), second edition, Section 33; also, Dirac, P. A. M. (1933). The Lagrangian in Quantum Mechanics. *Phys. Zeit. Sowjet.*, 3, 1, 64-72[; alternative formulation of quantum mechanics in terms of Lagrangian in place of Hamiltonian, *coordinates* and *velocities* instead of *coordinates* and *momenta*, allows *equations of motion* to be expressed as stationary property of *action function* which is the time-integral of the Lagrangian and *relativistic invariant*, there is no corresponding *action principle* in terms of *coordinates and momenta* in Hamiltonian theory, Lagrangian closely connected with theory of *contact transformations*, transformation functions have classical analogues expressible in terms of the Lagrangian, function of coordinates at time t and time t + dt rather than of the coordinates and velocities, applies to *field dynamics* using suitable field quantities or potentials as coordinates, "*many-time*" theory, each coordinate a function of four *space-time* variables instead of one time variable in particle theory, *generalized transformation function*].

In fact, we now see that to sufficient approximations the two quantities may be taken to be proportional to each other. *Dirac's remarks were the starting point of the present development*. The points he makes concerning the passage to the classical limit $h \to 0$ are very beautiful, and I may perhaps be excused for briefly reviewing them here.

First, we note that the wave function at x" at time t" can be obtained from that at x' at time t' by

$$\psi(x'', t'') = \text{Lim}_{\varepsilon\to 0} \int \ldots \int \times \exp [i/h \sum_{i=0}^{j-1} S(x_{i+1}, x_i)] \times \psi(x', t')\, dx_0/A\, dx_1/A \ldots dx_{j-1}/A, \quad (35)$$

where we put $x_0 \equiv x''$ and $x_j \equiv x''$ where $i\varepsilon = t'' - t'$ (between the times t' and t" we assume no restriction is being put on the region of integration). This can be seen either by repeated applications of (18)

$$[\psi(x_{k+1}, t + \varepsilon) = \int \exp [i/h\, S(x_{k+1}, x_k)]\, \psi(x_k, t) dx_k/A. \quad (18)]$$

or directly from Eq. (15)

$$[\psi(x_k, t) = \text{Lim}_{\varepsilon\to 0} \int_{R'} \exp [i/h \sum_{i=-\infty}^{k-1} S(x_{i+1}, x_i)] dx_{k-1}/A\, dx_{k-2}/A\ldots . \quad (15)]$$

Now we ask, as $h \to 0$ what values of the intermediate coordinates x_i, contribute most strongly to the integral? These will be the values most likely to be found by experiment and therefore will determine, in the limit, the classical path. If h is very small, *the exponent will be a very rapidly varying function of any of its variables x_i.* As x_i varies, the positive and negative contributions of the exponent nearly cancel. *The region at which x_i contributes most strongly is that at which the phase of the exponent varies least rapidly with x_i (method of stationary phase).*

> [Dirac (1933). The Lagrangian in Quantum Mechanics: "Now in the quantum theory we have
>
> $$(q_t \mid q_T) = \int (q_t \mid q_m)\, dq_m\, (q_m \mid q_{m-1})\, dq_{m-1} \cdots (q_2 \mid q_1)\, dq_1\, (q_1 \mid q_T), \quad (11)$$
>
> where q_k denotes q at the intermediate time t_k, (k = 1, 2 … m)
>
> [and $\exp [i \int_T^t Ldt/h]$ is the classical analogue of $(q_t \mid q_T)$. (8)]
>
> Equation (11) at first sight does not seem to correspond properly to equation (10)
>
> $$[\exp [i \int_T^t Ldt/h] = \exp [i \int_{tm}^t Ldt/h] \exp [i \int_{tm-1}^{tm} Ldt/h] \ldots \exp [i \int_1^2 Ldt/h], \exp [i \int_T^1 Ldt/h].$$
>
> Let us examine this discrepancy by seeing what becomes of (11) when we regard t as extremely small. From the results (8) and (9)
>
> [$(q_{t+dt} \mid q_t)$ corresponds to $\exp [iL\, dt/h]$ (9)]
>
> we see that the integrand in (11) must be of the form $e^{iF/h}$ where F is a function of $q_T, q_1, q_2 \ldots q_m, q_t$ which remains finite as h tends to zero. Let us now picture one of the intermediate q's say q_k, as varying continuously while the others are fixed. Owing to the smallness of h, we shall then in general have F|h varying extremely rapidly. This means that $e^{iF/h}$ will vary periodically with a very high frequency about the value zero, as a result of which its integral will be practically zero. The only important part in the domain of integration of q_k is thus that for which a comparatively large variation in q_k produces only a very small variation in F. This part is the neighborhood of a point for which F is stationary with respect to small variations in q_k.]

Call the sum in the exponent S;

$$S = \sum_{i=0}^{j-1} S(x_{i+1}, x_i). \qquad (36)$$

Then the *classical orbit* passes, approximately, through those points x_i at which the rate of change of S with x_i, is small, or in the limit of small h, zero, i.e., the classical orbit passes through the points at which $\partial S/\partial x_i$ for all x_i. Taking the limit $\varepsilon \to 0$, (36) becomes in view of (11)

$$S = \int_{t'}^{t''} L(\dot{x}(t), x(t))\, dt. \qquad (37)$$

We see then that *the classical path is that for which the integral (37) suffers no first-order change on varying the path.* This is Hamilton's principle and leads directly to the Lagrangian *equations of motion.*

8. *Operator Algebra*

Matrix Elements

Given the *wave function* and *Schroedinger's equation,* of course all of the machinery of operator or matrix algebra can be developed. It is, however, rather interesting to express these concepts in a somewhat different language more closely related to that used in stating the postulates. Little will be gained by this in elucidating operator algebra. In fact, the results are simply a translation of simple operator equations into a somewhat more cumbersome notation. On the other hand, the new notation and point of view are very useful in certain applications described in the introduction. Furthermore, the form of the equations permits natural extension to a wider class of operators than is usually considered (e.g., ones involving quantities referring to two or more different times). If any generalization to a wider class of action functionals is possible, the formulae to be developed will play an important role.

We discuss these points in the next three sections. This section is concerned mainly with definitions. *We shall define a quantity which we call a transition element between two states.* It is essentially a matrix element. But instead of being the matrix element between a state ψ and another χ corresponding to the same time, these two states will refer to different times. In the following section *a fundamental relation between transition elements will be developed from which the usual commutation rules between coordinate and momentum may be deduced.* The same relation also yields Newton's equation of motion in matrix form. Finally, *in Section 10 we discuss the relation of the Hamiltonian to the operation of displacement in time.*

We begin by defining a transition element in terms of the probability of transition from one *state* to another. More precisely, suppose we have a situation similar to that described in deriving (17)

> ["*the probability that a system in state ψ will be found by an experiment whose characteristic state is χ* is
>
> $$\left| \int \chi^*(x, t)\psi(x, t)\, dx \right|^2. \qquad (17)]$$

The region R consists of a region R' previous to t', all space between t' and t'' and the region R'' after t''. We shall study the probability that a system in region R' is later found in region R''. This is given by (17)

$$[| \int \chi^*(x, t)\psi(x, t)\, dx |^2. \qquad (17)]$$

We shall discuss in this section how it changes with changes in the form of the Lagrangian between t' and t''. In Section 10 we discuss how it changes with changes in the preparation R' or the experiment R''.

The *state* at time t' is defined completely by the preparation R'. It can be specified by a *wave function* $\psi(x', t')$ obtained as in (15)

$$[\psi(x_k, t) = \mathrm{Lim}_{\varepsilon\to 0} \int_{R'} \exp [i/h \sum_{i=-\infty}^{k-1} S(x_{i+1}, x_i)]dx_{k-1}/A\, dx_{k-2}/A\ldots, \quad (15)]$$

but containing only integrals up to the time t'. Likewise, the *state* characteristic of the experiment (region R'') can be defined by a function $\chi(x'', t'')$ obtained from (16)

$$[\chi^*(x_k, t) = \mathrm{Lim}_{\varepsilon\to 0} \int_{R''} \exp [i/h \sum_{i=k}^{\infty} S(x_{i+1}, x_i)]\, .\, 1/A\, dx_{k+1}/A\, dx_{k+2}/A\ldots (16)]$$

with integrals only beyond t''. The wave function $\psi(x'', t'')$ at time t'' can, of course, also be gotten by appropriate use of (15). It can also be gotten from $\psi(x', t')$ by (35)

$$[\psi(x'', t'') = \mathrm{Lim}_{\varepsilon\to 0} \int \ldots \int \times \exp [i/h \sum_{i=0}^{j-1} S(x_{i+1}, x_i)]$$
$$\times\, \psi(x', t')\, dx_0/A\, dx_1/A \ldots dx_{j-1}/A. \qquad (35)]$$

According to (17) with t'' used instead of t, the probability of being found in χ it prepared in ψ is the square of what we shall call the transition amplitude R $\chi^*(x'', t'')\psi(x'', t'')dx''$. We wish to express this in terms of χ at t'' and ψ at t'. This we can do with the aid of (35). Thus, the chance that a system prepared in *state* $\psi_{t'}$ at time t' will be found after t'' to be in a state $\chi_{t''}$ is the *square of the transition amplitude*

$$[<\chi_{t''}|1|\psi_{t'}>_S = \mathrm{Lim}_{\varepsilon\to 0} \int \ldots \int \chi^*(x'', t'') \times$$
$$\times \exp (iS/\hbar)\psi(x', t')\, dx_0/A\, dx_1/A \ldots dx_{j-1}/A\, dx_j, \qquad (38)$$

where we have used the abbreviation (36)

$$[S = \sum_{i=0}^{j-1} S(x_{i+1}, x_i). \qquad (36)]$$

In the language of ordinary quantum mechanics if the Hamiltonian, **H**, is constant, $\psi(x, t'') = \exp [-i(t'' - t')\mathbf{H}/\hbar]\psi(x, t')$ so that (38) is the matrix element of $\exp [-i(t''- t')H/\hbar]$ between *states* $\chi_{t''}$ and $\psi_{t'}$.

If F is any function of the coordinates x_i for $t' < t_i < t''$, we shall define the transition element of F between the *states* ψ at t' and χ at t'' for the *action* S as ($x'' \equiv x_j$, $x' \equiv x_0$):

$$<\chi_{t''}|F|\psi_{t'}> = \mathrm{Lim}_{\varepsilon\to 0} \int \ldots \int \times \chi^*(x'', t'')F(x_0, x_1, \ldots x_i) \cdot$$
$$\cdot \exp [i/\hbar \sum_{i=0}^{j-1} S(x_{i+1}, x_i)]\, \psi(x', t')\, dx_0/A \ldots dx_{j-1}/A\, dx_i. \qquad (39)$$

In the limit $\varepsilon \to 0$, F is a functional of the path x(t).

We shall see presently why such quantities are important. It will be easier to understand if we stop for a moment to find out what the quantities correspond to in conventional notation. Suppose F is simply x_k, where k corresponds to some time $t = t_k$. Then on the right-hand side of (39) the integrals from x_0 to x_{k-1} may be performed to produce $\psi(x_k, t)$ or $\exp [-i(t - t')\mathbf{H}/\hbar]\psi_{t'}$. In like manner the integrals on x_i for $j \geq i > k$ give $\chi^*(x_k, t)$ or $\{\exp [-i(t'' - t)\mathbf{H}/\hbar]\chi_{t''}$. Thus, the transition element of x_k,

$$<\chi_{t''}|F|\psi_{t'}>_S = \int \chi^*_{t''} e^{(i/\hbar)h(t''-t)}\, x e^{-(i/\hbar)\mathbf{H}(t-t')}\, \psi_{t'} dx = \int \chi^*(x, t) x \psi(x, t) dx \qquad (40)$$

is the matrix element of x at time $t = t_k$ between the *state* which would develop at time t from $\psi_{t'}$ at t' and the *state* which will develop from time t to $\chi_{t''}$ at t". It is, therefore, the matrix element of x(t) between these *states*.

Likewise, according to (39)

$$[<\chi_{t''}|F|\psi_{t'}> = \text{Lim}_{\varepsilon\to 0} \int \ldots \int \times\chi^*(x'', t'') F(x_0, x_1, \ldots x_i) \cdot$$
$$\cdot \exp\, [i/\hbar \textstyle\sum_{i=0}^{j-1} S(x_{i+1}, x_i)]\, \psi(x', t')\, dx_0/A \ldots dx_{j-1}/A\, dx_i. \qquad (39)]$$

with $F = x_{k+1}$, the transition element of x_{k+1} is the matrix element of $x(t + \varepsilon)$. The transition element of $F = (x_{k+1} - x_k)/\varepsilon$ is the matrix element of $(x(t + \varepsilon) - x(t))/\varepsilon$ or of $i(\mathbf{H}x - x\mathbf{H})/\hbar$, as is easily shown from (40). We can call this the matrix element of velocity $x^{\cdot}(t)$.

Suppose we consider a second problem which differs from the first because, for example, the potential is augmented by a small amount U(, xt). Then in the new problem the quantity replacing S is $S' = S + \sum_i \varepsilon U(x_i, t_i)$. Substitution into (38)

$$[<\chi_{t''}|1|\psi_{t'}>_S = \text{Lim}_{\varepsilon\to 0} \int \ldots \int \chi^*(x'', t'') \times$$
$$\times \exp\, (iS/\hbar) \psi(x', t')\, dx_0/A\, dx_1/A \ldots dx_{j-1}/A\, dx_j \qquad (38)]$$

leads directly to

$$<\chi_{t''}|1|\psi_{t'}>_{S'} = <\chi_{t''}|\exp\, i\varepsilon/\hbar \textstyle\sum_{i=1}^{j} U(x_i, t_i|\psi_{t'}>_S. \qquad (41)$$

Thus, transition elements such as (39) are important insofar as F may arise in some way from a change δS in an *action* expression. We denote, by observable *functionals*, those *functionals* F which can be defined, (possibly indirectly) in terms of the changes which are produced by possible changes in the *action* S. The condition that a *functional* be observable is somewhat similar to the condition that an operator be Hermitian. The observable *functionals* are a restricted class because the *action* must remain a quadratic function of velocities. From one observable *functional* others may be derived, for example, by

$$<\chi_{t''}|F|\psi_{t'}>_{S'} = <\chi_{t''}|F\, \exp\, i\varepsilon/\hbar \textstyle\sum_{i=1}^{j} U(x_i, t_i|\psi_{t'}>_S \qquad (42)$$

which is obtained from (39).

Incidentally, (41) leads directly to an important perturbation formula. If the effect of U is small the exponential can be expanded to first order in U and we find

$$<\chi_{t''}|1|\psi_{t'}>_{S'} = <\chi t''|1|\psi_{t'}>_S + i/\hbar <\chi_{t''}| \textstyle\sum_i \varepsilon U(x_i, t_i)|\psi_{t'}>. \qquad (43)$$

Of particular importance is the case that $\chi_{t''}$ is a *state* in which $\psi_{t'}$ would not be found at all were it not for the disturbance, U (i.e., $<\chi_{t''}|1|\psi_{t'}>S = 0$). Then

$$1/\hbar^2\, |<\chi_{t''}| \textstyle\sum_i \varepsilon U(x_i, t_i)|\psi_{t'}>_S|^2 \qquad (44)$$

is the probability of transition as induced to first order by the perturbation. In ordinary notation,

$$<\chi_{t''}| \textstyle\sum_i \varepsilon U(x_i, t_i)|\psi_{t'}>S = \int \{ \int \chi^*_{t''} e^{-(i/\hbar)\mathbf{H}(t''-t)}\, U e^{-(i/\hbar)\mathbf{H}(t-t')}\, \psi_{t'} dx\}\, dt$$

so that *(44) reduces to the usual expression*[18] *for time dependent perturbations.*

[18] Dirac, P. A. M. *The Principles of Quantum Mechanics*. (The Clarendon Press, Oxford, 1935), second edition, Section 47, Eq. (20).

9. *Newton's Equations*

The Commutation Relation

In this section we find that different *functionals* may give identical results when taken between any *two* states. This equivalence between *functionals* is the statement of operator equations in the new language.

If F depends on the various coordinates, we can, of course, define a new *functional* $\partial F/\partial x_k$ by differentiating it with respect to one of its variables, say x_k $(0 < k < j)$. If we calculate $<\chi_{t''}|\partial F/\partial x_k|\psi_{t'}>_S$ by (39)

$$[<\chi_{t''}|F|\psi_{t'}> = \text{Lim}_{\varepsilon\to 0} \int \ldots \int \times\chi*(x'', t'')F(x_0, x_1, \ldots x_i) \cdot \cdot\exp [i/\hbar \textstyle\sum_{i=0}^{j-1} S(x_{i+1}, x_i)]\, \psi(x', t')\, dx_0/A \ldots dx_{j-1}/A\, dx_i. \qquad (39)]$$

the integral on the right-hand side will contain $\partial F/\partial x_k$. The only other place that the variable x_k appears is in S. Thus, the integration on x_k can be performed by parts. The integrated part vanishes (assuming *wave functions* vanish at infinity) and we are left with the quantity $- F(\partial/\partial x_k) \exp(iS/\hbar)$ in the integral. However,

$$(\partial/\partial x_k) \exp(iS/\hbar) = (i/\hbar)(\partial S/\partial x_k) \exp(iS/\hbar),$$

so the right side represents the *transition element* of $- (i/\hbar)F(\partial S/\partial x_k)$, i.e.,

$$<\chi_{t''}|\partial F/\partial x_k|\psi_{t'}>_S = - i/\hbar <\chi_{t''}|F\ \partial S/\partial x_k|\psi_{t'}>_S. \qquad (45)$$

This very important relation shows that two different functionals may give the same result for the transition element between any two states. We say they are *equivalent* and symbolize the relation by

$$- \hbar/i\ \partial F/\partial x_k \leftrightarrow_S F\ \partial S/\partial x_k, \qquad (46)$$

the symbol $\leftrightarrow$ S emphasizing the fact that *functionals equivalent under one action may not be equivalent under another.* The quantities in (46) need not be observable. The equivalence is, nevertheless, true. Making use of (36)

$$[S = \textstyle\sum_{i=0}^{j-1} S(x_{i+1}, x_i). \qquad (36)]$$

one can write

$$- \hbar/i\ \partial F/\partial x_k \leftrightarrow_S F[\partial S(x_{k+1}, x_k)/\partial x_k + \partial S(x_k, x_{k-1})/\partial x_k]. \qquad (47)$$

This equation is true to zero and first order in ε and has as consequences the commutation relations of momentum and coordinate, as well as the Newtonian equations of motion in matrix form.

In the case of our simple one-dimensional problem, $S(x_{i+1}, x_i)$ is given by the expression (15)

$$[\psi(x_k, t) = \text{Lim}_{\varepsilon\to 0} \int_{R'} \exp [i/\hbar \textstyle\sum_{i=-\infty}^{k-1} S(x_{i+1}, x_i)]dx_{k-1}/A\ dx_{k-2}/A\ldots, \qquad (15)],$$

so that

$$\partial S(x_{k+1}, x_k)/\partial x_k = -\,m(x_{k+1} - x_k)/\varepsilon,$$

and

$$\partial S(x_k, x_{k-1})/\partial x_k = +\,m(x_k - x_{k-1})/\varepsilon - \varepsilon V'(x_k);$$

where we write V'(x) for the *derivative of the potential, or force*. Then (47)

$$[-\,\hbar/i\ \partial F/\partial x_k \leftrightarrow_S F[\partial S(x_{k+1}, x_k)/\partial x_k + \partial S(x_k, x_{k-1})/\partial x_k]. \qquad (47)]$$

becomes

$$-\,\hbar/i\ \partial F/\partial x_k \leftrightarrow_S F[-m\{(x_{k+1} - x_k)/\varepsilon - (x_k - x_{k-1})/\varepsilon\} - \varepsilon V'(x_k)]. \qquad (48)$$

If F does not depend on the variable x_k, this gives Newton's equations of motion. For example, if F is constant, say unity, (48) just gives (dividing by ε)

$$0 \leftrightarrow_S -\,m/\varepsilon\ \{(x_{k+1} - x_k)/\varepsilon - (x_k - x_{k-1})/\varepsilon\} - V'(x_k).$$

Thus, the transition element of mass times acceleration $[(x_{k+1} - x_k)/\varepsilon - (x_k - x_{k-1})/\varepsilon]/\varepsilon$ between any two states is equal to the transition element of force $-V'(x_k)$ between the same states. This is the matrix expression of Newton's law which holds in quantum mechanics.

What happens if F does depend upon x_k? For example, let $F = x_k$. Then (48) gives, since $\partial F/\partial x_k = 1$,

$$-\,\hbar/i \leftrightarrow_S x_k\ [-m\{(x_{k+1} - x_k)/\varepsilon - (x_k - x_{k-1})/\varepsilon\} - \varepsilon V'(x_k)]$$

or, neglecting terms of order ε,

$$m\{(x_{k+1} - x_k)/\varepsilon - (x_k - x_{k-1})/\varepsilon\}\ x_k \leftrightarrow_S \hbar/i. \qquad (49)$$

In order to transfer an equation such as (49) into conventional notation, we shall have to discover what matrix corresponds to a quantity such as $x_k x_{k+1}$. It is clear from a study of (39)

$$[\langle\chi_{t''}|F|\psi_{t'}\rangle = \mathrm{Lim}_{\varepsilon\to 0} \int \ldots \int \times\chi^*(x'', t'')F(x_0, x_1, \ldots x_i)\ \cdot$$
$$\cdot \exp\,[i/\hbar \textstyle\sum_{i=0}^{j-1} S(x_{i+1}, x_i)]\ \psi(x', t')\ dx_0/A \ldots dx_{j-1}/A\ dx_i. \quad (39)]$$

that if F is set equal to, say, $f(x_k)g(x_{k+1})$, the corresponding operator in (40)

$$[\langle\chi_{t''}|F|\psi_{t'}\rangle_S = \int \chi^*_{t''} e^{(i/\hbar)h(t''-t)}\ x e^{-(i/\hbar)\mathbf{H}(t-t')}\ \psi_{t'}dx = \int \chi^*(x, t)x\psi(x, t)dx \quad (40)]$$

is

$$e^{-(i/\hbar)(t''-t-\varepsilon)\mathbf{H}} g(x) e^{-(i/\hbar)\varepsilon\mathbf{H}} f(x) e^{-(i/\hbar)(t-t')\mathbf{H}},$$

the matrix element being taken between the states $\chi_{t''}$ and $\psi_{t'}$. The operators corresponding to functions of x_{k+1} will appear to the left of the operators corresponding to functions of x_k, i.e., *the order of terms in a matrix operator product corresponds to an order in time of the corresponding factors in a functional*. Thus, if the *functional* can be and is written in such a way that in each term factors corresponding to later times appear to the left of factors corresponding to earlier terms, the corresponding operator can immediately be written down if the order of the operators is kept the same as in the *functional*[19].

[19] Dirac has also studied operators containing quantities referring to different times. See reference 2. [Dirac, P. A. M. (April, 1945). On the Analogy Between Classical and

Quantum Mechanics. *Rev. Mod. Phys*., 17, 195; mathematical methods available for working with *non-commuting quantities* much weaker than those available for *commuting quantities* owing to the fact that the only functions of *non-commuting variables* that one has been able to define are those expressible algebraically, shows how this difficulty can be avoided in the case when the *non-commuting quantities* are *observables*, can set up a theory of functions of them (a generalization of the concept known as *well-ordered functions*) of almost the same degree of generality as the usual functions of *commuting variables*, can use this theory to make closer analogy between classical and quantum mechanics, enables discussion of trajectories for motion of a particle in quantum mechanics, method is given for defining *general functions of non-commuting observables*, method is developed to provide formal *probability* for *non-commuting observables* to have numerical values (in general a complex number), also enables the analogy between classical and quantum *contact transformations* to be set up on a more general basis].

Obviously, the order of factors in a *functional* is of no consequence. The ordering just facilitates translation into conventional operator notation. To write Eq. (49)

$[m\{(x_{k+1} - x_k)/\varepsilon - (x_k - x_{k-1})/\varepsilon\}\, x_k \leftrightarrow_S \hbar/i.$ (49)]

in the way desired for easy translation would require the factors in the second term on the left to be reversed in order. We see, therefore, that it corresponds to

$$\mathbf{px} - \mathbf{xp} = \hbar/i$$

where we have written **p** for the operator $m\mathbf{x}^{\bullet}$.

The relation between *functionals* and the corresponding operators is denned above in terms of the order of the factors in time. It should be remarked that this rule must be especially carefully adhered to when quantities involving velocities or higher derivatives are involved. The correct *functional* to represent the operator $(x^{\bullet})^2$ is actually $(x_{k+1} - x_k)/\varepsilon(x_k - x_{k-1})/\varepsilon$ rather than $[(x_{k+1}-x_k)/]$ 2. The latter quantity diverges as $1/\varepsilon$ as $\varepsilon\rightarrow 0$. This may be seen by replacing the second term in (49) by its value $x_{k+1}\cdot m(x_{k+1} - x_k)/\varepsilon$ calculated an instant later in time. *This does not change the equation to zero order in* ε. We then obtain (dividing by ε)

$$\{(x_{k+1} - x_k)/\varepsilon\}^2 \leftrightarrow_S -\hbar/im\varepsilon. \qquad (50)$$

This gives the result expressed earlier that the root mean square of the "velocity" $(x_{k+1} - x_k)/\varepsilon$ between two successive positions of the path is of order $\varepsilon^{-1/2}$.

It will not do then to write the *functional* for *kinetic energy*, say, simply as

$$\tfrac{1}{2} m[(x_{k+1} - x_k)/\varepsilon]^2 \qquad (51)$$

for *this quantity is infinite* as $\varepsilon\rightarrow 0$. In fact, it is not an observable functional.

One can obtain the *kinetic energy* as an observable *functional* by considering the first-order change in *transition amplitude* occasioned by a change in the *mass* of the particle. Let m be changed to $m(1 + \delta)$ for a short time, say ε, around t_k. The change in the *action* is $\tfrac{1}{2}\delta\varepsilon m\{(x_{k+1} - x_k)/\varepsilon\}^2$ the derivative of which gives an expression like (51). But the change in m changes the *normalization constant* 1/A corresponding to dx_k as well as the *action*. The constant is changed from $\{2\pi\hbar\varepsilon i/m\}^{-1/2}$ to $\{2\pi\hbar\varepsilon i/m(1 + \delta)\}^{-1/2}$ or by $\frac{1}{2}\,\delta(2\pi\hbar i/m)^{-1/2}$ to first order in δ. The total effect of the change in *mass* in Eq. (38)

$$[\langle\chi_{t''}|1|\psi_{t'}\rangle_S = \mathrm{Lim}_{\varepsilon\to 0} \int \ldots \int \chi^*(x'', t'') \times$$
$$\times \exp(iS/\hbar)\psi(x', t')\, dx_0/A\, dx_1/A \ldots dx_{j-1}/A\, dx_j \qquad (38)]$$

to the first order in δ is

$$\langle\chi_{t''}\,|\tfrac{1}{2}\,\delta\varepsilon i m[(x_{k+1} - x_k)/\varepsilon]^2/\hbar + \tfrac{1}{2}\,\delta|\,\psi_{t'}\rangle.$$

We expect the change of order δ lasting for a time ε to be of order δε. Hence, dividing by $\delta\varepsilon i/\hbar$, we can define the *kinetic energy functional* as

$$\text{K.E.} = \tfrac{1}{2}\,m[(x_{k+1} - x_k)/\varepsilon]^2 + \hbar/2\varepsilon i. \qquad (52)$$

This is finite as → 0 in view of (50)

$$[\{(x_{k+1} - x_k)/\varepsilon\}^2 \leftrightarrow_S - \hbar/im\varepsilon. \qquad (50)]$$

By making use of an equation which results from substituting $m(x_{k+1}-x_k)/\varepsilon$ for F in (48)

$$[-\hbar/i\; \partial F/\partial x_k \leftrightarrow_S F[-m\{(x_{k+1} - x_k)/\varepsilon - (x_k - x_{k-1})/\varepsilon\} - \varepsilon V'(x_k)]. \qquad (48)]$$

we can also show that the expression (52) is equal (to order ε) to

$$\text{K.E.} = \tfrac{1}{2}\,m\{(x_{k+1} - x_k)/\varepsilon - (x_k - x_{k-1})/\varepsilon\}. \qquad (53)$$

That is, *the easiest way to produce observable functionals involving powers of the velocities is to replace these powers by a product of velocities, each factor of which is taken at a slightly different time.*

10. *The Hamiltonian*

Momentum

The Hamiltonian operator is of central importance in the usual formulation of quantum mechanics. We shall study in this section the *functional* corresponding to this operator. We could immediately define the Hamiltonian functional by adding the *kinetic energy functional* (52) or (53) to the *potential energy*. This method is artificial and does not exhibit the important relationship of the Hamiltonian to time. *We shall define the Hamiltonian functional by the changes made in a state when it is displaced in time.*

To do this we shall have to digress a moment to point out that the subdivision of time into equal intervals is not necessary. Clearly, any subdivision into instants t_i will be satisfactory; the limits are to be taken as the largest spacing, $t_{i+1} - t_i$ approaches zero. *The total action S must now be represented as a sum*

$$S = \textstyle\sum_i S(x_{i+1}, t_{i+1}; x_i, t_i), \qquad (54)$$

where

$$S(x_{i+1}, t_{i+1}; x_i, t_i) = \int_{t_i}^{t_{i+1}} L\{\dot{x}(t), x(t)\}\, dt, \qquad (55)$$

the integral being taken along the classical path between x_i at t_i and x_{i+1} at t_{i+1}. For the simple one-dimensional example this becomes, with sufficient accuracy,

$$S(x_{i+1}, t_{i+1}; x_i, t_i) = [m/2 \{(x_{i+1} - x_i)/(t_{i+1} - t_i)\}^2 - V(x_{i+1})] (t_{i+1} - t_i); \quad (56)$$

the corresponding *normalization constant* for integration on dx_i is

$$A = \{2\pi\hbar i(t_{i+1} - t_i)/m\}^{-1/2}.$$

The relation of H to the change in a *state* with displacement in time can now be studied. Consider a *state* $\psi(t)$ defined by a space-time region R'. Now imagine that we consider another *state* at time t, $\psi_\delta(t)$, denned by another region R'_δ. Suppose the region R'_δ is exactly the same as R' except that it is earlier by a time δ, i.e., displaced bodily toward the past by a time δ. All the apparatus to prepare the system for R'_δ is identical to that for R' but is operated a time δ sooner. If L depends explicitly on time, it, too, is to be displaced, i.e., the state ψ_δ is obtained from the L used for state ψ except that the time t in L_δ is replaced by $t + \delta$. We ask how does the state ψ_δ differ from ψ? In any measurement the chance of finding the system in a fixed region R" is different for R' and R'_δ. Consider the change in the transition element $<\chi|1|\psi_\delta>_{S\delta}$ produced by the shift δ. We can consider this shift as effected by decreasing all values of t_i by δ for $i \leq k$ and leaving all t_i fixed for $i > k$, where the time t lies in the interval between t_{k+1} and t_k[20].

[20] From the point of view of mathematical rigor, if δ is finite, as $\to 0$ one gets into difficulty in that, for example, the interval $t_{k+1} - t_k$ is kept finite. This can be straightened out by assuming δ to vary with time and to be turned on smoothly before $t = t_k$ and turned off smoothly after $t = t_k$. Then keeping the rime variation of δ fixed, let $\to 0$. Then seek the first-order change as $\delta \to 0$. The result is essentially the same as that of the crude procedure used above.

This change will have no effect on $S(x_{i+1}, t_{i+1}; x_i, t_i)$ as defined by (55)

$$[S(x_{i+1}, t_{i+1}; x_i, t_i) = \int_{ti}^{ti+1} L\{x^\cdot(t), x(t)\}\, dt, \quad (55)]$$

as long as both t_{i+1} and t_i are changed by the same amount. On the other hand, $S(x_{k+1}, t_{k+1}; x_k, t_k)$ is changed to $S(x_{k+1}, t_{k+1}; x_k, t_k - \delta)$. The constant 1/A for the integration on dx_k, is also altered to $(2\pi\hbar i(t_{k+1} - t_k + \delta)/m)^{-1/2}$. The effect of these changes on the transition element is given to the first order in δ by

$$<\chi|1|\psi>_S - <\chi|1|\psi_\delta>_{S\delta} = i\delta/\hbar <\chi|H_k|\psi>_S, \quad (57)$$

here the Hamiltonian *functional* H_k is defined by

$$H_k = \partial S(x_{k+1}, t_{k+1}; x_k t_k)/\partial t_k + \hbar/2i(t_{k+1} - t_k). \quad (58)$$

The last term is due to the change in 1/A and serves to keep H_k finite as $\varepsilon \to 0$. For example, for the expression (56) this becomes

$$H_k = m/2 \{(x_{k+1} - x_k)/(t_{k+1} - t_k)\} + \hbar/2i(t_{k+1} - t_k) + V(x_{k+1}),$$

which is just *the sum of the kinetic energy functional* (52)

$$[K.E. = ½ m[(x_{k+1} - x_k)/\varepsilon]^2 + \hbar/2\varepsilon i. \quad (52)]$$

and that of the potential energy V (x_{k+1}).

The *wave function* $\psi_\delta(x, t)$ represents, of course, the same *state* as $\psi(x, t)$ will be after time δ, i.e., $\psi(x, t + \delta)$. Hence, (57) is intimately related to the *operator equation* (31)

$$[-\ h/i\ \partial\psi/\partial t = \mathbf{H}\psi. \qquad (31)]$$

One could also consider changes occasioned by a time shift in the final *state* χ. Of course, nothing new results in this way for it is only the relative shift of χ and ψ which counts. One obtains an alternative expression

$$H_k = -\ \partial S(x_{k+1}, t_{k+1}; x_k, t_k)/\partial t_{k+1} + \hbar/2i(t_{k+1} - t_k). \qquad (59)$$

This differs from (58) only by terms of order. The time rate of change of a *functional* can be computed by considering the effect of shifting both initial and final *state* together. This has the same effect as calculating the *transition element* of the functional referring to a later time. What results is the analog of the *operator equation*

$$\hbar/i\ \mathbf{f}^{\cdot} = \mathbf{Hf} - \mathbf{fH}.$$

The *momentum functional* pt can be defined in an analogous way by considering the changes made by displacements of *position*:

$$\langle\chi|1|\psi\rangle_S - \langle\chi|1|\psi_\Delta\rangle_{S\Delta} = i\Delta/\hbar\ \langle\chi|p_k|\psi\rangle_S.$$

The *state* ψ_Δ is prepared from a region R'_Δ which is identical to region R' except that it is moved a distance Δ in space. (The Lagrangian, if it depends explicitly on x, must be altered to $L\Delta = L(x^{\cdot}, x - \Delta)$ for times previous to t.) One finds[21]

$$p_k = \partial S(x_{k+1}, x_k)/\partial x_{k+1} = -\ \partial S(x_{k+1}, x_k)/\partial x_k. \qquad (60)$$

[21] We did not immediately substitute p_i from (60) into (47)
$[-\ \hbar/i\ \partial F/\partial x_k \leftrightarrow_S F[\partial S(x_{k+1}, x_k)/\partial x_k + \partial S(x_k, x_{k-1})/\partial x_k]. \qquad (47)]$
because (47) would then no longer have been valid to both zero order and the first order in ε. We could derive the *commutation relations*, but not the *equations of motion*. The two expressions in (60) represent the *momenta* at each end of the interval t_i to t_{i+1}. They differ by $V'(x_{k+1})$ because of the force acting during the time ε.

Since $\psi_\Delta(x, t)$ is equal to $\psi(x - \Delta, t)$, the close connection between p_k and the x–derivative of the wave function is established.

Angular momentum operators are related in an analogous way to rotations.

The derivative with respect to t_{i+1} of $S(x_{i+1}, t_{i+1}; x_i, t_i)$ appears in the definition of H_i. The derivative with respect to t_{i+1} defines p_i. But the derivative with respect to t_{i+1} of $S(x_{i+1}, t_{i+1}; x_i, t_i)$ is related to the derivative with respect to x_{i+1}, for the function $S(x_{i+1}, t_{i+1}; x_i, t_i)$ defined by (55)

$$[S(x_{i+1}, t_{i+1}; x_i, t_i) = \int_{ti}^{ti+1} L\{x^{\cdot}(t), x(t)\}\ dt \qquad (55)]$$

satisfies the Hamilton-Jacobi equation. Thus, *the Hamilton-Jacobi equation is an equation expressing H_i in terms of the p_i. In other words, it expresses the fact that time displacements of states are related to space displacements of the same states*. This idea leads directly to a derivation of the Schroedinger equation which is far more elegant than the one exhibited in deriving Eq. (30)

$$[-\ h/i\ \partial\psi/\partial t = 1/2m^{\cdot}\ (h/i\ \partial/\partial x)^2\ \psi + V\ (x)\psi. \qquad (30)]$$

11. *Inadequacies of the Formulation*

The formulation given here suffers from a serious drawback. The mathematical concepts needed are new. At present, it requires an unnatural and cumbersome subdivision of the time interval to make the meaning of the equations clear. Considerable improvement can be made through the use of the notation and concepts of the *mathematics of functionals*. However, it was thought best to avoid this in a first presentation. *One needs, in addition, an appropriate measure for the space of the argument functions x(t) of the functionals.*

It is also incomplete from the physical standpoint. One of the most important characteristics of quantum mechanics is its invariance under unitary transformations. These correspond to the *canonical transformations* of classical mechanics. Of course, the present formulation, being equivalent to ordinary formulations, can be mathematically demonstrated to be invariant under these transformations. However, it has not been formulated in such a way that *it is physically obvious that it is invariant*. This incompleteness shows itself in a definite way. *No direct procedure has been outlined to describe measurements of quantities other than position.* Measurements of *momentum*, for example, of one particle, can be defined in terms of measurements of *positions* of other particles. *The result of the analysis of such a situation does show the connection of momentum measurements to the Fourier transform of the wave function. But this is a rather roundabout method to obtain such an important physical result.* It is to be expected that the postulates can be generalized by the replacement of the idea of "*paths in a region of space-time R*" to "paths of class R", or "paths having property R". But which properties correspond to which physical measurements has not been formulated in a general way.

12. *A Possibility Generalization*

The formulation suggests an obvious generalization. There are interesting classical problems which satisfy a *principle of least action* but for which the *action* cannot be written as an integral of a function of *positions* and *velocities*. The *action* may involve *accelerations*, for example. Or, again, if interactions are not instantaneous, it may involve the product of coordinates at two different times, such as $\int x(t)x(t+T)\,dt$. The *action*, then, cannot be broken up into a sum of small contributions as in (10)

$$[S = \sum_i S(x_{i+1}, x_i). \qquad (10)]$$

As a consequence, no wave function is available to describe a state. Nevertheless, a transition probability can be defined for getting from a region R' into another R''. Most of the theory of the *transition elements* $\langle\chi_{t''}|F|\psi_{t'}\rangle_S$ can be carried over. One simply invents a symbol, such as $\langle R''|F|R'\rangle_S$ by an equation such as (39)

$$[\langle\chi_{t''}|F|\psi_{t'}\rangle = \mathrm{Lim}_{\varepsilon\to 0} \int \ldots \int \times\chi^*(x'', t'')F(x_0, x_1, \ldots x_i)\,\cdot$$
$$\cdot\exp\left[i/\hbar \sum_{i=0}^{j-1} S(x_{i+1}, x_i)\right] \psi(x', t')\, dx_0/A \ldots dx_{j-1}/A\, dx_i. \quad (39)]$$

but with the expressions (19) and (20) for ψ and χ substituted, and the more general action substituted for S. Hamiltonian and momentum functionals can be defined as in section 10. Further details may be found in a thesis by the author[23].

[23] The theory of electromagnetism described by Wheeler and Feynman can be expressed in a *principle of least action involving the coordinates of particles alone*. [Wheeler, J. A. & Feynman R. P. (April, 1945). Interaction with the Absorber as the Mechanism of Radiation. *Rev. Mod. Phys*., 17, 157-81[; the motive of the analysis was to clear the present quantum theory of interacting particles of those of its difficulties which have a purely classical origin, method was to define as closely as one can within the bounds of classical theory the proper use of the *field* concept in the description of nature, this paper represents the third part of the survey, an analysis of the *mechanism of radiation* believed to complete the last tie between *action at a distance* and *field theory*, only section now finished, difficulties to obtain a satisfactory account of the *field* generated by an accelerated charge at a remote point and to understand the source of the force experienced by the charge itself as a result of its motion, takes up suggestion by Tetrode (1922) that *the act of radiation should have some connection with the presence of an absorber*, develops this idea into the thesis that the force of radiative reaction arises from the action on the source owing to the half-advanced fields of the particles of the absorber, absorber response as the mechanism of radiative reaction, expresses in terms of *action at a distance*, *Wheeler–Feynman absorber theory, assumes solutions of the electromagnetic field equations* must be invariant under time-reversal transformation as are the field equations themselves, considers an accelerated charge located in the absorbing system as the source of radiation, *elementary particles not self-interacting*, a complete correspondence is established between *action at a distance* and the usual formulation of *field theory* in the case of a completely absorbing system].

It was an attempt to quantize this theory, without reference to the fields, which led the author to study the formulation of quantum mechanics given here. The extension of the ideas to cover the case of more general *action functions* was developed in his Ph.D. thesis, "*The principle of least action in quantum mechanics*" submitted to Princeton University, 1942 [Feynman, R. P. (1942). *The Principle of Least Action in Quantum Mechanics*. Ph.D. thesis, Princeton University (in Brown, L. M. (ed.) (2005). *Feynman's Thesis. A New Approach to Quantum Theory*. World Scientific); non-relativistic quantum mechanics, space-time viewpoint, *principle of least action*, functional, Lagrangian, action functions].

13. *Application to Eliminate Field Oscillators*

One characteristic of the present formulation is that it can give one a sort of bird's-eye view of the *space-time* relationships in a given situation. Before the integrations on the x_i are performed in an expression such as (39)

$$[\langle\chi_{t''}|F|\psi_{t'}\rangle = \mathrm{Lim}_{\varepsilon\to 0} \int \dots \int \times\chi^*(x'', t'')F(x_0, x_1, \dots x_i) \cdot$$
$$\cdot \exp [i/\hbar \textstyle\sum_{i=0}^{j-1} S(x_{i+1}, x_i)]\, \psi(x', t')\, dx_0/A \dots dx_{j-1}/A\, dx_i \quad (39)]$$

one has a sort of format into which various F *functionals* may be inserted. One can study how what goes on in the quantum-mechanical system at different times is interrelated. To make these vague remarks somewhat more definite, we discuss an example.

In classical electrodynamics the fields describing, for instance, the interaction of two particles can be represented as a set of oscillators. The *equations of motion* of these oscillators may be solved and the oscillators essentially eliminated (Lienard and Wiechert potentials). The interactions which result involve relationships of the motion of one particle at one time, and of the other particle at another time. In quantum electrodynamics the field is again represented as a set of oscillators. But the motion of the oscillators cannot be worked out and the oscillators eliminated. It is true that the oscillators representing longitudinal waves may be eliminated. The result is instantaneous *electrostatic interaction*. The electrostatic elimination is very instructive as it shows up the difficulty of *self-interaction* very distinctly. In fact, it shows it up so clearly that there is no ambiguity in deciding what term is incorrect and should be omitted. *This entire process is not relativistically invariant, nor is the omitted term. It would seem to be very desirable if the oscillators, representing transverse waves, could also be eliminated. This presents an almost insurmountable problem in the conventional quantum mechanics.* We expect that the motion of a particle *a* at one time depends upon the motion of *b* at a previous time, and vice versa. *A wave function* $\psi(x_a, x_b; t)$*, however, can only describe the behavior of both particles at one time*. There is no way to keep track of what *b* did in the past in order to determine the behavior of *a*. The only way is to specify the *state* of the set of oscillators at t, which serve to "remember" what *b* (and *a*) had been doing.

The present formulation permits the solution of the motion of all the oscillators and their complete elimination from the equations describing the particles. This is easily done. *One must simply solve for the motion of the oscillators before one integrates over the various variables* x_i*, for the particles.* It is the integration over x_i which tries to condense the past history into a single *state* function. This we wish to avoid. Of course, the result depends upon the initial and final *states* of the oscillator. If they are specified, the result is an equation for $<\chi_{t''}|1|\psi_{t'}>$ like (38)

$$[<\chi_{t''}|1|\psi_{t'}>_S = \text{Lim}_{\varepsilon\to 0} \int \ldots \int \chi^*(x'', t'') \times$$
$$\times \exp(iS/\hbar)\psi(x', t')\, dx_0/A\, dx_1/A \ldots dx_{j-1}/A\, dx_j, \qquad (38)]$$

but containing as a factor, besides exp (iS/ħ) another functional G depending only on the coordinates describing the paths of the particles.

We illustrate briefly how this is done in a very simple case. …

…

Proceeding in an analogous manner one finds that all of the oscillators of the electromagnetic field can be eliminated from a description of the motion of the charges.

[14.] *Statistical Mechanics*

Spin and Relativity

Problems in the theory of measurement and statistical quantum mechanics are often simplified when set up from the point of view described here. For example, *the influence of a perturbing measuring instrument can be integrated out in principle as we did in detail for the oscillator*. The *statistical density matrix* has a fairly obvious and useful generalization. It results from considering the square of (38). It is an expression similar to (38) but containing integrations over two sets of variables dx_i and dx'_i. The exponential is replaced by $\exp i(S - S')/\hbar$, where S' is the same function of the x'_i as S is of x_i. It is required, for example, to describe the result of the elimination of the field oscillators where, say, the final *state* of the oscillators is unspecified and one desires only the sum over all final *states* *m*.

Spin may be included in a formal way. The *Pauli spin equation* can be obtained in this way: One replaces the vector *potential interaction* term in $S(x_{i+1}, x_i)$,

$$e/2c\ (x_{i+1} - x_i) \cdot A(x_i) + e/2c\ (x_{i+1} - x_i) \cdot A(x_{i+1})$$

arising from expression (13)

$$[\exp [i/h \textstyle\sum_{i=k}^{\infty} S(x_{i+1}, x_i)] \cdot \exp [i/h \textstyle\sum_{i=-\infty}^{k-1} S(x_{i+1}, x_i)]. \qquad (13)]$$

by the expression

$$e/2c\ [\sigma \cdot \{x_{i+1} - x_i\}\{\sigma \cdot A(x_i)\}] + e/2c\ [\{\sigma \cdot A(x_{i+1})\}\{\sigma \cdot (x_{i+1} - x_i)\}].$$

Here A is the vector *potential,* x_{i+1} and x_i the vector *positions* of a particle at times t_{i+1} and t_i, and σ is Pauli's spin vector matrix. The quantity Φ must now be expressed as $\Pi_i \exp iS(x_{i+1}, x_i)/\hbar$ for this differs from the exponential of the sum of $S(x_{i+1}, x_i)$. *Thus,* Φ *is now a spin matrix.*

The *Klein Gordon relativistic equation can also be obtained formally by adding a fourth coordinate to specify a path.* One considers a "*path*" as being specified by four functions $x^{(\mu)}(\tau)$ of a parameter τ. The parameter τ now goes in steps as the variable t went previously. The quantities $x^{(1)}(t)$, $x^{(2)}(t)$, $x^{(3)}(t)$ are the space coordinates of a particle and $x^{(4)}(t)$ is a corresponding time. The Lagrangian used is

$$\textstyle\sum_{\mu=1}^{l} [(dx^{\mu}/d\tau)^2 + (e/c)(dx^{\mu}/d\tau)A_{\mu}],$$

where A_μ is the 4-vector *potential* and the terms in the sum for $\mu = 1, 2, 3$ are taken with reversed sign. If one seeks a *wave function* which depends upon τ periodically, one can show this must satisfy the Klein Gordon equation. *The Dirac equation results from a modification of the Lagrangian used for the Klein Gordon equation, which is analogous to the modification of the non-relativistic Lagrangian required for the Pauli equation.* What results directly is the square of the usual Dirac operator.

These results for spin and relativity are purely formal and add nothing to the understanding of these equations. There are other ways of obtaining the Dirac equation which offer some promise of giving a clearer physical interpretation to that important and beautiful equation.

The author sincerely appreciates the helpful advice of Professor and Mrs. H. C. Corben and of Professor H. A. Bethe. He wishes to thank Professor J. A. Wheeler for very many discussions during the early stages of the work.

Dirac, P. A. M. (May, 1948). Quantum Theory of Localizable Dynamical Systems.

[*Phys. Rev.*, 73, 9, 1092-103; https://doi.org/10.1103/PhysRev.73.1092.]

Received January 12, 1948.

Institute for Advanced Study, Princeton, New Jersey.

A dynamical system is called *localizable* if its *wave functions* can be expressed in terms of variables, each referring to physical conditions at only one point in *space-time*. These variables may be at points on any three-dimensional space-like surface in space-time. A general investigation is made of how the *wave function* varies when the surface is varied in any way. The variation of the *wave function* is given by equations of the Schrödinger type involving certain operators which play the role of Hamiltonians. The *commutation relations* for these operators are obtained. The theory works entirely with *relativistic* concepts and it provides the general pattern which any *relativistic* quantum theory must conform to, provided the dynamical system is *localizable*.

[Referenced in Schwinger (November, 1948) [Quantum Electrodynamics. I. A Covariant Formulation], page 1448.]

Abstract

A dynamical system is called *localizable* if its *wave functions* can be expressed in terms of variables, each referring to physical conditions at only one point in space-time. These variables may be at points on any three-dimensional space-like surface in space-time.

A general investigation is made of how the *wave function* varies when the surface is varied in any way. The variation of the *wave function* is given by equations of the Schrödinger type involving certain operators $H^n(u)$ which play the role of Hamiltonians. The *commutation relations* for these operators are obtained (Eqs. (50), (51)). The theory works entirely with *relativistic* concepts and it provides the general pattern which any *relativistic* quantum theory must conform to, provided the dynamical system is localizable.

Feynman, R. P. (October, 1948). A Relativistic Cut-Off for Classical Electrodynamics.

[*Phys. Rev.*, 74, 939-946; https://doi.org/10.1103/PhysRev.74.939.]

Received June 8, 1948.

Cornell University, Ithaca, New York.

In this paper a consistent classical theory is described which the author believes can be quantized, previous attempts to address *problem of infinite self-energy* that results from assuming *point electron* in *relativistic* theory met with considerable difficulties when attempt made to quantize them, the *potential* at a point in space at a given time depends on the charge at a distance r from the point at a time previous by t = r (taking the speed of light as unity), *relativistically* interaction occurs between events whose four-dimensional interval s defined by $s^2 = t^2 - r^2$ vanishes, this results in *infinite action of a point electron on itself, the present theory modifies this idea by assuming that substantial interaction exists as long as the interval s is time-like and less than some small length a of order of the electron radius*, reduces infinite *self-energy* to a finite value for accelerations which are not extreme, *action* of an electron on itself appears as *electromagnetic mass*, formulates in terms of *action at a distance*, satisfies Maxwell's equations, not the usual *retarded* solution for which there is no *self-force* but *half the retarded plus half the advanced solution*, effect of modification is to change slightly the field of one particle on another when they are very close and to add a *self-force, little reason to believe that the ideas used here to solve the divergences of classical electrodynamics will prove fruitful for quantum electrodynamics*.

> [Feynman, R. P. (December, 1965). Nobel Lecture. The Development of the Space-Time View of Quantum Electrodynamics: "It is most striking that most of the ideas developed in the course of this research were not ultimately used in the final result. For example, the *half-advanced and half-retarded potential* was not finally used, the *action* expression was not used, the idea that charges do not act on themselves was abandoned"].

Abstract

Ordinarily it is assumed that *interaction between charges* occurs along light cones, that is, only where the four-dimensional interval $s^2 = t^2 - r^2$ is exactly zero. We discuss the modifications produced if, as in the theory of F. Bopp, substantial interaction is assumed to occur over a narrow range of s^2 around zero. This has no practical effect on the *interaction of charges* which are distant from one another by several electron radii. *The action of a charge on itself is finite* and behaves as *electromagnetic mass* for accelerations which are not excessive. There also results a classical representation of the phenomena of pair production in sufficiently strong fields.

Quantum electrodynamics is built from a classical counterpart that already contains many difficulties which remain upon quantization. *It has been hoped that if a* [relativistic] *classical electro- dynamics could be devised which would not contain the difficulty of infinite self-energy, and this theory could be quantized*, then the problem of a self-consistent quantum electrodynamics would be solved. For this reason, many successful attempts have been made to produce such a classical theory. The field equations can be made non-linear,[1]

[1] Born, M. & Infeld, L. (March, 1934). Foundations of the New Field Theory. *Roy. Soc. Proc., A*, 144, 852, 425-51; https://doi.org/ 10.1098/rspa.1934.0059.

the fields produced by or acting on an electron can be redefined,[2,3]

[2] Dirac, P. A. M. (August, 1938). Classical Theory of Radiating Electrons. *Roy. Soc. Proc., A*, 167, 929, 148-69[; Dirac referenced Frenkel (1925) in describing *relativistic* form of classical theory of radiating electrons assuming electron to be a *point charge* with no volume, *an extended electron inconceivable in special theory of relativity due to intrinsic connection between space and time*, Lorentz *model of electron as possessing mass on account of the electromagnetic field around it fails* without further assumptions if the electromagnetic field varies too rapidly or if the acceleration of the electrons is too great, no natural way of introducing further assumptions has been discovered, discovery of neutron introduced a form of mass that is difficult to believe is of electromagnetic nature, theory of the positron in which positive and negative values of an electron play symmetrical roles cannot be fitted into idea of electromagnetic idea of mass which requires all mass to be positive, *departure from electromagnetic theory of mass removes main reason for believing in finite size of the electron*, results in difficulty that the *field* in the immediate neighborhood of the electron has an infinite mass, in quantum mechanics results in divergence in the solution of the equations that describe the interaction of an electron with an electromagnetic field and prevents its application to high-energy radiation processes, *Dirac chose to represent electron by point charge and to avoid the difficulties with the infinite energy by direct omission or subtraction of unwanted terms* as had been used in the *theory of the positron. new theory must be in agreement with special relativity* and the conservation of energy and momentum, first addressed problem of a *single electron moving in an electromagnetic field*, used equations for electromagnetic *potentials* in terms of the *charge-current density* vector, *assumed that the charge-current density vector vanishes everywhere except on the world-line of the electron where it is infinitely great*, derived *field* quantities from electromagnetic *potentials*, obtained solutions in terms of *retarded* and *advanced* potentials and potentials representing the incoming and outgoing radiation, calculated field of radiation produced by the electron, in agreement with classical theory but provided a value for the field of radiation throughout space-time, obtained exact *equations of motion* of the electron in a specified incident *field* from the laws of conservation of energy and momentum within limits of classical theory, imposed condition that solutions that occur in Nature when there is no incident field are those for which velocity in terms of proper time is constant, requires solutions of equations of motion *for which the final acceleration as well as the initial position and velocity are prescribed*, electron responds to pulse of electromagnetic radiation before it reaches the center of the electron, behaves as if it has a radius of order $1/a$ where $a = 3m/e^2$, implies that signal can be transmitted *faster than the speed of light* through the interior of an electron, *interior of electron is a region of failure of some of the elementary properties of space-time*, showed how theory can be extended to any number of electrons interacting with each other and

with a field of radiation]. An excellent discussion of these matters is given by Eliezer, C. J. (July, 1947). The Interaction of Electrons and an Electromagnetic Field. *Rev. Mod. Phys.* 19, 147; https://doi.org/10.1103/RevModPhys.19.147.

[3] Wheeler, J. A. & Feynman R. P. (April, 1945). Interaction with the Absorber as the Mechanism of Radiation. *Rev. Mod. Phys*., 17, 157-81[; the motive of the analysis was to clear the present quantum theory of interacting particles of those of its difficulties which have a purely classical origin, method was to define as closely as one can within the bounds of classical theory the proper use of the *field* concept in the description of nature, this paper represents the third part of the survey, an analysis of the *mechanism of radiation* believed to complete the last tie between *action at a distance* and *field theory*, only section now finished, difficulties to obtain a satisfactory account of the *field* generated by an accelerated charge at a remote point and to understand the source of the force experienced by the charge itself as a result of its motion, takes up suggestion by Tetrode (1922) that *the act of radiation should have some connection with the presence of an absorber*, develops this idea into the thesis that the force of radiative reaction arises from the action on the source owing to the half-advanced fields of the particles of the absorber, absorber response as the mechanism of radiative reaction, expresses in terms of *action at a distance*, *Wheeler–Feynman absorber theory, assumes solutions of the electromagnetic field equations* must be invariant under time-reversal transformation as are the field equations themselves, considers an accelerated charge located in the absorbing system as the source of radiation, *elementary particles not self-interacting*, a complete correspondence is established between *action at a distance* and the usual formulation of *field theory* in the case of a completely absorbing system].

or one may resort to some averaging of the fields over space or time.[4]

[4] There are many theories of this nature. The author's theory is essentially that of Bopp, F. (1942). *Ann. Physik*, 42, 573[; not found]. R. Peierls and H. McManus have developed a theory in which the electron is pictured as a rigid distribution of charge in both space and time. The theory can be shown to be exactly equivalent to the present one, at least for a class of f functions. Their physical ideas may offer advantages over the present one in which the function f is not so directly interpretable. I thank Dr. McManus for a copy of his thesis. For a summary of another theory of this type see Podolsky, B. & Schwed, P. (January, 1948). Review of a Generalized Electrodynamics. *Rev. Mod. Phys*., 20, 40; https://doi.org/10.1103/RevModPhys.20.40. A somewhat different type is that of Rosen, N. (August, 1947). Statistical Geometry and Fundamental Particles. *Phys. Rev*., 72, 298; https://doi.org/10.1103/PhysRev.72.298.

[These papers attempt to avoid or remove the infinite *self-energy* produced by assuming a *point electron*. From 1941 to 1947, Freidrich Bopp was a staff scientist at the Kaiser-Wilhelm Institut für Physik (KWIP, after World War II reorganized and renamed the Max Planck Institute for Physics), located in Berlin-Dahlem. He worked on the German nuclear energy project; collaborators on aspects of this project were for a time known collectively as the Uranverein (Uranium Club). In 1944, when most of the KWIP was evacuated to Hechingen in Southern Germany due to air raids on Berlin, he went there as the Institute's Deputy Director. When the American Alsos Mission evacuated Hechingen and Haigerloch, near the end of

World War II, French armed forces occupied Hechingen. Bopp did not get along with them and described the initial French policy objectives towards the KWIP as exploitation, forced evacuation to France, and seizure of documents and equipment. In order to put pressure on Bopp to evacuate the KWIP to France, the French Naval Commission imprisoned him for five days and threatened him with further imprisonment if he did not cooperate in the evacuation.]

These theories have, however, met with considerable difficulties when an attempt has been made to quantize them. *In this paper a consistent classical theory is described which the author believes can be quantized.* Some preliminary results of the quantization of this theory will be discussed in a future paper. Some of the physical ideas of the classical form of the theory are sufficiently interesting in themselves to warrant their discussion first in a separate paper.

The *potential* at a point in space at a given time depends on the charge at a distance r from the point at a time previous by t = r (taking the speed of light as unity). Speaking *relativistically*, interaction occurs between events whose four-dimensional interval, s, defined by $s^2 = t^2 - r^2$, vanishes. *There results, however, an infinite action of a point electron on itself. The present theory modifies this idea by assuming that substantial interaction exists as long as the interval s is time-like and less than some small length, a, of order of the electron radius.* When t is large since $\Delta(s^2) = 2t.\Delta t$ this means a spread in the time of arrival of a signal of amount of order $a^2/2t$. For charges separated by many electron radii there is, therefore, essentially no effect of the modification. For the action of an electron on itself, however, there is a considerable modification. *The result is to reduce the infinite self-energy to a finite value. For accelerations which are not extreme, the action of an electron on itself appears simply as an electromagnetic mass.* If desired in the classical theory, all the mass of an electron may be represented as electromagnetic. (In the quantum theory this cannot be done in a reasonable way as the *electromagnetic mass* comes out quite small under reasonable assumptions for *a*.) *We have, therefore, a consistent classical theory which does not disagree with classical experience.*

In the remainder of the paper, we formulate this idea mathematically, and draw one or two simple consequences. *We then discuss a curious feature of this theory. It can give a classical representation of the phenomena of pair production in sufficiently strong fields.* This is of interest because the physical ideas may possibly be carried over to give a clearer understanding of the *hole theory* of positrons.

The main result which is to be carried over to quantum problems is this: *In any process in which there is no permanent emission of quanta one must assume the field quanta to have a "density" $g(k_4, \mathbf{K})$ in frequency, and wave number space.* This replaces the usual assumption that the frequency k_4 equals the magnitude of the wave number, K, and that the density in wave numbers **K**, is uniform (corresponding to $g(k_4, \mathbf{K}) = \delta(k_4^2 - K^2)$). The properties $g(k_4, \mathbf{K})$ ought to have are discussed more fully below.

Mathematical Formulation

It is most convenient (but not necessary) to formulate these ideas in the language of *action at a distance*.[3] Hence a brief summary of that point of view is given here. We start with Fokker's action principle that the action

$$S = \sum_a m_a \int (da_\mu da_\mu)^{1/2} + \sum'_{a,b} e_a e_b \int\int \delta(s_{ab}^2) da_\mu db_\mu, \qquad (1)$$

is an extremum. Here a_μ represents, for $\mu = 1$ to 4, the three space coordinates and the time coordinate of a particle a of mass m_a, charge e_a. We shall later consider them as functions of a parameter α, say. The b_μ are corresponding quantities for a particle b, etc. The symbol $x_\mu y_\mu$ means $x_4y_4 - x_1y_1 - x_2y_2 - x_3y_3$ and $s_{ab}^2 = (a_\mu - b_\mu)(a_\mu - b_\mu)$. The δ is Dirac's delta function.

The integrals are taken over the trajectories of the particles. The $\sum'$ means the sum over all pairs a,b with $a \neq$ b. We consider varying the path $a_\mu(\alpha)$ of particle a. Defining

$$A_\mu^{(b)}(x) = e_b \int \delta(s_{ab}^2)\, db_\mu, \qquad (2)$$

where x stands for x_μ, a point in space time, we can write (1) as

$$S = \sum_a m_a \int (da_\mu da_\mu)^{\frac{1}{2}} + \sum_a \sum_{b \neq a} e_a \int A_\mu^{(b)}(a) da_\mu.$$

The result of seeking an extremum of this is to lead in the well-known way to the *equations of motion*,

$$m_a \frac{d}{d\tau_a}\left(\frac{da_\nu}{d\tau_a}\right) = e_a \frac{da_\mu}{d\tau_a} \sum_{b \neq a} F_{\mu\nu}^{(b)}(a), \qquad (3)$$

where we can call $F_{\mu\nu}^{(b)}(x)$ the field at x caused by particle b. It is given by

$$F_{\mu\nu}^{(b)}(x) = \partial A_\mu^{(b)}(x) - \partial A_\nu^{(b)}(x)/\partial x_\mu.$$

We have written $d\tau_a = (da_\mu da_\mu)^{1/2}$ for the proper time along the path of a.

...

... Thus $F_{\mu\nu}^{(b)}(x)$ satisfies Maxwell's equations. But the special solution (2) is not the usual retarded solution but is rather *half the retarded plus half the advanced solution* of Lienard and Wiechert[5] (since $\delta(t^2 - r^2) = (1/2r)\{\delta(t + r) + \delta(t - r)\}$). Thus, we may write (dots representing derivatives with respect to τ_a, and the fields being calculated at the point $x_\mu = a_\mu$),

$$m_a \ddot{a}_r = e_a \dot{a}_\mu \sum_{b \neq a} \{\tfrac{1}{2} F(b)_{\mu\nu\ \mathrm{ret}} + (\tfrac{1}{2} F(b)_{\mu\nu\ \mathrm{adv}}\}. \qquad (5)$$

[5] This use of *advanced and retarded potentials* is really unnecessary for an understanding of the modifications of electrodynamics which is the main point of the paper. It results from the author's desire to start with a *principle of least action*, for it is in this form that the transition to quantum theory can be made.

This can be compared to the usual theory which just uses *retarded effects* by writing it in the form

$$m_a a_r^{\cdot\cdot} = e_a a_\mu^{\cdot} [\textstyle\sum_{b\neq a} F^{(b)}{}_{\mu\nu\ ret} + \frac{1}{2}\sum_{all\ b} \{F^{(b)}{}_{\mu\nu\ adv} - F^{(b)}{}_{\mu\nu\ ret}\}$$

$$- \tfrac{1}{2}\{F^{(a)}{}_{\mu\nu\ adv} - F^{(a)}{}_{\mu\nu\ ret}\}] \qquad (6)$$

as in the paper by Wheeler and Feynman[3] [Eq. (39), page 169].

> [Wheeler, J. A. & Feynman R. P. (April, 1945). Interaction with the Absorber as the Mechanism of Radiation: "On the *a*th charge the entire field acting is given, according to the theory of *action at a distance*, by the sum
>
> $$\textstyle\sum_{k\neq a} (\frac{1}{2} F^{(k)}{}_{ret} + \frac{1}{2} F^{(k)}{}_{adv}) \qquad (38)$$
>
> This expression can be broken down into three parts:
>
> $$\textstyle\sum_{k\neq a} F^{(k)}{}_{ret} + (\frac{1}{2} F^{(k)}{}_{ret} - \frac{1}{2} F^{(k)}{}_{adv}) - \frac{1}{2}\sum_{all\ k} (F^{(k)}{}_{ret} - F^{(k)}{}_{adv}) \quad (39)]$$

As in that paper the first term is the *retarded field* of other charges, the second term vanishes in a world where all emitted light is eventually absorbed,[6] and the third term, depending only on the motion of *a*, is the force of radiative damping.

> [6] That the second term vanishes in these circumstances may be seen as follows. If a source radiates for a time, *at a very long time afterwards the total retarded field vanishes, for all the light is absorbed.* But also, *the total advanced field vanishes at this time* (for charges are no longer accelerating and the advanced field exists only at times previous to their motion). Hence, the difference vanishes everywhere at this time and, since it is a solution of Maxwell's homogeneous equations, at all times.

Thus (1) is equivalent to (6) and thus satisfactorily describes the known laws of classical electrodynamics. *There is no self-energy.*

According to the above, a particle does not act upon itself, as the term with a = b in the sum $\sum'_{a,b} e_a e_b \int \ldots$ in the *action* has been omitted. (Radiation resistance is pictured as in indirect effect of source on absorber and absorber on source.) The field of each particle must be kept separate in order to exclude, when asking for the force on a particle, the field of the particle itself.

There is no need to do so, but it is an interesting question to try to reinstate the idea of a universal field. This requires that a particle be allowed to act on itself and the term a = b to be included in the *action* sum. *This leads immediately to an infinite self-force.* This difficulty can be eliminated if the $\delta(s_{ab}{}^2)$ is replaced, as Bopp[4] has suggested, by some other function $f(s_{ab}{}^2)$ of the invariant $s_{ab}{}^2$, which behaves like $\delta(s_{ab}{}^2)$ for large dimensions but differs for small. (We shall discuss the properties of this function later, but as an example to keep in mind, consider

$f(s^2) = (1/2a^2) \exp(-|s|/a)$ for $s^2 > 0$, and $f(s^2) = 0$ for $s^2 < 0$ with a of order of the electron radius e^2/mc^2.)

…

The effect of the modification in the theory using *retarded fields* is therefore to change, slightly, the field of one particle on another when they are very close, and to add a *self-force …*

…

> [Error reported in Feynman (1949), Space-Time Approach to Quantum Electrodynamics, page 776: "This relation is given incorrectly in Feynman (October, 1948). A Relativistic Cut-Off for Classical Electrodynamics, equation just preceding 16."]

…

There are many interesting problems presented by these ideas. For example, will pairs be produced ad infinitum by the field, or only to that extent that we can guarantee that the positrons will be annihilated by electrons in the future? Again, in a weak field can a large number of pairs be created which separate slightly in the field (which is insufficient to tear the two apart) and thus produce a large polarization of that field? It is hoped that an application of these ideas to a study of positron hole theory will appear in a future paper.

I should like to thank Professor J. A. Wheeler for inoculating me with many ideas without which this work would not have been done.

Appendix

…

… *There is, therefore, little reason to believe that the ideas used here to solve the divergences of classical electrodynamics will prove fruitful for quantum electrodynamics.* Nevertheless, the corresponding modifications were attempted with quantum electrodynamics and appear to solve some of the divergence difficulties of that theory. This will be discussed in a future paper.

Feynman, R. P. (November, 1948). Relativistic Cut-Off for Quantum Electrodynamics.

Phys. Rev., 74, 1430-8; https://doi.org/10.1103/PhysRev.74.1430.]

Received July 12, 1948.

Cornell University, Ithaca, New York.

Describes model based on quantization of a classical theory for which all quantities automatically come out finite described in his previous paper [Feynman (October, 1948). A Relativistic Cut-Off for Classical Electrodynamics], contains an arbitrary function on which numerical results depend, only term that depends significantly (logarithmically) on the cut-off frequency is the *self-energy* which can be used to *renormalize* the electron mass, remaining terms are nearly independent of the function, applies only to results for processes in which virtual quanta are emitted and absorbed, *terms representing processes involving a pair production followed by annihilation of the same pair are infinite and not made convergent by the present scheme,* problems of *permanent emission* and the position of *positron theory* need to be addressed, *the present paper may be looked upon as presenting an arbitrary rule to cut off at high frequencies in a relativistically invariant manner the otherwise divergent integrals appearing in quantum field theories*, produces finite invariant *self-energy* for a free electron, but problem of *polarization of the vacuum* not solved, alternative cut-off procedure which eliminates high frequency intermediate states offers to solve *vacuum polarization* problems as well.

> [Oppenheimer, J. R. (1948). Electron Theory. *Report to the Solvay Conference for Physics at Brussels*, Belgium, September 27 to October 2, 1948, pages 6-7. (Reprint in Schwinger, J. (ed). (1958). *Selected Papers on Quantum electrodynamics*, Dover, New York, pages 150-1: "Now it is true that the fundamental equations of quantum-electrodynamics are gauge and Lorentz covariant. But they have in a strict sense no solutions expansible in powers of e. If one wishes to explore these solutions, bearing in mind that certain infinite terms will, in a later theory, no longer be infinite, one needs a covariant way of identifying these terms, and for that, not merely the field equations themselves, but *the whole method of approximation and solution must at all stages preserve covariance. This means that the familiar Hamiltonian methods, which imply a fixed Lorentz frame t = constant, must be renounced*; neither Lorentz frame nor gauge can be specified until after, in a given order in e, all terms have been identified, and those bearing on the definition of charge and mass recognized and relegated; then of course, in the actual calculation of transition probabilities and the reactive corrections to them, or in the determination of stationary states in fields which can be treated as static, and in the reactive corrections thereto, the introduction of a definite coordinate system and gauge for these no longer singular and completely well-defined terms can lead to no difficulty.

It is probable that, at least to order e^2, more than one covariant formalism can be developed. Thus, *Stueckelberg's four-dimensional perturbation theory*[26] would seem to offer a suitable starting point, *as also do the related algorithms of Feynman*[27].

[26] Stueckelberg, E. C. G. (September, 1934). Relativistisch invariante Störungstheorie des Diracschen Elektrons I. Teil: Streustrahlung und Bremsstrahlung. (Relativistically invariant perturbation theory of Dirac's electron Part I: scattered radiation and Bremsstrahlung.) *Ann. Phys.*, 413, 4, 367-89.

[27] Feynman, R. P. (November, 1948). Relativistic Cut-Off for Quantum Electrodynamics. *Phys. Rev.*, 74, 1430-8.

But a method originally suggested by Tomonaga[28], and independently developed and applied by Schwinger[22], would seem, apart from its practicality, to have the advantage of very great generality and a complete conceptual consistency.

[28] Tomonaga, S. (1943). On a Relativistically Invariant Formulation of the Quantum Theory of Wave Fields. *Bull. I. P. C. R. Riken-iho*, 22, 545 (in Japanese); translation Tomonaga, S. (August, 1946). *Prog. Theor. Phys.* 1, 2, 27-42; Koba, Z., Tati, T. & Tomonaga, S. (1947a). On a Relativistically Invariant Formulation of the Quantum Theory of Wave Fields. II. Case of Interacting Electromagnetic and Electron Fields. § 1-4. *Prog. Theor. Phys.*, 2, 101–116; Koba, Z., Tati, T. & Tomonaga, S. (1947b). On a Relativistically Invariant Formulation of the Quantum Theory of Wave Fields. III. Case of Interacting Electromagnetic and Electron Fields. § 5-7. *Prog. Theor. Phys.*, 2, 198–208.

[22] Schwinger, J. (November, 1948). Quantum Electrodynamics. I. A Covariant Formulation. *Phys. Rev.*, 74, 10, 1439-61, and in press.

It has also been shown by Dyson[29] how Feynman's algorithms can be derived from the Tomonaga equations.

[29] Dyson, F. J., *Phys. Rev.*, in press."]

Abstract

A *relativistic cut-off of high frequency quanta*, similar to that suggested by Bopp, is shown to produce a finite invariant *self-energy* for a free electron. The electromagnetic line shift for a bound electron comes out as given by Bethe and Weisskopf's wave packet prescription. The scattering of an electron in a potential, without radiation, is discussed. The cross section remains finite. *The problem of polarization of the vacuum is not solved.* Otherwise, the results will in general agree essentially with those calculated by the prescription of Schwinger. An alternative cut-off procedure analogous to one proposed by Wataghin, which eliminates high frequency intermediate states, is shown to do the same things but to offer to solve *vacuum polarization* problems as well.

The main problems of quantum electrodynamics have been essentially solved by the observations of Bethe[1] and of Weisskopf[2] that the divergent terms in the line shift problem can be thought to be contained in a *renormalization* of the mass of a free electron.

[1] Bethe, H. A. (August, 1947). The Electromagnetic Shift of Energy Levels. *Phys. Rev.*, 72, 339-41[; Lamb and Retherford results show that fine structure of second quantum state of hydrogen does not agree with *Dirac wave equation*, Schwinger, Weisskopf, and Oppenheimer suggest might be due to *shift of energy levels by interaction of electron with the radiation field*, this shift comes out infinite in all existing theories and has therefore always been ignored, possible to identify the most strongly (linearly) divergent term in the level shift with an *electromagnetic mass effect* which must exist for a bound as well as for a free electron, already included in the *observed mass* of the electron so should be subtracted, assumes *relativistic cut-off* in quantum energies (frequencies) of included atomic states, then calculation of Lamb shift for hydrogen atom using *non-relativistic* ordinary radiation theory gives shift of the levels due to *radiation interaction* in close agreement with observed value, removes discrepancy with Dirac theory, did not carried out *relativistic* calculations.]; Bethe, H. A. (1947). *Ibid.*, 73, 1271A [Not found].
[2] Schwinger J. & Weisskopf, V. (1948). *Phys, Rev.*, 73, 1272A [Not found].

That this principle applies as well to other problems was demonstrated by Lewis[3] in analyzing the radiationless scattering of an electron in a potential.

[3] Lewis, H. W. (January, 1948). On the Reactive Terms in Quantum Electrodynamics. *Phys. Rev.*, 73, 2, 173-6; https://doi.org/10.1103/PhysRev.73.173.

Ambiguities which remained in the subtraction procedures are removed by Schwinger[2,4].

[4] Schwinger, J. (February, 1948). On quantum-electrodynamics and the magnetic moment of the electron. *Phys. Rev.*, 73, 4, 416-7[; *electrodynamics unquestionably requires revision at ultra-relativistic energies,* desirable to isolate those aspects of the current theory that essentially involve high energies and are subject to modification by a more satisfactory theory, this goal has been achieved by transforming the Hamiltonian of current *hole theory* electrodynamics to exhibit explicitly the logarithmically divergent self-energy of a free electron which arises from the virtual emission and absorption of light quanta, the electromagnetic *self-energy* of a free electron can be ascribed to an *electromagnetic mass* which must be added to the mechanical mass of the electron, new Hamiltonian involves experimental *electron mass* rather than unobservable *mechanical mass*, electron now interacts with radiation field only in presence of external field such that only an accelerated electron can emit or absorb a light quantum, *interaction energy* of electron with external field subject to *finite* radiative correction, *polarization of the vacuum* still produces logarithmically divergent term proportional to *interaction energy* of electron in an external field, such term equivalent to altering value of *electron charge* by constant factor with only final value being identified with experimental charge, interaction between matter and radiation produces *renormalization* of electron charge and mass, all divergences contained in *renormalization* factors, radiative correction for energy of electron in external magnetic field corresponds to *additional magnetic moment associated with electron spin* of magnitude $\delta\mu/\mu = (\frac{1}{2}\pi)e^2/hc = 0.001162$, experimental measurements on hyperfine splitting of ground states of atomic hydrogen and deuterium larger than expected from

directly measured nuclear moments, finds additional *electron spin magnetic moment* to account for measured hydrogen and deuterium *hyperfine structures* to be $\delta\mu/\mu = 0.00126$ and $\delta\mu/\mu = 0.00131$ respectively, these discrepancies accounted for by additional spin magnetic moment to the electron of $\delta\mu/\mu = 0.0018 \pm 0.00003$, *values yielded by relativistic calculation of Lamb shift differ only slightly from those conjectured by Bethe on basis of non-relativistic calculation and are in good accord with experiment*].

He formulated, in a general way, which terms are to be identified in a future correct theory with rest mass, and hence should be omitted from a calculation which does not renormalize the mass. These results are remarkable because they solve the problem without the addition of any new fundamental lengths or dimensions.

The solution given by Schwinger does, however, assume that in some future theory the divergent self-energy terms will be finite. Therefore, it is of interest to point out that there is a model, a modification of ordinary electrodynamics, for which all quantities automatically do come out finite. With this model the ideas of Bethe, Oppenheimer, and Lewis and Schwinger can be directly confirmed.

The model results from the quantization of a classical theory described in a previous paper[5].

[5] Feynman, R. P. (October, 1948). A Relativistic Cut-Off for Classical Electrodynamics. *Phys. Rev.*, 74, 939-946.

In this paper we describe only the results for processes in which only virtual quanta are emitted and absorbed. *The problems of permanent emission and the position of positron theory must be more completely studied.* It is hoped that a complete physical theory may be published in the near future. Lacking such a complete picture, *the present paper may be looked upon merely as presenting an arbitrary rule to cut off at high frequencies in a relativistically invariant manner the otherwise divergent integrals appearing in quantum field theories*. For electrodynamics *the rule is to consider the (positive) frequency ω and wave number k of the field oscillators as independent* and to integrate them over the density function $g(\omega^2 - k^2)d\omega dk$ where

$$g(\omega^2 - k^2) = \int_0^\infty [\delta(\omega^2 - k^2) - \delta(\omega^2 - k^2 - \lambda^2)]G(\lambda)\, d\lambda \qquad (1)$$

Here $\delta(x)$ is Dirac's delta function and $G(\lambda)$ is some smooth function such that $\int_0^\infty G(\lambda)\, d\lambda = 1$ and for which the mean values of λ which are important are of order of the frequency $137\ mc^2/h$, or higher. Ordinary quantum electrodynamics replaces the function $g(\omega^2 - k^2)$ by $\delta(\omega^2 - k^2)$. According to (1), the density g is not everywhere positive[5]. Therefore, the model is essentially that due to Bopp[6].

[6] Bopp, F. (1942). *Ann. Physik*, 42, 573 [Not found.].

The model therefore contains an arbitrary function and the numerical results depend on the form of $G(\lambda)$. However, the only term that depends seriously (logarithmically) on the cut-off frequency is the *self-energy*, which can be used to renormalize the electron mass. After this is done, the remaining terms are nearly independent of the function $G(\lambda)$.

We shall illustrate these points by studying the particular examples of *self-energy* and radiationless scattering. *We shall then discuss an alternative cut-off procedure in which the density of electron states is cut off rather than that of the quanta*. This promises to solve problems of *vacuum polarization* which are not touched by the former procedure.

Self-energy

The *transverse self-energy* of a free electron, of mechanical mass μ, in state of momentum P_0 energy $E_0 = (\mu^2 + P_0^2)^{1/2}$ is given to the first order in e^2 by the second-order perturbation theory, *using the one-electron theory of Dirac*, by

$$\Delta E = \ldots . \qquad (2)$$

Here the intermediate state f arises from the initial state through emission of a quantum of momentum **k** and of energy $k = |\mathbf{k}|$ (the velocity of light is taken as unity, as is Planck's constant). Thus, in the intermediate state the electron has momentum $\mathbf{P}_f = \mathbf{P}_0 - \mathbf{k}$ and an energy of magnitude $E_f = +(\mu^2 + P_f^2)^{1/2}$ but which may be either plus or minus in sign. The sums indicate the sum over all such intermediate states (actually just two) for each sign of the energy. *The terms for positive and negative energy have been separated* and the sums are written $\sum_+$ and $\sum_-$ for these two cases. The $(f|\alpha_i|0)$ are the matrix elements of Dirac's α-matrices, the sum on i being over the two directions of polarization of the quanta. We shall henceforth write the integral $d\mathbf{k}/k$ over k space by its equivalent $2\int d\omega d\mathbf{k}\,\delta(\omega^2 - k^2)$, the integral being over all positive ω, and all wave numbers **k**. We shall also write ω for k in the energy denominators as we shall later wish to distinguish the energy of a quantum and the magnitude of the momentum change that its recoil represents....

...

According to the theory of holes, the last term, the transition to negative energy states, is to be left out; such transitions are prevented because the negative levels are already occupied. On the other hand, in the vacuum, electrons in state of energy $-E_f$ could make virtual transitions to positive energy state E_0. This is now prevented by the presence of an electron in the state E_0, so that, relative to the vacuum, the *transverse self-energy* is

$$\Delta E = \ldots . \qquad (3)$$

The treatment of the *longitudinal self-energy* is usually different, for the longitudinal oscillators are first eliminated from the Hamiltonian, their effect being the term e^2/r_{00} where r_{00} is the meaningless distance of the electron from itself. These terms must be expressed as integrals over oscillators and combined with (3) before the change suggested by (1) is to be performed. An additional point of confusion is that the longitudinal elimination assumes the intermediate states to form a complete set as they do in (2), but the situation in (3) is not so clear. Fortunately, all these points may be most easily circumvented by simply not eliminating the longitudinal oscillators from the field Hamiltonian at all. One need simply to specify that the sum on i in (3) now be interpreted to mean the sum over each of three perpendicular space directions minus a term for the time direction. We may write $\sum_i \alpha_i\Lambda\alpha_i = \boldsymbol{\alpha}.\Lambda.\boldsymbol{\alpha} - \Lambda$, which is a relativistic combination since $\alpha_4 = 1$. One does not need to be concerned about the gauge condition in a problem in which all quanta are virtual, for the quanta are created by a charge which is conserved. This solution automatically ensures the gauge condition just as the Lienard Wiechert classical solution of the Maxwell equations will automatically satisfy the gauge condition if the charge which produces the potential is conserved.

With this convention for $\sum_i$ Eq. (3) represents the *total self-energy*. It is easily calculated. …

We may use this result to show that the level shift for an electron in a bound state given in the present theory will be essentially that given by Weisskopf and Bethe according to their so-called wave-packet method. The change in energy of our electron in a bound state may be calculated in a straightforward manner according to the present formulation. One would simply start with Eq. (2) but with the wave functions and energies for states 0 and f being appropriate for the potential by which the electron is bound. Then one would integrate over $g(\omega^2 - k^2)$ rather than $\delta(\omega^2 - k^2)$ and obtain a definite finite result. The result would show a fairly large change in E_0 depending logarithmically on λ.

A good part of this change could be accounted for as simply due to the change in E_0 that would occur if the mass of the electron were altered from μ to $m = \mu + \Delta\mu$. We can define the true term shift, then, as the complete change in E_0, less $\Delta\mu(\partial E_0/\partial\mu)$, the change due to using μ instead of m in computing the energy with radiation absent. But $\partial E_0/\partial\mu$ is by perturbation theory the expected value $(\psi_0^*|\ \beta\ |\psi_0)$ of β for the state ψ_0 in question. From the result (12), however, this is also equivalent to computing the self-energy of a wave packet ψ_0, assuming the electron as free. But *Bethe*[1] *and Weisskopf*[2] *compute their term shift by just this prescription: the total effect less the self-energy of the free packet*. The only difference here is that we would compute the term shift integral on $g(\omega^2 - k^2)$ rather than $\delta(\omega^2 - k^2)$. But since the integral converges either way, the difference between the two results is very small, being of order of (μ^2/λ_0^2) times smaller than the result.

Radiationless scattering

We can study the *radiationless scattering* problem in a similar manner. *This problem is the correction to the scattering by a first-order potential due to the possibility of emission and absorption of a virtual quantum.* For example, this emission and absorption can occur at any time previous to the scattering. (It would, in this case, be nearly equivalent to a change in mass in the wave function of the electron arriving at the scatterer.) There will be a large change in cross section, which would be expected as the result of a change in mass of the electron plus a smaller change caused essentially by emissions previous to and absorptions subsequent to the scattering. As in the case of the self-energy in a field and, in fact, in all such problems, we will really be interested in those effects of radiation over and above that resulting from the change in mass. It is, therefore, simpler to compute the difference between the desired quantity calculated with no radiation and electrons of mass m, and the same quantity computed with the possibility of a virtual quantum emission and absorption with an electron of mass μ. This difference, which we shall call the radiative correction, can be looked upon as the result of perturbation due to the addition to the Hamiltonian of both the radiative interaction terms and a term $-\beta\Delta\mu$. The latter term can, as we have shown, be represented by the integral over oscillators of

... (13)

when acting on a free electron state of positive energy E_0 and momentum P_0. When acting on a state of negative energy $-E_0$, the term can be shown in a similar manner to be the expression (13) with the sign of E_0 changed in the denominator.

Terms like these are just the ones that Schwinger[4]

> [[4] Schwinger, J. (February, 1948). On quantum-electrodynamics and the magnetic moment of the electron.]

thought should be omitted from the Hamiltonian if one wishes to get meaningful results, so that the present model agrees with Schwinger's prescription.

When this process is applied to the scattering problem to obtain the radiative correction to the matrix elements, we are left with several residual terms. First, the emissions and absorptions previous to scattering are not exactly equivalent to a change in mass. If the emission occurs too close (in time) to the scattering, the absorption must occur in a restricted time, rather than at leisure as for a free electron forming $\beta\Delta\mu$. ...

...

The remaining terms are those for which the potential scattering occurs between the emission and absorption. They may be worked out as by Dancoff[10] (except that we include the longitudinal waves by summing i from 1 to 4).

> [10] Dancoff, S. M. (May, 1939). On Radiative Corrections for Electron Scattering. *Phys. Rev.*, 55, 959[; a *relativistic* treatment of the radiative correction of order $e^2/\hbar c$ to the elastic scattering cross section leads to the following results: (a) for the scattering in an electrostatic field of a particle described by the Pauli-Weisskopf theory, the correction is finite; (b) for the scattering of a Dirac electron in an electrostatic field, the correction diverges logarithmically and is positive; (c) the convergence or divergence of the correction depends critically on the type of scattering potential considered].

...

The integrals do, however, diverge logarithmically at the lower limit of small momentum transfer. This infra-red catastrophe has been completely cleared up by Bloch and Nordsieck[11].

> [11] Bloch, F. & Nordsieck, A. (July, 1937). Note on the Radiation Field of the Electron. *Phys. Rev.*, 52, 54; https://doi.org/10.1103/PhysRev.52.54.

They show that for very long wave-length quanta the amplitude for emission and reabsorption of more than one quantum is not negligible. Inclusion of these higher order terms, which is necessary only in the *non-relativistic* region, solves the problem. ...

...

… Thus, the terms up to second order can be represented by matrix elements of first and second space and time derivatives of the *potential*. That is, the *radiative correction to the scattering* in any potential is equivalent to the first order in e^2 and in the potential, to the scattering produced by a perturbation ΔH to the Dirac Hamiltonian. The perturbation up to terms of first and second derivatives of the vector potential **A** and the scalar potential φ is calculated in this manner to be

ΔH = … . (18)

The first term, where $\mathbf{B} = \nabla \times \mathbf{A}$ and $\mathbf{E} = -\nabla\varphi - (1/c)\,\partial\mathbf{A}/\partial t$, has the same effect as an alteration in the electron magnetic moment[13] by a fraction $e^2/2\pi\hbar c$.

[13] Pauli, W. (1933). *Handbuch der Physik*, Vol. 24/1, p. 233.

This effect was first discovered by Schwinger[4].

Line shift

The perturbation to H given here is useful not only for scattering problems but also for the *line-shift* problem. The actual motion of an electron in a binding potential can be visualized as simply a continued sequence of scatterings in this potential. For each scattering we can calculate the effect of virtual quanta in the way outlined above. However, it is possible, if the potential is strong, that two scatterings occur between the emission and reabsorption of the quantum, in which case the above formula for ΔH is incorrect. In hydrogen the potential over most of the atom is sufficiently weak that this does not occur with effective probability. The very long wave-length quanta do have a tendency to exist in the virtual state for long periods, but they have been eliminated by the cut-off λ_{min} at low frequencies.

In hydrogen, then, the *line shift* due to quanta above minimum wave number k_{min} is the expected value, for the state in question, of

ΔH = … . (19)

where $\varphi = e/r$, r being the distance to the proton, and we have used the relation

$\ln \lambda_{min} = \ln (2k_{min}) - 1$.

[Feynman (September, 1949), Space-Time Approach to Quantum Electrodynamics, page 777: "*Footnote 13*. "That the result given in B [Feynman (November, 1948) in Eq. (19)] *was in error* was repeatedly pointed out to the author, in private communication, by V. F. Weisskopf and J. B. French, as their calculation, completed simultaneously with the author's early in 1948, gave a different result. French has finally shown that *although the expression for the radiation-less scattering B, Eq. (18), or (24) above is correct, it was incorrectly joined onto Bethe's non-relativistic result*. He shows that the relation $\ln 2k_{max} - 1 = \ln\lambda_{min}$ used by the author should have been $\ln 2k_{max} - 5/6 = \ln\lambda_{min}$. This results in adding a term $-(1/6)$ to the logarithm in B, Eq. (19) so that the result now agrees with that of French & Weisskopf (April, 1949). [The Electromagnetic Shift of Energy Levels], and Kroll & Lamb (February, 1949). [On the self-energy of a bound electron]. *The author feels unhappily responsible for the very considerable delay in the publication of French's result occasioned by this error.*"]

The first term ensures that the fine structure separation correction will be that expected from the change in the electron's magnetic moment. The second may be combined with Bethe's non-relativistic calculation for quanta below k_{min}[14].

[14] Using Eq. (18), Professor Bethe finds 1050 megacycles for the separation between $2p_{1/2}$ and $2s_{1/2}$ in hydrogen. (Solvay Report.)

Application to other processes

The important problem of verifying that the self-energy will not diverge in higher-order approximations has not been carried to completion. It appears unlikely that trouble will arise here. If that is true the model probably gives sensible answers to all problems of quantum electrodynamics other than those involving Uehling polarization effects, discussed below. It has been found to give finite self-mass if we have, instead of a vector field, a scalar field or a pseudoscalar field, coupled to the electron in the simplest way possible without gradient operators. If the field quanta have mass M, $g(\omega^2 - k^2)$ is replaced by $g(\omega^2 - k^2 - M^2)$, and the values of λ of importance are chosen to be large compared to M.

The results for electrodynamics, then, after *mass renormalization*, depend only slightly on the form of $G(\lambda)$ and the size of λ_0. Since λ_0 may be taken to be extremely large without spoiling the smallness of $\Delta\mu$, there would appear to be good reason to drop the dependence on λ altogether. Thus, the $G(\lambda)$ appears only as a complicated scaffold which is removed after the calculation is done.

On the other hand, electrodynamics probably does break down somewhere and it is interesting to keep the terms in λ for various phenomena to see if one might be selected which is particularly sensitive to λ. This phenomenon would then be a promising one to study experimentally. The Moller interaction between two electrons is modified by the present theory. There is, of course, the *radiative correction*, but in addition to that there is simply a change due to the change in the density function for the quanta which can be exchanged. The Moller interaction ordinarily is proportional to $1/q^2$, where q is the magnitude of the momentum transferred from one electron to the other in the center of gravity system. The modification is only that this factor is changed to $\int_0^\infty \{1/q^2 - 1/(q^2 + \lambda^2)\}G(\lambda)\, d_\lambda$. This represents a decrease in cross section for hard collisions. If λ is of order 137 μc^2, we would need electrons in the center of gravity system of roughly 30 Mev to find a strong effect. This corresponds, however, to bombardment of stationary electrons by electrons of 3½ Bev[15].

[15] A more promising way to obtain processes with high momentum transfer would be wide-angle scattering of electrons from nuclei. But here deviations from expectations might be associated with uncertainties in the nuclear charge distributions rather than electrodynamics. Very wide angle pair production is a phenomena which does occur for

high energy incident γ-rays with large momentum transfer in a region not too close to the nucleus.

It is interesting to note that the Moller interaction can be viewed as simply a correction to *self-energy* due to the exclusion principle. The *self-energy* of two electrons, 1 and 2, is not the sum of the self-energy of each, for one of the virtual states that 2 could ordinarily enter by emission of a quantum is now occupied by 1. The difference between the *self-energy* of two electrons and the sum of the *self-energy* of each separately comes out to be just their *interaction energy*.

Vacuum polarization. Alternative cut-off procedures

In the above calculation, *terms of the type discussed by Uehling[16] have been omitted.*

[16] Uehling, E. A. (July, 1935). Polarization Effects in the Positron Theory. *Phys. Rev.*, 48, 55; https://doi.org/10.1103/PhysRev.48.55.

These terms *represent processes involving a pair production followed by annihilation of the same pair*. For example, a pair produced by the potential may annihilate again emitting a quantum. This quantum is then absorbed by the electron in state 1 transferring it to state 2. *These terms are infinite and are not made convergent by the present scheme.* There is some point, nevertheless, to solving problems at first without taking them into account. This is because their net effect is only to alter the effective *potential* in which the electron finds itself, for it may be scattered either directly or by the quantum produced by the Uehling terms. That is, if this problem of *polarization of the vacuum* is solved it will mean, if there is any effect, simply that the potential A, φ appearing in the Dirac equation and (to high order) in such terms as (18) should be replaced by new "polarized" potentials A', φ'.

These *polarization terms* can be characterized in a *relativistically* invariant manner. All the terms which have been calculated above contain matrix elements of operators between states in a sequence such as 1 to f, f to g, g to 2. The omitted *polarization terms* contain transitions like f to g, g to f, 1 to 2. For higher order processes the *polarization terms* are those which do not contain a continued sequence of transitions from the initial to the final state.

The *polarization terms* are not affected in any helpful way by the changes in the density of quanta. *It is likely that this problem will have its answer in a changed physical viewpoint*. However, there is a simple alternative procedure to produce finite self-energies which also makes convergent the integrals appearing in Serber's[17] treatment of the polarization problem.

[17] Serber, R. (July, 1935). Linear Modifications in the Maxwell Field Equations. *Phys. Rev.*, 48, 49; https://doi.org/10.1103/PhysRev.48.49.

(Since, however, this treatment of Serber already presupposes a partial subtraction procedure of Heisenberg and Dirac, the situation is not so clear here as in the *self-energy* problem.)

From the point of view of coordinate space, the reason that the electronic *self-energy* diverges appears to be this. A virtual light quantum emitted at one point spreads out as $\delta(t^2 - r^2)$ from the origin. The *wave packet* of the electron spreading out after the emission of the quantum has, as a consequence of Dirac's equation, a similar discontinuous value along the light cone. It is the continued coincidence of these singularities which makes the matrix element for the subsequent absorption of the quantum infinite. The method outlined above of changing $\delta(\omega^2 - k^2)$ to $g(\omega^2 - k^2)$ has the effect of changing $\delta(t^2 - r^2)$ to $f(t^2 - r^2)$ where $f(s^2)$ is everywhere finite and goes to zero rapidly for $| s2 | > 1/\lambda_0^2$. The quanta have been moved away from the electrons so that overlap on the light cone is reduced.

An obvious alternative procedure is to move the electron *wave function* away from the quanta. This is easily done in a very similar manner. We assume the density of electron *states* of energy E, momentum **P** to be $g(E^2 - P^2 - \mu^2)$ rather than $\delta(E^2 - P^2 - \mu^2)$[18].

[18] This is seen to be essentially the method proposed by Wataghin. [Wataghin, G. (January, 1934). Bemerkung über die Selbstenergie der Elektronen. (Remark on the self-energy of electrons.) *Zeit. Physik*, 88, 92-8; https://doi.org/10.1007/BF01352311.]

The quanta are conventional, $\omega = k$, density $d\mathbf{k}/k$. The *self-energy* integrals (2) can, of course, be expressed as an integral over the intermediate state momentum $\mathbf{P}_i$ rather than $\mathbf{k}$. Replacing $d\mathbf{P}_i/E_i$ by $g(E_i^2 - P_i^2 - \mu^2)dE_i d\mathbf{P}_i$, we find

$$\Delta E'_0 = \dots .$$

... The projection operators are unchanged since it is only the density of *states* which we wish to alter. They are still $\Lambda_i^{2\pm} = (E_i \pm \alpha.\mathbf{P}_i \pm \beta\mu)/2E_i$. The result of this calculation is to verify that $\Delta E'_0$ is finite, (depending logarithmically on λ_0). The other problems can be analyzed in the same way.

In the problem of *polarization of the vacuum*, the *wave functions* of both electron and positron ordinarily spread with a singularity on the light core. The matrix element for their subsequent annihilation is therefore infinite. With the modification here described these *wave functions* are made less singular and their overlap integral is finite. The polarization integrals in Serber's article[17] may now be integrated to yield finite results.

Other than terms which might be removed by a small *renormalization of charge* (depending logarithmically on λ_0), the net effect in (17) would be to change the – (3/8) in the last term of (17) to – (3/8) – (1/5). However, the real existence of such polarization corrections is, in the author's view, uncertain. These matters will be discussed in much more detail in future publications. *Also reserved for future publication is a more complete physical theory from which the results reported here may be directly deduced.* It yields much more powerful techniques for setting up problems and performing the required integrations.

The author would like to express his gratitude to Mr. P. V. C. Hough for assistance in the calculations and to Professor H. A. Bethe and Dr. F. Dyson and many others for useful discussions.

Schwinger, J. (November, 1948). Quantum Electrodynamics. I. A Covariant Formulation.

[*Phys. Rev.*, 74, 10, 1439-61; https://doi.org/10.1103/PHYSREV.74.1439.]

Received July 29, 1948.

Harvard University, Cambridge, MA.

Lack of convergence in current formulations of quantum electrodynamics indicates that revision of electrodynamic concepts at ultra-relativistic energies is necessary, elementary phenomenon in which divergences occur as a result of virtual transitions involving particles with unlimited energy are *polarization of the vacuum* and *self-energy of the electron* which express *the interaction of the electromagnetic and matter fields with their own vacuum fluctuations*, this alters the constants characterizing the properties of the individual fields and their mutual coupling by infinite factors, the question is whether all divergencies can be isolated in such unobservable *renormalization* factors, *this paper is occupied with the formulation of a completely covariant electrodynamics*, manifest covariance with respect to Lorentz and gauge transformations essential in a divergent theory, customary *canonical commutation relations* fail to exhibit the desired covariance since they refer to field variables at equal times and different points of space, *can be put in covariant form by replacing the four-dimensional surface t = const. by a space-like surface*, offers the advantage over the Schrodinger representation in which all operators refer to the same time providing distinct separation between *kinematical* and *dynamical* aspects, formulation that retains evident covariance of the Heisenberg representation but offers something akin to Schrodinger representation can be based on distinction between the properties of *non-interacting fields*, and the effects of *coupling between fields*, constructs a *canonical transformation* that changes the *field equations* in the *Heisenberg representation* into those of *non-interacting fields*, *supplementary condition* restricting the admissible states of the system and the *commutation relations* must be added to the *equations of motion*, describes the coupling between fields in terms of a varying state vector, then simple matter to evaluate commutators of *field* quantities at arbitrary *space-time* points, one thus obtains an obviously covariant and practical form of quantum electrodynamics expressed in a mixed Heisenberg-Schrodinger representation called the *interaction representation*, discusses *covariant* elimination of longitudinal field in which customary distinction between longitudinal and transverse fields is replaced by a suitable *covariant* definition, describes collision processes in terms of an invariant *collision operator* which is the unitary operator that determines the over-all change in state of a system as the result of interaction, notes that a *second paper* treats the problems of electron and photon *self-energy* together with the *polarization of the vacuum* and a *third paper* is concerned with the determination of the *radiative corrections* to the properties of an electron and the comparison with experiment [this was not addressed, it stated that "*radiative corrections to energy levels* will be treated in the next paper of the series" but *this did not appear, nor are there any references to it*].

[Oppenheimer, J. R. (1948). Electron Theory. *Report to the Solvay Conference for Physics at Brussels*, Belgium, September 27 to October 2, 1948, pages 6-7. (Reprint in Schwinger, J. (ed). (1958). *Selected Papers on Quantum electrodynamics*,

Dover, New York, pages 150-1: "Now it is true that the fundamental equations of quantum-electrodynamics are gauge and Lorentz covariant. But they have in a strict sense no solutions expansible in powers of e. If one wishes to explore these solutions, bearing in mind that certain infinite terms will, in a later theory, no longer be infinite, one needs a covariant way of identifying these terms, and for that, not merely the field equations themselves, but *the whole method of approximation and solution must at all stages preserve covariance. This means that the familiar Hamiltonian methods, which imply a fixed Lorentz frame t = constant, must be renounced*; neither Lorentz frame nor gauge can be specified until after, in a given order in e, all terms have been identified, and those bearing on the definition of charge and mass recognized and relegated; then of course, in the actual calculation of transition probabilities and the reactive corrections to them, or in the determination of stationary states in fields which can be treated as static, and in the reactive corrections thereto, the introduction of a definite coordinate system and gauge for these no longer singular and completely well-defined terms can lead to no difficulty.

It is probable that, at least to order e^2, more than one covariant formalism can be developed. Thus, *Stueckelberg's four-dimensional perturbation theory*[26] would seem to offer a suitable starting point, as also do the related algorithms of Feynman[27].

> [26] Stueckelberg, E. C. G. (September, 1934). Relativistisch invariante Störungstheorie des Diracschen Elektrons I. Teil: Streustrahlung und Bremsstrahlung. (Relativistically invariant perturbation theory of Dirac's electron Part I: scattered radiation and Bremsstrahlung.) *Ann. Phys.*, 413, 4, 367-89
>
> [27] Feynman, R. P. (November, 1948). Relativistic Cut-Off for Quantum Electrodynamics. *Phys. Rev.*, 74, 1430-8.

But a method originally suggested by Tomonaga[28], and *independently developed and applied by Schwinger*[22], would seem, apart from its practicality, to have the advantage of very great generality and a complete conceptual consistency.

> [28] Tomonaga, S. (1943). On a Relativistically Invariant Formulation of the Quantum Theory of Wave Fields. *Bull. I. P. C. R. Riken-iho*, 22, 545 (in Japanese); translation Tomonaga, S. (August, 1946). *Prog. Theor. Phys.* 1, 2, 27-42; Koba, Z., Tati, T. & Tomonaga, S. (1947a). On a Relativistically Invariant Formulation of the Quantum Theory of Wave Fields. II. Case of Interacting Electromagnetic and Electron Fields. § 1-4. *Prog. Theor. Phys.*, 2, 101–116; Koba, Z., Tati, T. & Tomonaga, S. (1947b). On a Relativistically Invariant Formulation of the Quantum Theory of Wave Fields. III. Case of Interacting Electromagnetic and Electron Fields. § 5-7. *Prog. Theor. Phys.*, 2, 198–208.
>
> [22] Schwinger, J. (November, 1948). Quantum Electrodynamics. I. A Covariant

Formulation. *Phys. Rev.*, 74, 10, 1439-61, and in press.

It has also been shown by Dyson[29] how Feynman's algorithms can be derived from the Tomonaga equations.

[29] Dyson, F. J., *Phys. Rev.*, in press."]

Abstract.

Attempts to avoid the divergence difficulties of quantum electrodynamics by mutilation of the theory have been uniformly unsuccessful. The lack of convergence does indicate that *a revision of electrodynamic concepts at ultra-relativistic energies is indeed necessary*, but no appreciable alteration of the theory for moderate *relativistic* energies can be tolerated. The elementary phenomena in which divergences occur, in consequence of virtual transitions involving particles with unlimited energy, are the *polarization of the vacuum* and the *self-energy of the electron*, effects *which essentially express the interaction of the electromagnetic and matter fields with their own vacuum fluctuations.* The basic result of these *fluctuation interactions* is to alter the constants characterizing the properties of the individual fields, and their mutual coupling, albeit *by infinite factors. The question is naturally posed whether all divergences can be isolated in such unobservable renormalization factors*; more specifically, *we inquire whether quantum electrodynamics can account unambiguously for the recently observed deviations from the Dirac electron theory*, without the introduction of fundamentally new concepts.

This paper, the first in a series devoted to the above question, *is occupied with the formulation of a completely covariant electrodynamics*. Manifest covariance with respect to Lorentz and gauge transformations is essential in a divergent theory since the use of a particular reference system or gauge in the course of calculation can result in a loss of covariance in view of the ambiguities that may be the concomitant of infinities.

[*Covariance* is the property of a function of retaining its form when the variables are linearly transformed. An equation is said to be Lorentz *covariant* if it can be written in terms of Lorentz *covariant quantities*. The key property of such equations is that if they hold in one inertial frame, then they hold in any inertial frame; this follows from the result that if all the components of a tensor vanish in one frame, they vanish in every frame.]

It is remarked, in the first section, that the customary *canonical commutation relations*, which fail to exhibit the desired covariance since they refer to field variables at equal times and different points of space, *can be put in covariant form by replacing the four-dimensional surface $t = const.$ by a space-like surface*. The latter is *such that light signals cannot be propagated between any two points on the surface*. In this manner, a formulation of quantum electrodynamics is constructed in the *Heisenberg representation*, which is obviously *covariant* in all its aspects. It is not entirely suitable, however, as a practical means of treating *electrodynamic* questions, since *commutators of field quantities at points separated by a time-like interval can be constructed only by solving the equations of* *motion*. This situation is to be contrasted with that of the *Schrodinger representation*, in which all operators refer to the same time, thus providing a distinct separation between kinematical and dynamical aspects. *A formulation that retains the evident covariance of the Heisenberg representation, and yet offers something akin to the advantage of the Schrodinger representation can be based on the distinction between the properties of non-interacting fields, and the effects of coupling between fields.*

In the second section, we construct a *canonical transformation* that changes the *field equations* in the *Heisenberg representation* into those of *non-interacting fields*, and therefore *describes the coupling between fields in terms of a varying state vector*. It is then a simple matter to evaluate commutators of *field* quantities at arbitrary *space-time* points. *One thus obtains an obviously covariant and practical form of quantum electrodynamics, expressed in a mixed Heisenberg-Schrodinger representation, which is called the interaction representation.*

The third section is devoted to a discussion of the *covariant* elimination of the longitudinal field, in which the customary distinction between longitudinal and transverse fields is replaced by a suitable *covariant* definition. *The fourth section is concerned with the description of collision processes in terms of an invariant collision operator, which is the unitary operator that determines the over-all change in state of a system as the result of interaction.* It is shown that the *collision operator* is simply related to the *Hermitian reaction operator*, for which a variational principle is constructed.

Introduction.

The predictions of quantum electrodynamics concerning higher order perturbation effects have long been discredited in view of the divergent nature of the results. Several attempts[1] have been made to arbitrarily remove supposedly objectionable features of the theory - the so-called "*subtraction physics*".

[1] Dirac, P. A. M. (March, 1934). Discussion of the infinite distribution of electrons in the theory of the positron. *Proc. Camb. Phil. Soc*., 30, 2, 150-63 [; attempt by Dirac to address problem with his *relativistic* 'hole' theory which implies an infinite number of negative-energy electrons (per unit volume) with energies extending continuously from $-mc^2$ to $-\infty$, when an electromagnetic field is present positive- and negative-energy states cannot be distinguished in *relativistically* invariant way, need to set up assumptions for production of *electromagnetic field* by the electron distribution such that any finite change in distribution produces a change in the field in agreement with Maxwell's equations and such that the infinite field which would be required by Maxwell's equations from an infinite density of electrons is in some way cut out, *assumes each electron has its own individual wave function in space-time* and each electron moves in an *electromagnetic field* which is the same for all electrons part coming from external causes and part from the electron distribution itself, introduces *relativistic density matrix* $\sum_{oc} \psi_{k'}(x' t') \psi^*_{k''}(x'' t'')$ referring to two points in space and two times, separates density distribution into two parts where one contains the singularities, and the other describes the *electric* and *current densities* physically present.]; Heisenberg, W. (March, 1934). Bemerkungen zur Diracschen Theorie des Positrons. (Remarks on the Dirac theory of positron.) *Zeit. Phys*., 90, 3-4, 209-31

[; https://doi.org/10.1007/ BF01333516[; Heisenberg's reconstruction of Dirac's *theory of the positron* in the formalism of quantum electrodynamics, demands that symmetry between positive and negative charge should be expressed in the basic equations from the outset, in addition to the well-known difficulties with the divergences no new infinities should appear in the formalism, theory should provide an approximation for the treatment of problems that have been treated by quantum electrodynamics up to now, Dirac [(1934). Discussion of the infinite distribution of electrons in the theory of the positron.] showed that a quantum mechanical system of many electrons that fulfill the Pauli principle and move in a given force field without back-reaction can be characterized by a *density matrix* (x′, t′, k′ | R | x″, t″, k″) = ∑n ψ*n (x′, t′, k′) ψ_n (x″, t″, k″) when ψ_n(x′, t′, k′) means the normalized eigenfunctions of the states that possess one electron, and x′, t′, k′ (x″, t″, k″, resp.) are position, time, and spin variables, all physically-important properties of quantum-mechanical systems like *charge density, current density*, etc., can be read off from the *density matrix*, the temporal change in the density matrix is determined by the Dirac differential equation, Dirac made different choice of *density matrix* representing external field resulting in a different energy and impulse density, by restricting oneself to an *intuitive analogue theory of matter fields the negative energy levels in the Dirac theory could be avoided by replacing the homogeneous Dirac differential equation with an inhomogeneous equation* where the inhomogeneity is indicative of pair creation, for most practical applications e.g. pair creation, annihilation, Compton scattering, etc. *the theory described here does not yield anything new compared to the formulation of the Dirac theory*, in the Maxwell theory a continuous charge distribution also leads to a finite self-energy; it is the "quantization" that leads to the infinite self-energy, *if one represents the quantization of the electromagnetic field by point-like light quanta then the infinitude of the self-energy also emerges in the intuitive theory of matter waves.*];
Heitler, W. & Peng, H. W. (1942). *Proc. Camb. Phil. Soc.*, 38, 296.

All such efforts have been fruitless; either failing in their avowed purpose, or lacking internal consistency[2].

[2] Serber, R. (April, 1936). A Note on Positron Theory and Proper Energies. *Phys. Rev.*, 49, 545; https://doi.org/10.1103/PhysRev.49.545; Bethe, H. A. & Oppenheimer, J. R. (October, 1946). Reaction of Radiation on Electron Scattering and Heitler's Theory of Radiation Damping. *Phys. Rev.*, 70, 451; https://doi.org/10.1103/PhysRev.70.451.

The unqualified success of quantum electrodynamics in applications involving the lowest order of perturbation theory indicates its essential validity for moderately *relativistic* particle energies. *The objectionable aspects of quantum electrodynamics are encountered in virtual processes involving particles with ultra-relativistic energies.* The two basic phenomena of this type are the *polarization of the vacuum* and the *self-energy of the electron.*

[*Vacuum polarization*: The possibility of creating electron-positron pairs with unlimited energy, through the virtual creation and annihilation of electron-positron pairs by an *electromagnetic field*, results in the generation of unlimited charge and current in the vacuum.]

The phrase "*polarization of the vacuum*" describes the modification of the properties of an *electromagnetic field* produced by its interaction with the charge fluctuations of the vacuum. In the language of perturbation theory, the phenomenon considered is *the generation of charge and current in the vacuum through the virtual creation and annihilation of electron-positron pairs by the electromagnetic field. If the electromagnetic field is that of a light quantum, the vacuum polarization effects are equivalent to ascribing a proper mass to the photon.* Previous calculations have yielded non-vanishing, divergent expressions for the light quantum *proper mass*. However, *the latter quantity must be zero in a proper gauge invariant theory*. The failure to obtain this result from a *gauge invariant* formulation can be ascribed only to a faulty application of the theory, rather than to an essential deficiency thereof. *When the electromagnetic field is that of a given current distribution, one obtains a logarithmically divergent contribution to the vacuum polarization current which is everywhere proportional to the given distribution.* This divergent result expresses the possibility, according to present theory, of creating electron-positron pairs with unlimited energy, a situation that presumably will be corrected in a more satisfactory theory. Thus, *the physically significant divergence arising from the vacuum polarization phenomenon occurs in a factor that alters the strength of all charges, a uniform renormalization that has no observable consequences other than the conflict with the empirical finiteness of charge*.

> [*Self-energy of an electron*: In a Lorentz invariant (relativistic) theory, the possibility of an electron emitting light quanta with unlimited energy results in the addition of an infinite *electromagnetic proper mass* to the electron's observable mechanical mass and to the infinite *self-energy* of the electron.]

The interaction between the electromagnetic field vacuum fluctuations and an electron, or more exactly, the electron-positron matter field, modifies the properties of the matter field and produces the self-energy of an electron. The mechanism here under discussion is commonly described as the *virtual emission and absorption of a light quantum by an otherwise free electron*, although an equally important effect is *the partial suppression, via the exclusion principle, of the coupled vacuum fluctuations of the electromagnetic and matter fields*.

In a Lorentz invariant theory, self-energy effects for a free electron can only result in the addition of an electromagnetic proper mass to the electron's mechanical proper mass. Calculations performed for a stationary electron[3] *have yielded a logarithmically divergent electromagnetic proper mass, a divergence that results from the possibility of emitting light quanta with unlimited energy.*

> [3] Weisskopf, V. F. (July, 1939). On the Self-energy and Electromagnetic Field of the Electron. *Phys. Rev.*, 56, 1, 72[; *the main purpose of this paper is to show the physical significance of the logarithmic divergence of the self-energy of the electron and to demonstrate the reason for its occurrence*, the *self-energy of the electron is its total energy in free space when isolated from other particles or light quanta*,
> $W = T + (1/8\pi) \int (H^2 + E^2)\, dr$ where T is the *kinetic energy* of the electron and H and E are

the *magnetic* and *electric field strengths* at point r, identifies three reasons why quantum theory of the electron results in infinite *self-energy* of the electron, claims that *quantum kinematics shows that the radius of the electron must be assumed to be zero*, resulting in infinite energy of the *electrostatic field*, the contributions of the electric and magnetic fields of the spin to the *self-energy* of the electron cancel one another, *quantum theory of the electromagnetic field postulates the existence of field strength fluctuations in empty space*, gives rise to an additional energy which diverges more strongly that the electrostatic self-energy, induces the electron to perform vibrations with energy that diverges quadratically *for an infinitely small radius*, Dirac's positron theory implies that the charge and magnetic dipole of the electron are extended over a finite region, explains why the *self-energy* is only *logarithmically infinite*, *divergences are consequence of assumption of point electron*].

It is here, as in the *vacuum polarization* problem, that modifications will be introduced in a more satisfactory theory. However, *the electromagnetic proper mass merely produces a renormalization of the electron mass that has no observable consequences, other than the conflict with the empirical finiteness of mass.*

It is evident that these *two phenomena* are quite analogous and *essentially describe the interaction of each field with the vacuum fluctuations of the other field. The effect of these fluctuation interactions is simply to alter the fundamental constants e and m, although by logarithmically divergent factors*. However, it may be argued that a future modification of the theory, *inhibiting the virtual creation of particles that possess energies many orders of magnitude in excess of* mc^2, will ascribe a value to these logarithmic factors not vastly different from unity. *The charge and mass renormalization factors will then differ only slightly from unity*, as befits a perturbation theory, in consequence of the small *coupling constant* for the *matter* and *electromagnetic* fields,

$$e^2/4\pi hc = 1/137.$$

We may now ask the fundamental question: *Are all the physically significant divergences of the present theory contained in the charge and mass renormalization factors*? Will the consideration of interactions more complicated than these simple vacuum fluctuation effects introduce new divergences; or will all further phenomena involve only moderate *relativistic* energies, and thus be comparatively insensitive to the high energy modifications that are presumably to be introduced in a more satisfactory theory? This series of papers represents an attempt to supply at least a partial answer to the question, which has acquired an immediate importance in view of recent conclusive evidence that the electromagnetic properties of the electron are not fully described by the *Dirac wave equation*. Fine structure measurements on hydrogen, deuterium[4], and ionized helium[5] have revealed energy level displacements that imply the existence of a weak, short range repulsive interaction between electron and proton.

[4] Lamb, Jr., W. E. & Retherford, R. C. (August, 1947). Fine Structure of the Hydrogen Atom by a Microwave Method. *Phys. Rev.*, 72, 3, 241-3[; the spectrum of hydrogen has a fine structure of the energy levels which according to the *Dirac wave equation* for an electron moving in a Coulomb field is due to the combined effects of *relativistic* variation of mass with velocity and spin-orbit coupling, according to this theory the $2^2S_{1/2}$ state

should exactly coincide in energy with the $2^2P_{1/2}$ state which is the lower of the two P states, previous attempts at measurement have alternated between finding confirmation and discrepancies of as much as eight percent, using microwave method depending on a novel property of the $2^2S_{1/2}$ level results indicate that contrary to the *Dirac wave equation* the $2^2S_{1/2}$ state is higher than the $2^2P_{1/2}$ by about 1000 Mc/sec].

[5] Mack, J. E. & Austern, N. (November, 1947). Newly Observed Structure in He II λ 4686. *Phys. Rev.*, 72, 972; https://doi.org/10.1103/PhysRev.72.972.

Experiments on the hyperfine structure of hydrogen and deuterium[6], together with electron g value determinations for several states of gallium and sodium[7], *prove that the electron possesses a small additional spin magnetic moment.*

[6] Nafe, J. E., Nelson, E. B. & Rabi, I. I. (June, 1947). The Hyperfine Structure of Atomic Hydrogen and Deuterium. *Phys. Rev.*, 71, 914; https://doi.org/10.1103/PhysRev.71.914; Nagle, D. E., Julian, R. S. & Zacharias, J. R. (November, 1947). The Hyperfine Structure of Atomic Hydrogen and Deuterium. *Phys. Rev.*, 72, 971; https://doi.org/10.1103/PhysRev.72.971.

[7] Kusch, P. & Foley, H. M. (December, 1947). Precision Measurement of the Ratio of the Atomic 'g Values' in the $^2P_{3/2}$ and $^2P_{1/2}$ States of Gallium. *Phys. Rev.*, 72, 1256; https://doi.org/10.1103/PhysRev.72.1256.2; Foley, H. M. & Kusch, P. (February, 1948). On the Intrinsic Moment of the Electron. *Phys. Rev.*, 73, 412; https://doi.org/10.1103/PhysRev.73.412, also in Schwinger, J. (ed). (1958). *Selected Papers on Quantum electrodynamics*. Dover, New York, page 135.

Immediately upon completion of the Lamb-Retherford experiment, *it was generally recognized*[8] *that the most probable explanation was to be found in higher order electrodynamic effects; the radiative corrections to the properties of a bound electron other than mass and charge renormalization.*

[8] Discussion at the Shelter Island Conference on the Foundations of Quantum Mechanics, June 1947.

A provisional *non-relativistic* calculation[9] lent support to this view.

[9] Bethe, H. A. (August, 1947). The Electromagnetic Shift of Energy Levels. *Phys. Rev.*, 72, 339-41[; Lamb and Retherford results show that fine structure of second quantum state of hydrogen does not agree with *Dirac wave equation*, Schwinger, Weisskopf, and Oppenheimer suggest might be due to *shift of energy levels by interaction of electron with the radiation field*, this shift comes out infinite in all existing theories and has therefore always been ignored, possible to identify the most strongly (linearly) divergent term in the level shift with an *electromagnetic mass effect* which must exist for a bound as well as for a free electron, already included in the *observed mass* of the electron so should be subtracted, assumes *relativistic cut-off* in quantum energies (frequencies) of included atomic states, then calculation of Lamb shift for hydrogen atom using *non-relativistic* ordinary radiation theory gives shift of the levels due to *radiation interaction* in close agreement with observed value, removes discrepancy with Dirac theory, did not carried out *relativistic* calculations].

However, *it required a completely relativistic treatment*[10] *to demonstrate that radiative corrections could account simultaneously for the two apparently unrelated deviations from the Dirac electron theory.*

[10] Schwinger, J. (February, 1948). On quantum-electrodynamics and the magnetic moment of the electron. *Phys. Rev.*, 73, 4, 416-7[; *electrodynamics unquestionably requires revision at ultra-relativistic energies,* desirable to isolate those aspects of the current theory that essentially involve high energies and are subject to modification by a more satisfactory theory, this goal has been achieved by transforming the Hamiltonian of current *hole theory* electrodynamics to exhibit explicitly the logarithmically divergent self-energy of a free electron which arises from the virtual emission and absorption of light quanta, the electromagnetic *self-energy* of a free electron can be ascribed to an *electromagnetic mass* which must be added to the mechanical mass of the electron, new Hamiltonian involves experimental *electron mass* rather than unobservable *mechanical mass*, electron now interacts with radiation field only in presence of external field such that only an accelerated electron can emit or absorb a light quantum, *interaction energy* of electron with external field subject to *finite* radiative correction, *polarization of the vacuum* still produces logarithmically divergent term proportional to *interaction energy* of electron in an external field, such term equivalent to altering value of *electron charge* by constant factor with only final value being identified with experimental charge, interaction between matter and radiation produces *renormalization* of electron charge and mass, all divergences contained in *renormalization* factors, radiative correction for energy of electron in external magnetic field corresponds to *additional magnetic moment associated with electron spin* of magnitude $\delta\mu/\mu = (\frac{1}{2}\pi)e^2/hc = 0.001162$, experimental measurements on hyperfine splitting of ground states of atomic hydrogen and deuterium larger than expected from directly measured nuclear moments, finds additional *electron spin magnetic moment* to account for measured hydrogen and deuterium *hyperfine structures* to be $\delta\mu/\mu = 0.00126$ and $\delta\mu/\mu = 0.00131$ respectively, these discrepancies accounted for by additional spin magnetic moment to the electron of $\delta\mu/\mu = 0.0018 \pm 0.00003$, *values yielded by relativistic calculation of Lamb shift differ only slightly from those conjectured by Bethe on basis of non-relativistic calculation and are in good accord with experiment*].

It is our major task to enlarge on this development.

In order to isolate the divergent aspects of quantum electrodynamics in a manner that is Lorentz and gauge invariant, it is necessary to employ a formulation of the theory that preserves these covariant features at all stages. The use of a particular reference system or gauge in the course of calculation can result in a loss of covariance in view of the ambiguities that may arise in a divergent theory. *The first paper is occupied with the development of a suitable covariant formulation.* In the *second paper* we treat the problems of electron and photon self-energy, together with the polarization of the vacuum.

[This appeared as Schwinger, J. (February, 1949). Quantum Electrodynamics. II. Vacuum Polarization and Self-Energy. *Phys. Rev.*, 75, 4, 651-79; below.]

The *third paper* is concerned with the major topic, the determination of the radiative corrections to the properties of an electron, and the comparison with experiment.

[The third paper appeared as Schwinger, J. (September, 1949). Quantum Electrodynamics. III. The Electromagnetic Properties of the Electron—Radiative Corrections to Scattering. *Phys. Rev.*, 76, 6, 790–817; below, but this did not address how *induction of a current in a vacuum by an electron results in an alteration in its electromagnetic properties.* It states that this paper is concerned with the computation of the second-order corrections to the current operator and the application to electron scattering. *Radiative corrections to energy levels* will be treated in the next paper of the series", but this did not appear.]

Scalar and vector matter fields will be discussed in a *fourth paper.* It is hoped that successive papers of this series will deal with such subjects as the corrections to the Klein-Nishina formula, the scattering of light by light, and by a Coulomb field.

1. *Covariance in the Heisenberg representation*

In this section, we employ the following notation: Greek subscripts assume values ranging from 1 to 4, and a repeated index is to be so summed. The coordinate vector of a four-dimensional point x is denoted by $x_\mu = (r, ict)$. The real time coordinate $x_0 = (1/i)x_4 = ct$ is also used. In particular, the four-dimensional element of volume is defined as $d\omega = dx_0dx_1dx_2dx_3$. The *four-vector potential* of the *electromagnetic field* is $A_\mu(x) = \{A(r, t), i\phi(r, t))$, while $\psi_\alpha(x)$ designates the four-component *Dirac spinor.* ...

... The notation γ_μ is used for the four Hermitian matrices that obey the *anticommutation relations*

$$\gamma_\mu\gamma_\nu + \gamma_\nu\gamma_\mu = 2\delta_{\mu\nu}. \tag{1.1}$$

The *adjoint spinor* $\psi^{-}_\alpha(x)$ is defined by

$$\psi^{-}_\alpha(x) = \psi^{+}(x)\gamma_4 = \gamma_4^{T}\psi^{+}(x), \tag{1.2}$$

where $\psi^{-}_\alpha(x)$ is the *Hermitian conjugate* of $\psi_\alpha(x)$. The so-called charge *conjugate spinor* $\psi'_\alpha(x)$ and its *adjoint* $\psi^{-\prime}_\alpha(x)$ are represented by

$$\psi'(x) = C\psi^{-}(x), \qquad \psi^{-\prime}(x) = C^{-1}\psi(x). \tag{1.3}$$

Here C is a matrix such that

$$\gamma_\mu^{T} = \gamma_\mu^{*} = -\,C^{-1}\gamma_\mu C, \tag{1.4}$$

which has the property of being skew-symmetric:

$$C^{T} = -\,C \tag{1.5}$$

and unitary:

$$C^{+}C = 1. \tag{1.6}$$

In the latter equation, $C^+ = C^{T*}$ is the *Hermitian conjugate matrix*. For the particular representation in which all elements of γ_4 are imaginary, while all elements of the other matrices are real, the conditions on C are satisfied with $C = -\gamma_4$. With this choice, $\psi'(x) = \psi^+(x)$; *charge* and *Hermitian conjugation* are equivalent. Finally,

$$\kappa_0 = m_0 c/\hbar \tag{1.7}$$

where m_0 is the *mechanical proper mass* of the electron.

The *equations of motion* of the coupled *electromagnetic* and *electron-positron matter fields* can be derived from the *variational principle*:

$$\delta \int L \, d\omega = 0, \tag{1.8}$$

where the *Lagrangian density L* is

$$L = \ldots \tag{1.9}$$

and is so constructed that it is *invariant with respect to Lorentz transformations, gauge transformations and charge conjugation*. The proof of *Lorentz invariance* follows the conventional treatment and need not be repeated. *Gauge invariance*, that is, *invariance under the combined transformations*

$$\begin{aligned} A_\mu(x) &\rightarrow A_\mu(x) - \partial\Delta(x)/\partial x_\mu \\ \psi(x) &\rightarrow \exp[-ie/hc\ \Delta(x)]/\psi(x) \\ \psi'(x) &\rightarrow \exp[ie/hc\ \Delta(x)]/\psi'(x) \end{aligned} \tag{1.10}$$

induced by a *scalar function of position*, $\Delta(x)$, would be generally valid were it not for the term in the *Lagrangian density* that refers to the *electromagnetic field* alone. The addition to *L* arising therefrom is

...

of which the first term has no effect on the *equations of motion*. Hence *gauge invariance* is restricted to the group of generating functions that obey

$$\partial^2\Delta(x)/\partial x_\mu^2 = \Box^2\Delta(x) = 0 \tag{1.11}$$

Invariance under charge conjugation expresses the complete symmetry between positive and negative charge. The interchange of $\psi(x)$ and $\psi'(x)$, together with +e and –e, evidently leaves the *Lagrangian density* unaltered.

In order to obtain the *equations of motion* for the *matter field*, it is necessary to express the *Lagrangian density* entirely in terms of $\psi(x)$ and $\psi^*(x)$, or alternatively, $\psi'(x)$ and $\psi^{*\prime}(x)$. By virtue of Eqs. (1.3), (1.4), and (1.5)

$$[\psi'(x) = C\psi^-(x), \qquad \psi^{-\prime}(x) = C^{-1}\psi(x), \tag{1.3}$$

$$\gamma_\mu^T = \gamma_\mu^* = -C^{-1}\gamma_\mu C, \tag{1.4}$$

$$C^T = -C, \tag{1.5)]}$$

the following relations hold

$$\psi^{-\prime}\gamma_\mu\psi' = C^{-1T}\gamma_\mu C\psi^- = \psi\gamma_\mu^T\psi^-$$

$$\psi^{-\prime}\psi' = \psi C^{-1T} C\psi^{-} = -\psi\psi^{-}, \tag{1.12}$$

...

We find, as the result of variation, apart from discarded divergencies,

$$\delta L = \ldots, \tag{1.13}$$

where

$$j_\mu(x) = iec/2\,[\psi^{-}(x)\gamma_\mu\psi(x) - \psi^{-\prime}(x)\gamma_\mu\psi'(x)] \tag{1.14}$$

represents the four-vector of *charge* and *current*; $j_\mu = (j_\mu, ic\rho)$. It is consistent with the form of the *commutation relations* imposed on the field quantities to infer that

$$\Box^2 A_\mu(x) = -\,1/c\; j_\mu(x), \tag{1.15}$$

and

$$[\gamma_\mu\{\partial/\partial x_\mu - ie/\hbar c\; A_\mu(x)\} + \kappa_0]\;\psi(x) = 0$$
$$[\gamma_\mu{}^T\{\partial/\partial x_\mu + ie/\hbar c\; A_\mu(x)\} - \kappa_0]\;\psi^{-}(x) = 0, \tag{1.16}$$

The *Dirac equations* for the *matter field* can also be cast in the *charge conjugate* form

$$[\gamma_\mu\{\partial/\partial x_\mu + ie/\hbar c\; A_\mu(x)\} + \kappa_0]\;\psi'(x) = 0$$
$$[\gamma_\mu{}^T\{\partial/\partial x_\mu - ie/\hbar c\; A_\mu(x)\} - \kappa_0]\;\psi^{-\prime}(x) = 0. \tag{1.17}$$

To the equations of motion must be added a supplementary condition, and the commutation relations. The *supplementary condition*

$$\partial A_\mu(x)/\partial x_\mu\; \Phi = 0 \tag{1.18}$$

restricts the *admissible states* of the system, as characterized by the constant vector Φ of our *Heisenberg representation*. The compatibility of (1.18) with the *equations of motion* is a consequence of the *charge conservation equation*

$$\partial j_\mu(x)/\partial x_\mu = 0. \tag{1.19}$$

The customary *Maxwell equations*, involving the *field strengths*

$$F_{\mu\nu} = \partial A_\nu(x)/\partial x_\mu - \partial A_\mu(x)/\partial x_\nu, \tag{1.20}$$

rather than the potentials, appear as derived supplementary conditions:

$$[\partial F_{\mu\nu}(x)/\partial x_\mu + 1/c\; j_\nu(x)]\Phi = 0. \tag{1.21}$$

The *commutation relations*, in their conventional canonical form, read

$$[A_\mu(r,t),\; 1/c\;\partial/\partial t\; A_\nu(r',t)] = ihc\delta_{\mu\nu}\delta(r - r') \tag{1.22a}$$
$$\{\psi_\alpha(r,t),\; (\psi^*(r',t)\gamma_4)_\beta\} = \delta_{\alpha\beta}\delta(r - r') \tag{1.22b}$$

where the bracket symbols signify the *commutator* and *anticommutator*, respectively:

$$[A,B] = AB - BA, \qquad \{A,B\} = AB + BA. \tag{1.23}$$

...

It should be noted that the particle field commutation relations are invariant with regard to charge conjugation. Thus,

... (1.24)

A further remark concerns the consistency of the *supplementary condition* and the *commutation relations*. Since (1.18) contains the arbitrary point x, one will obtain additional supplementary conditions by commutation, unless

$$[\partial A_\mu(x)/\partial x_\mu, \partial A_\nu(x')/\partial x_\nu'] = 0 \quad (1.25)$$

for arbitrary x and x'. In actuality, the *canonical commutation relations* are such as to yield (1.25). It must be realized that the *commutator*, considered as a function of x, obeys the *wave equation*, whence the validity of (1.25) is assured provided the *commutator* and its time derivative vanish for t = t'. This is easily verified.

The physical quantities characterizing the distribution of *energy* and *momentum* in the field are combined in the *canonical energy-momentum tensor*

$$T_{\mu\nu} = \ldots \quad (1.26)$$

...

which satisfies the *conservation equation*

$$\partial/\partial x_\mu \, T_{\mu\nu} = 0 \quad (1.27)$$

...

The *canonical tensor* can be replaced by a symmetrical *energy-momentum tensor*

$$\Theta_{\mu\nu} = \ldots \quad (1.29)$$

However, it is only the expectation value of $\Theta_{\mu\nu}$,

$$< \Theta_{\mu\nu}> = < \Phi,\Theta_{\mu\nu}\Phi>, \quad (1.30)$$

that satisfies the conservation equation

$$\partial/\partial x_\mu < \Theta_{\mu\nu}> = 0 \quad (1.31)$$

...

The symmetrical energy-momentum tensor is evidently invariant with respect to *gauge transformations* and *charge conjugation*. ...

...

The spatial volume integrals

$$P_\mu = - i/c \int T_{4\mu} \, d\upsilon \quad (1.35)$$

form a time independent four-vector that unites the *momentum* and *energy* integrals of the *equations of motion*; P_μ =(P,iW/c). ...

...

The operators P_μ form the *infinitesimal generators of the coordinate translation group*. ...

More generally, if F(x) is an arbitrary function of the field variables at the point x, but does not explicitly involve position coordinates,

$$i/\hbar\ [F(x)_\nu P_\nu] = \partial F(x)/\partial x_\nu, \qquad (1.38)$$

One can exploit this aspect of the operators P_μ to prove anew that they constitute constants of the motion, and to demonstrate that the *canonical commutation relations* are consistent with the *equations of motion.* In a similar way, one can introduce other operator constants of the motion which compose the *angular momentum* tensor. *These quantities form the infinitesimal generators of the Lorentz group, and with their aid the covariance of the canonical quantization scheme can be demonstrated.* However, it is at this point that we must deviate from the conventional development that here has so briefly been outlined.

The equations of motion and the supplementary condition are manifestly covariant; the canonical commutation relations lack this essential characteristic since a special Lorentz reference system is employed. The *commutation relations* involve field variables at two points of a four-dimensional surface characterized by t = const. *We shall achieve the desired covariance by replacing such surfaces with the invariant concept of a space-like surface.* The latter is such that light signals cannot be propagated between any two points on the surface. In terms of the position vectors of two points, x_μ and x_μ', it is required that

$$(x_\mu - x_\mu')^2 = (r - r')^2 - c^2(t - t')^2 > 0, \qquad (1.39)$$

which clearly involves no special reference system. Surfaces of the type t = const. form a special non-covariant class of plane *space-like surfaces.* The customary *commutation relations* are essentially an expression of the kinematical independence of field quantities at different points of space for a given time. *It is evident that the proper covariant description of this general property should involve field quantities at two space-time points that cannot be connected by light signals, that is, two points on a space-like surface. Accordingly, we endeavor thus to generalize the commutation relations into a manifestly covariant form.*

The simplest basis for a generalization of (1.22a)

$$[[A_\mu(r,t),\ 1/c\ \partial/\partial t\ A_\nu(r',t)] = ihc\delta_{\mu\nu}\delta(r - r') \qquad (1.22a)]$$

is provided by the two statements that express the properties of $\delta(r - r')$:

$$[A_\mu(r,t),\ 1/c\ \partial/\partial t\ A_\nu(r',t)] = 0, \qquad r \neq r' \qquad (1.40a)$$

$$\int [A_\mu(r,t),\ 1/c\ \partial/\partial t\ A_\nu(r',t)]\ dv' = ihc\delta_{\mu\nu} \qquad (1.40b)$$

in which the spatial volume integration is extended over an arbitrary region that includes the point r. The *proper generalization of (1.40a)*, together with the other vanishing *electromagnetic field* commutators, is simply

$$[A_\mu(x), A_\nu(x')] = 0, \quad (x_\mu - x_\mu')^2 > 0; \qquad (1.41)$$

that is, *the field quantities associated with two distinct points on a space-like surface commute.* In order to generalize (1.40b), it will prove convenient to define a four-vector differential surface area:

$d\sigma_\mu = (dx_2dx_3dx_0, dx_1dx_3dx_0, dx_1dx_2dx_0, dx_1dx_2dx_3/i).$ (1.42)

Considered as defining the direction of the normal to a *space-like surface*, $d\sigma_\mu$ must be a *time-like vector*, that is, $d\sigma_\mu^2 < 0$. It should be noted that our definitions of surface area and volume are such that the volume generated by the displacement δx_μ imparted to the surface area $d\sigma_\mu$ is $\delta\omega = d\sigma_\mu\delta x_\mu$. It is evident from the notation $dv' = id\sigma_4'$, $\partial A_\nu(r',t)/\partial ct = i\partial A_\nu(x')/\partial x_4'$, that the *proper covariant generalization of (1.40b) is*

$\int_\sigma [A_\mu(x), \partial/\partial x_\lambda' A_\nu(x')] d\sigma_\lambda' = hc/i\ \delta_{\mu\nu}$ (1.43)

in which the x' integration is extended over an arbitrary portion of a *space-like surface* σ that includes the point x.

In order to demonstrate the self-consistency of these and further *covariant commutation relations*, we must show that the values attributed to such surface integrals are unaltered as the *space-like surface* σ passing through the point x is varied; and, for a fixed surface relative to the point x, that the *commutation relations* are compatible with an arbitrary displacement of x. The latter requirement involves a detailed consideration of the *equations of motion* and will be discussed at an appropriate place. The verification of the first requirement is facilitated by introducing the notion of the *functional derivative*. The quantity occurring on the left side of (1.43)

[$\int_\sigma [A_\mu(x), \partial/\partial x_\lambda' A_\nu(x')] d\sigma_\lambda' = hc/i\ \delta_{\mu\nu}$ (1.43)]

involves the field variables at all points of the surface σ and is thus a functional of the *space-like surface* σ, say F[σ]. We may compare this with the functional of a neighboring *space-like surface* σ', F[σ'], which surface is such that it deviates from σ only in a neighborhood of the point x. If the volume enclosed between the and surfaces, δω, is allowed to approach zero, we obtain a definition of the functional derivative of F[σ] at the point x:

$\delta F[\sigma]/\delta\sigma(x) = \mathrm{Lim}_{\delta\omega\to 0}\ \{F[\sigma'] - F[\sigma]\}/\delta\omega,$ (1.44)

in which the notation emphasizes that we are considering the variation in F produced by a deformation of the surface σ at the point x. …

…

… It can be shown, as before, that the commutation relations are also valid for the charge conjugate matter fields. …

…

An obviously covariant definition of the energy-momentum four-vector, replacing Eq. (1.35)

[$P_\mu = - i/c \int T_{4\mu}\, d\upsilon$ (1.35)]

is

$P_\mu = - i/c \int_\sigma d\sigma_\lambda T_{\lambda\mu}(x),$ (1.55) in which the integration is extended over an entire space-like surface. The conservation laws now have their covariant expression in the statement that P_μ is independent of the surface σ. Thus

$$\delta/\delta\sigma(x)\, P_\mu c = \delta/\delta x_\lambda\, T_{\lambda\mu}(x) = 0. \qquad (1.56)$$

The conservation law for the total charge

$$Q = 1/c \int_\sigma d\sigma_\mu j_\mu(x)$$

has an analogous expression:

$$\delta/\delta\sigma(x)\, Qc = \delta/\delta x_\mu\, j_\mu(x) = 0. \qquad (1.57)$$

...

It can now be shown that the covariant *commutation relations* are consistent with the *equations of motion*. We examine the change in the *commutator* or *anticommutator* of two field variables associated with two points on a *space-like surface*, produced by a rigid displacement of the surface. In other words, we seek to evaluate

$$\delta/\delta x_\nu\, [F(x),G(x-\xi)] \text{ or } \delta/\delta x_\nu\, \{F(x), G(x-\xi)\},$$

where ξ_μ is a *space-like vector* and F, G are any two field variables. It is a consequence of elementary identities that if F and G obey the *equations of motion* (1.38)

$$[i/h\, [F(x)_\nu P_\nu] = \partial F(x)/\partial x_\nu, \qquad (1.38)]$$

so also do the brackets

$$[F(x),G(x-\xi)] \text{ and } \{F(x), G(x-\xi)\}.$$

Therefore, the specification of such brackets as ξ-dependent multiples of the unit operator is self-consistent, since both the derivative with respect to x_ν, and the commutator with P_ν, vanish.

The formulation of quantum mechanics that has now been developed is obviously covariant in all its aspects. However, it is not entirely suitable as a practical means of treating electrodynamic questions. In the course of application, it is often necessary to evaluate commutators of held quantities at points separated by a time-like interval. Such commutators are to be constructed by solving the *equations of motion* subject to boundary conditions on a *space-like surface*. *This jumbling of the kinematical and dynamical aspects of the situation is a detriment in the systematic discussion of electrodynamic problems.* At the opposite extreme is the Schrodinger picture, in which all operators are time independent, and the time development of the system is represented by a varying *state vector*; a procedure that is non-covariant in its aspect. *We now seek a formulation that enables us to retain the evident covariance of the Heisenberg representation, and yet offers something akin to the advantage of the Schrodinger representation, a distinct separation between kinematical and dynamical aspects.* The desired separation is to be found in that between the *elementary properties of non-interacting fields*, and the *modification of these properties by the coupling between fields*. For *non-interacting fields*, it is a simple matter to carry out the program previously mentioned, and construct commutation relations for field quantities at arbitrary *space-time* points. *In order to exploit this advantage, it is necessary to find a canonical transformation that changes the equations of motion for field quantities in the Heisenberg representation into those of non-interacting fields, and therefore describes the coupling between fields in terms of a varying state vector*. We shall perform this transformation in the next section, and thus obtain an obviously covariant and practical form of quantum electrodynamics, expressed in a *mixed Heisenberg-Schrodinger representation*, which may be called the *interaction representation*[11].

[11] The *interaction representation* can be regarded as a field generalization of the *many-time formalism*, from which point of view it has already been considered by Tomonaga. [Tomonaga, S. (1943). On a Relativistically Invariant Formulation of the Quantum Theory of Wave Fields. *Bull. I. P. C. R. Riken-iho*, 22, 545 (in Japanese); translation Tomonaga, S. (August, 1946). *Prog. Theor. Phys*. 1, 2, 27-42[; *existing formalism of quantum field theory not perfectly relativistic*, commutation relations refer to points in space at different times, Schrodinger equation for the vector representing the state of the system is a function of time, time variable plays a different role than space variables, *probability amplitude* not *relativistically invariant* in the space-time world, follows Dirac (1932) [Relativistic Quantum Mechanics] in *generalizing the notion of probability amplitude as far as is required by the theory of special relativity*, substitutes four-dimensional form of the commutation relations, *generalizes the Schrodinger equation* following the Dirac (1933) [The Lagrangian in Quantum Mechanics] *many-time* formalism, then introduces his *super many-time theory* [$\{H_{12}(P) + h/i\, \partial/\partial C_P\}\ \Psi[C] = 0$ at point P on surface C with infinitely many time variables representing the local time for each position in the space, results in *relativistic interaction representation*, three-dimensional manifold (space-like "surface") in four-dimensional space-time world, not necessary to also assume time-like surfaces for the variable surface as was required by Dirac, previous formalism built up in way too analogous to ordinary *non-relativistic* mechanics, theory was divided into one section giving the kinematical relations between various quantities at the same instant of time and another section determining the causal relations between quantities at different instants of time with the *commutation relations* belong to the first section and the *Schrodinger equation* to the second, this way of separating the theory into two sections very *unrelativistic*, "same instant of time" plays a distinct role, *new formalism consists of one section giving laws of behavior of the fields when they are left alone and the other giving the laws determining the deviation from this behavior due to interactions*, can be carried out *relativistically*, although theory in more satisfactory form no new contents added, *divergence difficulties inherited*, fundamental equations admit only catastrophic solutions *due to non-vanishing zero-point amplitudes of the fields which inheres in the operator* $H_{12}(P)$*, a more profound modification of the theory is required in order to remove this fundamental difficulty*].

Relativistic quantum theories have also been discussed recently by Dirac [Dirac, P. A. M. (May, 1948). Quantum Theory of Localizable Dynamical Systems. *Phys. Rev*., 73, 9, 1092-103; a dynamical system is called *localizable* if its *wave functions* can be expressed in terms of variables, each referring to physical conditions at only one point in *space-time*. These variables may be at points on any three-dimensional space-like surface in space-time. A general investigation is made of how the *wave function* varies when the surface is varied in any way. The variation of the *wave function* is given by equations of the

Schrödinger type involving certain operators which play the role of Hamiltonians. The *commutation relations* for these operators are obtained. The theory works entirely with *relativistic* concepts and it provides the general pattern which any *relativistic* quantum theory must conform to, provided the dynamical system is *localizable*].

2. *The interaction representation*

To alter the *equations of motion* in the above outlined manner, we introduce a *unitary operator* U[σ], defined for a *space-like* surface σ, and construct the *state vector* of the *interaction representation*

$$\Psi[\sigma] = U[\sigma]\Phi, \qquad (2.1)$$

which depends upon the surface σ, in contrast with the constant vector Φ of the *Heisenberg representation*. The *expectation value* of some *field variable* **F**(x) becomes (in this section, the operators of the *Heisenberg representation* will be denoted by bold face letters)

$$(\Phi,\mathbf{F}(x)\Phi) = (\Psi[\sigma],U[\sigma]\mathbf{F}(x)U^{-1}[\sigma]\Psi[\sigma]) = (\Psi[\sigma],F(x)\Psi[\sigma]) \qquad (2.2)$$

which defines the operators of the interaction representation in terms of those of the Heisenberg representation:

$$F(x) = U[\sigma]\mathbf{F}(x)U^{-1}[\sigma]. \qquad (2.3)$$

It is understood that σ is a *space-like* surface passing through the point x. In order, however, that F(x) depend only on the point x and not on the particular surface σ, *the form of U[σ] must be restricted*, as indicated by the following requirement:

$$\delta/\delta\sigma(x')F(x) = \ldots = \ldots = 0$$
$$(x_\mu - x_\mu')^2 > 0. \qquad (2.4)$$

This will be satisfied if

$$\delta U[\sigma]/\delta\sigma(x')\, U^{-1}[\sigma]$$

is an invariant function of the field operators at the point x', since the commutation properties on the surface σ are unaffected by the *unitary transformation*. If, further, the unitary character of U[σ] is to be preserved by its *equation of motion*, it is necessary that

$$i\, \delta U[\sigma]/\delta\sigma(x)\, U^{-1}[\sigma]$$

be a Hermitian operator. Therefore, on writing

$$ihc\, \delta U[\sigma]/\delta\sigma(x) = H(x)U[\sigma], \qquad (2.5)$$

we obtain a *covariant equation of motion* for U[σ], in which H(x) is a Hermitian operator, *an invariant function of the field quantities* at the point x, and has the dimensions of an *energy density*. The *equation of motion* for Ψ[σ] is, correspondingly,

$$ihc\, \delta\Psi[\sigma]/\delta\sigma(x) = H(x)\Psi[\sigma], \qquad (2.6)$$

We have obtained the conditions that must be satisfied by any canonical transformation.

It will now be shown that the special transformation desired is attained with H(x) chosen as the negative of the coupling term in the Lagrangian density, that is,

$$H(x) = -(1/c)j_\mu(x)A_\mu(x). \tag{2.7}$$

To construct the equations of motion in the interaction representation, we first note that the gradient of any field quantity can be exhibited as a *functional derivative*, through an obvious generalization of *Gauss' theorem*:

$$\partial F[x]/\partial x_\nu = \delta/\delta\sigma(x) \int_\sigma F(x')\, d\sigma_\nu'$$
$$= \delta/\delta\sigma(x)\, U[\sigma] \int_\sigma \mathbf{F}(x')\, d\sigma_\nu'\, U^{-1}[\sigma]. \tag{2.8}$$

The *functional derivative* in the latter form affects both the surface of integration and the operator U[σ], whence

$$\partial F[x]/\partial x_\nu = U[\sigma]\, \partial\mathbf{F}[x]/\partial x_\nu\, U^{-1}[\sigma]$$
$$+ \int_\sigma [\delta U[\sigma]/\delta\sigma(x)\, U^{-1}[\sigma], F(x')]\, d\sigma_\nu'$$
$$= U[\sigma]\, \partial\mathbf{F}[x]/\partial x_\nu\, U^{-1}[\sigma]$$
$$- i/\hbar c \int_\sigma [H(x), F(x')]\, d\sigma_\nu'. \tag{2.9}$$

If we first place $F(x) = A_\mu(x)$, it is immediately found from the covariant *commutation relations* on the *space-like* surface σ, that

$$\partial A_\mu(x)/\partial x_\nu = U[\sigma]\, \partial A_\mu(x)/\partial x_\nu\, U^{-1}[\sigma], \tag{2.10}$$

which, indeed, is necessary, in order that the electromagnetic field commutation relations retain their form under this canonical transformation. However, with $F(x) = \delta A_\mu(x)/\delta x_\nu$, one obtains

$$\Box^2 A_\mu(x) = \ldots = \ldots = 0; \tag{2.11}$$

the equations of motion for the electromagnetic field in the interaction representation are those of an isolated field. In addition, the *supplementary condition* is unchanged in form:

$$\partial A_\mu(x)/\partial x_\nu\, \Psi[\sigma] = 0 \tag{2.12}$$

provided the point x lies on the surface σ. Finally, if $F(x) = \gamma_\nu\psi(x)$,

$$(\gamma_\nu\, \partial/\partial x_\nu + \kappa_0)\psi(x) = \ldots . \tag{2.13}$$

But, according to (1.12) and (1.14)

$$[\psi^{-\prime}\gamma_\mu\psi' = C^{-1T}\gamma_\mu C\psi^- = \psi\gamma_\mu{}^T\psi^-$$
$$\psi^{-\prime}\psi' = \psi C^{-1T}C\psi^- = -\psi\psi^-, \tag{1.12}$$
$$j_\mu(x) = iec/2\, [\psi^-(x)\gamma_\mu\psi(x) - \psi^{-\prime}(x)\gamma_\mu\psi'(x)]] \tag{1.14}$$

$$_\mu(x) = iec/2\, [\psi^*(x)\gamma_\mu\psi(x) - \psi(x)\, \psi^*(x)\gamma_\mu], \tag{2.14}$$

and

$$[j_\mu(x), \gamma_\nu\psi(x')] = -iec\, \{\gamma_\nu\psi(x'), \psi^*{}_\alpha(x)\}(\gamma_\mu\psi(x')_\alpha, \tag{2.15}$$

so that

$$(\gamma_\nu\, \partial/\partial x_\nu + \kappa_0)\psi(x) = \ldots = 0; \qquad (2.16)$$

the *equations of motion* for the *matter field* in the *interaction representation* are those of an isolated field. *This completely proves the correctness of the choice (2.7)*

$$[H(x) = -(1/c)j_\mu(x)A_\mu(x). \qquad (2.7)]$$

for H(x).

We may now proceed to construct the general commutation laws for the field quantities in the new representation, by employing their elementary equations of motion. This process will be facilitated by introducing two *invariant functions of position*, D(x) and Δ(x), which are associated with the *electromagnetic* and *matter fields*, respectively, and have the following covariant definitions:

$$\Box^2 D(x) = 0; \quad D(x) = 0, \quad x_\mu^2 > 0$$
$$\int \partial D(x)/\partial x_\mu\, d\sigma_\mu = 1, \qquad (2.17)$$

$$\{\Box^2 D(x) - \kappa_0^2\}\Delta(x) = 0; \quad \Delta(x) = 0, \quad x_\mu^2 > 0$$
$$\int \partial \Delta(x)/\partial x_\mu\, d\sigma_\mu = 1, \qquad (2.18)$$

In these definitions, the surface integrations are to be extended over a *space-like* surface that includes the origin. It is easily verified that the constant value attributed to the surface integrals for arbitrary σ is consistent with the other equations. *The detailed construction of these and related functions will be postponed to the second paper of this series*; the properties contained in the equations of definition will suffice for our present purposes. It is easily deduced for example, that D and Δ are odd functions of the coordinates:

$$D(-x) = -D(x), \qquad \Delta(-x) = -\Delta(x). \qquad (2.19)$$

We note that

$$\ldots, \qquad (2.20)$$

which implies that the surface integral is independent of the particular surface σ. By choosing σ to be, successively, a *space-like* surface through the points x' and x", it is inferred that

$$\Delta(x'' - x') = -\Delta(x' - x'') \qquad (2.21)$$

which proves the second statement of (2.19). The proof for D(x) is identical.

The importance of these invariant functions stems from their utility in expressing the solutions of the equations of motion in terms of boundary values prescribed on some space-like surface. The *electromagnetic potentials* are uniquely determined if $A_\mu(x)$ and its normal derivative are specified on a surface σ. The explicit realization of this relation is provided by

$$A_\mu(x) = \int_\sigma [D(x - x')\, \partial/\partial x_\nu'\, A_\mu(x') - A_\mu(x')\, \partial/\partial x_\nu'\, D(x - x')]\, d\sigma_\nu'. \qquad (2.22)$$

To verify this statement, it is sufficient to observe that, analogously to Eq. (2.20), the right side of (2.22) is independent of σ, which can be specially chosen as a *space-like* surface through the point x, yielding

$$\int_\sigma A_\mu(x') \partial/\partial x_v' D(x - x')] d\sigma_v' = A_\mu(x)$$

as the value of the surface integral. The corresponding solution of the boundary value problem for the first order *Dirac equation* is provided by

$$\psi(x) = \int_\sigma S(x - x')\gamma_\mu\psi(x') d\sigma_\mu', \qquad (2.23)$$

where

$$S(x) = (\gamma_v \partial/\partial x_v - \kappa_0)\Delta(x)$$

Following the general pattern, we remark that the right side of (2.23) is independent of σ:

$$\delta/\delta\sigma(x') \int_\sigma S(x - x')\gamma_\mu\psi(x') d\sigma_\mu' = \ldots = \ldots = 0, \qquad (2.25)$$

since

$$(\gamma_\mu \partial/\partial x_\mu + \kappa_0)S(x) = \partial/\partial x_v S(x)\gamma_\mu + \kappa_0 S(x) = \{\Box^2 - \kappa_0^2\}\Delta(x) = 0; \quad (2.26)$$

and that an evaluation with a surface through the point x, gives

$$\int_\sigma \gamma_v\gamma_\mu \partial/\partial x_v \Delta(x - x')\psi(x') d\sigma_\mu' = \ldots = \psi(x)$$

with the aid of the lemma (1.58). The *adjoint equation*

$$\psi^-(x) = \int_\sigma d\sigma_\mu'\psi^-(x')\gamma_\mu S(x' - x) \qquad (2.27)$$

can be proved directly, or inferred from (2.23).

The construction of the general commutation relations is now trivial. To evaluate $[A_\mu(x), A_v(x')]$, for example, it is merely necessary to express $A_\mu(x)$in terms of the field variables on a *space-like* surface that includes the point x', and employ the *commutation relations* for such surfaces. Thus

$$[A_\mu(x), A_v(x')] = -\int_\sigma D(x - x'') \text{ x } [A_v(x'), \partial/\partial x_\lambda'' A_\mu(x'')] d\sigma_\lambda'',$$

whence

$$[A_\mu(x), A_v(x')] = i\hbar c\delta_{\mu v}D(x - x'). \qquad (2.28)$$

In a similar way

$$\{\psi_\alpha(x),\psi^-_\beta(x')\} = \int_{\alpha\gamma} S(x - x'') \text{ x } [\{\gamma_\mu\psi(x'')\}_\gamma,\psi^-_\beta(x')] d\sigma_\mu''$$

so that

$$\{\psi_\alpha(x),\psi^-_\beta(x')\} = 1/i\, S_{\alpha\beta}(x - x') = 1/i\, (\gamma_\mu \partial/\partial x_\mu - \kappa_0)_{\alpha\beta}\Delta(x - x'). \qquad (2.29)$$

All other *matter field* anti-commutators vanish. Of course, the *matter field commutation relations* are invariant with respect to *charge conjugation*.

Finally, we turn to the generalization of the *supplementary condition* (2.12)

$[\partial A_\mu(x)/\partial x_\nu\ \Psi[\sigma] = 0,$ (2.12)]

which consists of removing the restriction that x be situated on the surface σ. It follows from (2.22)

$[A_\mu(x) = \int_\sigma [D(x - x')\ \partial/\partial x_\nu'\ A_\mu(x') - A_\mu(x')\ \partial/\partial x_\nu'\ D(x - x')]\ d\sigma_\nu'.$ (2.22)]

that, for an arbitrary point x,

$$\partial A_\mu(x)/\partial x_\mu\ \Psi(\sigma) = \int_\sigma D(x - x'')\ x\ \partial/\partial x_\nu''\{\partial A_\mu(x'')/\partial x_\mu''\}d\sigma_\nu\ \Psi(\sigma). \quad (2.30)$$

However, according to (2.9)

$$[\partial F[x]/\partial x_\nu = U[\sigma]\ \partial \mathbf{F}[x]/\partial x_\nu\ U^{-1}[\sigma]$$
$$+ \int_\sigma [\delta U[\sigma]/\delta\sigma(x)\ U^{-1}[\sigma], F(x')]\ d\sigma_\nu'$$
$$= U[\sigma]\ \partial \mathbf{F}[x]/\partial x_\nu\ U^{-1}[\sigma]$$
$$- i/\hbar c \int_\sigma [H(x), F(x')]\ d\sigma_\nu', \quad (2.9)]$$

with $F = \partial A_\mu(x'')/\partial x_\mu''$,

$$\partial/\partial x_\nu''\{\partial A_\mu(x'')/\partial x_\mu''\}\Psi(\sigma) = \ldots = \ldots = \ldots . \quad (2.31)$$

In the last transformation, we have used the lemma (1.58) and the fact that the *electromagnetic field commutators* contain only the difference in coordinates of the two points involved. On introducing (2.31) into (2.30) and performing the x'' integration, we find without further difficulty that

$$[\partial A_\mu(x)/\partial x_\mu - \int_\sigma D(x - x')\ 1/c\ j_\mu(x')\ d\sigma_\mu']\ \Psi(\sigma) = 0, \quad (2.32)$$

which is the supplementary condition for the interaction representation. Although the consistency of the *supplementary condition* is guaranteed by the corresponding property in the *Heisenberg representation*, it is well to verify it directly. However, since the proof involves the commutation properties of the current four-vector, we digress briefly to derive the necessary theorems.

It is easy to deduce from the expression (2.14) for $j_\mu(x)$, and the anti-commutator (2.29), that

$$\ldots . \quad (2.33)$$

Of course, all components of j_μ commute at two distinct points on a *space-like* surface. However, the important statement is that a time-like component of j_μ commutes with all components of the *current* at the same point. We prove this by demonstrating that

$$\int_\sigma [j_\mu(x), j_\nu(x')]\ d\sigma_\nu' = 1/c\ [j_\mu(x), Q] = 0 \quad (2.34)$$

where σ is any *space-like* surface, which, in particular, can include the point x. The validity of this statement follows immediately from (2.23) and (2.27)

$$[\psi(x) = \int_\sigma S(x - x')\gamma_\mu\psi(x')\ d\sigma_\mu', \quad (2.23)$$
$$\psi^{-}(x) = \int_\sigma d\sigma_\mu'\psi^{-}(x')\gamma_\mu S(x' - x), \quad (2.27)]$$

since

$$\int_\sigma [j_\mu(x), j_\nu(x')]\ d\sigma_\nu' = \ldots = 0. \quad (2.35)$$

Indeed, *Eq. (2.34) is an expression of charge conservation*, for, according to (2.9)

$$[\partial F[x]/\partial x_\nu = U[\sigma]\, \partial \mathbf{F}[x]/\partial x_\nu\, U^{-1}[\sigma] + \int_\sigma [\delta U[\sigma]/\delta\sigma(x)\, U^{-1}[\sigma], F(x')]\, d\sigma_\nu' = U[\sigma]\, \partial \mathbf{F}[x]/\partial x_\nu\, U^{-1}[\sigma] - i/\hbar c \int_\sigma [H(x), F(x')]\, d\sigma_\nu', \qquad (2.9)]$$

with $F = j_\nu(x)$:

$$\partial j_\nu(x)/\partial x_\nu = i/\hbar c\ 1/c \int_\sigma [j_\mu(x), j_\nu(x')]\, d\sigma_\nu'\, A_\mu(x) = 1/\hbar c\ [j_\mu(x), Q] A_\mu(x). \qquad (2.36)$$

To prove the suitability of (2.32)

$$[[\partial A_\mu(x)/\partial x_\mu - \int_\sigma D(x - x')\ 1/c\ j_\mu(x')\, d\sigma_\mu'] \Psi(\sigma) = 0, \qquad (2.32)]$$

as a supplementary condition, we must show that it is consistent with the field equations of motion, the equation of motion for $\Psi(\sigma)$, and the commutation relations. In terms of the operator

$$\Omega[x,\sigma] = \partial A_\mu(x)/\partial x_\mu - \int_\sigma D(x - x')\ 1/c\ j_\mu(x')\, d\sigma_\mu' \qquad (2.37)$$

we must verify that

… , (2.38a)
… , (2.38b)
… , (2.38c)

The first statement is trivial. As to (2.38b), note that

…

while

…

in view of the property of $j_\mu(x)$ just established. Finally, the same property implies that

$$\Omega[x,\sigma], \Omega[x',\sigma] = \ \dots = -\,i\hbar c \Box^2 D(x - x') = 0.$$

Gauge invariance has a different aspect in the new representation from that of the Heisenberg representation, since the matter field equations do not involve the electromagnetic field. On introducing a change in *gauge*

$$A_\mu(x) \to A_\mu(x) - \partial \Delta(x)/\partial x_\mu,$$

where $\Delta(x)$ is a scalar function of position such that

$$\Box^2 \Delta(x) = 0,$$

the *supplementary condition*, *commutation relations* and *field equations of motion* are unaffected, but the *equation of motion* for $\Psi[\sigma]$ becomes

$$i\hbar c\ \delta\Psi[\sigma]/\delta\sigma(x) = [H(x) + \partial/\partial x_\mu\{1/c\ j_\mu(x)\Delta(x)\}]\ \Psi[\sigma], \qquad (2.39)$$

in which the *charge conservation equation* has been used. We shall show that it is possible to restore this equation to its original form, and thus prove *gauge invariance*, by a *canonical transformation* on $\Psi[\sigma]$. Indeed, the proper transformation is

$$\Psi[\sigma]\rightarrow e^{-iG[\sigma]}\Psi[\sigma] \quad (2.40)$$

where

$$G[\sigma] = 1/hc \int_\sigma 1/c\, j_\mu(x)\Delta(x)\, d\sigma_\mu. \quad (2.41)$$

The *equation of motion* for the new *state vector* is

$$\ldots, \quad (2.42)$$

as a consequence of the commutation properties of j,, on a *space-like* surface. We may now employ the simple expansion theorem

$$\ldots \quad (2\ 43)$$

to deduce that

$$i\hbar ciG\, \delta e^{-iG}/\delta\sigma(x) = \ldots = \partial/\partial x_\mu\{1/c\, j_\mu(x)\Delta(x)\},$$

in which *the commutability of* j_μ *with a time-like component of* j_μ *on the surface* σ *ensures that only the first term of the series survives*. We have thereby demonstrated the correctness of the transformation (2.40).

The form of the *energy-momentum* quantities, as well as their significance as displacement operators, *is altered by the canonical transformation that generates the interaction representation*. In the *Heisenberg representation*, the *functional derivative* of an operator is of immediate significance in computing the *functional derivative* of the expectation value of that operator:

$$\delta/\delta\sigma(x)(\Phi,F[\sigma]\Phi) = \{\Phi,\ \delta F[\sigma]/\delta\sigma(x)\ \Phi\}. \quad (2.44)$$

In the interaction representation, however, part of the change in the expectation value is accounted for by the variation in $\Psi[\sigma]$:

$$\delta/\delta\sigma(x)(\Psi[\sigma],F[\sigma]\Psi[\sigma]) = \{\ \Psi[\sigma],\ \delta F[\sigma]/\delta\sigma(x)\ \Psi[\sigma]\} + i/\hbar c\ \{\Psi[\sigma],[H(x),\ F[\sigma]\Psi[\sigma]. \quad (2.45)$$

Accordingly, it is natural to define the *total functional derivative* of an operator,

$$\Delta F[\sigma]/\Delta\sigma(x) = \ldots = U[\sigma]\ \delta F[\sigma]/\delta\sigma(x)\ U^{-1}[\sigma] \quad (2.46)$$

which is composed of the partial *functional derivatives, expressing the explicit coordinate variation and the implicit dynamical variation.* With this definition,

$$\delta/\delta\sigma(x)(\Psi[\sigma],F[\sigma]\Psi[\sigma]) = \{\Psi[\sigma],\Delta F[\sigma]/\Delta\sigma(x)\ \Psi[\sigma]\}. \quad (2.47)$$

If the *functional* is of the form

$$F[\sigma] = \int_\sigma F(x)\, d\sigma_\mu,$$

we are led to write

$$\Delta F[\sigma]/\Delta\sigma(x) = dF(x)/dx_\mu,$$

where

$$dF(x)/dx_\mu = \partial F(x)/\partial x_\mu + i/\hbar c\, [H(x), \int_\sigma F(x')\, d\sigma_\mu']$$

defines the *total coordinate derivative*. It should be clear that *the conservation theorem (1.56)*

$$[\delta/\delta\sigma(x)\, P_\mu c = \delta/\delta x_\lambda\, T_{\lambda\mu}(x) = 0. \qquad (1.56)]$$

and the equation of motion (1.38)

$$[i/\hbar\, [F(x),P_\nu] = \partial F(x)/\partial x_\nu, \qquad (1.38)]$$

in the interaction representation are to be written

$$\Delta P_\mu[\sigma]/\Delta\sigma(x) = \delta P_\mu[\sigma]/\delta\sigma(x) + i/\hbar c\, [H(x),P_\mu[\sigma]] = 0, \qquad (2.48)$$

and

$$i/\hbar c\, [F(x),P_\mu[\sigma]] = dF(x)/dx_\mu$$
$$= \partial F(x)/\partial x_\mu + i/\hbar c\, [H(x), \int_\sigma F(x')\, d\sigma_\mu'] \qquad (2.49)$$

Now the partial coordinate derivative $\partial F(x)/\partial x_\mu$ is that to be associated with the behavior of *non-interacting* fields, and can therefore be calculated from the *energy-momentum four-vector* of the isolated fields, $P_\mu^{(0)}$, according to

$$i/\hbar c\, [F(x), P_\mu^{(0)}] = dF(x)/dx_\mu. \qquad (2.50)$$

Therefore,

$$[F(x),P_\mu[\sigma] - P_\mu^{(0)}] = \ldots = -1/c \int_\sigma [F(x'), H(x')]\, d\sigma_\mu' \qquad (2.51)$$

in which we have used the fact that only the point x' = x will contribute to the surface integral. One may infer that

$$P_\mu[\sigma] = P_\mu^{(0)} - 1/c \int_\sigma H(x)]\, d\sigma_\mu, \qquad (2.52)$$

which, indeed, is compatible with the conservation theorem (2.48)

$$[\Delta P_\mu[\sigma]/\Delta\sigma(x) = \delta P_\mu[\sigma]/\delta\sigma(x) + i/\hbar c\, [H(x),P_\mu[\sigma]] = 0, \qquad (2.48)]$$

since

$$\Delta P_\mu[\sigma]/\Delta\sigma(x) = \delta P_\mu^{(0)}/\delta\sigma(x)$$
$$- 1/c\, \{\partial H(x)/\partial x_\mu - i/\hbar\, [H(x), P_\mu^{(0)}]\} = 0. \qquad (2.53)$$

The statement (2.52) can be confirmed by direct calculation. The appropriate transcription of the *Heisenberg operator* (1.64) involves the introduction of the total derivatives dA_λ/dx_ν and $d\psi/dx_\nu$. Only the latter differs from the explicit coordinate derivative. Now the operator $P_\nu^{(0)}$ is formally identical with (1.64), but expressed in terms of the *interaction representation* operators and their explicit coordinate derivatives. Therefore,

$$P_\nu[\sigma] = P_\nu^{(0)} + 1/2c \int_\sigma d\sigma_\mu'\, [\psi^-(x'),[H(x),\gamma_\mu\psi(x')]]\, d\sigma_\nu'. \qquad (2.54)$$

However,

$$[H(x),\gamma_\mu\psi(x')] = - i/c\, [j_\lambda(x), \gamma_\mu\psi(x')]A_\lambda(x) = \ldots, \qquad (2.55)$$

whence

$$P_\nu[\sigma] - P_\nu^{(0)} = \ldots = \ldots = 1/c^2 \int_\sigma j_\lambda(x) A_\lambda(x)\, d\sigma_\nu. \qquad (2.56)$$

which is the content of Eq. (2.52)

$$[P_\mu[\sigma] = P_\mu^{(0)} - 1/c \int_\sigma H(x)]\, d\sigma_\mu. \qquad (2.52)]$$

3. *Covariant elimination of the longitudinal field*

It is the function of the supplementary condition to ensure that the electromagnetic field contains no spinless light quanta, which have various unphysical properties. It is possible, indeed to eliminate the *scalar potential* and the longitudinal part of the *vector potential*, leaving only the transverse *vector potential* as the quantity truly descriptive of light waves. *Such conventional procedures suffer from a lack of covariance* which will be remedied in this section.

We shall show that one can replace the *electromagnetic field vector*, $A_\mu(x)$, by two scalar fields, $\Delta(x)$ and $\Delta'(x)$, together with a restricted vector field $G_\mu(x)$, in such a way that the *supplementary condition* involves only the scalar fields, while the *equation of motion* for $\Psi[\sigma]$ contains only $G_\mu(x)$, the sole physically significant part of the field. The decomposition will be conveniently expressed with the aid of an arbitrary time-like unit vector n_μ; $n_\mu^2 = -1$. The procedure of the customary theory corresponds to the special choice: $n_\mu = (0, 0, 0, i)$.

We decompose $A_\mu(x)$ into the gradient in the *time-like* direction specified by n_μ of a scalar operator $\Delta(x)$, the gradient in the *space-like* direction orthogonal to n_μ of a scalar operator $\Delta'(x)$ and the vector $\mathbf{G}_\mu(x)$ which has no component in the direction n_μ, and is divergence-less. …

…

We have thereby succeeded in constructing an *equation of motion* for $\Psi[\sigma]$ which no longer contains the *electromagnetic field* variables involved in the *supplementary condition. The additional term thus introduced is evidently the covariant generalization of the Coulomb interaction between charges.*

…

4. *The invariant collision operator*

While the *interactions* between fields and their vacuum fluctuations are conveniently regarded as modifying the properties of the *non-interacting* fields, other types of *interactions* are often best viewed as producing *transitions* among the *states* of the individual fields. We shall conclude this paper with a brief discussion of a *covariant* manner of describing such *transitions*. The change in *state* of several fields arising from their mutual *interaction* is described by the *equation of motion* (2.6)

$$[i\hbar c\, \delta\Psi[\sigma]/\delta\sigma(x) = H(x)\Psi[\sigma], \qquad (2.6)]$$

for the *state vector* $\Psi[\sigma]$. The question that must be answered in order to describe collisions between the particles associated with the quantized fields is: given the *state vector* on a surface σ_1, what is the *state vector* on the surface σ_2, in the limit as σ_1 and σ_2 recede into the remote and past and future, respectively? …

…

As a final remark, we observe that the representation of S as an integral extended over all *space-time* indicates that it is unaffected by a translation of the coordinate system, and therefore commutes with the operator $P_\mu^{(0)}$ (cf. Eq. (2.50)

[$i/\hbar c\ [F(x), P_\mu^{(0)}] = dF(x)/dx_\mu$. (2.50)]

$$[S, P_\mu^{(0)}] = 0. \quad (4.24)$$

This is the energy-momentum conservation law for collision processes, since, according to (4.10), the expectation value of $P_\mu^{(0)}$ is unchanged by the course of *interaction*, for an arbitrary initial *state*.

Freeman John Dyson (December 15, 1923 – February 28, 2020)

Dyson was an English-American theoretical physicist and mathematician known for his works in quantum field theory, astrophysics, random matrices, mathematical formulation of quantum mechanics, condensed matter physics, nuclear physics, and engineering. He was Professor Emeritus in the Institute for Advanced Study in Princeton and a member of the Board of Sponsors of the Bulletin of the Atomic Scientists.

Dyson was born on 15 December 1923, in Crowthorne in Berkshire, England. He was the son of Mildred (née Atkey) and George Dyson. His father, a prominent composer, was later knighted. His mother had a law degree, and after Dyson was born, she worked as a social worker. Dyson had one sibling, his older sister, Alice, who remembered him as a boy surrounded by encyclopedias and always calculating on sheets of paper. At the age of four he tried to calculate the number of atoms in the Sun. As a child, he showed an interest in large numbers and in the solar system, and was strongly influenced by the book Men of Mathematics by Eric Temple Bell. Politically, Dyson said he was "brought up as a socialist".

From 1936 to 1941 Dyson was a scholar at Winchester College, where his father was Director of Music. At the age of 17 he studied pure mathematics with Abram Besicovitch as his tutor at Trinity College, Cambridge, where he won a scholarship at age 15.

At the age of 19 he was assigned to war work in the Operational Research Section (ORS) of RAF Bomber Command, where he developed analytical methods for calculating the ideal density for bomber formations to help the Royal Air Force bomb German targets during World War II. After the war, Dyson was readmitted to Trinity College, where he obtained a BA degree in mathematics. From 1946 to 1949 he was a fellow of his college.

In 1947 Dyson published two papers in number theory. Friends and colleagues described him as shy and self-effacing, with a contrarian streak that his friends found refreshing but intellectual opponents found exasperating. "I have the sense that when consensus is forming like ice hardening on a lake, Dyson will do his best to chip at the ice", Steven Weinberg said of him. His friend the neurologist and author Oliver Sacks said: "A favorite word of Freeman's about doing science and being creative is the word 'subversive'. He feels it's rather important not only to be not orthodox, but to be subversive, and he's done that all his life."

On G. I. Taylor's advice and recommendation, Dyson moved to the United States in the fall of 1947 as a Commonwealth Fellow for postgraduate study *with Hans Bethe at Cornell University* (1947–1948). There he made the acquaintance of Richard Feynman. Dyson recognized the brilliance of the flamboyant American and over the course of that year he began meeting with Feynman, just at the time that Feynman was working out his new approach to quantum electrodynamics. Dyson and Feynman talked often during the spring of 1948 about Feynman's diagrams and how they could be used—conversations that continued in close quarters when the two drove across the country together that summer, just a few months after Feynman's Pocono Manor presentation. Later that summer, Dyson attended the summer school on theoretical physics at the University of Michigan, which featured detailed lectures by Julian Schwinger on his own, non-diagrammatic approach to renormalization. The summer school offered Dyson the opportunity to talk informally and at length with Schwinger in much the same way that he had already been talking with Feynman. By September 1948, Dyson had spent intense, concentrated time talking directly with both Feynman and Schwinger about their new techniques. At the end of the summer, he then moved to the Institute for Advanced Study (1948–1949) before returning to England (1949–51), where he was a research fellow at the University of Birmingham.

Shortly after his arrival in Princeton, Dyson submitted an article to the *Physical Review* that compared Feynman's and Schwinger and Tomonaga's methods.

> [Dyson was 24 years old when this paper was submitted; Feynman and Schwinger were both 30.]

This demonstrated the equivalence of two formulations of quantum electrodynamics, Feynman's diagrams and the operator method developed independently by Schwinger and Tomonaga. Dyson was the first person after their creator to appreciate the power of Feynman diagrams and his paper written in 1948 and published in 1949 was the first to make use of them. He said in that paper that Feynman diagrams were not just a computational tool but a physical theory and developed rules for the diagrams that completely solved the *renormalization* problem. Dyson's paper and also his lectures presented Feynman's theories of quantum electrodynamics in a form that other physicists could understand, facilitating the physics community's acceptance of Feynman's work. Oppenheimer, in particular, was persuaded by Dyson that Feynman's new theory was as valid as Schwinger's and Tomonaga's.

> [Murray Gell-Mann always referred to *Feynman diagrams* as *Stueckelberg diagrams*, after the Swiss physicist, Ernst Stueckelberg, who devised a similar notation many years earlier. [Stueckelberg, E.C.G. (1934), Relativistisch invariante Störungstheorie des Diracschen Elektrons I. Teil: Streustrahlung und Bremsstrahlung. (Relativistically invariant perturbation theory of Dirac's electron Part I: scattered radiation and Bremsstrahlung.) *Ann. Phys*., 413, 4, 367-389.]
>
> Stueckelberg was motivated by the need for a manifestly covariant formalism for quantum field theory, but did not provide as automated a way to handle symmetry factors and loops, although he was first to find the correct physical interpretation in terms of forward and backward in time particle paths, all without the path-integral.
>
> Historically, as a book-keeping device of covariant perturbation theory, the graphs were called *Feynman–Dyson diagrams* or *Dyson graphs*, because the path integral was unfamiliar when they were introduced, and Dyson's derivation from perturbation theory borrowed from the perturbative expansions in statistical mechanics was easier to follow for physicists trained in earlier methods.]

Dyson married his first wife, the Swiss mathematician Verena Huber, on 11 August 1950. They had two children, Esther and George, before divorcing in 1958.

Dyson joined the faculty at Cornell as a physics professor in 1951, though he still had no doctorate. In December 1952, Oppenheimer, the director of the Institute for Advanced Study in Princeton, New Jersey, offered Dyson a lifetime appointment at the Institute, "for proving me wrong", in Oppenheimer's words. Dyson remained at the Institute until the end of his career. In 1957 he became a US citizen.

In November 1958 he married Imme Jung, a German (born 1936), and they had four more children: Dorothy, Mia, Rebecca, and Emily Dyson.

From 1957 to 1961 Dyson worked on Project Orion, which proposed the possibility of space-flight using nuclear pulse propulsion. A prototype was demonstrated using conventional explosives, but the 1963 Partial Test Ban Treaty, in which Dyson was involved and which he supported, permitted only underground nuclear weapons testing, and the project was abandoned in 1965.

In 1966, independently of Elliott H. Lieb and Walter Thirring, Dyson and Andrew Lenard published a paper proving that the Pauli exclusion principle plays the main role in the stability of bulk matter. Hence it is not the electromagnetic repulsion between outer-shell orbital electrons that prevents two stacked wood blocks from coalescing into a single piece, but the exclusion principle applied to electrons and protons that generates the classical macroscopic normal force.

Dyson also did work in a variety of topics in mathematics, such as topology, analysis, number theory and random matrices. In 1973 the number theorist Hugh Lowell Montgomery was visiting the Institute for Advanced Study and had just made his pair correlation conjecture concerning the distribution of the zeros of the Riemann zeta function. He showed his formula to the mathematician Atle Selberg, who said that it looked like something in mathematical physics and that Montgomery should show it to Dyson, which he did. Dyson recognized the formula as the pair correlation function of the Gaussian unitary ensemble, which physicists have studied extensively. This suggested that there might be an unexpected connection between the distribution of primes (2, 3, 5, 7, 11, ...) and the energy levels in the nuclei of heavy elements such as uranium.

Around 1979 Dyson worked with the Institute for Energy Analysis on climate studies. This group, under Alvin Weinberg's direction, pioneered multidisciplinary climate studies, including a strong biology group. Also, during the 1970s, Dyson worked on climate studies conducted by the JASON defense advisory group.

Dyson retired from the Institute for Advanced Study in 1994. He won numerous scientific awards, but never a Nobel Prize. Nobel physics laureate Steven Weinberg said that the Nobel committee "fleeced" Dyson, but Dyson remarked in 2009, "I think it's almost true without exception if you want to win a Nobel Prize, you should have a long attention span, get hold of some deep and important problem and stay with it for ten years. That wasn't my style." Dyson was a regular contributor to The New York Review of Books, and published a memoir, *Maker of Patterns: An Autobiography Through Letters* in 2018.

Dyson disagreed with the scientific consensus on climate change. He believed that some of the effects of increased CO2 levels are favorable and not taken into account by climate scientists, such as increased agricultural yield, and further that the positive benefits of CO2 likely outweigh the negative effects. *He was skeptical about the simulation models used to predict climate change*, arguing that political efforts to reduce causes of climate change distract from other global problems that should take priority.

> [Dyson agreed that technically humans and additional CO2 emissions contribute to warming. However, he felt that the benefits of additional CO2 outweighed any associated negative effects. He said that in many ways increased atmospheric carbon dioxide is beneficial, and that it is increasing biological growth, agricultural yields and forests. He believed that existing simulation models of climate change fail to account for some important factors, and that the results thus contain too great a margin of error to reliably predict trends. He argued that political efforts to reduce the causes of climate change distract from other global problems that should take priority and compared acceptance of climate change as real to religion.
>
> Climate scientist James Hansen said that Dyson "doesn't know what he's talking about… If he's going to wander into something with major consequences for humanity and other life on the planet, then he should first do his homework – which he obviously has not done on global warming." Dyson replied that "[m]y objections to the global warming propaganda are not so much over the technical facts, about which I do not know much, but it's rather against the way those people behave and the kind of intolerance to criticism that a lot of them have." In a 2014 interview he said, "What I'm convinced of is that we don't understand climate… It will take a lot of very hard work before that question is settled."]

Dyson died on 28 February 2020 at a hospital near Princeton, New Jersey, from complications following a fall. He was 96.

Dyson, F. J. (February, 1949). The Radiation Theories of Tomonaga, Schwinger, and Feynman.

[*Phys. Rev.*, 75, 3, 486-502; https://doi.org/10.1103/PhysRev.75.486; also in Schwinger, J. (ed). (1958). *Selected Papers on Quantum electrodynamics*. Dover, New York, pages 275-91.]

Received October 6, 1948.

Institute for Advanced Study, Princeton, New Jersey.

The recent and independent formulations of quantum electrodynamics by Tomonaga, Schwinger, and Feynman have made two notable advances, the foundations and applications of the theory have been simplified by being presented in a *completely relativistic way* and the divergence difficulties *have been at least partially overcome*, the advantages of the Feynman formulation are simplicity and ease of application while those of Tomonaga-Schwinger are generality and theoretical completeness, this paper presents a unified development of quantum electrodynamics embodying the main features of Tomonaga-Schwinger and Feynman radiation theories, *emphasis on application of the theory*, aims to show how the Schwinger theory can be applied to specific problems in such a way as to incorporate the ideas of Feynman, the main results are general formulas from which radiative reactions on motions of electrons can be calculated, divides *energy-density* into two parts, *energy of interaction* of two fields with each other and energy produced by external forces, *interaction energy* alone is treated as a perturbation, important results of the paper are the *equation of motion* $ihc[\partial\Omega/\partial\sigma(x_0)] = \{S(\sigma)\}^{-1}H^e(x_0)S(\sigma)\Omega$ for the *state vector* $\Omega(\sigma)$ and the interpretation of the *state vector* Ω, simplifies Schwinger theory for using it for calculations, demonstrates equivalence of the theories within their common domain of applicability, in the *Schwinger theory* the aim is to calculate the matrix elements of the "*effective external potential energy*" between *states* specified by their *state vectors*, in the *Feynman theory* the basic principle is to "preserve symmetry between past and future" so the matrix elements of the operator are evaluated in a "*mixed representation*" in which the matrix elements are calculated between an *initial state* specified by its *state vector* and a *final state* specified by its *state vector,* a *graph* corresponding to a particular matrix element is used not merely as an aid to calculation but as a picture of the physical process which gives rise to that matrix element, derives fundamental formulas for the operator in the *equation of motion* for the *state vector* $\Omega(\sigma)$ which represents the interaction of a physical particle with an external field for both the Schwinger and the Feynman theories, derives set of rules by which matrix element of Feynman operator may be written down in form suitable for numerical evaluation, shows equivalence of the two theories, develops graphical representation of matrix elements, *the theory as a whole cannot be put into a finally satisfactory form so long as divergencies occur in it however skillfully these divergencies are circumvented*, present treatment should be regarded as justified by its success in applications rather than by its theoretical derivation, *paper suffers from a series of significant errors*.

[Judging by the errors reported in the *Notes in Proof*, the 24-year-old Dyson was a little too clever by half in what appears to be a rush to publish Schwinger and Feynman's theories before they published them themselves.]

A unified development of the subject of quantum electrodynamics is outlined, embodying the main features both of the Tomonaga-Schwinger and of the Feynman radiation theory. The theory is carried to a point further than that reached by these authors, in the discussion of higher order *radiative reactions* and *vacuum polarization* phenomena. However, the theory of these higher order processes is a program rather than a definitive theory, since *no general proof of the convergence of these effects is attempted.* The chief results obtained are (a) a *demonstration of the equivalence of the Feynman and Schwinger theories*, and (b) a *considerable simplification of the procedure involved in applying the Schwinger theory to particular problems*, the simplification being the greater the more complicated the problem.

I. *Introduction*

As a result of the recent and independent discoveries of Tomonaga[1], Schwinger[2], and Feynman[3], the subject of *quantum electrodynamics* has made two very notable advances.

[1] Tomonaga, S. (1946). On a Relativistically Invariant Formulation of the Quantum Theory of Wave Fields. *Prog. Theor. Phys*., 1, 27–42. Translated from the paper (1943). *Bull. I. P. C. R. Riken-iho*, 22, 545 (appeared originally in Japanese); Koba, Z., Tati, T. & Tomonaga, S. (1947). On a Relativistically Invariant Formulation of the Quantum Theory of Wave Fields. II. Case of Interacting Electromagnetic and Electron Fields. § 1-4. *Prog. Theor. Phys*., 2, 101–116; https://doi.org/10.1143/ PTP.2.101; Koba, Z., Tati, T. & Tomonaga, S. (1947). On a Relativistically Invariant Formulation of the Quantum Theory of Wave Fields. III. Case of Interacting Electromagnetic and Electron Fields. § 5-7. *Prog. Theor. Phys*., 2, 198–208; https://doi.org/10.1143/ PTP.2.198; Kanesawa, S. & Tomonaga, S. (1948). On a Relativistically Invariant Formulation of the Quantum Theory of Wave Fields. IV. Case of Interacting Electromagnetic and Meson Fields. § 1-4. *Prog. Theor. Phys.* 3, 1–13; https://doi.org/10.1143/PTP.3.1; Kanesawa, S. & Tomonaga, S. (1948). On a Relativistically Invariant Formulation of the Quantum Theory of Wave Fields. V. Case of Interacting Electromagnetic and Meson Fields. § 5-7. *Prog. Theor. Phys.* 3, 101–113; https://doi.org/10.1143/PTP.3.101; Tomonaga, S. & Oppenheimer, J. R. (July, 1948). On Infinite Field Reactions in Quantum Field Theory. *Phys. Rev*., 74, 224–225; https://doi.org/10.1103/PhysRev.74.224.

[2] Schwinger, J. (February, 1948). On quantum-electrodynamics and the magnetic moment of the electron. *Phys. Rev*., 73, 4, 416-7; Schwinger, J. (November, 1948). Quantum Electrodynamics. I. A Covariant Formulation. *Phys. Rev*., 74, 10, 1439-61. Several papers, giving a complete exposition of the theory, are in course of publication.

[3] Wheeler, J. A. & Feynman R. P. (April, 1945). Interaction with the Absorber as the Mechanism of Radiation. *Rev. Mod. Phys*. 17, 157 (1945); Feynman, R. P. (April, 1948). Space-Time Approach to *Non-Relativistic* Quantum Mechanics. *Rev. Mod. Phys*., 20, 367; [this paper originated whilst Dyson was at Cornell University, where he had met and worked with Feynman]; Feynman, R. P. (October, 1948). A *Relativistic* Cut-Off for Classical Electrodynamics. *Phys. Rev*., 74, 939; Feynman, R. P. (November, 1948). *Relativistic* Cut-Off for Quantum Electrodynamics. *Phys. Rev*., 74, 1430 [last two re-ordered]; These articles describe early stages in the development of Feynman's theory, little of which is yet published.

On the one hand, both the foundations and the applications of the theory have been simplified by being presented in a completely relativistic way; on the other, the divergence difficulties have been at least partially overcome. In the reports so far published, emphasis has naturally been placed on the second of these advances; the magnitude of the first has been somewhat obscured by the fact that the new methods have been applied to problems which were beyond the range of the older theories, so that the simplicity of the methods was hidden by the complexity of the problems. Furthermore, the theory of Feynman differs so profoundly in its formulation from that of Tomonaga and Schwinger, and so little of it has been published, that its particular advantages have not hitherto been available to users of the other formulations. *The advantages of the Feynman theory are simplicity and ease of application, while those of Tomonaga-Schwinger are generality and theoretical completeness.*

The present paper aims to show how the Schwinger theory can be applied to specific problems in such a way as to incorporate the ideas of Feynman. To make the paper reasonably self-contained it is necessary to outline the foundations of the theory, following the method of Tomonaga; but this paper is not intended as a substitute for the complete account of the theory shortly to be published by Schwinger. *Here the emphasis will be on the application of the theory*, and the major theoretical problems of gauge-invariance and of the divergencies will not be considered in detail. *The main results of the paper will be general formulas from which the radiative reactions on the motions of electrons can be calculated, treating the radiation interaction as a small perturbation, to any desired order of approximation.* These formulas will be expressed in Schwinger's notation, but are in substance identical with results given previously by Feynman. The contribution of the present paper is thus intended to be twofold: first, to simplify the Schwinger theory for the benefit of those using it for calculations, and second, to demonstrate the equivalence of the various theories within their common domain of applicability*.

*After this paper was written, the author was shown a letter, published in *Progress of Theoretical Physics*, 3, 205 (1948) by Z. Koba and G. Takeda. [Koba, Z. & Takeda, G. (June, 1948). Radiative Corrections in e^2 for an Arbitrary Process Involving Electrons. Positrons, and Light Quanta. *Prog. Theor. Phys*., 3, 2, 203-5, letter to the editor
[; investigated further the *radiation correction in collision processes between electrons, positrons and photons* in a general manner and confirmed modified Hamiltonian sufficient to eliminate divergence difficulty *as long as we are concerned with the first radiative correction, introduces "transition diagram method"* [similar to the Stueckelberg (1934) diagrams, later introduced by Feynman at Ponoco in spring 1948 and published in September 1949, and subsequently renamed *Feynman diagrams*] *to analyze complicated connection between initial and final states through a number of intermediate ones*, expresses *electrons* and *positrons* as *world lines* in *momentum space*, while *emission* and *absorption* of *photons* are "*leaps*" of these *world lines*, diagram acquires two additional *leaps* (the *emission* and *reabsorption* of a *virtual photon*) when *radiative corrections* are taken into account, two distinct ways of attaching them, existing *world lines* gain new *leaps* or additional closed *world line* is introduced and coupled with existing ones through *virtual photon*, first case implies "*mass-type*" or "*self-energy-type*" divergences when *emission* and

reabsorption of the *virtual photon* take place in succession which can be eliminated by *counter-self-energy term*, second case represents *vacuum polarization* corrections which can be subdivided according to number of *leaps* in additional closed *world line*, which afford quadratic and logarithmic divergences that are cancelled by the "*counter-terms*" and found not to contribute to divergence].

[The letter referred to "full account in later issue of this journal". This was not found online, nor were any references to it found.]

The letter is dated May 22, 1948, and briefly describes a method of treatment of radiative problems, similar to the method of this paper. Results of the application of the method to a calculation of the second-order radiative correction to the Klein-Nishina formula are stated. All the papers of Professor Tomonaga and his associates which have yet been published were completed before the end of 1946. The isolation of these Japanese workers has undoubtedly constituted a serious loss to theoretical physics.

[Dyson's claim regarding papers that had been published is not true. Only the series of five papers cited above, *On a Relativistically Invariant Formulation of the Quantum Theory of Wave Fields*, were completed before the end of 1946: Koba, Z. & Tomonaga, S. (published December, 1947 (???), letter dated December 30, 1947), Application of the "Self-Consistent" Subtraction Method to the Elastic Scattering of an Electron; and Koba, Z. & Tomonaga, S. (published September, 1948, received March 1, 1948), On Radiation Reactions in Collision Processes. I: Application of the "Self-Consistent" Subtraction Method to the Elastic Scattering of an Electron, *reference Bethe's August, 1947 paper*. Koba, Z. & Takeda, G. (published March, 1948, letter dated March 18, 1948), Radiative Corrections for Compton Scattering; and Tati, T. & Tomonaga, S. (published December, 1948, received May 23, 1948), A Self-Consistent Subtraction Method in the Quantum Field Theory, *reference Lewis and Epstein's January, 1948 papers*. Koba, Z. & Takeda, G. (published December, 1948, received May 23, 1948). Radiation Reaction in Collision Process, II: Radiative Corrections for Compton Scattering, *references the first part of this article which was received on the same day*. Koba, Z. & Takeda, G. (published June, 1948, letter dated May 22, 1948), Radiative Corrections in e^2 for an Arbitrary Process Involving Electrons. Positrons, and Light Quanta, *references their previous letter dated March 18, 1948*. Tomonaga, S. & Oppenheimer, J. R. (received June 1, 1948, published July, 1948). On Infinite Field Reactions in Quantum Field Theory, cited above, *was received on June 1, 1948*.

Nor is it true that Japanese scientists were isolated. In 1921, Nishina studied under Ernest Rutherford at the Cavendish Laboratory at Cambridge. In 1923, he moved to Copenhagen to work with Niels Bohr and they became good friends. He also visited Gottingen and Hamburg. Before he left Copenhagen in 1928, he published a paper on Compton scattering, a collaboration with Oscar Klein, from which the Klein–Nishina formula derives. After he returned home, he convinced Werner Heisenberg to visit Japan and give a number of lectures. Later, in 1937, Bohr and his wife also visited Nishina in Tokyo.

In 1931, he established his own laboratory at RIKEN, and invited some Western scholars to Japan including Heisenberg, Dirac and Bohr to stimulate Japanese physicists. In 1937, his lab constructed the first cyclotron outside the United States (and the second in the world). While he was constructing a larger cyclotron, he consulted Ernest Lawrence for help. Lawrence assisted him in buying an electromagnet from an American company, and hosted some of Nishina's assistants in 1940. Communication between the two physicists was cut off by the start of the war.

Nishina and his colleagues were busy between 1940 and June 1945, on research on nuclear fission and the development of an atomic bomb for the Imperial Japanese Army, as were his US, British and German counterparts on their countries' efforts.

In 1940, Nishina was asked by Lieutenant General Takeo Yasuda to research nuclear fission. In April 1941, the Imperial Japanese Army (IJA) officially authorized *Ni-Go*, the project to research an atomic bomb. Nishina's team's initial conclusion was that an atomic bomb was theoretically, but not technically, feasible, but they continued to pursue uranium enrichment via gaseous thermal diffusion (a process that had been abandoned by the Allies in 1941).

> *On December 7, 1941, the Imperial Japanese Navy Air Service launched a surprise military strike on the United States naval base at Pearl Harbor, which led to the formal entry of the US into World War II the next day.*

Nishina also participated in a committee convened by the Technical Research Institute of the IJN between 1942 and 1943. After several meetings, the committee disbanded with pessimistic conclusions. Meanwhile, Nishina continued his research for the IJA. In June 1945, a few months after *RIKEN* facilities were bombed, Nishina declared the project over, and resources were handed over to the Imperial Japanese Navy's (IJN) *F-Go* project.

By the end of July 1945, the Imperial Japanese Navy (IJN) had become incapable of conducting major operations and an Allied invasion of Japan was imminent. The day after the US dropped an atomic bomb on Hiroshima on August 9, 1945, Nishina received a copy of a restricted press release about the bomb that had come from President Truman. Nishina confirmed to the government the veracity of the claims about atomic weaponry.

The surrender of Japan in World War II was announced by Emperor Hirohito on August 15, and formally signed on September 2, 1945, bringing the war's hostilities to a close. In October 1945, Nishina asked the occupying forces for permission to continue using his remaining cyclotrons for biological and medical research. Permission was initially granted, then rescinded under orders from the Secretary of War. Every cyclotron in Japan was destroyed. The ones from *RIKEN* were disassembled and thrown into the Gulf of Tokyo.]

II. *Outline of Theoretical Foundations*

Notes added in proof; The argument of Section II is an over-simplification of the method of Tomonaga[1], and *is unsound*. There is an error in the derivation of (3); derivatives occurring in H(r) give rise to noncommutativity between H(r) and field quantities at r' when r is a point on σ infinitesimally distant from r'. The argument should be amended as follows. Φ is defined only for flat surfaces t(r) = t, and for such surfaces (3) and (6) are correct. Ψ is defined for general surfaces by (12) and (10), and is verified to satisfy (9). For a flat surface, Φ and Ψ are then shown to be related by (7). Finally, since H_1 does not involve the derivatives in H, the argument leading to (3) can be correctly applied to prove that for general σ the state-vector $\Psi(\sigma)$ will completely describe results of observations of the system on σ.

Relativistic quantum mechanics is a special case of *non-relativistic* quantum mechanics, and it is convenient to use the usual *non-relativistic* terminology in order to make clear the relation between the mathematical theory and the results of physical measurements. In *quantum electrodynamics* the dynamical variables are the *electromagnetic potentials* $A_\mu(r)$ and the spinor *electron-positron field* $\psi_\alpha(r)$; each component of each *field* at each point r of space is a separate variable. Each dynamical variable is, in the Schrodinger representation of quantum mechanics, a time-independent operator operating on the *state vector* Φ of the system. The nature of Φ (*wave function* or abstract vector) need not be specified; its essential property is that, given the Φ of a system at a particular time, the results of all measurements made on the system at that time are statistically determined. The variation of Φ with time is given by the *Schrodinger equation*

$$ih[\partial/\partial t]\Phi = \{\int H(r)d\tau\}\Phi, \qquad (1)$$

where H(r) is the operator representing the total *energy-density* of the system at the point r. The general solution of (1) is

$$\Phi(t) = \{[-it/h]\int H(r)d\tau\}\Phi_0, \qquad (2)$$

with Φ_0 any constant *state vector*.

Now in a *relativistic system*, the most general kind of measurement is not the simultaneous measurement of field quantities at different points of space. It is also possible to measure independently field quantities at different points of space at different times, provided that the points of space-time at which the measurements are made lie outside each other's light cones, so that the measurements do not interfere with each other. *Thus, the most comprehensive general type of measurement is a measurement of field quantities at each point r of space at a time t(**r**), the locus of the points (**r**, t(**r**)) in space-time forming a 3-dimensional surface σ which is space-like (i.e., every pair of points on it is separated by a space-like interval)*. Such a measurement will be called "an observation of the system on σ". It is easy to see what the result of the measurement will be. At each point r' the field quantities will be measured for a *state* of the system with *state vector* Φ(t(r')) given by (2). But all observable quantities at r' are operators which commute with the *energy-density* operator H(r) at every point r different from r', and it is a general principle of quantum mechanics that if B is a unitary operator commuting with A, then for any state Φ the results of measurements of A are the same in the *state* Φ as in the *state* $B\Phi$. Therefore, the results of measurement of the field quantities at r' in the *state* $\Phi(t(r'))$ are the same as if the *state* of the system were

$$\Phi(\sigma) = \{[-it/h] \int t(r)H(r)d\tau\}\Phi_0, \qquad (3)$$

which differs from $\Phi(t(r'))$ only by a unitary factor commuting with these field quantities. The important fact is that the *state vector* $\Phi(\sigma)$ depends only on σ and not on r'. The conclusion reached is that observations of a system on σ give results which are completely determined by attributing to the system the *state vector* $\Phi(\sigma)$ given by (3).

The Tomonaga-Schwinger form of the Schrodinger equation is a differential form of (3). Suppose the surface σ to be deformed slightly near the point r into the surface σ', the volume of *space-time* separating the two surfaces being V. Then the quotient

$$[\Phi(\sigma') - \Phi(\sigma)]/V$$

tends to a limit as $V \to 0$, which we denote by $\partial\Phi/\partial\sigma(r)$ and call the *functional derivative* of Φ with respect to σ at the point r. From (3) it follows that

$$ihc[\partial\Phi/\partial\sigma(r)] = H(r)\Phi, \qquad (4)$$

and (3) is, in fact, the general solution of (4).

The whole meaning of an equation such as (4) depends on the physical meaning which is attached to the statement "a system has a constant *state vector* Φ_0" In the present context, this statement means "results of measurements of field quantities at any given point of space are independent of time". *This statement is plainly non-relativistic, and so (4) is, in spite of appearances, a non-relativistic equation.*

The simplest way to introduce a new *state vector* Ψ which shall be a *relativistic invariant* is to require that the statement "a system has a constant *state vector* Ψ" shall mean "a system consists of photons, electrons, and positrons, traveling freely through space without interaction or external disturbance". For this purpose, let

$$H(r) = H_0(r) + H_1(r), \qquad (5)$$

where H_0 is the *energy-density* of the *free electromagnetic and electron fields*, and H_1 is that of their *interaction* with each other and with any external disturbing forces that may be present. A system with constant Ψ is, then, one whose H_1 is identically zero; by (3) such a system corresponds to a Φ of the form

$$\Phi(\sigma) = T(\sigma)\,\Phi_0,$$
$$T(\sigma) = \exp\{-it/h] \int t(r)H_0(r)d\tau\}. \qquad (6)$$

It is therefore consistent to write generally

$$\Phi(\sigma) = T(\sigma)\,\Psi(\sigma), \qquad (7)$$

thus defining the new *state vector* Ψ of any system in terms of the old Φ. The differential equation satisfied by Ψ is obtained from (4), (5), (6), and (7) in the form

$$ihc[\partial\Psi/\partial\sigma(r)] = \{T(\sigma)\}^{-1}H_1(r)T(\sigma)\Psi. \tag{8}$$

(5), (6), & (8) *are in error*; see Section IV below:
This is because $S(\infty)$ includes the effects of the *electromagnetic self-energy* of the electron, and this *self-energy* gives an expectation value to $S(\infty)$ *which is different from unity (and indeed infinite) in a one-electron state*, so that Eq. (21) cannot be satisfied. *The mistake that has been made occurred in trying to represent the observed electron with its electromagnetic self-energy by a wave field with the same characteristic rest-mass as that of the "bare" electron*. To correct the mistake, let δm denote the *electromagnetic mass* of the electron, i.e., *the difference in rest-mass between an observed and a "bare" electron*. Instead of (5), the division of the *energy-density* $H(\mathbf{r})$ should have taken the form

$$H(r) = \{H_0(r) + \delta mc^2\psi^*(r)\beta\psi(r)\} + \{H_1(r) - \delta mc^2\psi^*(r)\beta\psi(r)\}.$$

The first bracket on the right here represents the *energy-density* of the free electromagnetic and electron fields with the observed electron rest-mass, and should have been used instead of $H_0(r)$ in the definition (6) of $T(\sigma)$. Consequently, the second bracket should have been used instead of $H_1(r)$ in Eq. (8).

Now if $q(r)$ is any time-independent field operator, the operator

$$q(x_0) = \{T(\sigma)\}^{-1}q(r)T(\sigma)$$

is just the corresponding time-dependent operator as usually defined in *quantum electrodynamics*[4].

[4] See, for example, Gregor Wentzel, (1943). *Einführung in die Quantentheorie der Wellenfelder*. Franz Deuticke, Wien, pp. 18-26. (Translated by Charlotte Houtermans and J. M. Jauch, with an Appendix by J. M. Jauch. *Quantum Theory of Fields*. Interscience, 1949. Dover, 2003.)

It is a function of the point x_0 of *space-time* whose coordinates are $(r, ct(r))$, but is the same for all surfaces σ passing through this point, by virtue of the commutation of $H_1(r)$ with $H_0(r')$ for $r' \neq r$. Thus (8) may be written

$$ihc[\partial\Psi/\partial\sigma(x_0)] = H_1(x_0)\Psi, \tag{9}$$

where $H_1(x_0)$ is the time-dependent form of the *energy-density* of interaction of the two fields with each other and with external forces. *The left side of (9) represents the degree of departure of the system from a system of freely traveling particles and is a relativistic invariant*; $H_1(x_0)$ is also an invariant, and *thus is avoided one of the most unsatisfactory features of the old theories, in which the invariant H_1 was added to the non-invariant H_0. Equation (9) is the starting point of the Tomonaga-Schwinger theory.*

III. *Introduction of perturbation theory*

Notes added in proof

A *covariant perturbation theory* similar to that of Section III has previously been developed by Stueckelberg, E. C. G. (September, 1934). Relativistisch invariante Störungstheorie des Diracschen Elektrons. Teil I: Streustrahlung und Bremsstrahlung. (Relativistically invariant perturbation theory of Dirac's electron. Part I: Scattered radiation and Bremsstrahlung.) *Ann. d. Phys.*, (413, 4,) 21, 367-89; (January, 1944). An Unambiguous Method of Avoiding Divergence Difficulties in Quantum Theory. *Nature*, 153, 143-4; https://doi.org/10.1038/153143a0; "The classical theory of a point charge and the quantum theory of wave packets contain in their usual form well-known divergences. Dirac (August, 1938) [Classical Theory of Radiating Electrons], has elaborated a classical particle theory which avoids these difficulties, but his results cannot as yet be applied to quantum theory. On the other hand, Heisenberg (July, 1943) [Die beobachtbaren Größen in der Theorie der Elementarteilchen. (The "observable quantities" in the theory of elementary particles.)], has recently proposed a new formalism which permits the calculation of collision cross-sections in quantum theory without being disturbed by diverging terms. *However, no connecting link such as the correspondence principle has been given in order to apply this quantum formalism to a given problem such as Rutherford scattering, Compton effect or radiation damping*".

[*Bremsstrahlung* is electromagnetic radiation produced by the deceleration of a charged particle when deflected by another charged particle.]

Equation (9) can be solved explicitly. For this purpose, it is convenient to introduce a one-parameter family of space-like surfaces filling the whole of *space-time*, so that one and only one member, $\sigma(x)$, of the family passes through any given point x. Let $\sigma_0, \sigma_1, \sigma_2, \ldots$ be a sequence of surfaces of the family, starting with σ_0 and proceeding in small steps steadily into the past. By

$$\int_{\sigma 1}^{\sigma 0} H_1(x)dx$$

is denoted the integral of $H_1(x)$ over the 4- dimensional volume between the surfaces σ_1 and σ_0, similarly, by

$$\int_{-\infty}^{\sigma 0} H_1(x)dx, \qquad \int_{\sigma 0}^{\infty} H_1(x)dx$$

are denoted integrals over the whole volume to the past of σ_0 and to the future of σ_0, respectively.

Consider the operator

$$U = U(\sigma_0) = \{1 - [i/hc] \int_{\sigma 1}^{\sigma 0} H_1(x)dx\} \times \{1 - [i/hc] \int_{\sigma 2}^{\sigma 1} H_1(x)dx\} \ldots, \tag{10}$$

the product continuing to infinity and the surfaces $\sigma_0, \sigma_1, \ldots$ being taken in the limit infinitely close together. U satisfies the differential equation

$$ihc[\partial U/\partial\sigma(x_0)] = H_1(x_0)U, \tag{11}$$

and the general solution of (9) is

$$\Psi(\sigma) = U(\sigma)\,\Psi_0, \quad (12)$$

with Ψ_0 any constant vector.

Expanding the product (10) in ascending powers of H_1 gives a series

$$U = 1 + (-\,i/hc) \int_{-\infty}^{\sigma 0} H_1(x_1)dx_1 + (-\,i/hc)^2 \times \int_{-\infty}^{\sigma 0} dx_1 \int_{-\infty}^{\sigma(x1)} H_1(x_2)H_1(x_1)dx_2 + \ldots . \quad (13)$$

Further, U is by (10) obviously unitary, and

$$U^{-1} = U^{-} = 1 + (i/hc) \int_{-\infty}^{\sigma 0} H_1(x_1)dx_1 + (i/hc)^2 \times \int_{-\infty}^{\sigma 0} dx_1 \int_{-\infty}^{\sigma(x1)} H_1(x_2)H_1(x_1)dx_2 + \ldots . \quad (14)$$

It is not difficult to verify that U is a function of σ_0 alone and is independent of the family of surfaces of which σ_0 is one member. *The use of a finite number of terms of the series (13) and (14), neglecting the higher terms, is the equivalent in the new theory of the use of perturbation theory in the older electrodynamics.*

The operator $U(\infty)$, obtained from (10) by taking σ_0 in the infinite future, is a transformation operator transforming a *state* of the system in the infinite past (representing, say, converging streams of particles) into the same *state* in the infinite future (after the particles have interacted or been scattered into their final outgoing distribution), *This operator has matrix elements corresponding only to real transitions of the system, i.e., transitions which conserve energy and momentum. It is identical with the Heisenberg S matrix*[5].

[5] Heisenberg, W. (July, 1943). Die beobachtbaren Größen in der Theorie der Elementarteilchen. (The "observable quantities" in the theory of elementary particles.) *Zeit. Phys.*, 120, 513-38[; S-matrix theory as a principle of particle interactions]; *idem*, 673; and (1946). Der mathamatische Rahman der Quantentheorie der Wellnfelder. (The mathematical framework of quantum theory of wave fields.) *Zeits. Naturforschung*, A, 1, 11-12, 608-22; https://doi.org/10.1515/zna-1946-11-1202.

IV. *Elimination of the radiation interaction*

In most of the problems of electrodynamics, the *energy-density* $H_1(x_0)$ divides into two parts —

$$H_1(x_0) = H^i(x_0) + H^e(x_0), \quad (15)$$

$$H^i(x_0) = -\,[1/c]\, j_\mu(x_0)A_\mu(x_0). \quad (16)$$

the first part being the *energy of interaction* of the two fields with each other, and the second part the *energy produced by external forces*. It is usually not permissible to treat H* as a small perturbation as was done in the last section. Instead, H^i alone is treated as a perturbation, the aim being to eliminate H^i but to leave H^e in its original place in the *equation of motion* of the system.

Operators $S(\sigma)$ and $S(\infty)$ are defined by replacing H_1 by H^i in the definitions of $U(o)$ and $U(\infty)$. Thus $S(\sigma)$ satisfies the equation

$$ihc[\partial S/\partial\sigma(x_0)] = H^i(x_0)S. \quad (17)$$

Suppose now a new type of *state vector* $\Omega(\sigma)$ to be introduced by the substitution

$$\Psi(\sigma) = S(\sigma)\Omega(\sigma) \quad (18)$$

By (9), (15), (17), and (18) the *equation of motion* for $\Omega(\sigma)$ is

$$ihc[\partial\Omega/\partial\sigma(x_0)] = \{S(\sigma)\}^{-1}H^e(x_0)S(\sigma)\Omega. \quad (19)$$

The elimination of the radiation interaction is hereby achieved; only the question, "How is the new *state vector* $\Omega(\sigma)$ to be interpreted?", remains.

It is clear from (19) that a system with a constant Ω is a system of electrons, positrons, and photons, moving under the influence of their mutual interactions, but in the absence of external fields. In a system where two or more particles are actually present, their interactions alone will, in general, cause real transitions and scattering processes to occur. For such a system, it is rather "unphysical" to represent a state of motion including the effects of the interactions by a constant *state vector*; *hence, for such a system the new representation has no simple interpretation*. However, the most important systems are those in which only one particle is actually present, and its interaction with the vacuum fields gives rise only to virtual processes. In this case the particle, including the effects of all its interactions with the vacuum, appears to move as a free particle in the absence of external fields, and it is eminently reasonable to represent such a state of motion by a constant *state vector*. Therefore, it may be said that the operator,

$$H_T(x_0) = \{S(\sigma)\}^{-1}H^e(x_0)S(\sigma), \quad (20)$$

on the right of (19) *represents the interaction of a physical particle with an external field, including radiative corrections*. Equation (19) describes the extent to which the motion of a single physical particle deviates, in the external field, from the motion represented by a constant *state-vector*, i.e., from the motion of an observed "free" particle. If the system whose *state vector* is constantly Ω undergoes no real transitions with the passage of time, then the *state vector* Ω is called "steady". More precisely, Ω is steady if, and only if, it satisfies the equation

$$S(\infty)\Omega = \Omega. \quad (21)$$

As a general rule, one-particle *states* are steady and many-particle *states* unsteady. There are, however, two important qualifications to this rule.

First, the *interaction* (20) itself will almost always cause transitions from steady to unsteady *states*. For example, if the initial *state* consists of one electron in the field of a proton, H_T will have matrix elements for transitions of the electron to a new *state* with emission of a photon, and such transitions are important in practice. Therefore, although the interpretation of the theory is simpler for steady *states*, it is not possible to exclude unsteady *states* from consideration.

Second, if a one-particle *state* as hitherto defined is to be steady, the definition of $S(\sigma)$ must be modified. This is because $S(\infty)$ includes the effects of the *electromagnetic self-energy* of the electron, and this *self-energy* gives an expectation value to $S(\infty)$ *which is different from unity (and indeed infinite) in a one-electron state*, so that Eq. (21) cannot be satisfied. *The mistake that has been made occurred in trying to represent the observed electron with its electromagnetic self-energy by a wave field with the same characteristic rest-mass as that of the "bare" electron.* To correct the mistake, let δm denote the *electromagnetic mass* of the electron, i.e., *the difference in rest-mass between an observed and a "bare" electron.* Instead of (5), the division of the *energy-density* $H(\mathbf{r})$ should have taken the form

$$H(r) = \{H_0(r) + \delta mc^2\psi^*(r)\beta\psi(r)\} + \{H_1(r) - \delta mc^2\psi^*(r)\beta\psi(r)\}.$$

The first bracket on the right here represents the *energy-density* of the free *electromagnetic and electron fields* with the observed electron *rest-mass*, and should have been used instead of $H_0(r)$ in the definition (6) of $T(\sigma)$. Consequently, the second bracket should have been used instead of $H_1(r)$ in Eq. (8).

The definition of $S(\sigma)$ has therefore to be altered by replacing $H^i(x_0)$ by[6]

$$H^I(x_0) = H^i(x_0) + H^S(x_0) = H^i(x_0) - \delta mc^2\bar{\psi}(x_0)\psi(x_0). \quad (22)$$

[6] Here Schwinger's notation $\bar{\psi} = \psi^*\beta$ is used.

The value of δm can be adjusted so as to cancel out the self-energy effects in $S(\infty)$ (this is only a formal adjustment since the value is actually infinite), and then Eq. (21) will be valid for one-electron *states*. For the photon *self-energy*, no such adjustment is needed since, as proved by Schwinger, the photon *self-energy* turns out to be identically zero.

The foregoing discussion of the *self-energy* problem is intentionally only a sketch, but it will be found to be sufficient for practical applications of the theory. A fuller discussion of the theoretical assumptions underlying this treatment of the problem will be given by Schwinger in his forthcoming papers. *Moreover, it must be realized that the theory as a whole cannot be put into a finally satisfactory form so long as divergencies occur in it, however skillfully these divergencies are circumvented; therefore, the present treatment should be regarded as justified by its success in applications rather than by its theoretical derivation.*

The important results of the present paper up to this point are Eq. (19)
[the *equation of motion* for $\Omega(\sigma)$ is

$$ihc[\partial\Omega/\partial\sigma(x_0)] = \{S(\sigma)\}^{-1}H^e(x_0)S(\sigma)\Omega. \quad (19)]$$

and the interpretation of the *state vector* Ω. *The state vector Ψ of a system can be interpreted as a wave function giving the probability amplitude of finding any particular set of occupation numbers for the various possible states of free electrons, positrons, and photons. The state vector Ω of a system with a given Ψ on a given surface σ is, crudely* *speaking, the Ψ which the system would have had in the infinite past if it had arrived at the given Ψ on σ under the influence of the interaction $H^i(x_0)$ alone.*

The definition of Ω being unsymmetrical between past and future, a new type of *state vector* Ω' can be defined by reversing the direction of time in the definition of Ω. Thus, the Ω' of a system with a given Ψ on a given σ is the Ψ which the system would reach in the infinite future if it continued to move under the influence of $H^i(x_0)$ alone. More simply, Ω' can be defined by the equation

$$\Omega'(\sigma) = S(\infty)\Omega(\sigma). \qquad (23)$$

Since S(∞) is a unitary operator independent of σ, the *state vectors* Ω and Ω' are really only the same vector in two different representations or coordinate systems. Moreover, for any steady *state* the two are identical by (21).

V. *Fundamental formulas of the Schwinger and Feynman theories*

The Schwinger theory works directly from Eqs. (19) and (20), the aim being to calculate the matrix elements of the "effective external potential energy" H_T between states specified by their state vectors Ω. The states considered in practice always have Ω of some very simple kind, for example, Ω representing systems in which one or two free-particle states have occupation number one and the remaining free-particle states have occupation number zero. By analogy with (13), $S(\sigma_0)$ is given by

$$S(\sigma_0) = 1 + (-i/hc) \int_{-\infty}^{\sigma 0} H^I(x_1)dx_1 + (-i/hc)^2 \times \int_{-\infty}^{\sigma 0} dx_1 \int_{-\infty}^{\sigma(x1)} H^I(x_2)H^I(x_1)dx_2 + \dots , \qquad (24)$$

and $(S(\sigma_0))^{-1}$ by a corresponding expression analogous to (14). Substitution of these series into (20) gives at once

$$H_T(x_0) = \sum_{n=0}^{\infty} (i/hc)^n \int_{-\infty}^{\sigma(x0)} dx_1 \int_{-\infty}^{\sigma(x1)} dx_2 \dots \times \int_{-\infty}^{\sigma(xn-1)} dx_n \times H^I(x_n), [\dots, [H^I(x_2), [H^I(x_1), H^e(x_0)]]\dots]]. \qquad (25)$$

Notes added in proof

Schwinger's "effective potential" is not H_T given by (25), but is $H_T' = QH_TQ^{-1}$. Here Q is a "square root" of S(∞) obtained by expanding $(S(\infty))^{1/2}$ by the binomial theorem. The physical meaning of this is that Schwinger specifies states neither by Ω nor by Ω', but by an *intermediate state-vector* Ω" = QΩ = Q^{-1}Ω', whose definition is symmetrical between past and future. H_T' is also symmetrical between past and future. For one-particle states, H_T and H_T' are identical.

The repeated commutators in this formula are characteristic of the Schwinger theory, and their evaluation gives rise to long and rather difficult analysis. Using the first three terms of the series, Schwinger was able to calculate the second-order radiative corrections to the equations of motion of an electron in an external field, and obtained satisfactory agreement with experimental results. In this paper the development of the Schwinger theory will be carried no further; in principle the radiative corrections to the equations of motion of electrons could be calculated to any desired order of approximation from formula (25).

In the Feynman theory the basic principle is to preserve symmetry between past and future. Therefore, the matrix elements of the operator H_T are evaluated in a "mixed representation"; *the matrix elements are calculated between an initial state specified by its state vector Ω_1 and a final state* specified by its state vector Ω_2'. The matrix element of H_T between two such states in the Schwinger representation is

$$\Omega_2{}^*H_T\Omega_1 = \Omega_2{}'^*S(\infty)H_T\Omega_1, \tag{26}$$

and therefore, the operator which replaces H_T in the mixed representation is

$$H_F(x_0) = S(\infty)H_T(x_0) = S(\infty)(S(\sigma))^{-1}H^e(x_0)S(\sigma). \tag{27}$$

Going back to the original product definition of $S(\sigma)$ analogous to (10), it is clear that $S(\infty)$ x $(S(\sigma_0))^{-1}$ is simply the operator obtained from $S(\sigma)$ by interchanging past and future. Thus,

$$R(\sigma) = S(\infty)(S(\sigma))^{-1} = 1 + (-\,i/hc) \text{ x } \int^{\infty} H^I(x_1)dx_1$$
$$+ (-\,i/hc)^2 \int_{\sigma}^{\infty} dx_1 \text{ x } \int_{\sigma(x1)}{}^{\infty} H^I(x_2)H^I(x_1)dx_2 + \ldots, \tag{28}$$

The physical meaning of a mixed representation of this type is not at all recondite. In fact, a mixed representation is normally used to describe such a process as *bremsstrahlung* of an electron in the field of a nucleus when the Born approximation is not valid; the process of *bremsstrahlung* [electromagnetic radiation produced by the deceleration of a charged particle when deflected by another charged particle] is a radiative transition of the electron from a state described by a Coulomb wave function, with a plane ingoing and a spherical outgoing wave, to a state described by a Coulomb wave function with a spherical ingoing and a plane outgoing wave. The initial and final states here belong to different orthogonal systems of wave functions, and so the transition matrix elements are calculated in a mixed representation. In the Feynman theory the situation is analogous, "only the roles of the radiation interaction and the external (or Coulomb) field are interchanged; the radiation interaction is used instead of the Coulomb field to modify the state vectors (wave functions) of the initial and final states, and the external field instead of the radiation interaction causes transitions between these state vectors.

In the Feynman theory there is an additional simplification. For if matrix elements are being calculated between two states, either of which is steady (and this includes all cases so far considered), the mixed representation reduces to an ordinary representation. This occurs, for example, in treating a one-particle problem such as the radiative correction to the *equations of motion* of an electron in an external field; the operator $H_F(x_0)$, although in general it is not even Hermitian, can in this case be considered as an effective external *potential energy* acting on the particle, in the ordinary sense of the words.

This section will be concluded with the derivation of the fundamental formula (31) of the Feynman theory, which is the analog of formula (25) of the Schwinger theory.

$$[H_T(x_0) = \sum_{n=0}^{\infty} (i/hc)^n \int_{-\infty}^{\sigma(x0)} dx_1 \int_{-\infty}^{\sigma(x1)} dx_2 \ldots$$
$$\text{x} \int_{-\infty}^{\sigma(xn-1)} dx_n \text{ x } H^I(x_n), [\ldots, [H^I(x_2), [H^I(x_1), H^e(x_0)]]\ldots]].(25)$$

Notes added in proof

Schwinger's "effective potential" is not H_T given by (25), but is $H_T' = QH_TQ^{-1}$. Here Q is a "square root" of $S(\infty)$ obtained by expanding $(S(\infty))^{1/2}$ by the binomial theorem. The physical meaning of this is that Schwinger specifies states neither by Ω nor by Ω', but by an *intermediate state-vector* $\Omega'' = Q\Omega = Q^{-1}\Omega'$, whose definition is symmetrical between past and future. H_T' is also symmetrical between past and future. For one-particle states, H_T and H_T' are identical.

If

$$F_1(x_1), \ldots, F_n(x_n)$$

are any operators defined, respectively, at the points $x_1, \ldots, x_n$, of space-time, then

$$P(F_1(x_1), \ldots, F_n(x_n)) \quad (29)$$

will denote the product of these operators, taken in the order, reading from right to left, in which the surfaces $\sigma(x_1), \ldots, \sigma(x_n)$ occur in time. In most applications of this notation $F_i(x_i)$ will commute with $F_j(x_j)$ so long as x_i and x_j are outside each other's light cones; when this is the case, it is easy to see that (29) is a function of the points $x_1, \ldots, x_n$ only and is independent of the surfaces $\sigma(x_i)$. Consider now the integral

$$I_n = \int_{-\infty}^{\infty} dx_1 \ldots \int_{-\infty}^{\infty} dx_n \, P\{H^e(x_0), H^I(x_1), \ldots, H^I(x_n)\}.$$

Since the integrand is a symmetrical function of the points $x_1, \ldots, x_n$, the value of the integral is just n! times the integral obtained by restricting the integration to sets of points $x_1, \ldots, x_n$ for which $\sigma(x_i)$ occurs after $\sigma(x_{i+1})$ for each i. The restricted integral can then be further divided into (n+1) parts, the j'th part being the integral over those sets of points with the property that $\sigma(x_0)$ lies between $\sigma(x_{j-1})$ and $\sigma(x_j)$ (with obvious modifications for $j = 1$ and $j = n+1$). Therefore,

$$I_n = n! \sum_{j=1}^{n+1} \int_{-\infty}^{\sigma(x0)} dx_1 \ldots \int_{-\infty}^{\sigma(xn-1)} dx_n$$
$$\text{x} \int_{\sigma(x0)}^{\infty} dx_{j-1} \ldots \int_{\sigma(x2)}^{\infty} dx_1 \text{ x } H^I(x_1) \ldots H^e(x_0), H^I(x_j) \ldots H^I(x_n). \quad (30).$$

Now if the series (24) and (28) are substituted into (27), sums of integrals appear which are precisely of the form (30). Hence finally

$$H_F(x_0) = \sum_{n=0}^{\infty} (-i/hc)^n \, [1/n!] I_n$$
$$= \sum_{n=0}^{\infty} (-i/hc)^n \, [1/n!] \int_{-\infty}^{\infty} dx_1 \ldots \int_{-\infty}^{\infty} dx_n$$
$$\text{x } P\{H^e(x_0), H^I(x_j) \ldots, H^I(x_n)\}. \quad (31)$$

By this formula the notation $H_F(x_0)$ is justified, for this operator now appears as a function of the point x_0 alone and not of the surface σ. *The further development of the Feynman theory is mainly concerned with the calculation of matrix elements of (31) between various initial and final states.*

As a special case of (31) obtained by replacing H^e by the unit matrix in (27),

$$S(\infty) = \sum_{n=0}^{\infty} (-i/hc)^n \, [1/n!] \int_{-\infty}^{\infty} dx_1 \ldots \int_{-\infty}^{\infty} dx_n$$
$$\text{x } P\{H^I(x_j) \ldots, H^I(x_n)\}. \quad (32)$$

Notes added in proof

Equation (32) can most simply be obtained directly from the product expansion of $S(\infty)$.

VI. *Calculation of matrix elements*

In this section the application of the foregoing theory to a general class of problems will be explained. The ultimate aim is to obtain a set of rules by which the matrix element of the operator (31) between two given states may be written down in a form suitable for numerical evaluation, immediately and automatically. *The fact that such a set of rules exists is the basis of the Feynman radiation theory; the derivation in this section of the same rules from what is fundamentally the Tomonaga-Schwinger theory constitutes the proof of equivalence of the two theories.*

To avoid excessive complication, the type of matrix element considered will be restricted in two ways. First, it will be assumed that the external *potential energy* is

$$H^e(x_0) = -[1/c]\, j_\mu(x_0)A_\mu{}^e(x_0), \tag{33}$$

that is to say, the *interaction energy* of the electron-positron *field* with *electromagnetic potentials* $A_\mu{}^e(x_0)$ which are given numerical functions of space and time. Second, matrix elements will be considered only for transitions from a state A, in which just one electron and no positron or photon is present, to another state B of the same character. These restrictions are not essential to the theory, and are introduced only for convenience, in order to illustrate clearly the principles involved. The electron-positron *field operator* may be written

$$\psi_\alpha(x) = \sum_u \phi_{u\alpha}(x)a_u, \tag{34}$$

where the $\phi_{u\alpha}(x)$ are spinor wave functions of free electrons and positrons, and the a_u are annihilation operators of electrons and creation operators of positrons. Similarly, the *adjoint* operator

$$\psi^-{}_\alpha(x) = \sum_u \phi^-{}_{u\alpha}(x)a^-{}_u, \tag{35}$$

where $\alpha^-{}_u$ are annihilation operators of positrons and creation operators of electrons. The *electromagnetic field* operator is

$$A_\mu(x) = \sum_\upsilon (A_{\upsilon\mu}(x)b_\upsilon + A_{\upsilon\mu}{}^*(x)\, b^-{}_\upsilon, \tag{36}$$

where b_υ and $b^-{}_\upsilon$ are photon annihilation and creation operators, respectively. The charge-current 4-vector of the electron field is

$$j_\mu(x) = iec\psi^-(x)\psi(x); \tag{37}$$

strictly speaking, this expression ought to be anti-symmetrized to the form[7]

$$j_\mu(x) = iec\{\psi^-{}_\alpha(x)\psi_\beta(x) - \psi_\beta(x)\psi^-{}_\alpha(x)\}(\gamma_\mu)_{\alpha\beta}, \tag{38}$$

but it will be seen later that this is not necessary in the present theory.

[7] See Pauli, W. (1941). *Rev. Mod. Phys.*, 13, 203, Eq. (96), p. 224.

Consider the product P occurring in the n'th integral of (31); let it be denoted by P_n. From (16), (22), (33), and (37) it is seen that P_n is a sum of products of $(n + 1)$ operators ψ_α, $(n+1)$ operators ψ^-_α, and not more than n operators A_μ, multiplied by various numerical factors. By Q_n may be denoted a typical product of factors ψ_α, ψ^-_α, and A_μ, not summed over the indices such as α and μ, so that P_n is a sum of terms such as Q_n. Then Q_n will be of the form (indices omitted)

$$Q_n = \psi^-(x_{i0})\psi(x_{i0})\psi^-(x_{i1})\psi(x_{i1}) \ldots \psi^-(x_{in})\psi(x_{in})$$
$$\times A(x_{j1}) \ldots A(x_{jm}), \tag{39}$$

where $i_0, i_1, \ldots, i_n$ is some permutation of the integers $0, 1, \ldots, n$, and $j_1, \ldots, j_m$ are some, but not necessarily all, of the integers $1, \ldots, n$, in some order. Since none of the operators ψ^- and ψ commute with each other, it is especially important to preserve the order of these factors. Each factor of Q_n is a sum of creation and annihilation operators by virtue of (34), (35), and (36), and so Q_n itself is a sum of products of creation and annihilation operators.

Now consider under what conditions a product of creation and annihilation operators can give a non-zero matrix element for the transition A→B. Clearly, one of the annihilation operators must annihilate the electron in state A, one of the creation operators must create the electron in state B, and the remaining operators must be divisible into pairs, the members of each pair respectively creating and annihilating the same particle. Creation and annihilation operators referring to different particles always commute or anti-commute (the former if at least one is a photon operator, the latter if both are electron-positron operators). Therefore, if the two single operators and the various pairs of operators in the product all refer to different particles, the order of factors in the product can be altered so as to bring together the two single operators and the two members of each pair, without changing the value of the product except for a change of sign if the permutation made in the order of the electron and positron operators is odd. In the case when some of the single operators and pairs of operators refer to the same particle, it is not hard to verify that the same change in order of factors can be made, provided it is remembered that the division of the operators into pairs is no longer unique, and the change of order is to be made for each possible division into pairs and the results added together.

It follows from the above considerations that the matrix element of Q_n for the transition A→B is a sum of contributions, each contribution arising from a specific way of dividing the factors of Q_n into two single factors and pairs. A typical contribution of this kind will be denoted by M. The two factors of a pair must involve a creation and an annihilation operator for the same particle, and so must be either one ψ^- and one ψ or two A; the two single factors must be one ψ^- and one ψ. The term M is thus specified by fixing an integer k, and a permutation $r_0, r_1, \ldots, r_n$ of the integers $0, 1, \ldots, n$, and a division $(s_1,t_1), (s_2,t_2), \ldots, (s_h,t_h,)$ of the integers $j_1, \ldots, j_m$ into pairs; clearly $m = 2h$ has to be an even number; the term M is obtained by choosing for single factors $\psi^-(x_k)$ and $\psi(x_{rk})$, and for associated pairs of factors $(\psi^-(x_i),\psi(x_{ri}))$ for $i = 0, 1, \ldots, k-1, k+1, \ldots, n$ and $(A(x_{si}),A(x_{ti}))$ for $i = 1, \ldots, h$.

In evaluating the term M, the order of factors in Q_n is first to be permuted so as to bring together the two single factors and the two members of each pair, but without altering the order of factors within each pair; the result of this process is easily seen to be

$$Q_n' = \varepsilon P\{\psi^{-}(x_0),\psi(x_{r0})\} \ldots P\{\psi^{-}(x_n),\psi(x_{rn})$$
$$\times P\{A(x_{s1}), A(x_{t1})\} \ldots P\{A(x_{sh}), A(x_{th})\}, \qquad (40)$$

a factor ε being inserted which takes the value ± 1 according to whether the permutation of ψ^{-} and ψ factors between (39) and (40) is even or odd. Then in (40) each product of two associated factors (but not the two single factors) is to be independently replaced by the sum of its matrix elements for processes involving the successive creation and annihilation of the same particle.

Given a bilinear operator such as $A_\mu(x)A_\nu(y)$, the sum of its matrix elements for processes involving the successive creation and annihilation of the same particle is just what is usually called the "*vacuum expectation value*" of the operator, and has been calculated by Schwinger. This quantity is, in fact (note that Heaviside units are being used)

$$[A_\mu(x)A_\nu(y)]_0 = \tfrac{1}{2}\, hc\delta_{\mu\nu}\{D^{(1)} + iD\}(x-y),$$

where $D^{(1)}$ and D are Schwinger's invariant D functions. The definitions of these functions will not be given here, because it turns out that the vacuum expectation value of $P(A_\mu(x)A_\nu(y))$ takes an even simpler form. Namely,

$$[P[(A_\mu(x),A_\nu(y)]]_0 = \tfrac{1}{2}\, hc\delta_{\mu\nu}D_F(x-y), \qquad (41)$$

where D_F is the type of D function introduced by Feynman. $D_F(x)$ is an even function of x, with the integral expansion

$$D_F(x) = -[i/2\pi^2] \int_0^\infty \exp[i\alpha x^2]\, d\alpha, \qquad (42)$$

where x^2 denotes the square of the invariant length of the 4-vector x. In a similar way it follows from Schwinger's results that

$$[P[\psi^{-}_\alpha(x),\psi_\beta(x)]]_0 = \tfrac{1}{2}\, \eta(x,y)\, S_{F\beta\alpha}(x-y), \qquad (43)$$

where

$$S_{F\beta\alpha}(x) = -\{\gamma_\mu(\partial/\partial x_\mu) + \kappa_0\}_{\beta\alpha}\Delta_F(x), \qquad (44)$$

κ_0 is the reciprocal Compton wave-length of the electron, $\eta(x,y)$ is -1 or $+1$ according as $\sigma(x)$ is earlier or later than $\sigma(y)$ in time, and Δ_F is a function with the integral expansion

$$\Delta_F = -[i/2\pi^2] \int_0^\infty \exp[i\alpha x^2 - i\,\kappa_0^2/4\alpha]\, d\alpha. \qquad (45)$$

Substituting from (41) and (44) into (40), the matrix element M takes the form (still omitting the indices of the factors ψ^{-}, ψ, and A of Q_n)

$$M = \ldots \qquad (46)$$

The single factors $\psi^-(x_k)$ and $\psi(x_{rk})$ are conveniently left in the form of operators, since the matrix elements of these operators for effecting the transition A→B depend on the wave functions of the electron in the states A and B. Moreover, the order of the factors $\psi^-(x_k)$ and $\psi(x_{rk})$ is immaterial since they anti-commute with each other; hence it is permissible to write

$$P[\psi^-(x_k),\psi(x_{rk})] = \eta(x_k,x_{rk})\ \psi^-(x_k),\psi(x_{rk}).$$

Therefore (46) may be rewritten

$$M= \ldots \qquad (47)$$

with $\varepsilon' = \ldots$. (48)

Now the product in (48) is $(-1)^p$, where p is the number of occasions in the expression (40) on which the ψ of a P bracket occurs to the left of the ψ^-. Referring back to the definition of ε after Eq. (40), it follows that ε' takes the value + 1 or – 1 according to whether the permutation of ψ^- and ψ factors between (39) and the expression

$$\psi^-(x_0)\psi(x_{r0})\ldots\ \psi^-(x_n)\psi(x_{rn}) \qquad (49)$$

is even or odd. But (39) can be derived by an even permutation from the expression

$$\psi^-(x_0)\psi(x_0)\ldots\ \psi^-(x_n)\psi(x_n), \qquad (50)$$

and the permutation of factors between (49) and (50) is even or odd according to whether the permutation $r_0, r_1, \ldots, r_n$ of the integers 0, 1, …, n is even or odd. Hence, finally, ε' in (47) is + 1 or – 1 according to whether the permutation $r_0, r_1, \ldots, r_n$ is even or odd. It is important that ε' depends only on the type of matrix element M considered, and not on the points $x_0, x_1, \ldots, x_n$; therefore, it can be taken outside the integrals in (31).

One result of the foregoing analysis is to justify the use of (37), instead of the more correct (38), for the charge-current operator occurring in H^e and H^i. For it has been shown that in each matrix element such as M the factors ψ^- and ψ in (38) can be freely permuted, so that (38) can be replaced by (37), except in the case when the two factors form an associated pair. In the exceptional case, M contains as a factor the *vacuum expectation* value of the operator $j_\mu(x_i)$ at some point x_i; this expectation value is zero according to the correct formula (38), *though it would be infinite according to (37)*; thus, the matrix elements in the exceptional case are always zero. The conclusion is that *only those matrix elements are to be calculated for which the integer r_i differs from i for every $i \neq k$, and in these elements the use of formula (37) is correct.*

To write down the matrix elements of (31) for the transition A→B, it is only necessary to take all the products Q_n, replace each by the sum of the corresponding matrix elements M given by (47), reassemble the terms into the form of the P_n from which they were derived, and finally substitute back into the series (31). The problem of calculating the matrix elements of (31) is thus in principle solved. However, *in the following section it will be shown how this solution-in-principle can be reduced to a much simpler and more practical procedure.*

VII. *Graphical representation of matrix elements*

Let an integer n and a product P_n occurring in (31) be temporarily fixed. The points x_0, x_1, …, x_n may be represented by (n + 1) points drawn on a piece of paper. A type of matrix element M as described in the last section will then be represented graphically as follows. For each associated pair of factors ($\psi^-(x_i)$, $\psi(x_{ri})$) with $i \neq k$, draw a line with a direction marked in it from the point x_i to the point x_{ri}. For the single factors $\psi^-(x_i)$, $\psi(x_{ri})$, draw directed lines leading out from x_k to the edge of the diagram, and in from the edge of the diagram to x_{rk}. For each pair of factors ($A(x_{si})$,$A(x_{ti})$), draw an undirected line joining the points x_{si} and x_{ti}. The complete set of points and lines will be called the "*graph*" of M; clearly there is a one-to-one correspondence between types of matrix element and graphs, and the exclusion of matrix elements with r = i for $i \neq k$ corresponds to the exclusion of graphs with lines joining a point to itself. The directed lines in a graph will be called "*electron lines*", the undirected lines "*photon lines*".

Through each point of a graph pass two electron lines, and therefore the electron lines together form one open polygon containing the vertices x_k and x_{rk}, and possibly a number of closed polygons as well. The closed polygons will be called "*closed loops*", and their number denoted by *l*. Now the permutation r_0, r_1, …, r_n of the integers 0, 1, …, n is clearly composed of (*l* + 1) separate cyclic permutations. A cyclic permutation is even or odd according to whether the number of elements in it is odd or even. Hence the parity of the permutation r_0, r_1, …, r_n is the parity of the number of even-number cycles contained in it. But the parity of the number of odd-number cycles in it is obviously the same as the parity of the total number (n + 1) of elements. The total number of cycles being (*l* + 1), the parity of the number of even-number cycles is (*l* – 1). Since it was seen earlier that the ε' of Eq. (47) is determined just by the parity of the permutation r_0, r_1, …, r_n, the above argument yields the simple formula

$$\varepsilon' = (-1)^{l-n}. \qquad (51)$$

This formula is one result of the present theory which can be much more easily obtained by intuitive considerations of the sort used by Feynman. *In Feynman's theory the graph corresponding to a particular matrix element is regarded, not merely as an aid to calculation, but as a picture of the physical process which gives rise to that matrix element.* For example, an electron line joining x_1 to x_2 represents the possible creation of an electron at x_1 and its annihilation at x_2, together with the possible creation of a positron at x_2 and its annihilation at x_1. This interpretation of a graph is obviously consistent with the methods, and in Feynman's hands has been used as the basis for the derivation of most of the results, of the present paper. *For reasons of space, these ideas of Feynman will not be discussed in further detail here.*

To the product P_n correspond a finite number of graphs, one of which may be denoted by G; all possible G can be enumerated without difficulty for moderate values of n. To each G corresponds a contribution C(G) to the matrix element of (31) which is being evaluated.

It may happen that the graph G is disconnected, so that it can be divided into subgraphs, each of which is connected, with no line joining a point of one subgraph to a point of another. In such a case it is clear from (47) that C(G) is the product of factors derived from each subgraph separately. The subgraph G_1 containing the point x_0 is called the "*essential part*" of G, the remainder G_2 the "*inessential part*". There are now two cases to be considered, according to whether the points x_k and x_{rk} lie in G_2 or in G_1 (they must clearly both lie in the same subgraph). In the first case, the factor $C(G_2)$ of C(G) can be seen by a comparison of (31) and (32) to be a contribution to the matrix element of the operator $S(\infty)$ for the transition A→B. Now letting G vary over all possible graphs with the same G_1 and different G_2, the sum of the contributions of all such G is a constant $C(G_1)$ multiplied by the total matrix element of $S(\infty)$ for the transition A→B. But for one-particle states the operator $S(\infty)$ is by (21) equivalent to the identity operator and gives, accordingly, a zero matrix element for the transition A→B. Consequently, the disconnected G for which x_k and x_{rk} lie in G_2 give zero contribution to the matrix element of (31), and can be omitted from further consideration. When x_k and x_{rk} lie in G_1, again the C(G) may be summed over all G consisting of the given G_1 and all possible G_2, but this time the connected graph G_1 itself is to be included in the sum. The sum of all the C(G) in this case turns out to be just $C(G_1)$ multiplied by the expectation value in the vacuum of the operator $S(\infty)$. But the vacuum state, being a steady state, satisfies (21), and so the expectation value in question is equal to unity. Therefore, the sum of the C(G) reduces to the single term $C(G_1)$, and again the disconnected graphs may be omitted from consideration.

The elimination of disconnected graphs is, from a physical point of view, somewhat trivial, since these graphs arise merely from the fact that meaningful physical processes proceed simultaneously with totally irrelevant fluctuations of fields in the vacuum. However, similar arguments will now be used to eliminate a much more important class of graphs, namely, those involving self-energy effects. A "*self-energy part*" of a graph G is defined as follows; it is a set of one or more vertices not including x_0, together with the lines joining them, which is connected with the remainder of G (or with the edge of the diagram) only by two electron lines or by one or two photon lines. For definiteness it may be supposed that G has a self-energy part F, which is connected with its surroundings only by one electron line entering F at x_1, and another leaving F at x_2, the case of photon lines can be treated in an entirely analogous way. The points x_1 and x_2 may or may not be identical. From G a "*reduced graph*" G_0 can be obtained by omitting F completely and joining the incoming line at x_1, with the outgoing line at x_2 to form a single electron line in G_0, the newly formed line being denoted by λ. Given G_0 and λ, there is conversely a well determined set Γ of graphs G which are associated with G_0 and X in this way; G_0 itself is considered also to belong to Γ. It will now be shown that the sum $C(\Gamma)$ of the contributions C(G) to the matrix element of (31) from all the graphs G of Γ reduces to a single term $C'(G_0)$.

Suppose, for example, that the line λ in G_0 leads from a point x_3 to the edge of the diagram. Then C(Go) is an integral containing in the integrand the matrix element of

$$\psi^{-}_{\alpha}(x_3) \tag{52}$$

for creation of an electron into the state B. Let the *momentum-energy* 4-vector of the created electron be **p**; the matrix element of (52) is of the form

$$Y_{\alpha}(x_3) = a_{\alpha} \exp[-i(p.x_3)/h] \tag{53}$$

with a_{α} independent of x_3. Now consider the sum C(Γ). It follows from an analysis of (31) that C(Γ) is obtained from $C(G_0)$ by replacing the operator (52) by

$$\sum_{n=0}^{\infty} (-i/hc)^n [1/n!] \int_{-\infty}^{\infty} dy_1 \ldots \int_{-\infty}^{\infty} dy_n \\ \text{x } P\{\psi^{-}_{\alpha}(x_3), H^I(y_1), \ldots, H^I(y_n)\} \tag{54}$$

(This is, of course, a consequence of the special character of the graphs of Γ.) It is required to calculate the matrix element of (54) for a transition from the vacuum state 0 to the state B, i.e., for the emission of an electron into state B. This matrix element will be denoted by Z_{α}; C(Γ) involves Z_{α} in the same way that $C(G_0)$ involves (53). Now Z_{α} can be evaluated as a sum of terms of the same general character as (47); it will be of the form

$$Z_{\alpha} = \sum_i \int_{-\infty}^{\infty} K_i^{\alpha\beta} (y_i - x_3)\, Y_{\beta}(y_i)\, dy_i,$$

where the important fact is that K_i is a function only of the coordinate differences between y_i and x_3. By (53), this implies that

$$Z_{\alpha} = R_{\alpha\beta}(p)\, Y_{\beta}(x_3), \tag{55}$$

with R independent of x_3. *From considerations of relativistic invariance*, R must be of the form

$$\delta_{\beta\alpha} R_1(p^2) + (p_{\mu}\gamma_{\mu})_{\beta\alpha} R_2(p^2),$$

where p^2 is the square of the invariant length of the 4-vector **p**. But since the matrix element (53) is a solution of the Dirac equation,

$$p^2 = -h^2\kappa_0^2, \quad (p_{\mu}\gamma_{\mu})_{\beta\alpha} Y_{\beta} = ih\kappa_0 Y_{\alpha},$$

and so (55) reduces to

$$Z_{\alpha} = R_1 Y_{\alpha}(x_3),$$

with R_1 an absolute constant. Therefore, the sum C(Γ) is in this case just $C'(G_0)$, where $C'(G_0)$ is obtained from $C(G_0)$ by the replacement

$$\psi^{-}(x_3) \rightarrow R_1 \psi^{-}(x_3). \tag{56}$$

In the case when the line λ leads into the graph G_0 from the edge of the diagram to the point x_3, it is clear that C(Γ) will be similarly obtained from $C(G_0)$ by the replacement

$$\psi(x_3) \rightarrow R_1^* \psi(x_3). \tag{57}$$

There remains the case in which λ leads from one vertex x_3 to another x_4 of G_0. In this case $C(G_0)$ contains in its integrand the function

$$\tfrac{1}{2}\,\eta(x_3,x_4)\,S_{F\beta\alpha}(x_3 - x_4), \tag{58}$$

which is the vacuum expectation value of the operator

$$P\{\psi^{-}(x_3),\, \psi(x_4)\} \tag{59}$$

according to (43). Now in analogy with (54), C(Γ) is obtained from $C(G_0)$ by replacing (59) by

$$\sum_{n=0}^{\infty} (-\,i/hc)^n\,[1/n!] \int_{-\infty}^{\infty} dy_1 \ldots \int_{-\infty}^{\infty} dy_n$$
$$\times\, P\{\psi^{-}_{\alpha}(x_3),\, \psi(x_4),\, H^I(y_1),\, \ldots,\, H^I(y_n)\} \tag{60}$$

and the vacuum expectation value of this operator will be denoted by

$$\tfrac{1}{2}\,\eta(x_3,x_4)\,S'_{F\beta\alpha}(x_3 - x_4). \tag{61}$$

By the methods of Section VI, (61) can be expanded as a series of terms of the same character as (47); this expansion will not be discussed in detail here, but it is easy to see that it leads to an expression of the form (61), with $S_F'(x)$ a certain universal function of the 4- vector x. It will not be possible to reduce (61) to a numerical multiple of (58), as Z_α was in the previous case reduced to a multiple of Y_α. Instead, there may be expected to be a series expansion of the form

$$S_{F\beta\alpha}(x) = \{R_2 + a_1(\square^2 - \kappa_0^2) + a_2(\square^2 - \kappa_0^2)^2 + \ldots, \tag{62}$$

where $\square^2$ is the Dalembertian operator and the *a*, *b* are numerical coefficients.

Notes added in proof

Equation (62) is incorrect. The function S_F' is well-behaved, but its Fourier transform has a logarithmic dependence on frequency, which makes an expansion precisely of the form (62) impossible.

In this case C(Γ) will be equal to the $C'(G_0)$ obtained from $C(G_0)$ by the replacement

$$S_F(x_3 - x_4) \rightarrow S_F'(x_3 - x_4). \tag{63}$$

Applying the same methods to a graph G with a *self-energy* part connected to its surroundings by two photon lines, the sum C(Γ) will be obtained as a single contribution $C'(G_0)$ from the reduced graph G_0, $C'(G_0)$ being formed from $C(G_0)$ by the replacement

$$D_F(x_3 - x_4) \rightarrow D_F'(x_3 - x_4). \tag{64}$$

The function D_F' is defined by the condition that

$$\tfrac{1}{2}\,hc\,\delta_{\mu\nu}\,D_F'(x_3 - x_4) \tag{65}$$

is the vacuum expectation value of the operator

$$\sum_{n=0}^{\infty} (-\,i/hc)^n\,[1/n!] \int_{-\infty}^{\infty} dy_1 \ldots \int_{-\infty}^{\infty} dy_n$$
$$\times\, P\{A_\mu(x_3),\, A_\nu(x_4),\, H^I(y_1),\, \ldots,\, H^I(y_n)\}, \tag{66}$$

and may be expanded in a series

$$D_F'(x) = \{R_3 + c_1\Box^2 + c_2(\Box^2)^2 + \dots\}D_F(x). \quad (67)$$

Finally, it is not difficult to see that for graphs G with self-energy parts connected to their surroundings by a single photon line, the sum $C(\Gamma)$ will be identically zero, and so such graphs may be omitted from consideration entirely.

As a result of the foregoing arguments, the contributions C(G) of graphs with *self-energy* parts can always be replaced by modified contributions $C'(G_0)$ from a reduced graph G_0. A given G may be reducible in more than one way to give various G_0, but if the process of reduction is repeated a finite number of times a G_0 will be obtained which is "totally reduced", contains no *self-energy* part, and is uniquely determined by G. The contribution $C'(G_0)$ of a totally reduced graph to the matrix element of (31) is now to be calculated as a sum of integrals of expressions like (47), but with a replacement (56), (57), (63), or (64) made corresponding to every line in G_0. This having been done, the matrix element of (31) is correctly calculated by taking into consideration each totally reduced graph once and once only.

The elimination of graphs with *self-energy* parts is a most important simplification of the theory. For according to (22), H^I contains the subtracted part H^S, which will give rise to many additional terms in the expansion of (31). But if any such term is taken, say, containing the factor $H^S(x_i)$ in the integrand, every graph corresponding to that term will contain the point x_i joined to the rest of the graph only by two electron lines, and this point by itself constitutes a *self-energy* part of the graph. Therefore, all terms involving H^S are to be omitted from (31) in the calculation of matrix elements. *The intuitive argument for omitting these terms is that they were only introduced in order to cancel out higher order self-energy terms arising from H^i, which are also to be omitted; the analysis of the foregoing paragraphs is a more precise form of this argument.* In physical language, the argument can be stated still more simply; since δm is an unobservable quantity, it cannot appear in the final description of observable phenomena.

VIII. *Vacuum polarization and charge renormalization*

The question now arises: What is the physical meaning of the new functions D_F' and S_F', and of the constant R_1? In general terms, the answer is clear. The physical processes represented by the *self-energy* parts of graphs have been pushed out of the calculations, but these processes do not consist entirely of unobservable interactions of single particles with their self-fields, and so cannot entirely be written off as "*self-energy processes*". In addition, *these processes include the phenomenon of vacuum polarization, i.e., the modification of the field surrounding a charged particle by the charges which the particle induces in the vacuum*. Therefore, the appearance of D_F', S_F', and R_1 in the calculations may be regarded as an explicit representation of the *vacuum polarization* phenomena which were implicitly contained in the processes now ignored.

In the present theory there are two kinds of vacuum polarization, one induced by the external field and the other by the quantized electron and photon fields themselves; these will be called "external" and "internal", respectively. It is only the internal polarization which is represented yet in explicit fashion by the substitutions (56), (57), (63), (64); the external will be included later.

To form a concrete picture of the function D_F', it may be observed that *the function $D_F(y - z)$ represents in classical electrodynamics the retarded potential of the charge at y acting on the charge at z, together with the retarded potential of the charge at z acting on the charge at y*. Therefore, D_F may be spoken of loosely as "*the electromagnetic interaction between two point-charges*". In this semiclassical picture, *D_F' is then the electromagnetic interaction between two point-charges, including the effects of the charge-distribution which each charge induces in the vacuum.*

The complete phenomenon of vacuum polarization, as hitherto understood, is included in the above picture of the function D_F'. There is nothing left for S_F' to represent. Thus, *one of the important conclusions of the present theory is that there is a second phenomenon occurring in nature, included in the term vacuum polarization* as used in this paper, but additional to vacuum polarization in the usual sense of the word. The nature of the second phenomenon can best be explained by an example.

The scattering of one electron by another may be represented as caused by a potential energy (the Moller interaction) acting between them. If one electron is at y and the other at s, then, as explained above, the effect of vacuum polarization of the usual kind is to replace a factor D_F in this potential energy by D_F'. Now consider an analogous, but unorthodox, representation of the Compton effect, or the scattering of an electron by a photon. If the electron is at y and the photon at s, the scattering may be again represented by a potential energy, containing now the operator $S_F(y - z)$ as a factor; *the potential is an exchange potential, because after the interaction the electron must be considered to be at z and the photon at y*, but this does not detract from its usefulness. By analogy with the 4-vector charge-current density j_μ which interacts with the potential D_F, a spinor Compton-effect density u_α may be defined by the equation

$$u_\alpha(x) = A_\mu(x)(\gamma_\mu)_{\alpha\beta}\psi_\beta(x),$$

and an adjoint spinor by

$$u^-_\alpha(x) = \psi^-_\beta(x)(\gamma_\mu)_{\beta\alpha}A_\mu(x),$$

These spinors are not directly observable quantities, but the Compton effect can be adequately described as an *exchange potential*, of magnitude proportional to $S_F(y - z)$, acting between the Compton-effect density at any point y and the adjoint density at z. The second vacuum polarization phenomenon is described by a change in the form of this potential from S_F to S_F'. Therefore, the phenomenon may be pictured in physical terms as the inducing, by a given element of Compton-effect density at a given point, of additional Compton-effect density in the vacuum around it.

In both sorts of internal vacuum polarization, the functions D_F and S_F, in addition to being altered in shape, become multiplied by numerical (and actually divergent) factors R_3 and R_2, also the matrix elements of (31) become multiplied by numerical factors such as $R_1R_1^*$.

However, it is believed (this has been verified only for second-order terms) that all n'th-order matrix elements of (31) will involve these factors only in the form of a multiplier

$$(eR_2R_3^{1/2})^n;$$

this statement includes the contributions from the higher terms of the series (62) and (67). Here e is defined as the constant occurring in the fundamental interaction (16) by virtue of (37), Now the only possible experimental determination of e is by means of measurements of the effects described by various matrix elements of (31), and so the directly measured quantity is not e but $eR_2R_3^{1/2}$. Therefore, in practice the letter e is used to denote this measured quantity, and the multipliers R no longer appear explicitly in the matrix elements of (31); the change in the meaning of the letter e is called "*charge renormalization*", and is essential if e is to be identified with the observed electronic charge. As a result of the renormalization, the divergent coefficients R_1, R_2, and R_3 in (56), (57), (62), and (67) are to be replaced by unity, and the higher coefficients a, b, and c by expressions involving only the renormalized charge e.

The external vacuum polarization induced by the potential A_μ^e is, physically speaking, only a special case of the first sort of internal polarization; it can be treated in a precisely similar manner. Graphs describing external polarization effects are those with an "*external polarization part*", namely, a part including the point x_0 and connected with the rest of the graph by only a single photon line. Such a graph is to be "*reduced*" by omitting the polarization part entirely and renaming with the label x_0 the point at the further end of the single photon line. A discussion similar to those of Section VII leads to the conclusion that only reduced graphs need be considered in the calculation of the matrix element of (31), and that the effect of external polarization is explicitly represented if in the contributions from these graphs a replacement

$$A_\mu^e(x) \rightarrow A_\mu^{e\prime}(x) \tag{68}$$

is made. After a renormalization of the unit of potential, similar to the renormalization of charge, the modified potential $A_\mu^{e\prime}$ takes the form

$$A_\mu^{e\prime}(x) = (1 + c_1\square^2 + c_2(\square^2)^2 + \ldots)\, A_\mu^e(x), \tag{69}$$

where the coefficients are the same as in (67).

It is necessary, in order to determine the functions D_F', S_F', and $A_\mu^{e\prime}$, to go back to formulas (60) and (66). The determination of the vacuum expectation values of the operators (60) and (66) is a problem of the same kind as the original problem of the calculation of matrix elements of (31), and the various terms in the operators (60) and (66) must again be split up, represented by graphs, and analyzed in detail. However, since D_F' and S_F' are universal functions, this further analysis has only to be carried out once to be applicable to all problems.

It is one of the major triumphs of the Schwinger theory that it enables an unambiguous interpretation to be given to the phenomenon of vacuum polarization (at least of the first kind), and to the vacuum expectation value of an operator such as (66). *In making this interpretation, profound theoretical problems arise, particularly concerned with the gauge invariance of the theory, about which nothing will be said here*. For Schwinger's solution of these problems, the reader must refer to his forthcoming papers. Schwinger's argument can be transferred without essential change into the framework of the present paper.

Having overcome the difficulties of principle, Schwinger proceeded to evaluate the function D_F' explicitly as far as terms of order $\alpha = (e^2/4\pi hc)$ (Heaviside units). In particular, he found for the coefficient c_1 in (67) and (69) the value $(-\alpha/15\pi\kappa_0^2)$ to this order[8].

[8] Schwinger's results agree with those of the earlier, theoretically unsatisfactory treatment of vacuum polarization. The best account of the earlier work is Weisskopf, V. F. (1936). Uber die Elektrodynamik des Vakuums auf Grund der Quanten-theorie des Elektrons. (About the electrodynamics of the vacuum based on the quantum theory of the electron.) *Kgl. Danske Sels. Math.-Fys. Medd.*, 14, 6.

It is hoped to publish in a sequel to the present paper a similar evaluation of the function S_F', the analysis involved is too complicated to be summarized here.

IX. *Summary of results*

In this section the results of the preceding pages will be summarized, so far as they relate to the performance of practical calculations. In effect, this summary will consist of a set of rules for the application of the Feynman radiation theory to a certain class of problems.

Suppose an electron to be moving in an external field with interaction energy given by (33). Then the interaction energy to be used in calculating the motion of the electron, including radiative corrections of all orders, is

$$\begin{aligned} H_E(x_0) &= \sum_{n=0}^{\infty} (-i/hc)^n [1/n!]\, J_n \\ &= \sum_{n=0}^{\infty} (-i/hc)^n [1/n!] \int_{-\infty}^{\infty} dx_1 \ldots \int_{-\infty}^{\infty} dx_n \\ &\quad \times P\{H^e(x_0), H^i(x_1), \ldots, H^i(x_n)\}, \end{aligned} \tag{70}$$

with H^i given by (16), and the P notation as defined in (29).

To find the effective n'th-order radiative correction to the potential acting on the electron, it is necessary to calculate the matrix elements of J_n for transitions from one one-electron state to another. These matrix elements can be written down most conveniently in the form of an operator K_n bilinear in ψ^- and ψ, whose matrix elements for one-electron transitions are the same as those to be determined. In fact, the operator K_n itself is already the matrix element to be determined if the ψ^- and ψ contained in it are regarded as one-electron wave functions.

To write down K_n, the integrand P_n in J_n is first expressed in terms of its factors ψ^-, ψ, and A, all suffixes being indicated explicitly, and the expression (37) used for j_μ. All possible graphs G with (n + 1) vertices are now drawn as described in Section VII, omitting disconnected graphs, graphs with self-energy parts, and graphs with external vacuum polarization parts as defined in Section VIII. It will be found that in each graph there are at each vertex two electron lines and one photon line, with the exception of x_0 at which there are two electron lines only; further, such graphs can exist only for even n. K_n is the sum of a contribution K(G) from each G.

Given G, K(G) is obtained from J_n by the following transformations. First, for each photon line joining x and y in G, replace two factors $A_\mu(x)A_\nu(y)$ in P_n (regardless of their positions) by

$$\tfrac{1}{2}\, hc\delta_{\mu\nu}D_F'(x-y), \qquad (71)$$

with D_F' given by (67) with $R_3 = 1$, the function D_F being defined by (42). Second, for each electron line joining x to y in G, replace two factors $\bar{\psi}_\alpha(x)\psi_\beta(y)$ in P_n (regardless of positions) by

$$\tfrac{1}{2}\, S_{F\beta\alpha}'(x-y) \qquad (72)$$

with S_F' given by (62) with $R_2 = 1$, the function S_F being defined by (44) and (45). Third, replace the remaining two factors $P(\bar{\psi}_\gamma(z)\psi_\delta(w))$ in P_n by $\bar{\psi}_\gamma(z)\psi_\delta(w)$ in this order. Fourth, replace $A_\mu^e(x_0)$ by $A_\mu^{e\prime}(x_0)$ given by

$$A_\mu^{e\prime}(x) = A_\mu^e(x) - [\alpha/15\pi\kappa_0^2]\, \square^2 A_\mu^e(x) \qquad (73)$$

or, more generally, by (69). Fifth, multiply the whole by $(-1)^l$, where l is the number of closed loops in G as defined in Section VII.

The above rules enable K_n to be written down very rapidly for small values of n. It should be observed that if K_n is being calculated, and if it is not desired to include effects of higher order than the n'th, then D_F', S_F', and $A_\mu^{e\prime}$ in (71), (72), and (73) reduce to the simple functions D_F, S_F, and A_μ^e. Also, the integrand in J_n is a symmetrical function of $x_1, \ldots, x_n$; therefore, graphs which differ only by a relabeling of the vertices $x_1, \ldots, x_n$ give identical contributions to K_n and need not be considered separately.

The extension of these rules to cover the calculation of matrix elements of (70) of a more general character than the one-electron transitions hitherto considered presents no essential difficulty. All that is necessary is to consider graphs with more than two "loose ends", representing processes in which more than one particle is involved. This extension is not treated in the present paper, chiefly because it would lead to unpleasantly cumbersome formulas.

X. *Example — Second-order radiative corrections*

Notes added in proof

The term L still contains two divergent parts. One is an "infra-red catastrophe" removable by standard methods. The other is an "ultraviolet" divergence, and has to be interpreted as an additional charge-renormalization, or, better, cancelled by part of the charge-renormalization calculated in Section VIII.

As an illustration of the rules of procedure of the previous section, these rules will be used for writing down the terms giving second-order radiative corrections to the motion of an electron in an external field. Let the energy of the external field be …

…

The above expression L, is formally simpler than the corresponding expression obtained by Schwinger, but the two are easily seen to be equivalent. In particular, the above expression does not lead to any great reduction in the labor involved in a numerical calculation of the Lamb shift. Its advantage lies rather in the ease with which it can be written down.

In conclusion, the author would like to express his thanks to the Commonwealth Fund of New York for financial support, and to Professors Schwinger and Feynman for the stimulating lectures in which they presented their respective theories.

Fukuda, H., Miyamoto, Y. & Tomonaga, S. (March, 1949). A Self-Consistent Subtraction Method in the Quantum Field Theory. II-1.

[*Prog. Theor. Phys. (Kyoto)*, 4, 1, 47-59; https://doi.org/10.1143/PTP.4.47.]

Received September 23, 1948.

Physics Institutes, Tokyo University and Tokyo Bunrika Daigaku.

Tati and Tomonaga (December 1948) proposed a *subtraction procedure* to be used in the treatment of quantum-theoretical problems involving infinite field reactions, this consists in generalizing the method of *canonical transformation*, and by means of this transformation the electron and the radiation fields are separated into two parts, one bound with each other and the other unbound and propagating freely in the absence of an external field, then the infinite energies related to the infinite *self-energy* of the free electron and to the *vacuum polarization* of the radiation field are separated as the interaction energies between bound parts of the fields, then *these infinite terms in the energy are dropped off considering that only the remaining finite terms have physical meaning*, when an *external field is present* the unbound fields no longer propagate freely and transitions take place, the electron wave is able to change its state of propagation not only elastically by the external field but also by emitting or absorbing unbound photons, thus *an interaction appears between electron and radiation* which were free from interaction with each other in the absence of the external field, this interaction causes a radiation reaction upon the electron so that the motion of the electron will be modified, *in this paper we give an example of how one can calculate this reaction and obtained a finite result as the consequence of our subtraction procedure*, we are able to obtain a *finite radiative level-shift* of a bound electron in the external field and a *finite e^2-correction* to the *scattering cross section*, estimated Lamb shift for hydrogen atom as 1076 Mcycles by adding additional terms to Bethe's value, we have thus successfully obtained finite answers for these field reaction problems and they are of the magnitude agreeing with experimental results, but *we must nevertheless confess that the calculation carried out in this paper is still unsatisfactory because we had to make a non-relativistic treatment in the evaluation of the effective energies,* our work is therefore only of a provisionary character, *still problematic whether this procedure corresponds to the correct prescription.*

1. *Introduction.*

In the first paper of the same title[1]*

[1] Tati, T. & Tomonaga, S. (December, 1948). A Self-Consistent Subtraction Method in the Quantum Field Theory, I. *Prog. Theor. Phys.*, 3, 4, 391-406[; notes procedure previously used by Bloch and Nordsieck and then by Pauli and Fierz in the treatment of problems with the self-field of an electron by first separating the radiation field into part bound to the electron and part of unbound photons by means of a canonical transformation, they then obtained a term in the Hamiltonian that could be interpreted as the *interaction energy* of electron with the radiation field bound to it, this term though infinite gave rise to a modification of mass of electron so that it could be amalgamated into the mass term in the Hamiltonian for the free electron, the term was dropped off to obtain *observed mass* when electron mass reinterpreted as already including it, same result as Bethe for level-shift of

bound electron in *non-relativistic* treatment, Tati and Tomonaga considered it desirable to obtain *relativistic* generalization of this treatment using a canonical transformation, paper addresses *field reaction problem when no external field* to e^2 approximation, does not address infinity related to *vacuum polarization* effect that occurs in *relativistic* treatment, *omitted as cannot be amalgamated into equation for free radiation*, starts from *relativistic* formalism of quantum field theory proposed by Tomanaga [August, 1946. On a Relativistically Invariant Formulation of the Quantum Theory of Wave Fields] in which Schrodinger equation with Hamiltonian for interaction density between radiation and electron fields is $\{H(P) - i\, \partial/\partial C_P\}\ \Psi[C] = 0$ where H(P) is the *interaction energy density* between radiation and electron fields at the world point P, Ψ[C] is the generalized Schrodinger functional which is a functional of the space-like variable surface C in the four-dimensional world and $\partial/\partial C_P$ its partial functional differentiation at the point P, the point P being considered as lying on C, decomposes fields into parts oscillating with positive and negative corresponding to electron and positron, transforms Schrodinger functional retaining terms up to the order e^2, expands integrand into Fourier integral and defines 4-vector to obtain Hamiltonian for transformed equation that enables use of first order calculation, allows processes through intermediate states such as emission and reabsorption of virtual particles (self-energy of an electron) to be treated as direct processes, calculates the commutator and rearranges factors in each term into the correct order, identifies (1) terms to represent *interactions between electrons* caused by exchange of virtual photons including ordinary M¢ller interaction between two electrons (2) terms connected with *self-energy* of an electron (3) terms responsible for the *scattering of a photon* by a free electron, terms for *creation of a pair* by two photons (4) variables describing the radiation field only representing modification of the radiation field in vacuo due to the *polarization of the vacuum* that cause an infinite energy level shift of the vacuum itself and an infinite *self-energy* of a photon, and (5) a *mass-modifying term* with a logarithmic diverging quantity representing the electromagnetic mass of an electron, assumes *mass-modifying term* already included in free field equation so no level-shift caused by the interaction between the electron and radiation, subtraction method "*self-consistent*" in this sense, gives *relativistic* generalization of transformation that separates *radiation field* into field of "unbound" photons and field of photons bound to the electron, procedure similar to Hartree method of the *self-consistent* field in which *interaction* between electrons is considered as a perturbation but some part of its effect is already included].

* This first paper will be cited as I.

Tati and one of the present authors have proposed a *subtraction procedure* to be used in the treatment of quantum-theoretical problems involving infinite field reactions. This method consists in *generalizing the method of canonical transformation* used first by Bloch and Nordsieck[2] and then by Pauli and Fierz[3]

[2] Bloch, F. & Nordsieck, A. (July, 1937). Note on the Radiation Field of the Electron. *Phys. Rev.*, 52, 54.

[3] Pauli, W. & Fierz, M. (March, 1938). Zur Theorie der Emission langwelliger Lichtquanten. (On the theory of the emission of long-wave light quanta.) *Nuovo Cimento*, 15, 3, 167-88.

in the treatment of similar problems in a *non-relativistic* approximation, and *by means of this transformation the electron and the radiation fields are separated into two parts, one bound with each other and the other unbound and propagating freely in the absence of an external field***[4]

> ** *The usefulness of the canonical transformation in the field reaction problem was pointed out also by Schwinger*, reference (4).
>
> [4] Schwinger, J. (February, 1948). On quantum-electrodynamics and the magnetic moment of the electron. *Phys. Rev.*, 73, 4, 416-7[; *electrodynamics unquestionably requires revision at ultra-relativistic energies,* desirable to isolate those aspects of the current theory that essentially involve high energies and are subject to modification by a more satisfactory theory, this goal has been achieved by transforming the Hamiltonian of current *hole theory* electrodynamics to exhibit explicitly the logarithmically divergent self-energy of a free electron which arises from the virtual emission and absorption of light quanta, the electromagnetic *self-energy* of a free electron can be ascribed to an *electromagnetic mass* which must be added to the mechanical mass of the electron, new Hamiltonian involves experimental *electron mass* rather than unobservable *mechanical mass*, electron now interacts with radiation field only in presence of external field such that only an accelerated electron can emit or absorb a light quantum, *interaction energy* of electron with external field subject to *finite* radiative correction, *polarization of the vacuum* still produces logarithmically divergent term proportional to *interaction energy* of electron in an external field, such term equivalent to altering value of *electron charge* by constant factor with only final value being identified with experimental charge, interaction between matter and radiation produces *renormalization* of electron charge and mass, all divergences contained in *renormalization* factors, radiative correction for energy of electron in external magnetic field corresponds to *additional magnetic moment associated with electron spin* of magnitude $\delta\mu/\mu = (½\ \pi)e^2/hc = 0.001162$, experimental measurements on hyperfine splitting of ground states of atomic hydrogen and deuterium larger than expected from directly measured nuclear moments, finds additional *electron spin magnetic moment* to account for measured hydrogen and deuterium *hyperfine structures* to be $\delta\mu/\mu = 0.00126$ and $\delta\mu/\mu = 0.00131$ respectively, these discrepancies accounted for by additional spin magnetic moment to the electron of $\delta\mu/\mu = 0.0018 \pm 0.00003$, *values yielded by relativistic calculation of Lamb shift differ only slightly from those conjectured by Bethe on basis of non-relativistic calculation and are in good accord with experiment.*]

Then infinite energies, one related to the infinite self-energy of the free electron and the other to that giving rise to the vacuum polarization of the radiation field, are separated as the interaction energies between bound parts of the fields. *Then these infinite terms in the energy are dropped off considering that only the remaining finite terms have physical meaning.*

Now when an external field is present, the unbound fields do no longer propagate freely, and, by virtue of the external field, transitions take place. The electron wave is able to change its state of propagation not only elastically affected by the external field but also emitting or absorbing unbound photons; thus, an interaction appears between electron and radiation which were free from interaction with each other in the absence of the external field. This interaction causes now a radiation reaction upon the electron so that the motion of the electron will be modified. *We shall give in this paper an example how one can calculate this reaction and the result obtained is really finite as the consequence of our subtraction procedure introduced in I.* Thus, for example, we are able to obtain a *finite radiative level-shift* of a bound electron in the external field and a *finite e^2-correction* to the *scattering cross section*, the latter problem having been discussed recently by Koba and one of us[5].

[5] Koba, Z. & Tomonaga, S. (December, 1947). Application of the "Self-Consistent" Subtraction Method to the Elastic Scattering of an Electron. *Prog. Theor. Phys.*, 2, 4, 218, letter to the editor; https://doi.org/10.1143/ptp/2.4.218 [; applies Tati and Tomonaga's "self-consistent" subtraction method to *elastic scattering of electron by a fixed electrostatic potential* to adjust mass for perturbation energy, "counter self-energy" term gives rise to additional correction that cancels divergence in previous result]. Also in Koba, Z. & Tomonaga, S. (September, 1948). On Radiation Reactions in Collision Processes. I: Application of the "Self-Consistent" Subtraction Method to the Elastic Scattering of an Electron. *Progr. Theor. Phys.*, 3, 3, 290-303[; applies Tati and Tomonaga's "self-consistent" subtraction method to the elastic scattering of an electron by a fixed electrostatic potential, *formal infinity associated with mass-modifying term attributed to defect of current theory so empirical value substituted for theoretical value* (assuming it was already included in the free field equation), maintains total Hamilton describing the interaction of the electron and electromagnetic fields unaltered, requires interaction-energy part of the Hamiltonian to undergo corresponding change by inclusion of "counter-self-energy" term ("mass-type" correction to scattering cross-section), results in finite value for self-energy of electron in the e^2 approximation, when new formalism applied to elastic scattering of electron *effective cross-section* for scattering by a fixed potential in zeroth approximation has value in good agreement with experiment but as soon as reaction of electromagnetic field with electron is taken into account the correction becomes infinite, first infinite term in modified Hamiltonian eliminated using *subtraction hypothesis of positron theory*, second difficulty disappears by applying *"self-consistent" subtraction method* using modified Hamiltonian, first term can also be eliminated if one interprets the sum of the external potential and its infinite correction due the *vacuum polarization* effect as the physically observable potential and *substitute the empirical value for it* whilst interaction part of the Hamiltonian supplemented by an additional "counter-vacuum polarization" term, *method by no means give the real solution of the fundamental difficulty of the quantum electrodynamics* but reveals nature of various diverging terms and reduces them to two quantities - the *self-energy* and the *vacuum polarization*, in this way *it becomes possible in an unambiguous and consistent manner to treat the field reaction problem without touching the fundamental difficulty by employing the finite empirical values instead of the infinite "theoretical" values for these two quantities*].

As will be shown below *we can find finite effective interaction energies which describe the interaction between electron and radiation* as mentioned above. One of these interaction energies is first order in e and is essentially of the form $c\phi^{\dagger}OAU^{o}\phi$, U^{o} being the potential (multiplied by e) of the external field and O some operator containing Dirac matrices and operators upon $\phi^{\dagger}$, A, U^{o} and ϕ. This energy is responsible for the Bremsstrahlung of an electron in the field U^{o} or allied phenomena. Another interaction energy is of the second order in e and has the form $c^2\phi^\dagger O U^o \phi$. This energy as well as the trivial zero order term $\phi^\dagger \gamma U^o \phi$, give rise to elastic scattering of the electron in the field U^o, the second order term giving rise to the e^2-correction to the ordinary elastic scattering caused by the zero-order term. The famous *energy-level shift of a bound electron* observed by Lamb and Retherford[6]

> [6] Lamb, Jr., W. E. & Retherford, R. C. (August, 1947). Fine Structure of the Hydrogen Atom by a Microwave Method. *Phys. Rev*., 72, 3, 241-3[; the spectrum of hydrogen has a fine structure of the energy levels which according to the *Dirac wave equation* for an electron moving in a Coulomb field is due to the combined effects of *relativistic* variation of mass with velocity and spin-orbit coupling, according to this theory the $2^2S_{1/2}$ state should exactly coincide in energy with the $2^2P_{1/2}$ state which is the lower of the two P states, previous attempts at measurement have alternated between finding confirmation and discrepancies of as much as eight percent, using microwave method depending on a novel property of the $2^2S_{1/2}$ level results indicate that contrary to the *Dirac wave equation* the $2^2S_{1/2}$ state is higher than the $2^2P_{1/2}$ by about 1000 Mc/sec].

can be calculated as a combined effect of these two terms, namely as the first order modification of the energy-level due to the second order term combined with the second order effect due to the first order term.

Although we have thus successfully obtained finite answers for these field reaction problems and they are of the magnitude agreeing with experimental results, *we must nevertheless confess that the calculation carried out in this paper is still unsatisfactory because we had to make a non-relativistic treatment in the evaluation of the effective energies,* which have the form $\infty - \infty$. In the calculation with such improper expressions, from which we wish to draw a finite conclusion, the results are often affected by the way of calculation so that it will possibly occur that the approximate treatment would miss the essential point. But we think we have been able to confirm, at least, that the result converges by virtue of our subtraction procedure. *A method more satisfactory from the relativistic point of view is now being investigated.*

§ 2. *General Formulation.*

...

§ 3. *Evaluation of the Effective Energies.*

In evaluating the effective energies found in the preceding section *it is desirable to carry out the calculation relativistically as far as possible*. In applying our method to the calculation of the level shift or of the e^2-correction for the scattering of an electron with a *non-relativistic* velocity, it is allowed to make a *non-relativistic* approximation in some sense. But it is important to introduce this approximation first in the end stage of the calculation after we have obtained a converging expression. *This is because we are dealing with expressions of the form $\infty - \infty$, and the values obtained from such improper expressions depend so much on the way of evaluation that it is necessary to carry through the calculation on some definite prescription, which should be, of course, relativistically invariant.* In such a situation there is much fear of that the calculation does not correspond to the true prescription if one introduces the *non-relativistic* approximation in a too early stage. *We must confess, however, that we are yet unable to carry out the calculation relativistically throughout; we had to introduce a non-relativistic procedure from the beginning. Our work is therefore, only of a provisionary character.* We believe, however, that *we have been able to prove, at least, that our subtraction procedure is really effective in removing the divergency of the theory.*

The first *non-relativistic procedure* we had to introduce is to consider that the variable surface C is a plane parallel to the xyz-plane. By this process our calculation becomes much parallel with the ordinary perturbational calculation.

...

We omit here the calculation of (f) because this term represents the effect of vacuum polarization caused by the external field which should be dropped off with a similar reasoning as used in dropping off the mass correction term (8). (But see § 6)*. ...

Thus far our expressions are *non-relativistic* only in the sense that we have taken the surface C of a special form i.e. a plane parallel to the *xyz*-plane. We now introduce in the next section the essentially *non-relativistic* approximation assuming that the velocity of the electron under consideration is far smaller than the light velocity. (* *§§ [4,] 5 and 6 will appear in the next issue of this journal*).

Fukuda, H., Miyamoto, Y. & Tomonaga, S. (June, 1949). A Self-Consistent Subtraction Method in the Quantum Field Theory. II-2.

[*Prog. Theor. Phys. (Kyoto)*, 4, 2, 121-9; https://doi.org/10.1143/ptp/4.2.121.]

Received September 23, 1948.

Physics Institutes, Tokyo University and Tokyo Bunrika Daigaku.

§ 5[4]. *Approximate Treatment for the Electron with Non-Relativistic Velocity**.

* §§ 1, 2, 3 [and 4] of this appeared in (1949). *Prog. Theor. Phys.*, 4, 1, 47-59.

We now enter into the second stage of the *non-relativistic* approximation assuming that $p_1/\surd(p_1^2 + x^2) \ll 1$, $p_2/\surd(p_2^2 + x^2) \ll 1$. We thus expand the integrands in (V') in power series of $p_1/\surd(p_1^2 + x^2)$ and $p_2/\surd(p_2^2 + x^2)$ and retain terms up to the second order. ...
...

This fact shows that, although separately the terms (A), (B), (C), (E) and (F) diverge, the energy density (A) + (B) + ... + (F) as a whole converges on the ultra-violet side. Collecting together the terms diverging on the infra-red side we obtain, on the other hand,

$$- i\gamma\nu \ln 2\varepsilon.\{P_1/(e.P_1) - P_2/(e.P_2)\}^2. \tag{20}$$

It is to be noticed that the expression (20) has the same form with the expression for the difference between the *self-fields* of an electron with the *momentum* $\mathbf{p}_1$ and $\mathbf{p}_2$ respectively; the expression found by Bloch and Nordsieck[1] as well as by Pauli and Fierz[2] in the *non-relativistic* approximation.

[2] Bloch, F. & Nordsieck, A. (July, 1937). Note on the Radiation Field of the Electron. *Phys. Rev.*, 52, 54; https://doi.org/10.1103/PhysRev.52.54.

[3] Pauli, W. & Fierz, M. (March, 1938). Zur Theorie der Emission langwelliger Lichtquanten. (On the theory of the emission of long-wave light quanta.) *Nuovo Cimento*, 15, 3, 167-88; https://doi.org/10.1007/BF02958939.

This term diverges logarithmically when we go over to the limit $\varepsilon \rightarrow 0$.

Although our expression for the effective interaction energy contains thus a divergency on the infra-red side, this makes no difficulty. In applying our result to the scattering problem, for instance, this divergency can be eliminated if one takes into account the scattering process with the emission of a photon. In the calculation of the energy-level shift it is eliminated by the contribution from the second order effect caused by the term (III) because the latter effect has the same divergency with the opposite sign. This situation will be discussed in the next section.
...

As was mentioned above it is necessary to take into account also the processes involving the emission of a photon in order to eliminate the divergency on the infra-red side. For such processes the energy density (III) was responsible. The evaluation of (III) is carried out quite easily because it contains no divergency. …

…

In this way we have obtained the *energy densities* necessary for the calculation of the field reaction upon electron. *We shall now go over to the application of our result to some field-reaction problems.*

§ 5. *Application.*

(a) *Anomalous Magnetic Moment of an Electron.*

[The *anomalous magnetic moment* of a particle is a contribution of effects of quantum mechanics, expressed by Feynman diagrams with loops, to the magnetic moment of that particle. The "Dirac" magnetic moment, corresponding to tree-level Feynman diagrams (which can be thought of as the classical result), can be calculated from the Dirac equation. It is usually expressed in terms of the g-factor; the Dirac equation predicts g = 2. For particles such as the electron, this classical result differs from the observed value by a small fraction of a percent. The difference is the *anomalous magnetic moment*, denoted by *a* and defined as $a = (g - 2)/2$.

The one-loop contribution to the anomalous magnetic moment—corresponding to the first and largest quantum mechanical correction—of the electron is found by calculating the *vertex function* shown in the adjacent diagram.

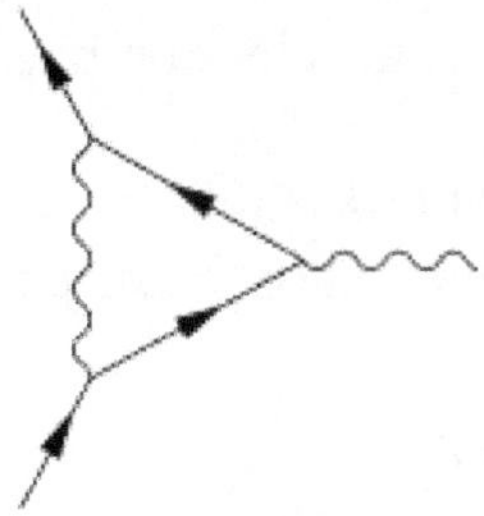

One-loop correction to a fermion's magnetic dipole moment.
(User:Harmaa, Public domain, via Wikimedia Commons.)

The calculation is relatively straightforward[1] and the one-loop result is:

[1] Peskin, M. E. & Schroeder, D. V. (1995). *An Introduction to Quantum Field Theory*. Addison-Wesley, Section 6.3.

$a_e = \alpha/2\pi \approx 0.0011614$

where α is the fine structure constant. This result was first found by Julian Schwinger in 1948 and is engraved on his tombstone. As of 2016, the coefficients of the quantum electrodynamics (QED) formula for the anomalous magnetic moment of the electron are known *analytically*[1] up to α^3

[1] Laporta, S. & Remiddi, E. (1996). The analytical value of the electron (g − 2) at order α^3 in QED. *Physics Letters* B, 379, 1–4, 283-91; arXiv:hep-ph/9602417; doi:10.1016/0370-2693(96)00439-X.

and have been *calculated* up to order α^5

a_e = 0.001159652181643(764).

The QED prediction agrees with the experimentally measured value to more than 10 significant figures, making the magnetic moment of the electron the most accurately verified prediction in the history of physics.

The current *experimental* value and uncertainty is:

a_e = 0.00115965218073(28).

According to this value, a_e is known to an accuracy of around 1 part in 1 billion (10^9). This required measuring g to an accuracy of around 1 part in 1 trillion (10^{12}).

An *empirical formula* has been reported for the *anomalous magnetic moment*[2]:

[2] Efimov, S.P. (2021). Relation of the Anomalous Magnetic Moment of the Electron with Proton and Neutron Masses. *Russ. Phys. J.*, 64, 6, 978-83; https://doi.org/10.1007/s11182-021-02417-z

$\delta\mu_e = eh/2c\ [3/\sqrt{\{2(m_p - \exp(1)\delta m)\}}]$,

where m_p is the proton mass and δm is difference between the neutron and proton masses ($m_n - m_p$). Therefore, the value a_e is equal to

$a_e = 3m_e/\sqrt{[2\{m_p - \exp(1)\delta m\}]}$ = 0.0011596523..,

where m_e is the electron mass.]

The last term of (VII) expresses that the electron interacts with the magnetic field as if it had an extra magnetic moment of the magnitude

$$\delta\mu = 1/2\pi\ (e^2/\hbar c)\ \mu, \qquad (23)$$

[Schwinger (February, 1948). On quantum-electrodynamics and the magnetic moment of the electron: "The simplest example of a radiative correction is that for the energy of an electron in an external magnetic field. The detailed application of the theory shows that the radiative correction to the magnetic interaction energy corresponds to an additional magnetic moment associated with the electron spin, of magnitude $\delta\mu/\mu = (\frac{1}{2}\pi)e^2/\hbar c$ = 0.001162".]

This anomaly in the *magnetic moment* will be a cause of the deviation of the hyperfine structure of line spectra from the prediction of the ordinary theory[3].

[3] Breit, G. (November, 1947). Does the electron have an intrinsic magnetic moment? *Phys. Rev.*, 72, 10, 984; https://doi.org/10.1103/PhysRev.72.984

[; Schwinger (February, 1948): "However, Breit has not correctly drawn the consequences of his empirical hypothesis. The effects of a nuclear magnetic field and a constant magnetic field do not involve different combinations of μ and δμ"];

Schwinger, J. (February, 1948). On quantum-electrodynamics and the magnetic moment of the electron. *Phys. Rev.*, 73, 4, 416-7[; *electrodynamics unquestionably requires revision at ultra-relativistic energies,* desirable to isolate those aspects of the current theory that essentially involve high energies and are subject to modification by a more satisfactory theory, this goal has been achieved by transforming the Hamiltonian of current *hole theory* electrodynamics to exhibit explicitly the logarithmically divergent self-energy of a free electron which arises from the virtual emission and absorption of light quanta, the electromagnetic *self-energy* of a free electron can be ascribed to an *electromagnetic mass* which must be added to the mechanical mass of the electron, new Hamiltonian involves experimental *electron mass* rather than unobservable *mechanical mass*, electron now interacts with radiation field only in presence of external field such that only an accelerated electron can emit or absorb a light quantum, *interaction energy* of electron with external field subject to *finite* radiative correction, *polarization of the vacuum* still produces logarithmically divergent term proportional to *interaction energy* of electron in an external field, such term equivalent to altering value of *electron charge* by constant factor with only final value being identified with experimental charge, interaction between matter and radiation produces *renormalization* of electron charge and mass, all divergences contained in *renormalization* factors, radiative correction for energy of electron in external magnetic field corresponds to *additional magnetic moment associated with electron spin* of magnitude $\delta\mu/\mu = (\frac{1}{2}\pi)e^2/hc = 0.001162$, experimental measurements on hyperfine splitting of ground states of atomic hydrogen and deuterium larger than expected from directly measured nuclear moments, finds additional *electron spin magnetic moment* to account for measured hydrogen and deuterium *hyperfine structures* to be $\delta\mu/\mu = 0.00126$ and $\delta\mu/\mu = 0.00131$ respectively, these discrepancies accounted for by additional spin magnetic moment to the electron of $\delta\mu/\mu = 0.0018 \pm 0.00003$, *values yielded by relativistic calculation of Lamb shift differ only slightly from those conjectured by Bethe on basis of non-relativistic calculation and are in good accord with experiment*].

In this problem also the term $-(1/6)\rho_2\sigma\boldsymbol{A}$ (VII) may play a role. This effect will be accounted for as a reinterpretation of the *nuclear moment*, but *its appearance is somewhat problematic* as will be mentioned in the next section*.

* *Note 2*. The problematic term was occurring also in the work of American authors, which could be eliminated by using Pauli's "regulator".

(b) *Energy-Level Shift of a Hydrogen-like Atom.*

The *interaction energy* (VII') causes a part of the level shift of the bound electron as the first order effect, which together with the second order effect due to the *interaction energy* (III') contributes to the totallevel shift. This part of the level shift can be calculated in the following way:

...

… Neglecting the small contribution from the second term in (III') and neglecting the retardation too, the calculation for this part can be carried out in a similar manner as was done by Bethe[4].

[4] Bethe, H. A. (August, 1947). The Electromagnetic Shift of Energy Levels. *Phys. Rev.*, 72, 339-41[; Lamb and Retherford results show that fine structure of second quantum state of hydrogen does not agree with *Dirac wave equation*, Schwinger, Weisskopf, and Oppenheimer suggest might be due to *shift of energy levels by interaction of electron with the radiation field*, this shift comes out infinite in all existing theories and has therefore always been ignored, possible to identify the most strongly (linearly) divergent term in the level shift with an *electromagnetic mass effect* which must exist for a bound as well as for a free electron, already included in the *observed mass* of the electron so should be subtracted, assumes *relativistic cut-off* in quantum energies (frequencies) of included atomic states, then calculation of Lamb shift for hydrogen atom using *non-relativistic* ordinary radiation theory gives shift of the levels due to *radiation interaction* in close agreement with observed value, removes discrepancy with Dirac theory, did not carried out *relativistic* calculations].

If we use Bethe's value

… = 1040 Mcycles, (28)

the total level shift for the $2S_{1/2}$ term is

… = 1059 Mcycles. (27')

For the $2P_{1/2}$ term, on the other hand, (26) gives

… = – 17 Mcycles. (26')

From (27) and (25') the observable line shift is found to be

$$\Delta E(2S_{1/2}) - \Delta E(2P_{1/2}) = e^2/\pi\hbar c\ Ze^2/6x^2a^2\ \{\ln x/(E_n - E_m)_{av} - \ln 2 + 23/24\} \quad (29)$$

$$= 1076 \text{ Mcycle.} \quad (29')$$

[where x is the mass of the electron.]

(c) *e^2-Correction to the Rutherford Formula*[5].

[5] Braunbeck, W. & Weimann, C. (1938). *Zeit. Phys.*, 110, 360; Endo, S., Kinoshita, T. & Koba, Z. *Prog. Theor. Phys.*, in press.

A correction to the cross section for the elastic scattering of an electron is caused by the first term of (VII'). …

§ 6. *Concluding Remarks.*

In concluding the paper, we shall add the following remarks. In our calculation we have dropped off altogether the term (f) in (IV_1) + (IV_2) which represents the effect of the vacuum polarization due to the external field U_1^o. *But it is still problematic whether this procedure corresponds to the correct prescription.* It is possible that only the diverging part in this expression is to be subtracted, thereby the separation of the diverging part from the converging remainder being performed by some suitable prescription. For instance, it is possible to set up a prescription in such a way that the converging part separated has the same form as that obtained by Serber[7] on the ordinary positron theory.

[7] Serber, R. (July, 1935). Linear Modifications in the Maxwell Field Equations. *Phys. Rev.*, 48, 49; https://doi.org/10.1103/PhysRev.48.49.

Then the effect of this converging remainder will be that which was calculated by Uehling[8],

[8] Uehling, E. A. (July, 1935). Polarization Effects in the Positron Theory. *Phys. Rev.*, 48, 55; https://doi.org/10.1103/PhysRev.48.55.

who found the level shift of – 17 Mcycle for the S-level. When we take account of this effect, the total shift for the $2S_{1/2}$-level becomes 1032 Mcycles, and the observable line shift becomes 1049 Mcycles*.

* *Note 1* [added in proof]. Corresponding to our value 1049 Mcycles for the line shift mentioned in the first line of page 128, they obtained 1051 Mcycles. This difference is due to the different values of ln $\{K/[E_n - E_m]$ used. If we use instead of 7.63 Bethe's new value 7.6876, we obtain the same value as theirs. In fact, the analytical expression for this line shift is $\Delta E(2S_{1/2}) - \Delta E(2P_{1/2}) = e^2/\pi\hbar c\ Ze^2/6x^2a^2\ \{\ln x/(E_n - E_m)_{av} - \ln 2 + 23/24 - 1/5\}$ [where x is the mass of the electron], which is obtained by adding Uehling's result to our (29), agrees with the corresponding expression of the cited authors.

[Bethe (August, 1947). The Electromagnetic Shift of Energy Levels: "Using this figure and $K = mc^2$, the logarithm has the value 7.63, and we find
$W_{ns}' = 136 \ln \{K/[E_n - E_m]\} = 1040$ megacycles." (12)]

The second fact we must emphasize is again about *the provisionary character of our calculation because of its imperfectness in view of the relativistic invariance. In evaluating the improper expression of the form ∞ – ∞ it is necessary to carry through the calculation in obedience to some definite prescription which must be, of course, relativistic invariant and free from any ambiguity arising from the pathological nature of the D-functions occurring in our formalism.* As a matter of fact, our method of calculating the integral over dX', first expanding the D-functions in Fourier integrals and then use the formula (7) of I, *often gives results depending on the order of integration, on the choice of the integration variables, on the reference system in which the integrations are performed and so on.*

Thus, for instance, in the calculation of the *electromagnetic mass correction* described in I we had to evaluate the integral according to the following disposition: we first transform the reference system into one in which the electron is at rest and perform the integral in this system then we go back to the original system making use of the *relativistically* invariant character of the formula. If one carried out here the calculation in a different way, one could not obtain the correct answer. *So, it may possibly occur that also the calculation in this paper would not give the correct answer because the calculation was carried out in a non-relativistic approximation from the beginning.* In fact, our result has a singular property that it has quite a different form for space and time component, and this seems to show the incorrectness of our way of calculation.

As one sees in (21), for instance, the expression for the energy has different form for $\nu \neq 4$ and for $\nu = 4$. *The appearance of the term* $-(1/6)p_2\sigma\boldsymbol{A}$ *in (VII) seems also to show the inadequateness of our calculation***

** Note 2 [added in proof]. The problematic term was occurring also in the work of American authors, which could be eliminated by using Pauli's "regulator",

because this term requires that the electron charge must be interpreted when it interacts with a magnetic field whereas no such reinterpretation is required when it interacts with an electric field. In the latest issue of the *Physical Review* we found that the unpublished result of Bethe and Schwinger of the same problem was cited by Fowler[9].

[9] Fowles, G. R. (July, 1948). The S-Level Shift in Ionized Helium. *Phys. Rev.*, 74, 219; https://doi.org/10.1103/PhysRev.74.219.

Comparing our result (29) with theirs, we find that in their formula 1/2 stands in the place where 23/24 stands in ours. This difference will probably be due to the difference of their treatment from ours*.

* See *Note 1* above.

It is desirable to develop a scheme of calculation which fits better for the use of the D-function than the imperfect scheme used by us, which perhaps will be done along the line suggested lately by Nambu[10]

[10] Nambu, Y., to be published in this journal.

It will be of much interest to compare our results with those obtained by the *cohesive force-field hypothesis* [abandoned *C-meson theory*]. Such a calculation is undertaken by the group of Nagoya University.

Notes added in proof. Several papers of American authors treating the same problem appeared meanwhile in the Physical Review (Kroll, N. M. & Lamb, Jr., W. E. (February, 1949). On the self-energy of a bound electron. *Phys. Rev.*, 75, 3, 388-98; https://doi.org/10.1103/PhysRev.75.388; French, J. B. & Weisskopf, V. F. (April, 1949). The Electromagnetic Shift of Energy Levels. *Phys. Rev.*, 75, 8, 1240-8; Schwinger, J. (February, 1949). Quantum Electrodynamics. II. Vacuum Polarization and Self-Energy. *Phys. Rev.*, 75, 4, 651-79) so we should like to add some remarks in order to relate our results to theirs. *Note 1. Note 2.*

Kanesawa, S, & Koba, Z. (September, 1949). A Remark on Relativistically Invariant Formulation of the Quantum Field Theory.

[*Prog. Theor. Phys*., 4, 3, 297-311; https://doi.org/10.1143/ptp/4.3.297.]

Received April 18, 1949.

New formalism of quantum field theory more satisfactory from stand-point of *relativistic invariance*, formulated in terms of *invariant space-time* concepts but employed *canonical formalism* in which time variable unnecessarily distinguished from space variables, *the purpose of the present paper is to show that the same formalism can be reached without referring to the canonical formulation*, in current theory the *interaction Hamilton density* coincides with *interaction Lagrange density* except the sign if surface-dependent part neglected, suggests to connect *generalized interaction Lagrange density* directly to generalized Schrodinger equation, guiding principle the integrability of the Tomonaga-Schwinger equation, shows how to apply new method to the general case when canonical description is impossible, *problematic aspects of theory such as universal length or ultraviolet divergencies were not addressed*, aim of paper lies in demonstrating the possibility of including more general kinds of interactions in field theory than those which allow the canonical description.

1. *Introduction and summary.*

In a series of papers[1,2]

[1] Tomonaga, S. (1943). On a Relativistically Invariant Formulation of the Quantum Theory of Wave Fields. *Bull. I. P. C. R. Riken-iho*, 22, 545 (in Japanese); translation Tomonaga, S. (August, 1946). *Prog. Theor. Phys*. 1, 2, 27-42[; *existing formalism of quantum field theory not perfectly relativistic*, commutation relations refer to points in space at different times, Schrodinger equation for the vector representing the state of the system is a function of time, time variable plays a different role than space variables, *probability amplitude* not *relativistically invariant* in the space-time world, follows Dirac (1932) [Relativistic Quantum Mechanics] in *generalizing the notion of probability amplitude as far as is required by the theory of special relativity*, substitutes four-dimensional form of the commutation relations, *generalizes the Schrodinger equation* following the Dirac (1933) [The Lagrangian in Quantum Mechanics] *many-time* formalism, then introduces his *super many-time theory* [$\{H_{12}(P) + h/i\ \partial/\partial C_P\}\ \Psi[C] = 0$ at point P on surface C with infinitely many time variables representing the local time for each position in the space, results in *relativistic interaction representation*, three-dimensional manifold (space-like "surface") in four-dimensional space-time world, not necessary to also assume time-like surfaces for the variable surface as was required by Dirac, previous formalism built up in way too analogous to ordinary *non-relativistic* mechanics, theory was divided into one section giving the kinematical relations between various quantities at the same instant of time and another section determining the causal relations between quantities at different instants of time with the *commutation relations* belong to the first section and the *Schrodinger equation* to the second, this way of separating the theory into two sections very *unrelativistic*, "same instant of time" plays a distinct role, *new formalism consists of one section giving laws of behavior of the fields when they are left alone and the other giving*

the laws determining the deviation from this behavior due to interactions, can be carried out *relativistically*, although theory in more satisfactory form no new contents added, *divergence difficulties inherited*, fundamental equations admit only catastrophic solutions *due to non-vanishing zero-point amplitudes of the fields which inheres in the operator $H_{12}(P)$, a more profound modification of the theory is required in order to remove this fundamental difficulty*]. This paper will be cited as I.

[2] Koba, Z., Tati, T. & Tomonaga, S. (1947). On a Relativistically Invariant Formulation of the Quantum Theory of Wave Fields. II. Case of Interacting Electromagnetic and Electron Fields. § 1-4. *Prog. Theor. Phys.*, 2, 3, 101-16; https://doi.org/10.1143/ PTP.2.101; *ibid.* On a Relativistically Invariant Formulation of the Quantum Theory of Wave Fields. III. Case of Interacting Electromagnetic and Electron Fields. § 5-7. *Prog. Theor. Phys.*, 2, 4, 198-208; https://doi.org/10.1143/PTP.2.198. This paper will be cited as II.

Kanesawa, S. & Tomonaga, S. (March, 1948). On a Relativistically Invariant Formulation of the Quantum Theory of Wave Fields. IV. Case of Interacting Electromagnetic and Meson Fields. § 1-4. *Prog. Theor. Phys.*, 3, 1, 1-13; https://doi.org/10.1143/ptp/3.1.1; *ibid.* On a Relativistically Invariant Formulation of the Quantum Theory of Wave Fields. V. Case of Interacting Electromagnetic and Meson Fields. § 5-7. *Prog. Theor. Phys.*, 3, 101-13. This paper will be cited as III.

Miyamoto, Y. (June, 1948). On the Interaction of the Meson and Nucleon Field in the Super-Many-Time Theory. *Prog. Theor. Phys.*, 3, 2, 124-40; https://doi.org/10.1143/ptp/ 3.2.124. This paper will be cited as IV.

Tomonaga and others developed a new formalism of the quantum field theory which was more satisfactory than the current one from the stand-point of the *relativistic invariance*. The theory was formulated in terms only of concepts possessed of *invariant space-time* meaning while in the current theory one had to use *non-relativistic* concepts borrowed from the ordinary quantum mechanics. In the process of derivation of this theory. however, *one employed canonical formalism in which time variable was unnecessarily distinguished from space variables. It is the purpose of the present paper to show that one can reach the same formalism without referring to the canonical formulation.*

According to the results obtained by applying our theory to several actual cases, the generalized Schrodinger equation in each physical system has the Lorentz invariant form despite of its *non-relativistic* way of derivation, and moreover *the interaction Hamilton density coincides with the interaction Lagrange density except the sign if we neglect its surface-dependent part. These facts suggest to connect directly the interaction Lagrange density to the generalized Schrodinger equation.*

In section 2 we show that the same result as the original formalism is obtained by this method without using the *canonical momenta*. In the fundamental equation of our new formalism

$$\{L_P[C] - \hbar/i\ \delta/\delta C_P\}\ \Psi[C] = 0 \qquad (1.1)$$

the generalized *interaction Lagrange density* $L_P[C]$ appears instead of the generalized *Hamilton density* $H_P[C]$ which is related to the former by means of

$$H_P[C] = - L_P[C], \qquad (1.2)$$

this function $L_P[C]$ is directly determined from the *interaction Lagrange den*sity L(P) as a special solution of the equation for the integrability of (1.1).

$$[L_X[C] - \hbar/i\ \delta/\delta C_X,\ L_{X'}[C] - \hbar/i\ \delta/\delta C_{X'},] = 0$$

or

$$\delta L_X[C]/\delta C_{X'} - \delta L_{X'}[C]/\delta C_X = -\ i/\hbar\ [L_X[C], L_{X'}[C]] \qquad (1.3)$$

[*the integrability condition*] where X and X' are the *two world points which lie in a space-like direction with respect to each other.* A general prescription for constructing this special solution $L_P[C]$ is formulated in section 3.

In section 4 we give a proof of the equivalence of our new method of formulation to the old one when the system under consideration is describable in a canonical form. Because the method proposed in section 3 is applicable also when the canonical description is impossible, our method can describe more general interactions than those occurring in the current one. In section 5 we shall give a simple illustration how one can formulate the quantum field theory in such a case.

Strictly speaking there is yet no justification of applying our formalism to this general case, since all we have proved is that our formalism coincides with the ordinary one when the latter is possible, as was kindly pointed out by Dr. Watanabe[3]

[3] At the annual meeting of the Physical Society of Japan, May 1948.

Indeed, there can be no criterion with regard to cases when the canonical formalism fails except agreement with experimental facts. *But we may be allowed to hope that our method of generalization, whose guiding principle is the integrability of the Tomonaga-Schwinger equation*, is perhaps a correct one, because according to our several experiences this condition is an essential feature of the current quantum field theory, so that it will hold in general as far as one is concerned only with a complicated type of coupling and *not with those problematic aspects of the theory such as universal length or ultraviolet divergencies*.

Further we must emphasize that our special solution obtained by the prescription described in section 3 is not the unique one which makes the fundamental equation (1.1) integrable[4].

[4] Tani, S. (April, 1949). Note on the Formal Solution of the Tomonaga-Schwinger Equation. *Prog. Theor. Phys*., 4, 209: "The fundamental equation of the Tomonaga-Schwinger theory[1,2]

[1] Tomonaga, S. (1946). *loc. cit.* and a series of works following this Part I.

[2] Schwinger, J. (November, 1948). Quantum Electrodynamics. I. A Covariant Formulation. *Phys. Rev.*, 74, 10, 1439–1461.

is of the form

$$\{H(x) - \hbar/i\ \delta/\delta\sigma(x)\}\ \Psi[\sigma] = 0 \qquad (1)$$

$$[\{L_P[C] - \hbar/i\ \delta/\delta C_P\}\ \Psi[C] = 0 \qquad (1.1)]$$

where the interaction density H(x) is subject to the condition

$$[H(x), H(x')] = O,\ \text{when}\ (x_\mu - x'_\mu)^2 > 0. \qquad (2)".$$

Thus, the aim of the present paper lies rather in demonstrating the possibility of including more general kind of interactions in the field theory than those which allow the canonical description. There arise then the questions *how the energy-momentum law will be formulated* and *what the generalization of the stress-energy tensor* in this general case. Such questions will be discussed in a later occasion.

2. *Some examples for the case where the canonical formulation is possible.*

Before entering into individual cases, we give the common assumptions which are made throughout all examples. They are as follows: the field variables contained in the interaction Lagrange density L(X) are solutions of the free field equations, consequently, satisfy the same commutation relations with those between free ones. Now, as was shown by Pauli[5]

[5] Confer W. Pauli, Solvay Berichte (1939),

the four-dimensional commutation relations between the vacuum field variables can be determined by the requirement that the energy and momentum of the field are quantized so that the field shows a particle nature.

Example I. Case of *electron field interacting with electro-magnetic field*[6].

[6] Confer the paper II [Koba, Z., Tati, T. & Tomonaga, S. (1947). On a Relativistically Invariant Formulation of the Quantum Theory of Wave Fields. II & III. Case of Interacting Electromagnetic and Electron Fields].

The *interaction Lagrange density* in our case is given by

$$L(X) = e\,(\phi^{*}\alpha\phi A), \qquad (2.1)$$

As was shown in the paper II, the equation

$$[L(X), L(X')] = O \qquad (2.2)$$

holds *when the world points X and X' lie space-like to each other.* According to our general prescription for constructing the $L_P[C]$ which will be given in the next section, it is found that in such a case we can use L(P) itself as $L_P[C]$, so that

$$\{L(P) - \hbar/i\ \delta/\delta C_P\}\ \Psi[C] = 0 \qquad (2.3)$$

$$[[\{L_P[C] - \hbar/i\ \delta/\delta C_P\}\ \Psi[C] = 0 \qquad (1.1)]$$

In this case it is obvious that

$$L(P) = -\,H(P) \qquad (2.4)$$

where H(P) means the *interaction Hamilton density of the fields*. Of course, the equation (2.3) coincides with the equation (III) in II.

Example 2. Case of scalar-(pseudoscalar-) meson field interacting with electromagnetic field[7].

[7] Confer the paper III [Kanesawa, S. & Tomonaga, S. (1948). On a Relativistically Invariant Formulation of the Quantum Theory of Wave Fields. IV & V. Case of Interacting Electromagnetic and Meson Fields].

...

Example 3. Case of vector-(pseudovector-) meson field interacting with electromagnetic field[7].

[7] *Loc. cit.*

...

In the above-mentioned examples, we have considered only the cases containing one kind of coupling constant *e* between two fields, but analogous statements can be extended easily to the cases where more than two coupling constants exist between the two interacting fields, for example to the case of nucleon field interacting with meson field[8].

[8] Confer the paper IV [Miyamoto, Y. (June, 1948). On the Interaction of the Meson and Nucleon Field in the Super-Many-Time Theory].

3. *The prescription for constructing the required special solution of the equation (1.3)*

Thus far we have considered only the system which is describable in canonical form, but our formalism can also be extended to the case where canonical variables cannot be defined in the ordinary sense. Let us suppose for simplicity that there are only two fields interacting with each other, and the total Lagrange density is given by

$$L_{total} = L_{I} + L_{II} + L_{III} \qquad (3.1)$$

where L_I and L_{II} mean respectively the Lagrange densities of the first and the second fields. L_{III} the interaction one. ...

...

(I) *Case where both fields are of integral spin.*

...

(II) *Case where both fields are of half odd integral spin.*

...

... It is to be noticed that $L_P[C]$ obtained in this way contains, in general. not only the unit normal of the surface C but its derivatives, so that it depends not only on the inclination of the surface C at the point P but on curvature etc. of C at this point.

4. *Proof of equality between the canonical formalism and ours.*

In this section we prove that our new formalism is perfectly equivalent to the old theory if the physical system under consideration can be written in a canonical form. For this purpose, it is sufficient to prove the following relation:

$$L_X[C_0] = -H(X) \quad (4.1)$$

where $L_X[C]$ means the *generalized interaction Lagrange density* which is derived from the interaction Lagrange density L(X) by the prescription described in section 3 so that (1.1)

$$[[\{L_P[C] - \hbar/i\ \delta/\delta C_P\}\ \Psi[C] = 0 \quad (1.1)]$$

satisfies the *integrability condition* (1.3)

$$[\delta L_X[C]/\delta C_{X'} - \delta L_{X'}[C]/\delta C_X = -\ i/\hbar\ [L_X[C], L_{X'}[C]], \quad (1.3)]$$

and C_0 is a special surface of C which is a plane parallel to the xyz-plane.

Now, for the first step of the proof of the equation (4.1) we will summarize the outline of the canonical formalism in the one-time theory. …

…

In this way we have furnished the proof of the equation (4.1).

5. *An example.*

We consider here the modified theory of β-decay introduced by Konopinski and Uhlenbeck[9] as an example of our generalized formulation.

[9] Konopinski, E. J. & Uhlenbeck, G. E. (July, 1935). On the Fermi Theory of β-Radioactivity. *Phys. Rev.*, 48, 7; https://doi.org/10.1103/PhysRev.48.7.

…

In this case the situation was especially simple so that the generalized Lagrange density depends only on the inclination of C but not on its curvature etc. although the system under consideration does not allow the ordinary canonical description.

In conclusion we express our cordial thanks to Professor S. Tomonaga for his kind guidance and encouragement.

Kroll, N. M.* & Lamb, Jr., W. E. (February, 1949). On the self-energy of a bound electron.

[*Phys. Rev.*, 75, 3, 388-98; https://doi.org/ 10.1103/PhysRev.75.388; also in Schwinger, J. (ed). (1958). *Selected Papers on Quantum electrodynamics*. Dover, New York, pages 414-24.]

Received October 7, 1948.

Columbia University, New York, New York.

*Now National Research Fellow at The Institute for Advanced Study.

Calculation of electromagnetic shift of energy levels of bound electron based on usual formulation of *relativistic* quantum electrodynamics and *positron theory* gave 1052 megacycles per second for $^{2}2S_{1/2} - {}^{2}2P_{1/2}$ shift in hydrogen in close agreement with the *non-relativistic* calculation of 1052 megacycles per second by Bethe.

Abstract

The electromagnetic shift of the energy levels of a bound electron has been calculated *on the basis of the usual formulation of relativistic quantum electrodynamics and positron theory*. The theory gives a finite result of 1052 megacycles per second for the shift $^{2}2S_{1/2} - {}^{2}2P_{1/2}$ in hydrogen, in close agreement with the *non-relativistic* calculation by Bethe.

French, J. B. & Weisskopf, V. F. (April, 1949). The Electromagnetic Shift of Energy Levels.

[*Phys. Rev.*[PW], 75, 8, 1240-8; https://doi.org/10.1103/PhysRev.75.1240.]

Received December 10, 1948.

Massachusetts Institute of Technology, Cambridge, MA.

Relativistic calculation of Lamb shift using conventional form of perturbation theory and removing infinite *self-energy* of the electron by subtracting "mass operator" from Hamiltonian gives 1051 megacycles per second for $2s_{1/2} - 2p_{1/2}$ separation in hydrogen, compared with *non-relativistic* calculation of 1040 megacycles per second by Bethe, and *relativistic* calculation of 1052 megacycles per second by Kroll & Lamb, results of the different calculations suggest that they are not dependent on whether they are *relativistic* or *non-relativistic*.

[The initial calculation of the correct Lamb shift by Bethe (August, 1947) was *non-relativistic*. Various *relativistic* calculations followed the next year. A comedy of errors ensued: *both Feynman and Schwinger made an incorrect patch between hard and soft photon processes, and so obtained identical, but incorrect, predictions for the Lamb shift,* and the weight of their reputations delayed the publication of the correct, if pedestrian, calculations by Kroll & Lamb and French & Weisskopf until February, 1949.

Date received		Calculated Lamb shift (mc/sec.):	
June 27, 1947	Bethe	1040	*Non-relativistic*
December 30, 1947	Schwinger	1016#	*Relativistic*
July 12, 1948	Feynman	1023#	*Relativistic*
September 23, 1948	Fukuda, Miyamoto, & Tomonaga	1076	*Relativistic*
October 7, 1948	Kroll & Lamb	1052	*Relativistic*
December 10, 1948	French & Weisskopf	1051	*Relativistic*

before correction; after correction agreed with French & Weisskopf.]

[Feynman (September, 1949). Space-Time Approach to Quantum Electrodynamics: *Footnote 13*. "That the result given in Feynman (1948). (Relativistic Cut-Off for Quantum Electrodynamics), in Eq. (19) was in error was repeatedly pointed out to the author, in private communication, by V. F. Weisskopf and J. B. French, as their calculation, completed simultaneously with the author's early in 1948, gave a different result. French has finally shown that *although the expression for the radiation-less scattering B, Eq. (18) or (24) above is correct, it was incorrectly joined onto Bethe's non-relativistic result*. He shows that the relation $\ln 2k_{max} - 1 = \ln\lambda_{min}$ used by the author should have been $\ln 2k_{max} - 5/6 = \ln\lambda_{min}$. This results in adding a term – (1/6) to the logarithm in B, Eq. (19) so that the result now agrees with that of French & Weisskopf (April, 1949). The Electromagnetic Shift of Energy Levels, and Kroll & Lamb (1949). On the self-energy of a bound electron. The author feels unhappily responsible for the very considerable delay in the publication of French's result occasioned by this error.”]

Abstract.

The effect of the *interaction with the radiation field* in changing the energy levels of an electron in an external field is calculated using the conventional form of perturbation theory. *The infinite self-energy of the electron which occurs in the same approximation is removed by subtracting from the Hamiltonian a "mass operator" M.* The criteria used in deriving M are that it should correctly give the *self-energy* for a free electron in the absence of an external field and that the amended Hamiltonian should give a properly covariant form for the level shift in an external field. It is pointed out that M is uniquely determined by these requirements.

The results give 1051 mc/sec. for the $2s_{1/2} - 2p_{1/2}$ *separation in hydrogen* and also show the surplus magnetic moment of the electron as $\alpha/2\pi$ Bohr magnetons, as first found by Schwinger. The amended Hamiltonian can be used for determining the radiative corrections for other processes.

Schwinger, J. (February, 1949). Quantum Electrodynamics. II. Vacuum Polarization and Self-Energy.

[*Phys. Rev.*, 75, 4, 651-79; https://doi.org/10.1103/ PHYSREV.75.651.]

Received November 1, 1948.

Harvard University, Cambridge, Massachusetts.

Interaction representation is applied to *polarization of the vacuum* and the *self-energies of the electron and photon*, in *first section* the vacuum of the non-interacting *electromagnetic* and *matter* fields is *covariantly* defined as *state* for which the eigenvalue of an arbitrary time-like component of the *energy-momentum four-vector* is an absolute minimum, covariant decomposition of field operators into positive and negative frequency components introduced to characterize vacuum *state vector*, shows that *state vector for electromagnetic vacuum* annihilated by positive frequency part of transverse four-vector potential and *state vector for matter vacuum* annihilated by positive frequency part of Dirac spinor and its charge conjugate, these properties of vacuum *state vector* employed in the calculation of the *vacuum expectation* values of quadratic field quantities, specifically the *energy-momentum tensors* of the independent *electromagnetic* and *matter fields* and the *current four-vector*, infers that the *electromagnetic energy-momentum tensor* and *current vector* must vanish in the vacuum, while the *matter field energy-momentum tensor* vanishes in the vacuum only by the addition of a suitable multiple of the unit tensor, *second section treats the induction of a current in the vacuum by an external electromagnetic field*, supposes that *external electromagnetic field* does not produce actual electron-positron pairs, considers only phenomenon of *virtual pair creation*, restriction is introduced by requiring that establishment and subsequent removal of the external field produce no net change in state for the *matter field*, *demonstrates that the induced current at a given space-time point involves the external current in the vicinity of that point and not the electromagnetic potentials*, this *gauge invariant* result shows that a light wave propagating at remote distances from its source induces no current in the vacuum and is therefore undisturbed in its passage through space, indicates *absence of a light quantum self-energy effect*, *current* induced at a point consists of two parts, a logarithmically divergent multiple of the *external current* at that point *which produces an unobservable renormalization of charge* and a more involved finite contribution which is the physically significant *induced current*, *third section considers the modification of the matter field properties arising from interaction with the vacuum fluctuations of the electromagnetic field*, analysis carried out with two alternative formulations, one employing the complete *electromagnetic potential* together with a *supplementary condition*, the other using the *transverse potential* with the variables of the *supplementary condition* eliminated, no real processes produced by first order coupling between the fields, alternative *equations of motion* for the *state vector* are constructed from which the first order interaction term has been eliminated and replaced by the *second order coupling* which it generates, this includes the *self-action* of individual particles and light quanta, the *interaction* of different particles, and a *coupling* between particles and light quanta which produces such effects as Compton scattering and two quantum pair annihilation, concludes from comparison of the alternative procedures that *for the treatment of virtual light quantum processes the separate consideration of longitudinal and transverse fields is an inadvisable complication*, light quantum *self-energy* term is shown to vanish, while that for a particle has the form for a change in *proper mass* but is logarithmically divergent in agreement with previous calculations, identification of *self-energy* effect with a change in *proper mass* is confirmed by removing this term from the state vector *equation of motion*, alters the *matter field equations of motion* in the expected manner, verifies that *the energy and momentum modifications produced by self-interaction effects are entirely accounted for by the addition of the electromagnetic proper mass to the mechanical proper mass—an unobservable mass renormalization*, an appendix is devoted to the construction of several invariant functions associated with the *electromagnetic* and *matter* fields.

> [Schwinger (November, 1948). Quantum Electrodynamics. I. A Covariant Formulation: "In the second paper we treat the problems of *electron and photon self-energy*, together with the *polarization of the vacuum*, ...".]

Abstract.

The *covariant formulation of quantum electrodynamics*, developed in a previous paper, is here applied to two elementary problems—the *polarization of the vacuum* and the *self-energies of the electron and photon.*

In the first section the vacuum of the non-interacting *electromagnetic* and *matter* fields is *covariantly* defined as that state for which the eigenvalue of an arbitrary time-like component of the *energy-momentum four-vector* is an absolute minimum. It is remarked that this definition must be compatible with the requirement that the vacuum expectation values of a physical quantity in various coordinate systems should be, not only *covariantly* related, but identical, since the vacuum has a significance that is independent of the coordinate system. *In order to construct a suitable characterization of the vacuum state vector, a covariant decomposition of the field operators into positive and negative frequency components is introduced*, and the properties of these associated fields developed. It is shown that the *state vector* for the *electromagnetic vacuum* is annihilated by the positive frequency part of the *transverse four-vector potential*, while that for the *matter vacuum* is annihilated by the positive frequency part of the *Dirac spinor* and of its *charge conjugate*. These defining properties of the vacuum *state vector* are employed in the calculation of the *vacuum expectation* values of quadratic field quantities, specifically the *energy-momentum tensors* of the independent *electromagnetic* and *matter fields*, and the *current four-vector*. It is inferred that the *electromagnetic energy-momentum tensor*, and the *current vector* must vanish in the vacuum, while the *matter field energy-momentum tensor* vanishes in the vacuum only by the addition of a suitable multiple of the unit tensor.

The second section treats the induction of a *current* in the vacuum by an *external electromagnetic field*. It is supposed that the latter does not produce actual electron-positron pairs; that is, we consider only the phenomenon of *virtual pair creation*. This restriction is introduced by requiring that the establishment and subsequent removal of the external field produce no net change in state for the *matter field*. *It is demonstrated, in a general manner, that the induced current at a given space-time point involves the external current in the vicinity of that point, and not the electromagnetic potentials*. This *gauge invariant* result shows that a light wave, propagating at remote distances from its source, induces no current in the vacuum and is therefore undisturbed in its passage through space. *The absence of a light quantum self-energy effect is thus indicated.* The *current* induced at a point consists, more precisely, of two parts: a logarithmically divergent multiple of the *external current* at that point, which produces an unobservable renormalization of *charge*, and a more involved finite contribution, which is the physically significant *induced current*. The latter agrees with the results of previous investigations.

The modification of the *matter field* properties arising from *interaction* with the vacuum fluctuations of the *electromagnetic field* is considered in the third section. The analysis is carried out with two alternative formulations, one employing the complete *electromagnetic potential* together with a *supplementary condition*, the other using the *transverse potential*, with the variables of the *supplementary condition* eliminated. It is noted that no real processes are produced by the first order coupling between the fields. Accordingly, alternative *equations of motion* for the *state vector* are constructed, from which the first order interaction term has been eliminated and replaced by the second order coupling which it generates. The latter includes the *self-action* of individual particles and light quanta, the *interaction* of different particles, and a *coupling* between particles and light quanta which produces such effects as Compton scattering and two quantum pair annihilation. It is concluded from a comparison of the alternative procedures that, *for the treatment of virtual light quantum processes, the separate consideration of longitudinal and transverse fields is an inadvisable complication.* The light quantum *self-energy* term is shown to vanish, while that for a particle has the anticipated form for a change in *proper mass*, although the latter is logarithmically divergent, in agreement with previous calculations. To confirm the identification of the *self-energy* effect with a change in *proper mass*, it is shown that the result of removing this term from the state vector *equation of motion* is to alter the *matter field equations of motion* in the expected manner. *It is verified, finally, that the energy and momentum modifications produced by self-interaction effects are entirely accounted for by the addition of the electromagnetic proper mass to the mechanical proper mass—an unobservable mass renormalization.* An appendix is devoted to the construction of several invariant functions associated with the *electromagnetic* and *matter* fields.

Dyson, F. J. (June, 1949). The S Matrix in Quantum Electrodynamics.

[*Phys. Rev.*, 75, 1736-55; https://doi.org/10.1103/PhysRev.75.1736; also in Schwinger, J. (ed). (1958). *Selected Papers on Quantum electrodynamics*. Dover, New York, pages 292-311.]

Received February 24, 1949.

Institute for Advanced Study, Princeton, New Jersey.

The covariant *quantum electrodynamics* of Tomonaga, Schwinger and Feynman is used as basis for a general treatment of scattering problems involving electrons, positrons, and photons, addresses relation between Schwinger and Feynman theories when restriction to one-electron problems is removed, in these more general circumstances the two theories appear as complementary rather than identical, *the Feynman method is essentially a set of rules for the calculation of the elements of the Heisenberg S matrix* corresponding to any physical process and can be applied directly to all kinds of scattering problems, *the Schwinger method evaluates radiative corrections by exhibiting them as extra terms appearing in the Schrodinger equation, practical usefulness of S matrix as connecting link between the Feynman technique of calculation and Hamiltonian formulation of quantum electrodynamics*, the *Feynman radiation theory* provides a set of rules for the calculation of matrix elements between *states* composed of any number of ingoing and outgoing free particles, it is thus an S matrix theory, scattering processes including the creation and annihilation of particles are completely described by Heisenberg's S matrix, the elements of this matrix are calculated by a consistent use of perturbation theory to any desired order, detailed rules given for carrying out such calculations, divergences arising from higher order radiative corrections removed from S matrix by consistent use of *mass and charge renormalization, the operators so calculated are divergence-free, the divergent parts at every stage of the calculation being explicitly dropped after being separated from the finite parts*, must be justified *a posteriori* by the fact that they ultimately lead to a clear separation of finite from infinite expressions, *such an a posteriori justification of dubious manipulations is an inevitable feature of any theory which aims to extract meaningful results from not completely consistent premises*, the perturbation theory of this paper is *applicable only to a restricted class of problems* when not only the radiation interaction but also the external potential is small enough to be treated as a perturbation, *does not give a satisfactory approximation either in problems involving bound states or in scattering problems at low energies,* in other situations the Schwinger theory will have to be used in its original form, *problems of extending treatment to include bound-state phenomena and of proving convergence of the theory as the order of perturbation itself tends to infinity not addressed*, analysis suggests that *the divergences of electrodynamics are directly attributable to the fact that the Hamiltonian formalism is based upon an idealized conception of measurability*, now no longer a compelling necessity for a future theory to abandon some essential features of the present electrodynamics, *the present electrodynamics is certainly incomplete, but is no longer certainly incorrect.*

[The *S-matrix* or scattering matrix relates the initial state and the final state of a physical system undergoing a scattering process. It is used in quantum mechanics, scattering theory and quantum field theory. *S-matrix theory* was proposed as a principle of particle interactions by Heisenberg in 1943, following Wheeler's 1937 introduction of the S-matrix. It avoided the notion of space and time by replacing it with abstract mathematical properties of the S-matrix. *In S-matrix theory, the S-matrix relates the infinite past to the infinite future in one step, without being decomposable into intermediate steps corresponding to time-slices.*

S-matrix theory was largely abandoned by physicists in the 1970s, as quantum chromodynamics was recognized to solve the problems of strong interactions within the framework of field theory. But in the guise of string theory, S-matrix theory is still a popular approach to the problem of quantum gravity.]

Abstract

The covariant *quantum electrodynamics* of Tomonaga, Schwinger, and Feynman is used as the basis for a general treatment of scattering problems involving electrons, positrons, and photons. Scattering processes, including the creation and annihilation of particles, are completely described by the *S matrix* of Heisenberg. It is shown that the elements of this matrix can be calculated, by a consistent use of perturbation theory, to any desired order in the fine-structure constant. Detailed rules are given for carrying out such calculations, and it is shown that divergences arising from higher order radiative corrections can be removed from the S matrix by a consistent use of the ideas of *mass and charge renormalization.*

Not considered in this paper are the problems of extending the treatment to include bound-state phenomena, and of proving the convergence of the theory as the order of perturbation itself tends to infinity.

I. *Introduction*

In a previous paper[1] (to be referred to in what follows as I) the radiation theory of Tonmnaga[2] and Schwinger[3] was applied in detail to the problem of the *radiative corrections to the motion of a single electron in a given external field.*

[1] Dyson, F. J. (February, 1949). The Radiation Theories of Tomonaga, Schwinger, and Feynman. *Phys. Rev.*, 75, 3, 486-502 [; the recent and independent formulations of quantum electrodynamics by Tomonaga, Schwinger, and Feynman have made two notable advances, the foundations and applications of the theory have been simplified by being presented in a *completely relativistic way* and the divergence difficulties *have been at least partially overcome*, the advantages of the Feynman formulation are simplicity and ease of application while those of Tomonaga-Schwinger are generality and theoretical completeness, this paper presents a unified development of quantum electrodynamics embodying the main features of Tomonaga-Schwinger and Feynman radiation theories, *emphasis on application of the theory*, aims to show how the Schwinger theory can be applied to specific problems in such a way as to incorporate the ideas of Feynman, the main results are general formulas from which radiative reactions on motions of electrons can be calculated, divides *energy-density* into two parts, *energy of interaction* of two fields with each other and energy produced by external forces, *interaction energy* alone is treated as a

perturbation, important results of the paper are the *equation of motion* $ihc[\partial\Omega/\partial\sigma(x_0)] = \{S(\sigma)\}^{-1}H^e(x_0)S(\sigma)\Omega$ for the *state vector* $\Omega(\sigma)$ and the interpretation of the *state vector* Ω, simplifies Schwinger theory for using it for calculations, demonstrates equivalence of the theories within their common domain of applicability, in the *Schwinger theory* the aim is to calculate the matrix elements of the "*effective external potential energy*" between *states* specified by their *state vectors*, in the *Feynman theory* the basic principle is to "preserve symmetry between past and future" so the matrix elements of the operator are evaluated in a "*mixed representation*" in which the matrix elements are calculated between an *initial state* specified by its *state vector* and a *final state* specified by its *state vector,* a *graph* corresponding to a particular matrix element is used not merely as an aid to calculation but as a picture of the physical process which gives rise to that matrix element, derives fundamental formulas for the operator in the *equation of motion* for the *state vector* $\Omega(\sigma)$ which represents the interaction of a physical particle with an external field for both the Schwinger and the Feynman theories, derives set of rules by which matrix element of Feynman operator may be written down in form suitable for numerical evaluation, shows equivalence of the two theories, develops graphical representation of matrix elements, *the theory as a whole cannot be put into a finally satisfactory form so long as divergencies occur in it however skillfully these divergencies are circumvented*, present treatment should be regarded as justified by its success in applications rather than by its theoretical derivation, *paper suffers from a series of significant errors*].

[2] Tomonaga, S. (1946). On a Relativistically Invariant Formulation of the Quantum Theory of Wave Fields. *Prog. Theor. Phys*., 1, 27–42. Translated from the paper (1943). *Bull. I. P. C. R. Riken-iho*, 22, 545 (appeared originally in Japanese); Koba, Z., Tati, T. & Tomonaga, S. (1947). On a Relativistically Invariant Formulation of the Quantum Theory of Wave Fields. II. Case of Interacting Electromagnetic and Electron Fields. § 1-4. *Prog. Theor. Phys*., 2, 101–116; https://doi.org/10.1143/ PTP.2.101; Koba, Z., Tati, T. & Tomonaga, S. (1947). On a Relativistically Invariant Formulation of the Quantum Theory of Wave Fields. III. Case of Interacting Electromagnetic and Electron Fields. § 5-7. *Prog. Theor. Phys*., 2, 198–208; https://doi.org/10.1143/ PTP.2.198; Kanesawa, S. & Tomonaga, S. (1948). On a Relativistically Invariant Formulation of the Quantum Theory of Wave Fields. IV. Case of Interacting Electromagnetic and Meson Fields. § 1-4. *Prog. Theor. Phys.* 3, 1–13; https://doi.org/10.1143/PTP.3.1; Kanesawa, S. & Tomonaga, S. (1948). On a Relativistically Invariant Formulation of the Quantum Theory of Wave Fields. V. Case of Interacting Electromagnetic and Meson Fields. § 5-7. *Prog. Theor. Phys.* 3, 101–113; https://doi.org/10.1143/PTP.3.101; Tomonaga, S. & Oppenheimer, J. R. (July, 1948). On Infinite Field Reactions in Quantum Field Theory. *Phys. Rev.*, 74, 224–225; https://doi.org/10.1103/PhysRev.74.224; Ito, D., Koba, Z. & Tomonaga, S. (1948). Corrections due to the Reaction of " Cohesive Force Field" to the Elastic Scattering of an Electron. *Prog. Theor. Phys*. 3, 3, 276-289; 3, 4, 325-337; [*abandoned C-meson field theory*]; Koba, Z. & Tomonaga, S. (September, 1948). On Radiation Reactions in Collision Processes. I: Application of the "Self-Consistent" Subtraction Method to the Elastic Scattering of an Electron. *Progr. Theor. Phys. (Kyoto)*, 3, 3, 290-303. [*Omitted* Tati, T. & Tomonaga, S. (December, 1948). A Self-Consistent Subtraction Method in the Quantum Field Theory, I. *Ibid*., 3, 4, 391-406.]

[3] Schwinger, J. (February, 1948). On quantum-electrodynamics and the magnetic moment of the electron. *Phys. Rev.*, 73, 4, 416-7; Schwinger, J. (November, 1948). Quantum

Electrodynamics. I. A Covariant Formulation. *Phys. Rev.*, 74, 10, 1439-61. Schwinger, J. (1949). Quantum Electrodynamics. II. Vacuum Polarization and Self-Energy. *Phys. Rev.*, 75, 4, 651-79.

It was shown that the rules of calculation for corrections of this kind *were identical with those which had been derived by Feynman[4] from his own radiation theory.*

[4] Feynman, R. P. (November, 1948). Relativistic Cut-Off for Quantum Electrodynamics. *Phys. Rev.*, 74, 1430-8 [; describes model based on quantization of a classical theory for which all quantities automatically come out finite described in his previous paper [Feynman (October, 1948). A Relativistic Cut-Off for Classical Electrodynamics], contains an arbitrary function on which numerical results depend, only term that depends significantly (logarithmically) on the cut-off frequency is the *self-energy* which can be used to *renormalize* the electron mass, remaining terms are nearly independent of the function, applies only to results for processes in which virtual quanta are emitted and absorbed, *terms representing processes involving a pair production followed by annihilation of the same pair are infinite and not made convergent by the present scheme,* problems of *permanent emission* and the position of *positron theory* need to be addressed, *the present paper may be looked upon as presenting an arbitrary rule to cut off at high frequencies in a relativistically invariant manner the otherwise divergent integrals appearing in quantum field theories*, produces finite invariant *self-energy* for a free electron, but problem of *polarization of the vacuum* not solved, alternative cut-off procedure which eliminates high frequency intermediate states offers to solve *vacuum polarization* problems as well].

For the one-electron problem the *radiative corrections* were fully described by an operator H_T (Eq. (20) of I) which appeared as the "effective potential" acting upon the electron, after the interactions of the electron with its own self-field had been eliminated by a *contact transformation.*

[Dyson (February, 1949). The Radiation Theories of Tomonaga, Schwinger, and Feynman: "the *equation of motion* for [state vector] $\Omega(\sigma)$ is

$$ihc[\partial\Omega/\partial\sigma(x_0)] = \{S(\sigma)\}^{-1}H^e(x_0)S(\sigma)\Omega. \qquad (19)$$

... Equation (19) describes the extent to which the motion of a single physical particle deviates, in the external field, from the motion represented by a constant *state-vector*, i.e., from the motion of an observed "free" particle.

... it may be said that the operator,

$$H_T(x_0) = \{S(\sigma)\}^{-1}H^e(x_0)S(\sigma), \qquad (20)$$

on the right of (19) *represents the interaction of a physical particle with an external field, including radiative corrections*"]

The difference between the Schwinger and Feynman theories lay only in the choice of a particular representation in which the matrix elements of H_T were calculated (Section V of I).

The present paper deals with the relation between the Schwinger and Feynman theories when the restriction to one-electron problems is removed. In these more general circumstances, the two theories appear as complementary rather than identical. *The*

Feynman method is essentially a set of rules for the calculation of the elements of the Heisenberg S matrix corresponding to any physical process, and can be applied with directness to all kinds of scattering problems[5].

[5] The idea of using standard electrodynamics as a starting point for an explicit calculation of the S matrix has been previously developed by Stueckelberg, E. C. G. (1941). The Interpretation of the Positron as a positive Energy Electron traveling backward in time. *Helv. Phys. Acta*, 14, 51-80; (1944). *Idem*, 17, 3; (1945). *Idem*, 18, 195; (1946). *Idem*, 19, 242; (January, 1944). An Unambiguous Method of Avoiding Divergence Difficulties in Quantum Theory. *Nature*, 153, 143-4; https://doi.org/10.1038/153143a0 [; "The classical theory of a point charge and the quantum theory of wave packets contain in their usual form well-known divergences, Dirac (August, 1938. Classical Theory of Radiating Electrons), has elaborated a classical particle theory which avoids these difficulties, but his results cannot as yet be applied to quantum theory. On the other hand, Heisenberg [Heisenberg (July, 1943). Die beobachtbaren Größen in der Theorie der Elementarteilchen. (The "observable quantities" in the theory of elementary particles.)] has recently proposed a new formalism which permits the calculation of collision cross-sections in quantum theory without being disturbed by diverging terms. However, no connecting link such as the correspondence principle has been given in order to apply this quantum formalism to a given problem such as Rutherford scattering, Compton effect or radiation damping"]; Stueckelberg, E. C. G. (1947). *Phys. Soc. Cambridge Conference Report*, 199; Rivier, D. & Stueckelberg, E. C. G. (July, 1948). A Convergent Expression for the Magnetic Moment of the Neutron. *Phys. Rev.*, 74, 218; https://doi.org/10.1103/ PhysRev.74.218.

Stueckelberg anticipated several features of the Feynman theory, in particular the use of the function D_F (in Stueckelberg's notation D^C) to represent retarded (i.e., causally transmitted) electromagnetic interactions. For a review of the earlier part of this work, see Wentzel, G. (January, 1947). *Rev. Mod. Phys*., 19, 1; https://doi.org/10.1103/ RevModPhys.19.1. The use of *mass renormalization* in scattering problems is due to Lewis, H. W. (January, 1948). On the Reactive Terms in Quantum Electrodynamics. *Phys. Rev*., 73, 173; https://doi.org/10.1103/PhysRev.73.173.

The Schwinger method evaluates radiative corrections by exhibiting them as extra terms appearing in the Schrodinger equation of a system of particles and is suited especially to bound-state problems. In spite of the difference of principle, the two methods in practice involve the calculation of closely related expressions; moreover, *the theory underlying them is in all cases the same*. The systematic technique of Feynman, the exposition of which occupied the second half of I and occupies the major part of the present paper, is therefore now available for the evaluation not only of the S matrix but also of most of the operators occurring in the Schwinger theory.

The prominent part which the S matrix plays in this paper is due to its practical usefulness as the connecting link between the Feynman technique of calculation and the Hamiltonian formulation of quantum electrodynamics. This practical usefulness remains, whether or not one follows Heisenberg in believing that the S matrix may eventually replace the Hamiltonian altogether. *It is still an unanswered question, whether the finiteness of the S matrix automatically implies the finiteness of all observable quantities, such as bound-state energy levels, optical transition probabilities, etc., occurring in electrodynamics*. An affirmative answer to the question is in no way essential to the arguments of this paper. Even if a finite S matrix does not of itself imply finiteness of other observable quantities, it is probable that all such quantities will be finite; to verify this, it will be necessary to repeat the analysis of the present paper, keeping all the time closer to the original Schwinger theory than has here been possible. There is no reason for attributing a more fundamental significance to the S matrix than to other observable quantities, nor was it Heisenberg's intention to do so. In the last section of this paper, tentative suggestions are made for a synthesis of the Hamiltonian and Heisenberg philosophies.

II. *The Feynman theory as an S matrix theory*

The S matrix was originally defined by Heisenberg in terms of the stationary solutions of a scattering problem. A typical stationary solution is represented by a time-independent wave function Ψ' which has a part representing ingoing waves which are asymptotically of the form Ψ_1', and a part representing outgoing waves which are asymptotically of the form Ψ_2'. The S matrix is the transformation operator S with the property that

$$\Psi_2' = S\Psi_1' \qquad (1)$$

for every stationary state Ψ'.

In Section III of I an operator $U(\infty)$ was defined and stated to be identical with the S matrix. Since $U(\infty)$ was defined in terms of time-dependent wave functions, a little care is needed in making the identification. In fact, the equation

$$\Psi_2 = U(\infty)\Psi_1 \qquad (2)$$

held, where Ψ_1 and Ψ_2 were the asymptotic forms of the ingoing and outgoing parts of a wave function Ψ in the Ψ-representation of I (the "*interaction representation*" of Schwinger[3]). Now the time-independent wave function Ψ' corresponds to a time-dependent wave function

$$\exp[(-i/h)Et]\,\Psi'$$

in the Schrodinger representation, where E is the *total energy* of the *state*; and this corresponds to a wave function in the *interaction representation*

$$\Psi = \exp[(+i/h)t(H_0 - E)]\,\Psi', \qquad (3)$$

where H_0 is the *total free particle Hamiltonian*. However, the asymptotic parts of the wave function Ψ', both ingoing and outgoing, represent freely traveling particles of total energy E, and are therefore eigenfunctions of H_0 with eigenvalue E. This implies, in virtue of (3), that the asymptotic parts Ψ_1 and Ψ_2 of Ψ are actually time-independent and equal, respectively, to Ψ_1' and Ψ_2'. Thus (1) and (2) are identical, and $U(\infty)$ is indeed the S matrix. Incidentally, $U(\infty)$ is also the "*invariant collision operator*" defined by Schwinger[3].

There is a series expansion of $U(\infty)$ analogous to (32) of I,

[As a special case of (31) obtained by replacing H^e by the unit matrix in (27),

$$S(\infty) = \sum_{n=0}^{\infty} (-i/hc)^n [1/n!] \int_{-\infty}^{\infty} dx_1 \ldots \int_{-\infty}^{\infty} dx_n$$
$$\times P\{H^I(x_j)\ldots, H^I(x_n)\}. \qquad (32)]$$

namely,

$$U(\infty) = \sum_{n=0}^{\infty} (-i/hc)^n [1/n!] \int_{-\infty}^{\infty} dx_1 \ldots \int_{-\infty}^{\infty} dx_n$$
$$\times P\{H^I(x_j)\ldots, H^I(x_n)\}. \qquad (4)$$

Here the P notation is as defined in Section V of I, and

$$H_1(x) = H^I(x) + H^e(x) \qquad (5)$$

is the sum of the *interaction energies* of the electron field with the photon field and with the external potentials. *The Feynman radiation theory provides a set of rules for the calculation of matrix elements of (4), between states composed of any number of ingoing and outgoing free particles.* Also, quantities contained in (4) are the only ones with which the Feynman rules can deal directly. *The Feynman theory is thus correctly characterized as an S matrix theory.*

One particular way to analyze $U(\infty)$ is to use (5) to expand (4) in a series of terms of ascending order in H^e. Substitution from (5) into (4) gives

$$U(\infty) = \sum_{m=0}^{\infty} \sum_{n=0}^{\infty} (-i/hc)^{m+n} [1/m!n!] \int_{-\infty}^{\infty} dx_1 \ldots \int_{-\infty}^{\infty} dx_n$$
$$\times \int_{-\infty}^{\infty} dx_{m+n}\, P\{H^e(x_1)\ldots, H^e(x_m), x\, H^I(x_{m+1})\ldots, H^I(x_{m+n})\}. \qquad (6)$$

ln this double series, the term of zero order in H^e is $S(\infty)$, given by (32) of I. The term of first order is

$$U_1 = (-i/hc) \int_{-\infty}^{\infty} H_F(x)dx, \qquad (7)$$

where H_F is given by (31) of I

$$[H_F(x_0) = \sum_{n=0}^{\infty} (-i/hc)^n [1/n!] \int_{-\infty}^{\infty} dx_1 \ldots \int_{-\infty}^{\infty} dx_n$$
$$\times P\{H^e(x_0), H^I(x_j)\ldots, H^I(x_n)\}. \qquad (31)]$$

Clearly, $S(\infty)$ is the S matrix representing scattering of electrons and photons by each other in the absence of an external potential; *U_1 is the S matrix representing the additional scattering produced by an external potential, when the external potential is treated in the first Born approximation*; higher terms of the series (6) would correspond to treating the external potential in the second or higher Born approximation. The operator H_F played a prominent part in I, where it was in no way connected with a Born approximation; however, it was there introduced in a somewhat unnatural manner, and its physical meaning is made clearer by its appearance in (7)

$$[U_1 = (-i/hc) \int_{-\infty}^{\infty} H_F(x)dx. \qquad (7)]$$

In fact, H_F may be defined by the statement that

$$(-i/h)(\delta t)(\delta \omega)H_F(x)$$

is the contribution to the S matrix that would be produced by an external potential of strength H^e, acting for a small duration δt and over a small volume $\delta\omega$ in the neighborhood of the space-time point x.

The remainder of this section will be occupied with a statement of the Feynman rules for evaluating $U(\infty)$. Proofs will not be given, because *the rules are only trivial generalizations of the rules which were given in I for the evaluation of matrix elements of H_F corresponding to one-electron transitions*.

In evaluating $U(\infty)$ we shall not make any distinction between the external and radiative parts of the *electromagnetic field*; this is physically reasonable since it is to some extent a matter of convention how much of the field in a given situation is to be regarded as "external". The *interaction energy* occurring in (4)

$$[U(\infty) = \sum_{n=0}^{\infty} (-i/hc)^n [1/n!] \int_{-\infty}^{\infty} dx_1 \ldots \int_{-\infty}^{\infty} dx_n \times P\{H^I(x_j)\ldots, H^I(x_n)\}. \qquad (4)]$$

is then

$$H_I(x) = - ieA_\mu(x)\psi^-(x)\gamma_\mu\psi(x) - \delta mc^2\psi^-(x)\psi(x), \qquad (8)$$

where A_μ is the *total electromagnetic field*, and the term in δm is included in order to allow for the fact that the *interaction representation* is defined in terms of the *total mass* of an electron including its "*electromagnetic mass*" δm (see Section IV of I). The first step in the evaluation of $U(\infty)$ is to substitute from (8) into (4), writing out in full the suffixes of the operators ψ^-_α, ψ_β which are concealed in the matrix product notation of (8). After such a substitution, (4) becomes

$$U(\infty) = \sum_{n=0}^{\infty} J_n, \qquad (9)$$

where J_n is an n-fold integral with an integrand which is a polynomial in ψ^-_α, ψ_β and A_μ operators.

The most general matrix element of J_n is obtained by allowing some of the ψ^-_α, ψ_β and A_μ operators to annihilate particles in the initial state, some to create particles in the final state, while others are associated in pairs to perform a successive creation and annihilation of intermediate particles. The operators which are not associated in pairs, and which are available for the real creation and annihilation of particles, are called "free"; a particular type of matrix element of J_n is specified by enumerating which of the operators in the integrand are to be free and which are to be associated in pairs. As described more fully in Section VII of I, each type of matrix element of J_n is uniquely represented by a "*graph*" G consisting of n points (bearing the labels $x_1, \ldots, x_n$) and various lines terminating at these points.

The relation between a type of matrix element of J_n and its graph G is as follows. For every associated pair of operators $(\psi^-(x), \psi(y))$, there is a directed line (electron line) joining x to y in G. For every associated pair of operators $(A(x), A(y))$, there is an undirected line (photon line) joining x and y in G. For every free operator $\psi^-(x)$, there is a directed line in G leading from x to the edge of the diagram. For every free operator $\psi(x)$, there is a directed line in G leading to x from the edge of the diagram. For every free operator $A(x)$, there is an undirected line in G leading from x to the edge of the diagram. Finally, for a particular type of matrix element of J_n it is specified that at each point x_i either the part of $H_I(x_i)$

containing $A_\mu(x_i)$ or the part containing δm is operating; correspondingly, at each vertex x_i of G there are either two electron lines (one ingoing and one outgoing) and one photon line, or else two electron lines only. Lines joining one point to itself are always forbidden.

In every graph G, the electron lines form a finite number m of open polygonal arcs with ends at the edge of the diagram, and perhaps in addition a number l of closed polygonal loops. The corresponding type of matrix element of J_n has m free operators ψ^- and m free operators ψ; the two end segments of any one open arc correspond to two free operators, one ψ^- and one ψ, which will be called a "free pair". The matrix elements of J_n are now to be calculated by means of an operator J(G), which is defined for each graph G of n vertices, and which is obtained from J_n by making the following five alterations.

First, at each point x_i, $H_I(x_i)$ is to be replaced by either the first or the second term on the right of (8)

[The *interaction energy* occurring in (4) is then

$$H_I(x) = -ieA_\mu(x)\psi^-(x)\gamma_\mu\psi(x) - \delta mc^2\psi^-(x)\psi(x), \qquad (8)$$

where A_μ is the *total electromagnetic field*, and the term in δm is included in order to allow for the fact that the *interaction representation* is defined in terms of the *total mass* of an electron including its "*electromagnetic mass*" δm,]

as indicated by the presence or absence of a photon line at the vertex x_i of G. Second, for every electron line joining a vertex x to a vertex y in G, two operators $\psi^-_\alpha(x)$ and $\psi_\beta(y)$ in J_n, regardless of their positions, are to be replaced by the function

$$\tfrac{1}{2}\, S_{F\beta\alpha}(x - y), \qquad (10)$$

as defined by (44) and (45) of I

[$$S_{F\beta\alpha}(x) = -\{\gamma_\mu(\partial/\partial x_\mu) + \kappa_0\}_{\beta\alpha}\Delta_F(x), \qquad (44)$$

where κ_0 is the reciprocal Compton wave-length of the electron, $\eta(x,y)$ is -1 or $+1$ according as $\sigma(x)$ is earlier or later than $\sigma(y)$ in time, and Δ_F is a function with the integral expansion

$$\Delta_F = -[i/2\pi^2]\int_0^\infty \exp[i\alpha x^2 - i\,\kappa_0^2/4\alpha]\,d\alpha. \qquad (45)]$$

Third, for every photon line joining two vertices x and y of G, two operators $A_\mu(x)$ and $A_\nu(y)$ in J_n, regardless of their positions, are to be replaced by the function

$$\tfrac{1}{2}\,\hbar c\delta_{\mu\nu}D_F(x - y), \qquad (11)$$

defined by (42) of I

[$$D_F(x) = -[i/2\pi^2]\int_0^\infty \exp[i\alpha x^2]\,d\alpha, \qquad (42)$$

where x^2 denotes the square of the invariant length of the 4-vector x.]

Fourth, all free operators in J_n are to be left unaltered, but the ordering by the P notation is to be dropped, and the order of the free ψ^- and ψ operators is to be arranged so that the two members of each free pair stand consecutively and in the order $\psi^-\psi$; the order of the free pairs among themselves, and of all free A_μ operators, is left arbitrary. Fifth, the whole expression J_n is to be multiplied by

$$(-1)^{n-l-m}. \qquad (12)$$

The *Feynman rules* for the evaluation of U(∞) are essentially contained in the above definition of the operators J(G). To each value of n correspond only a finite number of graphs G, and all possible matrix elements of U(∞) are obtained by substituting into (9)

$$[U(\infty) = \sum_{n=0}^{\infty} J_n, \qquad (9)]$$

for each J the sum of all the corresponding J(G). It is necessary only to specify how the matrix element of a given J(G) corresponding to a given scattering process may be written down.

The matrix element of J(G) for a given process may be obtained, broadly speaking, by replacing each free opera. tor in J(G) by the wave function of the particle which it is supposed to create or annihilate. More specifically, each free ψ^- operator may either create an electron in the final state or annihilate a positron in the initial state, and the reverse processes are performed by a free ψ operator. Therefore, for a transition from a state involving A electrons and B positrons to a state involving C electrons and D positrons, only operators J(G) containing (A + D) = (B + C) free pairs contribute matrix elements. For each such J(G), the (A + D) free ψ operators are to be replaced in all possible combinations by the A initial electron wave functions and the D final positron wave functions, and the (B + C) free ψ^- operators are to be similarly replaced by the initial positron and final electron wave functions, and the results of all such replacements added together, taking account of the anti-symmetry of the total wave functions of the system in the individual particle wave functions. In the case of the free A_μ operators, the situation is rather different, since each such operator may either create a photon in the final state, or annihilate a photon in the initial state, or represent merely the external potential. Therefore, for a transition from a state with A photons to a state with 8 photons, any J(G) with not less than (A + B) free A_μ operators may give a matrix element. If the number of free A_μ operators in J(G) is (A + B + C), these operators are to be replaced in all possible combinations by the (A + B) suitably normalized potentials corresponding to the initial and final photon states, and by the external potential taken C times, and the results of all such replacements added together, taking account now of the symmetry of the total wave functions in the individual photon states.

In practice cases are seldom likely to arise of scattering problems in which more than two similar particles are involved. The replacement of the free operators in J(G) by wave functions can usually be carried out by inspection, and the enumeration of matrix-elements of U(∞) is practically complete as soon as the operators J(G) have been written down.

The above rules for the calculation of U(∞) describe the state of affairs before any attempt has been made to identify and remove the various divergent parts of the expressions. In particular, contributions are included from all graphs G, even those which yield nothing but *self-energy* effects. For this reason, *the rules here formulated are superficially different from those given for the one electron problem in Section IX of I, which described the state of affairs after many divergencies had been removed.* Needless to say, *the rules are not complete until instructions have been supplied for the removal of all infinite quantities from the theory*; in Sections V-VII of this paper it will be shown how the formal structure of the S matrix makes such a complete removal of infinities *appear attainable*.

Another essential limitation is introduced into the S matrix theory by the use of the expansion (4)

$$[U(\infty) = \sum_{n=0}^{\infty} (-i/hc)^n [1/n!] \int_{-\infty}^{\infty} dx_1 \ldots \int_{-\infty}^{\infty} dx_n \times P\{H^I(x_j)\ldots, H^I(x_n)\}. \quad (4)]$$

All quantities discussed in this paper are expansions of this kind, in which *it is assumed that not only the radiation interaction but also the external potential is small enough to be treated as a perturbation*. It is well known that an expansion in powers of the external potential *does not give a satisfactory approximation, either in problems involving bound states or in scattering problems at low energies*. In particular, whenever a scattering problem allows the possibility of one of the incident particles being captured into a bound state, the capture process will not be represented in $U(\infty)$, since the initial and final states for processes described by $U(\infty)$ are always free-particle states. It is the expansion in powers of the external potential which breaks down when such a capture process is possible. Therefore, *it must be emphasized that the perturbation theory of this paper is applicable only to a restricted class of problems, and that in other situations the Schwinger theory will have to be used in its original form*.

III. *The S matrix in momentum space*

Both for practical applications to specific problems, and for general theoretical discussion, it is convenient to express the S matrix $U(\infty)$ in terms of *momentum* variables. For this purpose, it is enough to consider an expression which will be denoted by M, and which is a typical example of the units out of which all matrix elements of $U(\infty)$ are built up. A particular integer n and a particular graph G of n vertices being supposed fixed, the operator J(G) is constructed as in the previous section, and M is defined as the number obtained by substituting for each of the free operators in J(G) one particular free-particle *wave function*. More specifically, for each free operator $\psi(x)$ in J(G) there is substituted

$$\psi(k)e^{ik_\mu x_\mu}, \quad (13)$$

where k_μ is some constant 4 vector representing the *momentum* and *energy* of an electron, or minus the *momentum* and *energy* of a positron, and where $\psi(k)$ is a constant spinor. For each free operator $\psi^-(x)$ there is substituted

$$\psi^-(k')e^{-ik'_\mu x_\mu}, \quad (14)$$

where $\psi^-(k')$ is again a constant spinor. For each free operator $A_\mu(x)$ there is substituted

$$A_\mu(k'')\, e^{ik''_\mu x_\mu}, \quad (15)$$

where $A_\mu(k'')$ is a constant 4 vector which may represent the *polarization vector* of a quantum whose *momentum-energy 4-vector* is either plus or minus k''_μ; alternatively, $A_\mu(k'')$ may represent the Fourier component of the external *potential* with a particular *wave number and frequency* specified by the 4 vector k''. There is no loss of generality in splitting up the external *potential* into Fourier components of the form (15). When the substitutions (13), (14), (15) are made in J(G), the expression M which is obtained is still an n-fold integral over the whole of *space-time*, and in addition depends parametrically upon E constant 4 vectors in *momentum-space*, where E is the number of free operators in J(G).

The graph G will contain E external lines, i.e., lines with one end at a vertex and the other end at the edge of the diagram. To each of these external lines corresponds one constant 4 vector, which may be denoted by k_μ^i, i = 1, …, E, and one constant spinor or *polarization vector* appearing in M, either $\psi(k^i)$ or $\psi^-(k^i)$ or $A_\mu(k^i)$.

Suppose that G contains F internal lines, i.e., lines with both ends at vertices. To each of these lines corresponds a D_F or an S_F function in M, as specified by (11) or (10)

[$\frac{1}{2}\hbar c\delta_{\mu\nu}D_F(x-y)$, (11)

$\frac{1}{2}S_{F\beta\alpha}(x-y)$. (10)]

These functions have been expressed by Feynman as 4-dimensional Fourier integrals of very simple form, namely

$$D_F(x) = 1/4\pi^3 \int e^{-ip_\mu x_\mu}\,\delta_+(p^2)\,dp, \qquad (16)$$

$$S_F(x) = 1/4\pi^3 \int e^{-ip_\mu x_\mu}\,[+ip_\mu\gamma_\mu - \kappa_0] \times \delta_+(p^2 + \kappa_0^2)\,dp, \qquad (17)$$

where κ_0 is the electron reciprocal *Compton wavelength*,

$$p^2 = p_\mu p_\mu = p_1^2 + p_2^2 + p_3^2 - p_0^2, \qquad (18)$$

and the δ_+ function is defined by

$$\delta_+(a) = \tfrac{1}{2}\,\delta(a) + 1/2\pi i a = 1/2\pi \int_0^\infty e^{-iaz}\,dz. \qquad (19)$$

Substituting from (16) and (17) into M will introduce an F-fold integral over *momentum space*. Corresponding to each internal line of G, there will appear in M a 4-vector variable of integration, which may be denoted by p_μ^i, i = 1, …, F. However, after this substitution is made, the space-time variables $x_1, \ldots, x_n$ occur in M only in the exponential factors, and the integration over these variables can be performed. The result of the integration over x_i is to give

$$(2\pi)^4\delta(q_i), \qquad (20)$$

where the δ represents a simple 4-dimensional Dirac δ-function, and q_i is a 4-vector formed by taking an algebraic sum of the k^i and p^i 4 vectors corresponding to those lines of G which meet at x_i. The factor (20) in the integrand of M expresses the *conservation of energy and momentum* in the *interaction* occurring at the point x_i.

The transformation of M into terms of momentum variables is now complete. To summarize the results, M now appears as an F-fold integral over the variable 4 vectors p_μ^i in *momentum space*. In the integrand there appear, besides numerical factors;

(i) a constant spinor or polarization-vector, $\psi(k^i)$ or $\psi^{-}(k^i)$ or $A_\mu(k^i)$, corresponding to each *external line* of G;

(ii) a factor

$$D_F(p^i) = \delta_+\{(p^i)^2\} \tag{21}$$

corresponding to each internal *photon line* of G;

(iii) a factor

$$S_F(p^i) = [+ip^i_\mu\gamma_\mu - \kappa_0]\, \delta_+\{(p^i)^2 + \kappa_0^2)\} \tag{22}$$

corresponding to each internal *electron line* of G;

(iv) a factor

$$\delta(q_i) \tag{23}$$

corresponding to each *vertex* of G;

(v) a γ_μ operator, surviving from Eq. (8)

$$[H_I(x) = -ieA_\mu(x)\psi^{-}(x)\gamma_\mu\psi(x) - \delta mc^2\psi^{-}(x)\psi(x), \qquad (8)]$$

corresponding to each vertex of G at which there is *photon line.*

The important feature of the above analysis is that all the constituents of M are now localized and associated with individual lines and vertices in the graph G. It therefore becomes possible in an unambiguous manner to speak of "adding" or "subtracting" certain groups of factors in M, when G is modified by the addition or subtraction of certain lines and vertices. *As an example of this method of analysis, we shall briefly discuss the treatment in the S matrix formalism of the "Lamb shift" and associated phenomena.*

Suppose that a graph G, of any degree of complication, has a vertex x_1 at which two electron lines and a photon line meet. These three lines may be either internal or external, and the *momentum 4 vectors* associated with them in M may be either p^i or k^i; these 4 vectors are denoted by t^1, t^2, t^3 as indicated in Fig. 1. …

…

To summarize the results of this section, it has been shown that the S matrix formalism allows a wide variety of higher order radiative processes to be calculated *in the form of operators in momentum space*. Such operators appear as *radiative corrections* to the fundamental interaction between the photon and electron-positron fields, and need only to be calculated once to be applicable to the various special problems of electrodynamics.

IV. *Further reduction of the S matrix*

It was shown in Section VII of I that, for the one-electron processes there considered, only connected graphs needed to be taken into account. In constructing the S matrix in general, this is no longer the case; disconnected graphs give matrix elements of $U(\infty)$ representing two or more collision processes occurring simultaneously among separate groups of particles, and such processes have physical reality. It is only permissible to omit a disconnected graph when one of its connected components is entirely lacking in external lines; such a component without external lines will give rise only to a constant multiplicative phase factor in every matrix element of $U(\infty)$ and is therefore devoid of physical significance.

On the other hand, the treatment in Section VII of I of graphs with "*self-energy* parts" applies almost without change to the general S matrix formalism. A "*self-energy* part" of a graph is a connected part, consisting of vertices and internal lines only, which can be inserted into the middle of a single line of a graph G so as to give a meaningful graph G'. In Fig. 3 is shown an example of such an insertion made in one of the lines of Fig. 1. …
…

V. *Investigations of divergencies in the S matrix*

The δ+ function defined by (19)

$$[\delta_+(a) = \tfrac{1}{2}\,\delta(a) + 1/2\pi i a = 1/2\pi \int_0^\infty e^{-iaz}\,dz. \qquad (19)]$$

has the property that, if b is real and f(a) is any function analytic in the neighborhood of b, then

$$\int f(a)\delta_+(a-b)\,da = (1/2\pi i)\int f(a)\{1/(a-b)\}\,da \qquad (41)$$

where the first integral is along a stretch of the real axis including b, and the second integral is along the same stretch of the real axis but with a small detour into the complex plane passing underneath b. …
…

Thus every constituent part M of U(∞) can be written as an integral of a rational algebraic function of *momentum* variables, by using instead of (21) and (22)

$$[D_F(p^i) = \delta_+\{(p^i)^2\} \qquad (21)$$
$$S_F(p^i) = [+ip^i_\mu\gamma_\mu - \kappa_0]\,\delta_+\{(p^i)^2 + \kappa_0^2)\} \qquad (22)]$$

$$D_F(p^i) = 1/2\pi i\{(p^i)^2 \qquad (44)$$
$$S_F(p^i) = (ip^i_\mu\gamma_\mu - \kappa_0)/2\pi i\{(p^i)^2 + \kappa_0^2)\} \qquad (45)$$

This representation of D_F and S_F as rational functions in *momentum-space* has been developed and extensively used by Feynman (unpublished).

There may appear in M infinities of three distinct kinds. These are (i) singularities caused by the coincidence of two or more poles of the integrand, (ii) divergences at small *momenta* caused by a factor $D_F(p^i)$ in the integrand, (iii) divergences at large *momenta* due to insufficiently rapid decrease of the whole integrand at infinity.

In this paper no attempt will be made to explore the singularities of type (i). Such singularities occur, for example, when a many-particle scattering process may for special values of the particle momenta be divided into independent processes involving separate groups of particles. It is probable that all singularities of type (i) have a similarly clear physical meaning; these singularities have long been known in the form of vanishing energy denominators in ordinary perturbation theory, and have never caused any serious trouble.

A divergence of type (ii) is the so-called "infrared catastrophe", and is well known to be caused by the failure of an expansion in powers of α to describe correctly the radiation of low momentum quanta. It would presumably be possible to eliminate this divergence from the theory by a suitable adaptation of the standard Bloch-Nordsieck[7] treatment;

> [7] Bloch, F. & Nordsieck, A. (July, 1937). Note on the Radiation Field of the Electron. *Phys. Rev.*, 52, 54; https://doi.org/10.1103/PhysRev.52.54;

we shall not do this here. From a practical point of view, one may avoid the difficulty by arbitrarily writing instead of (44)

$$D_F(p^i) = 1/2\pi i\{(p^i)^2 + \lambda^2\}, \tag{46}$$

where λ is some non-zero *momentum*, smaller than any of the quantum *momenta* which are significant in the particular process under discussion[8].

> [8] The device of introducing λ in order to avoid infra-red divergences must be used with circumspection. Schwinger (unpublished) has shown that a long-standing discrepancy between two alternative calculations of the Lamb shift was due to careless use of λ in one of them.

It is the divergences of type (iii) which have always been the main obstacle to the construction of a consistent quantum electrodynamics, and which it is the purpose of the present theory to eliminate. In the following pages, attention will be confined to type (iii) divergences; *when the word "convergent" is used, the proviso "except for possible singularities of types (i) and (ii)" should always be understood.*

A divergent M is called "*primitive*" if, whenever one of the momentum 4 vectors in its integrand is held fixed, the integration over the remaining variables is convergent. Correspondingly, a *primitive divergent* graph is a connected graph G giving rise to divergent M, but such that, if any internal line is removed and replaced by two external lines, the modified G gives convergent M. *To analyze the divergences of the theory, it is sufficient to enumerate the primitive divergent M and G and to examine their properties.*
...

The possible *primitive divergent* graphs that have been found are all of a kind familiar to physicists. The case $E_e = 2$, $E_p = 0$ describes *self-energy effects of a single electron*; $E_e = 0$, $E_p = 2$ *self-energy effects of a single photon*; $E_e = 2$, $E_p = 1$ the *scattering of a single electron in an electromagnetic field*; and $E_e = 0$, $E_p = 4$ the *"scattering of light by light" or the mutual scattering of two photons*. Further, (55) shows that the divergence will never be more than logarithmic in the third and fourth cases, more than linear in the first, or more than quadratic in the second. Thus, it appears that, however far quantum electrodynamics is developed in the discussion of many-particle interactions and higher order phenomena, no essentially new kinds of divergence will be encountered. *This gives strong support to the view that "subtraction physics", of the kind used by Schwinger and Feynman, will be enough to make quantum electrodynamics into a consistent theory.*

VI. *Separation of divergencies in the S matrix*

First it will be shown that the "*scattering of light by light*" does not in fact introduce any divergence into the theory. The possible *primitive divergent* M in the case $E_e = 0$, $E_p = 4$ will be of the form

$$\delta(k^1 + k^2 + k^3 + k^4)A_\lambda(k^1)A_\mu(k^2)A_\nu(k^3)A_\rho(k^4)\ I_{\lambda\mu\nu\rho}, \qquad (56)$$

where $I_{\lambda\mu\nu\rho}$ is an integral of the type

$$\int R_{\lambda\mu\nu\rho}\ (k^1, k^2, k^3, k^4, p^i)\ dp^i, \qquad (57)$$

at most logarithmically divergent, and R is a certain rational function of the constant k^i and the variable p^i. ...

...

... To interpret this result physically, it is convenient to write (56) again in terms of *space-time* variables; this gives

$$M = \int I_{\lambda\mu\nu\rho}(0)A_\lambda(x)A_\mu(x)A_\nu(x)A_\rho(x)\ dx + N, \qquad (59)$$

where X is a convergent expression involving derivatives of the A(x) with respect to space and time. Now *the first term in (59) is physically inadmissible*; it is not *gauge-invariant*, and implies for example a scattering of light by an *electric field* depending on the absolute magnitude of the scalar *potential, which has no physical meaning. Therefore I(0) must vanish identically, and the whole expression (56) is convergent.*

The fact that the scattering of light by light is finite in the lowest order in which it occurs has long been known[9].

> [9] Euler, H. & Kockel, B. (1935). Über die Streuung von Licht an Licht nach der Diracschen Theorie. (On the Scattering of Light by Light in the Dirac Theory.) *Naturwiss.*, 23, 246-7; http://dx.doi.org/10.1007/BF01493898; Euler, H. (1936). Über die Streuung von Licht an Licht nach der Diracschen Theorie. (On the Scattering of Light by Light in the Dirac Theory.) *Ann. Phys.*, 26, 398-448; http://dx.doi.org/10.1002/andp.19364180503. In these early calculations of the scattering of light by light, the theory used is the Heisenberg electrodynamics, in which certain singularities are eliminated at the start by a procedure involving non-diagonal elements of the Dirac density matrix. In Feynman's calculation, on the other hand, a finite result is obtained without subtractions of any kind.

It has also been verified by Feynman by direct calculation, using his own theory as described in this paper. The graphs which give rise to the lowest order scattering are shown in Fig. 4. It is found that the divergent parts of the corresponding M exactly cancel when the three contributions are added, or, what comes to the same thing, when the function $R_{\lambda\mu\nu\rho}$ is symmetrized. It is probable that the absence of divergence in the scattering of light by light is in all cases due to a similar cancellation, and it should not be difficult to prove this by calculation and thus avoid making an appeal to gauge-invariance.

The three remaining types of primitive divergent M are, in fact, divergent. However, these are just the expressions which have been studied in Sections III and IV and shown to be completely described by the operators Λ_μ, Σ, and Π. …

…

Therefore, if some means can be found for isolating and removing the divergent parts from Λ_μ, Σ, and Π, the "irreducible" graphs defined in Section IV will not introduce any fresh divergences into the theory, and the rules of Section IV will lead to a divergence-free S matrix.

In considering Λ_μ, Σ, and Π in Section IV it was found convenient to divide vertex and self-energy parts themselves into the categories reducible and irreducible. An irreducible self-energy part W is required not only to have no vertex and *self-energy* parts inside itself; it is also required to be "proper", that is to say, it is not to be divisible into two pieces joined by a single line. …

…

VII. *Removal of divergences from the S matrix*

The task remaining is to complete the formulas (73), (78), and (82), which show how the infinite parts can be separated from the operators Γ_μ, S_F', and D_F', and to include the corrections introduced into these operators by the radiative reactions which they themselves describe. In other words, *we have to include radiative corrections to radiative corrections, and renormalizations of renormalizations, and so on ad infinitum*. This task. is not so formidable as it appears.

…

It is necessary finally to justify the dropping of the divergent terms. This will be done by showing that the "true" Γ_μ, S_F', and D_F', which are obtained if the divergent terms are not dropped, are only numerical multiples of those obtained by dropping divergences, and that the numerical multiples can themselves be eliminated from the theory *by a consistent use of the ideas of mass and charge renormalization. …*

…

It hardly needs to be pointed out that the arguments of this section have involved extensive manipulations of infinite quantities. These manipulations have only a formal validity, and must be justified *a posteriori* by the fact that they ultimately lead to a clear separation of finite from infinite expressions. *Such an a posteriori justification of dubious manipulations is an inevitable feature of any theory which aims to extract meaningful results from not completely consistent premises.*

…

VIII. *Summary of results*

The results of the preceding sections divide themselves into two groups. *On the one hand, there is a set of rules by which the element of the S matrix corresponding to any given scattering process may be calculated, without mentioning the divergent expressions occurring in the theory. On the other hand, there is the specification of the divergent expressions, and the interpretation of these expressions as mass and charge renormalization factors*.

The first group of results may be summarized as follows. *Given a particular scattering problem, with specified initial and final states, the corresponding matrix element of $U(\infty)$ is a sum of contributions from various graphs G as described in Section II.* A particular contribution M from a particular G is to be written down as an integral over momentum variables according to the rules of Section III; the integrand is a product of factors $\psi(k^i)$, $\psi^-(k^i)$, $A_\mu(k^i)$, $S_F(p')$, $D_F(p')$, $\delta(q_i)$, γ_μ, the factors corresponding in a prescribed way to the lines and vertices of G. According to Section IV, contributions M are only to be admitted from irreducible G; the effects of reducible graphs are included by replacing in M the factors ψ, ψ^-, A_μ, S_F, D_F, γ_μ by the corresponding expressions (37), (35), (36), (38). These replacements are then shown in Section VII to be equivalent to the following: each factor S_F in M is replaced by $S_{F1}'(e)$, each factor D_F by $D_{F1}'(e)$, each factor γ_μ by $\Gamma_{\mu 1}(e)$, each factor A_μ when it represents an external *potential* is replaced by

$$A_{\mu 1}(k^i) = 2\pi i\, D_{F1}'(e)\, (k^i)^2\, A_\mu(k^i), \qquad (102)$$

factors ψ, ψ^-, A_μ representing particle wave-functions are left unchanged, and finally e wherever it occurs in M is replaced by e_1. The definition of M is completed by the specification of $S_{F1}'(e)$, $D_{F1}'(e)$, $\Gamma_{\mu 1}(e)$; *it is in the calculation of these operators that the main difficulty of the theory lies*. The method of obtaining these operators is the process of successive substitution and integration explained in the first part of Section VII; *the operators so calculated are divergence-free, the divergent parts at every stage of the calculation being explicitly dropped after being separated from the finite parts by the method of Section VI.*

The above rules determine each contribution M to $U(\infty)$ as a divergence-free expression, which is a function of the *observed mass* m and the *observed charge* e_1 of the electron, *both of which quantities are taken to have their empirical values*. The divergent parts of the theory are irrelevant to the calculation of $U(\infty)$, being absorbed into the unobservable constants δm and e occurring in (8)

[The *interaction energy* …

$$H_1(x) = -\, ieA_\mu(x)\psi^-(x)\gamma_\mu\psi(x) - \delta mc^2\psi^-(x)\psi(x), \qquad (8)$$

where A_μ is the *total electromagnetic field*, and the term in δm is included in order to allow for the fact that the *interaction representation* is defined in terms of the *total mass* of an electron including its "*electromagnetic mass*" δm.]

A place where some ambiguity might appear in M is in the calculation of the operators $S_{F1}'(e)$, $D_{F1}'(e)$, $\Gamma_{\mu 1}(e)$, when the method of Section VI is used to separate out the finite parts $S(W, t^1)$, $D(W', t^1)$, $\Lambda_{\mu c}(V, t^1, t^2)$, from the expressions (67), (74), (80). Even in this place the rules of Section VI give unambiguous directions for making the separation; *only there is a question whether some alternative directions might be equally reasonable*. For example, it is possible to separate out a finite part from $\Sigma(W, t^1)$ according to (67), and not to make the further step of using (70) to separate out a finite part $S(W, t^1)$ which vanishes when (69) holds. Actually, it is easy to verify that such an alternative procedure will not change the value of M, but will only make its evaluation more complicated; it will lead to an expression for M in which one (infinite) part of the *mass* and *charge renormalizations* is absorbed into the constants δm and e, while other finite *mass and charge renormalizations* are left explicitly in the formulas. It is just these finite *renormalization* effects which the second step in the separation of $S(W, t^1)$ and $\Lambda_{\mu c}(V, t^1, t^2)$ is designed to avoid. *Therefore, it may be concluded that the rules of calculation of U(∞) are not only divergence-free but unambiguous.*

As anyone acquainted with the history of the Lamb shift[11] knows,

[11] Bethe, H. A. (1948). Electromagnetic Shift of Energy Levels, Report to Solvay Conference, Brussels.

the utmost care is required before it can be said that any particular rule of calculation is unambiguous. *The rules given in this paper are unambiguous, in the sense that each quantity to be calculated is an integral in momentum-space which is absolutely convergent at infinity; such an integral has always a well-defined value. However, the rules would not be unambiguous if it were allowed to split the integrand into several parts and to evaluate the integral by integrating the parts separately and then adding the results; ambiguities would arise if ever the partial integrals were not absolutely convergent.* A splitting of the integrals into conditionally convergent parts may seem unnatural in the context of the present paper, but occurs in a natural way when calculations are based upon a perturbation theory in which electron and positron states are considered separately from each other. The absolute convergence of the integrals in the present theory is essentially connected with the fact that the electron and positron parts of the electron-positron field are never separated; this finds its algebraic expression in the statement that the quadratic denominator in (45)

$$[S_F(p^i) = (ip^i_\mu\gamma_\mu - \kappa_0)/2\pi i\{(p^i)^2 + \kappa_0^2)\} \qquad (45)]$$

is never to be separated into partial fractions. Therefore *the absence of ambiguity in the rules of calculation of U(∞) is achieved by introducing into the theory what is really a new physical hypothesis, namely that the electron-positron field always acts as a unit and not as a combination of two separate fields.* A similar hypothesis is made for the *electromagnetic field*, namely that this field also acts as a unit and not as a sum of one part representing photon *emission* and another part representing photon *absorption.*

Finally, *it must be said that the proof of the finiteness and unambiguity of U(∞) given in this paper makes no pretense of being complete and rigorous.* It is most desirable that these general arguments should as soon as possible be supplemented by an explicit calculation of at least one fourth-order radiative effect, to make sure that no unforeseen difficulties arise in that order.

The second group of results of the theory is the identification of δm and e by (94) and (86). Although these two equations are strictly meaningless, both sides being infinite, yet it is a satisfactory feature of the theory that it determines the unobservable constants δm and e *formally as power series in the observable e_1, and not vice versa.* There is thus no objection in principle to identifying e_1 with the *observed electronic charge* and writing

$$(e_1^2/4\pi hc) = \alpha = 1/137. \qquad (103)$$

The constants appearing in (8)

$$[H_1(x) = -\, ieA_\mu(x)\psi^-(x)\gamma_\mu\psi(x) - \delta mc^2\psi^-(x)\psi(x), \qquad (8)]$$

are then, by (94) and (86),

$$\delta m = m(A1\alpha + A_1\alpha + A_2\alpha^2 + \ldots), \qquad (104)$$
$$e = e_1(1 + B_1\alpha + B_2\alpha^2 + \ldots), \qquad (105)$$

where the A_i and B_i are logarithmically divergent numerical coefficients, independent of m and e_1.

IX. *Discussion of further outlook*

The surprising feature of the S matrix theory, as outlined in this paper, is its success in avoiding difficulties. *Starting from the methods of Tomonaga, Schwinger and Feynman, and using no new ideas or techniques, one arrives at an S matrix from which the well-known divergences seem to have conspired to eliminate themselves.* This automatic disappearance of divergences is an empirical fact, which must be given due weight in considering the future prospects of electrodynamics. Paradoxically opposed to the finiteness of the S matrix is the second fact, that *the whole theory is built upon a Hamiltonian formalism with an interaction-function (8) which is infinite and therefore physically meaningless.*

The arguments of this paper have been essentially mathematical in character, being concerned with the consequences of a particular mathematical formalism. *In attempting to assess their significance for the future, one must pass from the language of mathematics to the language of physics.* One must assume provisionally that the mathematical formalism corresponds to something existing in nature, and then enquire to what extent the paradoxical results of the formalism can be reconciled with such an assumption. In accordance with this program, we interpret the contrast between the divergent Hamiltonian formalism and the finite S matrix as a contrast between two pictures of the world, seen by two observers having a different choice of measuring equipment at their disposal.

The first picture is of a collection of quantized fields with localizable interactions, and is seen by a fictitious observer whose apparatus has no atomic structure and whose measurements are limited in accuracy only by the existence of the fundamental constants c and h. This observer is able to make with complete freedom on a sub-microscopic scale the kind of observations which Bohr and Rosenfeld[12] employ in a more restricted domain in their classic discussion of the measurability of field-quantities; and he will be referred to in what follows as the "ideal" observer.

[12] Bohr, N. & Rosenfeld, L. (December, 1933). Zur Frage der Messbarkeit der elektromagnetischen Feldgrössen. (On the question of the measurability of electromagnetic

field quantities.) *Kgl. Dansk. Vid. Sels. Math.-Phys. Medd.*, 12, 8. A second paper by Bohr and Rosenfeld is to be published later, and is abstracted in a booklet by Pais, A. (1948). *Developments in the Theory of the Electron*. Princeton University Press, Princeton.

The second picture is of a collection of observable quantities (in the terminology of Heisenberg), and is the picture seen by a real observer, whose apparatus consists of atoms and elementary particles and whose measurements are limited in accuracy not only by c and h but also by other constants such as α and m. The real observer makes spectroscopic observations, and performs experiments involving bombardments of atomic systems with various types of mutually interacting subatomic projectiles, but to the best of our knowledge he cannot measure the strength of a single field undisturbed by the interaction of that field with others.

The ideal observer, utilizing his apparatus in the manner described in the analysis of the Hamiltonian formalism by Bohr and Rosenfeld[12]

[12] *Loc. cit.*

makes measurements of precisely this last kind, and it is in terms of such measurements that the commutation-relations of the fields are interpreted. The interaction-function (8)

$$[H_1(x) = -ieA_\mu(x)\bar{\psi}(x)\gamma_\mu\psi(x) - \delta mc^2\bar{\psi}(x)\psi(x), \qquad (8)]$$

will presumably always remain unobservable to the real observer, who is able to determine positions of particles only with limited accuracy, and who must always obtain finite results from his measurements. The ideal observer, however, using non-atomic apparatus whose location in space and time is known with infinite precision, is imagined to be able to disentangle a single field from its interactions with others, and to measure the interaction (8). In conformity with the Heisenberg uncertainty principle, it can perhaps be considered a physical consequence of the infinitely precise knowledge of location allowed to the ideal observer, that the value obtained by him when he measures (8) is infinite.

If the above analysis is correct, the divergences of electrodynamics are directly attributable to the fact that the Hamiltonian formalism is based upon an idealized conception of measurability. The paradoxical feature of the present situation does not then lie in the mere coexistence of a finite S matrix with an infinite interaction-function. The empirically found correlation, between expressions which are unobservable to a real observer and expressions which are infinite, is a physically intelligible and acceptable feature of the theory. The paradox is the fact that it is necessary in the present paper to start from the infinite expressions in order to deduce the finite ones. *Accordingly, what is to be looked for in a future theory is not so much a modification of the present theory which will make all infinite quantities finite, but rather a turning-round of the theory so that the finite quantities shall become primary and the infinite quantities secondary.*

One may expect that in the future a consistent formulation of electrodynamics will be possible, itself free from infinities and involving only the physical constants m and e_1, and such that a Hamiltonian formalism with interaction (8), with divergent coefficients δm and e, may in suitably idealized circumstances be deduced from it. The Hamiltonian formalism should appear as a limiting form of a description of the world as seen by a certain type of observer, the limit being approached more and more closely as the precision of measurement allowed to the observer tends to infinity.

The nature of a future theory is not a profitable subject for theoretical speculation. *The future theory will be built, first of all upon the results of future experiments, and secondly upon an understanding of the interrelations between electrodynamics and mesonic and nucleonic phenomena.* The purpose of the foregoing remarks is merely to point out that there is now no longer, as there has seemed to be in the past, a compelling necessity for a future theory to abandon some essential features of the present electrodynamics. *The present electrodynamics is certainly incomplete, but is no longer certainly incorrect.*

In conclusion, the author would like to express his profound indebtedness to Professor Feynman for many of the ideas upon which this paper is built, to Professor Oppenheimer for valuable discussions, and to the Commonwealth Fund of New York for financial support.

Richard John Eden (July 2, 1922 –September 25, 2021)

Eden was a British theoretical physicist who researched quantum field theory, nuclear theory and S-matrix theory in the 1950s and 1960s.

Eden was born in London on July 2, 1922. He received his doctorate in 1951 at Cambridge University under Paul Dirac (The Classical and Quantum Mechanics of Non-holonomic Systems). In the 1950s, Eden was a leading British exponent of analytic S-matrix studies in elementary particle physics. From 1964 to 1982 he was Reader in Theoretical Physics at the University of Cambridge.

In the 1950s, he attended the Institute for Advanced Study in Princeton. This experience led him to develop ideas for a College for Advanced Study in Cambridge. He was subsequently a founding Fellow of Clare Hall, Cambridge in 1966, and from 1987 to 1999 he was vice president of the college.

In 1972 he took up interdisciplinary energy studies. In 1974 he founded the Energy Research Group at the Cavendish Laboratory and from 1982 to 1989 was Professor of Energy Studies there.

He died in Poole on September 25, 2021, at the age of 99.

Eden, R. J. (September, 1949). Heisenberg's S Matrix for a System of Many Particles.

[*Proc. Roy. Soc., Series A*, 198, 1055, 540-559; https://doi.org/10.1098/rspa.1949.0118.]

Received March 12, 1949.

Peterhouse, University of Cambridge.

Communicated by P. A. M. Dirac, F.R.S.

This vignette is included as Dr. Eden provided lectures in Nuclear Physics and Quantum Theory to the author in 1963 as part of his M.A. Cantab. degree. Eden's thesis advisors were Dirac and Heisenberg. The author attended a joint lecture on May 23, 1963, by Heisenberg and Dirac (both speaking whilst writing and erasing on dueling pairs of blackboards at the same time) at the old Cavendish Laboratory.

Abstract

When creation and annihilation of particles is forbidden, the general features of a many-particle collision are illustrated by a scattering system containing two particles and a fixed scattering center. The term in the wave function, which is called by Moller *the 'outgoing' part, is shown to contain a term representing a totally outgoing wave, but also terms describing the interference effects between the incident wave of one particle and the outgoing wave of the other*. Corresponding to the latter terms, there exist singular eigenstates of the S matrix which are simultaneous eigenstates of the separate kinetic energies of the particles. The remaining eigenstates are called non-singular, and for these only the sum of the kinetic energies can be given a definite value. Analytic continuation of the non-singular eigenstates in the complex plane of total kinetic energy shows that the corresponding eigenvalues of S can be used to determine the energy levels of states with both particles bound to the center of force. The eigenvalues for the singular eigenstates will lead to the bound energy of a single particle in the scattering field of the other. The formalism is extended to include singular eigenstates which describe the scattering of one particle on a compound center made up of the other particle bound to the scatterer.

1. *Introduction*

The usual form of quantum theory contains many quantities which are not observable. These quantities are introduced as auxiliary variables for convenience in calculation and are only indirectly related to experimental results. Examples are wave functions or dynamical variables which are written as functions of *space-time* and therefore can be exactly localized. Such localizability would be invalid in a physical theory if, for instance, there is some universal constant which plays the part of a minimal length.

*Heisenberg (1949) * has suggested that the divergence difficulties of quantum-field theories are a direct consequence of the use of localizable variables.*

[* Heisenberg, W. (1949) *Two Lectures: 1. The Present Situation in the Theory of Elementary Particles; 2. Electron Theory of Superconductivity*. Cambridge University Press. In Blum, W., Dürr, HP., Rechenberg, H. (eds). (1984). *Scientific Review Papers, Talks, and Books. Wissenschaftliche Übersichtsartikel, Vorträge und Bücher. Gesammelte Werke. Collected Works, vol B*. Springer, Berlin, Heidelberg; https://doi.org/10.1007/978-3-642-61742-3_35.]

However, it seems essential that those parts of quantum theory which are directly related to experimentally observed quantities must also be incorporated in any new theory of elementary particles. *The theory of the characteristic matrix, or S matrix, is an attempt to set up a framework for a future theory which will contain none of the divergences but all the experimental results of present quantum mechanics.*

The S matrix is based on the idea of a stationary collision in a system of particles, and its matrix elements are closely connected with collision cross-sections for various processes. The original formulation of the theory was given by Heisenberg (1943a,b),

Heisenberg, W. (July, 1943). Die beobachtbaren Größen in der Theorie der Elementarteilchen. I (The "observable quantities" in the theory of elementary particles. I.) *Zeit. Phys.*, 120, 7–10, 513-38[; S-matrix theory as a principle of particle interactions]; (1943). Die beobachtbaren Größen in der Theorie der Elementarteilchen. II. (The "observable quantities" in the theory of elementary particles. II.) *Zeit. Phys.*, 120, 11–12, 673-702; https://doi.org/10.1007/BF01336936.

and the general properties of S have been studied by Moller (1945, 1946).

Moller, C. (1945). *K. danske vidensk. Selsk. Math-Fys. Medd.* 23, 1; (1946). *Ibid.*, 22, 19.

There are two main approaches to the development of S-matrix theory. *One is to study the properties of the S matrix obtained from non-relativistic quantum mechanics, and show that it will lead to all experimental results which have previously been obtained from a Hamiltonian.* In addition to cross-sections these include *the energy levels of bound states of two or more particles* and radio-active decay constants. *The second approach is to obtain properties, from non-relativistic quantum mechanics, which are restrictions on the S matrix itself.* Such properties must be independent of the particular form of Hamiltonian, and may then be assumed to apply even if no Hamiltonian exists. *The most important of these are the unitary condition which enables S to be interpreted as a transformation matrix*, and *the relativistic invariance of the eigenvalues of S.* Both these are general conditions; the outstanding requirement is a special condition, which depends on the particular dynamical system under consideration. If S can be determined from these conditions, aided by a need for simplicity, they would form the basis for a new theory.

In this paper the author will be concerned principally with *the first method of approach. The properties of a many-particle system are studied, with particular reference to the energy levels of bound states.*

The *Lorentz invariance* properties of S have been studied by Moller (1945) [*loc. cit.*] for a collision of two particles, and he has made a formal extension to a many-particle collision. *If one is not concerned with Lorentz invariance*, the co-ordinates of the center of gravity may be eliminated. The system is then mathematically equivalent to one in which particles are scattered on a fixed center of force, with conservation of energy but not of momentum. *Throughout this paper the author works in terms of a system of this type.*

In their papers on S-matrix theory, Heisenberg (1943a) and Moller (1945) [*loc. cit.*] *have assumed that the wave function for scattering may be put in a standard form having two parts, one representing the incident wave, and the other the 'outgoing' wave.* For a system of one particle and a fixed scattering center, Dirac (1948) [Dirac, P. A. M. (1948). *Principles of quantum mechanics*, 3rd ed. Oxford University Press] has given the justification for this standard form of the wave function. The 'outgoing' wave does in fact correspond to the particle moving outward in the asymptotic region of coordinate space. *It is shown in this paper, for a many-particle system, that the assumption of the standard form is related to the physical assumption that the particles interact only in a finite region of space.* This implies that no Coulomb forces are present. The 'outgoing' part of the wave function corresponds to a state in which at least one particle is moving outward in the asymptotic region of co-ordinate space.

For the scattering of one particle, Heisenberg (1944)

> Heisenberg, W. (1944). Die beobachtbaren Größen in der Theorie der Elementarteilchen. III. (The "observable quantities" in the theory of elementary particles. III.) *Zeit. Phys.*, 123, 1–2, 93-112; https://doi.org/10.1007/BF01375146.

has shown that analytic continuation of the eigenvalues of S, in the complex energy plane, will lead to the energy levels of closed states in which the particle is bound to the scattering center. *The main purpose of the present paper is to extend the work of Heisenberg to systems of two or more particles with a scattering center.* It is assumed throughout that the number of particles is a constant of the motion. *A system in which particles can be created and annihilated seems to present S-matrix theory with major difficulties* which can probably be overcome only by new assumptions. Some study of this problem has been made by Hu (1949) [Hu, N. (1948). Further Investigations on Heisenberg's Characteristic Matrix. *Proc. R. Irish Acad. A*, 52, 5, 51-68; https://www.jstor.org/stable/20488491], but his new assumptions lead to a *serious negative-energy difficulty.*

It is shown first that the wave function, for scattering of two particles with no mutual interaction on a fixed center, may be put into the standard form. The S matrix for the compound system is the product of the separate S matrices. This result can be extended to an arbitrary number of particles. After considering two particles having small mutual interaction, the interpretation of the 'outgoing' part of the wave function is examined. This is made up of interference terms between 'incident' waves of one particle and outgoing waves of the other, as well as the totally outgoing wave. The 'incident' wave of a particle is defined as the state in which this particle does not interact with the rest of the system; it will contain both incoming and outgoing parts. The product of the incident states of all the particles is called the *initial state* of the system.

In order to investigate bound states of a system containing two particles and a scattering center, a superposition of states is formed. The basic states of the superposition correspond to incident plane waves of equal energies. The weight function for the superposition can be chosen so that the total state is an eigenstate of the S matrix. Then the weight function can be called an eigenfunction of it determines the initial state of the system, which differs from the final state only by a phase factor. This phase factor is an eigenvalue of S.

Since the *total energy* W of the system commutes with S, every eigenfunction of S and W will contain a δ-function factor in the total energy of the two particles. Physically this means that, in a collision which is an eigenstate of S, one can know exactly the asymptotic value of the *total kinetic energy*. If, for a particular eigenstate, one knows the separate asymptotic values of the *kinetic energies* of the two particles, the eigenfunction will contain two δ-function factors, one for each *kinetic energy*. I call this a *singular eigenfunction*. It corresponds to a state in which the exchange of energy between the two particles is zero, so that asymptotically the separate *kinetic energies* are conserved. A necessary condition for the existence of singular eigenfunctions is that S contains terms having two δ-function factors in the separate *kinetic energies*. Such terms are always present in S, either as a product of incident waves, or as interference waves between one particle incident and the other outgoing. An eigenfunction which contains only the one δ-function factor in total kinetic energy is called a *non-singular eigenfunction*. The existence of such eigenfunctions is ensured if the separate *kinetic energies* do not commute with S.

By obtaining the asymptotic form of the wave function in co-ordinate space, we can investigate the conditions for bound states. For a non-singular eigenstate of S, analytic continuation of the corresponding eigenvalue, in the complex plane of *total kinetic energy*, leads to the *energy* levels of states with both particles bound to the center of force. For a singular eigenstate, keeping the *kinetic energy* of one particle fixed, analytic continuation of the eigenvalue leads to the *bound energy* of the other particle.

The important case, where the *fixed energy* value is the *bound energy* of one particle, can be included in this scheme. It corresponds to a state in which one particle is scattered on a compound center made up of the other particle bound to the scatterer. Since it is a singular eigenstate, ionization of the bound particle does not take place. When the total energy of the system is too low to allow ionization, the only possible states are a superposition of these singular eigenstates. With their corresponding eigenvalues these states define an S matrix which I call the *'partial' S matrix. It is not analytic and indicates new difficulties in the theory which are considered in more detail in another paper.*

The possibility of a mixed eigenfunction, containing both singular and non-singular parts, is considered. It is shown that these parts may be regarded as separate eigenfunctions of S, belonging to the same eigenvalue. This has the consequence that the singular part must correspond to a state in which one particle is bound. Finally, some aspects of the extension to an arbitrary number of particles are discussed. It seems unlikely that any new points of principle would arise in making this extension.

2. *Two-particle scattering with no mutual interaction*

Units are taken in which h = c = 1. Denote the *mass, momentum,* and *kinetic energy* of a particle by κ_n, k_n, W_n, for n = 1, 2. Then $W_n^2 = \kappa_n^2 + k_n^2$.

We consider two-particle scattering on a fixed center of force, *so that energy but not momentum must be conserved.* Since there is no mutual interaction, the system can be treated as the product of two single-particle scattering systems. In the momentum space of one system the incident wave can be represented by a ket vector $|k_n^A\rangle$, denoting a plane wave with momentum k_n^A. Then the wave function takes the form

$$< k_n \mid \psi_n \mid k_n^A> = \delta \ldots \qquad (2.1)$$

where $\delta\pm(x) = \ldots$ (2.2)

and δ(x) denotes Dirac's δ function.

...

... There will be many singular eigenfunctions, corresponding to kinetic energy conservation of any set of m particles where m < n. These eigenfunctions arise from the very complicated mixing between incident waves of some particles and outgoing waves of others, and also from the states in which some of the particles are bound to the scattering center. The concept of a partial S matrix will still be valid, but it will be discontinuous at a large number of energy values.

I would like to express my thanks to Professor P. A. M. Dirac for valuable advice and criticism, and to Professor W. Heisenberg for many helpful discussions.

Feynman, R. P. (September, 1949). The Theory of Positrons.

[*Phys. Rev.*, 76, 749-59; https://doi.org/10.1103/PhysRev.76.749; also in Schwinger, J. (ed). (1958). *Selected Papers on Quantum electrodynamics*. Dover, New York, pages 225-35.]

Received April 8, 1949.

Department of Physics, Cornell University, Ithaca, New York.

This is the first of a set of papers dealing with the solution of problems in *quantum electrodynamics*, main principle is to deal directly with *solutions* of the Hamiltonian differential equations rather than with equations themselves, this paper *addresses motion of electrons and positrons in given external potentials*, second paper considers *interactions* (*quantum electrodynamics*), problem of charges in a fixed potential usually treated by method of second quantization of the electron field using the *theory of holes*, here we replace Dirac's "hole theory" by reinterpreting *solutions* of the Dirac equation, results simplified by *following the charge rather than the particles* because *the number of particles is not conserved* whereas *charge is conserved*, in the approximation of classical *relativistic* theory the creation of an electron pair (electron and positron*)* represented by the start of two world lines from the point of creation, *following the charge rather than the particles corresponds to considering continuous world line as a whole rather than breaking it up into its pieces*, over-all *space-time* point of view, quantum mechanically the direction of the world lines is replaced by the direction of propagation of waves, quite different from Hamiltonian method which considers the future as developing continuously from the past, in a scattering problem this over-all view of the complete scattering process is similar to the *S-matrix view-point of Heisenberg*, temporal order of events during scattering which is analyzed in such detail by the Hamiltonian differential equation is irrelevant, development stemmed from idea that in *non-relativistic* quantum mechanics the *amplitude* for a given process can be considered as the sum of an *amplitude* for each *space-time* path available, in *relativistic* case the restriction that the paths must proceed always in one direction in time is removed, results more easily understood from more familiar physical viewpoint of *scattered waves* used in this paper, after the equations were worked out physically the proof of the equivalence to the *second quantization theory* was found, solution in terms of boundary conditions on *wave function* contains all possibilities of pair formation and annihilation together with the ordinary scattering processes, negative energy states appear in space-time as waves traveling away from the external potential backwards in time (as suggested in Stückelberg (December, 1942)), such a wave corresponds to a positron approaching the potential and annihilating the electron, a particle moving forward in time (electron) in a potential may be scattered forward in time (ordinary scattering) or backward (pair annihilation), when moving backward (positron) it may be scattered backward in time (positron scattering) or forward (pair production), *amplitude* for transition from an initial to a final state analyzed to any order in the *potential* by considering it to undergo a sequence of such scatterings, *amplitude* for a process involving many such particles is the product of the *transition amplitudes* for each particle, vacuum problems do not arise for charges which do not interact with one another, shows equivalent to the *theory of holes* in second quantization.

Abstract.

The problem of the behavior of positrons and electrons in given external potentials, neglecting their mutual interaction, is analyzed *by replacing the theory of holes by a reinterpretation of the solutions of the Dirac equation.* It is possible to write down a complete solution of the problem *in terms of boundary conditions on the wave function*, and this solution contains automatically all the possibilities of virtual (and real) pair formation and annihilation together with the ordinary scattering processes, including the correct relative signs of the various terms.

In this solution, *the "negative energy states" appear in a form which may be pictured (as by Stückelberg) in space-time as waves traveling away from the external potential backwards in time*. Experimentally, *such a wave corresponds to a positron approaching the potential and annihilating the electron.* A particle moving forward in time (electron) in a potential may be scattered forward in time (ordinary scattering) or backward (pair annihilation). When moving backward (positron) it may be scattered backward in time (positron scattering) or forward (pair production). For such a particle the *amplitude* for transition from an initial to a final state is analyzed to any order in the *potential* by considering it to undergo a sequence of such scatterings.

The amplitude for a process involving many such particles is the product of the transition amplitudes for each particle. The *exclusion principle* requires that antisymmetric combinations of *amplitudes* be chosen for those complete processes which differ only by exchange of particles. It seems that a consistent interpretation is only possible if the *exclusion principle* is adopted. The *exclusion principle* need not be taken into account in intermediate states. *Vacuum problems do not arise for charges which do not interact with one another, but these are analyzed nevertheless in anticipation of application to quantum electrodynamics.*

The results are also expressed in *momentum-energy* variables. *Equivalence to the second quantization theory of holes* is proved in an appendix.

1. *Introduction*

This is the first of a set of papers dealing with the solution of problems in *quantum electrodynamics*.

The main principle is to *deal directly with the solutions to the Hamiltonian differential equations rather than with these equations themselves. Here we treat simply the motion of electrons and positrons in given external potentials. In a second paper we consider the interactions of these particles, that is, quantum electrodynamics*.

The problem of charges in a fixed potential is usually treated by the method of second quantization of the electron field, using the ideas of the *theory of holes*. Instead, we show that by a suitable choice and interpretation of the solutions of *Dirac's equation* the problem may be equally well treated in a manner which is fundamentally no more complicated than Schrodinger's method of dealing with one or more particles. The various creation and annihilation operators in the conventional electron field view are required because *the number of particles is not conserved*, i.e., pairs may be created or destroyed. On the other hand, *charge is conserved* which suggests that *if we follow the charge, not the particle, the results can be simplified.*

In the approximation of classical *relativistic* theory, the creation of an electron pair (electron *A,* positron *B)* might be represented by the start of two world lines from the point of creation, 1. The world lines of the positron will then continue until it annihilates another electron, *C,* at a world point 2. Between the times t_1 and t_2 there are then three world lines, [two] before and after only one. However, the world lines of *C, B,* and *A* together form one continuous line albeit the "positron part" *B* of this continuous line is directed backwards in time. *Following the charge rather than the particles corresponds to considering this continuous world line as a whole rather than breaking it up into its pieces*. It is as though a bombardier flying low over a road suddenly sees three roads and it is only when two of them come together and disappear again that he realizes that he has simply passed over a long switchback in a single road.

This over-all *space-time* point of view leads to considerable simplification in many problems. One can take into account at the same time processes which ordinarily would have to be considered separately. For example, when considering the scattering of an electron by a potential one automatically takes into account the effects of virtual pair productions. The same equation, Dirac's, which describes the deflection of the world line of an electron in a field, can also describe the deflection (and in just as simple a manner) when it is large enough to reverse the time-sense of the world line, and thereby correspond to pair annihilation. *Quantum mechanically the direction of the world lines is replaced by the direction of propagation of waves.*

This view is quite different from that of the Hamiltonian method which considers the future as developing continuously from out of the past. Here we imagine the entire *space-time* history laid out, and that we just become aware of increasing portions of it successively. *In a scattering problem this over-all view of the complete scattering process is similar to the S-matrix view-point of Heisenberg.* The temporal order of events during the scattering, which is analyzed in such detail by the Hamiltonian differential equation, is irrelevant. The relation of these viewpoints will be discussed much more fully in the introduction to the second paper, in which the more complicated interactions are analyzed.

The development stemmed from the idea that in *non-relativistic* quantum mechanics the *amplitude* for a given process can be considered as the sum of an *amplitude* for each *space-time* path available.[1]

[1] Feynman, R. P. (April, 1948). Space-Time Approach to Non-Relativistic Quantum Mechanics. *Rev. Mod. Phys*., 20, 2, 367-87[; third formulation of *non-relativistic* quantum theory in addition to *differential equation of Schroedinger* and *matrix algebra of Heisenberg, path integral formulation* utilizing the *action* principle as suggested in Dirac (1933) [The Lagrangian in Quantum Mechanics] and Dirac (1945) [On the Analogy Between Classical and Quantum Mechanics], in quantum mechanics the probability of an

event which can happen in several different ways is the absolute square of a sum of complex contributions one from each alternative way $\varphi_{ac} = \sum_b \varphi_{ab}\varphi_{bc}$ where φ_{ab}, φ_{bc}, φ_{ac} are complex numbers such that $P_{ab} = |\varphi_{ab}|^2$, $P_{bc} = |\varphi_{bc}|^2$, and $P^q_{ac} = |\varphi_{ac}|^2$, and P_{ab} as the probability that if measurement A gave the result *a*, then measurement B will give the result *b*, and P^q_{ac} is the quantum mechanical probability that a measurement of C results in c when it follows a measurement of A giving a, the probability that a particle will be found to have a path lying somewhere within a region of space time is the absolute square of a sum of contributions - one from each path in the region, the contribution from a single path is postulated to be an exponential whose (imaginary) phase is the classical *action* for the path in question where *action* refers to time integral of Lagrangian along a path, restriction to finite time interval, the total contribution from all paths reaching x, t from the past is the *wave function* $\psi(x, t)$, shown to satisfy Schroedinger's equation, *probability amplitude for a space-time path associated with entire motion of particle as function of time rather than with position of particle at particular time*, establishes postulates that describe *non-relativistic quantum mechanics neglecting spin*, mathematically equivalent to Heisenberg and Schroedinger formulations, no fundamentally new results, *suffers serious drawbacks*, requires unnatural and cumbersome division of the time interval, not formulated so that it is physically obvious that it is invariant under unitary transformations, improvements could be made through use of notation and concepts of mathematics of functionals].

In view of the fact that in classical physics positrons could be viewed as electrons proceeding along world lines toward the past (reference 7) the attempt was made to remove, in the *relativistic* case, the restriction that the paths must proceed always in one direction in time.

[[7] The idea that positrons can be represented as electrons with proper time reversed relative to true time has been discussed by the author and others, particularly by Stueckelberg (1942). [Stueckelberg, E. C. C. (December, 1942). La mécanique du point matériel en théorie de relativité et en théorie des quanta. (Mechanics of the material point in relativity theory and quantum theory.) *Helv. Phys.,*15, 23-37; also in Lacki, J., Ruegg, H., Wanders, G. (eds) (2009). *E.C.G. Stueckelberg, An Unconventional Figure of Twentieth Century Physics*. Birkhäuser, Basel; https://doi.org/10.1007/978-3-7643-8878-2_20]; Feynman, R. P. (October, 1948). [A Relativistic Cut-Off for Classical Electrodynamics. *Phys. Rev*., 74, 939-946.] …]

It was discovered that the results could be even more easily understood from a more familiar physical viewpoint, that of scattered waves. This viewpoint is the one used in this paper. After the equations were worked out physically the proof of the equivalence to the second quantization theory was found. [2]

[2] The equivalence of the entire procedure (including photon interactions) with the work of Schwinger and Tomonaga has been demonstrated by Dyson, F. J. (February, 1949). The Radiation Theories of Tomonaga, Schwinger, and Feynman. *Phys. Rev*., 75, 3, 486-502 [; the recent and independent formulations of quantum electrodynamics by Tomonaga, Schwinger, and Feynman have made two notable advances, the foundations and applications of the theory have been simplified by being presented in a *completely relativistic way* and the divergence difficulties *have been at least partially overcome*, the advantages of the Feynman formulation are simplicity and ease of application while those

of Tomonaga-Schwinger are generality and theoretical completeness, this paper presents a unified development of quantum electrodynamics embodying the main features of Tomonaga-Schwinger and Feynman radiation theories, *emphasis on application of the theory*, aims to show how the Schwinger theory can be applied to specific problems in such a way as to incorporate the ideas of Feynman, the main results are general formulas from which radiative reactions on motions of electrons can be calculated, divides *energy-density* into two parts, *energy of interaction* of two fields with each other and energy produced by external forces, *interaction energy* alone is treated as a perturbation, important results of the paper are the *equation of motion* $ihc[\partial\Omega/\partial\sigma(x_0)] = \{S(\sigma)\}^{-1}H^e(x_0)S(\sigma)\Omega$ for the *state vector* $\Omega(\sigma)$ and the interpretation of the *state vector* Ω, simplifies Schwinger theory for using it for calculations, demonstrates equivalence of the theories within their common domain of applicability, in the *Schwinger theory* the aim is to calculate the matrix elements of the "*effective external potential energy*" between *states* specified by their *state vectors*, in the *Feynman theory* the basic principle is to "preserve symmetry between past and future" so the matrix elements of the operator are evaluated in a "*mixed representation*" in which the matrix elements are calculated between an *initial state* specified by its *state vector* and a *final state* specified by its *state vector,* a *graph* corresponding to a particular matrix element is used not merely as an aid to calculation but as a picture of the physical process which gives rise to that matrix element, derives fundamental formulas for the operator in the *equation of motion* for the *state vector* $\Omega(\sigma)$ which represents the interaction of a physical particle with an external field for both the Schwinger and the Feynman theories, derives set of rules by which matrix element of Feynman operator may be written down in form suitable for numerical evaluation, shows equivalence of the two theories, develops graphical representation of matrix elements, *the theory as a whole cannot be put into a finally satisfactory form so long as divergencies occur in it however skillfully these divergencies are circumvented*, present treatment should be regarded as justified by its success in applications rather than by its theoretical derivation, *paper suffers from a series of significant errors*].

First, we discuss the relation of the Hamiltonian differential equation to its solution, using for an example the Schrodinger equation. Next, we deal in an analogous way with the Dirac equation and show how the solutions may be interpreted to apply to positrons. The interpretation seems not to be consistent unless the electrons obey the *exclusion principle*. (Charges obeying the Klein-Gordon equations can be described in an analogous manner, but here consistency apparently requires Bose statistics.)[3]

[3] These are special examples of the general relation of spin and statistics deduced by Pauli, W. (October, 1940). The Connection Between Spin and Statistics. *Phys. Rev.* 58, 716; https://doi.org/10.1103/PhysRev.58.716; also in Schwinger, J. (ed). (1958). *Selected Papers on Quantum electrodynamics*. Dover, New York, pages 372-8[; "In the following paper we conclude for the relativistically invariant wave equation for free particles: From postulate (I), according to which the *energy must be positive*, the necessity of *Fermi-Dirac* statistics for particles with arbitrary half-integral spin; from postulate (II), according to which *observables on different space-time points with a space-like distance are commutable*, the necessity of *Einstein-Bose* statistics for particles with arbitrary integral spin. It has been found useful to divide the quantities which are irreducible against Lorentz

transformations into four symmetry classes which have a commutable multiplication like +1, −1, +ε, −ε with $\varepsilon^2=1$"].

A representation in *momentum and energy variables* which is useful for the calculation of matrix elements is described. A proof of the equivalence of the method to the *theory of holes* in second quantization is given in the Appendix.

2. *Green's function treatment of Schrodinger's Equation*

We begin by a brief discussion of the relation of the *non-relativistic wave equation* to its solution. The ideas will then be extended to *relativistic* particles, satisfying Dirac's equation, and finally in the succeeding paper to *interacting relativistic particles*, that is, quantum electrodynamics.

The Schrodinger equation

$$i\,\partial\psi/\partial t = \mathbf{H}\psi, \qquad (1)$$

describes the change in the wave function ψ *in an infinitesimal time* Δt *as due to the operation of an operator* $exp(-iH\Delta t)$. One can ask also, if $\psi(x_1, t_1)$ is the wave function at x_1 at time t_1, what is the wave function at time $t_2 > t_1$? It can always be written as

$$\psi\,(x_2, t_2) = \int K(x_2, t_2; x_1, t_1)\psi(x_1, t_1)d^2x_1, \qquad (2)$$

where K is a *Green's function* for the linear Eq. (1).

[In mathematics, a *Green's function* is the impulse response of an inhomogeneous *linear differential operator* defined on a domain with specified initial conditions or boundary conditions. This means that if L is the *linear differential operator*, then

- the *Green's function* G is the solution of the equation LG = δ, where δ is *Dirac's delta function*;
- the solution of the initial-value problem Ly = f is the *convolution* (G ∗ f), where G is the *Green's function.*

Convolution is a mathematical operation on two functions (f and g) that produces a third function (f*g) that expresses how the shape of one is modified by the other.

Green's functions are named after the British mathematician George Green, who first developed the concept in the 1820s. In the modern study of linear partial differential equations, Green's functions are studied largely from the point of view of fundamental solutions instead.

Under many-body theory, the term is also used in physics, specifically in *quantum field theory*, aerodynamics, aeroacoustics, electrodynamics, seismology and statistical field theory, *to refer to various types of correlation functions*, even those that do not fit the mathematical definition. In *quantum field theory*, Green's functions take the roles of *propagators*.

In *quantum mechanics* and *quantum field theory,* the *propagator* is a function that specifies the *probability amplitude* for a particle to travel from one place to another in a given period of time, or to travel with a certain *energy* and *momentum.*

In *Feynman diagrams*, which serve to calculate the rate of collisions in *quantum field theory*, virtual particles contribute their *propagator* to the rate of the scattering event described by the respective diagram.]

(We have limited ourselves to a single particle of coordinate x, but the equations are obviously of greater generality.)

If H is a constant operator having eigenvalues E_n, eigenfunctions ϕ_n so that $\psi(x, t_1)$ can be expanded as $\sum_n C_n\phi_n(x)$, then $\psi(x, t_2) = \exp\{-i E_n(t_2 - t_1)\} \times C\phi_n(x)$.

Since $C_n = \int \phi_n{}^*(x_1)\psi(x_1, t_1)d^2x_1$, one finds (where we write 1 for x_1, t_1 and 2 for x_2, t_2) in this case

$$K(2, 1) = \sum_n \phi_n(x_2)\phi_n{}^*(x_1) \exp\{-i E_n(t_2 - t_1)\}, \qquad (3)$$

for $t_2 > t_1$.

We shall find it convenient for $t_2 < t_1$ to define $K(2, 1) = 0$ (Eq. (2) is then not valid for $t_2 < t_1$). It is then readily shown that in general K can be defined by that solution of

$$(i\, \partial/\partial t_2 - H_2)K(2, 1) = i\delta(2, 1), \qquad (4)$$

which is zero for $t_2 < t_1$, where $\delta(2, 1) = \delta(t_2 - t_1)\delta(x_2 - x_1) \times \delta(y_2 - y_1)\delta(z_2 - z_1)$ and the subscript 2 on H_2 means that the operator acts on the variables of 2 of K(2, 1). When H is not constant, (2) and (4) are valid but K is less easy to evaluate than (3).[4]

[4] For a *non-relativistic* free particle, where $\phi_n = \exp\{ip\cdot x\}$, $E_n = p^2/2m$, (3) gives, as is well known

$$K_0(2, 1) = \int \exp[-(ip \bullet x_1 - ip\cdot x_2) - ip^2(t_2 - t_1)/2m]d^3p(2\pi)^{-3}$$
$$= (2\pi i m^{-1}(t_2 - t_1)i^{-3/2} \exp(\tfrac{1}{2}\, im(x_2 - x_1)^2(t_2 - t_1)i^{-1})$$

for $t_2 > t_1$ and $K_0 = 0$ for $t_2 < t_1$.

We can call K(2, 1) the *total amplitude* for arrival at x_2, t_2 starting from x_1, t_1. (It results from adding an amplitude exp iS, for each *space-time* path between these points, where S is the action along the path[1].) The *transition amplitude* for finding a particle in *state* $\chi(x_2, t_2)$ at time t_2, if at t_1 it was in $\psi(x_1, t_1)$, is

$$\int \chi^*(2)K(2, 1)\psi(1)\, d^2x_1 d^2x_2. \qquad (5)$$

A quantum mechanical system is described equally well by specifying the function K, or by specifying the Hamiltonian H from which it results. For some purposes the specification in terms of K is easier to use and visualize. We desire eventually to discuss *quantum electrodynamics* from this point of view.

To gain a greater familiarity with the K function and the point of view it suggests, we consider a simple perturbation problem. Imagine we have a particle in a weak *potential*

U(x, t), a function of position and time. We wish to calculate K(2, 1) if U differs from zero only for t between t_1 and t_2. We shall expand K in increasing powers of U:

$$K(2, 1) = K_0(2, 1) + K^{(1)}(2, 1) + K^{(2)}(2, 1) + \dots . \qquad (6)$$

To zero order in U, K is that for a free particle, $K_0(2, 1)$.[4]

To study the first order correction $K^{(1)}(2, 1)$, first consider the case that U differs from zero only for the infinitesimal time interval Δt_3 between some time t_3 and $t_3 + \Delta t_3$ ($t_1<t_3<t_2$). Then if $\psi(1)$ is the *wave function* at $\mathbf{x}_1$, t_1, the *wave function* at x_3, t_3 is

$$\psi(3) = \int K_0(3, 1)\psi(1)\, d^2x_1, \qquad (7)$$

since from t_1 to t_3 the particle is free. For the short interval Δt_3 we solve (1) as

$$\begin{aligned}\psi(x, t_3+\Delta t_3) &= \exp(-iH\Delta t_3)\, \psi(x, t_3)\\ &= (1 - iH_0\Delta t_3 - iU\Delta t_3)\psi(x, t_3),\end{aligned}$$

where we put $H = H_0 + U$, H_0 being the Hamiltonian of a free particle. Thus $\psi(x, t_3+\Delta t_3)$ differs from what it would be if the *potential* were zero (namely $(1 - iH_0\Delta t_3)\psi(x, t_3)$) by the extra piece

$$\Delta\psi = - iU(x, t_3) \bullet \psi(x_3, t_3)\Delta t_3, \qquad (8)$$

which we shall call the *amplitude scattered by the potential*. The *wave function* at 2 is given by

$$\psi(x_2, t_2) = \int K_0(x_2, t_2; x_3, t_3+\Delta t_3)\psi(x_3, t_3+\Delta t_3)\, d^3x_3,$$

since after $t_3+\Delta t_3$ the particle is again free. Therefore, the change in the wave function at 2 brought about by the *potential* is (substitute (7) into (8) and (8) into the equation for $\psi(x_2, t_2)$:

$$\Delta\psi(2) = - i \int K_0(2, 3)U(3)K_0(3, 1)\, \psi(1)\, d^3x_1 d^3x_3\Delta t_3.$$

In the case that the *potential* exists for an extended time, it may be looked upon as a sum of effects from each interval Δt_3 so that the total effect is obtained by integrating over t_3 as well as x_3. From the definition (2) of K then, we find

$$K^{(1)}(2, 1) = - i \int K_0(2, 3)U(3)K_0(3, 1)\, d\tau_3, \qquad (9)$$

where the integral can now be extended over all space and time, $d\tau_3 = d^3x_3\Delta t_3$. Automatically there will be no contribution if t_3 is outside the range t_1 to t_2 because of our definition, $K_0(2, 1) = 0$ for $t_2< t_1$.

We can understand the result (6), (9) this way. We can imagine that a particle travels as a free particle from point to point, but is scattered by the *potential* U. Thus, the *total amplitude* for arrival at 2 from 1 can be considered as the sum of the *amplitudes* for various alternative routes. It may go directly from 1 to 2 (*amplitude* $K_0(2, 1)$, giving the zero-order term in (6)). Or (see Fig. l(a)) it may go from 1 to 3 (*amplitude* $K_0(3, 1)$), get scattered there by the *potential* (scattering *amplitude* – iU(3) per unit volume and time) and then go from 3 to 2 (*amplitude* $K_0(2, 3)$). This may occur for any point 3 so that summing over these alternatives gives (9).

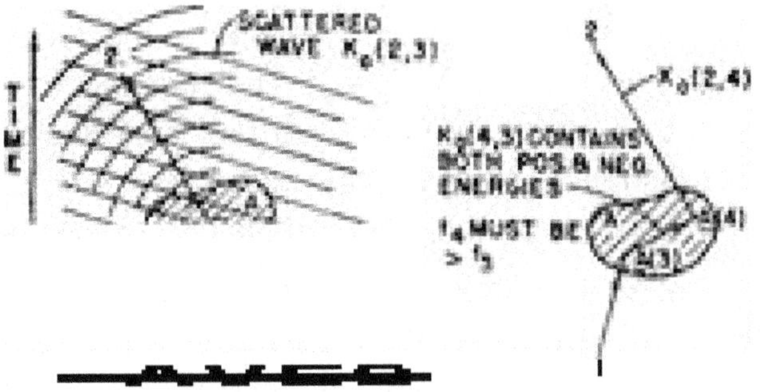

(a) First order eq. (9) (b) Second order eq. (10)

Fig. 1. *The Schrodinger (and Dirac) equation can be visualized as describing the fact that plane waves are scattered successively by a potential.* Figure 1 (a) illustrates the situation in first order. $K_0(2, 3)$ is the *amplitude* for a free particle starting at point 3 to arrive at 2. The shaded region indicates the presence of the *potential* **A** which scatters at 3 with *amplitude* – i**A**(3) per cm^3sec. (Eq. (9)). In (b) is illustrated the second order process (Eq. (10)), the waves scattered at 3 are scattered again at 4. However, *in Dirac one-electron theory $K_0(4, 3)$ would represent electrons both of positive and of negative energies* proceeding from 3 to 4. This is remedied by choosing a different scattering kernel $K_0(4, 3)$, Fig. 2.

Again, it may be scattered twice by the *potential* (Fig. 1(b)). It goes from 1 to 3 ($K_0(3, 1)$), gets scattered there (– iU(3)) then proceeds to some other point, 4, in *space-time* (*amplitude* $K_0(4, 3)$) is scattered again (– iU(4)) and then proceeds to 2 ($K_0(2, 4)$). Summing over all possible places and times for 3, 4 find that the second order contribution to the *total amplitude* $K^{(2)}(2, 1)$ is

$$(-i)^2 \iint K_0(2, 4)U(4)K_0(4, 3) \times U(3)K_0(3, 1)\, d\tau_3 d\tau_4. \qquad (10)$$

This can be readily verified directly from (1) just as (9) was. One can in this way obviously write down any of the terms of the expansion (6).[5]

[5] We are simply solving by successive approximations as an integral equation (deducible directly from (I) with $H = H_0 + U$, and (4) with $H = H_0$),

$$\psi(2) = -i \int K_0(2, 3)U(3)\psi(3)\, d\tau_3 + \int K_0(2, 1)\psi(1)\, d^3x_1,$$

where the first integral extends over all space and all times t_3 greater than the t_1 appearing in the second term, and $t_2>t_1$.

3. *Treatment of the Dirac equation*

We shall now extend the method of the last section to apply to the *Dirac equation.* All that would seem to be necessary in the previous equations is to consider H as the Dirac Hamiltonian, ψ as a symbol with four indices (for each particle). Then K_0 can still be defined by (3) or (4) and is now a 4-4 matrix which operating on the initial *wave function*, gives the final *wave function*. In (10), U(3) can be generalized to A4(3) – **α**·**A**(3) where A_4, **A** are the scalar and vector *potential* (times e, the electron charge) and **α** are Dirac matrices.

To discuss this, we shall define a convenient *relativistic* notation. We represent four-vectors like x,t by a symbol x_μ, where μ= 1, 2, 3, 4 and x_4 = t is real. Thus, the vector and scalar *potential* (times e) **A**, A_4 is A_μ. The four matrices β**α**, β can be considered as transforming as a four vector γ_μ (our γ_μ differs from Pauli's by a factor i for μ = 1, 2, 3). We use the summation convention $a_\mu b_\mu = a_4b_4 - a_1b_1 - a_2b_2 - a_3b_3 = a\cdot b$. In particular if a_μ is any four vector (but not a matrix) we write $\boldsymbol{a} = a_\mu\gamma_\mu$ so that $\boldsymbol{a}$ is a matrix associated with a vector ($\boldsymbol{a}$ will often be used in place of a_μ as a symbol for the vector). The γ_μ satisfy $\gamma_\mu\gamma_\nu + \gamma_\nu\gamma_\mu = 2\delta_{\mu\nu}$, where $\delta_{44} = +1$, $\delta_{11} = \delta_{22} = \delta_{33} = -1$, and the other $\delta_{\mu\nu}$ are zero. As a consequence of our summation convention $\delta_{\mu\nu}a_\nu = a_\mu$ and $\delta_{\mu\mu} = 4$. Note that $\boldsymbol{ab} + \boldsymbol{ba} = 2a\cdot b$ and that $\boldsymbol{a}^2 = a_\mu a_\mu = a\cdot a$ is a pure number. The symbol $\partial/\partial x_\mu$ will mean $\partial/\partial t$ for μ=4, and $-\partial/\partial x$, $-\partial/\partial y$, $-\partial/\partial z$ for μ = 1, 2, 3. Call $\nabla = \gamma_\mu\partial/\partial x_\mu = \beta\partial/\partial t + \beta\boldsymbol{\alpha}\cdot\boldsymbol{\nabla}$. We shall imagine hereafter, purely for *relativistic* convenience, that ϕ_n^* in (3) is replaced by its adjoint $\bar{\phi}_n = \phi_n^*\beta$.

Thus, *the Dirac equation for a particle, mass m, in an external field* $\boldsymbol{A} = A_\mu\gamma_\mu$ *is*

$$(i\nabla - m)\psi = \mathbf{A}\psi, \tag{11}$$

$[(i\gamma_\mu\partial/\partial x_\mu - m)\psi = \mathbf{A}\psi$.

Alternative forms of the *Dirac equation* include;

$\{p_0 + \rho_1 (\boldsymbol{\sigma}, \mathbf{p}) + \rho_3 mc\}\ \psi = 0$;

$\{i\Sigma\gamma_\mu p_\mu + mc\}\psi = 0$, ($\mu$ = 1, 2, 3, 4),

where $p_0 = ip_4$, $\rho_3 = \gamma_4$, $\rho_2\sigma_r = \gamma_r$, (r = 1, 2, 3);

$\{p_0 + e/c\ .\ A_0 + \rho_1 (\boldsymbol{\sigma}, \mathbf{p} + e/c\ \mathbf{A}) + \rho_3 mc\}\psi = 0$.

$\{p_0 + e/c\ A_0 + \alpha_1 (p_1 + e/c\ A_1) + \alpha_2 (p_2 + e/c\ A_2) + \alpha_3 (p_3 + e/c\ A_3) + \alpha_4 mc\}\psi = 0$.

$\{- i\hbar\gamma^\mu\delta_\mu + mc\}\psi = 0$,

where $\gamma^\mu\delta_\mu = (\beta^2/c\ \delta_t + \beta\alpha_1\delta_x + \beta\alpha_2\delta_y + \beta\alpha_3\delta_z) = \beta\ (\beta/c\ \delta_t + \sum^3_{n=1} \alpha_n\delta_n)$;

$(\beta mc^2 + c \sum^3_{n=1} \alpha_n p_n)\psi(x, t) = i\hbar\ \partial\psi(x, t)/\partial t$, or

$\{- i\hbar/c\ \partial\psi(x, t)/\partial t + \sum^3_{n=1} \alpha_n p_n)\psi(x, t) + \beta mc\psi(x, t)\} = 0$,

where,

$\psi = \psi(x, t)$ is the wave function for the electron of rest mass

m is the rest mass of the electron

x, t are the space-time coordinates

$p_n = -ih\ \partial/\partial x_n$, (n = 1, 2, 3) are the momentum components (momentum operator in the Schrödinger equation)

c is the speed of light

$\hbar$ is the reduced Planck constant

α_1, α_2, α_3 and β are 4 x 4 matrices] and Eq. (4)

$$[(i\,\partial/\partial t_2 - H_2)K(2, 1) = i\delta(2, 1), \qquad (4)]$$

determining the propagation of a free particle becomes

$$(i\nabla_2 - m)K_+(2, 1) = i\delta(2, 1), \qquad (12)$$

the index 2 on ∇_2 indicating differentiation with respect to the coordinates $x_{2\mu}$ which are represented as 2 in $K_+(2, 1)$ and $\delta(2, 1)$.

The function $K_+(2, 1)$ is defined in the absence of a field. If a *potential* **A** is acting a similar function, say $K_+^{(A)}(2, 1)$ can be defined. It differs from $K_+(2, 1)$ by a first order correction given by the analogue of (9)

$$[K^{(1)}(2, 1) = -i \int K_0(2, 3)U(3)K_0(3, 1)\, d\tau_3, \qquad (9)]$$

namely

$$K_+^{(1)}(2, 1) = -i \int K_+(2, 3)\mathbf{A}(3)K_+(3, 1)\, d\tau_3, \qquad (13)$$

representing the *amplitude* to go from 1 to 3 as a free particle, get scattered there by the *potential* (now the matrix **A**(3) instead of U(3)) and continue to 2 as free. The second order correction, analogous to (10)

$$[(-i)^2 \iint K_0(2, 4)U(4)K_0(4, 3) \times U(3)K_0(3, 1)\, d\tau_3 d\tau_4. \qquad (10)]$$

is

$$K_+^{(2)}(2, 1) = -i \iint K_+(2, 4)\mathbf{A}(4) \times K_+(4, 3)\mathbf{A}(3)K_+(3, 1)\, d\tau_4 d\tau_3, \qquad (14)$$

and so on. In general $K_+^{(A)}$ satisfies

$$(i\nabla_2 - \mathbf{A}(2) - m)K_+^{(A)}(2, 1) = i\delta(2, 1), \qquad (15)$$

and the successive terms (13), (14) are the power series expansion of the integral equation

$$K_+^{(A)}(2, 1) = K_+(2, 1) - i \int K_+(2, 3)\mathbf{A}(3)K_+(3, 1)\, d\tau_3, \qquad (16)$$

which it also satisfies.

We would now expect to choose, for the special solution of (12), $K_+ = K_0$ where $K_0(2, 1)$ vanishes for $t_2 < t_1$ and for $t_2 > t_1$ is given by (3)

$$[K(2, 1) = \textstyle\sum_n \phi_n(x_2)\phi_n^*(x_1) \exp\{-i E_n(t_2 - t_1)\}, \qquad (3)]$$

where ϕ_n and E_n are the eigenfunctions and *energy* values of a particle satisfying *Dirac's equation*, and ϕ_n^* is replaced by $\bar{\phi}_n$.

The formulas arising from this choice, however, suffer from the drawback that they apply to the one electron theory of Dirac rather than to the hole theory of the positron. For example, consider as in Fig. 1(a) an electron after being scattered by a *potential* in a small region 3 of space-time. The one electron theory says (as does (3) with $K_+ = K_0$)

$$[K(2, 1) = \textstyle\sum_n \phi_n(x_2)\phi_n^*(x_1) \exp\{-i E_n(t_2 - t_1)\}, \qquad (3)]$$

that the scattered *amplitude* at another point 2 will proceed toward positive times with both positive and negative *energies*, that is with both positive and negative rates of change of *phase*. No wave is scattered to times previous to the time of scattering. These are just the properties of $K_0(2, 3)$.

On the other hand, according to the *positron theory* negative energy states are not available to the electron after the scattering. Therefore, the choice $K_+ = K_0$ is unsatisfactory. But there are other solutions of (12)

$$[(i\nabla_2 - m)K_+(2, 1) = i\delta(2, 1). \qquad (12)]$$

We shall choose the solution defining $K_+(2, 1)$ so that $K_+(2, 1)$ *for $t_2 > t_1$ is the sum of (3) over positive energy slates only.* Now this new solution must satisfy (12) for all times in order that the representation be complete. It must therefore differ from the old solution K_0 by a solution of the *homogeneous Dirac equation*. It is clear from the definition that the difference $K_0 - K_+$ is the sum of (3) over all *negative energy states*, as long as $t_2 > t_1$. But this difference must be a solution of the *homogeneous Dirac equation* for all times and must therefore be represented by the same sum over *negative energy states* also for $t_2 < t_1$. Since $K_0 = 0$ in this case, it follows that our new kernel, $K_+(2, 1)$, *for $t_2 < t_1$ is the negative of the sum (3) over negative energy states.* That is,

$$K_+(2, 1) = \sum_{\text{POS En}} \phi_n(2)\, \bar{\phi}_n(1) \text{ x } \exp\{-iE_n(t_2 - t_1)\} \quad \text{for } t_2 > t_1$$
$$= -\sum_{\text{NEG En}} \phi_n(2)\, \bar{\phi}_n(1) \text{ x } \exp\{-iE_n(t_2 - t_1)\} \quad \text{for } t_2 < t_1. \qquad (17)$$

With this choice of K_+ our equations such as (13) and (14)

$$[K_+^{(1)}(2, 1) = -i \int K_+(2, 3)\mathbf{A}(3)K_+(3, 1)\, d\tau_3, \qquad (13)$$
$$K_+^{(2)}(2, 1) = -i \iint K_+(2, 4)\mathbf{A}(4) \text{ x } K_+(4, 3)\mathbf{A}(3)K_+(3, 1)\, d\tau_4 d\tau_3 \qquad (14)]$$

will now give results equivalent to those of the *positron hole theory*.

That (14), for example, is the correct second order expression for finding at 2 an electron originally at 1 according to the *positron theory* may be seen as follows (Fig. 2).

(a) First order eq. (13)

(b) Virtual Scattering: second order eq. (14)

Fig. 2. The *Dirac equation* permits another solution K+(2, 1) if one considers that waves scattered by the potential can proceed backwards in time as in Fig. 2 (a). This is interpreted in the second order processes (b), (c), by noting that there is now the possibility (c) of virtual pair production at 4, the positron going to 3 to be annihilated. This can be pictured as similar to ordinary scattering (b) except that the electron is scattered backwards in time from 3 to 4. The waves scattered from 3 to 2' in (a) represent the possibility of a positron arriving at 3 from 2' and annihilating the electron from 1. This view is proved equivalent to *hole theory*: electrons traveling backwards in time are recognized as positrons.

Assume as a special example that $t_2>t_1$ and that the *potential* vanishes except in interval $t_2 - t_1$ so that t_4 and t_2 both lie between t_1 and t_2. First suppose $t_4>t_2$ (Fig. 2(b)). Then (since $t_3 >t_1$) the electron assumed originally in a positive energy state propagates in that state (by $K_+(3, 1)$) to position 3 where it gets scattered (**A**(3)). It then proceeds to 4, which it must do as a *positive energy* electron. This is correctly described by (14)

$$[K_+^{(2)}(2, 1) = - i \iint K_+(2, 4)\mathbf{A}(4) \times K_+(4, 3)\mathbf{A}(3)K_+(3, 1)\, d\tau_4 d\tau_3 \qquad (14)]$$

for $K_+(4, 3)$ contains only *positive energy* components in its expansion, as $t_4>t_3$. After being scattered at 4 it then proceeds on to 2, again necessarily in a *positive energy state*, as $t_2>t_4$.

In positron theory there is an additional contribution due to the possibility of virtual pair production (Fig. 2(c)). A pair could be created by the potential **A**(4) at 4, the electron of which is that found later at 2. The positron (or rather, the hole) proceeds to 3 where it annihilates the electron which has arrived there from 1.

This alternative is already included in (14) as contributions for which $t_4< t_3$, and its study will lead us to an interpretation of $K_+(4, 3)$ for $t_4< t_3$. The factor $K_+(2, 4)$ describes the electron (after the pair production at 4) proceeding from 4 to 2. Likewise, $K_+(3, 1)$ represents the electron proceeding from 1 to 3. $K_+(4, 3)$ must therefore represent the propagation of the positron or hole from 4 to 3. That it does so is clear. The fact that in *hole theory* the hole proceeds in the manner of an electron of *negative energy* is reflected in the fact that $K_+(4, 3)$ for $t_4< t_3$ is (minus) the sum of only *negative energy* components. In hole theory the real *energy* of these intermediate *states* is, of course, positive. This is true here too, since in the *phases* $\exp\{-iE_n(t_4 - t_3)\}$ defining $K_+(4, 3)$ in (17)

$$[K_+(2, 1) = \sum_{\text{POS En}} \phi_n(2)\, \bar{\phi}_n(1) \times \exp\{-iE_n(t_2 - t_1)\} \text{ for } t_2>t_1$$
$$= - \sum_{\text{NEG En}} \phi_n(2)\, \bar{\phi}_n(1) \times \exp\{-iE_n(t_2 - t_1)\} \text{ for } t_2<t_1, \qquad (17)]$$

E_n is negative but so is $t_4 - t_3$. That is, the contributions vary with t_3 as $\exp\{-iE_n(t_4 - t_3)\}$ as they would if the *energy* of the intermediate *state* were $| E_n |$. The fact that the entire sum is taken as negative in computing $K_+(4, 3)$ is reflected in the fact that in hole theory the *amplitude* has its sign reversed in accordance with the Pauli principle and the fact that the electron arriving at 2 has been exchanged with one in the sea[6].

[6] It has often been noted that the *one-electron theory* apparently gives the same matrix elements for this process as does hole theory. The problem is one of interpretation, especially in a way that will also give correct results for other processes, e.g., *self-energy*.

To this, and to higher orders, all processes involving virtual pairs are correctly described in this way.

The expressions such as (14) can still be described as a passage of the electron from 1 to 3 (K+(3, 1)), scattering at 3 by A(3), proceeding to 4 (K+(4, 3)), scattering again, A(4), arriving finally at 2. The scatterings may, however, be toward both future and past times, an electron propagating backwards in time being recognized as a positron.

This therefore suggests that *negative energy* components created by scattering in a potential be considered as waves propagating from the scattering point toward the past, and that such waves represent the propagation of a positron annihilating the electron in the potential.[7]

[7] The idea that positrons can be represented as electrons with proper time reversed relative to true time has been discussed by the author and others, particularly by Stuckelberg (1942). [Stuckelberg, E. C. C. (December, 1942). La mécanique du point matériel en théorie de relativité et en théorie des quanta. (Mechanics of the material point in relativity theory and quantum theory.) *Helv. Phys.,* 15, 23-37; also in Lacki, J., Ruegg, H., Wanders, G. (eds) (2009). *E.C.G. Stueckelberg, An Unconventional Figure of Twentieth Century Physics*. Birkhäuser, Basel; https://doi.org/10.1007/978-3-7643-8878-2_20]; Feynman, R. P. (October, 1948). [A Relativistic Cut-Off for Classical Electrodynamics. *Phys. Rev*., 74, 939-946.] The fact that classically the *action* (proper time) increases continuously as one follows a trajectory is reflected in quantum mechanics in the fact that the *phase*, which is $|E_n||t_2 - t_1|$ always increases as the particle proceeds from one scattering point to the next.

With this interpretation real pair production is also described correctly (see Fig. 3). For example, in (13)

$$[K_+^{(1)}(2, 1) = -i \int K_+(2, 3)\mathbf{A}(3)K_+(3, 1)\, d\tau_3, \qquad (13)]$$

if $t_1 < t_3 < t_2$ the equation gives the *amplitude* that if at time t_1 one electron is present at 1, then at time t_2 just one electron will be present (having been scattered at 3) and it will be at 2. On the other hand, if t_2 is less than t_3, for example, if $t_2 = t_1 < t_3$, the same expression gives the *amplitude* that a pair, electron at 1, positron at 2 will annihilate at 3, and subsequently no particles will be present. Likewise if t_2 and t_1 exceed t_3 we have (minus) the *amplitude* for finding a single pair, electron at 2, positron at 1 created by **A**(3) from a vacuum. If $t_1 > t_3 > t_2$, (13) describes the scattering of a positron. All these *amplitudes* are relative to the *amplitude* that a vacuum will remain a vacuum, which is taken as unity. (This will be discussed more fully later.)

...

4. *Problems involving several charges*

We wish next to consider the case that there are two (or more) distinct *charges* (in addition to pairs they may produce in virtual *states*). In a succeeding paper we discuss the *interaction* between such charges. Here we assume that they do not interact. In this case each particle behaves independently of the other. ...

...

All the amplitudes are relative and their squares give the relative probabilities of the various phenomena. Absolute probabilities result if one multiplies each of the probabilities by P_υ, the true probability that if one has no particles present initially there will be none finally. This quantity P_υ can be calculated by normalizing the relative probabilities such that the sum of the probabilities of all mutually exclusive alternatives is unity. (For example, if one starts with a vacuum one can calculate the relative probability that there remains a vacuum (unity), or one pair is created, or two pairs, etc. The sum is P_υ^{-1}.) *Put in this form the theory is complete and there are no divergence problems.* Real processes are completely independent of what goes on in the vacuum.

When we come, in the succeeding paper, to deal with interactions between charges, however, the situation is not so simple. There is the possibility that virtual electrons in the vacuum may interact electromagnetically with the real electrons. For that reason, processes occurring in the vacuum are analyzed in the next section, in which an independent method of obtaining P_υ is discussed.

5. *Vacuum problems*

An alternative way of obtaining absolute *amplitudes* is to multiply all *amplitudes* by C_υ, the vacuum-to-vacuum *amplitude*, that is, the absolute *amplitude* that there be no particles both initially and finally. We can assume $C_\upsilon = 1$ if no *potential* is present during the interval, and otherwise we compute it as follows. It differs from unity because, for example, a pair could be created which eventually annihilates itself again. Such a path would appear as a closed loop on a *space-time* diagram. …

…

5. *Energy-momentum representation*

The practical evaluation of the matrix elements in some problems is often simplified by working with *momentum* and *energy* variables rather than *space* and *time*. This is because the function $K_+(2, 1)$ is fairly complicated but we shall find that its Fourier transform is very simple, …

…

The author has many people to thank for fruitful conversations about this subject, particularly H. A. Bethe and F. J. Dyson.

Feynman, R. P. (September, 1949). Space-Time Approach to Quantum Electrodynamics.

[*Phys. Rev.*, 76, 6, 769-89; https://doi.org/10.1103/PhysRev.76.769; also in Schwinger, J. (ed). (1958). *Selected Papers on Quantum electrodynamics*. Dover, New York, pages 236-56.]

Received May 9, 1949.

Department of Physics, Cornell University, Ithaca, New York.

In previous paper motion of electrons *neglecting interaction* was analyzed by dealing directly with the *solutions* of the Hamiltonian time differential equations rather than the equations, here the same technique is applied to include *interactions* to express in simple terms the solution of problems in *quantum electrodynamics* rather than the differential equations from which they come, (1) it is shown that *a considerable simplification can be attained by writing down matrix elements for complex processes in electrodynamics*, a physical point of view is available which permitted them to be written down directly. The simplification results from the fact that previous methods separated into individual terms processes that were closely related physically, (2) *electrodynamics is modified by altering the interaction of electrons at short distances*, all matrix elements now finite with the exception of those relating to *vacuum polarization*, latter evaluated in manner suggested by Pauli and Bethe to give finite results, phenomena directly observable insensitive to the details of the modification, electrodynamics viewed as direct *interaction* at a distance between *charges* rather than behavior of a *field* (Maxwell's equations), *field point of view*, which separates the production and absorption of light, most practical for problems involving *real quanta* while *interaction view* best for discussion of *virtual quanta* when dealing with close collisions of particles or their actions on themselves, Hamiltonian method not well adapted to represent direct *action at a distance* between *charges* because action is delayed, forces use of field viewpoint rather than interaction viewpoint, for *collisions* much easier to treat the process as a whole, effects of longitudinal and transverse waves can be combined, begins with solution in *space* and *time* of Schrodinger equation for particles interacting instantaneously, solution is expressed in terms of *matrix elements, derived from Lagrangian form of quantum mechanics*, then modifies in accordance with the requirements of the Dirac equation and phenomenon of pair creation, made easier by reinterpreting the *theory of holes*, generalizes to *delayed interactions* of *relativistic* electrons, for practical calculations expressions developed in a power series of $e^2/\hbar c$, *derivation will appear in a separate paper*, by forsaking Hamiltonian method, the wedding of *relativity* and *quantum mechanics* accomplished most naturally, *relativistic* invariance becomes self-evident but the *matrix elements for complex processes* and the *self-energy diverges* then can see how the matrix elements can be written down directly, then introduces modification in interaction between charges at short distances, *assumes substantial interaction exists as long as the four-dimensional interval is time-like and less than some small length of order of the electron radius* [Feynman (October, 1948). A Relativistic Cut-Off for Classical Electrodynamics.], *convergence factors* then introduced such that the integrals with their convergence factors now converge, *self-energy* now convergent, corresponds to a correction to the electron mass, *all matrix elements now finite* with exception of those relating to *vacuum polarization, strict physical basis for rules of convergence not known*, after *mass and charge renormalization* results are equivalent to those of Schwinger in which the terms corresponding to corrections in mass and charge are identified and removed from the expressions for real processes, *although in the limit the two methods agree neither method appears to be completely satisfactory theoretically*, practical advantage of new method is that ambiguities can be more easily resolved by direct calculation of otherwise divergent integrals, complete method therefore available for calculations of all processes involving electrons and photons, shows how matrix element for any process can be written down directly in *Feynman diagrams*, paper includes first published "*Feynman diagram*", *attempt to find a consistent modification of quantum electrodynamics is incomplete*, *not at all clear that the convergence factors do not upset the physical consistency of the theory.*

Abstract

In this paper two things are done. (1) It is shown that a *considerable simplification can be attained in writing down matrix elements for complex processes* in electrodynamics. Further, a physical point of view is available which permits them to be written down directly for any specific problem. *Being simply a restatement of conventional electrodynamics, however, the matrix elements diverge for complex processes.* (2) Electrodynamics is modified *by altering the interaction of electrons at short distances. All matrix elements are now finite, with the exception of those relating to problems of vacuum polarization.* The latter are evaluated in a manner suggested by Pauli and Bethe, which also gives finite results for these matrices. *The only effects sensitive to the modification are changes in mass and charge of the electrons.* Such changes could not be directly observed. *Phenomena directly observable are insensitive to the details of the modification used* (except at extreme energies). For such phenomena, a limit can be taken as the range of the modification goes to zero. The results then agree with those of Schwinger. A complete, unambiguous, and presumably consistent, method is therefore available for the calculation of all processes involving electrons and photons.

The simplification in writing the expressions results from an emphasis on the over-all *space-time view* resulting from a study of the solution of the equations of electrodynamics. The relation of this to the more conventional Hamiltonian point of view is discussed. *It would be very difficult to make the modification which is proposed if one insisted on having the equations in Hamiltonian form.*

The methods apply as well to *charges* obeying the Klein-Gordon equation, and to the various *meson theories* of nuclear forces. Illustrative examples are given. Although a modification like that used in electrodynamics can make all matrices finite for all of the meson theories, for some of the theories it is no longer true that all directly observable phenomena are insensitive to the details of the modification used.

The actual evaluation of integrals appearing in the matrix elements may be facilitated, in the simpler cases, by methods described in the appendix.

This paper should be considered as a direct continuation of a preceding one[1]

[1] Feynman, R. P. (September, 1949). The Theory of Positrons. *Phys. Rev.* 76, 749-59[; this is the first of a set of papers dealing with the solution of problems in *quantum electrodynamics*, main principle is to deal directly with *solutions* of the Hamiltonian differential equations rather than with equations themselves, this paper *addresses motion of electrons and positrons in given external potentials*, second paper considers *interactions* (*quantum electrodynamics*), problem of charges in a fixed potential usually treated by method of second quantization of the electron field using the *theory of holes*, here we replace Dirac's "hole theory" by reinterpreting *solutions* of the Dirac equation, results simplified by *following the charge rather than the particles* because *the number of particles is not conserved* whereas *charge is conserved*, in the approximation of classical *relativistic* theory the creation of an electron pair (electron and positron) represented by the start of two world lines from the point of creation, *following the charge rather than the particles corresponds to considering continuous world line as a whole rather than breaking it up into its pieces*, over-all *space-time* point of view, quantum mechanically the direction of the world lines is replaced by the direction of propagation of waves, quite different from Hamiltonian method which considers the future as developing continuously from the past, in a scattering problem this over-all view of the complete scattering process is similar to the *S-matrix view-point of Heisenberg*, temporal order of events during scattering which is analyzed in such detail by the Hamiltonian differential equation is irrelevant, development stemmed from idea that in *non-relativistic* quantum mechanics the *amplitude* for a given process can be considered as the sum of an *amplitude* for each *space-time* path available, in *relativistic* case the restriction that the paths must proceed always in one direction in time is removed, results more easily understood from more familiar physical viewpoint of *scattered waves* used in this paper, after the equations were worked out physically the proof of the equivalence to the *second quantization theory* was found, solution in terms of boundary conditions on *wave function* contains all possibilities of pair formation and annihilation together with the ordinary scattering processes, negative energy states appear in space-time as waves traveling away from the external potential backwards in time (as suggested in Stückelberg (December, 1942)), such a wave corresponds to a positron approaching the potential and annihilating the electron, a particle moving forward in time (electron) in a potential may be scattered forward in time (ordinary scattering) or backward (pair annihilation), when moving backward (positron) it may be scattered backward in time (positron scattering) or forward (pair production), *amplitude* for transition from an initial to a final state analyzed to any order in the *potential* by considering it to undergo a sequence of such scatterings, *amplitude* for a process involving many such particles is the product of the *transition amplitudes* for each particle, vacuum problems do not arise for charges which do not interact with one another, shows equivalent to the *theory of holes* in second quantization], hereafter called I.

in which the motion of electrons, *neglecting interaction*, was analyzed, by dealing directly with the *solution* of the Hamiltonian differential equations. *Here the same technique is applied to include interactions* and, in that way, to express in simple terms the solution of problems in *quantum electrodynamics*.

For most practical calculations in *quantum electrodynamics* the solution is ordinarily expressed in terms of a *matrix element*. The matrix is worked out as an expansion in powers of $e^2/\hbar c$, the successive terms corresponding to the inclusion of an increasing number of virtual quanta. *It appears that a considerable simplification can be achieved in writing down these matrix elements for complex processes.* Furthermore, each term in the expansion can be written down and understood directly from a physical point of view, similar to the *space-time view* in I. *It is the purpose of this paper to describe how this may be done. We shall also discuss methods of handling the divergent integrals which appear in these matrix elements.*

The simplification in the formulae results mainly from the fact that previous methods unnecessarily separated into individual terms processes that were closely related physically. For example, in the exchange of a quantum between two electrons there were two terms depending on which electron emitted and which absorbed the quantum. Yet, in the virtual states considered, timing relations are not significant. Only the order of operators in the matrix must be maintained. We have seen (I), that in addition, processes in which virtual pairs are produced can be combined with others in which only positive energy electrons are involved. Further, *the effects of longitudinal and transverse waves can be combined together.* The separations previously made were on an *unrelativistic* basis (reflected in the circumstance that apparently *momentum* but not *energy* is conserved in intermediate states). *When the terms are combined and simplified, the relativistic invariance of the result is self-evident.*

We begin by discussing the solution in *space* and *time* of the Schrodinger equation for particles interacting instantaneously. The results are immediately generalizable to delayed interactions of *relativistic* electrons and we represent in that way the laws of *quantum electrodynamics.* We can then see how the matrix element for any process can be written down directly. In particular, the *self-energy* expression is written down.

So far, nothing has been done other than a restatement of conventional electrodynamics in other terms. Therefore, the *self-energy* diverges. *A modification*[2] *in interaction between charges is next made, and it is shown that the self-energy is made convergent and corresponds to a correction to the electron mass.*

[2] For a discussion of this modification in classical physics see Feynman, R. P. (October, 1948). A Relativistic Cut-Off for Classical Electrodynamics. *Phys. Rev.*, 74, 939-46[; *in this paper a consistent classical theory is described which the author believes can be quantized*, previous attempts to address *problem of infinite self-energy* that results from assuming *point electron* in *relativistic* theory met with considerable difficulties when attempt made to quantize them, the *potential* at a point in space at a given time depends on the charge at a distance r from the point at a time previous by t = r (taking the speed of light as unity), *relativistically* interaction occurs between events whose four-dimensional interval s defined by $s^2 = t^2 - r^2$ vanishes, this results in *infinite action of a point electron on itself, the present theory modifies this idea by assuming that substantial interaction exists as long as the interval s is time-like and less than some small length a of order of the electron radius*, reduces infinite *self-energy* to a finite value for accelerations which are not extreme, *action* of an electron on itself appears as *electromagnetic mass*, formulates in terms of *action at a distance*, satisfies Maxwell's equations, not the usual *retarded* solution for which there is no *self-force* but *half the retarded plus half the advanced solution*, effect

of modification is to change slightly the field of one particle on another when they are very close and to add a *self-force*, *little reason to believe that the ideas used here to solve the divergences of classical electrodynamics will prove fruitful for quantum electrodynamics*], hereafter referred to as A.

After the *mass correction* is made, other real processes are finite and insensitive to the "width" of the cut-off in the interaction[3].

[3] A brief summary of the methods and results will be found in Feynman, R. P. (November, 1948). Relativistic Cut-Off for Quantum Electrodynamics. *Phys. Rev.* 74, 1430-8 [; describes model based on quantization of a classical theory for which all quantities automatically come out finite described in his previous paper [Feynman (October, 1948). A Relativistic Cut-Off for Classical Electrodynamics], contains an arbitrary function on which numerical results depend, only term that depends significantly (logarithmically) on the cut-off frequency is the *self-energy* which can be used to *renormalize* the electron mass, remaining terms are nearly independent of the function, applies only to results for processes in which virtual quanta are emitted and absorbed, *terms representing processes involving a pair production followed by annihilation of the same pair are infinite and not made convergent by the present scheme,* problems of *permanent emission* and the position of *positron theory* need to be addressed, *the present paper may be looked upon as presenting an arbitrary rule to cut off at high frequencies in a relativistically invariant manner the otherwise divergent integrals appearing in quantum field theories*, produces finite invariant *self-energy* for a free electron, but problem of *polarization of the vacuum* not solved, alternative cut-off procedure which eliminates high frequency intermediate states offers to solve *vacuum polarization* problems as well], hereafter referred to as B.

Unfortunately, the modification proposed is not completely satisfactory theoretically (it leads to some difficulties of conservation of energy). It does, however, seem consistent and satisfactory to define the matrix element for all real processes as the limit of that computed here as the cut-off width goes to zero. *A similar technique suggested by Pauli and by Bethe can be applied to problems of vacuum polarization (resulting in a renormalization of charge) but again a strict physical basis for the rules of convergence is not known.*

After *mass and charge renormalization*, the limit of zero cut-off width can be taken for all real processes. The results are then equivalent to those of Schwinger[4] who does not make explicit use of the convergence factors.

[4] Schwinger, J. (November, 1948). Quantum Electrodynamics. I. A Covariant Formulation. *Phys. Rev.*, 74, 10, 1439-61[; *lack of convergence in current formulations of quantum electrodynamics indicates that revision of electrodynamic concepts at ultra-relativistic energies is necessary*, elementary phenomenon in which divergences occur as a result of virtual transitions involving particles with unlimited energy are *polarization of the vacuum* and *self-energy of the electron* which express *the interaction of the electromagnetic and matter fields with their own vacuum fluctuations*, this alters the constants characterizing the properties of the individual fields and their mutual coupling by infinite factors, the question is whether all divergencies can be isolated in such unobservable *renormalization* factors, *this paper is occupied with the formulation of a completely covariant electrodynamics*, manifest covariance with respect to Lorentz and gauge transformations

essential in a divergent theory, customary *canonical commutation relations* fail to exhibit the desired covariance since they refer to field variables at equal times and different points of space, *can be put in covariant form by replacing the four-dimensional surface t = const. by a space-like surface*, offers the advantage over the Schrodinger representation in which all operators refer to the same time providing distinct separation between *kinematical* and *dynamical* aspects, formulation that retains evident covariance of the Heisenberg representation but offers something akin to Schrodinger representation can be based on distinction between the properties of *non-interacting fields*, and the effects of *coupling between fields*, constructs a *canonical transformation* that changes the *field equations* in the *Heisenberg representation* into those of *non-interacting fields*, *supplementary condition* restricting the admissible states of the system and the *commutation relations* must be added to the *equations of motion*, describes the coupling between fields in terms of a varying state vector, then simple matter to evaluate commutators of *field* quantities at arbitrary *space-time* points, one thus obtains an obviously covariant and practical form of quantum electrodynamics expressed in a mixed Heisenberg-Schrodinger representation called the *interaction representation*, discusses *covariant* elimination of longitudinal field in which customary distinction between longitudinal and transverse fields is replaced by a suitable *covariant* definition, describes collision processes in terms of an invariant *collision operator* which is the unitary operator that determines the over-all change in state of a system as the result of interaction, notes that a *second paper* treats the problems of electron and photon *self-energy* together with the *polarization of the vacuum* and a *third paper* is concerned with the determination of the *radiative corrections* to the properties of an electron and the comparison with experiment [this was not addressed, it stated that "*radiative corrections to energy levels* will be treated in the next paper of the series" but *this did not appear, nor are there any references to it*].

The method of Schwinger is to identify the terms corresponding to corrections in mass and charge and, previous to their evaluation, to remove them from the expressions for real processes. This has the advantage of showing that the results can be strictly independent of particular cut-off methods. On the other hand, many of the properties of the integrals are analyzed using formal properties of invariant propagation functions. But one of the properties is that *the integrals are infinite* and it is not clear to what extent this invalidates the demonstrations. *A practical advantage of the present method is that ambiguities can be more easily resolved; simply by direct calculation of the otherwise divergent integrals. Nevertheless, it is not at all clear that the convergence factors do not upset the physical consistency of the theory*. Although in the limit the two methods agree, *neither method appears to be thoroughly satisfactory theoretically*. Nevertheless, it does appear that we now have available a complete and definite method for the calculation of physical processes to any order in *quantum electrodynamics*.

Since we can write down the solution to any physical problem, we have a complete theory which could stand by itself. It will be theoretically incomplete, however, in two respects. First, although each term of increasing order in e^2/hc can be written down it would be desirable to see some way of expressing things in finite form to all orders in e^2/hc at once. Second, although it will be physically evident that the results obtained are equivalent to those obtained by conventional electrodynamics the mathematical proof of this is not included. *Both of these limitations will be removed in a subsequent paper* (see also Dyson[4]).

Briefly the genesis of this theory was this. The conventional electrodynamics was *expressed in the Lagrangian form* of quantum mechanics described in the *Reviews of Modern Physics*[5].

> [5] Feynman, R. P. (April, 1948). Space-Time Approach to Non-Relativistic Quantum Mechanics. *Rev. Mod. Phys.*, 20, 2, 367-87[; third formulation of *non-relativistic* quantum theory in addition to *differential equation of Schroedinger* and *matrix algebra of Heisenberg*, *path integral formulation* utilizing the *action* principle as suggested in Dirac (1933) [The Lagrangian in Quantum Mechanics] and Dirac (1945) [On the Analogy Between Classical and Quantum Mechanics], in quantum mechanics the probability of an event which can happen in several different ways is the absolute square of a sum of complex contributions one from each alternative way $\varphi_{ac} = \sum_b \varphi_{ab}\varphi_{bc}$ where φ_{ab}, φ_{bc}, φ_{ac} are complex numbers such that $P_{ab} = |\varphi_{ab}|^2$, $P_{bc} = |\varphi_{bc}|^2$, and $P^q_{ac} = |\varphi_{ac}|^2$, and P_{ab} as the probability that if measurement A gave the result *a*, then measurement B will give the result *b*, and P^q_{ac} is the quantum mechanical probability that a measurement of C results in c when it follows a measurement of A giving a, the probability that a particle will be found to have a path lying somewhere within a region of space time is the absolute square of a sum of contributions - one from each path in the region, the contribution from a single path is postulated to be an exponential whose (imaginary) phase is the classical *action* for the path in question where *action* refers to time integral of Lagrangian along a path, restriction to finite time interval, the total contribution from all paths reaching x, t from the past is the *wave function* $\psi(x, t)$, shown to satisfy Schroedinger's equation, *probability amplitude for a space-time path associated with entire motion of particle as function of time rather than with position of particle at particular time*, establishes postulates that describe *non-relativistic quantum mechanics neglecting spin*, mathematically equivalent to Heisenberg and Schroedinger formulations, no fundamentally new results, *suffers serious drawbacks*, requires unnatural and cumbersome division of the time interval, not formulated so that it is physically obvious that it is invariant under unitary transformations, improvements could be made through use of notation and concepts of mathematics of functionals].
> The application to electrodynamics is described in detail by Groenewold, H. J. (1949). *Koninklijke Nederlandsche Akademia van Weteschappen. Proceedings*, LII, 3, 226.

The motion of the field oscillators could be integrated out (as described in Section 13 of that paper), the result being an *expression of the delayed interaction of the particles*. Next the modification of the delta-function *interaction* could be made directly from the analogy to the classical case[2].

> [2] *Loc. cit.* Feynman, R. P. (October, 1948). A Relativistic Cut-Off for Classical Electrodynamics.

This was still not complete because the Lagrangian method had been worked out in detail only for particles obeying the *non-relativistic* Schrodinger equation. It was then modified *in accordance with the requirements of the Dirac equation and the phenomenon of pair creation*. This was made easier by the reinterpretation of the *theory of holes* (I). *Finally for practical calculations the expressions were developed in a power series in $e^2/\hbar c$.* It was apparent that each term in the series had a simple physical interpretation. *Since the result was easier to understand than the derivation, it was thought best to publish the results first in this paper.*

Considerable time has been spent to make these first two papers as complete and as physically plausible as possible without relying on the Lagrangian method, because it is not generally familiar. It is realized that such a description cannot carry the conviction of truth which would accompany the derivation. *On the other hand, in the interest of keeping simple things simple the derivation will appear in a separate paper.*

The possible application of these methods to the various meson theories is discussed briefly. The formulas corresponding to a charge particle of zero spin moving in accordance with the Klein Gordon equation are also given. In an Appendix a method is given for calculating the integrals appearing in the matrix elements for the simpler processes.

The point of view which is taken here of the *interaction of charges* differs from the more usual point of view of *field theory*. Furthermore, the familiar Hamiltonian form of quantum mechanics must be compared to the over-all *space-time view* used here. The first section is, therefore, devoted to a discussion of the relations of these viewpoints.

1. *Comparison with the Hamiltonian method*

Electrodynamics can be looked upon in two equivalent and complementary ways. One is as *the description of the behavior of a field (Maxwell's equations)*. The other is as *a description of a direct interaction at a distance (albeit delayed in time) between charges* (the solutions of Lienard and Wiechert). From the latter point of view *light is considered as an interaction of the charges in the source with those in the absorber*. This is an impractical point of view because many kinds of sources produce the same kind of effects. *The field point of view separates these aspects into two simpler problems, production of light, and absorption of light*. On the other hand, *the field point of view is less practical when dealing with close collisions of particles (or their action on themselves)*. For here the source and absorber are not readily distinguishable, there is an intimate exchange of quanta. The fields are so closely determined by the motions of the particles that it is just as well not to separate the question into two problems but to *consider the process as a direct interaction*. Roughly, *the field point of view is most practical for problems involving real quanta*, while *the interaction view is best for the discussion of the virtual quanta involved. We shall emphasize the interaction viewpoint in this paper*, first because it is less familiar and therefore requires more discussion, and second because the important aspect in the problems with which we shall deal is the effect of *virtual quanta*.

The Hamiltonian method is not well adapted to represent the direct action at a distance between charges because that action is delayed. The Hamiltonian method represents the future as developing out of the present. If the values of a complete set of quantities are known now, their values can be computed at the next instant in time. *If particles interact through a delayed interaction*, however, one cannot predict the future by simply knowing the present motion of the particles. One would also have to know what the motions of the particles were in the past in view of the interaction this may have on the future motions. This is done in the Hamiltonian electrodynamics, of course, by requiring that one specify besides the present motion of the particles, the values of a host of new variables (the coordinates of the field oscillators) to keep track of that aspect of the past motions of the particles which determines their future behavior. *The use of the Hamiltonian forces one to choose the field viewpoint rather than the interaction viewpoint.*

In many problems, for example, the close collisions of particles, we are not interested in the precise temporal sequence of events. It is not of interest to be able to say how the situation would look at each instant of time during a collision and how it progresses from instant to instant. Such ideas are only useful for events taking a long time and for which we can readily obtain information during the intervening period. *For collisions it is much easier to treat the process as a whole*[6].

> [6] This is the viewpoint of the theory of the S matrix of Heisenberg.
>
> [Heisenberg, W. (July, 1943). Die beobachtbaren Größen in der Theorie der Elementarteilchen. (The "observable quantities" in the theory of elementary particles.) *Zeitschrift für Physik* (in German), 120, 7–10, 513–538: S-matrix theory as a principle of particle interactions.
>
> *S-matrix theory* was proposed as a principle of particle interactions by Heisenberg in 1943, following Wheeler's 1937 introduction of the S-matrix; Wheeler, J. A. (December, 1937). On the Mathematical Description of Light Nuclei by the Method of Resonating Group Structure. *Phys. Rev.*, 52 (11): 1107–22; https://doi.org/10.1103/PhysRev.52.1107.
>
> *S-matrix theory* was a proposal for replacing local quantum *field theory* as the basic principle of elementary particle physics. It avoided the notion of space and time by replacing it with abstract mathematical properties of the S-matrix. In S-matrix theory, the S-matrix relates the infinite past to the infinite future in one step, without being decomposable into intermediate steps corresponding to time-slices. This path was mostly abandoned because the resulting equations, devoid of any space-time interpretation, were very difficult to understand and solve.]

The *Moller interaction matrix* for the collision of two electrons is not essentially more complicated than the *non-relativistic Rutherford formula*, yet the mathematical machinery used to obtain the former from quantum electrodynamics is vastly more complicated than Schrodinger's equation with the e^2/r_{12} interaction needed to obtain the latter. The difference is only that in the latter the action is instantaneous so that the Hamiltonian method requires no extra variables, while in the former *relativistic* case it is delayed and the Hamiltonian method is very cumbersome.

We shall be discussing the solutions of equations rather than the time differential equations from which they come. We shall discover that *the solutions, because of the over-all space-time view that they permit, are as easy to understand when interactions are delayed as when they are instantaneous*. As a further point, *relativistic invariance will be self-evident*. The Hamiltonian form of the equations develops the future from the instantaneous present. But for different observers in relative motion the instantaneous present is different, and corresponds to a different 3-dimensional cut of *space-time*. Thus, the temporal analyses of different observers is different and their Hamiltonian equations are developing the process in different ways. These differences are irrelevant, however, for the solution is the same in any *space-time* frame. *By forsaking the Hamiltonian method, the wedding of relativity and quantum mechanics can be accomplished most naturally*.

We illustrate these points in the next section by studying the solution of Schrodinger's equation for *non-relativistic* particles interacting by an instantaneous Coulomb potential (Eq. 2). When the solution is modified to include the effects of delay in the interaction and the *relativistic* properties of the electrons, we obtain an expression of the laws of *quantum electrodynamics* (Eq. 4).

2. *The interaction between charges*

We study by the same methods as in I, *the interaction of two particles* using the same notation as I. We start by considering the *non-relativistic* case described by the Schrodinger equation (I, Eq. 1).

$$[i\, \partial\psi/\partial t = \mathbf{H}\psi, \qquad (1)]$$

The *wave function* at a given time is a function $\psi(x_a, x_b, t)$ of the coordinates x_a and x_b of each particle. Thus call $K(x_a, x_b, t; x_a', x_b', t')$ the *amplitude* that particle *a* at x_a' at time t' will get to x_a at t while particle *b* at x_b' at t' gets to x_b at t. If the particles are free and do not interact this is

$$K(x_a, x_b, t; x_a', x_b', t') = K_{0a}(x_a, t; x_a', t')\, K_{0b}(x_b, t; x_b', t')$$

where K_{0a} is the K_0 function for particle a considered as free. In this case we can obviously define a quantity like K, but for which the time t need not be the same for particles *a* and *b* (likewise for t'); e.g.,

$$K_0(3, 4; 1, 2) = K_{0a}(3, 1)\, K_{0b}(4, 2) \qquad (1)$$

can be thought of as the *amplitude* that particle *a* goes from x_1 at t_1 to x_3 at t_3 and that particle *b* goes from x_2 at t_2 to x_4 at t_4.

When the particles do interact, one can only define the quantity K(3, 4; 1, 2) precisely if the interaction vanishes between t_1 and t_2 and also between t_3 and t_4. *In a real physical system such is not the case*. There is such an enormous advantage, however, to the concept that we shall continue to use it, imagining that we can neglect the effect of interactions between t_1 and t_2 and between t_3 and t_4. For practical problems this means choosing such long time-intervals $t_3 - t_1$ and $t_4 - t_2$ that the extra interactions near the end points have small relative effects. As an example, in a scattering problem it may well be that the particles are so well separated initially and finally that the interaction at these times is negligible. Again, *energy* values can be defined by the average rate of change of *phase* over such long time intervals that errors initially and finally can be neglected. Inasmuch as any physical problem can be defined in terms of scattering processes, we do not lose much in a general theoretical sense by this approximation. *If it is not made it is not easy to study interacting particles relativistically*, for there is nothing significant in choosing $t_1 = t_3$ if $x_1 \neq x_3$, *as absolute simultaneity of events at a distance cannot be defined invariantly*. It is essentially to avoid this approximation that the complicated structure of the older quantum electrodynamics has been built up. *We wish to describe electrodynamics as a delayed interaction between particles.* If we can make the approximation of assuming a meaning to K(3, 4; 1, 2) the results of this *interaction* can be expressed very simply.

To see how this may be done, imagine first that the *interaction* is simply that given by a Coulomb *potential* e^2/r where r is the distance between the particles. If this be turned on only for a very short time Δt_0 at time t_0, the first order correction to K(3, 4; 1, 2) can be worked out exactly as was Eq. (9) of I

$$[K^{(1)}(2, 1) = -i \int K_0(2, 3)U(3)K_0(3, 1)\, d\tau_3, \qquad (9)]$$

by an obvious generalization to two particles:

$$K^{(1)}(3, 4; 2, 1) = -ie^2 \iint K_{0a}(3, 5)K_{0b}(4, 6)r_{56}^{-1} \times K_{0a}(5, 1)K_{0b}(6, 2)\, d^3x_5 d^3x_6 \Delta t_0,$$

where $t_5 = t_6 = t_0$. If now the *potential* were on at all times (so that strictly K is not defined unless $t_4 = t_3$ and $t_1 = t_2$), the first-order effect is obtained by integrating on t_0, which we can write as an integral over both t_5 and t_6 if we include a delta-function $\delta(t_5 - t_6)$ to insure contribution only when $t_5 = t_6$. Hence, the first-order effect of interaction is (calling to $t_5 - t_6 = t_{56}$):

$$K^{(1)}(3, 4; 2, 1) = -ie^2 \iint K_{0a}(3, 5)K_{0b}(4, 6)r_{56}^{-1} \times \delta(t_{56})K_{0a}(5, 1)K_{0b}(6, 2)\, d\tau_5 d\tau_6, \qquad (2)$$

where $d\tau = d^3x dt$.

We know, however, in classical electrodynamics, that the Coulomb potential does not act instantaneously, but is delayed by a time r_{56}, taking the speed of light as unity. This suggests simply replacing $r_{56}^{-1}\,\delta(t_{56})$ in (2) by something like $r_{56}^{-1}\,\delta(t_{56}-r_{56})$ to represent the delay in the effect of *b* on *a*.

This turns out to be not quite right[7], for when this interaction is represented by photons they must be of only *positive energy*, while the Fourier transform of $\delta(t_{56}-r_{56})$ contains frequencies of both signs.

[7] It, and a like term for the effect of *a* on *b*, leads to a theory which, in the classical limit, exhibits interaction through half-advanced and half-retarded *potentials*. Classically, this is equivalent to purely retarded effects within a closed box from which no light escapes (e.g., see A, or Wheeler, J. A. & Feynman R. P. (April, 1945). Interaction with the Absorber as the Mechanism of Radiation. *Rev. Mod. Phys*., 17, 157-81[; the motive of the analysis was to clear the present quantum theory of interacting particles of those of its difficulties which have a purely classical origin, method was to define as closely as one can within the bounds of classical theory the proper use of the *field* concept in the description of nature, this paper represents the third part of the survey, an analysis of the *mechanism of radiation* believed

to complete the last tie between *action at a distance* and *field theory*, only section now finished, difficulties to obtain a satisfactory account of the *field* generated by an accelerated charge at a remote point and to understand the source of the force experienced by the charge itself as a result of its motion, takes up suggestion by Tetrode (1922) that *the act of radiation should have some connection with the presence of an absorber*, develops this idea into the thesis that the force of radiative reaction arises from the action on the source owing to the half-advanced fields of the particles of the absorber, absorber response as the mechanism of radiative reaction, expresses in terms of *action at a distance*, *Wheeler–Feynman absorber theory, assumes solutions of the electromagnetic field equations* must be invariant under time-reversal transformation as are the field equations themselves, considers an accelerated charge located in the absorbing system as the source of radiation, *elementary particles not self-interacting*, a complete correspondence is established between *action at a distance* and the usual formulation of *field theory* in the case of a completely absorbing system]).
Analogous theorems exist in quantum mechanics but it would lead us too far astray to discuss them now.

It should instead be replaced by $\delta_+(t_{56}-r_{56})$ where

$$\delta_+(x) = \int_0^\infty e^{-i\omega x}\, d\omega/\pi = \lim_{\varepsilon\to 0} (\pi i)^{-1}/(x - i\varepsilon) = \delta(x) + (\pi i x)^{-1}. \qquad (3)$$

This is to be averaged with $r_{56}^{-1}\, \delta_+(-t_{56}-r_{56})$ which arises when $t_5<t_6$ and corresponds to *a* emitting the quantum which *b* receives. Since

$$(2r)^{-1}\{\delta_+(t - r) + \delta_+(- t - r)\} = \delta_+(t^2 - r^2),$$

this means $r_{56}^{-1}\, \delta(t_{56})$ is replaced by $\delta_+(s_{56}^2)$ where $s_{56}^2 = t_{56}^2 - r_{56}^2$ is the square of the *relativistically* invariant interval between points 5 and 6. Since in classical electrodynamics there is also an interaction through the *vector potential*, the complete interaction (see A, Eq. (1)) should be $(1 - (\mathbf{v}_5.\mathbf{v}_6)\, \delta_+(s_{56}^2)$, or in the *relativistic* case,

$$(1 - \boldsymbol{\alpha}_a.\boldsymbol{\alpha}_b)\, \delta_+(s_{56}^2) = \beta_a\beta_b e\gamma_{a\mu}\gamma_{b\mu}\, \delta_+(s_{56}^2)$$

Hence, we have *for electrons obeying the Dirac equation,*

$$K^{(1)}(3, 4; 2, 1) = - ie^2 \iint K_{+a}(3, 5)K_{+b}(4, 6)\gamma_{a\mu}\gamma_{b\mu} \\ \times\, \delta_+(t_{56}^2)K_{+a}(5, 1)K_{+b}(6, 2)\, d\tau_5 d\tau_6, \qquad (4)$$

where $\gamma_{\alpha\mu}$ and $\gamma_{\beta\mu}$ are the *Dirac matrices* applying to the spinor corresponding to particles *a* and *b*, respectively (the factor $\beta_a\beta_b$ being absorbed in the definition, I Eq. (17), of K_+)

$$[K_+(2, 1) = \textstyle\sum_{\text{POS En}} \phi_n(2)\, \bar\phi_n(1) \times \exp\{-iE_n(t_2 - t_1)\} \quad \text{for } t_2>t_1$$
$$= - \textstyle\sum_{\text{NEG En}} \phi_n(2)\, \bar\phi_n(1) \times \exp\{-iE_n(t_2 - t_1)\} \quad \text{for } t_2<t_1. \quad (17)]$$

This is our fundamental equation for electrodynamics. It describes the effect of exchange of one quantum (therefore first order in e^2) between two electrons. It will serve as a prototype enabling us to write down the corresponding quantities involving the exchange of two or more quanta between two electrons or the *interaction* of an electron with itself. *It is a consequence of conventional electrodynamics. Relativistic* invariance is clear. Since one sums over μ it contains the effects of both longitudinal and transverse waves in a *relativistically* symmetrical way.

We shall now interpret Eq. (4)

$$[K^{(1)}(3, 4; 2, 1) = -ie^2 \iint K_{+a}(3, 5)K_{+b}(4, 6)\gamma_{a\mu}\gamma_{b\mu}$$
$$\times\ \delta_+(t_{56}^2)K_{+a}(5, 1)K_{+b}(6, 2)\, d\tau_5 d\tau_6, \qquad (4)]$$

in a manner which will permit us to write down the higher order terms. It can be understood (see Fig. 1 below) as saying that the *amplitude* for "*a*" to go from 1 to 3 and "*b*" to go from 2 to 4 is altered to first order *because they can exchange a quantum*. Thus, "*a*" can go to 5 (*amplitude* $K_+(5, 1)$) emit a quantum (longitudinal, transverse, or scalar $\gamma_{\alpha\mu}$) and then proceed to 3 ($K_+(3, 5)$). Meantime "*b*" goes to 6 ($K_+(6, 2)$), absorbs the quantum ($\gamma_{\beta\mu}$) and proceeds to 4 ($K_+(4, 6)$). The quantum meanwhile proceeds from 5 to 6, which it does with *amplitude* $\delta_+(s_{56}^2)$. We must sum over all the possible quantum polarizations μ and positions and times of *emission* 5, and of *absorption* 6. Actually, if $t_5 > t_6$ it would be better to say that "*a*" absorbs and "*b*" emits but no attention need be paid to these matters, as all such alternatives are automatically contained in (4).

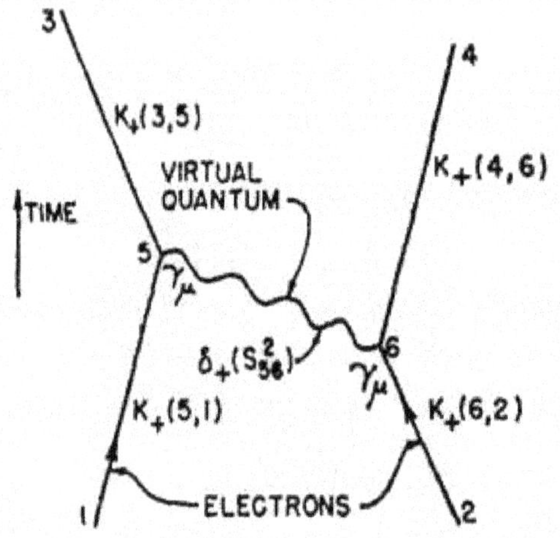

FIG. 1. The fundamental interaction Eq. (4). Exchange of one quantum between two electrons.

Fig. 1. The fundamental *interaction* Eq. (4). Exchange of one quantum between two electrons. [This is the first published "*Feynman diagram*".]

[This and Fig. 8 in the Appendix were reproduced in the *American Scientist* by permission of the American Physical Society. (Kaiser, D. (March-April, 2005). Physics and Feynman's Diagrams. *American Scientist*, 93, 156-65; http://www.americanscientist.org/IssueTOC/issue/701.) Fig. 1 is reproduced with the following text on page 158: "Figure 3. Electron-electron scattering is described by one of the earliest published Feynman diagrams (featured in "Sightings," September–October 2003). One electron (solid line at bottom right) shoots out a force-carrying particle—a virtual photon (wavy line)—which then smacks into the second electron (solid line at bottom left). The first electron recoils backward, while the second electron gets pushed off its original course. The diagram thus sketches a quantum-mechanical view of how particles with the same charge repel each other. As suggested by the term "Space-Time Approach" in the title of the article that accompanied this diagram, Feynman originally drew diagrams in which the dimensions were space and time; here the horizontal axis represents space. Today most physicists draw Feynman diagrams in a more stylized way, highlighting the topology of propagation lines and vertices."

This article also includes a photograph of Feynman, Wheeler, Schwinger, and other physicists gathered in June 1947 at Shelter Island, New York, several months before Feynman introduced his diagrams at the Pocono Manor Inn in March 1948.]

The correct terms of higher order in e^2 or involving larger numbers of electrons (interacting with themselves or in pairs) can be written down by the same kind of reasoning. They will be illustrated by examples as we proceed. *In a succeeding paper they will all be deduced from conventional quantum electrodynamics.*

Calculation, from (4)

$$[K^{(1)}(3, 4; 2, 1) = -ie^2 \iint K_{+a}(3, 5)K_{+b}(4, 6)\gamma_{a\mu}\gamma_{b\mu} \times \delta_+(t_{56}^2)K_{+a}(5, 1)K_{+b}(6, 2)\, d\tau_5 d\tau_6, \qquad (4)]$$

of the transition element between positive energy free electron *states* gives the Moller scattering of two electrons, when account is taken of the Pauli principle.

The *exclusion principle* for interacting charges is handled in exactly the same way as for non-interacting charges (I). For example, for two charges it requires only that one calculate K(3, 4; 1, 2) – K(4, 3; 1, 2) to get the net *amplitude* for arrival of charges at 3 and 4. It is disregarded in intermediate *states*. The interference effects for scattering of electrons by positrons discussed by Bhabha will be seen to result directly in this formulation. The formulas are interpreted to apply to positrons in the manner discussed in I.

As our primary concern will be for processes in which the quanta are virtual [*quanta* that have a known source and are eventually absorbed], we shall not include here the detailed analysis of processes involving real quanta in initial or final *state*, and shall content ourselves by only stating the rules applying to them[8].

[8] Although in the expressions stemming from (4) the quanta are *virtual*; this is not actually a theoretical limitation. One way to deduce the correct rules for real quanta from (4) is to note that in a closed system all quanta can be considered as *virtual* (i.e. they have a known source and are eventually absorbed) so that in such a system the present description is complete and equivalent to the conventional one. In particular, the relation of the Einstein A and B coefficients can be deduced. *A more practical direct deduction of the expressions for real quanta will be given in the subsequent paper.*
It might be noted that (4) can be rewritten as describing the action on *a*,

$$K^{(1)}(3,1) = i \int K_+(3,5) \times (A(5)K_+(5, 1)\, d\tau_5$$

of the potential

$$A_\mu(5) = e^2 \int K_+(4, 6)\, \delta_+(s_{56}^2)\, \gamma_\mu \times K_+(6, 2) d\tau_6$$

arising from Maxwell's equations $-\square^2 A_\mu = 4\pi j_\mu$ from a "current"

$j_\mu(6) = e^2K_+(4, 6)\gamma_\mu K_+(6, 2)$ produced by particle *b* in going from 2 to 4. This is virtue of the fact that δ_+ satisfies

$$-\square_2^2\,\delta_+(s_{21}^2) = 4\pi\delta(2, 1). \qquad (5)$$

The result of the analysis is, as expected, that they can be included by the same line of reasoning as is used in discussing the virtual processes, provided the quantities are normalized in the usual manner to represent single quanta. For example, the *amplitude* that an electron in going from 1 to 2 absorbs a quantum whose *vector potential*, suitably normalized, is $c_\mu \exp(-ik.x) = C_\mu(x)$ is just the expression (I, Eq. (13))

$$[K_+^{(1)}(2, 1) = -i\int K_+(2, 3)\mathbf{A}(3)K_+(3, 1)\, d\tau_3, \qquad (13)]$$

for scattering in a potential with **A** (3) replaced by **C** (3). Each quantum interacts only once (either in *emission* or in *absorption*), terms like (I, Eq. (14))

$$[K_+^{(2)}(2, 1) = -i\iint K_+(2, 4)\mathbf{A}(4) \times K_+(4, 3)\mathbf{A}(3)K_+(3, 1)\, d\tau_4 d\tau_3 \quad (14)]$$

occur only when there is more than one quantum involved. The Bose statistics of the quanta can, in all cases, be disregarded in intermediate *states*. The only effect of the statistics is to change the weight of initial or final *states*. If there are among quanta, in the initial *state*, some n which are identical then the weight of the *state* is (1/n!) of what it would be if these quanta were considered as different (similarly for the final *state*).

3. *The self-energy problem*

Having a term representing the mutual *interaction* of a pair of *charges*, *we must include similar terms to represent the interaction of a charge with itself.* For under some circumstances what appears to be two distinct electrons may, according to I, be viewed also as a single electron (namely in case one electron was created in a pair with a positron destined to annihilate the other electron). *Thus, to the interaction between such electrons must correspond the possibility of the action of an electron on itself*[9].

[9] These considerations make it appear unlikely that the contention of J. A. Wheeler and R. P. Feynman [Wheeler, J. A. & Feynman, R. P. (1945). Interaction with the Absorber as the Mechanism of Radiation. *Loc. cit.*], *that electrons do not act on themselves*, will be a successful concept in *quantum electrodynamics*.

[He does not mention the second part of this paper (Wheeler, J. A. & Feynman, R. P. (July, 1949). Classical Electrodynamics in Terms of Direct Interparticle Action. *Rev. Mod. Phys*., 21, 3, 425-33; https://doi.org/10.1103/RevModPhys. 21.425), which was published three months before this paper was published, and in which they argued "*Of quantum theories of fields and their possibilities we hardly know enough to demand on quantum grounds that such a direct self-interaction should exist*". They also conclude "that the theory of direct interparticle action, and the equivalent adjunct field theory, provide a physically reasonable and experimentally satisfactory account of the classical mechanical behavior of a system of point charges in electromagnetic interaction with one another, free of the ambiguities associated with the idea of a particle acting upon itself". It appears that Wheeler and Feynman may have had different views on this matter.]

This interaction is the heart of the self-energy problem.

Consider to first order in e^2 the action of an electron on itself in an otherwise force free region. The amplitude K(2, 1) for a single particle to get from 1 to 2 differs from $K_+(2, 1)$ to first order in e^2 by a term

$$K^{(1)}(2,1) = ie^2 \int\int K_+(2,4)\gamma_\mu K_+(4, 3)\gamma_\mu \text{ x } K_+(3, 1)\, d\tau_3 d\tau_4\, \delta_+(s_{43}{}^2). \quad (6)$$

It arises because the electron instead of going from 1 directly to 2, may go (Fig. 2) first to 3, ($K_+(3, 1)$), emit a quantum (γ_μ), proceed to 4, ($K_+(4, 3)$), absorb it (γ_μ), and finally arrive at 2 ($K_+(2, 4)$). The quantum must go from 3 to 4 ($\delta_+(s_{43}{}^2)$).

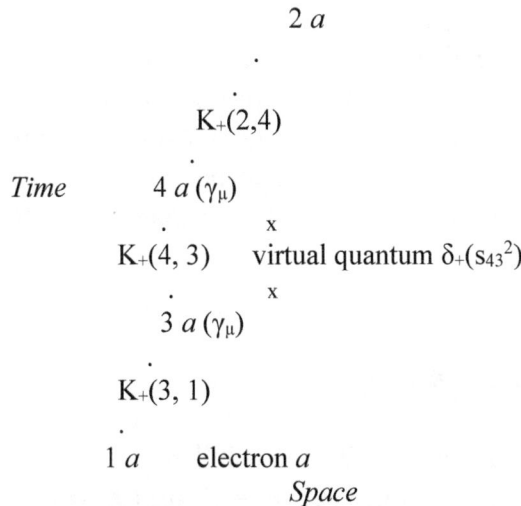

2 *a*

$K_+(2,4)$

Time 4 *a* (γ_μ)

x

$K_+(4, 3)$ virtual quantum $\delta_+(s_{43}{}^2)$

x

3 *a* (γ_μ)

$K_+(3, 1)$

1 *a* electron *a*

Space

Fig. 2. Interaction of an electron with itself, Eq. (6).

This is related to the self-energy of a free electron in the following manner. Suppose initially, time t_1, we have an electron in *state* f(1) which we imagine to be a *positive energy* solution of *Dirac's equation* for a free particle. After a long time $t_2 - t_1$ the perturbation will alter the *wave function*, which can then be looked upon as a superposition of free particle solutions (actually it only contains f). The *amplitude* that g(2) is contained is calculated as in (I, Eq. (21)). The diagonal element (g = f) is therefore

$$\iint f^*(1)\beta K^{(1)}(2, 1)\beta f(1) d^3x_1 d^3x_2. \quad (7)$$

The time interval $T = t_2 - t_1$ (and the spatial volume V over which one integrates) must be taken very large, for *the expressions are only approximate* (analogous to the situation for two interacting charges)[10].

[10] This is discussed in reference 5 [Feynman, R. P. (April, 1948). Space-Time Approach to Non-Relativistic Quantum Mechanics] in which it is pointed out that *the concept of a wave function loses accuracy if there are delayed self-actions.*

This is because, for example, *we are dealing incorrectly with quanta emitted just before t_2* which would normally be reabsorbed at times after t_2.

If $K^{(1)}(2,1)$ from (6)

$[K^{(1)}(2,1) = ie^2 \int\int K_+(2,4)\gamma_\mu K_+(4, 3)\gamma_\mu \text{ x } K_+(3, 1)\, d\tau_3 d\tau_4\, \delta_+(s_{43}{}^2).$ (6)]
is actually substituted into (7) the surface integrals can be performed as was done in obtaining I, Eq. (22) resulting in

$$- ie^2 \int\int f^-(4)\gamma_\mu K_+(4, 3)\gamma_\mu f(3)\delta_+(s_{43}{}^2)\, d\tau_3 d\tau_4. \qquad (8)$$

Putting for f(1) the plane wave u exp(– ip.x_1) where p_μ is the *energy* (p_4) and *momentum* of the electron ($\mathbf{p}^2 = m^2$), and u is a constant 4-index symbol, (8) becomes

$$- ie^2 \int\int\{u^-\gamma_\mu K_+(4, 3)\gamma_\mu, u\} \text{ x } \exp\{ip.(x_4 - x_3)\delta_+(s_{43}{}^2)\, d\tau_3 d\tau_4,$$

the integrals extending over the volume V and time interval T. Since $K_+(4, 3)$ depends only on the difference of the coordinates of 4 and 3, $x_{43\mu}$, the integral on 4 gives a result (except near the surfaces of the region) independent of 3. When integrated on 3, therefore, the result is of order VT. The effect is proportional to V, for the *wave functions* have been normalized to unit volume. If normalized to volume V, the result would simply be proportional to T. This is expected, for if the effect were equivalent to a change in *energy* ΔE, the *amplitude* for arrival in f at t_2 is altered by a factor exp(– iΔE($t_2 - t_1$)), or to first order by the difference – i(ΔE)T. Hence, we have

$$\Delta E = e^2 \iint\{u^-\gamma_\mu K_+(4, 3)\gamma_\mu, u\} \text{ x } \exp(ip.\, x_{43})\delta_+(s_{43}{}^2)\, d\tau_4, \qquad (9)$$

integrated over all space-time $d\tau_4$. *This expression will be simplified presently.* In interpreting (9) we have tacitly assumed that the *wave functions* are normalized so that (u^*u) = ($u^-\gamma_4 u$) = 1. The equation may therefore be made independent of the normalization by writing the left side as (ΔE) ($u^-\gamma_4 u$), or since ($u^-\gamma_4 u$) = (E/m) (u^-u) and mΔm= EΔE, as Δm(u^-u) where Δm is an equivalent change in mass of the electron. In this form invariance is obvious.

One can likewise obtain an expression for the energy shift for an electron in a hydrogen atom. Simply replace K_+ in (8)
$[- ie^2 \int\int f^-(4)\gamma_\mu K_+(4, 3)\gamma_\mu f(3)\delta_+(s_{43}{}^2)\, d\tau_3 d\tau_4,$ (8)]
by $K_+{}^{(V)}$, the exact kernel for an electron in the *potential*, $V= \beta e^2/r$, of the atom, and f by a *wave function* (of space and time) for an atomic state. In general, the ΔE which results is not real. The imaginary part is negative and in exp(–i(ΔE)T produces an exponentially decreasing *amplitude* with time. This is because we are asking for the *amplitude* that an atom initially with no photon in the field, will still appear after time T with no photon. If the atom is in a *state* which can radiate, this *amplitude* must decay with time. The imaginary part of ΔE when calculated does indeed give the correct rate of radiation from atomic *states*. It is zero for the *ground state a*nd for a free electron.

In the *non-relativistic region*, the expression for ΔE can be worked out as has been done by Bethe[11].

[11] Bethe, H. A. (August, 1947). The Electromagnetic Shift of Energy Levels. *Phys. Rev*., 72, 339-41[; Lamb and Retherford results show that fine structure of second quantum state of hydrogen does not agree with *Dirac wave equation*, Schwinger, Weisskopf, and

Oppenheimer suggest might be due to *shift of energy levels by interaction of electron with the radiation field*, this shift comes out infinite in all existing theories and has therefore always been ignored, possible to identify the most strongly (linearly) divergent term in the level shift with an *electromagnetic mass effect* which must exist for a bound as well as for a free electron, already included in the *observed mass* of the electron so should be subtracted, assumes *relativistic cut-off* in quantum energies (frequencies) of included atomic states, then calculation of Lamb shift for hydrogen atom using *non-relativistic* ordinary radiation theory gives shift of the levels due to *radiation interaction* in close agreement with observed value, removes discrepancy with Dirac theory, did not carried out *relativistic* calculations].

In the *relativistic region* (points 4 and 3 as close together as a Compton wave-length) the $K_+^{(V)}$ which should appear in (8) can be replaced to first order in V by K_+ plus $K_+^{(1)}(2, 1)$ given in I, Eq. (13)

$$[K_+^{(1)}(2, 1) = -\,i \int K_+(2, 3)\mathbf{A}(3)K_+(3, 1)\, d\tau_3. \qquad (13)]$$

The problem is then very similar to the radiationless scattering problem discussed below.

4. *Expression in momentum and energy space*

The evaluation of [the *self-energy*] (9)

$$[\Delta E = e^2 \int\int \{u^{-}\gamma_\mu K_+(4, 3)\gamma_\mu, u\} \text{ x } \exp(ip.\, x_{43})\delta_+(s_{43}{}^2)\, d\tau_4, \qquad (9)]$$

as well as all the other more complicated expressions arising in these problems, is very much simplified by working in the *momentum* and *energy* variables, rather than *space* and *time*. For this we shall need the Fourier Transform of $\delta_+(s_{21}{}^2)$ which is

$$\delta_+(s_{21}{}^2) = \pi^{-1} \int \exp(-ik.x_{21})\, \mathbf{k}^{-2}\, d^4k, \qquad (10)$$

which can be obtained from (3) and (5)

$$[\delta_+(x) = \int_0^\infty e^{-i\omega x}\, d\omega/\pi = \lim_{\varepsilon\to 0} (\pi i)^{-1}/(x - i\varepsilon) = \delta(x) + (\pi i x)^{-1}, \qquad (3)$$

$$-\,\Box_2{}^2\, \delta_+(s_{21}{}^2) = 4\pi\delta(2, 1) \qquad (5)]$$

or from I, Eq. (32) noting that $I_+(2, 1)$ for $m^2 = 0$ is $\delta_+(s_{21}{}^2)$ from I, Eq. (34). The $\mathbf{k}^{-2}$ means $(k.k)^{-1}$ or more precisely the limit as $\delta\to 0$ of $(k.k + i\delta)^{-1}$. Further d^4k means $(2\pi)^{-2}dk_1dk_2dk_3dk_4$. If we imagine that quanta are particles of zero *mass*, then we can make the general rule that all poles are to be resolved by considering the *masses* of the particles and quanta to have infinitesimal negative imaginary parts.

Using these results, we see that the *self-energy* (9)

$$[\Delta E = e^2 \int\int \{u^{-}\gamma_\mu K_+(4, 3)\gamma_\mu, u\} \text{ x } \exp(ip.\, x_{43})\delta_+(s_{43}{}^2)\, d\tau_4, \qquad (9)]$$

is the matrix element between u^- and u of the matrix

$$(e^2/\pi i) \int \gamma_\mu(\mathbf{p} - \mathbf{k} - m)^{-1}\gamma_\mu \mathbf{k}^{-2}\, d^4k, \qquad (11)$$

where we have used the expression (I, Eq. (31)) for the Fourier transform of K_+. This form for the *self-energy* is easier to work with than is (9).

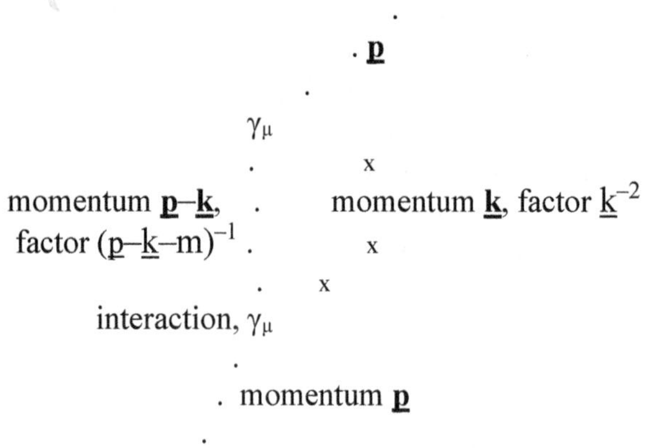

Momentum space

Fig. 3. Interaction of an electron with itself. Momentum space, Eq. (11).

The equation can be understood by imagining (Fig. 3) that the electron of *momentum* **p** emits (γ_μ) a quantum of *momentum* **k**, and makes its way now with *momentum* **p** – **k** to the next event (factor $(\mathbf{p}-\mathbf{k}-m)^{-1}$) which is to absorb the quantum (another γ_μ). The *amplitude* of propagation of quanta is $\mathbf{k}^{-2}$. (There is a factor $e^2/\pi i$ for each virtual quantum). One integrates over all quanta. The reason an electron of *momentum* **p** propagates as $1/(p-m)$ is that *this operator is the reciprocal of the Dirac equation operator, and we are simply solving this equation.* Likewise, *light* goes as $1/\mathbf{k}^2$, for this is the reciprocal D'Alembertian operator of the *wave equation of light*. The first γ_μ represents the *current* which generates the *vector potential*, while the second is the *velocity operator* by which this *potential* is multiplied in the *Dirac equation* when an external field acts on an electron.

Using the same line of reasoning, other problems may be set up directly in *momentum space*. For example, consider the *scattering in a potential* $\mathbf{A} = A_\mu\gamma_\mu$ varying in space and time as *a* exp(–iq.x). An electron initially in state of *momentum* $\mathbf{p}_1 = p_1\gamma_\mu$ will be deflected to *state* $\mathbf{p}_2$ where $\mathbf{p}_2 = \mathbf{p}_1 + \mathbf{q}$. The zero-order answer is simply the matrix element of *a* between *states* 1 and 2. We next ask for the first order (in e^2) radiative correction due to virtual radiation of one quantum. There are several ways this can happen.

See Feynman diagrams on page 775 of the article.

a. Eq. 12 b. Eq. 13. C. Eq. 14

Fig. 4. Radiative correction to scattering, momentum space

First for the case illustrated in Fig. 4(a), find the matrix:

$$(e^2/\pi i) \int \gamma_\mu(\mathbf{p}_2 - \mathbf{k} - m)^{-1}(\mathbf{p}_1 - \mathbf{k} - m)^{-1}\gamma_\mu\mathbf{k}^{-2}\, d^4k, \qquad (12)$$

For in this case, first[12]

[12] First, next, etc., here refer not to the order in true time but to the succession of events along the trajectory of the electron. That is, more precisely, to the order of appearance of the matrices in the expressions.

a quantum of *momentum* **k** is emitted (γ_μ), the electron then having *momentum* $\mathbf{p}_1 - \mathbf{k}$ and hence propagating with factor $(\mathbf{p}_1 - \mathbf{k} - \mathrm{m})^{-1}$. Next it is scattered by the *potential* (matrix ***a***) receiving additional *momentum* **q**, propagating on then (factor $((\mathbf{p}_2 - \mathbf{k} - \mathrm{m})^{-1}$ with the new *momentum* until the quantum is reabsorbed (γ_μ). The quantum propagates from *emission* to *absorption* (k^{-2}) and we integrate over all quanta (d^4k), and sum on *polarization* μ. When this is integrated on k_4, the result can be shown to be exactly equal to the expressions (16) and (17) given in B [Feynman, R. P. (1948). Relativistic Cut-Off for Quantum Electrodynamics. *Phys. Rev.* 74, 1430-8, (a relativistic cut-off of high frequency quanta is shown to produce a finite invariant *self-energy* for a free electron, but the problem of *polarization of the vacuum* is not solved) for the same process, the various terms coming from residues of the poles of the integrand (12)

$$[(e^2/\pi i) \int \gamma_\mu(\mathbf{p}_2 - \mathbf{k} - \mathrm{m})^{-1}(\mathbf{p}_1 - \mathbf{k} - \mathrm{m})^{-1}\gamma_\mu \mathbf{k}^{-2}\, d^4k. \qquad (12)]$$

Or again if the quantum is both *emitted* and *reabsorbed* before the scattering takes place one finds (Fig. 4(b))

$$(e^2/\pi i) \int \boldsymbol{a}(\mathbf{p}_1 - \mathrm{m})^{-1}\gamma_\mu(\mathbf{p}_1 - \mathbf{k} - \mathrm{m})^{-1}\gamma_\mu \mathbf{k}^{-2}\, d^4k, \qquad (13)$$

or if both *emission* and *absorption* occur after the scattering, (Fig. 4(c))

$$(e^2/\pi i) \int \gamma_\mu(\mathbf{p}_2 - \mathbf{k} - \mathrm{m})^{-1}(\mathbf{p}_2 - \mathrm{m})^{-1}\boldsymbol{a}\mathbf{k}^{-2}\, d^4k, \qquad (14)$$

These terms are discussed in detail below.

We have now achieved our simplification of the form of writing matrix elements arising from virtual processes. Processes in which a number of real quanta is given initially and finally offer no problem (assuming correct normalization).

See Feynman diagrams on page 775 of the article.

a. b.

Fig. 5. Compton scattering, Eq. (15).

For example, consider *the Compton effect* (Fig. 5(a)) in which an electron in *state* $\mathbf{p}_1$ absorbs a quantum of *momentum* $\mathbf{q}_1$, *polarization vector* $\mathbf{e}_{1\mu}$ so that its *interaction* is $e_{1\mu}\gamma_\mu = \mathbf{e}_1$, and emits a second quantum of *momentum* $-\mathbf{q}_2$, *polarization* $\mathbf{e}_2$ to arrive in final state of *momentum* $\mathbf{p}_2$. The matrix for this process is $\mathbf{e}_2(\mathbf{p}_1 + \mathbf{q}_1 - \mathrm{m})^{-1}\mathbf{e}_1$. The total matrix for the Compton effect is, then,

$$\mathbf{e}_2(\mathbf{p}_1 + \mathbf{q}_1 - \mathrm{m})^{-1}\mathbf{e}_1 + \mathbf{e}_1(\mathbf{p}_1 + \mathbf{q}_2 - \mathrm{m})^{-1}\mathbf{e}_2, \qquad (15)$$

the second. term arising because the *emission* of $\mathbf{e}_2$ may also precede the *absorption* of $\mathbf{e}_1$ (Fig. 5(b)). One takes matrix elements of this between initial and final electron *states* $(\mathbf{p}_1 + \mathbf{q}_1 = \mathbf{p}_2 - \mathbf{q}_2)$, to obtain the *Klein Nishina formula. Pair annihilation* with *emission* of two quanta, etc., are given by the same matrix, positron *states* being those with negative time component of **p**. Whether quanta are *absorbed* or *emitted* depends on whether the time component of **q** is positive or negative.

5. *The convergence of processes with virtual quanta*

These expressions are, as has been indicated, no more than a re-expression of conventional quantum electrodynamics. As a consequence, many of them are meaningless. For example, the self-energy expression (9) or (11) gives an infinite result when evaluated. The infinity arises, apparently, from the coincidence of the δ-function singularities in $K_+(4, 3)$ and $\delta_+(s_{43}^2)$. Only at this point is it necessary to make a real departure from conventional electrodynamics, a departure other than simply rewriting expressions in a simpler form.

We desire to make a modification of quantum electrodynamics analogous to the modification of classical electrodynamics described in a previous article, A.

> [Feynman (October, 1948). A Relativistic Cut-Off for Classical Electrodynamics, p. 776: "This requires that a particle be allowed to act on itself and the term a = b to be included in the *action* sum. *This leads immediately to an infinite self-force.* This difficulty can be eliminated if the $\delta(s_{ab}^2)$ is replaced, as Bopp[4] has suggested, by some other function $f(s_{ab}^2)$ of the invariant s_{ab}^2, which behaves like $\delta(s_{ab}^2)$ for large dimensions but differs for small].

There the $\delta(s_{12}^2)$ appearing in the action of interaction was replaced by $f(s_{12}^2)$ where f(x) is a function of small width and great height.

The obvious corresponding modification in the quantum theory is to replace the $\delta(s^2)$ appearing [in] the quantum mechanical *interaction* by a new function $f_+(s^2)$. *We can postulate that if the Fourier transform of the classical $f(s_{12}^2)$ is the integral over all k of $F(k^2)\exp(-ik.x)\,d^4k$, then the Fourier transform of $f_+(s^2)$ is the same integral taken over only positive frequencies k_4 for $t_2 > t_1$ and over only negative ones for $t_2 < t_1$ in analogy to the relation of $\delta_+(s^2)$ to $\delta(s^2)$.*

> [Page 770 (above): "*Unfortunately, the modification proposed is not completely satisfactory theoretically … a strict physical basis for the rules of convergence is not known.*"]

The function $f(s^2) = f(x.x)$ can be written* as

> * *This relation is given incorrectly in A* [Feynman, R. P. (October, 1948). A Relativistic Cut-Off for Classical Electrodynamics], equation just preceding 16.

$$f(x.x) = (2\pi)^{-2} \int_{k_4=0}^{\infty} \int \sin(k_4 \mid x_4 \mid) \times \cos(\mathbf{K.x})\, dk_4\, d^3\mathbf{K} g(k.k),$$

where g(k.k) is k_4^{-1} times the density of oscillators and may be expressed for positive k_4 as (A, Eq. (16))

$$g(\mathbf{k}^2) = \int_0^\infty (\delta(\mathbf{k}^2) - \delta(\mathbf{k}^2 - \lambda^2))\, G(\lambda)\, d\lambda,$$

where $\int_0^\infty G(\lambda)\, d\lambda = 1$ and G involves values of λ large compared to m. This simply means that the *amplitude* for propagation of quanta of *momentum* k is

$$-F_+(\mathbf{k}^2) = -\pi^{-1} \int_0^\infty (\mathbf{k}^{-2} - (\mathbf{k}^2 - \lambda^2)^{-1})\, G(\lambda)\, d\lambda,$$

rather than $\mathbf{k}^{-2}$. That is, writing $F_+(\mathbf{k}^2) = -\pi^{-1}\mathbf{k}^{-2}C(\mathbf{k}^2)$,

$$-f_+(s_{12}^2) = \pi^{-1} \int \exp(-ik.x12)\, \mathbf{k}^{-2}C(\mathbf{k}^2)\, d^4k. \qquad (16)$$

Every integral over an intermediate quantum which previously involved a factor $d^4k/\mathbf{k}^2$ *is now supplied with a convergence factor* $C(\mathbf{k}^2)$, where

$$C(\mathbf{k}^2) = \int_0^\infty -\lambda^2(\mathbf{k}^2 - \lambda^2)^{-1}G(\lambda)\, d\lambda. \qquad (17)$$

The poles are defined by replacing $\mathbf{k}^2$ by $\mathbf{k}^2 + i\delta$ in the limit $\delta \to 0$. That is λ^2 may be assumed to have an infinitesimal negative imaginary part.

The function $f_+(s_{12}^2)$ may still have a discontinuity in value on the light cone. This is of no influence for the Dirac electron. For a particle satisfying the Klein Gordon equation, however, the *interaction* involves gradients of the *potential* which reinstates the δ function if f has discontinuities. The condition that f is to have no discontinuity in value on the light cone implies $\mathbf{k}^{-2}C(\mathbf{k}^2)$ approaches zero as $\mathbf{k}^2$ approaches infinity. In terms of $G(\lambda)$ the condition is

$$\int_0^\infty \lambda^2 G(\lambda)\, d\lambda = 0. \qquad (18)$$

This condition will also be used in discussing the convergence of vacuum polarization integrals.

The expression for the *self-energy* matrix is now

$$(e^2/\pi i) \int \gamma_\mu (\mathbf{p} - \mathbf{k} - m)^{-1} \gamma_\mu\, \mathbf{k}^{-2}\, d^4k\, C(\mathbf{k}^2), \qquad (19)$$

which, since $C(\mathbf{k}^2)$ falls off at least as rapidly as $1/\mathbf{k}^2$, converges. For practical purposes we shall suppose hereafter that $C(\mathbf{k}^2)$ is simply $-\lambda^2/(\mathbf{k}^2 - \lambda^2)$ implying that some average (with weight $G(\lambda)d\lambda$ over values of λ may be taken afterwards. Since in all processes the quantum *momentum* will be contained in at least one extra factor of the form $(\mathbf{p} - \mathbf{k} - m)^{-1}$ *representing propagation of an electron while that quantum is in the field, we can expect all such integrals with their convergence factors to converge and that the result of all such processes will now be finite and definite (excepting the processes with closed loops,* discussed below, in which the diverging integrals are over the *momenta* of the electrons rather than the quanta).

The integral of (19) with $C(\mathbf{k}^2) = -\lambda^2(\mathbf{k}^2 - \lambda^2)^{-1}$ noting that $\mathbf{p}^2 = m^2$, $\lambda \gg m$ and dropping terms of order m/λ, is (see Appendix A)

$$(e^2/2\pi)[4m\{\ln(\lambda/m) + ½\} - \mathbf{p}\{\ln(\lambda/m) + 5/4\}]. \qquad (20)$$

When applied to a *state* of an electron of *momentum* **p** satisfying pu = mu, it gives for the *change in mass* (as in B [*Loc. cit.* Feynman (1948). Relativistic Cut-Off for Quantum Electrodynamics.], Eq. (9))

$$\Delta m = m(e^2/2\pi)\, \{3 \ln(\lambda/m) + ¾\}. \qquad (21)$$

6. *Radiative corrections to scattering*

We can now complete the discussion of the radiative corrections to scattering. In the integrals we include the *convergence factor* $C(\mathbf{k}^2)$, so that they converge for large $\mathbf{k}$. Integral (12)

$$[(e^2/\pi i) \int \gamma_\mu(\mathbf{p}_2 - \mathbf{k} - m)^{-1}(\mathbf{p}_1 - \mathbf{k} - m)^{-1}\gamma_\mu \mathbf{k}^{-2}\, d^4k, \qquad (12)]$$

is also not convergent because of the well-known *infra-red catastrophe*. For this reason, we calculate (as discussed in B) the value of the integral assuming the photons to have a small mass $\lambda_{min} \ll m \ll \lambda$. The integral (12) becomes

$$(e^2/\pi i) \int \gamma_\mu(\mathbf{p}_2 - \mathbf{k} - m)^{-1}(\mathbf{p}_1 - \mathbf{k} - m)^{-1} \times \gamma_\mu\, (\mathbf{k}^2 - \lambda_{min}{}^2)^{-1} d^4k(\mathbf{k}^2 - \lambda_{min}{}^2),$$

which when integrated (see Appendix B) gives $(e^2/2\pi)$ times

$$\ldots, \qquad (22)$$

where $(\mathbf{q}^2)^{1/2} = 2m \sin\theta$ and we have assumed the matrix to operate between states of *momentum* $\mathbf{p}_1$ and $\mathbf{p}_2 = \mathbf{p}_1 + \mathbf{q}$ and have neglected terms of order λ_{min}/m, m/λ, and q^2/λ^2. Here the only dependence on the convergence factor is in the term $r\boldsymbol{a}$, where

$$r = \ln(\lambda/m) + 9/4 - 2\ln(m/\lambda_{min}). \qquad (23)$$

As we shall see in a moment, the other terms (13), (14)

$$[(e^2/\pi i) \int \boldsymbol{a}(\mathbf{p}_1 - m)^{-1}\gamma_\mu(\mathbf{p}_1 - \mathbf{k} - m)^{-1}\gamma_\mu \mathbf{k}^{-2}\, d^4k, \qquad (13)$$

$$(e^2/\pi i) \int \gamma_\mu(\mathbf{p}_2 - \mathbf{k} - m)^{-1}(\mathbf{p}_2 - m)^{-1}\boldsymbol{a}\mathbf{k}^{-2}\, d^4k, \qquad (14)]$$

give contributions which just cancel the $r\boldsymbol{a}$ term. The remaining terms give for small $\mathbf{q}$,

$$(e^2/4\pi)\{1/2m\, (\mathbf{q}\boldsymbol{a} - \boldsymbol{a}\mathbf{q}) + 4\mathbf{q}^2/3m^2\, \boldsymbol{a}(\ln m/\lambda_{min} - 3/8)\}, \qquad (24)$$

which shows the change in *magnetic moment* and the *Lamb shift* as interpreted in more detail in B[13].

[13] That the result given in B [*Loc. cit.* Feynman (1948). Relativistic Cut-Off for Quantum Electrodynamics.] in Eq. (19) was in error was repeatedly pointed out to the author, in private communication, by V. F. Weisskopf and J. B. French, as their calculation, completed simultaneously with the author's early in 1948, gave a different result. French has finally shown that *although the expression for the radiation-less scattering B, Eq. (18) or (24) above is correct, it was incorrectly joined onto Bethe's non-relativistic result.* He shows that the relation $\ln 2k_{max} - 1 = \ln\lambda_{min}$ used by the author should have been $\ln 2k_{max} - 5/6 = \ln\lambda_{min}$. This results in adding a term – (1/6) to the logarithm in B, Eq. (19) so that the result now agrees with that of French, J. B. & Weisskopf, V. F. (April, 1949). The Electromagnetic Shift of Energy Levels. *Phys. Rev.*, 75, 8, 1240-8 and Kroll, N. M. & Lamb, Jr., W. E. (1949). On the self-energy of a bound electron. *Phys. Rev.*, 75, 388-98. The author feels unhappily responsible for the very considerable delay in the publication of French's result occasioned by this error. This footnote is appropriately numbered.

We must now study the remaining terms (13) and (14)

$$[(e^2/\pi i) \int \boldsymbol{a}(\mathbf{p}_1 - m)^{-1}\gamma_\mu(\mathbf{p}_1 - \mathbf{k} - m)^{-1}\gamma_\mu \mathbf{k}^{-2}\, d^4k, \qquad (13)$$

$$(e^2/\pi i) \int \gamma_\mu(\mathbf{p}_2 - \mathbf{k} - m)^{-1}(\mathbf{p}_2 - m)^{-1}\boldsymbol{a}\mathbf{k}^{-2}\, d^4k. \qquad (14)]$$

The integral on $\mathbf{k}$ in (13) can be performed (after multiplication by $C(\mathbf{k}^2)$) since it involves nothing but the integral (19)

$$[(e^2/\pi i) \int \gamma_\mu (\mathbf{p} - \mathbf{k} - m)^{-1} \gamma_\mu \mathbf{k}^{-2} d^4k\, C(\mathbf{k}^2), \qquad (19)]$$

for the *self-energy* and the result is allowed to operate on the *initial state* u_1, (so that $p_1u_1 = mu_1$). Hence the factor following $\boldsymbol{a}(\mathbf{p}_1 - m)^{-1}$ will be just Δm. But, if one now tries to expand $1/(\mathbf{p}_1 - m) = (\mathbf{p}_1 + m)/(\mathbf{p}_1{}^2 - m^2)$ one obtains an infinite result, since $\mathbf{p}_1{}^2 = m^2$. This is, however, just what is expected physically. For the quantum can be *emitted* and *absorbed* at any time previous to the scattering. Such a process has the effect of a change in *mass* of the electron in the *state* 1. It therefore changes the energy by ΔE and the *amplitude* to first order in ΔE by $- i\Delta E.t$ where t is the time it is acting, which is infinite. That is, the major effect of this term would be canceled by the effect of change of mass Δm.

The situation can be analyzed in the following manner. We suppose that the electron approaching the scattering *potential a* has not been free for an infinite time, but at some time far past suffered a scattering by a *potential b*. If we limit our discussion to the effects of Δm and of the virtual radiation of one quantum between two such scatterings each of the effects will be finite, though large, and their difference is determinate. The propagation from $\boldsymbol{b}$ to $\boldsymbol{a}$ is represented by a matrix

$$\boldsymbol{a}(\mathbf{p}' - m)^{-1}\boldsymbol{b}, \qquad (25)$$

in which one is to integrate possibly over $\mathbf{p}'$ (depending on details of the situation). (If the time is long between $\boldsymbol{b}$ and $\boldsymbol{a}$, the energy is very nearly determined so that $\mathbf{p}'^2$ is very nearly m^2.)

We shall compare the effect on the matrix (25) of the virtual quanta and of the change of mass Δm. The effect of a virtual quantum is

$$(e^2/\pi i) \int \boldsymbol{a}(\mathbf{p}' - m)^{-1}\gamma_\mu(\mathbf{p}' - \mathbf{k} - m)^{-1} \times \gamma_\mu(\mathbf{p}' - m)^{-1}\boldsymbol{b}\mathbf{k}^{-2}\, d^4k\, C(\mathbf{k}^2), \qquad (26)$$

while that of a *change of mass* can be written

$$\boldsymbol{a}(\mathbf{p}' - m)^{-1}\Delta m(\mathbf{p}' - m)^{-1}\boldsymbol{b}, \qquad (27)$$

and we are interested in the difference (26) – (27). A simple and direct method of making this comparison is just to evaluate the integral on $\mathbf{k}$ in (26) and subtract from the result the expression (27) where Δm is given in (21)

$$[\Delta m = m(e^2/2\pi)\, \{3 \ln(\lambda/m) + \tfrac{3}{4}\}. \qquad (21)]$$

The remainder can be expressed as a multiple $- r(\mathbf{p}'^2)$ of the unperturbed amplitude (25)

$$[\boldsymbol{a}(\mathbf{p}' - m)^{-1}\boldsymbol{b}; \qquad (25)]$$

$$- r(\mathbf{p}'^2)\boldsymbol{a}(\mathbf{p}' - m)^{-1}\boldsymbol{b}. \qquad (28)$$

This has the same result (to this order) as replacing the *potentials* $\boldsymbol{a}$ and $\boldsymbol{b}$ in (25) by

$(1 - \frac{1}{2} r(\mathbf{p}'^2)\boldsymbol{a}$ and $(1 - \frac{1}{2} r(\mathbf{p}'^2)\boldsymbol{b}$. In the limit, then, as $\mathbf{p}'^2 \to m^2$ the net effect on the scattering is $-\frac{1}{2}\, r\boldsymbol{a}$ where r, the limit of $r(\mathbf{p}'^2)$ as $\mathbf{p}'^2 \to m^2$ (assuming the integrals have an infrared cut-off), turns out to be just equal to that given in (23)

$$[r = \ln(\lambda/m) + 9/4 - 2\ln(m/\lambda_{min}). \qquad (23)]$$

An equal term $-\frac{1}{2}\, r\boldsymbol{a}$ arises from virtual transitions after the scattering (14)

$$[(e^2/\pi i)\int \gamma_\mu(\mathbf{p}_2 - \mathbf{k} - m)^{-1}(\mathbf{p}_2 - m)^{-1}\boldsymbol{a}\mathbf{k}^{-2}\, d^4k, \qquad (14)]$$

so that the entire $r\boldsymbol{a}$ term in (22) is canceled.

The reason that r is just the value of (12)

$$[(e^2/\pi i)\int \gamma_\mu(\mathbf{p}_2 - \mathbf{k} - m)^{-1}(\mathbf{p}_1 - \mathbf{k} - m)^{-1}\gamma_\mu\mathbf{k}^{-2}\, d^4k, \qquad (12)]$$

when $q^2 = 0$ can also be seen without a direct calculation as follows: Let us call **p** the vector of length m in the direction of **p'** so that if $\mathbf{p}'^2 = m(1+\varepsilon)^2$ we have $\mathbf{p}' = (1+\varepsilon)\mathbf{p}$ and we take ε as very small, being of order T^{-1} where T is the time between the scatterings $\boldsymbol{b}$ and $\boldsymbol{a}$. Since $(\mathbf{p}' - m)^{-1} = (\mathbf{p}' + m)/(\mathbf{p}'^2 - m^2) \approx (\mathbf{p} + m)/2m^2\varepsilon$, the quantity (25) is of order ε^{-1} or T. We shall compute corrections to it only to its own order (ε^{-1}) in the limit $\varepsilon \to 0$. The term (27)

$$[\boldsymbol{a}(\mathbf{p}' - m)^{-1}\Delta m(\mathbf{p}' - m)^{-1}\boldsymbol{b}, \qquad (27)]$$

can be written approximately[14] as

$$(e^2/\pi i)\int \boldsymbol{a}(\mathbf{p}' - m)^{-1}\gamma_\mu(\mathbf{p} - \mathbf{k} - m)^{-1} \times \gamma_\mu(\mathbf{p}' - m)^{-1}\boldsymbol{b}\mathbf{k}^{-2}\, d^4k\, C(\mathbf{k}^2),$$

$$[\text{cf} \quad (e^2/\pi i)\int \boldsymbol{a}(\mathbf{p}' - m)^{-1}\gamma_\mu(\mathbf{p}' - \mathbf{k} - m)^{-1} \times \gamma_\mu(\mathbf{p}' - m)^{-1}\boldsymbol{b}\mathbf{k}^{-2}\, d^4k\, C(\mathbf{k}^2), \qquad (26)]$$

using the expression (19)

$$[(e^2/\pi i)\int \gamma_\mu(\mathbf{p} - \mathbf{k} - m)^{-1}\gamma_\mu\,\mathbf{k}^{-2}\, d^4k\, C(\mathbf{k}^2) \qquad (19)]$$

for Δm.

[14] The expression is not exact because the substitution of Δm by the integral in (19) is valid only if **p** operates on a *state* such that **p** can be replaced by m. The error, however, is of order … so the net result is approximately $\boldsymbol{a}(\mathbf{p} - m)\,\boldsymbol{b}/4m^2$ and is not of order $1/\varepsilon$ but smaller, so that its effect drops out in the limit.

The net of the two effects is therefore approximately[15]

$$-(e^2/\pi i)\int \boldsymbol{a}(\mathbf{p}' - m)^{-1}\gamma_\mu(\mathbf{p} - \mathbf{k} - m)^{-1}\varepsilon\mathbf{p}(\mathbf{p} - \mathbf{k} - m)^{-1} \times \gamma_\mu(\mathbf{p}' - m)^{-1}\boldsymbol{b}\mathbf{k}^{-2}\, d^4k\, C(\mathbf{k}^2),$$

a term now of order $1/\varepsilon$ (since $(\mathbf{p}' - m)^{-1} \approx (\mathbf{p} + m)/2m^2\varepsilon$, and therefore the one desired in the limit.

[15] We have used, to first order, the general expansion (valid for any operators A, B)

$$(A + B)^{-1} = A^{-1} - A^{-1}BA^{-1} + A^{-1}BA^{-1}BA^{-1} - \ldots$$

with $A = \mathbf{p} - \mathbf{k} - m$ and $B = \mathbf{p}' - \mathbf{p} = \varepsilon\mathbf{p}$ to expand the difference of $(\mathbf{p}' - \mathbf{k} - m)^{-1}$ and $(\mathbf{p} - \mathbf{k} - m)^{-1}$.

Comparison to (28)

$$[-r(\mathbf{p}'^2)\boldsymbol{a}(\mathbf{p}' - m)^{-1}\boldsymbol{b}. \qquad (28)]$$

gives for r the expression

$$(\mathbf{p}_1 + m)/2m \int \gamma_\mu(\mathbf{p}_1 - \mathbf{k} - m)^{-1}(\mathbf{p}_1 m^{-1})(\mathbf{p}_1 - \mathbf{k} - m)^{-1} \times \gamma_\mu \mathbf{k}^{-2}\, d^4k\, C(\mathbf{k}^2). \qquad (29)$$

The integral can be immediately evaluated, since it is the same as the integral (12)

$$[(e^2/\pi i) \int \gamma_\mu(\mathbf{p}_2 - \mathbf{k} - m)^{-1}(\mathbf{p}_1 - \mathbf{k} - m)^{-1}\gamma_\mu \mathbf{k}^{-2}\, d^4k, \qquad (12)]$$

but with $\mathbf{q} = 0$, for $\boldsymbol{a}$ replaced by $\mathbf{p}_1/m$. The result is therefore $r.(\mathbf{p}_1/m)$ which when acting on the *state* u_1 is just r, as $\mathbf{p}_1 u_1 = m u_1$. For the same reason the term $(\mathbf{p}_1 + m)/2m$ in (29) is effectively 1 and we are left with – r of (23)[16]

$$[r = \ln(\lambda/m) + 9/4 - 2 \ln(m/\lambda_{min}). \qquad (23)]$$

[16] The *renormalization terms* appearing B, Eqs. (13), (14) [corrected]

$$(e^2/\pi i) \int \boldsymbol{a}(\mathbf{p}_1 - m)^{-1}\gamma_\mu(\mathbf{p}_1 - \mathbf{k} - m)^{-1}\gamma_\mu \mathbf{k}^{-2}\, d^4k, \qquad (13)$$

$$(e^2/\pi i) \int \gamma_\mu(\mathbf{p}_2 - \mathbf{k} - m)^{-1}(\mathbf{p}_2 - m)^{-1}\boldsymbol{a}\mathbf{k}^{-2}\, d^4k, \qquad (14)$$

when translated directly into the present notation do not give twice (29) but give this expression with the central $\mathbf{p}_1 m^{-1}$ factor replaced by $m\gamma_4/E_1$ where $E_1 = \mathbf{p}_1\mu$ for $\mu = 4$. When integrated it therefore gives … which gives just $r\boldsymbol{a}$, since $\mathbf{p}_1 u_1 = m u_1$.

In more complex problems starting with a free electron the same type of term arises from the effects of a virtual *emission* and *absorption* both previous to the other processes. They, therefore, simply lead to the same factor r so that the expression (23) may be used directly and these *renormalization integrals* need not be computed afresh for each problem.

In this problem of the radiative corrections to scattering the net result is insensitive to the cut-off. This means, of course, that by a simple rearrangement of terms previous to the integration we could have avoided the use of the *convergence factors* completely (see for example Lewis[17]).

[17] Lewis, H. W. (January, 1948). On the Reactive Terms in Quantum Electrodynamics. *Phys. Rev.*, 73, 173; https://doi.org/10.1103/PhysRev.73.173.

The problem was solved in the manner here in order to illustrate how the use of such *convergence factors*, even when they are actually unnecessary, may facilitate analysis somewhat by removing the effort and ambiguities that may be involved in trying to rearrange the otherwise divergent terms.

The replacement of δ_+ by f_+ given in (16), (17)

$$[-f_+(s_{12}^2) = \pi^{-1} \int \exp(-ik.x_{12})\, \mathbf{k}^{-2} C(\mathbf{k}^2)\, d^4k. \qquad (16)$$

$$C(\mathbf{k}^2) = \int_0^\infty -\lambda^2(\mathbf{k}^2 - \lambda^2)^{-1} G(\lambda)\, d\lambda. \qquad (17)]$$

is not determined by the analogy with the classical problem. In the classical limit only the real part of δ_+ (i.e., just δ) is easy to interpret. But by what should the imaginary part, $1/(\pi i s^2)$, of δ_+ be replaced? *The choice we have made here* (in defining, as we have, the location of the poles of (17)) *is arbitrary and almost certainly incorrect.* If the radiation resistance is calculated for an atom, as the imaginary part of (8)

$$[-ie^2 \int\int f^{*}(4)\gamma_\mu K_+(4, 3)\gamma_\mu f(3)\delta_+(s_{43}^2)\, d\tau_3 d\tau_4, \qquad (8)]$$

the result depends slightly on the function f_+. On the other hand, the light radiated at very large distances from a source is independent of f_+ The *total energy* absorbed by distant absorbers will not check with the *energy* loss of the source. We are in a situation analogous to that in the classical theory if the entire f function is made to contain only retarded contributions (see A, Appendix). One desires instead the analogue of $[F]_{ret}$ of A. *This problem is being studied.*

One can say therefore, that this attempt to find a consistent modification of quantum electrodynamics is incomplete (see also the question of closed loops, below). For *it could turn out that any correct form of f_+ which will guarantee energy conservation may at the same time not be able to make the self-energy integral finite. The desire to make the methods of simplifying the calculation of quantum electrodynamic processes more widely available has prompted this publication before an analysis of the correct form for f_+ is complete.* One might try to take the position that, since the energy discrepancies discussed vanish in the limit $\lambda \to\infty$, *the correct physics might be considered to be that obtained by letting $\lambda \to\infty$ after mass renormalization.* I have no proof of the mathematical consistency of this procedure, but the presumption is very strong that it is satisfactory. (It is also strong that a satisfactory form for f_+ can be found.)

7. *The problem of vacuum polarization*

In the analysis of the radiative corrections to scattering one type of term was not considered. The *potential* which we can assume to vary as $a_\mu \exp(-iq.x)$ creates a pair of electrons (see Fig. 6), *momenta* $\mathbf{p}_a$, $-\mathbf{p}_b$. This pair then re-annihilates, emitting a quantum $\mathbf{q} = \mathbf{p}_b - \mathbf{p}_a$, which quantum scatters the original electron from *state* 1 to *state* 2.

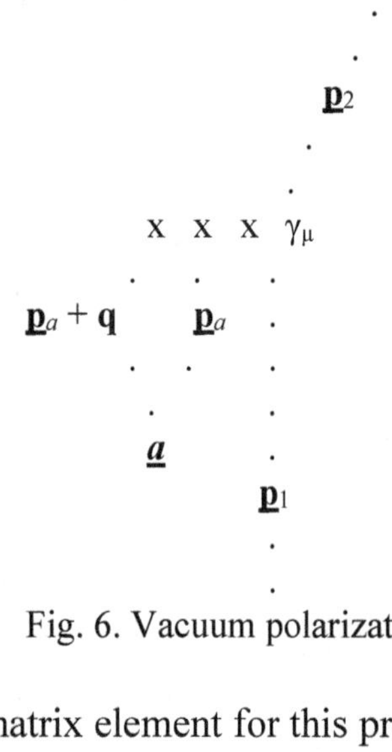

Fig. 6. Vacuum polarization effect on scattering, Eq. (30).

The matrix element for this process (and the others which can be obtained by rearranging the order in time of the various events) is

$$-(e^2/\pi i)(\bar{u}_2\, \gamma_\mu\, u_1) \int Sp[(\mathbf{p}_a + \mathbf{q} - m)^{-1} \times \gamma_\mu(\mathbf{p}_a - m)^{-1}\gamma_\mu]\, d^4p_a\, \mathbf{q}^{-2}\, C(\mathbf{q}^2)a_\nu. \quad (30)$$

This is because the potential produces the pair with *amplitude* proportional to $a_\nu\gamma_\nu$, the electrons of *momenta* $\mathbf{p}_a$ and $-(\mathbf{p}_a + \mathbf{q})$ proceed from there to annihilate, producing a quantum (factor γ_μ) which propagates (factor $\mathbf{q}^{-2}C(\mathbf{q}^2)$) over to the other electron, by which it is absorbed (matrix element of γ_μ between states 1 and 2 of the original electron ($\bar{u}_2\ \gamma_\mu\ u_1$)). All *momenta* $\mathbf{p}_a$ and *spin* states of the virtual electron are admitted, which means the spur and the integral on d^4p_a, are calculated.

One can imagine that the closed loop path of the positron-electron produces a *current*

$$4\pi j_\mu = J_{\mu\nu}a_\nu, \tag{31}$$

which is the source of the quanta which act on the second electron. The quantity

$$J_{\mu\nu} = -(e^2/\pi i) \int Sp[(\mathbf{p} + \mathbf{q} - m)^{-1} \times \gamma_\mu(\mathbf{p} - m)^{-1}\gamma_\mu]\, d^4p. \tag{32}$$

is then characteristic for this problem of *polarization of the vacuum.*

One sees at once that $J_{\mu\nu}$ diverges badly. The modification of δ to f alters the *amplitude* with which the *current* j_μ will affect the scattered electron, but *it can do nothing to prevent the divergence of the integral (32)* and of its effects.

One way to avoid such difficulties is apparent. From one point of view, we are considering all routes by which a given electron can get from one region of *space-time* to another, i.e., from the source of electrons to the apparatus which measures them. From this point of view *the closed loop path leading to (32) is unnatural*. It might be assumed that the only paths of meaning are those which start from the source and work their way in a continuous path (possibly containing many time reversals) to the detector. *Closed loops would be excluded.* We have already found that this may be done for electrons moving in a fixed *potential.*

Such a suggestion must meet several questions, however. *The closed loops are a consequence of the usual hole theory in electrodynamics. Among other things, they are required to keep probability conserved.* The probability that no pair is produced by a potential is not unity and its deviation from unity arises from the imaginary part of $J_{\mu\nu}$. Again, *with closed loops excluded, a pair of electrons once created cannot annihilate one another again, the scattering of light by light would be zero, etc. Although we are not experimentally sure of these phenomena, this does seem to indicate that the closed loops are necessary.* To be sure, it is always possible that these matters of probability conservation, etc., will work themselves out as simply in the case of interacting particles as for those in a fixed *potential. Lacking such a demonstration the presumption is that the difficulties of vacuum polarization are not so easily circumvented*[18].

[18] It would be very interesting to calculate the Lamb shift accurately enough to be sure that the 20 megacycles expected from vacuum polarization are actually present.

An alternative procedure discussed in B [*Loc. cit.* Feynman (1948). Relativistic Cut-Off for Quantum Electrodynamics.] is to *assume that the function $K_+(2, 1)$ used above is incorrect* and is to be replaced by a modified function K+' having no singularity on the light cone. The effect of this is to provide a *convergence factor* $C(\mathbf{p}^2 - m^2)$ for every integral over electron *momenta*[19].

[19] This technique also makes *self-energy* and radiationless scattering integrals finite even without the modification of' δ_+ to f_+ for the radiation (and the consequent *convergence factor* $C(k^2)$ for the quanta). See B. [*Loc. cit.* Feynman, R. P. (1948). Relativistic Cut-Off for Quantum Electrodynamics.]

This will multiply the integrand of (32)

$$[J_{\mu\nu} = -(e^2/\pi i) \int Sp[(\mathbf{p} + \mathbf{q} - m)^{-1} \times \gamma_\mu(\mathbf{p} - m)^{-1}\gamma_\mu]\, d^4p. \qquad (32)]$$

by $C(\mathbf{p}^2 - m^2)C((\mathbf{p} + \mathbf{q})^2 - m^2)$, since the integral was originally $\delta(\mathbf{p}_a - \mathbf{p}_b + q)\, d^4p_a\, d^4p_b$ and both $\mathbf{p}_a$ and $\mathbf{p}_b$ get *convergence factors. The integral now converges but the result is unsatisfactory*[20].

[20] Added to the terms given below (33) there is a term $\frac{1}{4}(\lambda^2 - 2\mu^2 + 1/3\,\mathbf{q}^2)\,\delta_{\mu\nu}$ for $C(\mathbf{k}^2) = -\lambda^2(\mathbf{k}^2 - \lambda^2)^{-1}$ *which is not gauge invariant.* (In addition, the *charge renormalization* has $-7/6$ added to the logarithm.)

One expects the *current* (31)

$$[4\pi j_\mu = J_{\mu\nu}a_\nu, \qquad (31)]$$

to be conserved, that t is $q_\mu j_\nu = 0$ or $q_\mu J_{\mu\nu} = 0$. Also, one expects no *current* if a_ν is a gradient, or $a_\nu = q_\nu$ times a constant. This leads to the condition $J_{\mu\nu}q_\nu = 0$ which is equivalent to $q_\mu J_{\mu\nu} = 0$ since $J_{\mu\nu}$ is symmetrical. But when the expression (32) is integrated with such *convergence factors* it does not satisfy this condition. *By altering the kernel from K to another, K', which does not satisfy the Dirac equation we have lost the gauge invariance, its consequent current conservation and the general consistency of the theory.*

One can see this best by calculating $J_{\mu\nu}q_\nu$ directly from (32). The expression within the spur becomes $(\mathbf{p} + \mathbf{q} - m)^{-1}\mathbf{q}(\mathbf{p} - m)^{-1}\gamma_\mu$ which can be written as the difference of two terms: $(\mathbf{p} - m)^{-1}\gamma_\mu - (\mathbf{p} + \mathbf{q} - m)^{-1}\gamma_\mu$. Each of these terms would give the same result if the integration d^4p were without a *convergence factor*, for the first can be converted into the second by a shift of the origin of $\mathbf{p}$, namely $\mathbf{p}' = \mathbf{p} + q$. This does not result in cancelation in (32) however, for the *convergence factor* is altered by the substitution.

A method of making (32) convergent without spoiling the gauge invariance has been found by Bethe and by Pauli. The *convergence factor* for light can be looked upon as the result of superposition of the effects of quanta of various masses (some contributing negatively). Likewise if we take the factor $C(\mathbf{p}^2 - m^2) = -\lambda^2(\mathbf{p}^2 - m^2 - \lambda^2)^{-1}$ so that $(\mathbf{p}^2 - m^2)^{-1}C(\mathbf{p}^2 - m^2) = (\mathbf{p}^2 - m^2)^{-1} - (\mathbf{p}^2 - m^2 - \lambda^2)^{-1}$ we are taking the difference of the result for electrons of *mass* m and *mass* $(\lambda^2 + m^2)^{-1}$. But we have taken this difference for *each* propagation between interactions with photons. They suggest instead that once created with a certain *mass* the electron should continue to propagate with this *mass* through all the potential interactions until it closes its loop. That is if the quantity (32), integrated over some finite range of $\mathbf{p}$, is called $J_{\mu\nu}(m^2)$ and the corresponding quantity over the same range of $\mathbf{p}$, but with m replaced by $(m^2 + \lambda^2)^{1/2}$, is $J_{\mu\nu}(m^2 + \lambda^2)$ we should calculate

$$J_{\mu\nu}{}^P = \int_0^\infty [J_{\mu\nu}(m^2) - J_{\mu\nu}(m^2 + \lambda^2)]G(\lambda)\, d\lambda, \qquad (32')$$

the function $G(\lambda)$ satisfying $\int_0^\infty G(\lambda)\, d\lambda = 1$ and $\int_0^\infty G(\lambda)\lambda^2\, d\lambda = 0$. Then in the expression for $J_{\mu\nu}{}^P$ the range of $\mathbf{p}$ integration can be extended to infinity as the integral now converges.

The result of the integration using this method is the integral on $d\lambda$ over $G(\lambda)$ of (see Appendix C)

$$J_{\mu\nu}{}^{P} = -\,e^2/\pi\,(q_\mu q_\nu - \delta_{\mu\nu}\mathbf{q}^2)\,\{-\,1/3\,(\ln(\lambda^2/m^2) - [(4m^2 + 2\mathbf{q}^2)/3\mathbf{q}^2\,(1 - \theta/\tan\theta) - 1/9]\} \qquad (33)$$

with $\mathbf{q}^2 = 4m^2 \sin^2\theta$.

The gauge invariance is clear, since $q_\mu(q_\mu q_\nu - \mathbf{q}^2\delta_{\mu\nu}) = 0$. Operating (as it always will) on a *potential* of zero divergence the $(q_\mu q_\nu - \mathbf{q}^2\delta_{\mu\nu})a_\nu$ is simply $-\,q^2 a_\nu$, the D'Alembertian of the potential, that is, *the current producing the potential*. The term $-\,1/3\,(\ln(\lambda^2/m^2))(q_\mu q_\nu - q^2\delta_{\mu\nu})$ therefore gives a *current* proportional to the *current* producing the *potential*. *This would have the same effect as a change in charge*, so that we would have a difference $\Delta(e^2)$ between e^2 and the experimentally observed *charge*, $e^2 + \Delta(e^2)$, analogous to the difference between m and the observed *mass*. *This charge depends logarithmically on the cut-off,* $\Delta(e^2)/e^2 = -\,(2e^2/3\pi)\ln(\lambda/m)$. *After this renormalization of charge is made, no effects will be sensitive to the cut-off.*

After this is done the final term remaining in (33), contains the usual effects[21] of *polarization of the vacuum*.

[21] Uehling, E. A. (July, 1935). Polarization Effects in the Positron Theory. *Phys. Rev.*, 48, 55; https://doi.org/10.1103/PhysRev.48.55; Serber, R. (July, 1935). Linear Modifications in the Maxwell Field Equations. *Phys. Rev.*, 48, 49; https://doi.org/10.1103/PhysRev.48.49.

It is zero for a free light quantum ($\mathbf{q}^2 = 0$). For small $\mathbf{q}^2$ it behaves as $(2/15)\mathbf{q}^2$ (adding $-\,1/5$ to the logarithm in the Lamb effect). For $\mathbf{q}^2 > (2m)^2$ it is complex, the imaginary part representing the loss in *amplitude* required by the fact that the probability that no quanta are produced by a *potential* able to produce pairs $((q^2)^{1/2} > 2m^2$ decreases with time. (To make the necessary analytic continuation, imagine m to have a small negative imaginary part, so that $(1 - \mathbf{q}^2/4m^2)^{1/2}$ becomes $-\,i(\mathbf{q}^2/4m^2 - 1)^{1/2}$ as $\mathbf{q}^2$ goes from below to above $4m^2$. Then $\theta = \pi/2 + iu$ where $\sinh u = +\,(\mathbf{q}^2/4m^2 - 1)^{1/2}$, and $-\,1/\tan\theta = i\tanh u = +\,i(\mathbf{q}^2 - 4m^2)^{1/2}(\mathbf{q}^2)^{-1}$. [sinh = hyperbolic sine.]

Closed loops containing a number of quanta or potential interactions larger than two produce no trouble. Any loop with an odd number of *interactions* gives zero (I, reference 9)

$$[K^{(1)}(2, 1) = -\,i\int K_0(2, 3)U(3)K_0(3, 1)\,d\tau_3. \qquad (9)]$$

Four or more potential *interactions* give integrals which are convergent even without a convergence factor as is well known. The situation is analogous to that for *self-energy*. *Once the simple problem of a single closed loop is solved there are no further divergence difficulties for more complex processes*[22].

[22] There are loops completely without external *interactions*. For example, a pair is created virtually along with a photon. Next, they annihilate, absorbing this photon. Such loops are disregarded on the grounds that they do not interact with anything and are thereby

completely unobservable. Any indirect effects they may have via the exclusion principle have already been included.

8. *Longitudinal waves*

In the usual form of quantum electrodynamics, the longitudinal and transverse waves are given separate treatment. Alternately the condition $(\partial A_\mu/\partial x_\mu)\psi = 0$ is carried along as a supplementary condition. In the present form no such special considerations are necessary for we are dealing with the solutions of the equation $-\Box^2 A_\mu = 4\pi j_\mu$ with a current j_μ which is conserved $\partial j_\mu/\partial x_\mu = 0$. That means at least $\Box^2(\partial A_\mu/\partial x_\mu) = 0$ and in fact our solution also satisfies $\partial A_\mu/\partial x_\mu = 0$.

To show that this is the case we consider the *amplitude* for *emission* (real or virtual) of a photon and show that the divergence of this *amplitude* vanishes. ...

...

This shows that unpolarized light is a relativistically invariant concept, and permits some simplification in computing cross sections for such light.

...

9. *Klein Gordon equation*

The methods may be readily extended to particles of spin zero satisfying the Klein Gordon equation ...

...

10. *Application to meson theories*

> [From theoretical considerations, in 1934 Hideki Yukawa predicted the existence and the approximate mass of the "*meson*" as the carrier of the nuclear force that holds atomic nuclei together. All mesons are unstable, with the longest-lived lasting for only a few hundredths of a microsecond. Heavier mesons decay to lighter mesons and ultimately to stable electrons, neutrinos and photons.]

The theories which have been developed to describe mesons and the *interaction* of nucleons can be easily expressed in the language used here. Calculations, to lowest order in the interactions can be made very easily for the various theories, *but agreement with experimental results is not obtained. Most likely all of our present formulations are quantitatively unsatisfactory*. We shall content ourselves therefore with a brief summary of the methods which can be used. ...
However, no good agreement with experiment results, when these are available, is obtained. Probably all of the formulations are incorrect. ...

The author is particularly indebted to Professor H. A. Bethe for his explanation of a method of obtaining finite and gauge invariant results for the problem of vacuum polarization. He is also grateful for Professor Bethe's criticisms of the manuscript, and for innumerable discussions during the development of this work. He wishes to thank Professor J. Ashkin for his careful reading of the manuscript.

Schwinger, J. (September, 1949). Quantum Electrodynamics. III. The Electromagnetic Properties of the Electron—Radiative Corrections to Scattering.

[*Phys. Rev.*, 76, 6, 790–817; https://doi.org/10.1103/PhysRev.76.790; also in Schwinger, J. (ed). (1958). *Selected Papers on Quantum electrodynamics*. Dover, New York, pages 169-96.]

Received May 26, 1949.

Harvard University, Cambridge, Massachusetts.

A covariant form of *quantum electrodynamics* has been developed and applied in the previous articles of this series to two elementary phenomena that are produced by the *vacuum fluctuations of the electromagnetic field*, these applications were the *polarization of the vacuum* expressing the modifications in the properties of an electromagnetic field arising from its interaction with the *matter field* vacuum fluctuations, and the *electromagnetic mass of the electron* embodying the corrections to the mechanical properties of the *matter field* in its single particle aspect, in these problems *the divergences that mar the theory are found to be concealed in unobservable charge and mass renormalization factors*, the previous paper was confined to consideration of *vacuum polarization* produced by the field of a prescribed *current* distribution, *we now consider how induction of current in a vacuum by an electron results in an alteration in its electromagnetic properties* revealed by scattering in Coulomb field and energy level displacements, *this paper is concerned with the computation of the second-order corrections to the current operator as modified by the coupling with the vacuum electromagnetic field and its application to electron scattering* by a Coulomb field, applies canonical transformation to *renormalize electron mass*, correction to *current operator* produced by coupling with electromagnetic field developed in power series, first- and second-order terms retained, results in second-order modifications in *current operator* of same general nature as the previously treated *vacuum polarization current* apart from contribution in form of *dipole current*, the latter implies fractional increase of $\alpha/2\pi$ in the *spin magnetic moment* of electron, the only flaw in second-order current correction is *logarithmic divergence attributable to infra-red catastrophe,* in the presence of an *external field* the first-order *current* correction will introduce a compensating divergence, thus the second-order corrections to particle electromagnetic properties cannot be completely stated without regard for the manner of exhibiting them by an *external field*, accordingly in the second section we consider the interaction of three systems - the *matter field*, the *electromagnetic field*, and a given *current distribution*, shows that this can be described in terms of an *external potential* coupled to the *current* operator as modified by the interaction with the *vacuum electromagnetic field*, applies to *scattering of an electron* by an *external field* which is regarded as a small perturbation, convenient to calculate the total rate at which collisions occur and then identify the *cross sections* for individual events, correction to the *cross section* for radiation-less scattering is determined by the second-order correction to the *current* operator, scattering that is accompanied by single quantum emission is a consequence of the first-order *current* correction, the final object of calculation is the *differential cross section* for scattering through a given angle with a prescribed maximum energy loss which is completely free of divergences, an *Appendix* is devoted to an alternative treatment of the *polarization of the vacuum* by an external field, *radiative corrections to energy levels will be treated in the next paper of the series* [*but this did not appear*],

> [Schwinger (November, 1948). Quantum Electrodynamics. I. A Covariant Formulation: "… a *third paper* is concerned with the major topic, the *determination of the radiative corrections to the properties of an electron, and the comparison with experiment*".
>
> Schwinger (September, 1949). Quantum Electrodynamics. III. The Electromagnetic Properties of the Electron—Radiative Corrections to Scattering, *see Abstract below*: "This paper is concerned with the computation of the second-order corrections to the current operator and the application to electron scattering. *Radiative corrections to energy levels will be treated in the next paper of the series*", but this did not appear.
>
> Nor has the *fourth paper* dealing with scalar and vector matter fields, or any subsequent papers dealing with the corrections to the Klein-Nishina formula, the scattering of light by light, and by a Coulomb field, been found. From his comments in his Nobel Lecture (below) which focused on his papers published in the early 1950's and early 1960's it appears that Schwinger lost interest in this formulation of *quantum electrodynamics* after this paper.]

Abstract.

The discussion of *vacuum polarization* in the previous paper of this series was confined to that produced by the field of a prescribed *current* distribution. *We now consider the induction of current in the vacuum by an electron*, which is a dynamical system and an entity indistinguishable from the particles associated with vacuum fluctuations. The additional *current* thus attributed to an electron implies an alteration in its *electromagnetic* properties which will be revealed by *scattering in a Coulomb field* and by *energy level displacements. This paper is concerned with the computation of the second-order corrections to the current operator and the application to electron scattering. Radiative corrections to energy levels will be treated in the next paper of the series.*

Following a canonical transformation *which effectively renormalizes the electron mass*, the correction to the *current* operator produced by the coupling with the *electromagnetic field* is developed in a power series, of which first- and second-order terms are retained. One thus obtains second-order modifications in the *current* operator which are of the same general nature as the previously treated *vacuum polarization* current, save for a contribution that has the form of a *dipole current. The latter implies a fractional increase of $\alpha/2\pi$ in the spin magnetic moment of the electron.*

The only flaw in the second-order current correction is a logarithmic divergence attributable to an infra-red catastrophe. It is remarked that, in the presence of an *external field*, the first-order *current* correction will introduce a compensating divergence. Thus, the second-order corrections to particle electromagnetic properties cannot be completely stated without regard for the manner of exhibiting them by an *external field*. Accordingly, we consider in the second section the interaction of three systems, the *matter field*, the *electromagnetic field*, and a given *current distribution*. It is shown that this situation can be described in terms of an *external potential* coupled to the *current* operator, as modified by the interaction with the *vacuum electromagnetic field*.

Application is made to the *scattering of an electron* by an *external field*, in which the latter is regarded as a small perturbation. It is found convenient to calculate the total rate at which collisions occur and then identify the *cross sections* for individual events. The correction to the *cross section* for radiation-less scattering is determined by the second-order correction to the *current* operator, while scattering that is accompanied by single quantum emission is a consequence of the first-order *current* correction. The final object of calculation is the *differential cross section* for scattering through a given angle with a prescribed maximum energy loss, *which is completely free of divergences*.

Detailed evaluations are given in two situations, the essentially *elastic scattering of an electron*, in which only a small fraction of the kinetic energy is radiated, and the *scattering of a slowly moving electron* with unrestricted energy loss.

The *Appendix* is devoted to an alternative treatment of the *polarization of the vacuum* by an external field. The conditions imposed on the *induced current* by the *charge conservation* and *gauge invariance* requirements are examined. It is found that the fulfillment of these formal properties requires the vanishing of an integral that is not absolutely convergent, but naturally vanishes for reasons of symmetry. This null integral is then used to simplify the expression for the *induced current* in such a manner that direct calculation yields a *gauge invariant* result. The *induced current* contains a logarithmically divergent multiple of the external current, which implies that a non-vanishing total charge, proportional to the external charge, is induced in the vacuum. The apparent contradiction with *charge conservation* is resolved by showing that a compensating charge escapes to infinity. Finally, the expression for the *electromagnetic mass of the electron* is treated with the methods developed in this paper.

A covariant form of *quantum electrodynamics* has been developed, and applied to two elementary vacuum fluctuation phenomena in the previous articles of this series[1].

[1] Schwinger, J. (November, 1948). Quantum Electrodynamics. I. A Covariant Formulation. *Phys. Rev.*, 74, 10, 1439-61[; *lack of convergence in current formulations of quantum electrodynamics indicates that revision of electrodynamic concepts at ultra-relativistic energies is necessary*, elementary phenomenon in which divergences occur as a result of virtual transitions involving particles with unlimited energy are *polarization of the vacuum* and *self-energy of the electron* which express *the interaction of the electromagnetic and matter fields with their own vacuum fluctuations*, this alters the constants characterizing the properties of the individual fields and their mutual coupling by infinite factors, the

question is whether all divergencies can be isolated in such unobservable *renormalization* factors, *this paper is occupied with the formulation of a completely covariant electrodynamics*, manifest covariance with respect to Lorentz and gauge transformations essential in a divergent theory, customary *canonical commutation relations* fail to exhibit the desired covariance since they refer to field variables at equal times and different points of space, *can be put in covariant form by replacing the four-dimensional surface t = const. by a space-like surface*, offers the advantage over the Schrodinger representation in which all operators refer to the same time providing distinct separation between *kinematical* and *dynamical* aspects, formulation that retains evident covariance of the Heisenberg representation but offers something akin to Schrodinger representation can be based on distinction between the properties of *non-interacting fields*, and the effects of *coupling between fields*, constructs a *canonical transformation* that changes the *field equations* in the *Heisenberg representation* into those of *non-interacting fields*, *supplementary condition* restricting the admissible states of the system and the *commutation relations* must be added to the *equations of motion*, describes the coupling between fields in terms of a varying state vector, then simple matter to evaluate commutators of *field* quantities at arbitrary *space-time* points, one thus obtains an obviously covariant and practical form of quantum electrodynamics expressed in a mixed Heisenberg-Schrodinger representation called the *interaction representation*, discusses *covariant* elimination of longitudinal field in which customary distinction between longitudinal and transverse fields is replaced by a suitable *covariant* definition, describes collision processes in terms of an invariant *collision operator* which is the unitary operator that determines the over-all change in state of a system as the result of interaction, notes that a *second paper* treats the problems of electron and photon *self-energy* together with the *polarization of the vacuum* and a *third paper* is concerned with the determination of the *radiative corrections* to the properties of an electron and the comparison with experiment [this was not addressed, it stated that "*radiative corrections to energy levels* will be treated in the next paper of the series" but *this did not appear, nor are there any references to it*];

Schwinger, J. (February, 1949). Quantum Electrodynamics. II. Vacuum Polarization and Self-Energy. *Phys. Rev.*, 75, 4, 651-79[; *interaction representation* is applied to *polarization of the vacuum* and the *self-energies of the electron and photon*, in *first section* the vacuum of the non-interacting *electromagnetic* and *matter* fields is *covariantly* defined as *state* for which the eigenvalue of an arbitrary time-like component of the *energy-momentum four-vector* is an absolute minimum, covariant decomposition of field operators into positive and negative frequency components introduced to characterize vacuum *state vector*, shows that *state vector for electromagnetic vacuum* annihilated by positive frequency part of transverse four-vector potential and *state vector for matter vacuum* annihilated by positive frequency part of Dirac spinor and its charge conjugate, these properties of vacuum *state vector* employed in the calculation of the *vacuum expectation* values of quadratic field quantities, specifically the *energy-momentum tensors* of the independent *electromagnetic* and *matter fields* and the *current four-vector*, infers that the *electromagnetic energy-momentum tensor* and *current vector* must vanish in the vacuum, while the *matter field energy-momentum tensor* vanishes in the vacuum only by the addition of a suitable multiple of the unit tensor, *second section treats the induction of a current in the vacuum by an external electromagnetic field*, supposes that *external electromagnetic field* does not produce actual electron-positron pairs, considers only phenomenon of *virtual pair creation*, restriction is introduced by requiring that establishment and subsequent removal of the external field produce no net change in state for the *matter field*, *demonstrates that the induced current at a given space-time point involves the external current in the vicinity of that point and not the electromagnetic potentials*, this *gauge invariant* result shows that a light wave propagating at remote distances from its source induces no current in the vacuum and is therefore undisturbed in its passage through space, indicates *absence of a light quantum self-energy effect*, *current* induced at a point consists of two parts, a logarithmically divergent multiple of the *external current* at that point *which produces an unobservable renormalization of charge* and a more involved finite contribution which is the physically significant *induced current*, *third section considers the modification of the matter field properties arising from interaction with the vacuum fluctuations of the electromagnetic field*, analysis carried out with two alternative formulations, one employing the complete *electromagnetic potential* together with a *supplementary condition*, the other using the *transverse potential* with the variables of the *supplementary condition* eliminated, no real processes produced by first order coupling between the fields, alternative *equations of motion* for the *state vector* are constructed from which the first order interaction term has been eliminated and replaced by the *second order coupling* which it generates, this includes the *self-action* of individual particles and light quanta, the *interaction* of different particles, and a *coupling* between particles and light quanta which produces such effects as Compton scattering and two quantum pair annihilation, concludes from comparison of the alternative procedures that *for the treatment of virtual light quantum processes the separate consideration of longitudinal and transverse fields is an inadvisable complication*, light quantum *self-energy* term is shown to vanish, while that for a particle has the form for a change in *proper mass* but is logarithmically divergent in agreement with previous calculations, identification of *self-energy* effect with a change in *proper mass* is confirmed by removing this term from the state vector *equation of motion*, alters the *matter field equations of motion* in the expected manner, verifies that *the energy and momentum modifications produced by self-interaction effects are entirely accounted for by the addition of the electromagnetic proper mass to the mechanical proper mass—an unobservable mass renormalization*, an appendix is devoted to the construction of several invariant functions associated with the *electromagnetic* and *matter* fields].

These applications were the *polarization of the vacuum*, expressing the modifications in the properties of an electromagnetic field arising from its interaction with the *matter field* vacuum fluctuations, and the *electromagnetic mass of the electron*, embodying the corrections to the mechanical properties of the *matter field*, in its single particle aspect, that are produced by the vacuum fluctuations of the electromagnetic field. In these problems, *the divergences that mar the theory are found to be concealed in unobservable charge and mass renormalization factors*.

The previous discussion of the *polarization of the vacuum* was concerned with a given current distribution, one that is not affected by the dynamical reactions of the electron-positron matter field. *We shall now consider the more complicated situation in which the original current is that ascribed to an electron or positron - a dynamical system - and an entity indistinguishable from the particles associated with the matter field vacuum fluctuations*. The changed electromagnetic properties of the particle will be exhibited in an external field, and may be compared with the experimental indications of deviations from the Dirac theory that were briefly discussed in I [Schwinger (November, 1948). Quantum Electrodynamics. I. A Covariant Formulation]. To avoid a work of excessive length, this discussion will be given in two papers. *In this paper we shall construct the current operator as modified, to the second order, by the coupling with the vacuum electromagnetic field. This will be applied to compute the radiative correction to the scattering of an electron by a Coulomb field*[2].

[2] A short account of the results has already been published, Schwinger, J. (1949). On Radiative Corrections to Electron Scattering. *Phys. Rev.*, 75, 5, 898; https://doi.org/10.1103/PhysRev.75.898; also in Schwinger, J. (ed). (1958). *Selected Papers on Quantum electrodynamics*. Dover, New York, pages 143-4.

The second paper will deal with the effects of radiative corrections on energy levels [but this did not appear].

[The first two papers Schwinger I and II correspond with what was covered in what is referred to in Feynman (November, 1950) [Mathematical Formulation of the Quantum Theory of Electromagnetic Interaction] (below), as Feynman I [Feynman (September, 1949). The Theory of Positrons], hereafter called I, and this and the promised second paper referred to above, which did not appear, correspond to what was covered in what was referred to as Feynman II [Feynman, (September, 1949). Space-Time Approach to Quantum Electrodynamics], hereafter called II.]

1. *Second-order corrections to the current operator.*

We shall evaluate the second-order modifications of the *current operator* produced by *the coupling between the matter and electromagnetic fields*. The latter is described by

$$i\hbar c\, \delta\Psi[\sigma]/\delta\sigma(x) = H(x)\Psi[\sigma],$$
$$H(x) = -(1/c)\, j_\mu(x)A_\mu(x). \qquad (1.1)$$

[Schwinger (November, 1948). Quantum Electrodynamics. I. A Covariant Formulation: "The equation of motion for $\Psi[\sigma]$ is, correspondingly,

$$i\hbar c\, \delta\Psi[\sigma]/\delta\sigma(x) = H(x)\Psi[\sigma], \qquad (2.6)$$

We have obtained the conditions that must be satisfied by any canonical transformation. It will now be shown that the special transformation desired is attained with H(x) chosen as the negative of the coupling term in the Lagrangian density, that is,

$$H(x) = -(1/c)\, j_\mu(x)A_\mu(x)." \qquad (2.7)]$$

Among the effects produced by this coupling is the *electromagnetic mass* of the electron, as contained in the *self-energy* operator $H_{t,0}(x)$. In order to describe the electron in terms of the experimental mass, we write (1.1) as

$$i\hbar c\, \delta\Psi[\sigma]/\delta\sigma(x) = \{H_{t,0}(x) + K(x)\}\Psi[\sigma], \qquad (1.2)$$

where

$$K(x) = H(x) - H_{t,0}(x). \qquad (1.3)$$

The canonical transformation

$\Psi[\sigma] \to W[\sigma]\Psi[\sigma]$,
$i\hbar c\ \delta W[\sigma]/\delta\sigma(x) = \{H_{t,0}(x)W[\sigma]$, (1.4)

then replace (1.2) with

$i\hbar c\ \delta\Psi[\sigma]/\delta\sigma(x) = W{-}1[\sigma]K(x)W[\sigma]\Psi[\sigma]$, (1.5)

while the operator representing the *current* becomes $W^{-1}[\sigma]j_\mu(x)W[\sigma]$. Now, as we have shown in II [Feynman (September, 1949). Space-Time Approach to Quantum Electrodynamics], the spinor $W^{-1}[\sigma]\psi(x)W[\sigma]$ obeys the Dirac equation for a particle of mass $m = m_0 + \delta m$, the *experimental mass* of the electron. Accordingly, the expectation value of the *current operator* can be computed as

$<j_\mu(x)> = \{\Psi[\sigma], j_\mu(x)\Psi[\sigma]\}$, (1.6)

where

$i\hbar c\ \delta\Psi[\sigma]/\delta\sigma(x) = K(x)\Psi[\sigma]$, (1.7)

with the understanding that the *experimental electron mass* is to be employed.

If a solution of the latter equation is constructed in the form

$\Psi[\sigma] = U[\sigma]\Psi_0$, (1.8)

the expectation value of the *current operator* becomes

$<j_\mu(x)> = \{\Psi_0, U^{-1}[\sigma]j_\mu(x)U[\sigma]\Psi_0\} = \{\Psi_0, j_\mu(x)\Psi_0\}$, (1.9)

in which the latter version describes the effect of the coupling between the fields by changing the *current operator* into

$j_\mu(x) = U^{-1}[\sigma]j_\mu(x)U[\sigma]$. (1.10)

The unitary operator $U[\sigma]$ obeys the equation of motion

$i\hbar c\ \delta U[\sigma]/\delta\sigma(x) = K(x)U[\sigma]$, (1.11)

which may be supplemented by the boundary condition

$U[-\infty] = 1$, (1.12)

in accordance with the supposition that coupling between the two fields is adiabatically established in the remote past.

The operator $j_\mu(x)$ can now be evaluated by remarking that

$j_\mu(x) = \ldots$ (1.13)

This process can be continued according to

… (1.14)

and yields $j_\mu(x)$ in the form of an infinite series,

$j_\mu(x) = \ldots$ (1.15)

An equivalent procedure, which exhibits $j_\mu(x)$ in a form that is more symmetrical between past and future, is based on the following observation,

… (1.16)

or

$j_\mu(x) = \ldots$ (1.17)

where

$S = U[\infty], \qquad U[-\infty] = 1,$ (1.18)

is the *collision operator* which describes the real transitions that permanently alter the state of the system. This process can be continued and finally yields

$j_\mu(x) = \ldots$ (1.19)

in which

… (1.20)

The collision operator S can be constructed in a similar manner. Thus,

$S - 1 = \ldots$ (1.21)

…

whence

$$(S - 1)/(S + 1) = -\,i/2\hbar c \int_{-\infty}^{\infty} [H_{2,0}(x) + H_{1,1}(x)]d\omega, \qquad (1.28)$$

which describes the real effects involving either two particles or one particle and a light quantum. Since we shall be concerned only with second-order effects referring to a single particle and the absence of light quanta, such real processes do not come into play and S is effectively unity. Consequently, the current operator is modified only by virtual processes, and is completely symmetrical between past and future. Thus, to the desired order of approximation,

$j_\mu(x) = \ldots$ (1.29)

The correction to the *current operator* may now be written

$j_\mu(x) - \mathcal{j}_\mu(x) = \ldots$ (1.30)

where

$\delta \mathcal{j}_\mu^{(1)}(x) = \ldots$ (1.31)

and

$(\delta \mathcal{j}_\mu^{(2)}(x)_{1,0} = \ldots$ (1.32)

are the first- and second-order corrections, respectively. …

…

A suitable representation of the solution of the Dirac equation with an altered mass parameter …, in terms of the actual spinor …, is provided by

… (1.55)

…

The resulting expression for the second-order correction to the *current operator* is

$$(\delta j_\mu^{(2)}(x))_{1,0} = \ldots \qquad (1.58)$$

...

Expressed in the same notation, the first term of (1.58) is (see II (2.44)),

$$\ldots \qquad (1.116)$$

from which we have omitted the *charge renormalization* term, with the understanding that the value of e is to be correspondingly altered. A rederivation of this result, employing methods akin to those presented in this paper, is given in the Appendix. Evidently the new contributions to the one particle *current operator*, as given in (1.114), are of the same general nature as the previously considered effect, (1.116), with the exception of the last term in (1.114). This is an addition to the *current vector* of the form

$$c(\partial/\partial x),\delta m_{\mu\nu}(x), \qquad (1.117)$$

where

$$\delta m_{\mu\nu}(x) = \alpha/2\pi \int F_0(x - x')\, m_{\mu\nu}(x')\, d\omega'. \qquad (1.118)$$

A *current vector* of this type can be interpreted as a dipole *current*, derived from an antisymmetrical dipole tensor $m_{\mu\nu}$ *which combines electric and magnetic dipole moment densities*. The tensor $m_{\mu\nu}$ is that characteristic of the Dirac theory, in which intrinsic dipole moments are related to the anti-symmetrical *spin tensor* $\sigma_{\mu\nu}$, the factor of proportionality being

$$\mu_0 = e/2\kappa = e\hbar/2mc, \qquad (1.119)$$

the Bohr magneton. According to (1.118), the correction to the *dipole tensor* at a point involves an average of $m_{\mu\nu}$ over the vicinity of that point. If all quantities are slowly varying, relative to $\hbar/mc$ and $\hbar/mc^2$ as units of length and time, an expansion in ascending powers of $\Box^2$ can be constructed, as in II (2.47). For this purpose, it is sufficient to expand the denominator in the first form of (1.112), thus obtaining

$$F_n(x) = \ldots \qquad (1.120)$$

Hence,

$$\delta m_{\mu\nu}(x) = \alpha/2\pi\, [m_{\mu\nu}(x) + 1/6\kappa^2\, \Box^2 m_{\mu\nu}(x) + \ldots], \qquad (1.121)$$

and, under conditions that permit the neglect of all but the first term in this series, an electron will act as though it possessed an additional *spin magnetic moment*[3]

$$\delta_\mu = (\alpha/2\pi)\, \mu_0. \qquad (1.122)$$

[3] This result was announced at the January, 1948 meeting of the American Physical Society. The formula is misprinted in a published note, Schwinger, J. (February, 1948). On quantum-electrodynamics and the magnetic moment of the electron. *Phys. Rev.*, 73, 4, 416-7[; *electrodynamics unquestionably requires revision at ultra-relativistic energies,* desirable to isolate those aspects of the current theory that essentially involve high energies and are subject to modification by a more satisfactory theory, this goal has been achieved

by transforming the Hamiltonian of current *hole theory* electrodynamics to exhibit explicitly the logarithmically divergent self-energy of a free electron which arises from the virtual emission and absorption of light quanta, the electromagnetic *self-energy* of a free electron can be ascribed to an *electromagnetic mass* which must be added to the mechanical mass of the electron, new Hamiltonian involves experimental *electron mass* rather than unobservable *mechanical mass*, electron now interacts with radiation field only in presence of external field such that only an accelerated electron can emit or absorb a light quantum, *interaction energy* of electron with external field subject to *finite* radiative correction, *polarization of the vacuum* still produces logarithmically divergent term proportional to *interaction energy* of electron in an external field, such term equivalent to altering value of *electron charge* by constant factor with only final value being identified with experimental charge, interaction between matter and radiation produces *renormalization* of electron charge and mass, all divergences contained in *renormalization* factors, radiative correction for energy of electron in external magnetic field corresponds to *additional magnetic moment associated with electron spin* of magnitude $\delta\mu/\mu = (\frac{1}{2}\pi)e^2/hc = 0.001162$, experimental measurements on hyperfine splitting of ground states of atomic hydrogen and deuterium larger than expected from directly measured nuclear moments, finds additional *electron spin magnetic moment* to account for measured hydrogen and deuterium *hyperfine structures* to be $\delta\mu/\mu = 0.00126$ and $\delta\mu/\mu = 0.00131$ respectively, these discrepancies accounted for by additional spin magnetic moment to the electron of $\delta\mu/\mu = 0.0018 \pm 0.00003$, *values yielded by relativistic calculation of Lamb shift differ only slightly from those conjectured by Bethe on basis of non-relativistic calculation and are in good accord with experiment*].

The misprint has unfortunately been copied by L. Rosenfeld in his book, *Nuclear Forces* (Interscience Publishers, Inc., New York, 1949), p. 438.

The comparison of this prediction with experiment will be discussed in the sequel to this paper.

[A sequel to this paper was not found on the internet.]

The final result for the *second-order correction* to the one particle current operator is

$$\{\delta j_\mu^{(2)}(x)\}_{1,0} = \ldots \tag{1.123}$$

...

It will be noted that the *total charge* computed from $\{\delta j_\mu^{(2)}(x)\}_{1,0}$ is zero, in agreement with evident *charge conservation* requirements, and the formal property that the operator of *total charge* commutes with all one-particle operators. *The apparent contradiction between these statements and the existence of the charge renormalization term is discussed in the Appendix, where it is shown that a compensating charge is created at infinity.*

Our result, (1.123), is marred only by the appearance of the logarithmic divergence associated with zero frequency quanta. It should be remarked, however, that $\{\delta j_\mu^{(2)}(x)\}_{1,0}$ is not a complete description of the *radiative corrections* under discussion. In order to measure the correction to the *current*, it is necessary to impose an external field. This will induce the emission of quanta, as described by $\delta j_\mu^{(1)}(x)$, among the effects of which is a *compensating low frequency divergence*. It will be apparent that, *as a consequence of the "infra-red catastrophe", the second-order corrections to particle electromagnetic properties cannot be completely stated without regard for the manner of exhibiting them by an external field.* We therefore turn to a discussion of the behavior of a single particle in an external field, as modified by the vacuum fluctuations of the electromagnetic field.

2. *Radiative corrections to electron scattering*

We shall now be concerned with the interaction of three systems - the matter field, the electromagnetic field, and a given current distribution. The latter may be associated with a nucleus or a macroscopic apparatus, two situations in which the reaction on the *current distribution* may have a negligible effect. A description of this state of affairs, in the *interaction representation*, is given by

$$i\hbar c\ \delta\Psi[\sigma]/\delta\sigma(x) = [-(1/c)\ \{j_\mu(x) + J_\mu(x)\}A_\mu(x)]\Psi[\sigma], \qquad (2.1a)$$
$$[\partial A_\mu(x')/\partial x_\mu' - (1/c \int_\sigma D(x - x')\{j_\mu(x) + J_\mu(x)\}d\sigma_\mu]\Psi[\sigma] = 0. \qquad (2.1b)$$

where $j_\mu(x)$ and $J_\mu(x)$ are the *current vectors* associated with the *matter field* and the *external system*, respectively. Both *current distributions* are coupled to the *electromagnetic field*, as characterized by $A_\mu(x)$. An equally valid way of stating matters is in terms of an *external electromagnetic field* acting on the *matter field current distribution*:

… (2.2a)

… (2.2b)

…

The functional $J[\sigma]$ is explicitly exhibited as

$$J[\sigma] = -1/\hbar c^2 \int_{-\infty}^{\sigma} J_\mu(x')\}A_\mu(x')]d\omega', \qquad (2.6)$$

in which the choice of lower limit corresponds to selecting the *retarded potentials* for the *electromagnetic field* generated by the given *current distribution*. …

…

… and the Dirac equation for $\psi(x)$ now involves the *experimental mass*. The further transformation

$$\Psi[\sigma] = U[\sigma]\Phi[\sigma], \qquad (2.21)$$

where

$$i\hbar c\ \delta U[\sigma]/\delta\sigma(x) = K(x)U[\sigma], \qquad U[-\infty] = 1, \qquad (2.22)$$

is the analog of (1.8)

$$[\Psi[\sigma] = U[\sigma]\Psi_0, \qquad (1.8)]$$

save that $\Phi[\sigma]$ varies in the presence of an external field,

$$i\hbar c\ \delta\Phi[\sigma]/\delta\sigma(x) = U^{-1}[\sigma]K^{(\sigma)}(x)U[\sigma]\Phi[\sigma],$$
$$= -(1/c)\ j_\mu(x)A_\mu^{(\sigma)}(x)]\Phi[\sigma], \qquad (2.23)$$

in response to the coupling with the current operator $j_\mu(x)$. *The latter contains the modifications produced by the vacuum electromagnetic field.* The *supplementary condition* (2.2b) appears as

… (2.24)

in consequence of these transformations. However,

… (2.25)

so that the *supplementary condition* associated with (2.23) is simply

$$\partial A_\mu(x')/\partial x_\mu' \, \Phi[\sigma] = 0. \quad (2.26)$$

As the first application of (2.23), we shall consider the *scattering of an electron produced by its interaction with an external field*, in which the latter is regarded as a small perturbation[4].

[4] Radiative corrections to scattering have been discussed by many authors. That a finite correction is obtained after a *renormalization of charge and mass* was independently observed by Koba, Z. & Tomonaga, S. (September, 1948). On Radiation Reactions in Collision Processes. I: Application of the "Self-Consistent" Subtraction Method to the Elastic Scattering of an Electron. *Progr. Theoret. Phys. (Kyoto)*, 3, 3, 290-303[; applies Tati and Tomonaga's "self-consistent" subtraction method to the elastic scattering of an electron by a fixed electrostatic potential, *formal infinity associated with mass-modifying term attributed to defect of current theory so empirical value substituted for theoretical value* (assuming it was already included in the free field equation), maintains total Hamilton describing the interaction of the electron and electromagnetic fields unaltered, requires interaction-energy part of the Hamiltonian to undergo corresponding change by inclusion of "counter-self-energy" term ("mass-type" correction to scattering cross-section), results in finite value for self-energy of electron in the e^2 approximation, when new formalism applied to elastic scattering of electron *effective cross-section* for scattering by a fixed potential in zeroth approximation has value in good agreement with experiment but as soon as reaction of electromagnetic field with electron is taken into account the correction becomes infinite, first infinite term in modified Hamiltonian eliminated using *subtraction hypothesis of positron theory*, second difficulty disappears by applying *"self-consistent" subtraction method* using modified Hamiltonian, first term can also be eliminated if one interprets the sum of the external potential and its infinite correction due the *vacuum polarization* effect as the physically observable potential and *substitute the empirical value for it* whilst interaction part of the Hamiltonian supplemented by an additional "counter-vacuum polarization" term, *method by no means give the real solution of the fundamental difficulty of the quantum electrodynamics* but reveals nature of various diverging terms and reduces them to two quantities - the *self-energy* and the *vacuum polarization*, in this way *it becomes possible in an unambiguous and consistent manner to treat the field reaction problem without touching the fundamental difficulty by employing the finite empirical values instead of the infinite "theoretical" values for these two quantities*];
Lewis, H. W. (January, 1948). On the Reactive Terms in Quantum Electrodynamics. *Phys. Rev.*, 73, 173; https://doi.org/10.1103/ PhysRev.73.173; and Schwinger, J. (February, 1948). On quantum-electrodynamics and the magnetic moment of the electron. *Loc. cit.* See also Feynman, R. P. (November, 1948). Relativistic Cut-Off for Quantum

Electrodynamics. *Phys. Rev.*, 74, 1430-8[; describes model based on quantization of a classical theory for which all quantities automatically come out finite described in his previous paper [Feynman (October, 1948). A Relativistic Cut-Off for Classical Electrodynamics], contains an arbitrary function on which numerical results depend, only term that depends significantly (logarithmically) on the cut-off frequency is the *self-energy* which can be used to *renormalize* the electron mass, remaining terms are nearly independent of the function, applies only to results for processes in which virtual quanta are emitted and absorbed, *terms representing processes involving a pair production followed by annihilation of the same pair are infinite and not made convergent by the present scheme,* problems of *permanent emission* and the position of *positron theory* need to be addressed, *the present paper may be looked upon as presenting an arbitrary rule to cut off at high frequencies in a relativistically invariant manner the otherwise divergent integrals appearing in quantum field theories*, produces finite invariant *self-energy* for a free electron, but problem of *polarization of the vacuum* not solved, alternative cut-off procedure which eliminates high frequency intermediate states offers to solve *vacuum polarization* problems as well].

We shall restrict the *external potential* to be that of a time independent field, which will eventually be specialized to the Coulomb field of a stationary nucleus.

A solution of (2.23)

$$[i\hbar c\, \delta\Phi[\sigma]/\delta\sigma(x) = U^{-1}[\sigma]K^{(\sigma)}(x)U[\sigma]\Phi[\sigma],$$
$$= -(1/c)\, j_\mu(x)A_\mu^{(\sigma)}(x)]\Phi[\sigma], \qquad (2.23)]$$

can be constructed in the form

$$\Phi[\sigma] = R[\sigma]\Phi 1, \qquad (2.27)$$

where

$$i\hbar c\, \delta\Phi[\sigma]/\delta\sigma(x) = H(x)R[\sigma], \qquad (2.28)$$

and

$$R[\sigma] \to 1, \qquad \sigma \to -\infty \qquad (2.29)$$

The *state vector* Φ_1 characterizes the *initial state* of the system, composed of one electron with definite energy and momentum, and no light quanta. The total probability, per unit time, that a scattering process occurs, can be obtained by evaluating the time rate of decrease of the probability that the system remains in the *initial state*,

$$\omega = -c \int d\upsilon\, \delta/\delta\sigma(x)\, |(\Phi 1,\Phi[\sigma])|^2$$
$$= -c \int d\upsilon\, \delta/\delta\sigma(x)\, |(\Phi 1,R[\sigma]\Phi 1)|^2. \qquad (2.30)$$

The integration is extended over the surface t = const., with $d\upsilon$ the three-dimensional volume element. …

…

This result is perfectly equivalent to the more conventional perturbation formula in which the rate of transition from the *initial state* is expressed as a sum of transition rates to all possible *final states* of equal *energy*. *The energy conservation law is here expressed by the time integration*, and the summation over all states other than the original is provided for by the removal from H(x) of the diagonal matrix element. *Our basic formula for calculating the transition rate for scattering of a particle by a time independent potential* is thus

$$\omega = \ldots . \tag{2.39}$$

We have not indicated that the diagonal matrix element is to be subtracted from $j_\mu(x)$, since it is sufficient to remove, in the final result, those transitions in which no change of state occurs.

…

We shall first consider the simple situation in which the emitted radiation exerts a negligible reaction on the electron. That is to say, we shall treat the essentially *elastic scattering of an electron*, in which only a small fraction of the electron kinetic energy is radiated. …

…

… It will be noted that the *infrared catastrophe*, as characterized by k_{min} has disappeared. However, it is possible, in principle, to consider the limit $\Delta E \rightarrow 0$, which would make δ *diverge logarithmically. It is well known that this difficulty stems from the neglect of processes involving more than one low frequency quantum*[5].

[5] Bloch, F. & Nordsieck, A. (July, 1937). Note on the Radiation Field of the Electron. *Phys. Rev.*, 52, 54; https://doi.org/10.1103/PhysRev.52.54.

Actually, the essentially *elastic scattering cross section* must approach zero as $\Delta E \rightarrow 0$; that is, *it never happens that a scattering event is unaccompanied by the emission of quanta.* This is described by replacing the radiative correction factor $1-\delta$ with $e^{-\delta}$ which has the proper limiting behavior as $\Delta E \rightarrow 0$. The further terms in the series expansion of $e^{-\delta}$ express the effects of higher order processes involving the multiple emission of soft quanta. However, for practical purposes, such a refinement is unnecessary. The accuracy with which the energy of a particle can be measured is such that the limit $\Delta E \rightarrow 0$ cannot be realized, and δ will be small in comparison with unity under presently accessible circumstances.

For a slowly moving particle,

$$\beta \ll 1: \delta(\vartheta, \Delta E \ll W) = 8\alpha/3\pi\, \beta^2 \sin^2 \vartheta/2[\log(mc^2/2\Delta E) + 19/30] \tag{2.103}$$

according to (2.95), the limiting form of H and the corresponding limiting form of F_n:

$$\lambda \ll 1; \quad F_n = 1/(2n+1). \tag{2.104}$$

The radiative correction thus increases linearly with the kinetic energy of the particle. In the extreme *relativistic* region, on the other hand,

$$(p_0/\kappa) \sin \vartheta/2 \gg 1:$$
$$\delta(\vartheta, \Delta E \ll W) = 4\alpha/\pi\, [\{\log(E/\Delta E) - 13/12\}\{\log(2p_0/K) \sin \vartheta/2 - 1/2\} + 17/72 + \tfrac{1}{2} \sin^2 \vartheta/2\, f(\vartheta)], \tag{2.105}$$

which has a logarithmic dependence on the particle energy. The asymptotic form (2.105) is quite accurate for even moderate energies. Thus, with $\vartheta = \pi/2$, $\Delta E = 10$ Kev and $W = 3.1$ Mev, which corresponds to $(p_0/\kappa) \sin \vartheta/2 = 5$, the value of δ computed from (2.105) differs from the correct value,

$$\delta = 8.6\ 10^{-2}, \qquad (2.106)$$

by only a fraction of a percent. It is evident from this numerical result that *radiative corrections to scattering cross sections can be quite appreciable.*

For the particular conditions chosen, ΔE can be materially increased (but still subject to $\Delta E \ll W$), without seriously impairing δ. Thus, with $\Delta E = 40$ Kev, $\delta = 6.3$ 10–2, while $\Delta E = 80$ Kev yields $\delta = 5.1\ 10^{-2}$. As to the energy dependence of δ, we remark that with a given accuracy in the determination of the energy, $\Delta E/E$, δ varies linearly with the logarithm of the energy. Thus, with $\Delta E/E = 0.04/3.6 = 1.1\ 10^{-2}$, an increase in the total energy by a factor of four produces an addition of $4.4\ 10^{-2}$ to δ, whence $\delta = 11\ 10^{-2}$ for W= 14 Mev, and $\delta = 15\ 10^{-2}$ for W=57 Mev.

The angular dependence of δ at *relativistic* energies is not fully described by the asymptotic formula (2.105)

$$[(p_0/\kappa) \sin \vartheta/2 \gg 1:$$

$$\delta(\vartheta, \Delta E \ll W) = 4\alpha/\pi\ [\{\log(E/\Delta E) - 13/12\}\{\log(2p_0/K) \sin \vartheta/2 - 1/2\}$$
$$+ 17/72 + \tfrac{1}{2} \sin2\ \vartheta/2\ f(\vartheta)], \qquad (2.105)]$$

since the underlying condition, $(p_0/\kappa) \sin \vartheta/2 \ll 1$, cannot be maintained with diminishing ϑ. Indeed, δ is proportional to $\sin^2 \vartheta/2$ at angles such that $(p_0/\kappa) \sin \vartheta/2 \ll 1$. However (2.105) can be used over a wide angular range, even at moderate energies. Thus, with W=3.1 Mev, $\Delta E = 40$ Kev, and $\vartheta = \pi/4$, which corresponds to $(p_0/\kappa) \sin \vartheta/2 = 2.7$, the value of δ deduced from (2.105) exceeds by only 2 percent the correct value, $\delta = 4.2\ 10^{-2}$. We may note that under the same energy conditions, but with $\vartheta = 3\pi/4$, $\delta = 7.2\ 10^{-2}$. *The angular dependence of δ may be particularly suitable for an experimental test of these predictions*, which involve the *relativistic* aspects of the radiative corrections to the electromagnetic properties of the electron.

> [No report of such a test has been found. If it had been successful, it would have been publicized.]

We have thus far considered only the essentially elastic scattering of an electron, in which radiative corrections arise primarily from virtual processes. If we wish to compute the *differential cross section for scattering* with an arbitrary maximum energy loss ΔE, it is only necessary to augment the essentially elastic cross section, in which the maximum energy loss is $\Delta E' \ll W$, by the *cross section for scattering* with the emission of a light quantum in the energy range from $\Delta E'$ to ΔE. The latter process involves the well-known *bremsstrahlung cross section* which, of course, is the content of (2.79)

$$[K = \Delta E/\hbar c. \qquad (2.79)]$$

This will be illustrated by the calculation of the *differential cross section* for the scattering of a slowly moving electron, irrespective of the final energy. The *differential cross section* per unit solid angle for scattering of an electron through the angle ϑ, in which a light quantum is emitted in the energy range from $\Delta E'$ to W, is

… (2.107)

according to the *non-relativistic* limit of (2.78). Here $d\omega$ is an element of solid angle associated with the direction of the unit vector $n = k/k_0$, and

$$|q| = (p^2 - 2\kappa k_0)^{1/2}. \quad (2.108)$$

On performing the integration over all emission directions of the light quantum, and introducing the new variable of integration, $x = |q|/|p|$, (2.107) becomes

… (2.109)

Thus, the contribution to δ produced by emission of quanta with energies in the range from $\Delta E'$ to W is

… . (2.110)

On adding this to $\delta(\vartheta,\Delta E')$, as given by (2.103)

[$\beta \ll 1$: $\delta(\vartheta,\Delta E \ll W) = 8\alpha/3\pi\ \beta^2 \sin^2 \vartheta/2[\log(mc^2/2\Delta E) + 19/30]$, (2.103)]

we obtain the desired result:

$$\beta \ll 1: \delta(\vartheta,W) = 8\alpha/3\pi\ \beta^2 \sin^2 \vartheta/2[\log(1/4\beta^2) + 19/30 + (\pi - \vartheta)\tan \vartheta/2 + \cos \vartheta/\cos^2 \vartheta/2 \log \csc \vartheta/2]. \quad (2.111)$$

It may be remarked, finally, that the analogous meso-nuclear phenomenon, the radiative correction to nucleon-nucleon scattering associated with virtual meson emission, will be a relatively more significant effect in view of the stronger couplings involved. This may well be the explanation of the discrepancy between the observed neutron-proton scattering cross section for high energy neutrons and the larger theoretical values computed from various assumed interaction potentials[6].

[6] A summary is given by Rosenfeld, L. (1949). *Nuclear Forces*. Interscience Publishers, Inc., New York, pp. 450, 454.

Appendix

In this section, we shall first give an alternative treatment of the *polarization of the vacuum by an external field*, employing the methods developed in the preceding pages. It is desired to compute the expectation value of $j_\mu(x)$,

$$\langle j_\mu(x)\rangle = \{\Psi[\sigma], j_\mu(x)\Psi[\sigma]\}, \quad (A.1)$$

[the expectation value of the *current operator* can be computed as

$\langle j_\mu(x)\rangle = \{\Psi[\sigma], j_\mu(x)\Psi[\sigma]\}$ (1.6)]

where $\Psi[\sigma]$ obeys

$$i\hbar c\ \delta\Psi[\sigma]/\delta\sigma(x) = -(1/c)\ j_\mu(x)A_\mu(x)\Psi[\sigma], \quad (A.2)$$

[$i\hbar c\ \delta\Phi[\sigma]/\delta\sigma(x) = -(1/c)\ j_\mu(x)A_\mu^{(\sigma)}(x)\Phi[\sigma]$ (2.23)]

and $A_\mu(x)$ is the *potential* or a prescribed *current distribution*. …

…

We may conclude that in the process of *vacuum polarization*, a non-vanishing, and indeed divergent charge is attached to the original charge distribution, and a compensating charge is created at infinity.

We shall finally apply the computational methods of this paper to evaluate the invariant expression for the *electromagnetic mass*,

…

…

$$\delta m/m = 3\alpha/2\pi \, \mathrm{Lim}_{K\to\infty}\{\log\,(K_0 + K)/\kappa - \tfrac{1}{4}\}, \qquad \text{(A.66)}$$

… .

Feynman, R. P. (November, 1950). Mathematical Formulation of the Quantum Theory of Electromagnetic Interaction.

[*Phys. Rev.*, 80, 3, 440; https://doi.org/10.1103/ PhysRev.80.440; also in Schwinger, J. (ed). (1958). *Selected Papers on Quantum electrodynamics*. Dover, New York, pages 257-74.]

Received June 8, 1950.

Department of Physics, Cornell University, Ithaca, New York*.

* Now at the California Institute of Technology, Pasadena California.

In two previous papers rules were given for calculation of matrix element for any process in electrodynamics, no complete proof of equivalence of these rules to conventional electrodynamics was given, nor was a closed expression given valid to all orders in $e^2/\hbar c$, this paper addresses these formal omissions, *gives the derivations of the formulas of Feynman (September 1949).* [*Space-Time Approach to Quantum Electrodynamics.*] *by means of the form of quantum mechanics given in Feynman (April 1948).*[*Space-Time Approach to Non-Relativistic Quantum Mechanics*], derivation of rules using Lagrangian form of quantum mechanics permits motion of any part to be solved first and the results to be used in solution of motion of other parts, *electromagnetic field* a simple system, *interaction of matter (electrons and positrons) and field analyzed by first solving for the behavior of the field in terms of the coordinates of the matter*, integration over field oscillator coordinates eliminates field variables from *equations of motion* of electrons, then address behavior of electrons, all of the rules given in second paper derived, *restricted to cases in which the particle's motion is non-relativistic* but the transition of the final formulas to the relativistic case is direct and the proof could have been kept relativistic throughout, *generalized formulation unsatisfactory because for situations of importance it gives divergent results, problems of divergences are not discussed, present theory assumes that substantial interaction exists as long as the interval is time-like and less than some small length of order of the electron radius.*

> [*Problems with infinities*, page 451: "... *the formulation is unsatisfactory* because for situations of importance it gives divergent results, even if $T_0[B]$ is finite. The modification proposed in II [Feynman (September, 1949). Space-Time Approach to Quantum Electrodynamics] of replacing $\delta_+(s_{12}^2)$ in (45), (48) by $f_+(s_{12}^2)$ is not satisfactory owing to the loss of the theorems of conservation of energy or probability discussed in II at the end of Sec. 6. ... *The choice we have made here ... is arbitrary and almost certainly incorrect*"."]

Abstract

The validity of the rules given in previous papers for the solution of problems in *quantum electrodynamics* is established. Starting with Fermi's formulation of the *field* as a set of harmonic oscillators, the effect of the oscillators is integrated out in the Lagrangian form of quantum mechanics. There results an expression for the effect of all virtual photons valid to all orders in $e^2/\hbar c$. It is shown that evaluation of this expression as a power series in $e^2/\hbar c$ gives just the terms expected by the aforementioned rules.

In addition, a relation is established between the *amplitude* for a given process in an arbitrary unquantized *potential* and in a quantum electrodynamical *field*. This relation permits a simple general statement of the laws of *quantum electrodynamics*.

A description, in Lagrangian quantum-mechanical form, of particles satisfying the Klein-Gordon equation is given in an Appendix. It involves the use of an extra parameter analogous to proper time to describe the trajectory of the particle in four dimensions.

A second Appendix discusses, in the special case of photons, the problem of finding what real processes are implied by the formula for virtual processes.

Problems of the divergences of electrodynamics are not discussed.

1. *Introduction*

In two previous papers[1]

[1] Feynman, R. P. (September, 1949). The Theory of Positrons. *Phys. Rev.*, 76, 749-59 [; this is the first of a set of papers dealing with the solution of problems in *quantum electrodynamics*, main principle is to deal directly with *solutions* of the Hamiltonian differential equations rather than with equations themselves, this paper *addresses motion of electrons and positrons in given external potentials*, second paper considers *interactions* (*quantum electrodynamics*), problem of charges in a fixed potential usually treated by method of second quantization of the electron field using the *theory of holes*, here we replace Dirac's "hole theory" by reinterpreting *solutions* of the Dirac equation, results simplified by *following the charge rather than the particles* because *the number of particles is not conserved* whereas *charge is conserved*, in the approximation of classical *relativistic* theory the creation of an electron pair (electron and positron*)* represented by the start of two world lines from the point of creation, *following the charge rather than the particles corresponds to considering continuous world line as a whole rather than breaking it up into its pieces*, over-all *space-time* point of view, quantum mechanically the direction of the world lines is replaced by the direction of propagation of waves, quite different from Hamiltonian method which considers the future as developing continuously from the past, in a scattering problem this over-all view of the complete scattering process is similar to the *S-matrix view-point of Heisenberg*, temporal order of events during scattering which is analyzed in such detail by the Hamiltonian differential equation is irrelevant, development stemmed from idea that in *non-relativistic* quantum mechanics the *amplitude* for a given process can be considered as the sum of an *amplitude* for each *space-time* path available, in *relativistic* case the restriction that the paths must proceed always in one direction in time is removed, results more easily understood from more familiar physical viewpoint of *scattered waves* used in this paper, after the equations were worked out physically the proof of the equivalence to the *second quantization theory* was found, solution in terms of boundary conditions on *wave function* contains all possibilities of pair formation and annihilation together with the ordinary scattering processes, negative energy states appear in space-time as waves traveling away from the external potential backwards in time (as suggested in Stückelberg (December, 1942)), such a wave corresponds to a positron

approaching the potential and annihilating the electron, a particle moving forward in time (electron) in a potential may be scattered forward in time (ordinary scattering) or backward (pair annihilation), when moving backward (positron) it may be scattered backward in time (positron scattering) or forward (pair production), *amplitude* for transition from an initial to a final state analyzed to any order in the *potential* by considering it to undergo a sequence of such scatterings, *amplitude* for a process involving many such particles is the product of the *transition amplitudes* for each particle, vacuum problems do not arise for charges which do not interact with one another, shows equivalent to the *theory of holes* in second quantization], hereafter called I,

and Feynman, R. P. (September, 1949). Space-Time Approach to Quantum Electrodynamics. *Phys. Rev.*, 76, 6, 769-89 [; in previous paper motion of electrons *neglecting interaction* was analyzed by dealing directly with the *solutions* of the Hamiltonian time differential equations rather than the equations, here the same technique is applied to include *interactions* to express in simple terms the solution of problems in *quantum electrodynamics* rather than the differential equations from which they come, (1) it is shown that *a considerable simplification can be attained by writing down matrix elements for complex processes in electrodynamics*, a physical point of view is available which permitted them to be written down directly. The simplification results from the fact that previous methods separated into individual terms processes that were closely related physically, (2) *electrodynamics is modified by altering the interaction of electrons at short distances*, all matrix elements now finite with the exception of those relating to *vacuum polarization*, latter evaluated in manner suggested by Pauli and Bethe to give finite results, phenomena directly observable insensitive to the details of the modification, electrodynamics viewed as direct *interaction* at a distance between *charges* rather than behavior of a *field* (Maxwell's equations), *field point of view*, which separates the production and absorption of light, most practical for problems involving *real quanta* while *interaction view* best for discussion of *virtual quanta* when dealing with close collisions of particles or their actions on themselves, Hamiltonian method not well adapted to represent direct *action at a distance* between *charges* because action is delayed, forces use of field viewpoint rather than interaction viewpoint, for *collisions* much easier to treat the process as a whole, effects of longitudinal and transverse waves can be combined, begins with solution in *space* and *time* of Schrodinger equation for particles interacting instantaneously, solution is expressed in terms of *matrix elements*, *derived from Lagrangian form of quantum mechanics*, then modifies in accordance with the requirements of the Dirac equation and phenomenon of pair creation, made easier by reinterpreting the *theory of holes*, generalizes to *delayed interactions* of *relativistic* electrons, for practical calculations expressions developed in a power series of $e^2/\hbar c$, *derivation will appear in a separate paper*, by forsaking Hamiltonian method, the wedding of *relativity* and *quantum mechanics* accomplished most naturally, *relativistic* invariance becomes self-evident but the *matrix elements for complex processes* and the *self-energy diverges* then can see how the matrix elements can be written down directly, then introduces modification in interaction between charges at short distances, *assumes substantial interaction exists as long as the four-dimensional interval is time-like and less than some small length of order of the electron radius* [Feynman (October, 1948). A Relativistic Cut-Off for Classical Electrodynamics.], *convergence factors* then introduced such that the integrals with their convergence factors now converge, *self-energy* now convergent, corresponds to a correction to the electron mass, *all matrix elements now finite* with exception of those relating to *vacuum polarization*, *strict physical basis for rules of convergence not known*, after *mass and charge renormalization* results are equivalent to those of Schwinger in which the terms corresponding to corrections in mass and charge are identified and removed from the expressions for real processes, *although in the limit the two methods agree neither method appears to be completely satisfactory theoretically*, practical advantage of new method is that ambiguities can be more easily resolved by direct calculation of otherwise divergent integrals, complete method therefore available for calculations of all processes involving electrons and photons, shows how matrix element for any process can be written down directly in *Feynman diagrams*, paper includes first published "*Feynman diagram*", *attempt to find a consistent modification of quantum electrodynamics is incomplete, not at all clear that the convergence factors do not upset the physical consistency of the theory*], hereafter called II.

rules were given for the calculation of the matrix element for any process in electrodynamics, to each order in $e^2/\hbar c$. No complete proof of the equivalence of these rules to the conventional electrodynamics was given in these papers. Secondly, no closed expression was given valid to all orders in $e^2/\hbar c$. In this paper these formal omissions will be remedied[2].

[2] See in this connection also the papers of S. Tomonaga; Tomonaga, S. & Oppenheimer, J. R.] (July, 1948). On Infinite Field Reactions in Quantum Field Theory. *Phys. Rev.*, 74, 224–225; https://doi.org/10.1103/PhysRev.74.224; Kanesawa, S. & Tomonaga, S. (1948). On a Relativistically Invariant Formulation of the Quantum Theory of Wave Fields. V. Case of Interacting Electromagnetic and Meson Fields. § 5-7. *Prog. Theor. Phys.* 3, 101–113; Schwinger, J. (September, 1949). Quantum Electrodynamics. III. The Electromagnetic Properties of the Electron—Radiative Corrections to Scattering. *Phys. Rev.*, 76, 6, 790–817; Dyson, F. J. (June, 1949). The S Matrix in Quantum Electrodynamics. *Phys. Rev.*, 75, 1736-55; Pauli, W. & Villars, F. (July, 1949). On the Invariant Regularization in Relativistic Quantum Theory. *Rev. Mod. Phys.*, 21, 434; https://doi.org/10.1103/ RevModPhys.21.434, also in Schwinger, J. (ed). (1958). *Selected Papers on Quantum electrodynamics*. Dover, New York, pages 198-208. The papers cited give references to previous work [The references for Tomonaga, Schwinger, and Dyson are the last papers in a series.]

In paper II [Feynman (September 1949). Space-Time Approach to Quantum Electrodynamics] it was pointed out that for many problems in electrodynamics the Hamiltonian method is not advantageous, and might be replaced by the *over-all space-time point of view* of a direct particle interaction. It was also mentioned that the Lagrangian form of quantum mechanics[3] was useful in this connection.

[3] Feynman, R. P. (April, 1948). Space-Time Approach to Non-Relativistic Quantum Mechanics. *Rev. Mod. Phys.*, 20, 2, 367-87[; third formulation of *non-relativistic* quantum theory in addition to *differential equation of Schroedinger* and *matrix algebra of Heisenberg*, *path integral formulation* utilizing the *action* principle as suggested in Dirac (1933) [The Lagrangian in Quantum Mechanics] and Dirac (1945) [On the Analogy Between Classical and Quantum Mechanics], in quantum mechanics the probability of an event which can happen in several different ways is the absolute square of a sum of complex contributions one from each alternative way $\varphi_{ac} = \sum_b \varphi_{ab}\varphi_{bc}$ where φ_{ab}, φ_{bc}, φ_{ac} are

> complex numbers such that $P_{ab} = |\varphi_{ab}|^2$, $P_{bc} = |\varphi_{bc}|^2$, and $P^q_{ac} = |\varphi_{ac}|^2$, and P_{ab} as the probability that if measurement A gave the result *a*, then measurement B will give the result *b*, and P^q_{ac} is the quantum mechanical probability that a measurement of C results in c when it follows a measurement of A giving a, the probability that a particle will be found to have a path lying somewhere within a region of space time is the absolute square of a sum of contributions - one from each path in the region, the contribution from a single path is postulated to be an exponential whose (imaginary) phase is the classical *action* for the path in question where *action* refers to time integral of Lagrangian along a path, restriction to finite time interval, the total contribution from all paths reaching x, t from the past is the *wave function* $\psi(x, t)$, shown to satisfy Schroedinger's equation, *probability amplitude for a space-time path associated with entire motion of particle as function of time rather than with position of particle at particular time*, establishes postulates that describe *non-relativistic quantum mechanics neglecting spin*, mathematically equivalent to Heisenberg and Schroedinger formulations, no fundamentally new results, *suffers serious drawbacks*, requires unnatural and cumbersome division of the time interval, not formulated so that it is physically obvious that it is invariant under unitary transformations, improvements could be made through use of notation and concepts of mathematics of functionals], *hereafter called C*.

The rules given in paper II were, in fact, first deduced in this form of quantum mechanics. We shall give this derivation here.

The advantage of a Lagrangian form of quantum mechanics is that in a system with interacting parts it permits a separation of the problem such that the motion of any part can be analyzed or solved first, and the results of this solution may then be used in the solution of the motion of the other parts. This separation is especially useful in *quantum electrodynamics* which represents the interaction of matter with the electromagnetic field. The electromagnetic field is an especially simple system and its behavior can be analyzed completely. *What we shall show is that the net effect of the field is a delayed interaction of the particles*. It is possible to do this easily only if it is not necessary at the same time to analyze completely the motion of the particles. *The only advantage in our problems of the form of quantum mechanics in C* [Feynman (April, 1948). Space-Time Approach to Non-Relativistic Quantum Mechanics] *is to permit one to separate these aspects of the problem*. There are a number of disadvantages, however, such as a lack of familiarity, the apparent (but not real) necessity for dealing with matter in *non-relativistic* approximation, and at times a cumbersome mathematical notation and method, as well as the fact that a great deal of useful information that is known about operators cannot be directly applied.

It is also possible to separate the field and particle aspects of a problem in a manner which uses operators and Hamiltonians in a way that is much more familiar. One abandons the notation that the order of action of operators depends on their written position on the paper and substitutes some other convention (such that the order of operators is that of the time to which they refer). The increase in manipulative facility which accompanies this change in notation makes it easier to represent and to analyze the formal problems in electrodynamics. *The method requires some discussion, however, and will be described in a succeeding paper*. [Feynman (October 1951). An Operator Calculus Having

Applications in Quantum Electrodynamics.]

In this paper we shall give the derivations of the formulas of II [Feynman (September 1949). Space-Time Approach to Quantum Electrodynamics] by means of the form of quantum mechanics given in C [Feynman (April 1948). Space-Time Approach to Non-Relativistic Quantum Mechanics].

The problem of interaction of matter and field will be analyzed by first solving for the behavior of the *field* in terms of the coordinates of the *matter*, and finally discussing the behavior of the *matter* (by *matter* is actually meant the electrons and positrons). That is to say, we shall first eliminate the *field* variables from the *equations of motion* of the electrons and then discuss the behavior of the electrons. In this way all of the rules given in the paper II will be derived.

Actually, the straightforward elimination of the *field* variables will lead at first to an expression for the behavior of an arbitrary number of Dirac electrons. Since the number of electrons might be infinite, this can be used directly to find the behavior of the electrons according to *hole theory* by imagining that nearly all the negative energy states are occupied by electrons. But, at least in the case of motion in a fixed *potential*, it has been shown that *this hole theory picture is equivalent to one in which a positron is represented as an electron whose space-time trajectory has had its time direction reversed.* To show that this same picture may be used in *quantum electrodynamics* when the *potentials* are not fixed, a special argument is made based on a study of the relationship of *quantum electrodynamics* to motion in a fixed *potential*. Finally, it is pointed out that this relationship is quite general and might be used for a general statement of the laws of *quantum electrodynamics*.

Charges obeying the Klein-Gordon equation can be analyzed by a special formalism given in Appendix A. A fifth parameter is used to specify the four-dimensional trajectory so that the Lagrangian form of quantum mechanics can be used. Appendix B discusses in more detail the relation of real and virtual photon emission. An equation for the propagation of a self-interacting electron is given in Appendix C.

In the demonstration which follows *we shall restrict ourselves temporarily to cases in which the particle's motion is non-relativistic*, but the transition of the final formulas to the relativistic case is direct, and the proof could have been kept *relativistic* throughout.

The transverse part of the electromagnetic field will be represented as an assemblage of independent harmonic oscillators each interacting with the particles, as suggested by Fermi.[4] We use the notation of Heitler[5].

[4] Fermi, E. (January, 1932). Quantum Theory of Radiation. *Rev. Mod. Phys.*, 4, 87–132; https://doi.org/10.1103/RevModPhys.4.87.

[5] Heitler, W. (1944). *The Quantum Theory of Radiation*, second edition, Oxford University Press, London.

2. *Quantum electrodynamics in Lagrangian form*

The Hamiltonian for a set of *non-relativistic* particles interacting with radiation is, classically $H = H_p + H_t + H_c + H_{tr}$, where $H_p + H_t = \sum_n \frac{1}{2} m_n^{-1} (\mathbf{p}_n - e_n\mathbf{A}^{tr}(\mathbf{x}_n))^2$ is the Hamiltonian of the particles of mass m_n, charge e_n, coordinate $\mathbf{x}_n$ and momentum $\mathbf{p}_n$ and their interaction with the *transverse part* of the electromagnetic field. This field can be expanded into plane waves

$$\mathbf{A}^{tr}(x) = (8\pi)^{1/2} \sum_{\mathbf{K}}[\mathbf{e}_1(q_{\mathbf{K}}^{(1)}\cos(\mathbf{K} \cdot \mathbf{x}) + q_{\mathbf{K}}^{(3)}\sin(\mathbf{K} \cdot \mathbf{x})) + \mathbf{e}_2(q_{\mathbf{K}}^{(2)}\cos(\mathbf{K} . \mathbf{x}) + q_{\mathbf{K}}^{(4)}\sin(\mathbf{K} . \mathbf{x}))] \quad (1)$$

where $\mathbf{e}_1$ and $\mathbf{e}_2$ are two orthogonal polarization vectors at right angles to the propagation vector $\mathbf{K}$, magnitude k. ... The Hamiltonian of the *transverse* field represented as oscillators is

$$H_{tr} = \frac{1}{2} \sum_{\mathbf{K}} \sum_{r=1}^{4} [(p_{\mathbf{K}}^{(r)})^2 + k^2(q_{\mathbf{K}}^{(r)})^2]$$

where $p_{\mathbf{K}}^{(r)}$ is the momentum conjugate to $q_{\mathbf{K}}^{(r)}$. The *longitudinal* part of the field has been replaced by the Coulomb interaction[6],

$$H_c = \frac{1}{2} \sum_n \sum_m e_n e_m / r_{nm}$$

where $r_{nm}^2 = (x_n - x_m)^2$.

[6] The term in the sum for n = m is obviously infinite but must be included for *relativistic* invariance. Our problem here is to re-express the usual (and divergent) form of electrodynamics in the form given in II [Feynman (September, 1949). Space-Time Approach to Quantum Electrodynamics]. Modifications for dealing with the divergences are discussed in II and we shall not discuss them further here.

As is well known[4], when this Hamiltonian is quantized, one arrives at the usual theory of quantum electrodynamics. To express these laws of quantum electrodynamics one can equally well use the Lagrangian form of quantum mechanics to describe this set of oscillators and particles. *The classical Lagrangian equivalent to this Hamiltonian is* $L = L_p + L_l + L_c + L_{tr}$ where

$$L_p = \frac{1}{2} \sum_n m_n \mathbf{x}'^2_n \quad (2a)$$
$$L_l = \sum_n e_n x'_n . \mathbf{A}^{tr}(\mathbf{x}_n) \quad (2b)$$
$$L_{tr} = \frac{1}{2} \sum_{\mathbf{K}} \sum_r [(q'_{\mathbf{K}}^{(r)})^2 - k^2(q_{\mathbf{K}}^{(r)})^2] \quad (2c)$$
$$L_c = -\frac{1}{2} \sum_n \sum_m e_n e_m / r_{mn}. \quad (2d)$$

When this Lagrangian is used in the Lagrangian forms of quantum mechanics of C [Feynman (April 1948). Space-Time Approach to Non-Relativistic Quantum Mechanics], what it leads to is, of course, mathematically equivalent to the result of using the Hamiltonian H in the ordinary way, and is therefore equivalent to the more usual forms of *quantum electrodynamics* (at least for *non-relativistic* particles). We may, therefore, proceed by using this Lagrangian form of *quantum electrodynamics*, with the assurance that the results obtained must agree with those obtained from the more usual Hamiltonian form.

…

In the Lagrangian form it is possible to eliminate the *transverse* oscillators as is discussed in C, Section 13. One must specify, however, the initial and final *state* of all oscillators. We shall first choose the special, simple case that all oscillators are in their *ground states* initially and finally, so that all photons are *virtual*. Later we do the more general case in which real quanta are present initially or finally. We ask, then, for the amplitude for finding no quanta present and the particles in *state* $\chi_{t''}$ at time t'', if at time t' the particles were in *state* $\psi_{t'}$ and no quanta were present.

The method of eliminating field oscillators is described in Section 13 of C. We shall simply carry out the elimination here using the notation and equations of C. To do this, for simplicity, we first consider in the next section the case of a particle or a system of particles interacting with a single oscillator, rather than the entire assemblage of the electromagnetic field.

3. *Forced harmonic oscillator*

We consider a harmonic oscillator, coordinate q, Lagrangian $L = \frac{1}{2}(\dot{q}^2 - \omega^2 q^2)$ interacting with a particle or system of particles, *action* S_p, through a term in the Lagrangian $q(t)\gamma(t)$ where $\gamma(t)$ is a function of the coordinates (symbolized as x) of the particle. …

…

… Thus, the complex *action* is additive, being the sum of the contributions like (15) for each of the several oscillators.

4. *Virtual transitions in the electromagnetic field*

We can now apply these results to eliminate the *transverse field oscillators* of the Lagrangian (2). At first, we can limit ourselves to the case of purely *virtual* transitions in the electromagnetic field, so that there is no photon in the field at t' and t''. That is, all of the *field oscillators* are making transitions from ground state to ground state.

…

The total complex *action* of the system is then[10] $S_p + R$.

[10] The classical action for this problem is just $S_p + R'$ where R' is the real part of the expression (24). In view of the generalization of the Lagrangian formulation of quantum mechanics suggested in Section 12 of C [Feynman (April, 1948). Space-Time Approach to Non-Relativistic Quantum Mechanics], one might have anticipated that R would have been simply R'. This corresponds, however, to boundary conditions other than no quanta present in past and future. It is harder to interpret physically. For a system enclosed in a light tight box, however, it appears likely that both R and R' lead to the same results.

Or, what amounts to the same thing; to obtain *transition amplitudes* including the effects of the *field* we must calculate the *transition element* of exp(iR):

$$\langle \chi_{t''} \mid \exp(iR) \mid \psi_{t'} \rangle_{S_p} \tag{25}$$

under the action S_p of the particles, excluding interaction. Expression (24) for R must be considered to be written in the usual manner as a Riemann sum and the expression (25) interpreted as defined in C [Feynman (April, 1948). Space-Time Approach to Non-Relativistic Quantum Mechanics.], Eq. (39). Expression (6) must be used for $\mathbf{x}'_n$ at time t.

Expression (25), with (24), then contains all the effects of *virtual* quanta on a (at least *non-relativistic*) system according to quantum electrodynamics. It contains the effects to all orders in $e^2/\hbar c$ in a single expression. *If expanded in a power series in $e^2/\hbar c$, the various terms give the expressions to the corresponding order obtained by the diagrams and methods of II.* We illustrate this by an example in the next section.

5. *Example of application of expression (25)*

We shall not be much concerned with the *non-relativistic* case here, as the relativistic case given below is as simple and more interesting. It is, however, very similar and at this stage it is worth giving an example to show how expressions resulting from (25) are to be interpreted according to the rules of C. For example, consider the case of a single electron, coordinate x, either free or in an external given *potential* (contained for simplicity in S_p, not in R)[11].

[11] One can show from (25) how the correlated effect of many atoms at a distance produces on a given system the effects of an external potential. Formula (24) yields the result that this potential is that obtained from Lienard and Wiechert by retarded waves arising from the charges and currents resulting from the distant atoms making transitions. Assume the wave functions χ and ψ if, can be split into products of wave functions for system and distant atoms and expand exp(iR) assuming the effect of any individual distant atom is small. Coulomb potentials arise even from nearby particles if they are moving slowly.

Its interaction with the *field* produces a reaction back on itself given by R as in (24) but in which we keep only a single term corresponding to m = n. Assume the effect of R to be small and expand exp(iR) as 1 + iR. ...

...

... We shall not be interested in *non-relativistic* quantum electrodynamics in detail. The situation is simpler for Dirac electrons. ...

6. *Extension to Dirac particles*

Expressions (24) and (25) and their proof can be readily generalized to the *relativistic* case according to the *one electron theory* of Dirac. We shall discuss the *hole theory* later. In the *non-relativistic* case, we began with the proposition that the *amplitude* for a particle to proceed from one point to another is the sum over paths of exp(iSp), that is, we have for example for a *transition element*

$$<\chi | 1 | \psi> = \ldots \quad (34)$$

...

7. *Extension to positron theory*

Since in (39) we have an arbitrary number of electrons, we can deal with the *hole theory* in the usual manner by imagining that we have an infinite number of electrons in negative energy states.

On the other hand, in paper I [Feynman (September 1949). The Theory of Positrons.] on the theory of positrons, it was shown that the results of the *hole theory* in a system with a given *external potential* A_μ were equivalent to those of the Dirac one electron theory if one replaced the propagation kernel, K_0, by a different one, K_+, and multiplied the resultant *amplitude* by factor C, involving A_μ. We must now see how this relation, derived in the case of external *potentials*, can also be carried over in electrodynamics to be useful in simplifying expressions involving the infinite sea of electrons.

To do this we study in greater detail the relation between a problem involving *virtual* photons and one involving purely external *potentials*. ...

...

8. *Generalized formulation of quantum electrodynamics*

The relation implied by (45) between the formal solution for the *amplitude* for a process in an arbitrary *unquantized external potential* to that in a *quantized field* appears to be of much wider generality. We shall discuss the relation from a more general point of view here (still limiting ourselves to the case of no photons in initial or final *state*).

In earlier sections we pointed out that as a consequence of the Lagrangian form of quantum mechanics the aspects of the particles' motions and the behavior of the field could be analyzed separately. *What we did was to integrate over the field oscillator coordinates first.* We could, in principle, have integrated over the particle variables first. That is, we first solve the problem with the *action* of the particles and their interaction with the *field* and then multiply by the exponential of the *action* of the field and integrate over all the *field oscillator* coordinates. (For simplicity of discussion let us put aside from detailed special consideration the questions involving the separation of the *longitudinal* and *transverse parts* of the *field.*) Now the integral over the particle coordinates for a given process is precisely the integral required for the analysis of the motion of the particles in an unquantized *potential*. With this observation we may suggest a generalization to all types of systems.

Let us suppose the formal solution for the *amplitude* for some given process with *matter* in an external *potential* $B_\mu(1)$ is some numerical quantity T_0. We mean *matter* in a more general sense now, for the motion of the *matter* may be described by the Dirac equation, or by the Klein-Gordon equation, or may involve charged or neutral particles other than electrons and positrons in any manner whatsoever. The quantity T_0 depends of course on the *potential* function $B_\mu(1)$; that is, it is a functional $T_0[B]$ of this *potential*. ...

...

Equation (45) or its solution (46), (47), (48) constitutes a very general and convenient formulation of the laws of *quantum electrodynamics* for *virtual processes*. Its *relativistic* invariance is evident if it is assumed that the unquantized theory giving $T_0[B]$ is invariant. It has been proved to be equivalent to the usual formulation for Dirac electrons and positrons (for Klein-Gordon particles see Appendix A). It is suggested that it is of wide generality. It is expressed in a form which has meaning even if it is impossible to express the matter system in Hamiltonian form; in fact, it only requires the existence of an *amplitude* for fixed *potentials* which obeys the *principle of superposition of amplitudes. …* The limitation to virtual quanta is removed in the next section.

On the other hand, *the formulation is unsatisfactory because for situations of importance it gives divergent results*, even if $T_0[B]$ is finite. The modification proposed in II

> [Feynman (September 1949). Space-Time Approach to Quantum Electrodynamics, p. 776: "*We desire to make a modification of quantum electrodynamics analogous to the modification of classical electrodynamics described in a previous article*, A.
>
> > [Feynman (October, 1948). A Relativistic Cut-Off for Classical Electrodynamics, p. 941: "This requires that a particle be allowed to act on itself and the term a = b to be included in the *action* sum. *This leads immediately to an infinite self-force*. This difficulty can be eliminated if the $\delta(s_{ab}^2)$ is replaced, as Bopp[4] has suggested, by some other function $f(s_{ab}^2)$ of the invariant s_{ab}^2, which behaves like $\delta(s_{ab}^2)$ for large dimensions but differs for small".]
>
> There the $\delta(s_{12}^2)$ appearing in the action of interaction was replaced by $f(s_{12}^2)$ where f(x) is a function of small width and great height.
>
> The obvious corresponding modification in the quantum theory is to replace the $\delta(s^2)$ appearing [in] the quantum mechanical *interaction* by a new function $f_+(s^2)$. *We can postulate that if the Fourier transform of the classical $f(s_{12}^2)$ is the integral over all k of $F(k^2)$ exp(–ik.x) d^4k, then the Fourier transform of $f_+(s^2)$ is the same integral taken over only positive frequencies k_4 for $t_2 > t_1$ and over only negative ones for $t_2 < t_1$ in analogy to the relation of $\delta_+(s^2)$ to $\delta(s^2)$.*]
>
> [pp. 769-70: "*Unfortunately, the modification proposed is not completely satisfactory theoretically … a strict physical basis for the rules of convergence is not known.*"]

of replacing $\delta_+(s_{12}^2)$ in (45), (48) by $f_+(s_{12}^2)$ is not satisfactory owing to the loss of the theorems of conservation of energy or probability discussed in II at the end of Sec. 6.

> [Feynman (September 1949). Space-Time Approach to Quantum Electrodynamics, p. 778: "The replacement of δ_+ by f_+ given in (16), (17)
>
> $$[-f_+(s_{12}^2) = \pi^{-1} \int \exp(-ik.x_{12})\, \mathbf{k}^{-2} C(\mathbf{k}^2)\, d^4k. \quad (16)$$
> $$C(\mathbf{k}^2) = \int_0^\infty -\lambda^2(\mathbf{k}^2 - \lambda^2)^{-1} G(\lambda)\, d\lambda. \quad (17)]$$

is not determined by the analogy with the classical problem. In the classical limit only the real part of δ_+ (i.e., just δ) is easy to interpret. But by what should the imaginary part, $1/(\pi i s^2)$, of δ_+ be replaced? *The choice we have made here* (in defining, as we have, the location of the poles of (17)) *is arbitrary and almost certainly incorrect*".]

There is the additional difficulty in positron theory that even $T_0[B]$ is infinite to begin with (vacuum polarization). Computational ways of avoiding these troubles are given in II and in the references of footnote 2.

9. *Case of real photons*

The case in which there are real photons in the initial or the final state can be worked out from the beginning in the same manner[17].

[17] For an alternative method starting directly from the formula (24) for virtual photons, see Appendix B.

We first consider the case of a system interacting with a single oscillator. From this result the generalization will be evident. This time we shall calculate the *transition element* between an *initial state* in which the particle is in *state* $\psi_{t'}$ and the oscillator is in its nth eigenstate (i.e., there are n photons in the field) to a final state with particle in $\chi_{t''}$, oscillator in mth level. …

…

APPENDIX A. THE KLEIN-GORDON EQUATION

In this Appendix we describe a formulation of the equations for a particle of spin zero which was first used to obtain the rules given in II [Feynman (September, 1949). Space-Time Approach to Quantum Electrodynamics] for such particles. The complete physical significance of the equations has not been analyzed thoroughly so that it may be preferable to derive the rules directly from the second quantization formulation of Pauli and Weisskopf. This can be done in a manner analogous to the derivation of the rules for the Dirac equation given in I [Feynman (September 1949). The Theory of Positrons.] or from the Schwinger-Tomonaga formulation[2] in a manner described, for example, by Rohrlich[18].

[18] Rohrlich, F. (to be published).

The formulation given here is therefore not necessary for a description of spin zero particles but is given only for its own interest as an alternative to the formulation of second quantization.

…

APPENDIX B. THE RELATION OF REAL AND VIRTUAL PROCESSES

If one has a general formula for all virtual processes, he should be able to find the formulas and states involved in real processes. That is to say, we should be able to deduce the formulas of Section 9 directly from the formulation (24), (25) (or its generalized equivalent such as (46), (48)) without having to go all the way back to the more usual formulation. We discuss this problem here.

That this possibility exists can be seen from the consideration that what looks like a real process from one point of view may appear as a virtual process occurring over a more extended time. For example, if we wish to study a given real process, such as the scattering of light, we can, if we wish, include in principle the source, scatterer, and eventual absorber of the scattered light in our analysis. We may imagine that no photon is present initially, and that the source then emits light (the energy coming say from kinetic energy in the source). The light is then scattered and eventually absorbed (becoming kinetic energy in the absorber). From this point of view the process is virtual; that is, we start with no photons and end with none. Thus, we can analyze the process by means of our formula for virtual processes, and obtain the formulas for real processes by attempting to break the analysis into parts corresponding to emission, scattering, and absorption[23].

> [23] The formulas for real processes deduced in this way are strictly limited to the case in which the light comes from sources which are originally dark, and that eventually all light emitted is absorbed again. We can only extend it to the case for which these restrictions do not hold by hypothesis, namely, that the details of the scattering process are independent of these characteristics of the light source and of the eventual disposition of the scattered light. The argument of the text gives a method for discovering formulas for real processes when no more than the formula for virtual processes is at hand. But with this method belief in the general validity of the resulting formulas must rest on the physical reasonableness of the above-mentioned hypothesis.

…

We have therefore found rules for real photons in terms of those for virtual. The real photons are a way of representing and keeping track of those aspects of the past behavior which may influence the future.

If one starts from a theory involving an arbitrary modification of the direct interaction δ_+ (or in more general situations) it is possible in this way to discover what kinds of states and physical entities will be involved if one tries to represent in the present all the information needed to predict the future. With the Hamiltonian method, which begins by assuming such a representation, it is difficult to suggest modifications of a general kind, for one cannot formulate the problem without having a complete representation of the characteristics of the intermediate states, the particles involved in interaction, etc. It is quite possible (in the author's opinion, it is very likely) that we may discover that in nature the relation of past and future is so intimate for short durations that no simple representation of a present may exist. In such a case a theory could not find expression in Hamiltonian form.

An exactly similar analysis can be made just as easily starting with the general forms (46), (48). Also, a *coordinate* representation of the photons could have been used instead of the familiar *momentum* one. One can deduce the rules (60), (61). Nothing essentially different is involved physically, however, so we shall not pursue the subject further here. Since they imply all the rules for real photons, Eqs. (46), (47), (48) constitute a compact statement of all the laws of quantum electrodynamics. *But they give divergent results.* Can the result after *charge and mass renormalization* also be expressed to all orders in $e^2/\hbar c$ in a simple way?

APPENDIX C. DIFFERENTIAL EQUATION FOR ELECTRON PROPAGATION

An attempt has been made to find a *differential wave equation* for the propagation of an electron interacting with itself, analogous to the Dirac equation, but containing terms representing the *self-action*. Neglecting all effects of closed loops, one such equation has been found, but not much has been done with it. It is reported here for whatever value it may have.

An electron acting upon itself is, from one point of view, a complex system of a particle and a field of an indefinite number of photons. To find a differential law of propagation of such a system we must ask first what quantities known at one instant will permit the calculation of these same quantities an instant later. Clearly, a knowledge of the position of the particle is not enough. We should need to specify: (1) the amplitude that the electron is at x and there are no photons in the field, (2) the amplitude the electron is at x and there is one photon of such and such a kind in the field, (3) the amplitude there are two photons, etc. That is, a series of functions of ever-increasing numbers of variables. Following this view, we shall be led to the *wave equation* of the theory of second quantization. …

The author appreciates his opportunities to discuss these matters with Professor H. A. Bethe and Professer J. Ashkin, and the help of Mr. M. Baranger with the manuscript.

Feynman, R. P. (October, 1951). An Operator Calculus Having Applications in Quantum Electrodynamics.

[*Phys. Rev.*, 84, 1, 108; https://doi.org/10.1103/PhysRev.84.108.]

Received May 23, 1951.

California Institute of Technology, Pasadena, California*

* Absent on leave at the University of Brazil, Rio de Janeiro, Brazil.

Suggests alteration in mathematical notation for handling operators, new notation permits considerable increase in ease of manipulation of complicated expressions involving operators, *no results which are new are obtained in this way, the mathematics is not completely satisfactory, no attempt has been made to maintain mathematical rigor, it is believed that to put the present methods on a rigorous basis may be quite a difficult task, beyond the abilities of the author.*

Abstract

An alteration in the notation used to indicate the order of operation of non-commuting quantities is suggested. Instead of the order being defined by the position on the paper, an ordering subscript is introduced so that $A_sB_{s'}$ means AB or BA depending on whether s exceeds s′ or *vice versa.* Then A_s can be handled as though it were an ordinary numerical function of s. An increase in ease of manipulating some operator expressions results. Connection to the theory of functionals is discussed in an appendix. Illustrative applications to quantum mechanics are made. In *quantum electrodynamics* it permits a simple formal understanding of the interrelation of the various present day theoretical formulations.

The operator expression of the Dirac equation is related to the author's previous description of positrons. An attempt is made to interpret the operator ordering parameter in this case as a fifth coordinate variable in an extended Dirac equation. Fock's parametrization, discussed in an appendix, seems to be easier to interpret.

In the last section a summary of the numerical constants appearing in formulas for transition probabilities is given.

In this paper we suggest an alteration in the mathematical notation for handling operators. This new notation permits a considerable increase in the ease of manipulation of complicated expressions involving operators. *No results which are new are obtained in this way*, but it does permit one to relate various formulas of operator algebra in quantum mechanics in a simpler manner than is often available. In particular, it is applied to *quantum electrodynamics* to permit an easier way of seeing the relationships among the conventional formulations, that of Schwinger and Tomanaga[1], and that of the author[2].

[1] See Schwinger, J. (September, 1949). Quantum Electrodynamics. III. The Electromagnetic Properties of the Electron—Radiative Corrections to Scattering. *Phys.*

Rev., 76, 6, 790–817, and Tomonaga, S. & Oppenheimer, J. R. (July, 1948). On Infinite Field Reactions in Quantum Field Theory. *Phys. Rev.*, 74, 224–225, [the last in each series of papers] where additional references to previous work may be found.

[2] The author's previous papers will hereafter be designated as follows:
Feynman, R. P. (April, 1948). Space-Time Approach to Non-Relativistic Quantum Mechanics. *Rev. Mod. Phys.*, 20, 2, 367-87 - C;
Feynman, R. P. (September, 1949). The Theory of Positrons. *Phys. Rev.*, 76, 749-59 - I;
Feynman, R. P. (September, 1949). Space-Time Approach to Quantum Electrodynamics. *Phys. Rev.*, 76, 6, 769-89 - II; and
Feynman, R. P. (November, 1950). Mathematical Formulation of the Quantum Theory of Electromagnetic Interaction. *Phys. Rev.*, 80, 3, 440-57 - III.

These relationships have already been discussed by many people, particularly Dyson[3].

[3] Dyson, F. J. (February, 1949). The Radiation Theories of Tomonaga, Schwinger, and Feynman. *Phys. Rev.*, 75, 3, 486-502; Dyson, F. J. (June, 1949). The S Matrix in Quantum Electrodynamics. *Phys. Rev.*, 75, 1736-55.

The connection was shown by means of a re-ordering of operators in each term of a perturbation power series. Here, the same end is achieved in much the same way without having to resort to such an expansion.

It is felt, *in the face of daily experimental surprises for meson theory*, that it might be worthwhile to spend one's time expressing *electrodynamics* in every physical and mathematical way possible. *There may be some hope that a thorough understanding of electrodynamics might give a clue as to the possible structure of the more complete theory to which it is an approximation.* This is one reason that this paper is published, even though *it is little more than a mathematical re-expression of old material.* A second reason is the desire to describe a mathematical method which may be useful in other fields.

The mathematics is not completely satisfactory. No attempt has been made to maintain mathematical rigor. The excuse is not that it is expected that rigorous demonstrations can be easily supplied. Quite the contrary, *it is believed that to put the present methods on a rigorous basis may be quite a difficult task, beyond the abilities of the author*.

The mathematical ideas are described and are illustrated with simple applications to quantum mechanics, in the first four sections. Some possible mathematical relations between the operator calculus described here and the theory of functionals is described in Appendix A, with further specific mathematical applications in Appendixes B and C. Section 5, and more particularly Secs. 6 to 9, apply specifically to *quantum electrodynamics* and may be omitted without loss by those whose interest is limited to mathematical questions. The use of a fifth variable to parametrize the Dirac equation is discussed in Secs. 8 and 9. … Section 10 gives a summary of the rules for computing matrix elements.

Dirac, P. A. M. (October, 1962). The Conditions for a Quantum Field Theory to be Relativistic.

[*Rev. Mod. Phys.*, 34, 592; https://doi.org/10.1103/RevModPhys.34.592.]

St. John's College, Cambridge, England

Abstract

A quantum field theory in agreement with *special relativity* can be built up from the infinitesimal operators of translation and rotation. These operators are expressible in terms of a *momentum density* and an *energy density*. The momentum density is determined by the geometrical properties of the fields concerned. The energy density has to satisfy commutation relations for which certain conditions hold.

Tomonaga, S. (May 6, 1966). Nobel Lecture. ***Development of Quantum Electrodynamics - 1966: Personal recollections***

[https://www.nobelprize. org/prizes/physics/1965/tomonaga/lecture/.]

> Sin-Itiro Tomonaga was jointly awarded the 1965 Nobel Prize in Physics for his work on *quantum electrodynamics*, along with Julian Swinger and Richard Feynman "for their fundamental work in *quantum electrodynamics*, with deep-ploughing consequences for the physics of elementary particles".
>
> Following the establishment of the theory of relativity and quantum mechanics, an initial *relativistic* theory was formulated for the interaction between charged particles and electromagnetic fields. The theory had to be reformulated, however, partly due to the observation of the Lamb shift in 1947, in which the supposed single energy level within a hydrogen atom was instead proven to be two similar levels. Sin-Itiro Tomonaga solved this problem in 1948 through a "*renormalization*" and thereby contributed to a new quantum electrodynamics.
> [Sin-Itiro Tomonaga – Facts. NobelPrize.org. <https://www.nobelprize.org/prizes/physics/1965/tomonaga/facts/.]

Tomonaga's Nobel Prize lecture describes the evolution of his work and that of others since 1932 when he started his research career up until 1948.

Development of Quantum Electrodynamics - 1966: Personal recollections.

(1) In 1932, when I started my research career as an assistant to Nishina, Dirac published a paper in the *Proceedings of the Royal Society, London*[1].

> [1] Dirac, P. A. M. (May, 1932). Relativistic Quantum Mechanics. *Roy. Soc. Proc., A*, 136, 829, 453-64.

In this paper, he discussed the formulation of *relativistic* quantum mechanics, especially that of *electrons interacting with the electromagnetic field*. At that time a comprehensive theory of this interaction had been formally completed by Heisenberg and Pauli[2],

> [2] Heisenberg, W. & Pauli, W. (July, 1929). Zur Quantendynamik der Wellenfelder. (On the quantum dynamics of wave fields.) *Zeit. Phys.*, 56, 1-61.

but Dirac was not satisfied with this theory and tried to construct a new theory from a different point of view. *Heisenberg and Pauli regarded the (electromagnetic) field itself as a dynamical system amenable to the Hamiltonian treatment*; its interaction with particles could be described by an *interaction energy*, so that the usual method of Hamiltonian quantum mechanics could be applied. On the other hand, *Dirac thought that the field and the particles should play essentially different roles. That is to say, according to him, "the role of the field is to provide a means for making observations of a system of particles" and therefore "we cannot suppose the field to be a dynamical system on the same footing as the particles and thus be something to be observed in the same way as the particles".*

Based on such a philosophy, Dirac proposed a new theory, the so-called *many-time theory*, which, besides being a concrete example of his philosophy was of much more satisfactory and beautiful form than other theories presented up to then.

> [Dirac's *many-time theory* appeared in Dirac (1933). The Lagrangian in Quantum Mechanics; *Phys. Zeit. Sowjetunion*, 3, 1, 64-72.]

In fact, from the *relativistic* point of view, these other theories had a common defect which was inherent in their Hamiltonian formalism. *The Hamiltonian dynamics was developed on the basis of non-relativistic concepts which make a sharp distinction between time and space*. It formulates a physical law by describing how the state of a dynamical system changes with time. Speaking quantum-mechanically, it is a formalism to describe how the *probability amplitude* changes with *time* t. Now, as an example, let us consider a system composed of N particles, and let the coordinates of each particle be $r_1, r_2, \ldots, r_N$. Then the *probability amplitude* of the system is a function of the N variables $r_1, r_2, \ldots, r_N$, and in addition, of the time t to which the *amplitude* is referred. Thus, this function contains only one time variable in contrast to N *space* variables. In the theory of relativity, however, *time* and *space* must be treated on an entirely equal footing so that the above imbalance is not satisfactory. On the other hand, *in Dirac's theory which does not use the Hamiltonian formalism, it becomes possible to consider different time variables for each particle, so that the probability amplitude can be expressed as a function of* $r_1, t_1, r_2\ t_2, \ldots, r_N\ t_N$. *Accordingly, the theory satisfies the requirement of the principle of relativity that time and space be treated with complete equality*. The reason why the theory is called the *many-time theory* is because N distinct time variables are used in this way.

This paper of Dirac's attracted my interest because of the novelty of its philosophy and the beauty of its form. Nishina also showed a great interest in this paper and suggested that I investigate the possibility of predicting some new phenomena by this theory.

Then I started computations to see whether the Klein-Nishina formula could be derived from this theory or whether any modification of the formula might result.

> [The Klein–Nishina formula gives the differential cross section (i.e. the "likelihood" and angular distribution) of photons scattered from a single free electron, calculated in the lowest order of quantum electrodynamics. It was first derived in 1928 by Oskar Klein and Yoshio Nishina, constituting one of the first successful applications of the Dirac equation. The formula describes both the Thompson scattering of low energy photons (e.g. visible light) and the Compton scattering of high energy photons (e.g. x-rays and gamma-rays), showing that the total cross section and expected deflection angle decrease with increasing photon energy.]

I found out immediately however, without performing the calculation through to the end, that it would yield the same answer as the previous theory. *This new theory of Dirac's was in fact mathematically equivalent to the older Heisenberg-Pauli theory*, and I realized during the calculation that *one could pass from one to the other by a unitary transformation*. The equivalence of these two theories was also discovered by Rosenfeld[3] and by Dirac-Fock-Podolsky[4] and was soon published in their papers.

[3] Rosenfeld, L. (November, 1932). Über eine mögliche Fassung des Diracschen Programms zur Quantenelektrodynamik und deren formalen Zusammenhang mit der Heisenberg-Paulischen Theorie. (On a possible version of Dirac's program on quantum electrodynamics and its formal connection with the Heisenberg-Paulian theory.) *Zeit. Phys.*, 76 (1932) 729-34; https://doi.org/10.1007/BF01341566.
[4] Dirac, P. A. M., Fock, V. A., Podolsky, B. (1932). On quantum electrodynamics. *Phys. Zeit. Sowjetunion*, 2, 468.

Though Dirac's *many-time* formalism turned out to be equivalent to the Heisenberg-Pauli theory, *it had the advantage that it gave us the possibility of generalizing the former interpretation of the probability amplitude*. Namely, while one could calculate the probability of finding particles at points with coordinates r_1, r_2,..., r_N, all at the time t according to the previous theory, one could now compute more generally the probability that the first particle is at r_1 at time t_1, the second at r_2 at time t_2, ... and the N-th at r_N at time t_N. This was first discussed by Bloch[5] in 1943.

[5] Bloch, F. (1943). Die physikalische Bedeutung mehrerer Zeiten in der Quantenelektrodynamik. (The physical significance of multiple times in quantum electrodynamics.) *Phys. Zeit. Sowjet.*, 5, 301-15.

(2) *In this many-time theory developed by Dirac, electrons were treated according to the particle picture. Alternatively, in quantum theory, any particle should be able to be treated according to the wave picture*. As a matter of fact, electrons were also treated as waves in the Heisenberg-Pauli theory, and it was well known that this wave treatment was frequently more convenient than the particle treatment. So, *the question arose as to whether one could reformulate the Heisenberg-Pauli theory in a way which would be more satisfactory relativistically, when electrons were treated as waves as well as the electromagnetic field.*

As Dirac already pointed out, the Heisenberg-Pauli theory is built upon the Hamiltonian formalism and therefore the *probability amplitude* contains only one *time* variable. That is to say, the *probability amplitude* is given as a function of the *field strength* at different *space* points and of one common *time* variable. However, the concept of a common *time* at different *space* points does not have a *relativistically* covariant meaning.

Around 1942, Yukawa[6] wrote a paper emphasizing this unsatisfactory aspect of the quantum field theory.

[6] Yukawa, H. (1942). *Kagaku* (Science), 7, 249; 8, 282; 9, 322 (in Japanese).

[In 1935, Hideki Yukawa had proposed the existence of a new kind of particle, the meson, in order to explain how protons and neutrons in the nucleus interact, for which he received the Nobel Prize in Physics in 1949 "for his prediction of the existence of mesons on the basis of theoretical work on nuclear forces."]

He thought it necessary to use the idea of the g.t.f. (*generalized transformation function*) proposed by Dirac[7] to correct this defect of the theory.

[7] Dirac, P. A. M. (1933). The Lagrangian in Quantum Mechanics. *Phys. Zeit. Sowjet.*, 3, 1, 64-72.

Here I shall omit talking about the g.t.f., but, briefly, Yukawa's idea was to introduce as the basis of a new theory a concept which generalized the conventional conception of the *probability amplitude*. However, as pointed out also by Yukawa, we encounter the difficulty that, in doing this, cause and effect cannot be clearly separated from each other. According to Yukawa, the inseparability of cause and effect would be an essential feature of quantum field theory, and without abandoning the causal way of thinking which strictly separates cause and effect, it would not be possible to solve various difficulties appearing in quantum field theory about which I will talk later. *I thought however, that it might be possible (without introducing such a drastic change as Yukawa and Dirac tried to do) to remedy the unsatisfactory, unpleasant aspect of the Heisenberg-Pauli theory of having a common time at different space points*. In other words, it should be possible, I thought, to define a *relativistically* meaningful *probability amplitude* which would be manifestly covariant, without being forced to give up the causal way of thinking. In having this expectation, I was recalling Dirac's *many-time theory* which had enchanted me 10 years before.

When there are N particles in Dirac's *many-time theory*, we assign a time t_1 to the first particle, t_2 to the second, and so on, thus introducing N different times, $t_1, t_2, \ldots, t_N$, instead of the one common time t. Similarly, I tried in quantum field theory to see whether it was possible to assign different *times*, instead of one common *time*, to each *space* point. And in fact, I was able to show that this was possible[8].

[8] Tomonaga, S. (1943). On a Relativistically Invariant Formulation of the Quantum Theory of Wave Fields. *Bull. I. P. C. R. Riken-iho*, 22, 545 (in Japanese); translation Tomonaga, S. (August, 1946). *Prog. Theor. Phys.* 1, 2, 27-42; Koba, Z., Tati, T. & Tomonaga, S. (1947a). On a Relativistically Invariant Formulation of the Quantum Theory of Wave Fields. II. Case of Interacting Electromagnetic and Electron Fields. § 1-4. *Prog. Theor. Phys.*, 2, 101-16; Koba, Z., Tati, T. & Tomonaga, S. (1947b). On a Relativistically Invariant Formulation of the Quantum Theory of Wave Fields. III. Case of Interacting Electromagnetic and Electron Fields. § 5-7. *Prog. Theor. Phys.*, 2, 198-208; Kanesawa, S. & Tomonaga, S. (1948a). On a Relativistically Invariant Formulation of the Quantum Theory of Wave Fields. IV. Case of Interacting Electromagnetic and Meson Fields. § 1-4. *Prog. Theor. Phys.* 3, 1-13; Kanesawa, S. & Tomonaga, S. (1948b). On a Relativistically Invariant Formulation of the Quantum Theory of Wave Fields. V. Case of Interacting Electromagnetic and Meson Fields. § 5-7. *Prog. Theor. Phys.* 3, 101-13.

As there are an infinite number of *space* points in *field theory* in contrast to the *finite* number of *particles* in *particle theory*, the number of *time* variables appearing in the *probability amplitude* became infinite. But it turned out that no essential difficulty appeared. An interpretation quite analogous to the one discussed by Bloch in connection with Dirac's *many-time theory* could be given to our *probability amplitude* containing an infinite number of *time* variables. Further, it was found that the theory thus formulated was completely covariant and that this covariant formulation was equivalent in its whole content to the Heisenberg-Pauli theory: it was shown, just as in the case of the *many-time theory*, that we could pass from one to the other by a *unitary transformation. I began this work about 1942, and completed it in 1946.*

(3) As I mentioned a little while ago, there are many difficulties in the quantum mechanics of fields. In particular, *infinite quantities always arise which are associated with the presence of field reactions in various processes*. The first phenomenon which attracted our attention as a manifestation of field reactions was the *electromagnetic mass* of the electron. The electron, having a charge, produces an electromagnetic field around itself. In turn, this field, the so-called *self-field* of the electron, interacts with the electron. We call this interaction the field reaction. *Because of the field reaction the apparent mass of the electron differs from the original mass.* The excess mass due to this field reaction is called the *electromagnetic mass* of the electron and the experimentally observed mass is the sum of the original mass and this *electromagnetic mass*. The concept of the *electromagnetic mass* had already appeared in the classical theory of the electron by Lorentz, who computed the *electromagnetic mass* by *applying the classical theory and obtained the result that the mass becomes infinite for the point (zero size) electron*. On the other hand, the *electromagnetic mass* was computed in quantum theory by various people, and here I mention particularly the work of Weisskopf[9].

[9] Weisskopf, V. F. (July, 1939). On the Self-energy and Electromagnetic Field of the Electron. *Phys. Rev.*, 56, 1, 72-85[; *the main purpose of this paper is to show the physical significance of the logarithmic divergence of the self-energy of the electron and to demonstrate the reason for its occurrence*, the *self-energy of the electron is its total energy in free space when isolated from other particles or light quanta*,
$W = T + (1/8\pi) \int (H^2 + E^2)\, dr$ where T is the *kinetic energy* of the electron and H and E are the *magnetic* and *electric field strengths* at point r, identifies three reasons why quantum theory of the electron results in infinite *self-energy* of the electron, claims that *quantum kinematics shows that the radius of the electron must be assumed to be zero*, resulting in infinite energy of the *electrostatic field*, the contributions of the electric and magnetic fields of the spin to the *self-energy* of the electron cancel one another, *quantum theory of the electromagnetic field postulates the existence of field strength fluctuations in empty space*, gives rise to an additional energy which diverges more strongly that the electrostatic self-energy, induces the electron to perform vibrations with energy that diverges quadratically *for an infinitely small radius*, Dirac's positron theory implies that the charge and magnetic dipole of the electron are extended over a finite region, explains why the *self-energy* is only *logarithmically infinite*, *divergences are consequence of assumption of point electron*].

According to him, the quantum-mechanical *electromagnetic mass* turned out to be infinite, and although the order of the divergence was much weaker than in the case of the Lorentz theory, the *observed mass*, which included this additional mass, would be infinite. This would be, of course, contrary to experiment.

In order to overcome the difficulty of an infinitely large *electromagnetic mass, Lorentz considered the electron not to be point-like but to have a finite size. It is very difficult, however, to incorporate a finite sized electron into the framework of relativistic quantum theory*. Many people tried various means to overcome this problem of infinite quantities, but nobody succeeded.

In connection with *field reactions*, the next problem which attracted the attention of physicists was determining *what kind of influence the field reaction exerts in electron-scattering processes*. Let us consider, as a concrete example, a problem in which an electron is scattered by an *external field*. In the ordinary treatment, we neglect the effect of *field reactions* on the scattered electron, assuming that it is negligibly small. Then the behavior of the scattering obtained by calculation (e.g. the Rutherford formula) fits very well with experiment. But what will happen if the influence of field reaction is taken into account? This theoretical problem was examined *non-relativistically* by Braunbeck-Weinmann[10] and Pauli-Fierz[11] and *relativistically* by Dancoff[12].

[10] Braunbeck, W. & Weinman, E. (May, 1938). Die Rutherford-Streuung mit BerÜcksichtigung der Ausstrahlung. (The Rutherford scattering with consideration of the radiance.) Zeit. Phys., 110, 5-6, 360-372; https://doi.org/10.1007/BF01341441.

[11] Pauli, W. & Fierz, M. (March, 1938). Zur Theorie der Emission langwelliger Lichtquanten. (On the theory of the emission of long-wave light quanta.) *Nuovo Cimento*, 15, 3, 167-88.

[12] Dancoff, S. M. (May, 1939). On Radiative Corrections for Electron Scattering. *Phys. Rev*., 55, 959.

While Dancoff applied an approximation method, the perturbation method, in his relativistic calculation, Pauli and Fierz treated the problem in such a way that the most important part of the field reaction was first separated out exactly by employing a *contact transformation* method which was similar to the one which Bloch-Nordsieck[13] had published a year before.

[13] Bloch, F. & Nordsieck, A. (July, 1937). Note on the Radiation Field of the Electron. *Phys. Rev*., 52, 54; https://doi.org/10.1103/PhysRev.52.54.

Since Pauli and Fierz adopted a *non-relativistic* model, and further simplified the problem by using the so-called dipole approximation, their calculation was especially transparent. At any rate, *both non-relativistic and relativistic calculations exhibited several infinities in the scattering processes**.

* The main purpose of the work of Bloch-Nordsieck and Pauli-Fierz was to solve the so-called *infrared catastrophe* which was one of a number of divergences. Since this difficulty

was resolved in their papers, we confine ourselves here to a discussion of the other divergences which are of the so-called *ultraviolet type*.

The conclusions of these people were fatal to the theory. That is, the influence of the *field reaction* becomes infinite in this problem. The effect of *field reaction* on a quantity called the *scattering cross section*, which expresses quantitatively the behavior of the scattering, rather than becoming negligibly small, becomes infinitely large. This does not, of course, agree with experiment.

This discouraging state of affairs generated in many people a strong distrust of *quantum field theory*. There were even those with the extreme view that the concept of *field reaction* itself had nothing to do at all with the true law of nature.

On the other hand, there was also the view that the *field reaction* might not be altogether meaningless but would play an essential role in the scattering processes, though the appearance of divergences revealed a defect of the theory. Heisenberg[14], in his paper published in 1939,

[14] Heisenberg, W. (1939). Zur Theorie der explosiven Schauer in der kosmischen Strahlung. (On the theory of explosive showers in cosmic rays. II.) *Zeit. Phys*., 113, 61-86; https://doi.org/10.1007/BF01371656.

emphasized that the *field reaction* would be crucial in *meson-nucleon scattering*. Just at that time I was studying at Leipzig, and I still remember vividly how Heisenberg enthusiastically explained this idea to me and handed me galley proofs of his forthcoming paper. Influenced by Heisenberg, *I came to believe that the problem of field reactions far from being meaningless was one which required a frontal attack*.

Thus, after coming back to Japan from Leipzig, I began to examine the nature of the infinities appearing in scattering processes at the same time that I was engaged in the above-mentioned work of formulating a covariant field theory. What I wanted to know was what kind of relationship exists between the infinity associated with the *scattering process* and that associated with the *mass*. If you read the above-mentioned papers of Bloch-Nordsieck and Pauli-Fierz, you will see that one of the terms containing infinite quantities is first separated out by a *contact transformation* and this term turns out to be just the term modifying the *mass*. Besides this kind of infinity there appeared, according to Pauli-Fierz, another kind of infinity characteristic of the *scattering process*. I further investigated a couple of simple models which were not realistic, but could be solved exactly. What was understood from these models, was that the most strongly divergent terms in the *scattering process* had the same form as the expression giving the modification of the particle *mass* due to field reactions, and therefore both should be manifestations of the same effect. In other words, at least a portion of the infinities appearing in the *scattering process* could be amalgamated into the infinity associated with the particle *mass*, leaving infinities proper to the *scattering process* alone. These turned out to be more weakly divergent than the infinity associated with the *mass*.

Since these conclusions were derived from *non-relativistic* or unrealistic models, it was still doubtful whether the same thing would occur in the case of *relativistic electrons interacting with the electromagnetic field.* Dancoff tried to answer this question. He calculated *relativistically* the infinities appearing in the *scattering process* and determined which of them could be amalgamated into the *mass* and which remained as infinities proper to the *scattering process* alone. He found that there remained, in the latter group of infinite terms, one which was at least as divergent as the infinity of the *mass*, a finding which differed from the conclusion based on fictitious models.

Actually, there are two kinds of field reactions in the case of the relativistic electron and electromagnetic field. One of them ought to be called "of *mass type*" and the other "of *vacuum polarization type*". *The field reaction of mass type changes the apparent electronic mass from its original value by the amount of the electromagnetic mass* as was calculated by Weisskopf. On the other hand, *the field reaction of vacuum polarization type changes the apparent electronic charge from its original value*. As was discussed in further papers by Weisskopf[15] and others, infinite terms appear in the apparent *electronic charge* if the effect of *vacuum polarization* is taken into account.

> [15] Weisskopf, V. F. (1936). Uber die Elektrodynamik des Vakuums auf Grund der Quanten-theorie des Elektrons. (About the electrodynamics of the vacuum based on the quantum theory of the electron.) *Kgl. Danske Sels. Math.-Fys. Medd.*, 14, 6, 1-36.

However, in this talk, for simplicity, I will mention only briefly the divergence of the *vacuum polarization* type.

(4) In the meantime, in 1946, Sakata[16] proposed a promising method of eliminating the divergence of the electron *mass* by introducing the idea of a *field* of *cohesive force*.

> [16] Sakata, S. & Hara, O. (April 8, 1947). The Self-Energy of the Electron and the Mass Difference of Nucleons. *Progr. Theoret. Phys. (Kyoto)*, 2, 30-1[; abandoned *C-meson field theory*].
>
> [The quoted reference, Sakata, *Progr. Theoret. Phys. (Kyoto)*, I (1946) 143, is incorrect. Presumably Tomonaga meant Sakata S. & Inoue, T. (1946). On the Correlations between Mesons and Yukawa Particles. *Progr. Theoret. Phys. (Kyoto)*, 1, 143-9. Report read at the Meson Symposium held at Tokyo in September, 1943, and published after World War II on being translated into English), but this also appears to be the wrong paper as it does not address the electron, but is focused on meson theory and does not refer to a *cohesive force field.*
>
> Presumably he intended to refer to Sakata, S. & Hara, O. (April 8, 1947). The Self-Energy of the Electron and the Mass Difference of Nucleons. *Progr. Theoret. Phys. (Kyoto)*, 2, 30-1; which sets out the C-meson theory but does not use the terms *cohesive force field* or *coupling constants* (see below), but these were used by Tomonaga in referring to this paper in Ito, D., Koba, Z. & Tomonaga, S. (December 1, 1947). Correction due to the Reaction of "Cohesive Force Field" for the Elastic

Scattering of an Electron. *Progr. Theoret. Phys. (Kyoto)*, 2, 4, 216-7, in a letter to the editor dated November 1, 1947:

> "Recently Sakata and Hara introduced a scalar neutral *cohesive force field* (the so-called C-meson field) which is assumed to interact with charged particles, and by imposing a relation
>
> $F^2 = 2e^2$ (1)
>
> on the *coupling constants* e, f of the latter with *electromagnetic* and *cohesive field* respectively, they succeeded in obtaining a non-diverging value for the self-energy of an electron and also suggested a possibility of explaining the mass difference between proton and neutron."

The reference in this letter, "S. Sakata and O. Hara, O. *Progr. Theoret. Phys. (Kyoto)*, 2 (1949), 30", is also incorrect. It should read 1947.]

It was the idea that there exists unknown field, of the type of the *meson field* which interacts with the electron in addition to the *electromagnetic field.* Sakata named this field the *cohesive force field*, because the apparent electronic *mass* due to the interaction of this field and the electron, though infinite, is negative and therefore the existence of this field could stabilize the electron in some sense.

Sakata pointed out the possibility that the electromagnetic mass and the negative new mass cancel each other and that the infinity could be eliminated by suitably choosing the coupling constant between this field and the electron. Thus, the difficulty which had troubled people for a long time seemed to disappear insofar as the *mass* was concerned. (It was found later that Pais[17] proposed the same idea in the U.S. independently of Sakata.)

> [17] Pais, A. (1945). On the theory of the electron and of the nucleon. *Phys. Rev.*, 68, 227-8 [; the reference in the publication of this lecture Pais, *Phys. Rev.*, 73, (1946) 173 was clearly incorrect as this journal number is for a different year and different author. It appears to have come from the reference (see below) for Lewis, H. W. (January, 1948). On the Reactive Terms in Quantum Electrodynamics. *Phys. Rev.*, 73, 173]

Then what concerned me most was whether the infinities appearing in the electron-scattering process could also be removed by the idea of a plus-minus cancellation.

An example of a computation of how the field reaction influences the scattering process was already given by Dancoff. *What we had to do was just to replace the electromagnetic field by the cohesive force field in Dancoff's calculation.* I mobilized young people around me and we performed the computation together[18].

> [18] Ito, D., Koba, Z. & Tomonaga, S. (September, 1948). Corrections due to the Reaction of "Cohesive Force Field" to the Elastic Scattering of an Electron. I. *Prog. Theor. Phys.*, 3, 3, 276-89; https://doi.org/10.1143/ptp/3.3.276[; abandoned C-meson field theory in which Sakata and Hara introduced a hypothetical *scalar particle* (called *C-meson*) in order to solve the divergence problem]; Koba, Z. & Takeda, S. (December, 1948). Radiation Reaction in Collision Process, II: Radiative Corrections for Compton Scattering. *Prog. Theor. Phys.*, 3, 4, 407-21; https://doi.org/10.1143/ptp/3.4.407.

Infinities with negative sign actually appeared in the *scattering cross-section* as was expected. However, when we compared these with the infinities with positive sign which Dancoff calculated for the *electromagnetic field*, the two infinities did not cancel each other completely. That is to say, according to our result, *the Sakata theory led to the cancellation of infinities for the mass but not for the scattering process*. It was also known that the infinity of *vacuum polarization* type was not cancelled by the introduction of the *cohesive force field*.

Unfortunately, Dancoff did not publish the detailed calculations in his paper, and while we were engaged in the above considerations, we felt it necessary to do Dancoff's calculation over again for ourselves in parallel with the computation of the influence of the *cohesive force field*. At the same time, I happened to discover a simpler method of calculation.

This new method of calculation was to use the technique of *contact transformations* based on the previously mentioned formalism of the covariant field theory and was in a sense a relativistic generalization of the Pauli-Fierz method. *This method had the advantage of separating the electromagnetic mass from the beginning*, just as was shown in their paper.

Our new method of calculation was not at all different in its contents from Dancoff's *perturbation method*, but had the advantage of making the calculation clearer. In fact, what took a few months in the Dancoff type of calculation could be done in a few weeks. And it was by this method that *a mistake was discovered in Dancoff's calculation*; we had also made the same mistake in the beginning. Owing to this new, more lucid method, we noticed that, among the various terms appearing in both Dancoff's and our previous calculation, *one term had been overlooked. There was only one missing term, but it was crucial to the final conclusion.* Indeed, if we corrected this error, *the infinities appearing in the scattering process of an electron due to the electro-magnetic and cohesive force fields cancelled completely, except for the divergence of vacuum polarization type*.

(5) When this unfortunate error of Dancoff's was discovered, we had to re-examine his conclusions concerning the relation between the divergence of the *scattering process* and the divergence of the *mass*, in particular, the conclusion that there remained a portion of the infinities of the scattering process which could not be amalgamated into the modification of the *mass*. In fact, *it turned out that after correcting the error, the infinity of mass type appearing in the scattering process could be reduced completely to the modification of the mass*, and the remaining field reaction belonging to the scattering proper was not divergent[19].

[19] Koba, Z. & Tomonaga, S. (1948). On Radiation Reactions in Collision Processes. I: Application of the "Self-Consistent" Subtraction Method to the Elastic Scattering of an Electron. *Progr. Theoret. Phys. (Kyoto)*, 3, 3, 290-303; Tati, T. & Tomonaga, S. (1948). A Self-Consistent Subtraction Method in the Quantum Field Theory, I. *Ibid.*, 3, 4, 391-406].

In other words, *the highest divergence part of the infinities appearing in the scattering process, in the relativistic as well as in the non-relativistic case, could be attributed to the infinity of mass*. The reason why the remaining part became finite in the relativistic case was due to the fact that the order of the highest divergence was only log co [???], and *after amalgamating the divergence into the mass term, the remainder was convergent.* The great value of this method of *contact transformations* was that *once the infinity of the mass was separated out, we obtained a divergence-free theoretical framework.*

In this way the nature of various infinities became fairly clear. Though I did not describe here the *infinity of vacuum polarization type*, this too appears in the scattering process, as mentioned earlier. However, Dancoff had already discovered that *this infinity could be amalgamated into an apparent change in the electronic charge*. To state the conclusion, therefore, *all infinities appearing in the scattering process can be attributed either to the infinity of the electromagnetic mass or to the infinity appearing in the electronic charge* – there are no other divergences in the theory.

It is a very pleasant thing that no divergence is involved in the theory *except for the two infinities of the electronic mass and charge. We cannot say that we have no divergences in the theory, since the mass and charge are in fact infinite*. It is to be noticed, however, that if we reduce the infinities appearing in the *scattering process* to modifications of *mass* and *charge*, the remaining terms all become finite. Further, if we examine the structure of the theory, after the infinities are amalgamated into the *mass* and *charge* terms, *we see that the only mass and charge appearing in the theory are the values modified by field reactions* – the original values and excess ones due to field reactions never appear separately.

This situation gives rise to the following possibility. *The theory does not of course yield a resolution of the infinities*. That is, since those parts of the modified mass and charge due to field reactions contain divergence, it is impossible to calculate them by the theory. However, *the mass and charge observed in experiments are not the original mass and charge but the mass and charge as modified by field reactions, and they are finite*. On the other hand, the *mass* and *charge* appearing in the theory are, as I mentioned above, after all the values modified by *field reactions*. Since this is so, and particularly since the theory is unable to calculate the modified *mass* and *charge*, *we may adopt the procedure of substituting experimental values for them phenomenologically.* When a theory is incompetent in part, it is a common procedure to rely on experiment for that part. *This procedure is called the renormalization of mass and charge*, and our method has brought the possibility that the theory will lead to finite results by the *renormalization* even if it contains defects.

The idea of *renormalization* is far from new. Many people used explicitly or implicitly this idea, and we find the word *renormalization* already in Dancoff's paper. In his calculation it appeared, because of an error that there still remained a divergence in the *scattering* even after the *renormalization* of the electron *mass*. This error was very unfortunate; if he had performed the calculation correctly, the history of *renormalization theory* would have been completely different.

(6) This period, around 1946-1948, was soon after the second world war, and it was quite difficult in Japan to obtain information from abroad. But soon we got the news that in the

U.S., Lewis and Epstein[20] found Dancoff's mistake and gave the same conclusions as ours, Schwinger[21] constructed a covariant field theory similar to ours, and he was probably performing various calculations making use of it.

[20] Lewis, H. W. (January, 1948). On the Reactive Terms in Quantum Electrodynamics. *Phys. Rev.*, 73, 173; https://doi.org/10.1103/PhysRev.73.173; Epstein, S. T. (January, 1948). Remarks on H. W. Lewis' Paper "On the Reactive Terms in Quantum Electrodynamics". *Ibid.*, 73, 177; https://doi.org/10.1103/ PhysRev.73.177.
[21] Schwinger, J. S. (November, 1948). On quantum-electrodynamics and the magnetic moment of the electron. *Phys. Rev.*, 73, 4, 416–417; Quantum Electrodynamics. I. A Covariant Formulation. *Ibid.*, 74, 10, 1439–1461; (February, 1949). Quantum Electrodynamics. II. Vacuum Polarization and Self-Energy. *Ibid.*, 75, 4, 651-79; (September, 1949). Quantum Electrodynamics. III. The Electromagnetic Properties of the Electron—Radiative Corrections to Scattering. *Ibid.*, 76, 6, 790–817.

In particular, little by little news arrived that the so-called Lamb-shift was discovered[22] as a manifestation of the *electromagnetic field reaction* and that Bethe[23] was calculating it theoretically.

[22] Lamb, Jr., W. E. & Retherford, R. C. (August, 1947). Fine Structure of the Hydrogen Atom by a Microwave Method. *Phys. Rev.*, 72, 3, 241-3.
[23] Bethe, H. A. (August, 1947). The Electromagnetic Shift of Energy Levels. *Phys. Rev.*, 72, 339-41.

The first information concerning the Lamb-shift was obtained not through the *Physical Review*, but through the popular science column of a weekly U.S. magazine. This information about the Lamb-shift prompted us to begin a calculation more exactly than Bethe's tentative one.

The Lamb-shift is a phenomenon in which the energy levels of a hydrogen atom show some shifts from the levels given by the Dirac theory. Bethe thought that the *field reactions* were primarily responsible for this shift. According to his calculation, *field reactions* give rise to an infinite *level shift*, but he thought that it should be possible to make it finite by a *mass renormalization* and a tentative calculation yielded a value almost in agreement with experiments.

This problem of the *level shift* is different from the *scattering process*, but it was conceivable that the *renormalization* which was effective in avoiding infinities in the *scattering process* would be workable in this case as well. In fact, the *contact transformation* method of Pauli and Fierz devised to solve the *scattering problem* could be applied to this case, clarifying Bethe's calculation and justifying his idea. Therefore, the method of covariant *contact transformations*, by which we did Dancoff's calculation over again would also be useful for the problem of performing the *relativistic* calculation for the Lamb-shift. This was our prediction.

The calculation of the Lamb-shift was done by many people in the U.S.[24].

[24] Kroll, N. M. & Lamb, Jr., W. E. (February, 1949). On the self-energy of a bound electron. *Phys. Rev.*, 75, 388-98; French, J. B. & Weisskopf, V. F. (April, 1949). The Electromagnetic Shift of Energy Levels. *Ibid.*, 75, 8, 1240-8.

Among others, Schwinger, commanding powerful mathematical techniques, and by making thorough use of the method of *covariant contact transformations*, very skillfully calculated not only the Lamb-shift but other quantities such as the anomalous *magnetic moment* of the electron. After long, laborious calculations, less skillful than Schwinger's, we[25] obtained a result for the Lamb-shift which was in agreement with Americans'.

[25] Fukuda, H., Miyamoto, Y. & Tomonaga, S. (March, 1949). A Self-Consistent Subtraction Method in the Quantum Field Theory. II-1. *Prog. Theor. Phys. (Kyoto)*, 4, 1, 47-59; (June, 1949). *Idem.*, II-2, 24, 2, 121-9.

Furthermore, Feynman[26] devised a convenient method based on an ingenious idea which could be used to extend the approximation of Schwinger and ours to higher orders,

[26] Feynman, R. P. (November, 1948). Relativistic Cut-Off for Quantum Electrodynamics. *Phys. Rev.*, 74, 1430-8; (September, 1949). The Theory of Positrons. *Phys. Rev.*, 76, 749-59.

[The latter was submitted for publication on April 8, 1949, after the first Dyson paper had been published (February) and after the second Dyson paper had been submitted (February 24) ff.].

and Dyson[27] showed that all infinities appearing in quantum electrodynamics could be treated by the *renormalization* procedure to an arbitrarily high order of approximation.

[27] Dyson, F. J. (February, 1949). The Radiation Theories of Tomonaga, Schwinger, and Feynman. *Phys. Rev.*, 75, 3, 486-502; (June, 1949). The S Matrix in Quantum Electrodynamics. *Ibid.*, 75, 1736-55.

Furthermore, this method devised by Feynman and developed by Dyson was shown by many people to be applicable not only to quantum electrodynamics, but to statistical mechanics and solid-state physics as well, and provided a new, powerful method in these fields. However, these matters *will probably be discussed by Schwinger and Feynman themselves* and need not be explained by me. So far, I have told you the story of how I played a tiny, partial role in the recent development of quantum electrodynamics, and here I would like to end my talk.

Feynman, R. P. (December 11, 1965). Nobel Lecture. *The Development of the Space-Time View of Quantum Electrodynamics.*

[https://www.nobelprize.org/prizes/physics/1965/feynman/lecture/.]

> Richard Feynman shared the 1965 Nobel Prize in Physics with Sin-Itiro Tomonaga and Julian Schwinger "for their fundamental work in *quantum electrodynamics*, with deep-ploughing consequences for the physics of elementary particles"
>
> Following the establishment of the theory of relativity and quantum mechanics, an initial *relativistic* theory was formulated for the interaction between charged particles and electromagnetic fields. This needed to be reformulated, however. In 1948 in particular, Richard Feynman contributed to creating a new quantum electrodynamics by introducing Feynman diagrams: graphic representations of various interactions between different particles. These diagrams facilitate the calculation of interaction probabilities. [Richard P. Feynman – Facts. NobelPrize.org. https://www.nobelprize.org/prizes/physics/1965/feynman/facts/]

Feynman's Nobel Prize lecture describes the sequence of ideas which occurred and by which he finally came out the other end with an unsolved problem for which he ultimately received the Nobel prize, he represented conventional electrodynamics with *retarded interaction*, not his half-advanced and half-retarded theory, the *action* expression was not used, the idea that *charges* do not act on themselves was abandoned, the *path-integral formulation* of quantum mechanics was useful for guessing at final expressions and at formulating the general theory of electrodynamics in new ways though was not absolutely necessary, the rest of his work was simply to improve the techniques then available for calculations, *making diagrams to help analyze perturbation theory quicker*, "*I don't think we have a completely satisfactory relativistic quantum-mechanical model*, … Therefore, I think that the *renormalization theory* is simply a way to sweep the difficulties of the divergences of electrodynamics under the rug".

> ["I never used all that machinery which I had cooked up to solve a single *relativistic* problem. … the half-advanced and half-retarded *potential* was not finally used, … I was representing the conventional electrodynamics with *retarded interaction*, and not my half-advanced and half-retarded theory, … the *action* expression was not used, the idea that *charges* do not act on themselves was abandoned. The *path-integral formulation* of quantum mechanics was useful for guessing at final expressions and at formulating the general theory of electrodynamics in new ways – although, strictly it was not absolutely necessary. The same goes for the idea of the positron being a backward moving electron, it was very convenient, but not strictly necessary for the theory because it is exactly equivalent to the negative energy sea point of view."
>
> "The rest of my work was simply to improve the techniques then available for calculations, making diagrams to help analyze perturbation theory quicker."

> “At this stage, I was urged to publish this because everybody said it looks like an easy way to make calculations, and wanted to know how to do it. I had to publish it, missing two things; one was proof of every statement in a mathematically conventional sense. … the second thing that was missing when I published the paper, an *unresolved difficulty*. … the calculations would give results which were not “unitary”, that is, for which the sum of the probabilities of all alternatives was not unity. … as far as I know, nobody has yet been able to resolve this question.”
>
> “*I believe there is really no satisfactory quantum electrodynamics*, but I'm not sure. And, I believe, that one of the reasons for the slowness of present-day progress in understanding the strong interactions is that there isn't any *relativistic* theoretical model, from which you can really calculate everything. ... *I don't think we have a completely satisfactory relativistic quantum-mechanical model*, … Therefore, I think that the *renormalization theory* is simply a way to sweep the difficulties of the divergences of electrodynamics under the rug”.]

We have a habit in writing articles published in scientific journals to make the work as finished as possible, to cover all the tracks, to not worry about the blind alleys or to describe how you had the wrong idea first, and so on. So, there isn't any place to publish, in a dignified manner, what you actually did in order to get to do the work, although, there has been in these days, some interest in this kind of thing. Since winning the prize is a personal thing, I thought I could be excused in this particular situation, if I were to talk personally about my relationship to *quantum electrodynamics*, rather than to discuss the subject itself in a refined and finished fashion. Furthermore, since there are three people who have won the prize in physics, if they are all going to be talking about *quantum electrodynamics* itself, one might become bored with the subject. So, what I would like to tell you about today are the sequence of events, really the sequence of ideas, which occurred, and by which I finally came out the other end with *an unsolved problem* for which I ultimately received a prize.

I realize that a truly scientific paper would be of greater value, but such a paper I could publish in regular journals. So, I shall use this Nobel Lecture as an opportunity to do something of less value, but which I cannot do elsewhere. I ask your indulgence in another manner. I shall include details of anecdotes which are of no value either scientifically, nor for understanding the development of ideas. They are included only to make the lecture more entertaining.

I worked on this problem about eight years until the final publication in 1947.

> [It is not clear what he is referring to here. The papers relating to his award of the Nobel prize were submitted and published between 1948 and 1950. *Feynman diagrams* were introduced at Ponoco in April 1948 and first published in 1949.]

The beginning of the thing was at the Massachusetts Institute of Technology, when I was an undergraduate student reading about the known physics, learning slowly about all these things that people were worrying about, and realizing ultimately that *the fundamental problem of the day was that the quantum theory of electricity and magnetism was not completely satisfactory*. This I gathered from books like those of Heitler and Dirac. I was inspired by the remarks in these books; not by the parts in which everything was proved and demonstrated carefully and calculated, because I couldn't understand those very well. At the young age what I could understand were the remarks about the fact that this doesn't make any sense, and the last sentence of the book of Dirac I can still remember, "It seems that some essentially new physical ideas are here needed." So, I had this as a challenge and an inspiration. I also had a personal feeling, that since they didn't get a satisfactory answer to the problem I wanted to solve, I don't have to pay a lot of attention to what they did do.

I did gather from my readings, however, that two things were the source of the difficulties with the *quantum electrodynamical* theories. The first was an *infinite energy of interaction of the electron with itself.* And this difficulty existed even in the classical theory. The other difficulty came from some *infinites which had to do with the infinite numbers of degrees of freedom in the field.* As I understood it at the time (as nearly as I can remember) this was simply the difficulty that if you quantized the harmonic oscillators of the field (say in a box) each oscillator has a ground state energy of $(\frac{1}{2})\hbar\omega$ and there is an infinite number of modes in a box of every increasing frequency ω, and therefore there is an infinite energy in the box. I now realize that that wasn't a completely correct statement of the central problem; it can be removed simply by changing the zero from which energy is measured. At any rate, I believed that the difficulty arose somehow from a combination of the electron acting on itself and the infinite number of degrees of freedom of the field.

Well, it seemed to me quite evident that the idea that a particle acts on itself, that the electrical force acts on the same particle that generates it, is not a necessary one – it is a sort of a silly one, as a matter of fact. And, *so I suggested to myself, that electrons cannot act on themselves, they can only act on other electrons. That means there is no field at all.* You see, if all charges contribute to making a single common field, and if that common field acts back on all the charges, then each charge must act back on itself. Well, that was where the mistake was, there was no field. It was just that when you shook one charge, another would shake later. *There was a direct interaction between charges, albeit with a delay.* The law of force connecting the motion of one charge with another would just involve a delay. Shake this one, that one shakes later. The sun atom shakes; my eye electron shakes eight minutes later, because of a direct interaction across.

Now, this has the attractive feature that it solves both problems at once. First, I can say immediately, I don't let the electron act on itself, I just let this act on that, *hence, no self-energy*! Secondly, *there is not an infinite number of degrees of freedom in the field. There is no field at all*; or if you insist on thinking in terms of ideas like that of a field, this field is always completely determined by the action of the particles which produce it. You shake this particle, it shakes that one, but if you want to think in a field way, the field, if it's there, would be entirely determined by the matter which generates it, and therefore, the field does not have any *independent* degrees of freedom and the infinities from the degrees of freedom would then be removed. As a matter of fact, when we look out anywhere and see light, we can always "see" some matter as the source of the light. We don't just see light (except recently some radio reception has been found with no apparent material source).

You see then that *my general plan was to first solve the classical problem, to get rid of the infinite self-energies in the classical theory, and to hope that when I made a quantum theory of it*, everything would just be fine.

That was the beginning, and the idea seemed so obvious to me and so elegant that I fell deeply in love with it. And, like falling in love with a woman, it is only possible if you do not know much about her, so you cannot see her faults. The faults will become apparent later, but after the love is strong enough to hold you to her. So, I was held to this theory, in spite of all difficulties, by my youthful enthusiasm.

Then I went to graduate school [Princeton, 1939] and somewhere along the line *I learned what was wrong with the idea that an electron does not act on itself.* When you accelerate an electron, it radiates energy and you have to do extra work to account for that energy. The extra force against which this work is done is called the *force of radiation resistance*. The origin of this extra force was identified in those days, following Lorentz, as the *action* of the electron itself. *The first term of this action, of the electron on itself, gave a kind of inertia* (not quite *relativistically* satisfactory). *But that inertia-like term was infinite for a point-charge*. Yet the next term in the sequence gave an energy loss rate, which for a point-charge agrees exactly with the rate you get by calculating how much energy is radiated. So, the force of *radiation resistance*, which is absolutely necessary for the conservation of energy would disappear if I said that a charge could not act on itself.

So, I learned in the interim when I went to graduate school the glaringly obvious fault of my own theory. But I was still in love with the original theory, and was still thinking that with it lay the solution to the difficulties of *quantum electrodynamics*. So, I continued to try on and off to save it somehow. *I must have some action develop on a given electron when I accelerate it to account for radiation resistance.* But if I let electrons only act on other electrons the only possible source for this action is another electron in the world. So, one day, when I was working for Professor Wheeler and could no longer solve the problem that he had given me, I thought about this again and I calculated the following. Suppose I have two charges – I shake the first charge, which I think of as a source and this makes the second one shake, but the second one shaking produces an effect back on the source. And so, I calculated how much that effect back on the first charge was, hoping it might add up the force of radiation resistance. It didn't come out right, of course, but I went to *Professor Wheeler* and told him my ideas. He said, – yes, but the answer you get for the problem with the two charges that you just mentioned will, unfortunately, depend upon the charge and the mass of the second charge and will vary inversely as the square of the distance R, between the charges, while the force of radiation resistance depends on none of these things. I thought, surely, he had computed it himself, but now having become a professor, I know that one can be wise enough to see immediately what some graduate student takes several weeks to develop. He also pointed out something that also bothered me, that if we had a situation with many charges all around the original source at roughly uniform density and if we added the effect of all the surrounding charges the inverse R square would be compensated by the R^2 in the volume element and we would get a result proportional to the thickness of the layer, which would go to infinity. That is, one would have an infinite total effect back at the source. And, finally he said to me, and you forgot something else, when you accelerate the first charge, the second acts later, and then the reaction back here at the source would be still later. In other words, the action occurs at the wrong time. I suddenly realized what a stupid fellow I am, for *what I had described and calculated was just ordinary reflected light, not radiation reaction*.

But, as I was stupid, so was *Professor Wheeler* that much more clever. For he then went on to give a lecture as though he had worked this all out before and was completely prepared, but he had not, he worked it out as he went along. First, *he said, let us suppose that the return action by the charges in the absorber reaches the source by advanced waves as well as by the ordinary retarded waves of reflected light; so that the law of interaction acts backward in time, as well as forward in time*. I was enough of a physicist at that time not to say, "Oh, no, how could that be?" For today all physicists know from studying Einstein and Bohr, that sometimes an idea which looks completely paradoxical at first, if analyzed to completion in all detail and in experimental situations, may, in fact, not be paradoxical. So, it did not bother me any more than it bothered Professor Wheeler to use advance waves for the back reaction – a solution of Maxwell's equations, which previously had not been physically used.

Professor Wheeler used advanced waves to get the reaction back at the right time and then he suggested this: If there were lots of electrons in the absorber, there would be an index of refraction n, so, the retarded waves coming from the source would have their wave lengths slightly modified in going through the absorber. Now, if we shall assume that the advanced waves come back from the absorber without an index – why? I don't know, let's assume they come back without an index – then, there will be a gradual shifting in phase between the return and the original signal so that we would only have to figure that the contributions act as if they come from only a finite thickness, that of the first wave zone. (More specifically, up to that depth where the phase in the medium is shifted appreciably from what it would be in vacuum, a thickness proportional to $1/(n\text{-}1)$.) Now, the less the number of electrons in here, the less each contributes, but the thicker will be the layer that effectively contributes because with less electrons, the index differs less from 1. The higher the charges of these electrons, the more each contribute, but the thinner the effective layer, because the index would be higher. And when we estimated it, (calculated without being careful to keep the correct numerical factor) sure enough, it came out that the action back at the source was completely independent of the properties of the charges that were in the surrounding absorber. Further, it was of just the right character to represent radiation resistance, but we were unable to see if it was just exactly the right size. He sent me home with orders to figure out exactly how much advanced and how much retarded wave we need to get the thing to come out numerically right, and after that, figure out what happens to the advanced effects that you would expect if you put a test charge here close to the source? For if all charges generate advanced, as well as retarded effects, why would that test not be affected by the advanced waves from the source?

I found that you get the right answer if you use half-advanced and half-retarded as the field generated by each charge. That is, one is to use the solution of Maxwell's equation which is symmetrical in time and that the reason we got no advanced effects at a point close to the source in spite of the fact that the source was producing an advanced field is this. Suppose the source is surrounded by a spherical absorbing wall ten light seconds away, and that the test charge is one second to the right of the source. Then the source is as much as eleven seconds away from some parts of the wall and only nine seconds away from other parts. The source acting at time t = 0 induces motions in the wall at time t = +10. Advanced effects from this can act on the test charge as early as eleven seconds earlier, or at t = –1. This is just at the time that the direct advanced waves from the source should reach the test charge, and it turns out the two effects are exactly equal and opposite and cancel out! At the later time t = +1 effects on the test charge from the source and from the walls are again equal, but this time are of the same sign and add to convert the half-retarded wave of the source to full retarded strength.

Thus, it became clear that there was the possibility that if we assume all actions are via half-advanced and half-retarded solutions of Maxwell's equations and assume that all sources are surrounded by material absorbing all the light which is emitted, then we could account for radiation resistance as a direct action of the charges of the absorber acting back by advanced waves on the source.

Many months were devoted to checking all these points. I worked to show that everything is independent of the shape of the container, and so on, that the laws are exactly right, and that the advanced effects really cancel in every case. We always tried to increase the efficiency of our demonstrations, and to see with more and more clarity why it works. I won't bore you by going through the details of this. Because of our using advanced waves, we also had many apparent paradoxes, which we gradually reduced one by one, and saw that there was in fact no logical difficulty with the theory. It was perfectly satisfactory.

We also found that we could reformulate this thing in another way, and that is by a principle of least action. Since *my original plan was to describe everything directly in terms of particle motions,* it was my desire to represent this new theory without saying anything about fields. It turned out that we found a form for an *action* directly involving the *motions* of the *charges* only, which upon variation would give the *equations of motion* of these *charges*. The expression for this action *A* is

$$A = \Sigma_i m_i \int \left(\dot{X}^i_\mu \dot{X}^i_\mu\right)^{\frac{1}{2}} d\alpha_i + \frac{1}{2} \sum_{\substack{ij \\ i\neq j}} e_i e_j \int\int \delta(I_{ij}^2)\, \dot{X}^i_\mu(\alpha_i)\, \dot{X}^j_\mu(\alpha_j)\, d\alpha_i d\alpha_j \quad (1)$$

where

$$I_{ij}^2 = [X^i_\mu(\alpha_i) - X^j_\mu(\alpha_j)]\,[X^i_\mu(\alpha_i) - X^j_\mu(\alpha_j)]$$

where $\dot{X}^i_\mu(\alpha_i)$ is the four-vector position of the *i*th particle as a function of some parameter α_i, $\dot{X}^i_\mu(\alpha_i)$ is $dX^i_\mu(\alpha)/d\alpha_i$. The first term is the integral of proper time, *the ordinary action of relativistic mechanics of free particles of mass* m_i. (We sum in the usual way on the repeated index μ.) The second term represents the *electrical interaction* of the *charges*. It is summed over each pair of *charges* (the factor ½ is to count each pair once, *the term* $i = j$ *is omitted to avoid self-action*). The *interaction* is a double integral over a delta function of the square of *space-time* interval I^2 between two points on the paths. Thus, *interaction* occurs only when this interval vanishes, that is, along light cones.

The fact that the interaction is exactly one-half advanced and half-retarded meant that we could write such a principle of least action, whereas interaction *via* retarded waves alone cannot be written in such a way.

So, all of classical electrodynamics was contained in this very simple form. It looked good, and therefore, it was undoubtedly true, at least to the beginner. *It automatically gave half-advanced and half-retarded effects and it was without fields. By omitting the term in the sum when* $i = j$, *I omit self-interaction and no longer have any infinite self-energy.* This then was the hoped-for solution to the problem of ridding classical electrodynamics of the infinities.

It turns out, of course, that you can reinstate fields if you wish to, but you have to keep track of the field produced by each particle separately. This is because to find the right field to act on a given particle, you must exclude the field that it creates itself. A single universal field to which all contribute will not do. This idea had been suggested earlier by Frenkel and so we called these Frenkel fields. *This theory which allowed only particles to act on each other was equivalent to Frenkel's fields using half-advanced and half-retarded solutions.*

There were several suggestions for interesting modifications of electrodynamics. We discussed lots of them, but I shall report on only one. It was to *replace this delta function in the interaction by another function, say,* $f(I^2_{ij})$, *which is not infinitely sharp*. Instead of having the action occur only when the interval between the two *charges* is exactly zero, we would replace the delta function of I^2 by a narrow-peaked thing. Let's say that $f(Z)$ is large only near $Z = 0$ width of order a^2. *Interactions* will now occur when $T^2 - R^2$ is of order a^2 roughly where T is the time difference and R is the separation of the *charges*. This might look like it disagrees with experience, but if a is some small distance, like 10^{-13} cm, it says that the time delay T in action is roughly $\sqrt{(R^2 \pm a^2)}$ or approximately, – if R is much larger than a, $T = R \pm a^2/2R$. This means that the deviation of time T from the ideal theoretical time R of Maxwell, gets smaller and smaller, the further the pieces are apart. Therefore, all theories involving in analyzing generators, motors, etc., in fact, all of the tests of electrodynamics that were available in Maxwell's time, would be adequately satisfied if were 10^{-13} cm. If R is of the order of a centimeter this deviation in T is only 10^{-26} parts. So, it was possible, also, to change the theory in a simple manner and to still agree with all observations of classical electrodynamics. You have no clue of precisely what function to put in for *f*, but it was an interesting possibility to keep in mind when developing quantum electrodynamics.

It also occurred to us that if we did that (replace δ by *f*) we could not reinstate the term i = j in the sum because this would now represent in a *relativistically* invariant fashion a *finite action of a charge on itself*. In fact, it was possible to prove that if we did do such a thing, the main effect of the *self-action* (for not too rapid accelerations) would be to produce a *modification of the mass*. In fact, there need be no mass m_i, term, all the *mechanical mass* could be *electromagnetic self-action*. So, if you would like, we could also have another theory with a still simpler expression for the action A. In expression (1) only the second term is kept, the sum extended over all i and j, and some function replaces δ. *Such a simple form could represent all of classical electrodynamics, which aside from gravitation is essentially all of classical physics.*

Although it may sound confusing, I am describing several different alternative theories at once. The important thing to note is that at this time we had all these in mind as different possibilities. There were several possible solutions of the difficulty of classical electrodynamics, any one of which might serve as a good starting point to the solution of the difficulties of quantum electrodynamics.

I would also like to emphasize that by this time I was becoming used to a physical point of view different from the more customary point of view. In the customary view, things are discussed as a function of time in very great detail. For example, you have the *field* at this moment, a differential equation gives you the *field* at the next moment and so on; a method, which I shall call *the Hamilton method, the time differential method.* We have, instead (in (1) say) a thing *that describes the character of the path throughout all of space and time.* The behavior of nature is determined by saying her whole *spacetime path* has a certain character. For an *action* like (1) the equations obtained by variation (of $X^i\mu$ (a_i)) are no longer at all easy to get back into Hamiltonian form. If you wish to use as variables only the coordinates of particles, then you can talk about the property of the paths – but the path of one particle at a given time is affected by the path of another at a different time. If you try to describe, therefore, things *differentially*, telling what the present conditions of the particles are, and how these present conditions will affect the future you see, it is impossible with particles alone, because something the particle did in the past is going to affect the future.

Therefore, you need a lot of bookkeeping variables to keep track of what the particle did in the past. These are called *field variables*. You will, also, have to tell what the *field* is at this present moment, if you are to be able to see later what is going to happen. From the overall *space-time* view of the *least action principle*, the *field* disappears as nothing but bookkeeping variables insisted on by the Hamiltonian method.

As a by-product of this same view, I received a telephone call one day at the graduate college at Princeton from Professor Wheeler, in which he said, "Feynman, I know why all electrons have the same charge and the same mass" "Why?" "Because, they are all the same electron!" And, then he explained on the telephone, "suppose that the world lines which we were ordinarily considering before in time and space – instead of only going up in time were a tremendous knot, and then, when we cut through the knot, by the plane corresponding to a fixed time, we would see many, many world lines and that would represent many electrons, except for one thing. If in one section this is an ordinary electron world line, in the section in which it reversed itself and is coming back from the future we have the wrong sign to the proper time – to the proper four velocities – and that's equivalent to changing the sign of the charge, and, therefore, that part of a path would act like a positron." "But, Professor", I said, "there aren't as many positrons as electrons." "Well, maybe they are hidden in the protons or something", he said. I did not take the idea that all the electrons were the same one from him as seriously as I took the observation that *positrons could simply be represented as electrons going from the future to the past in a back section of their world lines*. That, I stole!

To summarize, when I was done with this, as a physicist I had gained two things. One, *I knew many different ways of formulating classical electrodynamics*, with many different mathematical forms. I got to know how to express the subject every which way. Second, I had a point of view – *the overall space-time point of view* – and a disrespect for the Hamiltonian method of describing physics.

I would like to interrupt here to make a remark. The fact that electrodynamics can be written in so many ways – the differential equations of Maxwell, various minimum principles with fields, minimum principles without fields, all different kinds of ways, was something I knew, but I have never understood. It always seems odd to me that the fundamental laws of physics, when discovered, can appear in so many different forms that are not apparently identical at first, but, with a little mathematical fiddling you can show the relationship. An example of that is the Schrödinger equation and the Heisenberg formulation of quantum mechanics. I don't know why this is – it remains a mystery, but it was something I learned from experience. There is always another way to say the same thing that doesn't look at all like the way you said it before. I don't know what the reason for this is. I think it is somehow a representation of the simplicity of nature. A thing like the inverse square law is just right to be represented by the solution of Poisson's equation, which, therefore, is a very different way to say the same thing that doesn't look at all like the way you said it before. I don't know what it means, that nature chooses these curious forms, but maybe that is a way of defining simplicity. Perhaps a thing is simple if you can describe it fully in several different ways without immediately knowing that you are describing the same thing.

I was now convinced that since we had solved the problem of classical electrodynamics (and completely in accordance with my program from M.I.T., *only direct interaction between particles, in a way that made fields unnecessary*) that everything was definitely going to be all right. *I was convinced that all I had to do was make a quantum theory analogous to the classical one and everything would be solved.*

So, the problem is only to make a quantum theory, which has as its classical analog, this expression (1). Now, there is no unique way to make a quantum theory from classical mechanics, although all the textbooks make believe there is. What they would tell you to do, was find the *momentum variables* and replace them by $(\hbar/i)(\partial/\partial x)$, but I couldn't find a *momentum variable*, as there wasn't any.

The character of quantum mechanics of the day was to write things in the famous Hamiltonian way – in the form of a *differential equation*, which described how the wave function changes from instant to instant, and in terms of an operator, H. If the classical physics could be reduced to a Hamiltonian form, everything was all right. Now, *least action does not imply a Hamiltonian form if the action is a function of anything more than positions and velocities at the same moment*. If the *action* is of the form of the *integral of a function*, (usually called the Lagrangian) of the *velocities* and *positions* at the same time

$$S = \int L(\dot{x},x)\, dt$$

then you can start with the Lagrangian and then create a Hamiltonian and work out the quantum mechanics, more or less uniquely. But this thing (1) involves the key variables, *positions* at two different times, and therefore, it was not obvious what to do to make the quantum-mechanical analogue.

I tried – I would struggle in various ways. One of them was this; if I had harmonic oscillators interacting with a delay in time, I could work out what the normal modes were and guess that the quantum theory of the normal modes was the same as for simple oscillators and kind of work my way back in terms of the original variables. I succeeded in doing that, but I hoped then to generalize to other than a harmonic oscillator, but I learned to my regret something, which many people have learned. The harmonic oscillator is too simple; very often you can work out what it should do in quantum theory without getting much of a clue as to how to generalize your results to other systems.

So that didn't help me very much, but when I was struggling with this problem, I went to a beer party in the Nassau Tavern in Princeton. There was a gentleman, newly arrived from Europe (Herbert Jehle) who came and sat next to me. *Europeans are much more serious than we are in America because they think that a good place to discuss intellectual matters is a beer party*. So, he sat by me and asked, "what are you doing" and so on, and I said, "I'm drinking beer." Then I realized that he wanted to know what work I was doing and I told him I was struggling with this problem, and I simply turned to him and said, "listen, do you know any way of doing quantum mechanics, starting with *action* – where the *action integral* comes into the quantum mechanics?" "No", he said, "*but Dirac has a paper in which the Lagrangian, at least, comes into quantum mechanics*. I will show it to you tomorrow."

[Dirac, P. A. M. (1933). The Lagrangian in Quantum Mechanics. *Phys. Zeit. Sowjet.*, 3, 1, 64-72.]

Next day we went to the Princeton Library, they have little rooms on the side to discuss things, and he showed me this paper. What Dirac said was the following: There is in quantum mechanics a very important quantity which carries the wave function from one time to another, besides the differential equation but equivalent to it, a kind of a kernel, which we might call K(x', x), which carries the wave function j(x) known at time t, to the wave function j(x') at time, t + ε, Dirac points out that this function K was *analogous* to the quantity in classical mechanics that you would calculate if you took the exponential of *i*e, multiplied by the Lagrangian $L(\dot{x},x)$ imagining that these two positions x, x' corresponded t and t + ε. In other words,

K(x', x) is analogous to $e^{i\varepsilon L\{(x'-x)/\varepsilon\,,x\}/\hbar}$

Professor Jehle showed me this, I read it, he explained it to me, and I said, "what does he mean, they are analogous; what does that mean, *analogous*? What is the use of that?" He said, "you Americans! You always want to find a use for everything!" I said, that I thought that Dirac must mean that they were equal. "No", he explained, "he doesn't mean they are equal." "Well,", I said, "let's see what happens if we make them equal."

So, I simply put them equal, taking the simplest example where the Lagrangian is ½Mx² – V(x) but soon found I had to put a constant of proportionality A in, suitably adjusted. *When I substituted* $Ae^{i\varepsilon L/\hbar}$ *for K to get*

$$\psi(x',t+\varepsilon)=\int A\exp\left[\frac{i\varepsilon}{\hbar}L\left(\frac{x'-x}{\varepsilon},x\right)\right]\psi(x,t)\,dx$$

and just calculated things out by Taylor series expansion, out came the Schrödinger equation. So, I turned to Professor Jehle, not really understanding, and said, "well, you see Professor Dirac meant that they were proportional". Professor Jehle's eyes were bugging out – he had taken out a little notebook and was rapidly copying it down from the blackboard, and said, "no, no, this is an important discovery. You Americans are always trying to find out how something can be used. That's a good way to discover things!" So, I thought I was finding out what Dirac meant, but, as a matter of fact, had made the discovery that *what Dirac thought was analogous, was, in fact, equal. I had then, at least, the connection between the Lagrangian and quantum mechanics, but still with wave functions and infinitesimal times.*

It must have been a day or so later when I was lying in bed thinking about these things, that I imagined what would happen if I wanted to calculate the wave function at a finite interval later.

I would put one of these factors $e^{i\varepsilon L}$ in here, and that would give me the wave functions the next moment, t + ε and then I could substitute that back into (3) to get another factor of $e^{i\varepsilon L}$ and give me the wave function the next moment, t + 2ε and so on and so on. In that way I found myself thinking of a large number of integrals, one after the other in sequence.

In the integrand was the product of the exponentials, which, of course, was the exponential of the sum of terms like εL. Now, *L is the Lagrangian and ε is like the time interval dt, so that if you took a sum of such terms, that's exactly like an integral*. That's like Riemann's formula for the integral ∫ Ldt, you just take the value at each point and add them together. We are to take the limit as $\varepsilon \rightarrow 0$, of course. Therefore, the connection between the *wave function* of one instant and the *wave function* of another instant a finite time later could be obtained by an infinite number of integrals, (because ε goes to zero, of course) of exponential (iS/ħ) where *S* is the *action* expression (2). *At last, I had succeeded in representing quantum mechanics directly in terms of the action S.*

This led later on to the idea of the *amplitud*e for a path; that for each possible way that the particle can go from one point to another in *space-time*, there's an *amplitude*. That *amplitude* is e to the i/ħ times the *action* for the path. *Amplitudes* from various paths superpose by addition. *This then is another, a third way, of describing quantum mechanics, which looks quite different than that of Schrödinger or Heisenberg, but which is equivalent to them.*

Now immediately after making a few checks on this thing, what I wanted to do, of course, was to substitute the action (1) for the other (2). *The first trouble was that I could not get the thing to work with the relativistic case of spin one-half. However, although I could deal with the matter only non-relativistically,* I could deal with the light or the photon interactions perfectly well by just putting the interaction terms of (1) into any action, replacing the mass terms by the non-relativistic $(Mx^2/2)dt$. When the *action* has a delay, as it now had, and involved more than one time, I had to lose the idea of a *wave function*. That is, *I could no longer describe the program as; given the amplitude for all positions at a certain time to compute the amplitude at another time*. However, that didn't cause very much trouble. It just meant developing a new idea. Instead of *wave functions* we could talk about this; that if a *source* of a certain kind emits a particle, and a detector is there to receive it, we can give the *amplitude* that the *source* will emit and the detector receive. We do this without specifying the exact instant that the source emits or the exact instant that any detector receives, without trying to specify the state of anything at any particular time in between, but by just finding the amplitude for the complete experiment. And, then we could discuss how that *amplitude* would change if you had a scattering sample in between, as you rotated and changed angles, and so on, without really having any *wave functions*.

It was also possible to discover what the old concepts of *energy* and *momentum* would mean with this *generalized action*. And, so I believed that *I had a quantum theory of classical electrodynamics* – or rather of this new classical electrodynamics described by *action* (1). I made a number of checks. If I took the *Frenkel field* point of view, which you remember was more *differential*, I could convert it directly to quantum mechanics in a more conventional way. *The only problem was how to specify in quantum mechanics the classical boundary conditions to use only half-advanced and half-retarded solutions.* By some ingenuity in defining what that meant, *I found that the quantum mechanics with Frenkel fields, plus a special boundary condition, gave me back this action, (1) in the new form of quantum mechanics with a delay*. So, various things indicated that there wasn't any doubt I had everything straightened out.

It was also easy to guess how to modify the electrodynamics, if anybody ever wanted to modify it. I just changed the delta to an *f*, just as I would for the classical case. So, it was very easy, a simple thing. To describe the old retarded theory without explicit mention of fields I would have to write probabilities, not just amplitudes. I would have to square my amplitudes and that would involve double path integrals in which there are two S's and so forth. Yet, as I worked out many of these things and studied different forms and different boundary conditions. *I got a kind of funny feeling that things weren't exactly right. I could not clearly identify the difficulty and in one of the short periods during which I imagined I had laid it to rest, I published a thesis and received my Ph.D.*

> [Feynman, R. P. (1942). *The Principle of Least Action in Quantum Mechanics*. Ph.D. thesis, Princeton University (in Brown, L. M. (ed.) (2005). *Feynman's Thesis. A New Approach to Quantum Theory*. World Scientific); *non-relativistic* quantum mechanics, space-time viewpoint, *principle of least action*, functional, Lagrangian, action functions.]

During the war, I didn't have time to work on these things very extensively, but wandered about on buses and so forth, with little pieces of paper, and struggled to work on it and *discovered indeed that there was something wrong, something terribly wrong. I found that if one generalized the action from the nice Langrangian forms (2) to these forms (1) then the quantities which I defined as energy, and so on, would be complex*. The energy values of *stationary states* wouldn't be real and probabilities of events wouldn't add up to 100%. That is, if you took the probability that this would happen and that would happen – everything you could think of would happen, it would not add up to one.

Another problem on which I struggled very hard, was to represent relativistic electrons with this new quantum mechanics. I wanted to do a unique and different way – and not just by copying the operators of Dirac into some kind of an expression and using some kind of Dirac algebra instead of ordinary complex numbers. I was very much encouraged by the fact that *in one space dimension, I did find a way of giving an amplitude to every path by limiting myself to paths, which only went back and forth at the speed of light.* The amplitude was simple ($i\varepsilon$) to a power equal to the number of velocity reversals where I have divided the time into steps and I am allowed to reverse velocity only at such a time. *This gives (as approaches zero) Dirac's equation in two dimensions – one dimension of space and one of time*

$$(\hbar = M = c = 1).$$

Dirac's wave function has four components in four dimensions, but in this case, it has only two components and this rule for the *amplitude* of a path automatically generates the need for two components. Because if this is the formula for the *amplitudes* of a path, it will not do you any good to know the *total amplitude* of all paths, which come into a given point to find the *amplitude* to reach the next point. This is because for the next time, if it came in from the right, there is no new factor iε if it goes out to the right, whereas, if it came in from the left there was a new factor iε. So, to continue this same information forward to the next moment, it was not sufficient information to know the total amplitude to arrive, but you had to know the amplitude to arrive from the right and the amplitude to arrive to the left, independently. If you did, however, you could then compute both of those again independently and thus you had to carry two *amplitudes* to form a *differential equation* (first order in time).

And, so I dreamed that if I were clever, I would find a formula for the *amplitude* of a path that was beautiful and simple for three dimensions of space and one of time, which would be equivalent to the *Dirac equation*, and for which the four components, matrices, and all those other mathematical funny things would come out as a simple consequence – I *have never succeeded in that either*. But I did want to mention some of the unsuccessful things on which I spent almost as much effort, as on the things that did work.

To summarize the situation a few years after the war, I would say, I had much experience with *quantum electrodynamics*, at least in the knowledge of many different ways of formulating it, in terms of path integrals of *actions* and in other forms. One of the important by-products, for example, of much experience in these simple forms, was that it was easy to see how to combine together what was in those days called the *longitudinal* and *transverse fields*, and in general, to see clearly the *relativistic* invariance of the theory. Because of the need to do things differentially there had been, in the standard quantum electrodynamics, a complete split of the *field* into two parts, one of which is called the *longitudinal part* and the other mediated by the photons, or *transverse waves*. The *longitudinal part* was described by a Coulomb *potential* acting instantaneously in the Schrödinger equation, while the *transverse part* had entirely different description in terms of quantization of the *transverse waves*. This separation depended upon the *relativistic* tilt of your axes in *spacetime*. People moving at different velocities would separate the same field into longitudinal and transverse fields in a different way. Furthermore, *the entire formulation of quantum mechanics insisting, as it did, on the wave function at a given time, was hard to analyze relativistically*. Somebody else in a different coordinate system would calculate the succession of events in terms of wave functions on differently cut slices of *space-time*, and with a different separation of longitudinal and transverse parts. *The Hamiltonian theory did not look relativistically invariant, although, of course, it was*. One of the great advantages of the overall point of view, was that you could see the *relativistic invariance* right away – or as Schwinger would say – the covariance was manifest. *I had the advantage, therefore, of having a manifestedly covariant form for quantum electrodynamics* with suggestions for modifications and so on. I had the disadvantage that if I took it too seriously – I mean, if I took it seriously at all in this form, – *I got into trouble with these complex energies and the failure of adding probabilities to one and so on*. I was unsuccessfully struggling with that.

Then Lamb did his experiment, measuring the separation of the $2S½$ and $2P½$ levels of hydrogen, finding it to be about 1000 megacycles of frequency difference. Professor Bethe, with whom I was then associated at Cornell, is a man who has this characteristic: If there's a good experimental number you've got to figure it out from theory. So, he forced the quantum electrodynamics of the day to give him an answer to the separation of these two levels. He pointed out that the *self-energy* of an electron itself is infinite, so that the calculated energy of a bound electron should also come out infinite. *But, when you calculated the separation of the two energy levels in terms of the corrected mass instead of the old mass, it would turn out, he thought, that the theory would give convergent finite answers*. He made an estimate of the splitting that way and found out that it was still divergent, but he guessed that was probably due to the fact that he used an *unrelativistic* theory of the *matter*. *Assuming it would be convergent if relativistically treated, he estimated he would get about a thousand megacycles for the Lamb-shift, and thus, made the most important discovery in the history of the theory of quantum electrodynamics*. He worked this out on the train from Ithaca, New York to Schenectady and telephoned me excitedly from Schenectady to tell me the result, which I don't remember fully appreciating at the time.

Returning to Cornell, he gave a lecture on the subject, which I attended. He explained that it gets very confusing to figure out exactly which infinite term corresponds to what in trying to make the correction for the infinite change in mass. If there were any modifications whatever, he said, even though not physically correct, (that is not necessarily the way nature actually works) but any modification whatever at high frequencies, which would make this correction finite, then there would be no problem at all to figuring out how to keep track of everything. You just calculate the finite *mass correction* Dm to the *electron mass* m_o, substitute the numerical values of m_o + Dm for m in the results for any other problem and all these ambiguities would be resolved. If, in addition, this method were *relativistically* invariant, then we would be absolutely sure how to do it without destroying *relativistically* invariant.

After the lecture, I went up to him and told him, "I can do that for you, I'll bring it in for you tomorrow." I guess I knew every way to modify quantum electrodynamics known to man, at the time. So, I went in next day, and explained what would correspond to the modification of the delta-function to *f* and asked him to explain to me how you calculate the *self-energy* of an electron, for instance, so we can figure out if it's finite.

I want you to see an interesting point. I did not take the advice of Professor Jehle to find out how it was useful. *I never used all that machinery which I had cooked up to solve a single relativistic problem*. I hadn't even calculated the *self-energy* of an electron up to that moment, and was studying the difficulties with the conservation of probability, and so on, without actually doing anything, except discussing the general properties of the theory.

But now I went to Professor Bethe, who explained to me on the blackboard, as we worked together, how to calculate the *self-energy* of an electron. Up to that time when you did the integrals they had been logarithmically divergent. I told him how to make the *relativistically* invariant modifications that I thought would make everything all right. We set up the integral which then diverged at the sixth power of the *frequency* instead of logarithmically!

So, I went back to my room and worried about this thing and went around in circles trying to figure out what was wrong because I was sure physically everything had to come out finite, *I couldn't understand how it came out infinite*. I became more and more interested and finally realized I had to learn how to make a calculation. So, ultimately, I taught myself how to calculate the self-energy of an electron working my patient way through the terrible confusion of those days of negative energy states and holes and longitudinal contributions and so on. *When I finally found out how to do it and did it with the modifications I wanted to suggest, it turned out that it was nicely convergent and finite, just as I had expected.* Professor Bethe and I have never been able to discover what we did wrong on that blackboard two months before, but apparently, we just went off somewhere and we have never been able to figure out where. It turned out, that what I had proposed, if we had carried it out without making a mistake would have been all right and would have given a finite correction. Anyway, it forced me to go back over all this and to convince myself physically that nothing can go wrong. At any rate, the correction to *mass* was now finite, proportional to ln $(ma/\hbar)$ where a is the width of that function f which was substituted for δ. *If you wanted an unmodified electrodynamics, you would have to take a equal to zero, getting an infinite mass correction.* But that wasn't the point. *Keeping a finite, I simply followed the program outlined by Professor Bethe and showed how to calculate all the various things, the scatterings of electrons from atoms without radiation, the shifts of levels and so forth, calculating everything in terms of the experimental mass*, and noting that the results as Bethe suggested, were not sensitive to a in this form and even had a definite limit as ag0.

The rest of my work was simply to improve the techniques then available for calculations, making diagrams to help analyze perturbation theory quicker. Most of this was first worked out by guessing – you see, I didn't have the *relativistic theory of matter*. For example, it seemed to me obvious that the velocities in *non-relativistic* formulas have to be replaced by Dirac's matrix α or in the more *relativistic* forms by the operators γ_μ. I just took my guesses from the forms that I had worked out using path integrals for *non-relativistic matter*, but *relativistic* light. It was easy to develop rules of what to substitute to get the *relativistic* case. I was very surprised to discover that it was not known at that time, that every one of the formulas that had been worked out so patiently by separating *longitudinal* and *transverse* waves could be obtained from the formula for the *transverse* waves alone, if instead of summing over only the two perpendicular polarization directions you would sum over all four possible directions of polarization. It was so obvious from the action (1) that I thought it was general knowledge and would do it all the time. I would get into arguments with people, because I didn't realize they didn't know that; but it turned out that all their patient work with the *longitudinal* waves was always equivalent to just extending the sum on the two *transverse* directions of polarization over all four directions. This was one of the amusing advantages of the method. *In addition, I included diagrams for the various terms of the perturbation series, improved notations to be used, worked out easy ways to evaluate integrals, which occurred in these problems, and so on, and made a kind of handbook on how to do quantum electrodynamics.*

But one step of importance that was physically new was involved with the negative energy sea of Dirac, which caused me so much logical difficulty. I got so confused that I remembered Wheeler's old idea about the positron being, maybe, the electron going backward in time. Therefore, *in the time dependent perturbation theory that was usual for getting self-energy, I simply supposed that for a while we could go backward in the time, and looked at what terms I got by running the time variables backward.* They were the same as the terms that other people got when they did the problem a more complicated way, using holes in the sea, except, possibly, for some signs. These, I, at first, determined empirically by inventing and trying some rules.

I have tried to explain that all the improvements of *relativistic* theory were at first more or less straightforward, semi-empirical shenanigans. Each time I would discover something, however, I would go back and I would check it so many ways, compare it to every problem that had been done previously in electrodynamics (and later, in weak coupling meson theory) to see if it would always agree, and so on, until I was absolutely convinced of the truth of the various rules and regulations which I concocted to simplify all the work.

During this time, people had been developing meson theory, a subject I had not studied in any detail. I became interested in the possible application of my methods to perturbation calculations in meson theory. But what was meson theory? All I knew was that meson theory was something analogous to electrodynamics, except that particles corresponding to the photon had a mass. It was easy to guess the δ-function in (1), which was a solution of d'Alembertian equals zero, was to be changed to the corresponding solution of d'Alembertian equals m^2. Next, there were different kind of mesons – the one in closest analogy to photons, coupled *via* $\gamma_\mu\gamma_\mu$, are called vector mesons – there were also scalar mesons. Well, maybe that corresponds to putting unity in place of the γ_μ, I would here then speak of "pseudo vector coupling" and I would guess what that probably was. *I didn't have the knowledge to understand the way these were defined in the conventional papers because they were expressed at that time in terms of creation and annihilation operators, and so on, which, I had not successfully learned.* I remember that when someone had started to teach me about *creation* and *annihilation* operators, that this operator creates an electron, I said, "how do you create an electron? It disagrees with the conservation of charge", and in that way, I blocked my mind from learning a very practical scheme of calculation. Therefore, I had to find as many opportunities as possible to test whether I guessed right as to what the various theories were.

One day a dispute arose at a Physical Society meeting as to the correctness of a calculation by Slotnick of the interaction of an electron with a neutron using pseudo scalar theory with pseudo vector coupling and also, pseudo scalar theory with pseudo scalar coupling. He had found that the answers were not the same, in fact, by one theory, the result was divergent, although convergent with the other. Some people believed that the two theories must give the same answer for the problem. This was a welcome opportunity to test my guesses as to whether I really did understand what these two couplings were. So, I went home, and during the evening I worked out the electron neutron scattering for the pseudo scalar and pseudo vector coupling, saw they were not equal and subtracted them, and worked out the difference in detail. The next day at the meeting, I saw Slotnick and said, "Slotnick, I worked it out last night, I wanted to see if I got the same answers you do. I got a different answer for each coupling – but, I would like to check in detail with you because I want to make sure of my methods." And, he said, "what do you mean you worked it out last night, it took me six months!" And, when we compared the answers, he looked at mine and he asked, "what is that Q in there, that variable Q?" (I had expressions like ($\tan^{-1}Q$) /Q etc.). I said, "that's the *momentum* transferred by the electron, the electron deflected by different angles." "Oh", he said, "no, I only have the limiting value as Q approaches zero; the *forward scattering*." Well, it was easy enough to just substitute Q equals zero in my form and I then got the same answers as he did. *But it took him six months to do the case of zero momentum transfer, whereas, during one evening I had done the finite and arbitrary momentum transfer.* That was a thrilling moment for me, like receiving the Nobel Prize, because that convinced me, at last, I did have some kind of method and technique and understood how to do something that other people did not know how to do. That was my moment of triumph in which I realized I really had succeeded in working out something worthwhile.

At this stage, I was urged to publish this because everybody said it looks like an easy way to make calculations, and wanted to know how to do it. I had to publish it, missing two things; one was proof of every statement in a mathematically conventional sense. Often, even in a physicist's sense, I did not have a demonstration of how to get all of these rules and equations from conventional electrodynamics. But I did know from experience, from fooling around, that everything was, in fact, equivalent to the regular electrodynamics and had partial proofs of many pieces, although, I never really sat down, like Euclid did for the geometers of Greece, and made sure that you could get it all from a single simple set of axioms. *As a result, the work was criticized, I don't know whether favorably or unfavorably, and the "method" was called the "intuitive method".* For those who do not realize it, however, I should like to emphasize that there is a lot of work involved in using this "intuitive method" successfully. Because no simple clear proof of the formula or idea presents itself, it is necessary to do an unusually great amount of checking and rechecking for consistency and correctness in terms of what is known, by comparing to other analogous examples, limiting cases, etc. In the face of the lack of direct mathematical demonstration, one must be careful and thorough to make sure of the point, and one should make a perpetual attempt to demonstrate as much of the formula as possible. Nevertheless, a very great deal more truth can become known than can be proven.

It must be clearly understood that in all this work, *I was representing the conventional electrodynamics with retarded interaction, and not my half-advanced and half-retarded theory* corresponding to (1). I merely use (1) to guess at forms. And, one of the forms I guessed at corresponded to changing δ to a function f of width a^2, so that I could calculate finite results for all of the problems. This brings me to *the second thing that was missing* *when I published the paper, an unresolved difficulty. With δ replaced by f the calculations would give results which were not "unitary", that is, for which the sum of the probabilities of all alternatives was not unity.* The deviation from unity was very small, in practice, if a was very small. In the limit that I took a very tiny, it might not make any difference. And, so the process of the *renormalization* could be made, you could calculate everything in terms of the *experimental mass* and then take the limit and *the apparent difficulty that the unitary is violated temporarily seems to disappear. I was unable to demonstrate that, as a matter of fact, it does.*

It is lucky that I did not wait to straighten out that point, for *as far as I know, nobody has yet been able to resolve this question.* Experience with meson theories with stronger couplings and with strongly coupled vector photons, although not proving anything, convinces me that if the coupling were stronger, or if you went to a higher order (137th order of perturbation theory for electrodynamics), this difficulty would remain in the limit and there would be real trouble. That is, *I believe there is really no satisfactory quantum electrodynamics*, but I'm not sure. And, I believe, that one of the reasons for the slowness of present-day progress in understanding the *strong* interactions is that *there isn't any relativistic theoretical model, from which you can really calculate everything.* Although, it is usually said, that the difficulty lies in the fact that strong interactions are too hard to calculate, I believe, it is really because strong interactions in field theory have no solution, have no sense they're either infinite, or, if you try to modify them, the modification destroys the unitarity. *I don't think we have a completely satisfactory relativistic quantum-mechanical model*, even one that doesn't agree with nature, but, at least, agrees with the logic that the sum of probability of all alternatives has to be 100%. Therefore, *I think that the renormalization theory is simply a way to sweep the difficulties of the divergences of electrodynamics under the rug.* I am, of course, not sure of that.

This completes the story of the development of the *space-time* view of quantum electrodynamics. I wonder if anything can be learned from it. I doubt it. *It is most striking that most of the ideas developed in the course of this research were not ultimately used in the final result.* For example, *the half-advanced and half-retarded potential was not finally used, the action expression (1) was not used, the idea that charges do not act on themselves was abandoned.* The *path-integral formulation* of quantum mechanics was useful for guessing at final expressions and at formulating the general theory of electrodynamics in new ways – although, strictly it was not absolutely necessary. The same goes for *the idea of the positron being a backward moving electron, it was very convenient, but not strictly necessary for the theory because it is exactly equivalent to the negative energy sea point of view.*

We are struck by the very large number of different physical viewpoints and widely different mathematical formulations that are all equivalent to one another. *The method used here, of reasoning in physical terms, therefore, appears to be extremely inefficient.* On looking back over the work, I can only feel a kind of regret for the enormous amount of physical reasoning and mathematically re-expression which ends by merely re-expressing what was previously known, although in a form which is much more efficient for the calculation of specific problems. Would it not have been much easier to simply work entirely in the mathematical framework to elaborate a more efficient expression? This would certainly seem to be the case, but it must be remarked that although the problem actually solved was only such a reformulation, *the problem originally tackled was the (possibly still unsolved) problem of avoidance of the infinities of the usual theory*. Therefore, a new theory was sought, not just a modification of the old. Although *the quest was unsuccessful*, we should look at the question of the value of physical ideas in developing a *new* theory.

Many different physical ideas can describe the same physical reality. Thus, classical electrodynamics can be described by a field view, or an action at a distance view, etc. Originally, Maxwell filled space with idler wheels, and Faraday with fields lines, but somehow the Maxwell equations themselves are pristine and independent of the elaboration of words attempting a physical description. The only true physical description is that describing the experimental meaning of the quantities in the equation – or better, the way the equations are to be used in describing experimental observations. This being the case perhaps the best way to proceed is to try to guess equations, and disregard physical models or descriptions. For example, McCullough guessed the correct equations for light propagation in a crystal long before his colleagues using elastic models could make head or tail of the phenomena, or again, *Dirac obtained his equation for the description of the electron by an almost purely mathematical proposition. A simple physical view by which all the contents of this equation can be seen is still lacking.*

Therefore, *I think equation guessing might be the best method to proceed to obtain the laws for the part of physics which is presently unknown.* Yet, when I was much younger, I tried this equation guessing and I have seen many students try this, but it is very easy to go off in wildly incorrect and impossible directions. I think the problem is not to find the *best* or most efficient method to proceed to a discovery, but to find any method at all. Physical reasoning does help some people to generate suggestions as to how the unknown may be related to the known. Theories of the known, which are described by different physical ideas may be equivalent in all their predictions and are hence scientifically indistinguishable. However, they are not psychologically identical when trying to move from that base into the unknown. For different views suggest different kinds of modifications which might be made and hence are not equivalent in the hypotheses one generates from them in ones attempt to understand what is not yet understood. I, therefore, think that a good theoretical physicist today might find it useful to have a wide range of physical viewpoints and mathematical expressions of the same theory (for example, of *quantum electrodynamics*) available to him. This may be asking too much of one man. Then new students should as a class have this. If every individual student follows the same current fashion in expressing and thinking about electrodynamics or field theory, then the variety of hypotheses being generated to understand strong interactions, say, is limited. Perhaps rightly so, for possibly the chance is high that the truth lies in the fashionable direction. But, on the off-chance that it is in another direction – a direction obvious from an unfashionable view of field theory – who will find it? Only someone who has sacrificed himself by teaching himself *quantum electrodynamics* from a peculiar and unusual point of view; one that he may have to invent for himself. I say sacrificed himself because he most likely will get nothing from it, because the truth may lie in another direction, perhaps even the fashionable one.

But, if my own experience is any guide, the sacrifice is really not great because if the peculiar viewpoint taken is truly experimentally equivalent to the usual in the realm of the known there is always a range of applications and problems in this realm for which the special viewpoint gives one a special power and clarity of thought, which is valuable in itself. Furthermore, in the search for new laws, you always have the psychological excitement of feeling that possible nobody has yet thought of the crazy possibility you are looking at right now.

So, what happened to the old theory that I fell in love with as a youth? Well, I would say it's become an old lady, that has very little attractive left in her and the young today will not have their hearts pound anymore when they look at her. But we can say the best we can for any old woman, that she has been a very good mother and she has given birth to some very good children. And, I thank the Swedish Academy of Sciences for complimenting one of them. Thank you.

Julian Schwinger – Nobel Lecture, December 11, 1965. *Relativistic Quantum Field Theory*.

[https://www.nobelprize.org/prizes/physics/1965/schwinger/lecture/.]

> Schwinger was jointly awarded the Nobel Prize in Physics in 1965 for his work on *quantum electrodynamics*, along with Sin-Itiro Tomonaga and Richard Feynman "for their fundamental work in *quantum electrodynamics*, with deep-ploughing consequences for the physics of elementary particles"
>
> Following the establishment of the theory of relativity and quantum mechanics, an initial relativistic theory was formulated for the interaction between charged particles and electromagnetic fields. However, partly because the electron's *magnetic moment* proved to be somewhat larger than expected, the theory had to be reformulated. Julian Schwinger solved this problem in 1948 through "*renormalization*" and thereby contributed to a new quantum electrodynamics. [Julian Schwinger – Facts. NobelPrize.org. https://www.nobelprize.org/prizes/physics/1965/schwinger/facts/.]

Schwinger does not mention any of his 1948 and 1949 papers in his Nobel Prize lecture for which he was awarded the Nobel Prize, instead he focuses on papers he wrote between 1951 and 1965, he starts by describing the logical foundations of *quantum field theory* or *relativistic quantum mechanics* which he defines *as the synthesis of quantum mechanics with relativity*, he notes improvements in the formal presentation of quantum mechanical principles by himself described in a series of six papers on the theory of quantized fields published in *Physical Review* between June 1951 and June 1954 and by Feynman described in his April 1948 Space-Time Approach to Non-Relativistic Quantum Mechanics utilizing the concept of *action* based on a study of Dirac concerning the correspondence between the quantum *transformation function* and the classical *action*, he identifies two distinct formulations of quantum mechanics - a *differential formulation* utilizing a *differential* version of the *action* principle and the *path integral formulation* of Feynman, he claims that his *differential* version transcends the *correspondence principle* and incorporates on the same footing the two different kinds of quantum dynamical variable that are demanded empirically by the two known varieties of particles obeying Bose-Einstein or Fermi-Dirac statistics, the *quantum action principle is a differential statement about time transformation functions, all quantum-dynamical aspects of the system are derived from a single dynamical principle*, he also claims that *quantum field theory* had failed no significant test nor could any decisive confrontation be anticipated in the near future contrary to Feynman's reservations and Schwinger's statements in the preface to his 1958 book, *classical mechanics is a determinate theory*, knowledge of the *state* at a given time permits precise prediction of the result of measuring any property of the system, *quantum mechanics is only statistically determinate*, it is the *probability* of attaining a particular result on measuring any property of the system not the outcome of an individual microscopic observation that is predictable from knowledge of the *state*, *relativistic* structure of the *action principle* is completed by demanding that it present the same form independently of the particular partitioning of *space-time* into space and time, facilitated by the appearance of the *action operator* - the time integral of the Lagrangian - as the *space-time* integral of the Lagrange function, he also discusses a further eleven papers that he had published in *Physical Review* between July 1962 and October 1965 which attempted to extend his theory to gravitational fields and develop a *field theory of matter.*

> [From the middle of the 1950s, some of his theories began to be challenged. In the 1960s, Schwinger formulated and analyzed what is now known as the *Schwinger model*, quantum electrodynamics in one space and one time dimension, the first example of a confining theory. He attempted to formulate a theory of quantum electrodynamics with point *magnetic monopoles*, a program which met with limited success because monopoles are strongly interacting when the quantum of charge is small.
>
> From 1966, he began to develop his '*source theory*', which was another reformulation of *quantum electrodynamics* theory. However, the theory was not accepted by many of colleagues at Harvard.
>
> Starting in the 1980s, Schwinger began a series of papers on the Thomas-Fermi model of atoms. Then from 1989, he began to take interest in the non-mainstream research of cold fusion. However, his papers were not accepted for publication.]

The *relativistic* quantum theory of fields was born some thirty-five years ago [1930] through the paternal efforts of Dirac, Heisenberg, Pauli and others. It was a somewhat retarded youngster, however, and first reached adolescence seventeen years later [1947], an event which we are gathered here to celebrate. *But it is the subsequent development and more mature phase of the subject that I wish to discuss briefly today.*

I shall begin by describing to you the logical foundations of *relativistic quantum field theory.* No dry recital of lifeless « axioms » is intended but, rather, an outline of its organic growth and development *as the synthesis of quantum mechanics with relativity.* Indeed, *relativistic quantum mechanics* - the union of the complementarity principle of Bohr with the relativity principle of Einstein - is *quantum field theory.* I beg your indulgence for the mode of expression I must often use. Mathematics is the natural language of theoretical physics. It is the irreplaceable instrument for the penetration of realms of physical phenomena far beyond the ordinary experience upon which conventional language is based.

Improvements in the formal presentation of quantum mechanical principles, utilizing the concept of *action*, have been interesting by-products of work in *quantum field theory.* Both my efforts in this direction[1]

> [1] Some references are: Schwinger, J. (June, 1951). The Theory of Quantized Fields. I. *Phys. Rev.*, 82, 6, 914; https://doi.org/10.1103/PhysRev.82.914; also in Schwinger, J. (ed). (1958). *Selected Papers on Quantum electrodynamics*. Dover, New York, pages 342-55 [; "The conventional correspondence basis for quantum dynamics is here replaced by a self-contained quantum dynamical principle from which the *equations of motion* and the

commutation relations can be deduced"]; Schwinger, J. (August, 1953). The Theory of Quantized Fields. II. *Phys. Rev.*, 91, 3, 713; https://doi.org/10.1103/ PhysRev.91.713, also in Schwinger, J. (ed). (1958). *Selected Papers on Quantum electrodynamics*. Dover, New York, pages 356-71 [; "The arguments leading to the *differential* formulation of the *action principle* for a general field are presented, the general field is decomposed into two sets, which are identified with Bose-Einstein and Fermi-Dirac fields"]; Schwinger, J. (June, 1960). Unitary transformations and the action principle. PNAS, 46, 883-97; http://www.pnas.org/content/46/6/883.full.pdf [; "We shall consider, in particular, infinitesimal unitary transformations and their composition to form finite transformations. In the special example of a particular type of quantum degree of freedom, for which the complementary pair of properties have continuous spectra, this leads to a powerful tool for the construction of transformation functions - the *action principle*"].

and those of Feynman[2] (which began earlier)

[2] Feynman, R. P. (April, 1948). Space-Time Approach to Non-Relativistic Quantum Mechanics. *Rev. Mod. Phys.*, 20, 2, 367-87; Feynman, R. & Hibbs, A. (1965). *Quantum Mechanics and Path Integrals*, McGraw-Hill, New York, 1965.

were based on a study of Dirac concerning the correspondence between the quantum transformation function and the classical action. We followed quite different paths, however, and two distinct formulations of quantum mechanics emerged which can be distinguished as *differential* and *integral* viewpoints.

In order to suggest the conceptual advantages of these formulations, I shall indicate how the *differential* version transcends the *correspondence principle* and *incorporates, on the same footing, two different kinds of quantum dynamical variable*. It is just these two types *that are demanded empirically by the two known varieties of particle statistics*. The familiar properties of the variables q_k, p_k, $k= 1 \ldots n$, of the conventional quantum system enable one to derive the form of the *quantum action principle*. It is a *differential* statement about *time transformation functions*,

$$\delta\langle t_1|t_2\rangle = (i/\hbar)\langle t_1|\delta[\int_{t_2}^{t_1} dt\, L]|t_2\rangle \qquad (1)$$

which is valid for a certain class of kinematical and dynamical variations. The quantum Lagrangian operator of this system can be given the very symmetrical form

$$L = \sum_{k=1}^{n} \tfrac{1}{4} (p_k\, dq_k/dt - q_k\, dp_k/dt + dq_k/dt\, p_k - dp_k/dt\, q_k) - H(q,p,t) \qquad (2)$$

The symmetry is emphasized by collecting all the variables into the 2n-component Hermitian vector z(t) and writing

$$L = \tfrac{1}{4} (za\, dz/dt - dz/dt\, az) - H(z,t) \qquad (3)$$

where a is a real anti-symmetrical matrix, which only connects the complementary pairs of variables.

The *transformation function* depends explicitly upon the choice of *terminal states* and implicitly upon the dynamical nature of the system. If the latter is held fixed, any alteration of the *transformation function* must refer to changes in the *states*, as given by

$$\delta\langle t_1| = (i/\hbar)\langle t_1|G_1 \qquad \delta|t_2\rangle = -(i/\hbar)\, G_2|t_2\rangle \qquad (4)$$

where G_1 and G_2, are infinitesimal Hermitian operators constructed from *dynamical variables* of the system at the specified times. For a given dynamical system, then,

$$\delta[\int_{t_2}^{t_1} dt\, L] = G_1 - G_2 \qquad (5)$$

which is the *quantum principle of stationary action*, or *Hamilton's principle*, since there is no reference on the right-hand side to variations at intermediate times. The *stationary action principle* implies *equations of motion* for the *dynamical variables* and supplies explicit expressions for the *infinitesimal operators* $G_{1,2}$. The interpretation of these operators as generators of transformations on *states*, and on the *dynamical variables*, implies *commutation relations. In this way, all quantum-dynamical aspects of the system are derived from a single dynamical principle*. The specific form of the *commutation relations* obtained from the symmetrical treatment of the usual quantum system is given by the matrix statement

$$[z(t),z(t)] = i\hbar a^{-1} \qquad (6)$$

Note particularly how the anti-symmetry of the *commutator* matches the anti-symmetry of the matrix *a*.

We may now ask whether this general form of *Lagrangian operator*,

$$L = \tfrac{1}{4}\,(xA\, dx/dt - dx/dt\, Ax) - H(x,t) \qquad (7)$$

also describes other kinds of quantum systems, if the properties of the matrix A and of the Hermitian variables x are not initially assigned. There is one general restriction on the matrix A, however. It must be skew-Hermitian, as in the realization by the real, anti-symmetrical matrix *a*. Only one other simple possibility then appears, that of an imaginary, symmetrical matrix. We write that kind of matrix as $i\alpha$, where α is real and symmetrical, and designate the corresponding variables collectively by $\xi(t)$. The replacement of the antisymmetrical *a* by the symmetrical *a* requires that the anti-symmetrical *commutators* which characterize z(t) be replaced by symmetrical anti-*commutators* for $\xi(t)$, and indeed

$$\{\xi(t), \xi(t)\} = \hbar a^{-1} \qquad (8)$$

specifies the quantum nature of this second class of quantum variable. It has no classical analogue. The consistency of various aspects of the formalism *requires only that the Lagrangian operator be an even function of this second type of quantum variable*.

Time appears in quantum mechanics as a continuous parameter which represents an abstraction of the dynamical role of the measurement apparatus. The requirement of relativistic invariance invites the extension of this abstraction to include space and time coordinates. The implication that *space-time* localized measurements are a useful, if practically unrealizable idealization may be incorrect, but it is a grave error to dismiss the concept on the basis of *a priori* notions of measurability. *Microscopic measurement has no meaning apart from a theory*, and the idealized measurement concepts that are implicit in a particular theory must be accepted or rejected in accordance with the final success or failure of that theory to fulfill its avowed aims. *Quantum field theory has failed no significant test, nor can any decisive confrontation be anticipated in the near future.*

> [On pages xv-xvii of the Preface dated 1956 to Schwinger, J. (ed). (1958). *Selected Papers on Quantum electrodynamics*, Schwinger qualified this statement. He questioned whether *renormalization* simply corrected a mathematical error that causes the divergencies, or whether *there is a serious flaw in the structure of field theory*. He concluded that "the observational basis of quantum electrodynamics is self-contradictory" and that "a convergent theory cannot be formulated consistently within the framework of present space-time concepts" … "It can never explain the observed value of the dimensionless coupling constant measuring the electron charge … a full understanding of the electron charge can exist only when the theory of elementary particles has come to a stage of perfection that is presently unimaginable".]

Classical mechanics is a determinate theory. Knowledge of the *state* at a given time permits precise prediction of the result of measuring any property of the system. In contrast, *quantum mechanics is only statistically determinate*. It is the *probability* of attaining a particular result on measuring any property of the system, not the outcome of an individual microscopic observation, that is predictable from knowledge of the *state*. But both theories are causal - a knowledge of the *state* at one time implies knowledge of the *state* at a later time. A *quantum state* is specified by particular values of an optimum set of compatible physical properties, which are in number related to the number of degrees of freedom of the system. *In a relativistic theory, the concepts of « before » and « after » have no intrinsic meaning for regions that are in space-like relation*. This implies that measurements individually associated with different regions in *space-like* relation are causally independent, or compatible. Such measurements can be combined in the complete specification of a *state*. But since there is no limit to the number of disjoint spatial regions that can be considered, *a relativistic quantum system has an infinite number of degrees of freedom*.

The latter statement, incidentally, contains an implicit appeal to a general property that the mathematics of physical theories must possess - *the mathematical description of nature is not sensitive to modifications in physically irrelevant details*. An infinite total spatial volume is an idealization of the finite volume defined by the macroscopic measurement apparatus. Arbitrarily small volume elements are idealizations of cells with linear dimensions far below the level of some least distance that is physically significant. Thus, *it would be more accurate, conceptually, to assert that a relativistic quantum system has a number of degrees of freedom that is extravagantly large, but finite*.

The distinctive features of relativistic quantum mechanics flow from the idea that each small element of three-dimensional space at a given time is physically independent of all other such volume elements. Let us label the various degrees of freedom explicitly - by a point of three-dimensional space (in a limiting sense), and by other quantities of finite multiplicity. The dynamical variables then appear as

$$\chi_{a,x} \equiv \chi_a\,(t = x^0, x) \qquad (9)$$

which are a finite number of Hermitian operator functions of *space-time* coordinates, or quantum fields. The dynamical independence of the individual volume elements is expressed by a corresponding additivity of the Lagrangian operator

$$L = \int (dx)\, \mathscr{L} \qquad (10)$$

where the Lagrange function $\mathscr{L}$ *describes the dynamical situation in the infinitesimal neighborhood of a point*. The characteristic time derivative or kinematical part of L appears analogously in $\mathscr{L}$ in terms of the variables associated with the specified spatial point. The *relativistic* structure of the *action principle* is completed by demanding that it present the same form, independently of the particular partitioning of *space-time* into space and time. This is facilitated by the appearance of the *action operator*, the time integral of the Lagrangian, as the *space-time* integral of the Lagrange function. Accordingly, we require, as a sufficient condition, that the latter be a scalar function of its *field variables*, which implies that the known form of the *time derivative* term is supplemented by similar *space derivative* contributions. This is conveyed by

$$\mathscr{L} = \tfrac{1}{4}\,(\chi A^{\mu}\delta_{\mu}\chi - \delta_{\mu}\chi A^{\mu}\chi) - \mathscr{H}(\chi) \qquad (11)$$

where the A^{μ} are a set of four finite skew-Hermitian matrices. *A specific physical field is associated with submatrices of the* A^{μ}, *which are real and anti-symmetrical for a field* φ *that obeys Bose-Einstein statistics, or imaginary and symmetrical for a field* ψ *obeying Fermi-Dirac statistics.* Finally, the boundaries of the four-dimensional integration region, formed by three-dimensional space at the terminal times, are described by the invariant concept of the *space-like* surface σ, a three-dimensional manifold such that every pair of points is in *space-like* relation. The ensuing invariant form of the *action principle* of *relativistic quantum field theory* is (we now use atomic units, in which ħ = c = 1)

$$\delta\langle\sigma_1|\sigma_2\rangle = i\langle\sigma_1|\delta[\textstyle\int_{\sigma_2}^{\sigma_1}(dx)\,\mathscr{L}]|\sigma_2\rangle \qquad (12)$$

Relativity is a statement of equivalence within a class of descriptions associated with similar but different measurement apparatus. Space-time coordinates are an abstraction of the role that the measurement apparatus plays in defining a *space-time* frame of reference. The empirical fact, that all connected *space-time* locations and orientations of the measurement apparatus supply equivalent descriptions, is interpreted by the mathematical requirement of invariance under the group of proper orthochronous inhomogeneous Lorentz transformations, applied to the continuous numerical coordinates.

There is another numerical element in the quantum-mechanical description that has a measure of arbitrariness and expresses an aspect of *relativity*. I am referring to *the quantum-mechanical use of complex numbers* and of the mathematical equivalence of the two square roots of -1, $\pm i$. What general property of any measurement apparatus is subject to our control, in principle, but offers only the choice of two alternatives? The answer is clear - *a macroscopic material system can be constructed of matter, or of antimatter*! But let us not conclude too hastily that a *matter* apparatus and an *antimatter* apparatus are completely equivalent. It is characteristic of quantum mechanics that the dividing line between apparatus and system under investigation can be drawn somewhat arbitrarily, as long as the measurement apparatus always possesses the classical aspects required for the unambiguous recording of an observation. To preserve this feature, the interchange of *matter* and *antimatter* must be made on the whole assemblage of macroscopic apparatus and microscopic system. Since *the observational label of this duality is the algebraic sign of electric charge*, the microscopic interchange must reverse the vector of *electric current* j^{μ}, while maintaining the tensor $T^{\mu\nu}$ that gives the flux of *energy* and *momentum*. But this is just the effect of the coordinate transformation that reflects all four coordinates.

It is indeed true that the *action principle* does not retain its general form under either of the two transformations, the replacement of i with $-i$, and the reflection of all coordinates, but does preserve it under their combined influence. In more detail, *the effect of complex conjugation is equivalent to the reversal of operator multiplication, which distinguishes fields with the two types of statistics*. The reflection of all coordinates, a proper transformation, can be generated by rotations in the attached Euclidean space obtained by introducing the imaginary time coordinate $x_4 = ix^0$. *This transformation alters reality properties, distinguishing fields with integral and half-integral spin.* The combination of the two transformations replaces the original Lagrange function

$$\mathscr{L}(\varphi_{\text{int}}, \varphi_{1/2\,\text{int}}, \psi_{\text{int}}, \psi_{1/2\,\text{int}}) \tag{13}$$

with

$$\mathscr{L}(\varphi_{\text{int}}, i\varphi_{1/2\,\text{int}}, i\psi_{\text{int}}, \psi_{1/2\,\text{int}}). \tag{14}$$

If only fields of the types φ_{int}, $\Psi_{1/2\,\text{int}}$ are considered, which is the empirical connection between spin and statistics, the *action principle* is unaltered in form. *This invariance property of the action principles expresses the relativity of matter and antimatter*. That is the content of the so-called *TCP theorem*.

> [The *CPT theorem* appeared for the first time, implicitly, in Schwinger 1951, The Theory of Quantized Fields I, to prove the connection between spin and statistics. Charge, parity, and time reversal (CPT) symmetry is a fundamental symmetry of physical laws under the simultaneous transformations of charge conjugation (C), parity transformation (P), and time reversal (T). The CPT theorem says that CPT symmetry holds for all physical phenomena, or more precisely, that any Lorentz invariant local quantum field theory with a Hermitian Hamiltonian must have CPT symmetry.]

The anomalous response of the field types φ_{int}, $\Psi_{1/2\ int}$ is also the basis for the theoretical rejection of these possibilities as contrary to general physical requirements of positiveness, namely, the positiveness of *probability*, and the positiveness of *energy*.

The concept of *space-like* surface is not limited to plane surfaces. According to the *action principle*, an infinitesimal deformation of the *space-like* surface on which a *state* is specified changes that *state* by

$$\delta\langle\sigma| = i\langle\sigma| \int d\sigma_\mu\, T^{\mu\nu}\delta x^\nu \qquad (15)$$

which is the infinitely multiple relativistic generalization of the Schrodinger equation

$$\delta\langle t| = i\langle t|\, H(-\,\delta t). \qquad (16)$$

This set of *differential* equations must obey *integrability conditions*, which are *commutator* statements about the elements of the tensor $T^{\mu\nu}$. Since rigid displacements and rotations can be produced from arbitrary local deformations, the operator expressions of the group properties of Lorentz transformations must be a consequence of these *commutator conditions*. Foremost among the latter are the equal-time *commutators* of the energy density T^{00}, which suffice to convey all aspects of *relativistic invariance* that are not of a three-dimensional nature. A system that is invariant under three-dimensional translations and rotations will be Lorentz invariant if, at equal times[4]

$$-\,i[T^{00}(x), T^{00}(x')] = -\,i\{T^{0k}(x) + T^{0k}(x')\}\, \partial k\delta(x - x'). \qquad (17)$$

[4] Schwinger, J. (July, 1962). Non-Abelian Gauge Fields. Relativistic Invariance. *Phys. Rev.*, 127, 1, 324; https://doi.org/10.1103/PhysRev.127.324 [; "A simple criterion for Lorentz invariance in quantum field theory is stated as a *commutator condition* relating the *energy density* to the *momentum density*. With its aid a *relativistically* invariant radiation-gauge formulation is devised for a non-Abelian vector-gauge field coupled to a spin-½ Fermi field"].

This is a sufficient condition. Additional terms with higher derivatives of the delta function will occur, in general. But there is a distinguished class of physical system, which I shall call *local*, for which no further term appears. The phrase « *local system* » can be given a physical definition within the framework we have used or, alternatively, by viewing the *commutator condition* as a measurability statement about the property involved *in the response of a system to a weak external gravitational field*[5]

[5] Schwinger, J. (April, 1963). Commutation Relations and Conservation Laws. *Phys. Rev.*, 130, 1, 406; https://doi.org/10.1103/PhysRev.130.406[; "The response of a physical system to external *electromagnetic* and *gravitational* fields, as embodied in the *electric current* and *stress tensor* conservation laws, is used to derive the equal-time *commutation relations* for *charge density* and *energy density*"].

Only the external *gravitational potential* g_{00} is relevant here. A physical system is *local* if the operators $T^{\mu\nu}$, which may be explicit functions of g_{00} at the same time, do not depend upon time derivatives of g_{00}. The class of *local* systems is limited[6] to *fields* of spin 0, ½,1.

[6] Schwinger, J. (April, 1963). Energy and Momentum Density in Field Theory. *Phys. Rev.*, 130, 2, 800; https://doi.org/10.1103/PhysRev.130.800[; "It is shown that the *energy density commutator condition* in its simplest form is valid for interacting spin 0, ½, 1 field systems, but not for higher spin fields. The *action principle* is extended, for this purpose, to arbitrary coordinate frames. ... As the fundamental equation of *relativistic* quantum field theory, the *commutator condition* makes explicit the greater physical complexity of higher spin fields"].

Such *fields* are distinguished by their physical simplicity in comparison with fields of higher spin. *One may even question whether consistent relativistic quantum field theories can be constructed for non-local systems.*

The *energy density commutator condition* is a very useful test of *relativistic invariance*. Only a month or so ago I employed it to examine whether a *relativistic quantum field theory* could be devised to describe *magnetic* as well as *electric charge*. *Dirac pointed out many years ago that the existence of magnetic charge would imply a quantization of electric charge, in the sense that the product of two elementary charges, eg/ħc, could assume only certain values. According to Dirac, these values are any integer or half-integer*. In recent years, the theoretical possibility of *magnetic charge* has been attacked from several directions. *The most serious accusation is that the concept is in violation of Lorentz invariance*. This is sometimes expressed in the language of field theory by the remark that no manifestly scalar Lagrange function can be constructed for a system composed of *electromagnetic field* and *electric and magnetic charge-bearing fields*. Now *it is true that there is no relativistically invariant theory for arbitrary* e *and* g, *so that no formally invariant version could exist*. Indeed, the unnecessary assumption that $\mathscr{L}$ is a scalar must be relinquished in favor of the more general possibilities that are compatible with the *action principle*. But the *energy commutator condition* can still be applied. I have been able to show that *energy* and *momentum density operators* can be exhibited which satisfy the *commutator condition*, together with the three-dimensional requirements, provided *eg/ħc* possesses one of a discrete set of values. These values are integers, which is more restrictive than Dirac's *quantization condition*. *Such general considerations shed no light on the empirical elusiveness of magnetic charge*. They only emphasize that this novel theoretical possibility should not be dismissed lightly.

The physical systems that obey the *commutator* statement of *locality* do not include the *gravitational field*. But, this field, like the *electromagnetic field*, requires very special consideration. And these considerations make full use of the *relativistic field concept*. The dynamics of the *electromagnetic field* is characterized by invariance under *gauge transformations*, in which the phase of every *charge-bearing field* is altered arbitrarily, but continuously, at each *space-time* point while *electromagnetic potentials* are transformed inhomogeneously. The introduction of the *gravitational field* involves, not only the use of general coordinates and coordinate transformations, but the establishment at each point of an independent Lorentz frame. The *gravitational field gauge transformations* are produced by the arbitrary reorientation of these local coordinate systems at each point while *gravitational potentials* are transformed linearly and inhomogeneously. The formal extension of the *action principle* to include the *gravitational field* can be carried out[7]

[7] Schwinger, J. (May, 1963). Quantized Gravitational Field. *Phys. Rev.*, 130, 3, 1253; https://doi.org/10.1103/PhysRev.130.1253[; "A *gravitational action operator* is constructed that is invariant under *general coordinate transformations* and *local Lorentz (gauge) transformations*. To interpret the formalism the arbitrariness in description must be restricted by introducing *gauge conditions* and *coordinate conditions*. … *The question of Lorentz invariance is left undecided since the energy density operator is only given implicitly*"],

together with the verification of consistency conditions analogous to the *energy density commutator condition*[8]

[8] Schwinger, J. (November, 1963). Quantized Gravitational Field. II. *Phys. Rev.* 132, 3, 1317; https://doi.org/10.1103/PhysRev.132.1317[; "A consistent formulation is given for the *quantized gravitational field* in interaction with integer spin fields. Lorentz transformation equivalence within a class of physically distinguished coordinate systems is verified"].

To appreciate this tour de force, one must realize that the operator in the role of energy density is a function of the gravitational field, which is influenced by the energy density. Thus, the object to be tested is only known implicitly. It also appears that the detailed specification of the spatial distribution of *energy* lacks physical significance when gravitational phenomena are important. Only integral quantities or equivalent asymptotic field properties are physically meaningful in that circumstance. *It is in the further study of such boundary conditions that one may hope to comprehend the significance of the gravitational field as the physical mediator between the worlds of the microscopic and the macroscopic, the atom and the galaxy.*

I have now spoken at some length about *fields*. But it is in the language of *particles* that observational material is presented. How are these concepts related? Let us turn for a moment to the early history of our subject. *The quantized field appears initially as a device for describing arbitrary numbers of indistinguishable particles. It was defined as the creator or annihilator of a particle at the specified point of space and time*. This picture changed some-what as a consequence of the developments in quantum electrodynamics to which Feynman, Tomonaga, myself, and many others contributed.

It began to be appreciated that the observed properties of so-called elementary particles are partly determined by the effect of interactions. The *fields* used in the dynamical description were then associated with noninteracting or bare particles, but there was still a direct correspondence with physical particles. The weakness of *electromagnetic interactions*, as measured by the small value of the fine structure constant $e^2/\hbar c$ is relevant here, for *the same view-point failed disastrously when extended to strongly interacting nucleons and mesons*. The resulting widespread disillusionment with *quantum field theory* is an unhappy chapter in the history of high-energy theoretical physics, although it did serve to direct attention toward various useful *phenomenological** calculation techniques.

[* *relating to the science of phenomena as distinct from that of the nature of being*.]

The great qualitative difference between *weakly interacting* and *strongly interacting* systems was impressed upon me by a particular consideration which I shall now sketch for you[9]

[9] Schwinger, J. (January, 1962). Gauge Invariance and Mass. *Phys. Rev*., 125, 1, 397; https://doi.org/10.1103/PhysRev.125.397; (December, 1962). Gauge Invariance and Mass. II. *Phys. Rev*. 128, 5, 2425; https://doi.org/10.1103/PhysRev.128.2425[; "It is argued that the *gauge invariance* of a vector field does not necessarily imply zero *mass* for an associated particle if the *current vector coupling* is sufficiently strong. This situation may permit a deeper understanding of *nucleonic charge conservation* as a manifestation of a *gauge invariance*, without the obvious conflict with experience that a massless particle entails"].

In the absence of interactions there is an immediate connection between the quantized Maxwell field and a physical particle of zero mass, the photon. The null mass of the photon is the particle transcription of a field property, that electromagnetism has no well-defined range but weakens geometrically. Now one of the most important interaction aspects of quantum electrodynamics is the phenomenon of *vacuum polarization*. A variable *electromagnetic field* induces secondary *currents*, even in the absence of actual particle creation. *In particular, a localized charge creates a counter charge in its vicinity, which partially neutralizes the effect of the given charge at large distances. The implication that physical charges are weaker than bare charges by a universal factor is the basis for charge renormalization.* But once the idea of a partial neutralization of charge is admitted one cannot exclude the possibility of total charge neutralization. This will occur if the interaction exceeds a certain strength such that an oppositely charged particle combination, of the same nature as the photon, becomes so tightly bound that the corresponding mass diminishes to zero. Under these circumstances no long-range fields would remain and the massless particle does not exist. *We learn that the connection between the Maxwell field and the photon is not an a priori one, but involves a specific dynamical aspect, that electromagnetic interactions are weaker than the critical strength.* It is a natural speculation that another such *field* exists which couples more strongly than the critical amount to *nucleonic charge*, the property carried by all heavy fermions. *That hypothesis would explain the absolute stability of the proton*, in analogy with the electromagnetic explanation of electron stability, without challenging the uniqueness of the photon.

A *field operator* is a localized excitation which, applied to the vacuum state, generates all possible *energy-momentum*, or equivalently, *mass states* that share the other distinguishing properties of the *field*. The products of *field operators* widen and ultimately exhaust the various classes of *mass states*. If an isolated *mass value* occurs in a particular product, the *state* is that of a stable particle with corresponding characteristics. Should a small neighborhood of a particular *mass* be emphasized, the situation is that of an unstable particle, with a proper lifetime which varies inversely as the *mass width* of the excitation. The quantitative properties of the stable and unstable particles that may be implied by a given dynamical field theory cannot be predicted with presently available calculation techniques. In these matters, to borrow a phrase of Ingmar Bergman, and St. Paul, we see through a glass, darkly. Yet, in the plausible qualitative inference that a substantial number of particles, stable and unstable, will exist for sufficiently strong interactions among a few *fields* lies the great promise of *relativistic quantum field theory.*

Experiment reveals an ever-growing number and variety of unstable particles, which seem to differ in no essential way from the stable and long-lived particles with which they are grouped in tentative classification schemes. Surely one must hope that this bewildering complexity is the dynamical manifestation of a conceptually simpler substratum, which need not be directly meaningful on the observational level of particles. *The relativistic field concept is a specific realization of this general groping toward a new conception of matter.*

There is empirical evidence in favor of such simplification at a deeper dynamical level. Strongly interacting particles have been rather successfully classified with the aid of a particular *internal symmetry group*. It is the unitary group SU_3. The dimensionalities of particle multiplets that have been identified thus far are 1, 8, and 10. But the fundamental multiplet of dimensionality 3 is missing. It is difficult to believe in the physical significance of some transformation group without admitting the existence of objects that respond to the transformations of that group. Accordingly, I would describe the *observed situation* as follows. There are sets of *fundamental fields* that form triplets[10] with respect to the group U_3

[10] Schwinger, J. (August, 1964). Field Theory of Matter. *Phys. Rev.* 135, 3B, B816; https://doi.org/10.1103/PhysRev.135.B816[; "A speculative *field theory of matter* is developed. Simple computational methods are used in a preliminary survey of its consequences. The theory exploits the known properties of leptons by means of a *principle of symmetry* between *electrical* and *nucleonic charge*. There are *fundamental fields* with spins 0, ½, 1. The spinless field is neutral. Spin ½ and 1 fields can carry both *electrical* and *nucleonic* charge"]; (December, 1964). Field Theory of Matter. II. *Phys. Rev.*, 136, 6B, 1821; https://doi.org/10.1103/PhysRev.136.B1821[; "A qualitative dynamical interpretation is given for observed regularities of nonleptonic phenomena in *strong, electromagnetic, and weak interactions*".

The excitations produced by these fields are very massive and highly unstable. The low-lying mass excitations of mesons and baryons are generated by products of the fundamental *fields*. If these *fields* are assigned spin ½, as a specific model, it is sufficient to consider certain products of two and three fields to represent the general properties of mesons and baryons, respectively.

The cogency of this picture is emphasized by its role in clarifying a recent development in symmetry classification schemes. That is the provocative but somewhat mysterious suggestion that the *internal symmetry group* SU_3 be combined with space-time spin transformations to form the larger unitary group SU_6. *This idea, with its relativistic generalizations, has had some striking numerical successes but there are severe conceptual problems in reconciling Lorentz invariance with any union of internal and space-time transformations, as long as one insists on immediate particle interpretation*. The situation is different if one can refer to the *space-time* localizability that is the hallmark of the *field* concept[11].

> [11] Schwinger, J. (October, 1965). Field Theory of Matter. IV. *Phys. Rev.*, 140, 1B, 158; https://doi.org/10.1103/PhysRev.140.B158[; "The *relativistic* dynamics of 0^- and 1^- mesons in the idealization of U_3 symmetry is derived from the hypothesis that a compact group of transformations on *fundamental fields* induces a predominantly local and linear transformation of the *phenomenological fields* that are associated with particles. The physical picture of *phenomenological fields* as highly localized functions of *fundamental fields* implies that the *interaction* term of the *phenomenological* Lagrange function can have symmetry properties, expressed by invariance under the compact *transformation group*, that have no significance for the remainder of the Lagrange function, which describes the propagation of the physical excitations"].

Let us assume that the *interactions* among the *fundamental fields* are of such strength that *field products* at practically coincident points suffice to describe the excitation of the known relatively low-lying particles. The resulting quasi-local structures are in some sense *fields* that are associated with the physical particles. I call these *phenomenological fields*, as distinguished from the *fundamental fields* which are the basic *dynamical variables* of the system. Linear transformations on the *fundamental fields* can simulate the effect of external probes, which may involve both unitary and spin degrees of freedom. If these external perturbations are sufficiently gentle, the structure of the particles will be maintained and the *phenomenological field* transformed linearly with indefinite multiplets. It is not implausible that the highly localized interactions among the *phenomenological fields* will exhibit a corresponding symmetry. Thus, combined spin and unitary transformations appear as a device for characterizing some gross features of the unknown inner field dynamics of physical particles, as it operates in the neighborhood of a specific point. But these transformations can have no general significance for the transfer of excitations from point to point, and only lesser symmetries will survive in the final particle description.

Phenomenological fields are the basic concept in formulating the practical calculation methods of *strong interaction field theory*. They serve to isolate the formidable problem of the dynamical origin of physical particles from the more immediate questions referring to their properties and interactions. *In somewhat analogous circumstances, those of non-relativistic many-particle physics, the methods and viewpoint of quantum field theory*[12] *have been enormously successful.*

> [12] The general theory is described by Martin, P. C. & Schwinger, J. (September, 1959). Theory of Many-Particle Systems. I. *Phys. Rev.*, 115, 6, 1342; https://doi.org/10.1103/PhysRev.115.1342[; "This is the first of a series of papers dealing with *many-particle systems* from a unified, nonperturbative point of view. …"].

They have clarified the whole range of cooperative phenomena, while employing relatively simple approximation schemes. I believe that *phenomenological relativistic quantum field theory* has a similar future, and will replace the algorithms that were introduced during the period of revolt from field theory. *But the intuition that serves so well in non-relativistic contexts does not exist for these new conditions*. One has still to appreciate the precise rules of *phenomenological relativistic field theory*, which must supply a self-consistent description of the residual *interactions*, given that the strong *fundamental interactions* have operated to compose the various physical particles. *And when this is done, how much shall we have learned, and how much will remain unknown*, about the mechanism that builds matter from more primitive constituents? Are we not at this moment,

> *. . . like stout Cortez when with eagle eyes*
> *He star'd at the Pacific - and all his men*
> *Look'd at each other with a wild surmise –*
> *Silent, upon a peak in Darien.*

And now it only remains for me to say: *Tack sa mycket for uppmarksamheten*. [Thank you very much for your attention.]